Encyclopedia of the Life Course and Human Development

Encyclopedia of the Life Course and Human Development

VOLUME 1
CHILDHOOD AND ADOLESCENCE

Deborah Carr

EDITOR IN CHIEF

MACMILLAN REFERENCE USA
A part of Gale, Cengage Learning

GALE
CENGAGE Learning™

Detroit • New York • San Francisco • New Haven, Conn • Waterville, Maine • London

Encyclopedia of the Life Course and Human Development
Deborah Carr, Editor in Chief

For product information and technology assistance, contact us at
Gale Customer Support, 1-800-877-4253
For permission to use material from this text or product,
submit all requests online at **www.cengage.com/permissions**
Further permissions questions can be emailed to
permissionrequest@cengage.com

Since this page cannot legibly accommodate all copyright notices, the acknowledgments constitute an extension of the copyright notice.

While every effort has been made to ensure the reliability of the information presented in this publication, Gale, a part of Cengage Learning, does not guarantee the accuracy of the data contained herein. Gale accepts no payment for listing; and inclusion in the publication of any organization, agency, institution, publication, service, or individual does not imply endorsement of the editors or publisher. Errors brought to the attention of the publisher and verified to the satisfaction of the publisher will be corrected in future editions.

Library of Congress Cataloging-in-Publication Data

Encyclopedia of the life course and human development / Deborah Carr, editor in chief.
  v. ; cm.
  Includes bibliographical references and index.
  ISBN 978-0-02-866162-9 (set : alk. paper) – ISBN 978-0-02-866163-6 (vol. 1 : alk. paper) – ISBN 978-0-02-866164-3 (vol. 2 : alk. paper) – ISBN 978-0-02-866165-0 (vol. 3 : alk. paper)
  1. Social evolution—Encyclopedias. 2. Human evolution—Encyclopedias. I. Carr, Deborah S.

HM626.E538 2008
305.203—dc22   2008027490

Gale
27500 Drake Rd.
Farmington Hills, MI 48331-3535

ISBN-13: 978-0-02-866162-9 (set)   ISBN-10: 0-02-866162-1 (set)
ISBN-13: 978-0-02-866163-6 (vol. 1)   ISBN-10: 0-02-866163-X (vol. 1)
ISBN-13: 978-0-02-866164-3 (vol. 2)   ISBN-10: 0-02-866164-8 (vol. 2)
ISBN-13: 978-0-02-866165-0 (vol. 3)   ISBN-10: 0-02-866165-6 (vol. 3)

This title is also available as an e-book.
ISBN-13: 978-0-02-866166-7 ISBN-10: 0-02-866166-4
Contact your Gale sales representative for ordering information.

Editorial Board

Editorial and Production Staff

PROJECT EDITORS

Deirdre S. Blanchfield
Andrew G. Specht

EDITORIAL TECHNICAL SUPPORT

Mark Drouillard

MANUSCRIPT EDITORS

Judith Clinebell
Laurie J. Edwards
Jessica Hornik Evans
Christine Kelley
Eric Linderman
Eric Lowenkron
Kari Lucke

Raymond Lukens
David E. Salamie

PROOFREADER

John Krol

INDEXER

Laurie Andriot

PRODUCT DESIGN

Pamela A.E. Galbreath

IMAGING

Lezlie Light

GRAPHIC ART

Pre-PressPMG

PERMISSIONS

Leitha Etheridge-Sims

COMPOSITION

Evi Seoud

MANUFACTURING

Wendy Blurton

DIRECTOR, NEW PRODUCT DEVELOPMENT

Leigh Ann Cusack

PUBLISHER

Jay Flynn

Contents

Preface

Why are people so fascinated by their high school reunions? Alumni look forward to reuniting with former classmates for the very same reason that life course scholars do research: *to find out how people's lives turned out*. Life course and human development scholars seek to discover *how* and *why* human lives unfold as they do. Why do some young people succeed in high school, college, and in the workplace while others struggle to earn good grades, find rewarding jobs, and stay out of legal trouble? Why do brothers and sisters from the very same families often follow divergent paths as adults? How do sweeping social changes and historical events—like the Depression era of the 1930s, World War II in the 1940s, the sexual revolution of the 1960s, the stagflation of the 1970s, and the internet explosion in the 2000s—affect the goals, values and opportunities facing each new generation of young people? What factors predict whether old age is marked by physical and cognitive impairment, or vigor and mental acuity?

These are just few examples of the countless questions that life course scholars investigate—but finding answers requires a more systematic investigation than striding up to a former classmate at a high school reunion and asking "what's new?" Life course scholars have developed a sophisticated theoretical paradigm to understand human lives. Four key assumptions guide their selection of research questions, and their ways of thinking about human lives: (a) lives are embedded in and shaped by historical context; (b) the meaning and impact of a life transition is contingent on when it occurs; (c) lives are intertwined through social relationships; and (d) individuals construct their own lives through their choices and actions, yet within the constraints of historical and social circumstances.

Life course scholars also rely on rigorous research methods and data sources—including national censuses, sample surveys, in-depth interviews, and historical records—to document human lives. Because a key question of life course research is "how does historical time and place shape lives?" researchers often compare data obtained at different points in time, from different birth cohorts (i.e., individuals born at different points in history), and from different national and cultural contexts. Researchers also rely heavily on longitudinal data, or data obtained from the same person at multiple points in time, so they can track continuity and change within a single life.

ABOUT THE ENCYCLOPEDIA

We created the *Encyclopedia of the Life Course and Human Development* to introduce life course theory, research, and methods to seasoned social scientists, graduate students, undergraduate students taking their very first sociology course, and anyone who has ever wondered "what makes people turn out the way they do?" We hope that this volume conveys our enthusiasm for the creativity, breadth, and both theoretical and practical contributions of life course research.

This project was the brainchild of Barbara Rader, an acquisitions editor with Thomson Gale, who retired shortly after setting the wheels in motion for the encyclopedia. Barbara recognized that the core questions of life course sociology and human development are an essential part of most college-level social science courses, yet there existed no single compendium that brought together the latest research and theory on the life course for a broad audience. I agreed that this encyclopedia was an absolute necessity, and happily signed on as editor-in-chief. I promptly invited three esteemed colleagues to serve as the associate editors for each of the three volumes that comprise the encyclopedia: Robert Crosnoe, University of Texas-Austin (Volume 1, Childhood and Adolescence); Mary Elizabeth Hughes, Johns Hopkins University (Volume 2, Adulthood); and Amy Pienta, University of Michigan (Volume 3, Later Life).

The four-person editorial board then tackled the Herculean task of developing a list of roughly 500 topics to be covered in the encyclopedia. Each board member individually developed their own wish list of topics, and the team then discussed and debated each and every suggestion. Was a topic substantial enough to warrant its own entry? Which classic books were considered sufficiently influential to be included in the annotated bibliography? Which of the hundreds of eminent life course and human development scholars should be celebrated with a biography entry? Should the encyclopedia cover theory and research findings only, or should it also point readers to data sources so that they could conduct their own research? After much deliberation, we narrowed down our list from roughly 1,000 topics to around 450, and invited authors to write the entries. We had two goals in inviting authors. First, we wanted to feature authors who are widely recognized as experts in their field. Second, we wanted to include authors who represented the full range of the professional life course, from outstanding graduate students to eminent emeriti faculty. We are absolutely delighted that such an accomplished group of authors agreed to write for us, and are grateful for the care, creativity, and thoughtfulness that they invested in each and every entry.

Deciding on topics and authors was not the only challenge we faced. The editorial board also thought long and hard about their organizational framework. Should the three volumes reflect separate (yet clearly overlapping and mutually influential) life course stages? Or should it be organized by important life course domains, such as work, family, and health? Or, should there be no organizational framework imposed, and the entries simply run from A to Z? We ultimately decided to divide the encyclopedia into three volumes according to the life course stages that are typically the foci of undergraduate social science courses, such as Adolescent Development, and Sociology of Aging. Yet the debates didn't stop there. In which volume should we place each entry? Some choices were obvious; nearly all of the 34 entries on education-related issues were assigned to Volume 1 (with the exceptions of "Continuing Education," "Educational Attainment," and "Lifelong Learning"). Yet other decisions were less clear-cut. For instance, "Child Custody and Support" describes the impact of custody decisions on children, their adult parents, and their aged grandparents. In the end, we placed entries in the volume that we believed would be the most intuitive "home" for an entry, in the eyes of our readers.

We also questioned whether a topic should be covered in one volume only, or in each of the three. For more than a dozen topics—ranging from parent-child relations to health differentials/disparities to cultural images to genetic influences—we decided that the core research findings, theories and implications were sufficiently distinct for each life course stage as to warrant a unique entry for each volume. Our intellectual deliberations underscore key assumptions of the life course paradigm—life domains are intertwined, the fates of generations are linked, and no single period of life can be understood in isolation from one's earlier experiences and future aspirations. As readers peruse the encyclopedia, they will become keenly aware of the fuzziness of boundaries demarcating life course stages and domains. However, to help readers easily locate the topics they're seeking, we've provided a thematic outline that classifies individual entries by broad topical subheading and volume number.

Each of the three volumes follows an A-to-Z format that mixes long "composite" (i.e., multi-part) entries on broad topics like aging, childbearing, health behaviors, mental health, and stages of schooling, and short sidebars on up-to-the-minute and controversial topics like "Bankruptcy," "Mommy Wars/Images of Motherhood," "Hurried Child Syndrome," "Third Age," and "Virginity Pledges." Entries range in length from a brief 250 words to more than 7,000 words. Each entry defines key terms, explains why a concept is important for understanding human lives, reviews classic and contemporary works, documents sub-group patterns such as age, race, gender, or cross-national differences, identifies gaps and controversies in research, highlights the implications of life course research for policy and practice, and points to avenues for future study. Most entries are followed by a bibliography, cross-references to related entries, illustrations, and statistical charts and graphs.

Of course, we recognize that we could not cover every possible topic but we hope that the encyclopedia encompasses the topics, theories, scholars, data sources, and research methods that are most central to the study of the life course and human development. We believe that now is the perfect time to unveil this encyclopedia. Ph.D. level social scientists and graduate students have long understood the appeal and importance of adopting a life course approach when studying human lives. In the past decade, nearly a dozen excellent edited volumes and monographs have brought together world-class scholars to synthesize, critique, and extend research and theory on the life course (e.g., Mortimer & Shanahan 2003; Settersten 1999, 2003; Shanahan & Macmillan 2008). For the most part, these sophisticated analyses are targeted toward graduate students and seasoned researchers. However, we know of no other volume that provides a general readership with concise yet cutting-edge statements on nearly 500 topics that represent the range and breadth of life course research.

A BRIEF OVERVIEW OF THE LIFE COURSE PARADIGM

Upon their initial foray into studying the life course and human development, students sometimes ask the cynical question, "so, is life course sociology the study of *everything*?" Researchers working in the life course tradition *do* study a broad range of topics, ranging from childbearing to criminality, political participation to parenting styles, genetic influences to gender roles. Life course research also incorporates ideas from a broad range of academic disciplines, including biology, economics, epidemiology, genetics, gerontology, history, medicine, psychiatry, psychology, political science, and even statistics. However, scholars working in this tradition approach their work by taking a clear-eyed and highly focused lens on human behavior. The life course paradigm was first articulated by sociologist Glen H. Elder, Jr. (see Elder, 1994 for a review) as an approach to studying human lives that pays equal attention to *individual-level* biographies and *macrosocial* influences.

Time is a core component of the life course paradigm, and encompasses both *personal time* (one's own age and aging) as well as *historical time* (historical events and patterns). As such, life course scholars recognize the importance of looking at whole lives in context, rather than isolated stages, such as adolescence (Riley, Johnson, & Foner 1972). The emphasis on *whole lives* or "life trajectories" has its conceptual and methodological roots in the classic *Polish Peasant in Europe and America* study, conducted by W. I. Thomas and Florian Znaniecki (1918–1920). Thomas and Znaniecki studied life histories and also encouraged other researchers to "explore many types of individuals with regard to their experiences in various past periods of life in different situations and follow groups of individuals into the future, getting a continuous record of their experiences as they occur" (Volkart 1951, p. 593). The authors featured in the encyclopedia have heeded this call.

In the following sections, we revisit the four core life course themes mentioned in the opening paragraph of this Introduction, and provide a brief description of the historical roots and contemporary research relevant to each such theme. We hope this brief overview provides readers a foundation for understanding the vast and diverse range of subject matter addressed in this encyclopedia.

Historical Time and Place The life course of individuals is embedded in and shaped by the historical times and places they experience over their life time. Think about how your life is different from the lives of your parents, grandparents, or great-grandparents. The choice of one's occupation, plans for when (and whether) to marry and have children, the ability to purchase a home, whether one's schooling is interrupted by war or a family's financial crises, and one's life expectancy are just a few of the many life course experiences that have changed drastically in the past century. The notion that human lives are shaped by social and historical context is a core theme of the life course paradigm, and dates back to the writings of C. Wright Mills. In his classic book *The Sociological Imagination*, Mills (1959) proposed that to understand human behavior, scholars must consider both one's "biography" and "history." Mills noted that "the sociological imagination enables it possessor to understand the larger historical scene in terms of its meaning for the inner life and external career of a variety of individuals" (Mills 1959, p. 7).

The impact of history on individual lives is most evident during periods of rapid social change (Mannheim, 1928/1952). For example, during the latter half of the 20th century, women's social roles changed dramatically, as educational and occupational opportunities expanded in the wake of the Women's Movement. White middle-class women who were stay-at-home mothers in their 1950s witnessed their daughters grow up to have successful careers as lawyers, bankers, doctors, and other careers that historically were considered "men's" domain. Although mothers and daughters share many similarities, including genetic background, ethnicity, religion, and (often) social class, historical changes created a seismic divide in the life choices made by these two generations of women (Carr, 2004).

The impact of history also operates in very different ways, based on one's age when a major *historical trend* unfolds. Young people who were in elementary school when the internet explosion occurred can't remember life before e-mail, and are technologically-savvy computer whizzes. Older adults, by contrast, often struggled to become comfortable surfing the net, emailing, and text messaging. The effects of *specific historical events* also vary based on one's age when the event occurred. Research by Elder and colleagues showed that World War II had vastly different impacts on soldiers, based on their age during the war years. Young entrants had no family or work responsibilities when they shipped off to Japan or Europe, yet older soldiers were abandoning jobs and marriages when they headed overseas. While the young soldiers returned home to new adventures in work, family, and education (due in part to the educational benefits provided by the G.I. Bill), the older soldiers often came home to find their marriages were strained, or their former jobs were no longer available (see MacLean & Elder 2007 for a review). Throughout the encyclopedia, authors highlight the many ways that history shapes individual lives—through processes like economic restructuring and the loss of man-ufacturing jobs, the 1960s Civil Rights movement that expanded opportunities for ethnic and racial minorities in the United States, and technological and medical advances that enable older adults to live longer and healthier lives than ever before.

Place also affects how individual lives unfold. "Place" can be defined as broadly as one's nation, or as narrowly as one's neighborhood or city block. Nation-level characteristics, such its the level of economic development or "modernization" can profoundly influence its citizens' attitudes, values, gender roles, childbearing behavior, educational opportunities and even personality (Inkeles & Levinson, 1969). One's local social context also matters. Neighborhood characteristics like "social capital" or the social cohesiveness and integration of a city block (Coleman 1990), and "social disorganization" or the level of instability, poverty, and crime in one's neighborhood (Shaw & McKay 1942) can affect residents' educational prospects, physical and mental health, occupational opportunities, and even one's life span. Place also exposes individuals to potentially life-altering public policies. For example, persons living in nations with restrictive population policies, like China's one-child policy, have little choice over their childbearing. Encyclopedia entries take special care to highlight the ways that life course trajectories differ across neighborhoods, states, nations, and even continents. Although geography is hardly "destiny" it does play an important role in shaping life trajectories.

Timing in Lives The developmental impact of a succession of life transitions or events is contingent on when they occur in a person's life. How might your life have turned out if you married at age 16? Or if you waited until age 35 to wed? If you married at age 16, you might not have completed high school and might have gone on to have many children, or to hold a poorly paying job that did not require a high school diploma. If you marry for the first time at age 35, you probably have already completed your education, perhaps earning a graduate degree, and having spent many years in the paid work force prior to marrying. Yet marrying at age 35 may mean that one will have only one or two children, given that the likelihood of conceiving a child declines steadily for women after age 35.

These examples illustrate the importance of "social timing." Social timing refers to the ways that age shapes whether, when, how, and to what end one experiences important social roles and transitions between roles. The timing of life transitions reflects a broad range of biological, social, and political forces. For example, the age at which one can physically bear a child is contingent upon the *biological transition* to menarche (that is, a girl's first menstrual period). *Social norms* also provide guidelines for the "appropriate" time for making transitions. Life course sociologist Bernice Neugarten (1965) has observed that people are expected to comply with a "social clock." This refers to "age norms and age expectations [that] operate as prods and brakes upon behavior, in some instances hastening behavior and in some instances delaying it" (Neugarten et al. 1965, p. 710).

Neugarten and her colleagues conducted surveys that reveal Americans generally agree that there is a "right" age to marry, start a job, and find one's own home (Neugarten & Datan 1973). Norms dictating the "right" age for life transitions change over historical time, however. For example, marrying at age 19 and having one's first child at 20 was normal and even desirable for women, in the late 1950s. By contrast, few college students today (or their parents!) would endorse marrying at such a young age.

Popular culture reinforces the belief that there is a "best" time to make important life transitions. Recent Hollywood films like *Failure to Launch* and *Stepbrothers* are an obvious indicator that Americans believe that life transitions that occur "off-time" or later than is typical, are a sign of poor adjustment. Both films depict men in their late 30s who still live with their parents, and who are portrayed as incompetent. Similarly, the 30-something women portrayed in the hit television show *Sex & the City* bemoaned the fact that they were single, and often were made to feel like failures by their married peers. They also worried that they might not be able to have biological children, if they waited until their 40s to marry. These cultural messages carry another important theme of life course research: "mistimed" transitions—or transitions that occur earlier or later than one's peers often create psychological stress, difficult challenges, and social disapproval.

Although cultural norms *informally prescribe* the appropriate timing of life course transitions, public policies *mandate* the timing of many important transitions. Although state laws vary, the law typically dictates that children must stay in school until age 16, and that young people cannot marry until they are 18 years old unless they obtain parental permission. Likewise, the age that one can vote, drive, drink legally, serve in the military, retire with full Social Security benefits, or become President of the United States is dictated by federal or state law. Laws, like social norms, also change over historical time. While young children labored on farms and in factories in past centuries, child labor was banned in the United States by the Fair Labor Act of 1938, and strict rules now dictate the age at which children can work for pay.

Life course scholars recognize that legal, biological, and social time tables may be out of sync, which may cause difficulties as individuals negotiate their life choices. For instance, boys and girls may be physically able to bear a child at age 13, yet they may not be emotionally prepared to enter the role of parent. Public polices encourage (and in some cases, mandate) workers to retire at age 65, although many older employees are perfectly healthy and willing to remain in the work force for another decade. Married couples may not feel "financially ready" to have a child until both have their careers and finances in place, yet if they delay childbearing too long, it may become biologically unfeasible.

The concept of timing weaves through many entries in this encyclopedia. Entries on important life transitions such as marriage, childbearing, school-to-work transition, home ownership, relocation, widowhood, mortality, and retirement reveal the ways that the timing of such events is shaped by one's past, yet also sets the foundation for one's future. Many entries also show that the timing of events varies not only by one's birth cohort, but by factors such as race, ethnicity, nativity status, and social class. In doing so, the encyclopedia authors reveal the importance of heterogeneity in the life course.

Linked Lives Lives are lived interdependently, and social and historical influences are expressed through this network of shared relationships. The third theme of the life course paradigm is the notion of "linked lives." "Linked lives" refers to the way that each individual's life is embedded in a large network of social relationships—with parents, children, siblings, friends, coworkers, in-laws, romantic partners, and others. The notion that social relationships matter dates back to Émile Durkheim's (1997 [1897]) classic writings on social integration. Durkheim famously found that persons with tight-knit social networks had lower rates of suicide than those with weaker social ties. Married persons had lower suicide rates than the unmarried, Catholics fared better than Protestants, and parents revealed lower suicide rates than the childless. Since the publication of Durkheim's work, social scientists have sought to uncover why and how social relationships affect the life course.

Authors in the encyclopedia explore a wide variety of types of social relationships and document patterns such as the impact of parent-child relationships on child outcomes, how peer relationships among school-age persons affect psychological and educational success, and the ways that older persons' health is protected by spouses, friends, siblings, and significant others. The protective effects of social relationships even extent to studies of crime and deviance over the life course. For example, John Laub and Robert Sampson (2006) found that social relationships, especially marrying, having children, and securing a job, were considered the key pathways out of deviant careers for young men whose early years were spent in reform school.

The concept of linked lives also refers to the ways that *generations* are linked to one another. A core theme of life course research is intergenerational transmission; parents' transmit their values, attitudes, and even their socioeconomic achievements and intellectual resources to their offspring (e.g., Sewell & Hauser 1975; Baumrind 1987; Furstenberg, Brooks-Gunn, & Morgan 1987). Parents socialize their children, and thus teach their children how to become well-adjusted members of society. Although classic studies of socialization revealed the ways that children became like their parents, researchers also have focused on identifying why and how children turn out differently from their parents—often placing emphasis on the many other social relationships and social contexts that a child experiences. For example, James Coleman's (1961) classic study *Adolescent Society* shows how high schools students socialize their peers to hold values that are in opposition to the values held by their parents.

Life course sociologists also recognize that *life domains* are linked. Even within a single individual, work and family choices affect one another; working full-time may preclude one from being a stay-at-home parent, or intensive parenting demands may prevent one from working as many hours as one would like. Likewise, economic standing and physical health are mutually influential; poverty exposes people to health risks such as poor nutrition, yet poor health compromises one's ability to work full-time. Moreover, life course influences can occur both cross-person and cross-domain. A spouse's work strain may affect one's own psychological health, while a parent's job loss (and loss of health insurance). may affect a child's health.

The classic example of cross-generation, cross-domain linkages is Elder's (1974) *Children of the Great Depression* study. He found that when fathers lost their job during the Great Depression, mothers and young sons were forced to seek work and earn money, while young daughters often did household chores that would have otherwise fallen to the mother. Thus, the imbalance of power between mother and father weakened the father's ability to exert control, and shaped the children's gender role views. The family's income loss also created marital strain between the parents, and compromised their ability to supervise their children. Elder reveals the cascading consequences of this economic strain across generations and across domains. The complex interplay among generations, social relationships, and life domains is evident in nearly every encyclopedia entry and underscores the key life course theme that lives are embedded in rich networks of roles and relationships.

Agency Individuals construct their own life course through their choices and actions, within the opportunities and constraints of historical and social circumstances. Thus far, we have highlighted the ways that forces external to the individual, like history, birth cohort, wars, and one's social relationships shape individual biographies. Yet personal agency—or one's choices, aspirations, ambitions, and attitudes—also shape the life course. Whereas classic sociological research and theory has traditionally emphasized the ways that social structures constrain (or benefit) individuals, the life course paradigm views human behavior as a reflection of *both personal agency and structural constraint*. Individuals select social roles and opportunities that are consistent with their own personal preferences, traits, resources, and even genetic predispositions (e.g., Scarr & McCartney 1983)—yet freedom of choice is not distributed evenly throughout the population. Persons with fewer economic resources have fewer opportunities to seek out and pursue desirable options, while characteristics such as age, race, gender, physical ability status, sexual orientation and religion may create obstacles for some individuals—at least at certain points in history.

A compelling example of the ways that agency and structure influence life course trajectories can be found in John Clausen's (1993) book *American Lives*. Clausen tracked a cohort of men and women who were born in the early 20th century, and followed their lives for more than 60 years. He found that an important cluster of traits— "planful competence" — was a powerful predictor of successful careers and marriages, and good health more than five decades after the adolescents had graduated high school. "Planful competence," according to Clausen, encompasses self-confidence, intellectual investment, and dependability. These attributes, in turn, were associated with better academic performance in school, well-thought out plans for post-secondary schooling, and focus when selecting one's career. Clearly, planful competence encompasses one's own ambition, aspirations for the future, and conscientiousness in pursuing one's goals. At the same time, however, children from more advantaged social and economic backgrounds were more likely than their less well-off peers to enjoy high levels of competence.

Many encyclopedia entries offer compelling descriptions of the ways that psychological traits, like age identity, personality, self-esteem, and even body image, shape one's future orientations and accomplishments, yet also reflect the resources and relationships one has enjoyed in the past. Psychological factors like agency do not only exert *direct* effects on life chances, they also may *buffer against* the negative effects of early adversity, or exacerbate the difficulties that accompany critical life stressors. The interplay between the social and the psychological, micro- and macro-level phenomena, and biography and history, make up the foundation of life course research and scholarship. The entries to follow reveal the vitality and nuance that characterizes this field of study today.

BIBLIOGRAPHY

Baumrind, D. (1991.) The influence of parenting style on adolescent competence and substance use. *Journal of Early Adolescence* 11, 56–95.

Carr, D. (2004). "My daughter has a career—I just raised babies": Women's intergenerational social comparisons. *Social Psychology Quarterly* 67, 132–54.

Clausen, J. (1993). *American lives: Looking back at the children of the Great Depression.* New York: Free Press.

Coleman, J. S. (1961). *Adolescent society: The social life of teenagers and its impact on education.* New York: Free Press.

Coleman, J. S. (1990). *Foundations of social theory.* Cambridge, MA: Harvard University Press.

Durkheim, É. (1997). *Suicide.* New York: The Free Press. (Original work published 1897).

Elder, G.H., Jr. (1974). *Children of the Great Depression: Social change in life experience.* Chicago: University of Chicago Press.

Elder, G. H., Jr. (1994). Time, human agency, and social change: Perspectives on the life course. *Social Psychology Quarterly, 57,* 4–15.

Furstenberg, F. F., Jr., Brooks-Gunn, J., & Morgan, S.P. (1987). *Adolescent mothers in later life.* New York: Cambridge University Press.

Inkeles, A., & D. Levinson. (1969). National character: the study of modal personality and sociocultural systems. In G. Lindzey & E. Aronson (Eds.), *The Handbook of Social Psychology,* 2nd edition, Volume 4 (pp. 418–507). Reading, MA: Addison-Wesley.

Laub, J. H., & R. J. Sampson. (2006). *Shared beginnings, divergent lives: delinquent boys to age 70.* Cambridge, MA: Harvard University Press.

MacLean, A., & G. H. Elder, Jr. (2007). Military service in the life course. *Annual Review of. Sociology, 33,* 175-96.

Mannheim, K. (1928/1952). "The problem of generations." In K Mannheim, *Essays on the Sociology of Knowledge* (pp. 276-322). London: Routledge & Keagan Paul.

Mills, C. W. (1959). *The sociological imagination.* New York: Oxford University Press.

Mortimer, J. T., & M. J. Shanahan (Eds). (2003). *Handbook of the life course.* New York: Kluwer Academic/ Plenum Publishers.

Neugarten, B. L., M. W. Moore, & J. C. Lowe. (1965). Age norms, age constraints, and adult socialization. *American Journal of Sociology, 70,* 710–17.

Neugarten, B. L., & N. Datan.(1973). Sociological perspectives on the life cycle. In P. B. Baltes, & K. W. Schaie (Eds.), *Life-Span developmental psychology: personality and socialization* (pp. 53–69). New York: Academic Press.

Riley, M. W., M. E. Johnson, & A. Foner (Eds). 1972. *Aging and society: a sociology of age stratification* (Volume 3). New York: Russell Sage.

Scarr, S., & K. McCartney. (1983). How people make their own environments: a theory of genotype-environment effects. *Child Development, 54,* 424–35.

Settersten, R. A., Jr. (1999). *Lives in time and place: the problems and promises of developmental science.* Amityville, NY: Baywood Publishing Co.

Settersten, R. A., Jr., (Ed.) 2003. *Invitation to the life course: toward new understandings of later life.* Amityville, NY: Baywood Publishing.

Sewell, W. H., & R.M. Hauser. (19750. *Education, occupation, and earnings: achievements in the early career.* New York: Academic Press.

Shanahan, M. J., & R. Macmillan. (2007). *Biography and the sociological imagination: contexts and contingencies.* New York: W. W. Norton.

Shaw, C. R., & H. McKay. (1942). *Juvenile delinquency and urban areas.* Chicago: University of Chicago Press.

Thomas, W. I., & F. Znaniecki. (1974/1918–1920). *The Polish peasant in Europe and America (Volumes 1 & 2).* Urbana, Illinois: University of Illinois Press.

Volkart, E. H. (1951). *Social behavior and personality: contributions of W. I. Thomas to theory and social research.* New York: Social Science Research Council.

Deborah Carr
Editor in Chief

Introduction to Volume 1, Childhood and Adolescence

Although childhood and adolescence are familiar concepts, they are deceptively ambiguous. Each of these two stages of the life course can be internally divided into qualitatively different substages. Moreover, the boundaries of each stage are notoriously fuzzy. Two and half centuries ago, for example, childhood was a very short period of life, and the term "adolescence" was not even coined until the turn of the 20th century (by G. Stanley Hall in 1904). Today, the outer edge of adolescence is drifting upward, as powerful macro-level forces (e.g., economic globalization, declining mortality rates) have added a period of "adultolescence" to what has traditionally been thought to constitute the early life course (Settersten, Furstenberg, & Rumbaut 2005).

When designing this encyclopedia, the other editors and I were occasionally confused about whether many entries should go into Volume 1 or 2 and over whether other entries were primarily about children or their parents. For the sake of clarity, it was decided that childhood and adolescence involve the portions of the life course between conception and the early 20s and the experiences that young people have during this time as well as what happens to or directly affects them. This is about as precise as one can be with life stages that tend to be a little slippery across time and place. Whatever the exact definitions are, the importance of this volume lies in the *foundational* role of childhood and adolescence in the life course. By foundational, I mean that what happens during childhood and adolescence sets the stage for adulthood and beyond—when positive and negative trajectories begin that carry individuals through life, when experiences happen that can be built on or torn down by later experiences. As such, the entries included in Volumes 2 and 3 are prefaced by the entries in this volume.

To illustrate this foundational role of childhood and adolescence, consider work from two leading social scientists who study the early life course. First, the Nobel Prize-winning economist James Heckman has advocated an early action strategy for public policy by documenting how public investments in pre-school have bigger long-term returns than investments in later stages of schooling precisely because young children are more malleable and because initial experiences serve as building blocks for the rest of life. This idea has become a guiding force in policy intervention, such as universal pre-Kindergarten programs in several states that guarantee publicly funded pre-school education for all children as a means of reducing demographic disparities in school readiness (see the child care/early education entry by Gordon in this volume). Second, Glen H. Elder, Jr.—perhaps the most important architect of the life course perspective—began his career tracing the life trajectories of children growing up in historical eras of economic hardship, such as the Great Depression of the 1930s. The early experiences of these children left an imprint on their lives. Some never overcame their initial hardship. Most, however, were resilient because later experiences (in school, in the armed forces) provided turning points that counterbalanced these early hardships.

The pioneering work of Heckman, Elder, and others demonstrates the need to view what happens in childhood and adolescence as part of a larger sequence of inter-related events, experiences, and roles, as links in the same chain. Given that this volume mixes entries on macro-level and micro-level phenomena, the life course approach to understanding childhood and adolescence can be seen on the macro-level of the population (e.g., the evolving society) and on the micro-level of the individual (e.g., the developing person).

CHILDHOOD AND ADOLESCENCE IN POPULATION PERSPECTIVE

The size of the United States school population is at an all-time high (U.S. Census 2008). This milestone is but one manifestation of the incredible growth of the child/adolescent population in the United States and in many other parts of the world, which is just one trend driving home the point that the size and composition of the youth cohort can change dramatically over time with great consequences for society at large.

This growth in the youth population of the United States is a result of many things. Most importantly, it represents a natural progression of the Baby Boom generation—a succession of large birth cohorts from 1945–1965—entering their prime childbearing years. What has also had an impact is the uptick in immigration after the reform of U.S. immigration laws in the 1960s (see "Immigration, Childhood and Adolescence" by Galindo & Durham). By lifting national origin quotas for immigrants, these laws set into motion another population trend pertinent to understanding childhood and adolescence today. Specifically, the growing predominance of Latin American and Asian countries in the "sending" stream, coupled with differential fertility rates across race/ethnic groups, has fueled the increasing heterogeneity of American youth. When I was born in the early 1970s, for example, 4 of 5 youth in the United States were non-Hispanic white. Now, just over half are.

Two other population trends also warrant discussion. The first is that, because of dramatic increases in divorce, cohabitation, and non-marital fertility throughout the late 20th century, only a minority of American youth grow up in homes with continuously married parents, the family arrangement most probabilistically related to optimal child outcomes. This trend, in turn, varies considerably by race/ethnicity and social class (see "Family and Household Structure, Childhood and Adolescence" by Cavanagh). The second trend is that, although concepts of equal opportunity and social mobility are ingrained in American culture, socioeconomic inequality is at its highest point in modern history. For the most part, this stratification has occurred because gains in income and wealth have primarily been concentrated in the upper tail of the socioeconomic distribution. Since the oldest of today's youth cohort was being born in the 1980s, the fortunes of the upper and middle class have diverged while those of the middle and lower class have converged (Fischer and Hout 2006).

These population trends have varied in other parts of the world. Europe and Japan have low fertility rates, which means smaller youth cohorts, but this is not the case in Africa or Latin America. As a historically diverse society and major immigration destination, the United States has a more heterogeneous youth population than many nations. Moreover, because the United States lacks a large social safety net relative to other Western countries, inequalities here may be more problematic than in other countries. Still, although it is not always representative, the United States provides a clear case of how the demography of the youth population can be shape the future of a society.

To see how demographic trends link the present to the future, consider crime, education, and marriage/fertility. First, the years when one is most likely to commit crime are in late adolescence and young adulthood (see "Crime, Criminal Activity in Childhood and Adolescence" by Macmillan). As a result, societies often experience crime surges when the number of adolescents and young adults is large, a pattern suggesting that an upswing in crime is to be expected. Second, the large number of students in the education system means bigger schools and greater competition for courses, grades, and college slots. This trend is likely to fuel socioeconomic disparities in educational attainment at a time when economic restructuring is prioritizing technological innovation and, in turn, penalizing societies for not pushing the best candidates into the right positions (Fischer and Hout 2006). Third, the increasing race/ethnic diversity of today's youth cohort means that they will experience a more diverse marriage market, with increased opportunities to meet and interact with members of other groups meaning that interracial marriage rates could be headed upward (see "Transition to Marriage" by Sweeney). From a policy perspective, these examples of how the demography of today's child/adolescent population is linked to the future of American society suggests a need to foresee consequences of demographic change and craft policies accordingly, such as building schools, hiring teachers, and increasing college capacity. This strategy seems self-evident, but, as the current Social Security crisis illustrates, it is one not often employed.

CHILDHOOD AND ADOLESCENCE IN DEVELOPMENTAL PERSPECTIVE

Together, childhood and adolescence represent the period of most rapid change in the entire life course. Just think about how different a newborn is from a 20 year old—how they look, what they do, how they think and feel, what their worlds include. How this change unfolds—and on what schedule—results in a basic "package" of a person, in terms of appearance, personality, and skills that they will carry forward into the rest of life, that can be modified and altered in various ways but is rarely fundamentally reconstitued (Furstenberg 2000).

Consider physical, neurological, and psychosocial development. Over the course of childhood and adolescence, young people gain weight and height at often exponential rates, and their bodies gradually evolve in form to fit the functions they need to perform. This physical change is clearly encapsulated in puberty. Puberty is not a discrete event, but rather a transitional period that links childhood to adolescence. A *biological* process, it has *social* meaning, and is viewed as the dividing line between young and old (see "Puberty" by Cavanagh). Brain scanning has shed valuable light on neurological development, by, for example, mapping out the explosion of brain growth that occurs in early childhood and the ultimate pruning of neurological pathways as childhood progresses. Such scanning has also documented how the different rates of development in parts of the brain regulating social and emotional maturity from those regulating critical thinking and impuslivity over the course of adoelscence provide a biological explanation for the risk-taking and defiance that often occurs during the high school years (Steinberg 2008). Turning to other psychosocial trajectories, as young people move through the various substages of childhood and adolescence, they gradually gain the ability to take others' perspectives, self reflect, and and connect action to consequence. At the same time, their relationship systems expand from dyadic interactions to diffuse interpersonal networks, and they try on multiple identities before gradually integrating the different pieces of the self (Hartup and Stevens 1997). In turn, all of this physical, neurological, and psychosocial development occurs as young people move through a series of institutional structures, including early child care, school, and the health care system, in which the take on new roles, gain credentials, and learn about their place in the larger society.

To borrow images from the life course perspective, then, the life course is a tapestry of developmental trajectories, convoys of social relations, and institutional pathways. We can see why the structure and timing of this tapestry in childhood and adolescence is consequential for the future by considering crime, education, and marriage/fertility. First, the increase in impulsivity and susceptibility to peer influence that occurs in adolescence, which comes about because of both neurological and psychosocial changes, explains why rates of crime and sexual risk-taking rise during this period and fade thereafter. Because of the cumulative nature of the educational system, however, this rite of passage can have long-term consequences even after this period is over—when psychosocial maturity has caught up to physical maturity, and individuation from parents is complete. These long-term consequences can occur if what happens during adolescence throws off trajectories through school in some way that cannot be easily undone (see "Academic Achievement" by Grodsky). As examples, getting arrested can foreclose many future work and schooling opportunities, substance use can lower chances of going to college at a time when a college education is especially important to one's future, and an unplanned pregnancy can lead to truncated educational attainment and early marriage or cohabitation. Because adolescence is not isolated from the rest of life, the consequences of what happens during adolescence are not necessarily confined to those years. The same goes for childhood.

From a policy perspective, the intertwined nature of developmental trajectories, social convoys, and institutional pathways in childhood and adolescence means that efforts to target any one must be grounded in a holistic view of the person and with an eye towards long-term consequences. As with the discussion of the population level, this approach seems self-evident. Consider Eccles' work on the disconnect between the structure of middle schools in the United States and the developmental needs of early adolescence and the implications of this disconnect for healthy development and educational attainment long after the end of middle school. A reading of this work reveals how an understanding of the developmental processese of childhood and adolescence is not incorporated into the structure and organization of institutions serving developing children and adolescents.

PUTTING THE PIECES TOGETHER

The role of childhood and adolescence in the life course on the population level is inextricably tied to its same role on the developmental level. Popualtions, after all, are aggregates of developing individuals, and what goes on in a population is a setting of individual development. This interplay between the macro- and micro-levels is what the concept of the life course is all about—the intersection of societal history and personal biography (Mills 1959).

The biography metaphor is a good one for understanding childhood and adolescence. If the life course is a biography on a book store shelf, childhood and adolescence are the opening chapters. By the end of the book, the memory of those early chapters may have faded, but the reader cannot understand how the book ends if he or she skips Chapters 1 and 2.

BIBLIOGRAPHY

Eccles, J. S., Midgely, C., Wigfield, A., Buchanan, C.M., Reuman, D., & MacIver, D. (1993). "Development during adolescence: The impact of stage-environment fit on young adolescents' experiences in schools and in families." *American Psychologist* 48: 90–101.

Fischer, C. S., & Hout, M. (2006). *Century of Difference: How America Changed in the Last One Hundred Years*. New York: Russell Sage.

Furstenberg, F. F. (2000). "The sociology of adolescence and youth in the 1990s: A critical commentary." *Journal of Marriage and Family* 62:896–910.

Hall, G. S. (1904). *Adolescence: Its Psychology and Its Relations to Physiology, Anthropology, Sociology, Sex, Crime, Religion, and Education*. New York, Appleton.

Hartup, W., & Stevens, N. (1997). "Friendships and adaptation in the life course." *Psychological Bulletin* 121: 355–370.

Heckman, James. (2006). "Skill formation and the economics of investing in disadvantaged children." *Science* 312(5782): 1900–1902.

Mills, C. Wright. (1959). *The Sociological Imagination*. New York: Oxford University Press

Settersten, R. A., Furstenberg, F. F., Jr., & Rubén, G. R. (2005). *On the Frontier of Adulthood: Theory, Research, and Public Policy*. Chicago: University of Chicago Press.

Steinberg, L. (2008). "A social neuroscience perspetive on adolescent risk-taking." *Developmental Review* 28: 78–106.

U.S. Census Bureau, Housing and Household Economic Statistics Division, Education & Social Stratification Branch. (2008). *Historical Table A-1. School Enrollment of the Population Three Years Old and Over, by Level and Control of School, Race, and Hispanic Origin: October 1955 to 2005*. Retrieved on May 2, 2008, from http://www.census.gov/population/www/socdemo/school.html

Robert Crosnoe
Associate Editor

List of Articles

Contributors

Caitlin Abar
Human Development and Family
Studies
Pennsylvania State University
DRINKING, ADOLESCENT

jimi adams
Robert Wood Johnson Health and
Society Scholars Program
Columbia University
SOCIAL NETWORKS

Jennifer A. Ailshire
Doctoral Candidate
Department of Sociology
University of Michigan
EPIDEMIOLOGIC TRANSITION

James W. Ainsworth
Associate Professor
Department of Sociology
Georgia State University
VOCATIONAL TRAINING AND
EDUCATION

Karen Albright
Center for Culture and Health,
Semel Institute for Neuroscience and
Human Behavior
University of California, Los Angeles
SOCIAL CLASS

Katherine Allen
Professor
Department of Human
Development
Virginia Tech
SEXUAL ACTIVITY, ADULTHOOD

Karen Anderson
Department of Communication
Studies
University of North Texas
CULTURAL IMAGES, LATER LIFE
CULTURAL VALUES AND CHINESE
MEDIA

Georgia Anetzberger
Doctor
Health Care Administration
Program
Cleveland State University
ELDER ABUSE AND NEGLECT

William S. Aquilino
Professor
University of Wisconsin-Madison
NONCUSTODIAL PARENTS

Richard Arum
Professor of Sociology and Education
New York University
STAGES OF SCHOOLING: HIGH
SCHOOL

Myra A. Aud
Associate Professor
Sinclair School of Nursing
University of Missouri
AGING IN PLACE

Tamara A. Baker
Assistant Professor
School of Aging Studies
University of South Florida
ARTHRITIS

Karlene Ball
Doctor
Department of Psychology
University of Alabama at
Birmingham
OLDER DRIVERS

Albert Bandura
David Starr Jordan Professor of
Social Science in Psycholog
Stanford University
AGENCY

Jacqueline M. Baron
Department of Psychology
University of Florida
WISDOM

**Jenni M. Bernadette
Bartholomew**
Vice President of Community
Impact
United Way of Central
New York
Syracuse, New York
SOCIAL SECURITY

Jason Beckfield
Assistant Professor
Department of Sociology
Harvard University
INCOME INEQUALITY

Victoria Hilkevitch Bedford
Professor
School of Psychological Sciences

University of Indianapolis
SIBLING RELATIONSHIPS,
ADULTHOOD

Emily Beller
Independent Scholar
SOCIAL MOBILITY

Vern Bengtson
Research Professor of Gerontology
Andrus Gerontology Center
University of Southern California
THEORIES OF AGING

Kate M. Bennett
Doctor
School of Psychology
University of Liverpool, UK
WIDOWHOOD

Lars R. Bergman
Professor
Department of Psychology
Stockholm University
PERSON-ORIENTED APPROACHES

Amy L. Best
Associate Professor
Department of Sociology and
Anthropology
George Mason University
YOUTH CULTURE

Julia D. Betensky
Doctoral Candidate
Department of Psychology
Rutgers, The State University of New
Jersey
CARDIOVASCULAR DISEASE

David S. Bickham
Doctor
Center on Media and Child Health
Children's Hospital Boston
MEDIA AND TECHNOLOGY USE,
CHILDHOOD AND ADOLESCENCE

Kira Birditt
Assistant Professor
Institute for Social Research
University of Michigan
SOCIAL SUPPORT, LATER LIFE

Rosemary Blieszner
Alumni Distinguished Professor
Department of Human
Development
Virginia Polytechnic Institute and
State University
FRIENDSHIP, ADULTHOOD

Hans-Peter Blossfeld
Doctor
Department of Social and Economic
Sciences
Bamberg University, Germany
GLOBALIZATION

Susan Bluck
Associate Professor
Department of Psychology
University of Florida
WISDOM

Linda M. Blum
Associate Professor
Sociology and Women's Studies
University of New Hampshire
BREASTFEEDING

Jason D. Boardman
Associate Professor of Sociolog
University of Colorado, Boulder
GENETIC INFLUENCES, EARLY LIFE
NATURE VS. NURTURE DEBATE

Kathrin Boerner
Senior Research Scientist
Research Institute on Aging
Jewish Home Lifecare, New York
BALTES, MARGRET AND PAUL

Darya D. Bonds
Prevention Research Center
Arizona State University
RESILIENCE, CHILDHOOD AND
ADOLESCENCE

Axel Börsch-Supan
Doctor
Mannheim Research Institute for the
Economics of Aging
DATA SOURCES, LATER LIFE: SURVEY
OF HEALTH AGEING AND
RETIREMENT IN EUROPE (SHARE)

Michelle J. Boyd
Doctoral Student
Eliot-Pearson Department of Child
Development
Tufts University
DEVELOPMENTAL SYSTEMS THEORY

Robert Bozick
Research Scientist
RTI International
COLLEGE ENROLLMENT
JUNIOR/COMMUNITY COLLEGE

Peter D. Brandon
Broom Professor of Social
Demography

Department of Sociology and
Anthropology
Carleton College
POVERTY, CHILDHOOD AND
ADOLESCENCE

R. Mara Brendgen
Associate Professor
Département de Psychologie
Université du Québec à Montréal
AGGRESSION, CHILDHOOD AND
ADOLESCENCE

Gilbert Brenes-Camacho
Researcher in Demography
Central American Center for
Population
University of Costa Rica
DATA SOURCES, LATER LIFE:
LONGITUDINAL STUDY OF AGING
(LSOA)

Inge Bretherton
Professor Emerita
Department of Human
Development and Family Studies
University of Wisconsin-Madison
BOWLBY, JOHN

Diane R. Brown
Institute for Elimination of Health
Disparities
University of Medicine and
Dentistry of New Jersey
HEALTH DIFFERENTIALS/
DISPARITIES, LATER LIFE

Ivana Brown
Doctoral Candidate
Department of Sociology
Rutgers University
MOMMY WARS/IMAGES OF
MOTHERHOOD
MOTHERHOOD

Tony N. Brown
Associate Professor of Sociology
Vanderbilt University
SOCIALIZATION, RACE

Christopher R. Browning
Associate Professor
Department of Sociology
Ohio State University
NEIGHBORHOOD CONTEXT,
ADULTHOOD
NEIGHBORHOOD CONTEXT,
CHILDHOOD AND ADOLESCENCE

Kristen Bub
CHILD CARE AND EARLY EDUCATION

Mara Buchbinder
Doctor
Department of Anthropology
University of California, Los Angeles
ILLNESS AND DISEASE, CHILDHOOD
AND ADOLESCENCE

Michelle J. Budig
Associate Professor
Department of Sociology
University of Massachusetts
SELF-EMPLOYMENT

Regina M. Bures
Department of Sociology
University of Florida
AGE SEGREGATION

Sarah Burgard
Assistant Professor
Department of Sociology
University of Michigan
JOB CHARACTERISTICS AND JOB STRESS

Jeffrey A. Burr
Professor
Department of Gerontology
University of Massachusetts Boston
PRODUCTIVE AGING
VOLUNTEERING, LATER LIFE

Lori A. Burrington
Assistant Professor
Department of Sociology and Crime,
Law, and Justice
Pennsylvania State University
NEIGHBORHOOD CONTEXT,
ADULTHOOD
NEIGHBORHOOD CONTEXT,
CHILDHOOD AND ADOLESCENCE

Allison M. Burton
Post-Doctoral Research Fellow
The University of Texas MD
Anderson Cancer Center
HOSPICE AND PALLIATIVE CARE

Kay Bussey
Department of Psychology
Macquarie University
BANDURA, ALBERT

Rebecca Callahan
Doctor
Department of Language and
Literacy Education
University of Georgia
BILINGUAL EDUCATION

Carole A. Campbell
Doctor
Department of Sociology

California State University, Long
Beach
AIDS

Nichole M. Campbell
Design Studies Department
University of Wisconsin-Madison
RETIREMENT COMMUNITIES

Andrew Caparaso
Department of Sociology
University of Maryland
TIME USE, LATER LIFE

William Carbonaro
Professor
Department of Sociology
University of Notre Dame
INTERGENERATIONAL CLOSURE

Deborah Carr
Associate Professor
Department of Sociology and
Institute for Health, Health Care and
Aging Research
Rutgers University
DEATH AND DYING
DURKHEIM, ÉMILE
RESEARCH METHODS:
INTRODUCTION AND OVERVIEW
RESEARCH METHODS:
QUANTITATIVE ANALYSIS

Denise C. Carty
Doctoral Candidate
University of Michigan School of
Public Health
RACISM/RACE DISCRIMINATION

Thomas F. Cash
Emeritus Professor of Psychology
Old Dominion University
BODY IMAGE, ADULTHOOD

George Cavalletto
Adjunct Professor of Sociology
Hunter and Brooklyn Colleges, City
University of New York
FREUD, SIGMUND

Shannon E. Cavanagh
Assistant Professor
Department of Sociology and
Population Research Center
University of Texas at Austin
AGE NORMS
FAMILY AND HOUSEHOLD
STRUCTURE, CHILDHOOD AND
ADOLESCENCE
PUBERTY

Feinian Chen
Associate Professor
Department of Sociology and
Anthropology
North Carolina State University
RESEARCH METHODS, CONCEPTS:
CORRELATION VS. CAUSATION
RESEARCH METHODS, CONCEPTS:
VARIABLES

Zeng-yin Chen
Associate Professor
Department of Sociology
California State University, San
Bernardino
PARENT-CHILD RELATIONSHIPS,
CHILDHOOD AND ADOLESCENCE
PARENTING STYLE

Noelle Chesley
Doctor
Department of Sociology
University of Wisconsin-Milwaukee
MOEN, PHYLLIS

Juanita J. Chinn
University of Texas at Austin
HEALTH DIFFERENTIALS/
DISPARITIES, ADULTHOOD

Victor Cicirelli
Professor of Developmental and
Aging Psychology
Department of Psychological
Sciences
Purdue University
SIBLING RELATIONSHIPS, LATER LIFE

Verena R. Cimarolli
Lighthouse International
Arlene R. Gordon Research Institute
SENSORY IMPAIRMENTS

Andreana Clay
Doctor
Department of Sociology
San Francisco State University
CIVIC ENGAGEMENT, CHILDHOOD
AND ADOLESCENCE

Tim Clydesdale
Professor of Sociology
The College of New Jersey
RELIGION AND SPIRITUALITY,
CHILDHOOD AND ADOLESCENCE

Gene D. Cohen
Director
Center on Aging, Health and
Humanities
The George Washington University
CREATIVITY, LATER LIFE
PRAGMATIC CREATIVITY AND AGING

Bertram J. Cohler
William Rainey Harper Professor
The University of Chicago
ERIKSON, ERIK
GAYS AND LESBIANS, ADULTHOOD
SAME-SEX MARRIAGE

Ed Collom
Associate Professor of Sociology
University of Southern Maine
HOME SCHOOLING

Soria Colomer
Department of Language and
Literacy Education
University of Georgia
BILINGUAL EDUCATION

Carol McDonald Connor
Associate Professor of Education
Florida State University and the
Florida Center for Reading Research
STAGES OF SCHOOLING:
ELEMENTARY SCHOOL

Richard J. Contrada
Professor
Department of Psychology
Rutgers, The State University of New
Jersey
CARDIOVASCULAR DISEASE

Daniel Thomas Cook
Associate Professor
Department of Childhood Studies
Rutgers University-Camden
CONSUMPTION, CHILDHOOD AND
ADOLESCENCE

Elizabeth C. Cooksey
Department of Sociology
Ohio State University
SEXUAL ACTIVITY, ADOLESCENT

Carey E. Cooper
Post-doctoral Research Associate
Center for Research on Child
Wellbeing
Princeton University
FAMILY PROCESS MODEL

Jennifer Cornman
Professor
School of Public Health
UMDNJ
ASSISTIVE TECHNOLOGIES

William A. Corsaro
Professor
Department of Sociology
Indiana University, Bloomington
INTERPRETIVE THEORY
SOCIALIZATION

James E. Côté
Professor
Department of Sociology
University of Western Ontario,
Canada
CULTURAL IMAGES, ADULTHOOD

Donald Cox
Professor
Boston College
INTERGENERATIONAL TRANSFERS

Eileen M. Crimmins
Davis School of Gerontology
University of Southern California
ACTIVE LIFE EXPECTANCY
AGING
LIFE TABLE
THIRD AGE

Sarah R. Crissey
Population Research Center
University of Texas at Austin
DATING AND ROMANTIC
RELATIONSHIPS, CHILDHOOD
AND ADOLESCENCE

Robert Crosnoe
Associate Professor of Sociology
University of Texas at Austin
DATA SOURCES, CHILDHOOD AND
ADOLESCENCE: BALTIMORE STUDY
DATA SOURCES, CHILDHOOD AND
ADOLESCENCE: BENNINGTON
WOMEN'S STUDY
DATA SOURCES, CHILDHOOD AND
ADOLESCENCE: BERKELEY
GUIDANCE STUDY AND THE
OAKLAND GROWTH STUDY
DATA SOURCES, CHILDHOOD AND
ADOLESCENCE: EARLY
CHILDHOOD LONGITUDINAL
STUDIES
DATA SOURCES, CHILDHOOD AND
ADOLESCENCE: GENERAL ISSUES
DATA SOURCES, CHILDHOOD AND
ADOLESCENCE: NATIONAL
EDUCATIONAL LONGITUDINAL
STUDY
DATA SOURCES, CHILDHOOD AND
ADOLESCENCE: NATIONAL
LONGITUDINAL STUDY OF
ADOLESCENTDATA SOURCES,
CHILDHOOD AND ADOLESCENCE:
NATIONAL LONGITUDINAL
SURVEY OF YOUTH
DATA SOURCES, CHILDHOOD AND
ADOLESCENCE: STANFORD-
TERMAN STUDY
ELDER, GLEN H., JR.
SCHOOL CULTURE
STAGES OF SCHOOLING: OVERVIEW
AND INTRODUCTION

Mick Cunningham
Department of Sociology
Western Washington University
HOUSEWORK

Stephen J. Cutler
Professor
Department of Sociology
University of Vermont
MEDIA AND TECHNOLOGY USE,
LATER LIFE

K. Thomas D'Amuro
Department of Sociology
University at Buffalo, State
University of New York
POLICY, LATER LIFE WELL-BEING

Judith C. Daniluk
Professor
Department of Educational and
Counselling Psychology
University of British Columbia
INFERTILITY
WOMEN WAITING TOO LONG TO
BECOME MOTHERS

Jacinda K. Dariotis
Assistant Scientist, Faculty
Department of Population, Family,
and Reproductive Health
Johns Hopkins, Bloomberg School
of Public Health
CHILDLESSNESS

Rosalyn Benjamin Darling
Professor
Department of Sociology
Indiana University of Pennaylvania
DISABILITY, CHILDHOOD AND
ADOLESCENCE

Aniruddha Das
Department of Sociology
University of Chicago
SEXUAL ACTIVITY, LATER LIFE

Scott Davies
Professor
Department of Sociology
McMaster University
PRIVATE SCHOOLS

Kelly D. Davis
Department of Human
Development and Family Studies
Pennsylvania State University
GENDER IN THE WORKPLACE
OPTING OUT

Brian de Vries
Professor

Gerontology Program
San Francisco State University
GAYS AND LESBIANS, LATER LIFE

Gini Deibert
Doctor
Department of Sociology
Suffolk University
THEORIES OF DEVIANCE

Cindy Dell Clark
Associate Professor
Department of Human
Development and Family Studies
Pennsylvania State University,
Brandywine
ILLNESS AND DISEASE, CHILDHOOD
AND ADOLESCENCE

Lindsay Demers
Graduate Student
Department of Psychology
University of Massachusetts Amherst
MEDIA EFFECTS

Sarah Desai
Department of Sociology
University at Buffalo, State
University of New York
LONG-TERM CARE

Neha Deshpande-Kamat
Department of Human
Development and Family Studies
Iowa State University
COGNITIVE FUNCTIONING AND
DECLINE

Lane Destro
Graduate Student
Department of Sociology
Duke University
WEALTH

Norman Dolch
Professor Emeritus
Department of Social Science
Louisiana State University in
Shreveport
INSTITUTIONS, SOCIAL

Thurston Domina
Assistant Professor
Department of Education
University of California, Irvine
PARENTAL INVOLVEMENT IN
EDUCATION

Brent Donnellan
Department of Psychology
Michigan State University
PERSONALITY

Julia A. Rivera Drew
Department of Sociology
Brown University
HEALTH CARE USE, CHILDHOOD
AND ADOLESCENCE

Pamela Dubyak
University of Florida
SLEEP PATTERNS AND BEHAVIOR

Ruth E. Dunkle
School of Social Work
University of Michigan
OLDEST OLD

Rachel E. Durham
Assistant Research Scientist
Johns Hopkins University
IMMIGRATION, CHILDHOOD AND
ADOLESCENCE

Adam Matthew Easterbrook
Doctoral Candidate
Department of Sociology
University of British Columbia
HOME OWNERSHIP/HOUSING

Donna Eder
Professor
Department of Sociology
Indiana University
STAGES OF SCHOOLING: MIDDLE
SCHOOL

David J. Ekerdt
Professor of Sociology
Gerontology Center
University of Kansas
CONSUMPTION, ADULTHOOD AND
LATER LIFE

Amanda Floetke Elliott
Department of Ophthalmology
University of Alabama at
Birmingham
PAIN, ACUTE AND CHRONIC

Cynthia Fuchs Epstein
Distinguished Professor
Graduate Center, City University of
New York
ROSSI, ALICE

Lance D. Erickson
Department of Sociology
Brigham Young University
MENTORING

Mary Ann Erickson
Gerontology Institute
Ithaca College
ROLES
TIME BIND

Jennifer J. Esala
Doctor
University of New Hampshire
BREASTFEEDING

Jay Fagan
Professor
School of Social Administration
Temple University
FATHERHOOD
STAY AT HOME DADS

Robert Faris
Assistant Professor
Department of Sociology
University of California, Davis
BULLYING AND PEER
VICTIMIZATION

George Farkas
Professor
Department of Education
University of California, Irvine
HUMAN CAPITAL

Danielle Farrie
Professor
Department of Sociology
Temple University
FATHERHOOD
STAY AT HOME DADS

Erika Felts
Department of Sociology
University of California, Davis
ACADEMIC ACHIEVEMENT
GRADE RETENTION VS. SOCIAL
PROMOTION

Julie Fennell
Doctor
Brown University
BIRTH CONTROL

Rudy Fenwick
Associate Professor of Sociology
University of Akron
FLEXIBLE WORK ARRANGEMENTS

Jessica Fields
Associate Professor
Center for Research on Gender and
Sexuality
San Francisco State University
SEX EDUCATION/ABSTINENCE
EDUCATION
VIRGINITY PLEDGE

Nancy Foner
Distinguished Professor of Sociology
of Sociology

Hunter College and Graduate
Center, City University of New York
FONER, ANNE

Angela Fontes
University of Wisconsin
INVESTMENT
SAVING

Patricia A. Francis
Independent Scholar
MEAD, MARGARET

Erica Frankenberg
Doctor
The Civil Rights Project
University of California, Los Angeles
SEGREGATION, SCHOOL

Thomas N. Friemel
Assistant Professor
Institute of Mass Communication
and Media Research
University of Zurich, Switerland
MEDIA AND TECHNOLOGY USE,
ADULTHOOD
THE DIGITAL DIVIDE

W. Parker Frisbie
Professor of Sociology
Population Research Center
University of Texas at Austin
INFANT AND CHILD MORTALITY

Michelle L. Frisco
Department of Sociology and Crime,
Law and Justice
Pennsylvania State University
OBESITY, CHILDHOOD AND
ADOLESCENCE

Sylvia Fuller
Assistant Professor
Department of Sociology
University of British Columbia
JOB CHANGE

Elizabeth Gage
Department of Cancer Prevention
and Population Sciences
Roswell Park Cancer Institute
POLICY, HEALTH

Claudia Galindo
Postdoctoral Fellow
Johns Hopkins University
IMMIGRATION, CHILDHOOD AND
ADOLESCENCE

Leonid Gavrilov
Center on Aging at NORC
University of Chicago
GENETIC INFLUENCES, LATER LIFE

Natalia Gavrilova
Center on Aging at NORC
University of Chicago
GENETIC INFLUENCES, LATER LIFE

Sarah Gehlert
Doctor
Department of Comparative Human
Development
University of Chicago
CANCER, ADULTHOOD AND LATER
LIFE

Linda K. George
Professor of Sociology
Duke University
RELIGION AND SPIRITUALITY, LATER
LIFE
RILEY, MATILDA WHITE
SOCIAL SUPPORT, ADULTHOOD

Kerstin Gerst
Gerontology Institute and
Department
University of Massachusetts Boston
POVERTY, LATER LIFE

Heather J. Gibson
Department of Tourism, Recreation
and Sport Management
University of Florida
LEISURE AND TRAVEL, ADULTHOOD
LEISURE AND TRAVEL, LATER LIFE

Peggy C. Giordano
Doctor
Department of Sociology
Bowling Green State University
FRIENDSHIP, CHILDHOOD AND
ADOLESCENCE

Jennifer L. Glanville
Department of Sociology
University of Iowa
SOCIAL CAPITAL

Frances Goldscheider
Doctor
Department of Family Science
University of Maryland
FAMILY AND HOUSEHOLD
STRUCTURE, ADULTHOOD

Rachel A. Gordon
Associate Professor
Department of Sociology and
Institute of Government and Public
Affairs
University of Illinois at Chicago
POLICY, CHILD WELL-BEING

Kurt Gore
Doctor
The University of Texas at Austin
SOCIALIZATION, GENDER

Derek M. Griffith
Assistant Professor
Department of Health Behavior and
Health Education
University of Michigan School of
Public Health
RACISM/RACE DISCRIMINATION

Eric Grodsky
Assistant Professor of Sociology
University of Minnesota
ACADEMIC ACHIEVEMENT
GRADE RETENTION VS. SOCIAL
PROMOTION

Matthias Gross
Doctor
Helmholtz Centre for
Environmental Research - UFZ
Leipzig, Germany
SIMMEL, GEORG

Joseph G. Grzywacz
Associate Professor
Department of Family and
Community Medicine
Wake Forest University School of
Medicine
WORK-FAMILY CONFLICT

Diana Guelespe
Doctoral Candidate
Department of Sociology
Loyola University Chicago
LOPATA, HELENA

Michel Guillot
Associate Professor
Department of Sociology
University of Wisconsin-Madison
BIOLOGICAL LIMITS TO HUMAN LIFE
LIFE EXPECTANCY

Guang Guo
Carolina Population Center,
Carolina Center for Genome
Sciences
University of North Carolina Chapel
Hill
GENETIC INFLUENCES, ADULTHOOD

Anne E. Haas
Assistant Professor
Department of Sociology
Kent State University, Stark
ATTRACTIVENESS, PHYSICAL
COSMETIC SURGERY

Aaron T. Hagedorn
Doctor
Davis School of Gerontology
University of Southern California
ACTIVE LIFE EXPECTANCY
LIFE TABLE

John M. Hagedorn
Professor of Criminology, Law, and
Justice
University of Illinois-Chicago
GANGS

Elzbieta Halas
Professor
Institute of Sociology
University of Warsaw, Poland
ZNANIECKI, FLORIAN

Matthew Hall
Department of Sociology
Pennsylvania State University
RESIDENTIAL MOBILITY,
ADULTHOOD

Jenifer Hamil-Luker
Assistant Professor
Department of Sociology
University of North Carolina at
Greensboro
LIFELONG LEARNING

Madlene Hamilton
University of Texas at Austin
HIGH-STAKES TESTING

Rachel Hammel
Graduate assistant
Department of Sociology
Case Western Reserve University
STRESS IN LATER LIFE

Wen-Jui Han
Associate Professor
Columbia University School of
Social Work
MATERNAL EMPLOYMENT

Angel L. Harris
Assistant Professor of Sociology
Princeton University
OPPOSITIONAL CULTURE

Jake Harwood
Professor
Department of Communication
University of Arizona
CULTURAL IMAGES, LATER LIFE
CULTURAL VALUES AND CHINESE
MEDIA

Pamela Healy
Doctoral Student
Psychological and Brain Sciences
University of Louisville
DEMENTIAS

Gloria D. Heinemann
Independent Scholar
SHANAS, ETHEL

Kathleen Hentz
Professor
San Francisco State University
SEX EDUCATION/ABSTINENCE
EDUCATION
VIRGINITY PLEDGE

Melissa R. Herman
Department of Sociology
Dartmouth College
BIRACIAL YOUTH/MIXED RACE
YOUTH

Masa Higo
Department of Sociology
Boston College
RETIREMENT

Heather D. Hill
Assistant Professor
School of Social Service
Administration
University of Chicago
DATA SOURCES, ADULTHOOD:
CURRENT POPULATION STUDY
(CPS)

Steffen Hillmert
Professor
Department of Sociology
University of Tübingen, Germany
MAYER, KARL ULRICH

Christine L. Himes
Professor
Department of Sociology
Syracuse University
AGE STRUCTURE
FEMINIZATION OF OLD AGE
OBESITY, ADULTHOOD
POPULATION PYRAMID

Steven Hitlin
Assistant Professor
Department of Sociology
University of Iowa
CLAUSEN, JOHN
IDENTITY DEVELOPMENT

Melissa Hodges
Department of Sociology
University of Massachusetts Amherst
SELF-EMPLOYMENT

Randy Hodson
Professor of Sociology
Ohio State University
EMPLOYMENT, ADULTHOOD

Lynette Hoelter
Research Investigator
ICPSR
University of Michigan
DIVORCE AND SEPARATION

John P. Hoffmann
Department of Sociology
Brigham Young University
DRUG USE, ADOLESCENT

Jennifer Jellison Holme
Assistant Professor
Department of Educational
Administration
University of Texas at Austin
POLICY, EDUCATION

Rachel M. Holmes
Doctoral Candidate
Department of Psychology
University of Tennessee
DATING AND ROMANTIC
RELATIONSHIPS, ADULTHOOD

Anthony C. Holter
Alliance for Catholic Education
Leadership Program
University of Notre Dame
MORAL DEVELOPMENT AND
EDUCATION

Ann Horgas
Associate Professor
University of Florida
College of Nursing
PAIN, ACUTE AND CHRONIC

Edward M. Horowitz
Associate Professor
School of Communication
Cleveland State University
POLITICAL SOCIALIZATION

Erin McNamara Horvat
Associate Professor
Urban Education
Temple University
CULTURAL CAPITAL

Allan V. Horwitz
Professor of Sociology
Rutgers University
MENTAL HEALTH, ADULTHOOD

Stephanie Howells
Doctoral Candidate

Department of Sociology
McMaster University
PRIVATE SCHOOLS

Amy Hsin
Postdoctoral Fellow
University of Michigan
COGNITIVE ABILITY

Mary Elizabeth Hughes
Department of Population, Family
and Reproductive Health
Johns Hopkins, Bloomberg School
of Public Health
BABY BOOM COHORT

Robert A. Hummer
Professor and Chairperson
Department of Sociology
University of Texas at Austin
HEALTH DIFFERENTIALS/
DISPARITIES, ADULTHOOD
INFANT AND CHILD MORTALITY

Stephen Hunt
The Department of Sociology and
Criminology
University of the West of England,
Bristol, UK
SOCIAL STRUCTURE/SOCIAL SYSTEM

Kathryn Hynes
Assistant Professor
Department of Human
Development and Family Studies
Pennsylvania State University
GENDER IN THE WORKPLACE
OPTING OUT

Ellen L. Idler
Institute for Health, Health Care
Policy, and Aging Research, and
Department of Sociology
Rutgers University
SELF-RATED HEALTH

Michelle Inderbitzin
Associate Professor of Sociology
Oregon State University
JUVENILE JUSTICE SYSTEM

Margot I. Jackson
Office of Population Research and
Center for Health and Well-Being,
Princeton University
Department of Sociology, Brown
University
RESIDENTIAL MOBILITY, YOUTH

Thomas B. Jankowski
Associate Director
Institute of Gerontology

Wayne State University
AMERICAN ASSOCIATION OF
RETIRED PERSONS (AARP)
POLITICAL BEHAVIOR AND
ORIENTATIONS, LATER LIFE

James M. Jasper
Professor of Sociology
Graduate Center, City University of
New York
SOCIAL MOVEMENTS

Krista Jenkins
Assistant Professor of Political
Science
Department of Social Sciences and
History
Fairleigh Dickinson University
POLITICAL BEHAVIOR AND
ORIENTATIONS, ADULTHOOD

Kristi Rahrig Jenkins
Research Area Specialist
Addiction Research Center and
Research Investigator, Institute for
Social Research
University of Michigan
HEALTH BEHAVIORS, ADULTHOOD
HEALTH BEHAVIORS, LATER LIFE

Shane R. Jimerson
Doctor
University of California, Santa
Barbara
HIGH SCHOOL DROPOUT

Richard W. Johnson
Principal Research Associate
Urban Institute
PENSION BENEFIT GUARANTY
CORPORATION (PBGC)
PENSIONS

Daniela Jopp
Doctor
Institute of Gerontology
University of Heidelberg, Germany
BALTES, MARGRET AND PAUL

Kara Joyner
Associate Professor of Sociology
Bowling Green State University
TRANSITION TO PARENTHOOD

Boaz Kahana
Professor of Psychology
Cleveland State University
MENTAL HEALTH, LATER LIFE
STRESS IN LATER LIFE

Eva Kahana
Department of Sociology

Case Western Reserve University
MENTAL HEALTH, LATER LIFE
STRESS IN LATER LIFE

Howard B. Kaplan
Distinguished Professor
Department of Sociology
Texas A&M University
SELF-ESTEEM

Lisa A. Keister
Professor
Department of Sociology
Duke University
WEALTH

Christopher P. Kelley
Doctoral Candidate
Department of Sociology
University of Iowa
CLAUSEN, JOHN

Jessica Kelley-Moore
Associate Professor
Department of Sociology
Case Western Reserve University
CHRONIC ILLNESS, ADULTHOOD
AND LATER LIFE

Erin Kelly
Department of Sociology and
Minnesota Population Center
University of Minnesota
FAMILY AND MEDICAL LEAVE ACT
POLICY, FAMILY

M. Alexis Kennedy
Department of Criminal Justice
University of Nevada, Las Vegas
CHILD ABUSE

Corey L.M. Keyes
Department of Sociology
Emory University
COMMUNITARIANISM

Eric R. Kingson
Professor of Social Work and Public
Administration
School of Social Work
College of Human Ecology, Syracuse
University
SOCIAL SECURITY

James B. Kirby
Senior Social Scientist
Agency for Healthcare Research and
Quality
COMPLEMENTARY AND
ALTERNATIVE MEDICINE
HEALTH CARE USE, ADULTHOOD

Monica Kirkpatrick Johnson
Associate Professor of Sociology
Washington State University
EMPLOYMENT, YOUTH

Richard Kitchener
Doctor
Department of Philosophy
Colorado State University
PIAGET, JEAN

Pamela Koch
Department of Sociology
University of South Carolina
BIRTH ORDER
SIBLING RELATIONSHIPS,
CHILDHOOD AND
ADOLESCENCE

Emily Koert
Department of Educational and
Counselling Psychology
University of British Columbia
INFERTILITY
WOMEN WAITING TOO LONG TO
BECOME MOTHERS

Tanya Koropeckyj-Cox
Associate Professor
Department of Sociology
University of Florida
LONELINESS, LATER LIFE
SINGLEHOOD

Jennie Jacobs Kronenfeld
Medical Sociologist
Arizona State Univeristy
HEALTH DIFFERENTIALS/DISPARITIES,
CHILDHOOD AND ADOLESCENCE

Nancy Kutner
Professor
Departments of Rehabilitation
Medicine and Sociology
Emory University
CENTENARIANS

Annette M. La Greca
Distinguished Professor
Psychology and Pediatrics
University of Miami
SOCIAL DEVELOPMENT

Donna A. Lancianese
Department of Sociology
University of Iowa
IDENTITY DEVELOPMENT

Abby Larson
Department of Sociology
New York University
STAGES OF SCHOOLING: HIGH
SCHOOL

Edward O. Laumann
Department of Sociology
University of Chicago
SEXUAL ACTIVITY, LATER LIFE

Nathanael Lauster
Assistant Professor
Department of Sociology
University of British Columbia
HOME OWNERSHIP/HOUSING

Felicia LeClere
Institute for Social Research
University of Michigan
DATA SOURCES, ADULTHOOD:
GENERAL ISSUES
NEIGHBORHOOD CONTEXT, LATER
LIFE

Barrett A. Lee
Department of Sociology and
Population Research Institute
Pennsylvania State University
RESIDENTIAL MOBILITY,
ADULTHOOD

Jennifer Catherine Lee
Assistant Professor
Department of Sociology
Indiana University-Bloomington
SOCIOECONOMIC INEQUALITY IN
EDUCATION

Mal Leicester
Emeritis Professor
Department of Education
University of Nottingham, UK
CONTINUING EDUCATION

Erica Leifheit-Limson
Doctoral Candidate
Department of Epidemiology and
Public Health
Yale University
AGEISM/AGE DISCRIMINATION

Patricia Lengermann
Department of Sociology
The George Washington
University
SOCIOLOGICAL THEORIES

Janel M. Leone
Department of Child and Family
Studies
Syracuse University
DOMESTIC VIOLENCE

Richard Lerner
Bergstrom Chair in Applied
Developmental Science

Institute for Applied Research in
Youth Development
Tufts University
DEVELOPMENTAL SYSTEMS THEORY

Chase L. Lesane-Brown
Research Assistant Professor
Psychology and Human
Development
Vanderbilt University
SOCIALIZATION, RACE

Ron Lesthaeghe
Visiting Professor
Departments of Sociology
University of Michigan and
University of California Irvine
DEMOGRAPHIC TRANSITION
THEORIES

Elaine Leventhal
Professor of Medicine
Department of Medicine
Robert Wood Johnson Medical
School/UMDNJ
CARDIOVASCULAR DISEASE

Becca Levy
Associate Professor of Epidemiology
and Psychology
Yale University
AGEISM/AGE DISCRIMINATION

Katherine C. Little
Doctoral Candidate
Department of Psychology
University of Tennessee
DATING AND ROMANTIC
RELATIONSHIPS, ADULTHOOD

Guangya Liu
Department of Sociology and
Anthropology
North Carolina State University
RESEARCH METHODS: VARIABLES

Andrew S. London
Professor
Department of Sociology and Center
for Policy Research
Syracuse University
G.I. BILL
MILITARY SERVICE

Kyle C. Longest
Department of Sociology
University of North Carolina Chapel
Hill
SPORTS AND ATHLETICS

Charles Longino
Doctor

Department of Sociology
Wake Forest University
RESIDENTIAL MOBILITY, LATER LIFE

Samuel R. Lucas
Department of Sociology
University of California, Berkeley
EDUCATIONAL ATTAINMENT
SCHOOL TRACKING

Alicia Doyle Lynch
Doctoral Student
Eliot-Pearson Department of Child
Development
Tufts University
DEVELOPMENTAL SYSTEMS THEORY

Eleanor Race Mackey
Post-doctoral Fellow
Department of Psychiatry
Children's National Medical Center
SOCIAL DEVELOPMENT

Ross Macmillan
Doctor
Department of Sociology and Life
Course Center
University of Minnesota
CRIME AND VICTIMIZATION,
ADULTHOOD
CRIME, CRIMINAL ACTIVITY IN
CHILDHOOD AND ADOLESCENCE
INDIVIDUATION/STANDARDIZATION
DEBATE

Jennifer L. Maggs
Associate Professor of Human
Development and Family Studies
Pennsylvania State University
DRINKING, ADOLESCENT

Joseph L. Mahoney
Associate Professor
Department of Education
University of California, Irvine
ACTIVITY PARTICIPATION,
CHILDHOOD AND ADOLESCENCE
HURRIED CHILD SYNDROME

Christine A. Mair
Department of Sociology and
Anthropology
North Carolina State University
RESEARCH METHODS:
CORRELATION VS. CAUSATION

Jennifer Margrett
Department of Human
Development and Family Studies
Iowa State University
COGNITIVE FUNCTIONING AND
DECLINE

Rachel Marks
Doctor
UT MD Anderson Cancer Center
HOSPICE AND PALLIATIVE CARE

MG Marmot
Professor of Epidemiology
Department of Epidemiology &
Public Health
University College London, UK
DATA SOURCES, LATER LIFE:
ENGLISH LONGITUDINAL STUDY
OF AGEING (ELSA)

Barbara L. Marshall
Professor
Department of Sociology
Trent University
ANDROPAUSE/MALE MENOPAUSE

Linda G. Martin
Senior Fellow
RAND Corporation
DISABILITY AND FUNCTIONAL
LIMITATION, LATER LIFE

Steven Martin
Doctor
Department of Sociology
University of Maryland
PATTERNS OF TV VIEWING
TIME USE, ADULTHOOD

Amir Marvasti
Assistant Professor of Sociology
Penn State Altoona
THOMAS, W.I.

Benjamin T. Mast
Associate Professor
Psychological and Brain Sciences
University of Louisville
DEMENTIAS

Ryan J. McCammon
Department of Sociology
University of Michigan
COHORT VERSUS GENERATION
COHORT

Christina McCrae
Associate Professor
Department of Clinical and Health
Psychology
University of Florida
SLEEP PATTERNS AND BEHAVIOR

Anne McDaniel
Doctoral Candidate
Sociology Department
Ohio State University
GENDER AND EDUCATION
SINGLE-SEX SCHOOLS

Nancy L. McElwain
Assistant Professor
Department of Human and
Community Development
University of Illinois at Urbana-
Champaign
ATTACHMENT THEORY

Sarah McKinnon
Department of Sociology
University of Texas at Austin
INFANT AND CHILD MORTALITY

Anne McMunn
Senior Research Fellow
Department of Epidemiology &
Public Health
University College London, UK
DATA SOURCES, LATER LIFE:
ENGLISH LONGITUDINAL STUDY
OF AGEING (ELSA)

James W. McNally
Director
NACDA Program on Aging
University of Michigan
ASSISTED LIVING FACILITIES
DATA SOURCES, LATER LIFE:
GENERAL ISSUES
QUALITY OF LIFE
SUCCESSFUL AGING

Ann Meier
Assistant Professor
Department of Sociology
University of Minnesota
RELIGION AND SPIRITUALITY,
ADULTHOOD

Deborah Merrill
Doctor
Department of Sociology
Clark University
PARENT-CHILD RELATIONSHIPS,
LATER LIFE

Barret Michalec
Department of Sociology
Emory University
COMMUNITARIANISM

Steven J. Miller
Sinclair School of Nursing
University of Missouri
AGING IN PLACE

Steven Mintz
Doctor
Columbia University
CULTURAL IMAGES, CHILDHOOD
AND ADOLESCENCE

Phyllis Moen
Department of Sociology
University of Minnesota
CAREERS
DUAL CAREER COUPLES

Kathryn Dawson Moneysmith
Center for Human Resource
Research
DATA SOURCES, ADULTHOOD:
NATIONAL LONGITUDINAL
SURVEYS OF MATURE MEN AND
WOMEN

Bethany Morgan
Department of Sociology
University of Essex, UK
GIDDENS, ANTHONY

S. Philip Morgan
Department of Sociology
Duke University
CHILDBEARING
PRONATALIST POLICIES

Pamela A. Morris
Co-Director
Policy Area on Family Well-Being
and Children's Development
MDRC
BRONFENBRENNER, URIE

Daniel G. Morrow
Human Factors Division and
Beckman Institute of Advanced
Technology
University of Illinois at Urbana-
Champaign
HEALTH LITERACY

Jeylan T. Mortimer
Professor of Sociology
Department of Sociology
University of Minnesota
SCHOOL TO WORK TRANSITION

Mark Motivans
Department of Sociology
The George Washington University
CRIME AND VICTIMIZATION, LATER
LIFE

Anna S. Mueller
Doctoral Candidate
Department of Sociology and the
Population Research Center
University of Texas at Austin
BODY IMAGE, CHILDHOOD AND
ADOLESCENCE

Christopher Muller
Department of Sociology

Harvard University
INCARCERATION, ADULTHOOD

Jan E. Mutchler
Professor
Department of Gerontology
University of Massachusetts Boston
FAMILY AND HOUSEHOLD
STRUCTURE, LATER LIFE
POVERTY, LATER LIFE

Scott M. Myers
Associate Professor
Department of Sociology
Montana State University
SOCIAL INTEGRATION/ISOLATION,
ADULTHOOD

Michelle Napierski-Prancl
Associate Professor
Department of Sociology and
Criminal Justice
Russell Sage College
EATING DISORDERS

Darcia Narváez
Department of Psychology
University of Notre Dame
MORAL DEVELOPMENT AND
EDUCATION

A. Rebecca Neal
Assistant Professor of Psychology
University of Texas at Austin
AUTISM

Belinda L. Needham
Doctor
University of California, San
Francisco and Berkeley
GENDER DIFFERENCES IN
DEPRESSION

Harold W. Neighbors
Professor and Director
Center for Research on Ethnicity
Culture and Health
University of Michigan School of
Public Health
RACISM/RACE DISCRIMINATION

Shelley L. Nelson
Doctoral Candidate
Department of Sociology
Indiana University
SOCIOECONOMIC INEQUALITY IN
EDUCATION

Gillian Niebrugge
Department of Sociology
American University
SOCIOLOGICAL THEORIES

Dan Nyaronga
University of California Berkeley,
School of Public Health
HEALTH BEHAVIORS, CHILDHOOD
AND ADOLESCENCE

Kirsten Bengtson O'Brien
Department of Sociology
University of Minnesota
RELIGION AND SPIRITUALITY,
ADULTHOOD

Mary Beth Ofstedal
Institute for Social Research
University of Michigan
DATA SOURCES, LATER LIFE: HEALTH
AND RETIREMENT STUDY (HRS)

Diane Graves Oliver
Assistant Professor of Psychology
University of Michigan-Dearborn
ETHNIC AND RACIAL IDENTITY

Randall Olsen
Professor of Economics
Ohio State University
DATA SOURCES, ADULTHOOD:
NATIONAL LONGITUDINAL
SURVEYS OF MATURE MEN AND
WOMEN

Jacqueline Olvera
Assistant Professor
Department of Sociology
Barnard College
ECONOMIC RESTRUCTURING

Daphna Oyserman
Edwin J. Thomas Collegiate
Professor
School of Social Work, Department
of Psychology
University of Michigan
ETHNIC AND RACIAL IDENTITY

Fred Pampel
Department of Sociology
University of Colorado, Boulder
RELATIVE COHORT SIZE HYPOTHESIS

Maria E. Parente
Department of Education
University of California, Irvine
ACTIVITY PARTICIPATION,
CHILDHOOD AND ADOLESCENCE
HURRIED CHILD SYNDROME

Megan E. Patrick
Department of Human
Development and Family Studies
Pennsylvania State University
DRINKING, ADOLESCENT

Desmond U. Patton
Doctoral Student
Social Service Administration
University of Chicago
SCHOOL VIOLENCE

Chuck W. Peek
Department of Sociology
University of Florida
POPULATION AGING
SEX RATIO

M. Kristen Peek
Associate Professor
Department of Preventive Medicine
and Community Health
University of Texas Medical Branch
MARRIAGE IN LATER LIFE

Anthony Daniel Perez
Post-doctoral Research Associate
Center for Studies in Demography
and Ecology
University of Washington
BIRACIAL/MIXED RACE ADULTS

Carolyn C. Perrucci
Professor of Sociology
Purdue University
UNEMPLOYMENT

Robert Perrucci
Professor of Sociology
Purdue University
UNEMPLOYMENT

Julie Phillips
Associate Professor
Department of Sociology
Institute for Health, Health Care
Policy and Aging Research, Rutgers
University
CRIME, CRIMINAL ACTIVITY IN
ADULTHOOD

Jane Pilcher
Department of Sociology
University of Leicester, UK
MANNHEIM, KARL

Stephen B. Plank
Department of Sociology
Johns Hopkins University
COLEMAN, JAMES

Jason L. Powell
Senior Lecturer in Sociology
School of Sociology and Social Policy
University of Liverpool, UK
GLOBAL AGING
POLICY, EMPLOYMENT

Tetyana Pudrovska
Assistant Professor of Sociology
Population Research Center
University of Texas at Austin
MIDLIFE CRISES AND TRANSITIONS

Norella M. Putney
Research Assistant Professor
Department of Sociology
University of Southern California
THEORIES OF AGING

Zhenchao Qian
Professor of Sociology
The Ohio State University
MATE SELECTION

Amélie Quesnel-Vallée
Assistant Professor
Department of Epidemiology,
Biostatistics and Occupational
Health and Department of McGill
University
HEALTH INSURANCE

Barry T. Radler
Institute on Aging
University of Wisconsin-Madison
DATA SOURCES, ADULTHOOD:
MIDLIFE IN THE UNITED STATES
(MIDUS)

R. Kelly Raley
Associate Professor
Department Sociology
University of Texas at Austin
COHABITATION

Mark R. Rank
Herbert S. Hadley Professor of Social
Welfare
Washington University in St. Louis
POVERTY, ADULTHOOD

Marilyn Rantz
Professor
Sinclair School of Nursing
University of Missouri
AGING IN PLACE

Eric N. Reither
Assistant Professor of Sociology
Utah State University
ALLOSTATIC LOAD

Dorothy P. Rice
Professor Emeritus
Institute for Health and Aging, Social
and Behavioral Sciences
University of California, San
Francisco
HEALTH CARE USE, LATER LIFE

Barbara Riddick
Ph.D. Programme Director
School of Education
Durham University, UK
LEARNING DISABILITY

Carol L. Roan
Assistant Scientist
Department of Sociology
University of Wisconsin
DATA SOURCES, LATER LIFE:
WISCONSIN LONGITUDINAL
STUDY

John P. Robinson
Professor
Department of Sociology
University of Maryland
PATTERNS OF TV VIEWING
TIME USE, ADULTHOOD
TIME USE, LATER LIFE

Karen S. Rook
Professor
School of Social Ecology
University of California, Irvine
FRIENDSHIP, LATER LIFE

Judith E. Rosenstein
Department of Sociology and
Anthropology
Wells College
SEXISM/SEX DISCRIMINATION

Erin Ruel
Assistant Professor of Sociology
Georgia State University
INHERITANCE

Louise B. Russell
Research Professor
Institute for Health, Health Care
Policy, and Aging Research
Rutgers University
HEALTH CARE USE, LATER LIFE

Stephen T. Russell
Professor
Norton School of Family &
Consumer Sciences
University of Arizona
GAYS AND LESBIANS, YOUTH AND
ADOLESCENCE

Steven Rytina
Professor
Department of Sociology
McGill University, Canada
OCCUPATIONS

Michael Sadowski
Assistant Professor

Master of Arts in Teaching Program
Bard College
SCHOOL READINESS

Jarron M. Saint Onge
Assistant Professor
Department of Sociology
University of Houston
MORTALITY

Irwin Sandler
Regents' Professor
Department of Psychology
Arizona State University
RESILIENCE, CHILDHOOD AND
ADOLESCENCE

Erica Scharrer
Associate Professor
Department of Communication
University of Massachusetts Amherst
MEDIA EFFECTS

Kathryn Schiller
Associate Professor
Department of Educational
Administration and Policy Studies
University at Albany, State
University of New York
SCHOOL TRANSITIONS

Bob Schoeni
Doctor
ISR, Public Policy, and Economics
University of Michigan
DATA SOURCES, ADULTHOOD:
PANEL STUDY OF INCOME
DYNAMICS (PSID)

Christopher L. Seplaki
Assistant Professor
Department of Population, Family
and Reproductive Health
Johns Hopkins Bloomberg School of
Public Health and Johns Hopkins
Center on Aging and FRAILTY AND
ROBUSTNESS

Joyce Serido
Research Scientist
Norton School of Family and
Consumer Sciences
University of Arizona
LIFE EVENTS
STRESS IN ADULTHOOD

Richard A. Settersten, Jr.
Professor
Human Development and Family
Sciences
Oregon State University
NEUGARTEN, BERNICE

Michael J. Shanahan
Associate Professor of Sociology
University of North Carolina,
Chapel Hill
GENETIC INFLUENCES, EARLY LIFE
NATURE VS. NURTURE DEBATE

Benjamin A. Shaw
Doctor
Department of Health Policy,
Management, and Behavior
University at Albany, School of
Public Health
TRAUMA

Corey S. Shdaimah
Assistant Professor
School of Social Work
University of Maryland, Baltimore
MILLS, C. WRIGHT

Kristen M. Shellenberg
Doctoral Candidate
Department of Population, Family
and Reproductive Health
Johns Hopkins, Bloomberg School
of Public Health
ABORTION

Diane S. Shinberg
Assistant Professor
Department of Sociology
University of Memphis
MENOPAUSE

Kim M. Shuey
Department of Sociology
University of Western Ontario,
Canada
RISK

Merril Silverstein
Professor of Gerontology and
Sociology
University of Southern California
BENGTSON, VERN
CAREGIVING
DATA SOURCES, ADULTHOOD:
LONGITUDINAL STUDY OF
GENERATIONS (LSOG)
GRANDCHILDREN
PARENT-CHILD RELATIONSHIPS,
ADULTHOOD

Moira Sim
Associate Professor
School of Nursing, Midwifery, and
Post-Graduate Medicine
Edith Cowan University, Australia
END OF LIFE DECISION-MAKING

Cassandra Simmel
Professor
School of Social Work
Rutgers University
ADOPTED CHILDREN

Julie Smart
Professor
Department of Special Education
and Rehabilitation
Utah State University
AMERICANS WITH DISABILITIES ACT
DISABILITY, ADULTHOOD

Jacqui Smith
Professor
Department of Psychology and
Institute for Social Research
University of Michigan
DATA SOURCES, LATER LIFE: BERLIN
AGING STUDY
SELF

Julia B. Smith
Associate Professor of Education
Educational Leadership Department
Oakland University
HIGH SCHOOL ORGANIZATION

Tom W. Smith
National Opinion Research Center
University of Chicago
DATA SOURCES, ADULTHOOD:
GENERAL SOCIAL SURVEY (GSS)

Dana Sohmer
Doctor
University of Chicago
CANCER, ADULTHOOD AND LATER
LIFE

Donna Spencer
Doctoral Candidate
Department of Sociology
University of Minnesota
FAMILY AND MEDICAL LEAVE ACT
POLICY, FAMILY

Steven Stack
Doctor
Departments of Psychiatry and
Criminal Justice
Wayne State University
SUICIDE, ADULTHOOD
SUICIDE, LATER LIFE

Jeremy Staff
Assistant Professor
Department of Sociology
Pennsylvania State University
SCHOOL TO WORK TRANSITION

Cynthia K. Stanton
Department of Population, Family
and Reproductive Health
Johns Hopkins, Bloomberg School
of Public Health
MATERNAL MORTALITY

Lala Carr Steelman
Professor
Department of Sociology
University of South Carolina
BIRTH ORDER
SIBLING RELATIONSHIPS,
CHILDHOOD AND ADOLESCENCE

Gillian Stevens
Professor of Sociology
University of Illinois, Urbana-
Champaign
IMMIGRATION, ADULTHOOD

Jean Stockard
Professor Emerita
Department of Planning, Public
Policy, and Management
University of Oregon
SUICIDE, ADOLESCENCE

Charles E. Stokes
Department of Sociology
University of Texas at Austin
COHABITATION

Kathy Stolley
Assistant Professor
Department of Sociology and
Criminal Justice
Virginia Wesleyan College
ADOPTIVE PARENTS

Jeanne M. Stolzer
Associate Professor
University of Nebraska-Kearney
ATTENTION DEFICIT/
HYPERACTIVITY DISORDER
(ADHD)

Margaret R. Stone
Family Studies and Human
Development
University of Arizona
PEER GROUPS AND CROWDS

Debra Street
Associate Professor
Department of Sociology
University at Buffalo, State
University of New York
LONG-TERM CARE
NATIONAL HEALTH INSURANCE
POLICY, HEALTH
POLICY, LATER LIFE WELL-BEING

Kate W. Strully
Assistant Professor
Department of Sociology
University at Albany, State
University of New York
BARKER HYPOTHESIS
BIRTH WEIGHT

Teresa A. Sullivan
Department of Sociology
University of Michigan
BANKRUPTCY
DEBT

Christopher Swann
Department of Economics
University of North Carolina
Greensboro
FOSTER CARE

Megan M. Sweeney
Associate Professor of Sociology
University of California, Los Angeles
REMARRIAGE
TRANSITION TO MARRIAGE

Sarah L. Szanton
Assistant Professor
Johns Hopkins University School of
Nursing
FRAILTY AND ROBUSTNESS

Maximiliane E. Szinovacz
Professor of Gerontology
University of Massachusetts Boston
GRANDPARENTHOOD

Jennifer Tanner
Visiting Assistant Research Professor
Institute for Health, Health Care
Policy and Aging Research
Rutgers University
TRANSITION TO ADULTHOOD

Beth Tarasawa
Doctoral Candidate
Department of Sociology
Emory University
RACIAL INEQUALITY IN EDUCATION
STEREOTYPE THREAT

John R. Thelin
University Research Professor
Educational Policy Studies
University of Kentucky
COLLEGE CULTURE

Deborah Thorne
Assistant Professor
Ohio University
BANKRUPTCY
DEBT

Russell Toomey
Doctor
University of Arizona
GAYS AND LESBIANS, YOUTH AND
ADOLESCENCE

Casey Totenhagen
Division of Family Studies and
Human Development
University of Arizona
STRESS IN ADULTHOOD

Angela Valenzuela
Professor of Curriculum and
Instruction and Department of
Educational Administration
University of Texas at Austin
HIGH-STAKES TESTING

Theo van Tilburg
Professor
Department of Sociology
VU University, Amsterdam
SOCIAL INTEGRATION/ISOLATION,
LATER LIFE

Zoua Vang
Doctor
Department of Sociology and
Population Studies Center
University of Pennsylvania
SEGREGATION, RESIDENTIAL

Sarinnapha Vasunilashorn
Doctoral Candidate
University of Southern California
Davis School of Gerontology
AGING

Brandon Wagner
Sociology Department
University of North Carolina
GENETIC INFLUENCES, ADULTHOOD

Linda J. Waite
Professor of Sociology
University of Chicago
MARRIAGE
SEXUAL ACTIVITY, LATER LIFE

Douglas M. Walker
Associate Professor
Department of Economics and
Finance
College of Charleston
GAMBLING

David F. Warner
Assistant Professor
Department of Sociology

Case Western Reserve University
SOCIAL SELECTION-CAUSATION
DEBATE

Amber Watts
University of Southern California
THIRD AGE

Dan Weinles
Senior Research Associate
Greater Philadelphia Urban Affairs
Coalition
CULTURAL CAPITAL

Deborah Welsh
Professor
Department of Psychology
University of Tennessee
DATING AND ROMANTIC
RELATIONSHIPS, ADULTHOOD

Amy Wendling
Doctor
Creighton University
MARX, KARL

Gerben J. Westerhof
Associate Professor of Psychology
University of Twente, The
Netherlands
AGE IDENTITY

Les B. Whitbeck
Bruhn Professor of Sociology
University of Nebraska-Lincoln
HOMELESS, YOUTH AND
ADOLESCENTS

K. A. S. Wickrama
Professor
Department of Human
Development and Family Studies
Iowa State University
HEALTH BEHAVIORS, CHILDHOOD
AND ADOLESCENCE

Christopher Wildeman
Robert Wood Johnson Health and
Society Scholars Program
University of Michigan
INCARCERATION, ADULTHOOD

Anne M. Wilkinson
Professor
Cancer Council WA Chair in
Palliative and Supportive Care
Edith Cowan University, Australia
END OF LIFE DECISION-MAKING

Kristi Williams
Doctor
Department of Sociology
Ohio State University
DATA SOURCES, ADULTHOOD:
AMERICANS' CHANGING LIVES
(ACL)

John B. Williamson
Department of Sociology
Boston College
RETIREMENT

Janet M. Wilmoth
Professor
Department of Sociology and Center
for Policy Research
Syracuse University
MILITARY SERVICE

John Wilson
Professor
Department of Sociology
Duke University
VOLUNTEERING, ADULTHOOD

Hal Winsborough
Professor Emeritus in Sociology
Center for Demography and
Ecology
University of Wisconsin-Madison
DATA SOURCES, ADULTHOOD: U.S.
(DECENNIAL) CENSUS
RYDER, NORMAN

Emily B. Winslow
Prevention Research Center,
Department of Psychology
Arizona State University
RESILIENCE, CHILDHOOD AND
ADOLESCENCE

Judith Wittner
Professor
Department of Sociology

Loyola University Chicago
LOPATA, HELENA

John S. Wodarski
Professor, College of Social Work
The University of Tennessee
MENTAL HEALTH, CHILDHOOD AND
ADOLESCENCE

Michael E. Woolley
Doctor
School of Social Service
Administration
University of Chicago
SCHOOL VIOLENCE

Linda A. Wray
Associate Professor of Biobehavioral
Health
Department of Biobehavioral Health
Pennsylvania State University
DIABETES, ADULTHOOD AND LATER
LIFE

Yang Yang
Assistant Professor
Department of Sociology
University of Chicago
AGE, PERIOD, COHORT EFFECTS

Laura Zettel-Watson
Assistant Professor
Department of Psychology
California State University, Fullerton
FRIENDSHIP, LATER LIFE

Min Zhou
Professor of Sociology & Asian
American Studies
University of California, Los Angeles
ASSIMILATION

Kathleen M. Ziol-Guest
Harvard School of Public Health
CHILD CUSTODY AND SUPPORT

Carole B. Zugazaga
Associate Professor
Department of Sociology,
Anthropology and Social Work
Auburn University
HOMELESS, ADULTS

Thematic Outline

The following classification of articles arranged thematically gives an overview of the variety of entries and the breadth of subjects treated in the encyclopedia. Along with the index and the arrangement of the encyclopedia, the thematic outline should aid in the location of topics. It is our hope that it will do more, that it will direct the reader to articles that may not have been the object of a search, that it will facilitate a kind of browsing that invites the reader to discover new articles, new topics, related, perhaps tangentially, to those originally sought.

1. BIOGRAPHIES

VOLUME 1

Bandura, Albert
Bowlby, John
Bronfenbrenner, Urie
Clausen, John
Coleman, James
Elder, Glen H., Jr.
Erikson, Erik
Freud, Sigmund
Mead, Margaret
Piaget, Jean
Thomas, W. I.

VOLUME 2

Durkheim, Èmile
Giddens, Anthony
Mannheim, Karl
Marx, Karl
Mills, C. Wright
Simmel, Georg
Znaniecki, Florian

VOLUME 3

Baltes, Margret and Paul
Bengtson, Vern
Foner, Anne
Lopata, Helena
Mayer, Karl Ulrich
Moen, Phyllis
Neugarten, Bernice
Riley, Matilda White
Rossi, Alice
Ryder, Norman
Shanas, Ethel

2. BIOLOGICAL INFLUENCES

VOLUME 1

Birth Weight
Genetic Influences, Early Life
Puberty

VOLUME 2

Andropause/Male Menopause
Attractiveness, Physical
Genetic Influences, Adulthood
Menopause

VOLUME 3

Allostatic Load
Frailty and Robustness
Genetic Influences, Later Life
Life Expectancy

3. CORE CONCEPTS

VOLUME 1

Age Norms
Socialization

VOLUME 3

Gays and Lesbians, Later Life
Sexual Activity, Later Life

28. SOCIAL RELATIONSHIPS

VOLUME 1

Dating and Romantic Relationships,
    Childhood and Adolescence
Friendship, Childhood and Adolescence
Intergenerational Closure
Peer Groups and Crowds

VOLUME 2

Dating and Romantic Relationships,
    Adulthood
Friendship, Adulthood
Social Integration/Isolation, Adulthood
Social Networks
Social Support, Adulthood

VOLUME 3

Caregiving
Friendship, Later Life
Loneliness, Later Life
Social Integration/Isolation, Later Life
Social Support, Later Life

29. SOCIAL ROLES AND BEHAVIORS

VOLUME 1

Activity Participation, Childhood and
    Adolescence

Civic Engagement, Childhood and
    Adolescence
Political Socialization
Sports and Athletics

VOLUME 2

Communitarianism
Housework
Leisure and Travel, Adulthood
Political Behavior and Orientations,
    Adulthood
Roles
Social Movements
Time Use, Adulthood
Volunteering, Adulthood

VOLUME 3

Leisure and Travel, Later Life
Older Drivers
Political Behavior and Orientations,
    Later Life
Time Use, Later Life
Volunteering, Later Life

30. TECHNOLOGY

VOLUME 1

Media and Technology Use,
    Childhood and Adolescence

VOLUME 2

Media and Technology Use, Adulthood

VOLUME 3

Assistive Technologies
Media and Technology Use, Later Life

31. THEORY

VOLUME 1

Attachment Theory
Developmental Systems Theory
Family Process Model
Interpretive Theory

VOLUME 2

Relative Cohort Size Hypothesis
Sociological Theories

VOLUME 3

Demographic Transition Theories
Social Selection-Causation Debate
Theories of Aging

32. WORK AND WORKPLACE ISSUES

VOLUME 2

Careers
Dual Career Couples
Flexible Work Arrangements
Gender in the Workplace
Job Characteristics and Job Stress
Work-Family Conflict

A

ACADEMIC ACHIEVEMENT

The primary function of primary and most secondary schools in the United States is to imbue students with core competencies in the academic subjects, including English, mathematics, science, history, and social studies. Achievement refers to students' demonstrated command over these subjects and can be measured in a variety of ways. Although related to both cognitive ability and learning, academic achievement is distinct from each. Learning refers to a change in academic achievement whereas cognitive ability may contribute to the capacity of students to learn or their pace of learning. Achievement, in contrast, captures the knowledge or skills that students possess at a single point in time. Students' academic achievement has direct and indirect effects on their educational and occupational opportunities, physical and mental health, and general well-being over the life course.

MAIN THEMES AND THEORIES IN RESEARCH ON ACADEMIC ACHIEVEMENT

Sociologists have drawn on four major bodies of theoretical knowledge to guide their investigations of the determinants and outcomes of academic achievement: status attainment theory, reproduction and resistance theories, organizational theory, and human capital. In this section, each of these theories is reviewed briefly.

Status Attainment According to the status attainment model, academic achievement is a critical pathway through which social origins (or the social class of one's parents) affect one's own occupational achievements. This model originally measured occupational and educational attainment as a function of parental education and occupation alone (Blau & Duncan, 1967). William Sewell, Archibald Haller, and Alejandro Portes (1969) changed the model by adding measures of cognitive skills, academic achievement, and social psychological measures. They found that academic achievement in high school is significantly affected by cognitive skills but not of cognitive skills, independent of socioeconomic origins. Sewell et al. also showed that academic performance is a key predictor of educational and occupational outcomes. One avenue through which academic performance affects outcomes is through its effect on students' own educational and occupational aspirations as well as the aspirations their parents and teachers hold for them.

Reproduction and Resistance Critics of the status attainment model contend that it ignores structural barriers, such as discrimination, tracking, or variation in school quality, across groups that block even ambitious and academically able students from achieving in school. Whereas reproduction and resistance theorists, like status attainment researchers, view academic achievement as a key factor in explaining the relationship between social origins and destinations, they contend that academic achievement serves only as a means of reinforcing preexisting social distinctions among children. In one variant of reproduction theory, Pierre Bourdieu (1930–2002) described how students and parents from elite social backgrounds draw on their *cultural capital* to help them navigate the educational process. According to Bourdieu, schools and teachers validate the cultural capital of middle- and upper-class students both formally (through grades) and informally (through their daily interactions), whereas denigrating the cultural capital of working-class students. In contrast to Sewell, empirical work in this tradition suggests that student cultural capital, as measured by involvement in art, music, and literature, has a significant effect on high school grades, independent of social background and ability (DiMaggio, 1982).

Taking a more Marxian approach to social reproduction, Samuel Bowles and Herbert Gintis (2002) described how schools mirror the labor force in structure, stratification, and the social characteristics or personality traits it rewards. According to Bowles and Gintis, teachers nurture critical thinking skills in their middle- and upper-class students while rewarding punctuality and conformity in their working-class students. Thus grades reflect academic achievement in part but also reflect the adoption of class-appropriate personality attributes. Although these assertions regarding the importance of noncognitive over cognitive skills in the workplace enjoy substantial empirical support, research on the relationship between social class origins and grades fails to support the contention that teachers differentially reward their students based on class-appropriate personality characteristics (Olneck & Bills, 1980).

Some critics of reproduction theories argue that these theories overlook the degree to which working-class and poor students work to subvert the educational system. For example, Paul Willis (1977) showed how working-class boys in Britain resist engaging in mental labor in favor of manual labor as the latter "carries with it—though not intrinsically—the aura of the real adult world" (p. 103). According to Willis, the boys in his study made "partial penetrations" into the stratification system, rightly seeing academic achievement as a means of legitimating the social class hierarchy but failing to understand their own roles in perpetuating that system. Likewise, Giroux (1983) argued that working-class subordination through schooling is part of the process of self-formation within the working class itself.

Organizations of Schools The sociology of organizational theory views academic achievement as heavily influenced by the technology of instruction and structure of the school organization. These two perspectives come together to inform the substantial research on tracking and ability grouping in education. In the primary school grades, Rebecca Barr and Robert Dreeben (1983) found few differences in academic aptitude across classrooms but substantial differences in aptitude across reading groups within classrooms. Teachers teach to the mean of each ability group, and, as a result, academic achievement increases more sharply for some reading groups than it does for others. This observation, which is borne out across middle and high schools as well, led Alan Kerckhoff and Elizabeth Glennie (1999) to characterize the relationship between initial academic achievement and learning as the *Matthew effect*, whereby academic advantage and disadvantage are compounded over time. The Matthew effect tends to reinforce the relationship between social origins and academic achievement, as children of less educated and less affluent parents, as well as Latino and African American children, are more likely to be enrolled in lower-track classes than their more advantaged peers (Gamoran & Mare, 1989).

Human Capital Like the status attainment and reproduction theories, human capital theories view academic achievement as both an outcome of social origins and a determinant of an individual's occupational outcomes. Human capital can be defined as knowledge and skills that contribute to worker productivity and thus increase earnings and wealth. Parents invest in their children's human capital and that investment generally boosts the child's academic achievement. Research shows that these investments are neither random nor equally distributed among children, however. Parental investment is a function of their number of children, the child's apparent cognitive endowment, and the child's luck (Becker & Tomes, 1976).

Human capital theory also shows how achievement in school is related to occupational outcomes. Achievement in kindergarten through high school contributes to general human capital—that is, any basic knowledge and skills that can be used to excel in an occupation. Achievement in secondary school and beyond that begins to be tailored to an individual's specific occupational interests may pay off even more.

COMMON INDICATORS OF ACHIEVEMENT AND HOW THEY ARE USED

Like other unobserved characteristics, such as attitudes, cognitive ability, and self-esteem, measuring academic achievement is a challenge to educators and researchers alike. Academic achievement may be assessed at any point in the learning process. Teachers often measure academic achievement early in the semester to help guide their instructional practice. Several times over the course of the term they will monitor learning, or change in achievement (formative assessment), and at the end of the term they evaluate student mastery (summative assessment). School districts and states compel schools to assess students for accountability purposes, whereas colleges, universities, and graduate schools assess potential students prior to offering them admission. Assessment instruments vary widely depending on the education system, the subject, and the type of skill or knowledge to be assessed.

Indicators The most commonly used type of assessment for comparative purposes is the standardized assessment. Standardized assessments consist of a uniform prompt given to all examinees in a uniform setting. Standardized exams may consist of multiple-choice, short-answer, and essay questions. Standardized assessments seek to systematically eliminate sources of variation in assessment performance outside of student achievement (including the influence of teachers or examiners, the use of notes or readings, parent intervention, and so forth). Although multiple choice questions are the most common and least costly questions used in standardized assessments, constructed response items (such as the writing section of the scholastic aptitude test [SAT]) are also common. There are many alternatives to standardized assessment. For example, in portfolio assessment, evaluators review a collection of each student's work, or portfolio. Performance assessments measure student achievement by requiring the student to execute or complete a specific task or process that directly reflects his or her academic achievement. An example of this might be asking students to complete a lab experiment in a chemistry or physics course. Portfolio and performance assessments grew in popularity in the early 1990s when some states required that students be evaluated with alternative forms of assessment. Although potentially effective, the time-intensive nature of grading these assessments may be impractical for large classes.

How Indicators Are Used Measures of academic achievement are used for a number of different purposes in American education. Within classrooms, teachers measure academic achievement to monitor student learning, tailor instruction to student needs, and (in primary schools) assign students to ability groups, primarily in reading. Teachers and counselors rely on measures of academic achievement (including standardized test scores, grades, and teacher judgments) to assign students to academic classes that vary in their content coverage and pace or to targeted instructional programs, including classes for English-language learners, remedial classes, and classes for gifted and talented students.

Schools rely increasingly on high-stakes tests to make grade retention and promotion decisions at the primary school level and to award high school diplomas to those who have satisfied other requirements for high school

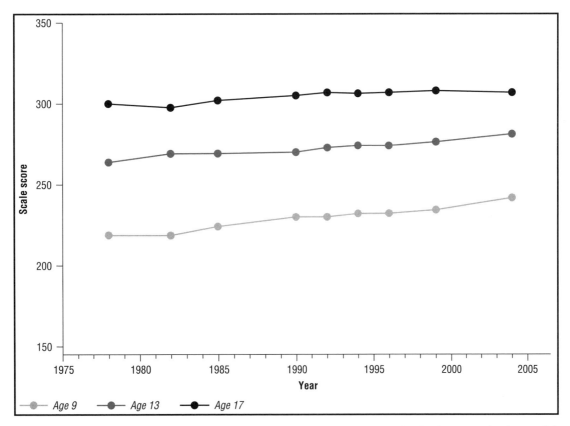

Figure 1. *This NAEP chart shows mathematics achievement scores over the last few decades, broken out by the age of the students.* **CENGAGE LEARNING, GALE.**

graduation. In most cases, individual schools are compelled to use tests in this manner by school districts or states. In 2007 more than two-thirds of all aspiring high school graduates were required to pass a high school exit exam to earn their diplomas. Alternatively, students who fail to pass the high school exit exam or to satisfy other requirements needed to attain a high school diploma may choose to take an alternative test of academic achievement in hopes of earning a general equivalency diploma.

State and federal policy makers also compel schools to assess student achievement to hold schools accountable for student learning. In 2001 the federal No Child Left Behind (NCLB) Act strengthened the call for accountability in core academic subjects by requiring that elementary, middle, and secondary schools present data each year on their students' academic achievement. Consequences for schools failing to meet that which is considered adequate yearly progress include financial sanction and, at the extreme, shift of personnel from failing schools to other schools.

At the postsecondary level, most baccalaureate-granting institutions and graduate and professional schools require that applicants take an entrance exam such as the SAT reasoning test, the American college test (ACT) for baccalaureate programs, the graduate record examination (GRE), or other professional exam such as the graduate management admission test (GMAT) for business or the medical college admission test (MCAT). Baccalaureate programs also may require one or more subject-specific achievement tests. Many colleges and universities use their own placement tests to determine what courses students must take in mathematics or composition. Although college instructors assess student achievement in the context of individual courses, and some colleges have summative assessments of student learning (such as an undergraduate thesis or comprehensive exam), most colleges in the past have not evaluated how much students learn over their undergraduate careers. However, as a result of pressures exerted by organizations such as the Spellings Commission, many colleges and universities are now considering assessments of achievement with which they could evaluate how much students learn during college (Secretary of Education's Commission on the Future of Higher Education, 2006). Finally, many professions (including those in law and medicine) regulate entrance, in part, with a standardized exam of student achievement.

TRENDS AND DISPARITIES IN ACADEMIC ACHIEVEMENT

Longitudinal Trends The National Assessment of Educational Progress (NAEP) provides a longitudinal measure of achievement that allows for direct comparison across years. National averages of student test scores have been fairly steady over the past 30 years. Figures 1 and 2 show mathematics and reading achievement scores on the NAEP, over the past few decades, broken out by the age of the students. Whereas 9-year-olds show improvement in their mathematics achievement, 13- and 17-year-olds have fairly consistent mathematics achievement scores. Reading scores changed little between 1971 and 2004, although there is more variation across time for the youngest students (U.S. Department of Education, 2004).

High school grade point averages (GPAs), by contrast, are less stable over time. The average GPA earned by high school graduates has increased steadily from 1987 to 2000 in every academic subject (U.S. Department of Education, 2005). Whether this shows that academic achievement has increased or simply reflects grade inflation is a topic of debate.

Racial and Ethnic Variation Racial and ethnic gaps in standardized test scores occur in almost all large, nationally representative databases. The magnitude of these gaps varies by academic subject, student age, and cohort. Gaps persist in standardized exams administered to high school students (the SAT and ACT) and college graduates (the GRE, GMAT, law school admission test [LSAT], and MCAT).

Inequalities also appear across racial and ethnic groups in high school GPA, with Black and Hispanic students having lower GPAs than White students and Asian students having higher GPAs than White students. Racial and ethnic differences in grades are similar in math, science, English, and social studies. Between 1987 and 2000 the Black-White GPA gap in all four subjects increased, whereas the Hispanic-White gap grew slightly in math and science but declined in English and social studies. The advantage Asian American or Pacific Islander students held over White students has decreased since 1987 (U.S. Department of Education, 2005).

Socioeconomic Variation Standardized test scores also consistently vary across measures of socioeconomic status. Differences in socioeconomic status across racial and ethnic groups account for some, but not all, of the gaps described above. Whether measured by parental education, occupation, or family income, less advantaged children have lower average levels of academic achievement than more advantaged children across a range of measures.

Sex Variation Gender differences in academic achievement vary over time and across subjects. Girls consistently have outscored boys in reading and writing on the NAEP but have not always had a clear advantage in mathematics and science. Girls enjoy a clear advantage

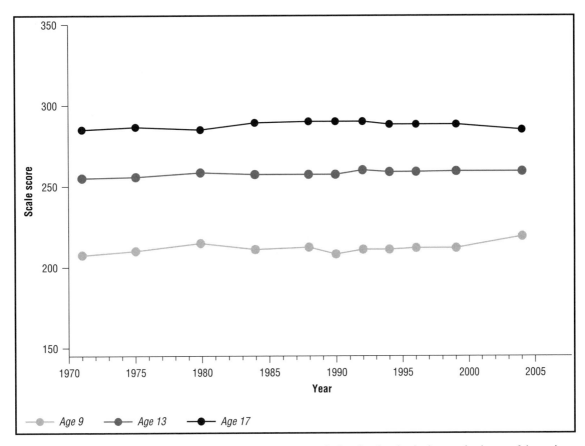

Figure 2. *This NAEP chart shows reading achievement scores over the last few decades, broken out by the age of the students.*
CENGAGE LEARNING, GALE.

in grades, however. In math, science, English, and social studies, between 1987 and 2000, male students never had a higher average GPA than female students (U.S. Department of Education, 2005).

OUTCOMES

Academic achievement contributes to a range of important adult outcomes. Secondary school academic achievement is among the most powerful predictors of attending college and of the selectivity of the various colleges students attend. Most people consider the primary benefits of academic achievement and higher education to be financial. However, in addition to economic returns, academic achievement and educational attainment confer a range of nonfinancial rewards.

Labor Market Achievement test scores in high school increase the likelihood that adults will be employed and, among those who are employed, make substantively and statistically significant contributions to earnings many years after students leave high school. Academic achievement measured by high school grades may also

affect earnings, although the evidence is thin compared to test scores. In the United States, James Rosenbaum and Takehiko Kariya (1991) found that high school grades do not have a significant effect on earnings for men, but they do for women. Returns to academic achievement are not limited to students who complete high school. A study conducted by John Tyler, Richard Murnane, and John Willett (1999) showed significant, positive effects of GED scores on earnings for White and non-White female and non-White male high school dropouts. College grades also appear to have a significant, positive effect on earnings for both men and women (Jones & Jackson, 1990). Although the academic achievement of college graduates varies substantially, their average level of academic achievement is greater than that of otherwise similar adults without a college degree. For example, in 2003 the average levels of prose and document literacy of adults with a bachelor's degree were about four-fifths of a standard deviation higher than those of adults with a high school diploma (Kutner et al., 2007). Because the baccalaureate signals a broad level of academic achievement, the literature on the economic returns to academic achievement as measured by educational attainment is briefly reviewed.

GRADE RETENTION VS. SOCIAL PROMOTION

Schooling in the United States is age graded with students progressing as they age, at least through the primary and middle grades. "Social promotion" has been criticized for passing students through the educational system without requiring that they are equipped with the knowledge and skills to succeed in the next grade level. This concern was heightened after the Ronald Reagan Administration's publication of *A Nation at Risk: The Imperative for Educational Reform* (1983), which declared that America's schools were failing to remain competitive with other countries.

Some studies show positive effects of grade-retention (Alexander, Entwisle, & Dauber, 1994), but the majority of research on outcomes of retention show that at best there is no difference between retained and promoted students, and at worst there are significant negative consequences (Heubert & Hauser, 1999). One study examined the outcomes of 44 different studies on grade retention and showed that the academic achievement of students that were retained was, on average, 0.4 standard deviations below that of promoted students (Holmes & Matthews, 1984). Research also shows that students retained in the eighth grade are significantly more likely to drop out of high school than are similar students who were not retained (Jacob & Lefgren, 2007).

College graduates make significantly more money than students who complete only high school. In 2005 the median income for adults (ages 25 to 64) who do not finish high school was $16,678 compared to $25,389 for those who complete only high school and $42,584 for adults with a bachelor's degree (U.S. Census Bureau, 2007).

Estimating just how much students benefit from attending college is difficult because there are other characteristics of students that complete only high school that differentiate them from college graduates. Several studies using innovative techniques have found a significant effect of college-going, yet the effect is smaller than the bivariate association between education and income (Angrist & Krueger, 1991).

Other Benefits Individuals who excel in academic work benefit not only from increased income but from a variety of other outcomes. Mark Kutner et al. (2007) showed that

adults with greater levels of literacy are more likely to vote; utilize print and Internet sources for information about current events, public affairs, and the government; and volunteer. Also, individuals who pursue more schooling have a greater understanding of current events and political issues, benefit from greater control over their work environment, and are healthier (Pallas, 2000).

WHERE THE STUDY OF ACADEMIC ACHIEVEMENT IS MOVING

Research on academic achievement will continue to focus largely on schools' contributions to the production of academic achievement and the effects that student academic achievement has on adult outcomes. Within primary and secondary schools, researchers and policy makers are increasingly interested in evaluating student growth, rather than achievement, at one point in time. In part as a response to potential sanctions imposed by the federal government under NCLB, many states have proposed using growth models to measure student proficiency rather than relying on changes in mean student achievement. More recently, interest in measuring learning has spread to higher education. As part of its effort to increase the quality of information available to students choosing among colleges, the U.S. Department of Education is advocating for a common database to which colleges report how much their graduates learn in college (Secretary of Education's Commission on the Future of Higher Education, 2006).

Research will also continue to explore how academic achievement shapes both occupational and nonoccupational outcomes. Some argue that the occupational returns to academic achievement are higher than ever as a result of rapid changes in technology and the availability of blue-collar jobs, whereas others suggest that personal attributes correlated with academic achievement, such as communication and interpersonal skills, may be more powerful predictors of labor market returns than academic achievement itself (Bowles & Gintis, 2002).

Sociologists of education have effectively shown how academic achievement is a factor throughout the life course as it is both a function of social origins and a predictor of future outcomes. Unfortunately this research also shows significant disparities in academic achievement between different racial and ethnic groups and students from varying levels of socioeconomic status. One hopes that as research on academic achievement progresses, researchers will gain greater insight into the processes that result in these inequalities and will be better equipped to form policies that can reduce disparities in academic achievement across groups.

SEE ALSO Volume 1: *Cognitive Ability; Gender and Education; High-Stakes Testing; Policy, Education;*

Racial Inequality in Education; School Culture; Socioeconomic Inequality in Education; Stages of Schooling.

BIBLIOGRAPHY

Alexander, K. L., Entwisle, D. R., & Dauber, S. L. (1994). *On the success of failure: A reassessment of the effects of retention in the primary.* New York: Cambridge University Press.

Angrist, J. D., & Krueger, A. B. (1991). Does compulsory school attendance affect schooling and earnings? *The Quarterly Journal of Economics, 106*(4), 979–1014.

Barr, R., & Dreeben, R. (1983). *How schools work.* Chicago: University of Chicago Press.

Becker, G. S., & Tomes, N. (1976). Child endowments and the quantity and quality of children. *The Journal of Political Economy, 84*(4), S143–S162.

Blau, P. M., & Duncan, O. D. (1967). *The American occupational structure.* New York: Wiley.

Bowles, S., & Gintis, H. (2002). Schooling in capitalist America revisited. *Sociology of Education, 75*(2), 1–18.

DiMaggio, P. (1982). Cultural capital and school success: The impact of status culture participation on the grades of U.S. high school students. *American Sociological Review, 47,* 189–210.

Gamoran, A., & Mare, R. D. (1989). Secondary school tracking and educational inequality: Compensation, reinforcement, or neutrality? *American Journal of Sociology, 94,* 1146–1183.

Giroux, H. A. (1983). Theories of reproduction and resistance in the new sociology of education: A critical analysis. *Harvard Educational Review, 53*(3), 257–293.

Hearn, J. C. (1991). Academic and nonacademic influences on the college destinations of 1980 high school graduates. *Sociology of Education, 64*(3), 158–171.

Heubert, J. P., & Hauser, R. M. (Eds.). (1999). *High stakes: Testing for tracking, promotion, and graduation.* Washington, DC: National Academy Press.

Holmes, C. T., & Matthews, K. M. (1984). The effects of nonpromotion on elementary and junior high school pupils: A meta-analysis. *Review of Educational Research, 54*(2), 225–236.

Jacob, B., & Lefgren, L. (2007). *The effect of grade retention on high school completion.* Cambridge, MA: National Bureau of Economic Research.

Jones, E. B., & Jackson, J. D. (1990). College grades and labor market rewards. *The Journal of Human Resources, 25*(2), 253–266.

Karen, D. (2002). Changes in access to higher education in the United States: 1980–1992. *Sociology of Education, 75*(3), 191–210.

Kerckhoff, A. C., & Glennie, E. (1999). The Matthew effect in American education. *Sociology of Education and Socialization, 12,* 35–66.

Kutner, M., Greenberg, E., Jin, Y., Boyle, B., Hsu, Y., et al. (2007). *Literacy in everyday life: Results from the 2003 national assessment of adult literacy.* Washington, DC: National Center for Education Statistics. Retrieved May 15, 2008, from http://nces.ed.gov/pubs2007

National Commission on Excellence in Education (1983). *A nation at risk: the imperative for educational reform.* Washington, DC: Author.

Olneck, M. R., & Bills, D. B. (1980). What makes Sammy run? An empirical assessment of the Bowles-Gintis correspondence theory. *American Journal of Education, 89*(1), 27–61.

Pallas, A. M. (2000). The effects of schooling on individual lives. In M. T. Halinan (Ed.), *Handbook of sociology and education.* New York: Kluwer; Plenum.

Rosenbaum, J. E., & Kariya, T. (1991). Do school achievements affect the early jobs of high school graduates in the United States and Japan? *Sociology of Education, 64*(2), 78–95.

Secretary of Education's Commission on the Future of Higher Education. (2006). *A test of leadership: Charting the future of U.S. higher education.* Washington, DC: U.S. Department of Education. Retrieved May, 21, 2008, from http://www.ed.gov/about

Sewell, W. H., Haller, A. O., & Portes, A. (1969). The educational and early occupational attainment process. *American Sociological Review, 31*(2), 159–168.

Tyler, J. H., Murnane, R. J., & Willet, J. B. (1999). *Do the cognitive skills of school dropouts matter in the labor market?* Cambridge, MA: National Bureau of Economic Research. Retrieved May 21, 2008, from http://ideas.repec.org

U.S. Census Bureau, Current Population Survey. (2007). *Annual social and economic supplement.* Retrieved May 31, 2008, from http://www.census.gov/population

U.S. Department of Education. (2004). *Institute of Education Sciences, National Center for Education Statistics, National Assessment of Educational Progress (NAEP), 1978-2004 long-term trend reading and mathematics assessments.* Retrieved May 31, 2008, from http://nces.ed.gov/nationsreportcard

U.S. Department of Education (2005). *Institute of Education Sciences, National Center for Education Statistics, High School Transcript Study (HSTS), selected years 1987–2000.* Retrieved May 31, 2008, from http://nces.ed.gov/nationsreportcard

Willis, P.E. (1977). *Learning to labor: How working-class kids get working-class jobs.* Farnborough, England: Saxon House.

Erika Felts
Eric Grodsky

ACADEMIC MOTIVATION

SEE Volume 1: *Academic Achievement.*

ACTIVITY PARTICIPATION, CHILDHOOD AND ADOLESCENCE

Extracurricular activities, or organized activities, are defined as voluntary, structured, school- or community-based activities in which school-age children and adolescents (ages 6 to 17) can participate outside of normal school hours. They include, but are not limited to, athletics, academic clubs (often called cocurricular activities), fine arts, musical activities, lessons, student government, and after-school programs. These are in contrast to alternative, unstructured, free time activities (i.e., hanging out with friends, watching television, and playing games). Research on adolescent leisure time indicates that a majority of children and youths participate in extracurricular activities. Accordingly, these activities are increasingly regarded as normative developmental contexts for the American youth. A review of participation in extracurricular activities carried out by Amy Feldman and Jennifer Matjasko (2005) suggests that extracurricular activities are central developmental settings for school-age children that are associated with many positive developmental outcomes.

The heightened interest in structured, out-of-school contexts stems from a number of factors. First, considerable increases in maternal employment and youth leisure time have created a gap in supervision between school and parental work hours. Investigators regard extracurricular activities as contexts that can limit the time a youth spends in unsupervised and unstructured activities outside of school, which have both been linked to negative developmental outcomes. Second, extracurricular activities are seen as opportunities to combat increases in academic underachievement, especially for at-risk youth. Last, extracurricular activities are considered to be contexts that promote mastery of skills not traditionally taught in the classroom, such as leadership, organization, and social problem solving. These influences have contributed to an increase in the popularity and availability of extracurricular activities since the early 1980s.

PARTICIPATION IN ORGANIZED ACTIVITIES

Approximately half of school-age children's time is devoted to their leisure activities. For many youths, most of these activities are school-sponsored extracurricular activities. In 2003, according to U.S. census data, 70% of school-age children (ages 6 to 17) were involved in at least one extracurricular activity (Dye & Johnson, 2007). Similarly, data from the National Longitudinal Study of Adolescent Health suggest that more than 70% of

Band Practice. *Approximately half of school-age children's time is spent in leisure activities.* AP IMAGES.

adolescents (ages 12 to 17) participate in at least one extracurricular activity (Feldman & Matjasko, 2005). Approximately 7 million children are enrolled in after-school programs. Overall, these percentages represent increases in extracurricular participation since the early 1970s (National Center for Youth Statistics, 1996).

School-age children initiate and continue to participate in organized activities for a variety of reasons; these reasons include, but are not limited to, developing and learning new skills, involvement in competition, developing one's identity, having fun, being with friends, and passing time. Additionally, school-age children are more likely to participate in extracurricular activities if they feel they are competent in the particular activity type. However, youths are not always the initiators of their participation in these activities. Parental encouragement affects participation decisions, and this is especially true for younger children who are more likely to be involved in activities valued by their parents.

There are considerable barriers to extracurricular participation—the most basic of which are prerequisites to involvement, including previous participation in the particular activity, academic performance requirements, and minimum skill level in the activities. Substantial barriers to participation also exist for families living in poverty, for whom availability of and access to extracurricular activities remain considerably lower than their counterparts at higher income levels. Inner-city and rural schools generally offer fewer opportunities for extracurricular participation, where economic resources tend to be fewer. In addition, lower-income families may have difficulty paying for supplemental fees that accompany some organized activities. Other obstacles include language barriers, difficulty finding transportation to and from the activity setting, and youth responsibilities that limit the amount of time available for participation (i.e., caring for younger siblings, working, and so forth).

Such factors may explain the lower rates of participation for the poorer youth. Data from the National Survey of America's Families, a nationally representative study of more than 42,000 households, suggest that youths living below the poverty line are half as likely to be involved in extracurricular activities as compared to those children whose family income is at least twice the income level defined as the federal poverty line (Casey, Ripke, & Huston, 2005). Notably, the National Center for Youth Statistics (1996) found that overall rates of high school senior extracurricular participation increased between 1972 and 1992 but not for those in the lowest income brackets. This conclusion is especially disheartening when considering that low-income children may garner the most benefit from participation.

HURRIED CHILD SYNDROME

Most research suggests that organized activity participation is beneficial for youth, yet there is mounting concern in the popular media that participation has become excessive. According to the over-scheduling hypothesis, youth are under considerable parental pressure to participate in numerous activities in order to increase chances for long-term educational success. This pressure and resulting stress and time commitment are thought to be harmful to youth and family functioning.

To evaluate the scientific basis of the over-scheduling hypothesis, Joseph Mahoney, Angel Harris, and Jacquelynne Eccles (2006) analyzed activity participation for a nationally representative sample of youth in the Panel Study of Income Dynamics (PSID). They conclude that many of the tenets of the over-scheduling hypothesis are not supported. First, most youth reported intrinsic (i.e., enjoyment) over extrinsic (i.e., parental pressure) motives for participation. Second, a majority of youth spent fewer than 10 hours per week participating in organized activities, with higher levels of participation associated with psychosocial adjustment and academic achievement. For a small percentage (less than 5%) participating more than 20 hours per week, only a few associated negative impacts were detected. By contrast, negative outcomes were consistently associated with no participation. Overall, the authors found minimal support for the over-scheduling hypothesis, but further support for increasing the availability of youth organized activities.

BIBLIOGRAPHY

Mahoney, J. L., Harris, A. L., & Eccles, J. S. (2006). Organized activity participation, positive youth development, and the over-scheduling hypothesis. *SRCD Social Policy Report, 20*(4),–31.

Rosenfeld, A., & Wise, N. (2000). *The over-scheduled child: Avoiding the hyper-parenting trap.* New York: St. Martin's; Griffin.

Involvement in extracurricular activities is also dependent on a number of other characteristics beyond income status, including age, gender, race, and neighborhood context. First, studies suggest differing rates and type of extracurricular involvement according to a child's age. Children are more likely to participate as they progress through elementary school, and this participation usually peaks in adolescence. Younger children are more likely to participate in lessons (i.e., music, dance), whereas adolescents are more likely to participate in sports.

Second, gender may influence the type of extracurricular involvement. Boys are more likely than girls to be involved in unorganized activities (i.e., hanging out with friends), whereas girls report more participation in clubs and lessons (Dye & Johnson, 2007). Studies exploring gender differences in extracurricular participation for school-age children find that girls are more likely to explore a wider variety of extracurricular activities as compared to boys. This difference may stem from boys being overrepresented in sports participation, a pattern found in a number of other investigations. These trends may reflect internal or external pressure to conform to traditional gender stereotypes. Interestingly, the gender of siblings and parents can affect activity choice. Children are more likely to be involved in stereotypically male activities when they have male brothers or when they are involved in more activities with their father as opposed to their mother.

Third, the amount and type of participation varies across race and ethnicity. According to U.S. census data (2003), European American school-age children are more likely to participate in extracurricular activities than traditionally defined minority students (Dye & Johnson, 2007). Hispanic adolescents, especially those of Mexican origin, report the least amount of participation in sports as compared to their European American, African American, and Asian counterparts. Furthermore, Hispanic students may also be at risk for lower extracurricular participation in general.

Finally, neighborhood context may impact organized activity participation. Findings from Reed Larson, Maryse Richards, Belinda Sims, and Jodi Dworkin (2001) suggest that urban youth spend significantly less time than their suburban counterparts in structured activities outside of school, taking into account a number of individual- and family-level factors. One explanation for these results may be the increased likelihood for restrictive parenting in more dangerous communities. Additionally, as mentioned previously, barriers to participation may be high for youths in urban, low-income settings, and the availability of extracurricular activities may be low.

The above differences in extracurricular participation illustrate that these activities are not isolated from other developmental contexts in which they are situated. Accordingly, differences in rates and type of participation are dependent on a number of child-, family-, and community-level factors. Researchers argue that further work is needed to elucidate the patterns of activity participation across different ages, genders, races, and poverty levels—including the amount and type of participation at any given time during the school years. They also contend that further work is required to investigate the developmental outcomes associated with participation and the possible mechanisms underlying these relations.

DEVELOPMENTAL CONSEQUENCES OF ORGANIZED ACTIVITY PARTICIPATION

Given that a majority of school-age children are involved in extracurricular activities, it is imperative to examine the consequences of this participation. Researchers studying the impact of extracurricular involvement on American youth tend to find benefits for child and adolescent adjustment. On the whole, partaking in extracurricular activities may provide adolescents with the social and human capital necessary to make a successful transition into American adulthood; in addition, this involvement will likely provide more opportunities for positive social, emotional, and academic development.

In regards to social adjustment, extracurricular contexts are associated with the development of relationships with peers and mentors and seem to increase the opportunity to observe and partake in prosocial group norms and behavior. In addition, extracurricular involvement is linked to forming friendships with academically oriented peers who are more apt to abide by conventional group norms, such as not skipping school and avoiding drug use. Accordingly, involvement in extracurricular activities is correlated with fewer behavioral problems among adolescent youth. Unstructured and unsupervised activity settings, by contrast, are related to behavior problems and delinquency, especially for low-income youth, who are already at heightened risk for these issues. Taken together, these findings suggest that extracurricular participation may be a way to address social adjustment disparities between low-income youth and their counterparts of higher socioeconomic status.

Research also suggests a link between extracurricular participation and positive psychological and emotional functioning. For example, some adolescents involved in extracurricular activities are less likely to experience a depressed mood, less likely to have anxiety problems, and are more likely to have high self-esteem than those involved in unsupervised and unstructured out-of-school activities. In addition, extracurricular activity contexts may support positive identity development. Bonnie Barber, Margaret Stone, James Hunt, and Jacquelynne Eccles (2005) argue that extracurricular activities promote positive identity development through opportunities for undertaking leadership positions, exploring and expressing one's identity in a social context, fostering relationships with peers and mentors, and developing individual interests and skills. However, it must be noted that negative identity development can also occur in these contexts, although research has generally linked extracurricular participation with positive impacts on identity.

Last, academic performance and achievement are also associated with organized activity participation. These include higher GPAs and lower rates of grade retention and school dropout. Presumably, improvements in academic functioning can be attributed to opportunities in extracurricular settings for positive social, psychological, and emotional development. Additionally, extracurricular activities are associated with increased cognitive stimulation, school engagement, and school connectedness and heightened motivation for learning, each of which may contribute to positive academic outcomes. Longitudinal research has also revealed long-term positive adjustment associated with extracurricular participation, including improved future employment and increased adult civic involvement, even after accounting for a number of child- and family-level factors.

FACTORS THAT INFLUENCE THE IMPACT OF ORGANIZED ACTIVITIES

An abundance of research has shown that extracurricular participation among youths is associated with numerous positive developmental outcomes; however, such benefits are not inevitable. A variety of factors impact the degree to which extracurricular participation is linked to beneficial outcomes. For example, developmental outcomes may depend on the consistency of participation. Jonathan Zaff, Kristin Moore, Angela Papillo, and Stephanie Williams (2003) found that consistent adolescent activity participation across the high school years was associated with more positive developmental outcomes than either occasional or no participation, even after controlling for individual- and family-level characteristics. Outcomes may also vary depending on the breadth and intensity of a youth's extracurricular involvement. Some studies have found that participation in a variety of extracurricular activities is more beneficial to youths, perhaps because it provides students with a variety of skills and experiences, allows for greater practice of these skills across multiple contexts, and provides additional resources that may buffer negative experiences occurring in other activities or unsupervised contexts.

High-intensity participation (i.e., more frequent participation in a fewer number of organized activities) has also been associated with positive outcomes, though findings are somewhat mixed. Youths participating in fewer organized activities may be able to invest more time and

effort into these activities, which may lead to greater knowledge and skill mastery. However, other studies (Busseri, Rose-Krasnor, Willoughby, & Chalmers, 2006) have found that participation in a variety of activities is associated with more positive outcomes than higher intensity extracurricular participation. Finally, some have suggested that too much time invested in extracurricular activities and other organized activities can lead to having an overly demanding schedule, which may be associated with negative outcomes. However, research generally suggests that greater extracurricular involvement is associated with improved youth adjustment.

The associated impacts of extracurricular involvement on development may also depend on the type of activity in which youths are involved. For example, studies investigating high school sports suggest both positive and negative consequences associated with sports participation. Beneficial impacts of sports participation include increases in initiative, educational aspirations, positive attitudes toward school, and high school completion. Sports involvement also offers opportunities to build skills such as problem solving, goal setting, managing time and emotions, teamwork, and maintaining physical health. However, other studies indicate negative associations with sports activity. For example, longitudinal work by Eccles and Barber (1999) found that sports involvement may increase the likelihood of alcohol use among adolescents (during the high school years), primarily because of specific peer networks in these settings. These contradictory findings suggest that beneficial impacts of extracurricular involvement may depend on the type of activity the youth is involved in, as well as the specific processes and peer associations occurring in these contexts.

INVESTIGATING UNDERLYING DEVELOPMENTAL PROCESSES

Unfortunately, few studies have examined the developmental processes taking place in the extracurricular settings that promote youth development. However, an initiative put forth by the National Research Council and Institute of Medicine examined features of organized activity contexts that are related to positive developmental outcomes—these include setting safety, structure, and prosocial norms, and opportunities for feelings of belongingness, supportive relationships, and skill and self-efficacy building. Extracurricular settings that meet these features are likely to contribute to positive youth development.

Many researchers are calling for further exploration of the underlying causes of positive outcomes for youths involved in extracurricular activities and the examination of specific processes in these settings. One possibility that would tie this work together is examining participation with a more holistic approach. In their review of extracurricular involvement, Feldman and Matjasko (2005) highlighted the need for investigations focusing on the patterns of youth extracurricular participation and their developmental implications. They contend that a paucity of research has compared involvement in more than a single extracurricular activity and that different patterns and profiles of participation may lead to different outcomes.

POLICY IMPLICATIONS
AND RESEARCH DIRECTIONS

Overall, most research on organized activity settings has shown positive consequences of participation for social, emotional, and academic development. This fact has fueled funding initiatives in both the public and private sectors to expand the availability and accessibility to extracurricular activities. Most notable is the 21st Century Community Learning Centers program, a federal initiative supporting funding of out-of-school youth programs.

Unfortunately, financial support for extracurricular activities is often pitted against funding for traditional school academic initiatives that have taken even greater precedence since the initiation of No Child Left Behind in 2001. School districts making budget concessions are more likely to cut funding for extracurricular activities before other list items. Additionally, U.S. Supreme Court decisions have actually increased barriers to participation. The Court has ruled in favor of (a) allowing schools to limit participation based on funding problems and (b) drug testing for all students partaking in extracurricular activities. Schools may also require students to meet minimum academic standings in order to participate in an organized activity. Unfortunately, these policies may limit extracurricular involvement for precisely those students who stand to benefit the most from them, as it could promote a return to positive developmental trajectories.

Despite the increased support and interest in extracurricular activities, substantial barriers to participation exist for youths from low-income families. Researchers call for policy makers to decrease barriers to participation for these youths, as extracurricular activities may address the socioeconomic disparities found in school achievement and overall adjustment. One possibility is to increase monies to low-income families in order to facilitate increased extracurricular involvement. However, David Casey, Marika Ripke, and Aletha Huston (2005) maintained that policies such as welfare reform have done little to increase family income. Family resources may be augmented using subsidies that cover fees and transpor-

tation costs for extracurricular activities. In fact, studies offering monetary subsidies and monetary assistance to families, such as the New Hope Evaluation in Milwaukee, saw increases in youth organized activity participation. Another implication is to increase funding for extracurricular activities at the school level, especially in low-income areas where availability is limited. A substantial way to guarantee funding is to include extracurricular involvement as part of a child's legal entitlement for a minimally adequate education. On the whole, researchers are calling for sustained funding for existing programs and an increase in the availability of extracurricular activities, especially for children in low-income communities.

Past work also suggests that improving the quality of extracurricular activities offered to youths should be a target for policy makers. James Quinn (2005) asserts that relatively few activity programs meet quality standards suggested by research. In addition, the focus of many programs has turned to raising test scores rather than concentrating on other important aspects of development. This change has resulted from an increased attention on standardized test scores since the initiation of No Child Left Behind.

Further research is necessary to inform policy practices suggested above. Researchers are increasingly moving toward investigations that offer a deeper understanding of extracurricular activities and their developmental impacts. Investigations such as those by Larson and colleagues (2005) demonstrates how research can elucidate the underlying processes that may underlie the impact of extracurricular involvement on youth adjustment. Some scholars suggest that a person-environment fit model be used when investigating whether and how extracurricular participation benefits young people. Indeed, Eccles (2005) suggests that future research is necessary to explicate the specific characteristics of the activity settings and the specific adult and student behaviors that influence participants. Further research is also necessary for explaining who participates in various extracurricular activities, who continues to participate, and what influences the type of organized activity youths decide to become involved in.

SEE ALSO Volume 1: *Data Sources, Childhood and Adolescence; Academic Achievement; Drinking, Adolescent; Drug Use, Adolescent; Identity Development; Self-Esteem; Social Capital; Sports and Athletics.*

BIBLIOGRAPHY

Barber, B. L. Stone, M. R., Hunt, J. E., & Eccles, J. S. (2005). Benefits of activity participation: The roles of identity affirmation and peer group norm sharing. In J. L. Mahoney, R. W. Larson, & J. S. Eccles (Eds.), *Organized activities as contexts of development: Extracurricular activities, after-school and community programs.* Mahwah, NJ: Lawrence Erlbaum.

Busseri, M. A., Rose-Krasnor, L., Willoughby, T., & Chalmers, H. (2006). A longitudinal examination of breadth and intensity of youth involvement and successful development. *Developmental Psychology, 42*(6),1313–1326.

Casey, D. M., Ripke, M. N., & Huston, A. C. (2005). Activity participation and the well-being of children and adolescents in the context of welfare reform. In J. L. Mahoney, R. W. Larson, & J. S. Eccles (Eds.), *Organized activities as contexts of development: Extracurricular activities, after-school and community programs.* Mahwah, NJ: Lawrence Erlbaum.

Dye, J. L., & Johnson, T. (2007). *A child's day: 2003 (selected indicators of child well-being).* Washington, DC: U.S. Department of Commerce. Retrieved March 19, 2008, from http://www.census.gov/prod/2007pubs

Eccles, J. S. (2005). The present and future of research on activity settings as developmental contexts. In J. L. Mahoney, R. W. Larson, & J. S. Eccles (Eds.), *Organized activities as contexts of development: Extracurricular activities, after-school and community programs.* Mahwah, NJ: Lawrence Erlbaum.

Eccles, J. S., & Barber, B. L. (1999). Student council, volunteering, basketball, or marching band: What kind of extracurricular involvement matters? *Journal of Adolescent Research, 14*(1), 10–43.

Feldman, A. F. & Matjasko, J. L. (2005). The role of school-based extracurricular activities in adolescent development: A comprehensive review and future directions. *Review of Educational Research, 75*(2), 159–210.

Heath, S. B. (1999). Dimensions of language development: Lessons from older children. In A. S. Masten (Ed.), *Cultural processes in child development: The Minnesota symposia on child psychology.* Mahwah, NJ: Lawrence Erlbaum.

Larson, R. W., Hansen, D. M., & Walker, K. (2005). Everybody's gotta give: Development of initiative and teamwork within a youth program. In J. L. Mahoney, R. W. Larson, & J. S. Eccles (Eds.), *Organized activities as contexts of development: Extracurricular activities, after-school and community programs.* Mahwah, NJ: Lawrence Erlbaum.

Larson, R. W., Richards, M. H., Sims, B., & Dworkin, J. (2001). How urban African-American young adolescents spend their time: Time budgets for locations, activities, and companionship. *American Journal of Community Psychology, 29*(4), 565–597.

National Center for Youth Statistics. (1996). *Youth Indicators, 1996: Trends in the well-being of American youth.* Washington, DC: Author. Retrieved March 18, 2008, from http://nces.ed.gov/pubs98

Quinn, J. (2005). Building effective practices and policies for out-of-school time. In J. L. Mahoney, R. W. Larson, & J. S. Eccles (Eds.), *Organized activities as contexts of development: Extracurricular activities, after-school and community programs.* Mahwah, NJ: Lawrence Erlbaum.

Zaff, J., Moore, K., Papillo, A., & Williams, S. (2003). Implications of extracurricular activity participation during adolescence on positive outcomes. *Journal of Adolescent Research, 18*(6), 599–630.

Maria E. Parente
Joseph L. Mahoney

ADOPTED CHILDREN

Adoption is one way that family formation occurs. The term *adoption* has a fairly uncomplicated connotation, but the process itself is multifaceted and encompasses several types of adoptive arrangements, each with its own set of administrative protocols, legal regulations, as well as developmental implications for the children. Although there are many similarities across the various types of adoption in how this process affects lifespan development, recent research has uncovered a more refined understanding of the discrete differences among diverse populations of adoptees.

THE ADOPTION PROCESS

In the early 21st century, formal adoption refers to a "legal procedure through which a permanent family is created for a child whose birth parents are unable, unwilling, or legally prohibited from caring for the child" (Triseliotis, Shireman, & Hundleby, 1997, p. 1). Over the past several decades, this "legal procedure" has become increasingly popular, yet there is no nationwide governmental authority that oversees all types of adoptive placements, despite the fact that all adoptions entail some sort of legal involvement. As such, determining the exact number of adoptions that occur for every type of adoptive arrangement is difficult. Currently, research estimates that there are about 1.5 million adopted children in the United States (Evan B. Donaldson Adoption Institute, 2002) and that about 2–4% of families include an adopted child. There are two exceptions to the imprecise collection of statistics on adoption, however: international adoptions and the adoption of children from the U.S. foster-care system (public adoptions). Due to legislative changes and other societal transformations, both of these forms of adoption have become increasingly widespread (U.S. Department of Health and Human Services, 2008). Not only are there expanding notions of who is qualified to be an adoptive parent, but potential adoptees are no longer limited to "healthy infants." The population of adoptees has become increasingly diverse in terms of age, race, country of origin, and developmental background, whereas contemporary potential adoptive parents encompass a wide range of demographic characteristics.

In the United States, adoptions can occur with the assistance of a public or private agency, or independently from a government-certified agency. Public agency adoptions are those that involve youth in the child welfare system (foster care or other type of out-of-home care). Such adoptions require the voluntary or involuntary termination of (biological) parental rights and include a range of children both in terms of demographic characteristics and developmental histories. Independent adoptions indicate that the arrangement occurred without the

National Adoption Day. *Adopted children from left, Christopher Futschik, 10; Jessica Sherman, 8; Achaunti Strong, 5; and Johane Strong, 7; stand with Connecticut Attorney General Richard Blumenthal to celebrate National Adoption Day in Hartford, Connecticut.* AP IMAGES.

assistance of an official agency but instead through a third party such as a clergy, doctor, or attorney who mediates the agreement between the birthparent and adoptive caregivers, with a judge authorizing the final agreement (Stolley, 1993). Private adoptions refer to the assistance of a nonpublic adoption agency in negotiating the matter; infants and children from a variety of backgrounds are placed through private adoption. Within these agency or independent entities, adoptions may be domestic or international, and may also involve transracial placements.

Although the practice of transracial adoptions has a tumultuous history—and was actively discouraged by adoption practitioners for many years—federal child welfare legislation from the 1990s has helped break down the barriers to this practice. In part due to the fact that so many ethnic minority children were residing in foster care indefinitely, the Multiethnic Placement Act of 1994 and its amended provisions, the Interethnic Placement Provisions of 1997, were passed by Congress; both prohibit the delay or denial of an adoptive placement on the basis of race, color, or national origin of either the

potential adoptive parent or the adoptee (U.S. Department of Health and Human Services, 1997). Simply put, adoption practitioners cannot legally factor in these characteristics when making decisions about the potential adoptive placement. Whereas there was once concern that the psychological adjustment of transracial adoptees would be irrevocably harmed, several longitudinal studies of children and young adults have significantly disputed this notion (Brooks & Barth, 1999). Yet transracial placement needs to be handled with sensitivity and careful forethought. This process may still engender difficulties for young adoptees in terms of development of racial identity and acculturation, while the adoptive family may struggle with feelings of isolation in the community (Lee, 2003).

In contrast to most of the other categories of adoptions, international adoptions are fairly well documented by the federal government. International or intercountry adoptions refer to the adoption of nonnative children and typically involve infants. This form of adoption has become increasingly popular in recent decades due to worldwide changes in countries' economic and political circumstances (e.g., Romania, Russia) as well as demographic and social transformations in the United States. For example, the number of native infants available for adoption in the United States has decreased due to increases in the utilization of abortion and in growing acceptance of single-parent households (Bartholet, 1993). Around the world, international adoptions have increased, including the adoption of American children abroad. Increasing in popularity since the Korean and Vietnam Wars, this practice has become especially widespread since the mid-1980s. Between 2002 and 2007, Americans adopted 100,000 children from other countries (U. S. Department of State, 2008). In 1990, for example, the number of international adoptions by American parents was just over 7,000 children; by 2006, this figure had tripled. Furthermore, children are adopted from dozens of countries, with the greatest number in 2006 originating from China, Guatemala, Russia, and South Korea.

In recent years, the face of adoptive parents has evolved dramatically. Whereas adoption was once primarily undertaken by childless, two-parent couples, in the early 21st century gay and lesbian couples, single parents, and low-income families are actively seeking to adopt children (Howard, 2006). It was only as recently as the early 1970s that single, unmarried men were able to adopt children on their own (Dorris, 1989). For foster children, it is not as uncommon as once thought for *kin adoption* to occur (i.e., a grandmother or aunt chooses to adopt a relative who has been placed in foster care; U.S. Department of Health and Human Services, 2002).

HISTORY OF ADOPTION

The practice of adoption has been evident, both formally and informally, for centuries, albeit with motives that are distinct from the contemporary American practice of adoption (Sokoloff, 1993; Triseliotis et al., 1997). For a number of centuries, maintaining the familial lineage was one of the ultimate goals of adoption. In ancient Rome, adoptions were performed for the sake of the adoptive family sustaining its familial lineage. Adult males were preferred for adoption in order to provide heirs to Roman emperors (Sokoloff, 1993; Triseliotis et al., 1997). In 17th-century England, parentless or dependent children were tended to in a manner that obviated the need for adoptive homes. The de facto policy for tending to unwanted children included placing them in almshouses or relegating them to positions as labor apprentices, indentured servants, or domestic help (Sokoloff, 1993).

In later centuries in the United States, a controversial method for tending to dependent children developed. Over the course of several decades, these children were shipped via trains from urban areas (primarily New York City) to farms in the Midwest, a movement referred to as the *orphan trains*. This permanent relocation was spearheaded by Charles Loring Brace under the auspices of the Children's Aid Society (also founded by Brace) because he felt that children deserved a better upbringing than living in institutions, or on the streets (O'Connor, 2001). Upon arriving at their destinations, the young children were put on display to the public, who then selected the children they wanted, often as farm labor. Although Brace's intention was to place unwanted children in stable, permanent adoptive families, the formal adoption of these children did not often occur. Moreover, many of these children were technically not orphans and had living biological parents in their former residences. However, the fact that many of these children may have benefited from this move west into secure families cannot be overlooked (O'Connor, 2001).

By the early 20th century in the United States, legal guidelines for adoptions emerged—mostly directed at protecting the privacy and secrecy of both the adoptee and the biological parent—and adoption has fluctuated in popularity since that time. The demand for infants, however, has remained fairly steady since World War I (Sokoloff, 1993).

HOW ADOPTION AFFECTS INDIVIDUALS' PSYCHOSOCIAL DEVELOPMENT

Do adopted children encounter more identity and developmental difficulties than nonadopted children? For many decades, researchers and practitioners assumed that adopted children were at a higher risk of poor outcomes in numerous interpersonal and developmental domains (Brodzinsky, Schecter, Braff, & Singer, 1984; Elonen & Schwartz, 1969; Sharma, McGue, & Benson, 1998). In part, theories for such problematic outcomes have vacillated from biological issues, such as the genetic inheritance of behavior problems (i.e., impulsivity; Deutsch et al., 1982), to psychodynamic issues related to long-term confusion on behalf of an adopted child about his or her familial origins, often referred to as the *adopted child syndrome* (Kirschner, 1996), to difficulties stemming from having resided in troubled international regions (Tizard, 1991). Alternatively, some researchers suggest that tainted perceptions on the part of adoptive parents and/or mental health professionals—who perhaps unwittingly look for signs of difficulties in adopted youth—fuel the reported higher rates of problem behaviors (Warren, 1992; Wegar, 1995).

Yet until the late 1990s, some of this research was flawed methodologically in that all types of adopted children were clustered together in many analyses, causing the entire population of adoptees to appear

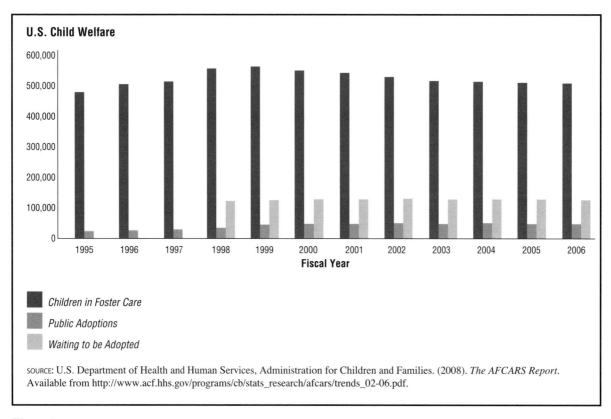

U.S. Child Welfare

- Children in Foster Care
- Public Adoptions
- Waiting to be Adopted

SOURCE: U.S. Department of Health and Human Services, Administration for Children and Families. (2008). *The AFCARS Report.* Available from http://www.acf.hhs.gov/programs/cb/stats_research/afcars/trends_02-06.pdf.

Figure 1. CENGAGE LEARNING, GALE.

troubled. More recent research, by contrast, has focused on specific subgroups of adoptees and thus has teased apart some of this heterogeneity. Such studies conclude that with so much variability in the types of adoptive placements, the developmental trajectory of the adoptee is partly contingent on how the child is placed in the home (Haugaard, 1998).

ADOPTION OF FOSTER CHILDREN

The foster-care population consists of children burdened by particularly harsh early developmental histories, primarily due to circumstances of physical and/or sexual abuse and neglect. For these reasons, and because foster children frequently already have biological families from whom they must separate, the adoption of foster children can be precarious. That is, public adoptions, unlike other types of adoptions (e.g., private adoption of infants), mostly entail the placement of toddlers and children who, because of their early histories, may possess emotional and behavioral difficulties that can engender unstable transitions into their new adoptive homes. Not only do many foster children have histories of maltreatment, but they are generally older and thus may have rather distinct memories—or ongoing relationships—with their biological parents and families, as well as with siblings

who are also placed in foster care. Although parents who undergo a public adoption are well-intentioned, these types of adoptions have inherent risks that may not be present in other types of adoptions. Research on adopted foster children indicates that both male and female adoptees may evidence internalizing and externalizing behavior problems, such as depression and anxiety and disruptive behavior disorders for extensive periods of time—especially if no services are provided (Simmel, 2007). Providing accurate preparation and ongoing services to the adoptee and the adoptive family is critical for cultivating and sustaining healthy psychosocial development (Simmel, 2007).

Recognizing that foster children face decreased odds of entering into an adoptive arrangement as they mature, the U.S. Congress passed legislation in 1997 (the Adoption and Safe Families Act) that partially addressed this challenge. Foster children beyond the age of 5 years are not likely to be adopted. If they do not reunify with their biological families, they remain in foster care or other institutional arrangements until age 18. Thus, for children in the child welfare system, adoption is one of the most desirable placement outcomes as it involves a sense of permanency for children, as opposed to the "temporary" option of residing in a succession of foster and group homes. The passage of the Adoption and Safe Families Act in 1997 ushered in a revived focus on the permanency and safety of foster children, and consequently the number of public adoptions increased in the United States (U.S. Department of Health and Human Services, 2008).

RECENT CHANGES IN ADOPTION PRACTICE

The potentially problematic outcomes for adopted children are not limited to foster children. Infants adopted at birth—including through private adoptions—are not necessarily free of problems. Difficulties in acquiring a sense of belonging and developing a personal identity may haunt some adoptees regardless of when or how they came into the adoptive home. Anecdotally, two recent personal memoirs shared insights about the authors' personal histories of adoption, the developmental impact of being an adoptee, and surrendering an infant to adoption. A. M. Homes (2007), adopted at infancy, discussed her search for and eventual reunion with her biological parents and the complexities of this process on her identity development, particularly when this new relationship became problematic. In contrast, Meredith Hall (2007) wrote about the painful process of being a pregnant teen in the 1960s and the extensive efforts to cover up her pregnancy and the adoption of her infant son. She eloquently described the impact on her own identity development, her reunion with her biological son, and how they both navigated her new role as a mother figure in his life as a young adult.

An increasingly common practice is open adoption, or maintaining some form of contact among the child adoptee, his or her adoptive family, and the biological parent or family who surrendered the child. This contact may take place through in-person contact, letters, phone calls, or e-mails. The extent of these contacts varies. For some it may be an annual letter from the adoptive parents; for others it may be monthly visits with the biological parent. As children begin to comprehend the nature of what it means to be adopted, usually between the ages of 7 and 11, allowing access to biological families holds enormous potential for facilitating positive growth. During adolescence, a time when struggling with identity issues is paramount, having some continuity with one's biological heritage could be instrumental in fomenting one's sense of self (Brodzinsky, 2005). This practice demonstrates a striking procedural change from traditional adoption practice, whereby all information about both parties was strictly concealed (Sokoloff, 1993). Reflecting societal changes in the 1970s pertaining to personal liberties (e.g., the civil rights movement, the women's movement), as well as an increase in both adult adoptees and birthparents seeking information about and/or reunification with one another, open adoption was initially a revolutionary idea that has gradually evolved into greater acceptance (Brokzinsky, 2005).

In addition, the use of open adoptions is becoming evident in some public adoptions in the United States, Canada, and Europe (Brodzinsky, 2005). It is not clear whether this practice is uniformly beneficial to the adoptee. Research shows that adoptive families may curtail the amount of contact with biological families after sustaining contact with them for the first few years (Brooks, Simmel, Barth, & Wind, 2005). However, as noted by David Brodzinsky (2005), open adoption is a "fluid process" and may fluctuate again as children reach the adolescent phase. Yet Brodzinsky remarks that having had an open adoption and maintaining open communication with adopted parents may have a positive developmental influence on adoptees' psychological development in both childhood and adulthood.

THE FUTURE OF ADOPTION RESEARCH

Within the broad practice of adoption, many different types of adoptive placements exist, affecting infants and children from many different age groups who possess diverse developmental backgrounds and demographic characteristics. To apply a set of research findings to the entire population of adoptees does a disservice to understanding the explicit and subtle differences among them. Whereas recent empirical investigations are beginning to hone in on the specific characteristics of unique types of adoptive arrangements, the next wave of research will be enhanced by taking this a step further. For example, given the potential vulnerability of children adopted from foster care, what are the long-term strengths and challenges faced by these families and can effective services be introduced to mitigate the stressors? For international adoptees, closer examination of the regional differences in how the sending countries care for their young orphans and the ultimate developmental impact of this early care is necessary.

Similarly, on a broader policy level, tighter regulations of independent, for-profit adoption agencies (that may be involved in domestic placements as well) are needed so that the rights of potential adoptive parents and international adoptees are carefully safeguarded. Finally, with respect to the advent of openness in adoption proceedings, how might this process help young adoptees reconcile potential identity issues with their biological heritage? Understanding the immediate (childhood) and long-term (early adulthood) impact of this process is an important step.

SEE ALSO Volume 1: *Foster Care; Parent-Child Relationships, Childhood and Adolescence;* Volume 2: *Adoptive Parents; Parent-Child Relationships, Adulthood.*

BIBLIOGRAPHY

Bartholet, E. (1993). International adoption: Current status and future prospects. *The Future of Children, 3,* 89–103.

Brodzinsky, D. M. (2005). Reconceptualizing openness in adoption. In D. Brodzinsky & J. Palacios (Eds.), *Psychological issues in adoption: Research and practice* (pp. 145–166). Westport, CT: Praeger.

Brodzinsky, D. M., Schecter, D. E., Braff, A. M., & Singer, L. M. (1984). Psychological and academic adjustment in adopted children. *Journal of Consulting and Clinical Psychology, 52,* 582–590.

Brooks, D., & Barth, R. P. (1999). Adjustment outcomes of adult transracial and inracial adoptees: Effects of race, gender, adoptive family structure, and placement history. *American Journal of Orthopsychiatry, 69,* 87–102.

Brooks, D., Simmel, C., Wind, L., & Barth, R. P. (2005). Contemporary adoptive families and implications for the next wave of adoption research. In D. Brodzinsky & J. Palacios (Eds.), *Psychological issues in adoption: Research and practice* (pp. 1–26). Westport, CT: Praeger.

Deutsch, C. K., Swanson, J. M., Bruell, J. H., Cantwell, D. P., Weinberg, F., & Baren, M. (1982). Overrepresentation of adoptees in children with the attention deficit disorder. *Behavior Genetics, 12,* 231–238.

Dorris, M. (1989). *The broken cord.* New York: Harper & Row.

Elonen, A. S., & Schwartz, E. M. (1969). A longitudinal study of emotional, social, and academic functioning of adopted children. *Child Welfare, 48,* 72–78.

Evan B. Donaldson Adoption Institute. (2002). *Overview of adoption in the United States.* New York: Author.

Hall, M. (2007). *Without a map: A memoir.* Boston: Beacon Press.

Haugaard, J. J. (1998). Is adoption a risk factor for the development of adjustment problems? *Clinical Psychology Review, 18,* 47–69.

Homes, A. M. (2007). *The mistress's daughter.* New York: Viking.

Howard, J. (2006). *Expanding the resources for children: Is adoption by gays and lesbians part of the answer for boys and girls who need homes?* New York: Evan B. Donaldson Adoption Institute. Retrieved May 13, 2008, from http://www.adoptioninstitute.org/publications

Kirschner, D. (1996). Adoption psychopathology and the "adopted child syndrome." In *The Hatherleigh guide to child and adolescent therapy* (pp. 103–123). New York: Hatherleigh Press.

Lee, R. M. (2003). The transracial adoption paradox: History, research, and counseling implications of cultural socialization. *The Counseling Psychologist, 31*(6), 711–744.

O'Connor, S. (2001). *Orphan trains: The story of Charles Loring Brace and the children he saved and failed.* Boston: Houghton Mifflin.

Peters, B. R., Atkins, M. S., & McKay, M. M. (1999). Adopted children's behavior problems: A review of five explanatory models. *Clinical Psychology Review, 19,* 297–328.

Sharma, A. R., McGue, M. K., & Benson, P. L. (1998). The psychological adjustment of United States adopted adolescents and their nonadopted siblings. *Child Development, 69,* 791–802.

Simmel, C. (2007). Risk and protective factors contributing to the longitudinal psychosocial well-being of adopted foster children. *Journal of Emotional and Behavioral Disorders, 15,* 237–249.

Sokoloff, B. Z. (1993). Antecedents of American adoption. *The Future of Children, 3,* 17–25.

Stolley, K. S. (1993). Statistics on adoption in the United States. *The Future of Children, 3,* 26–42.

Tizard, B. (1991). Intercountry adoption: A review of the evidence. *Journal of Child Psychology and Psychiatry and Allied Disciplines, 32,* 743–756.

Triseliotis, J., Shireman, J., & Hundleby, M. (1997). *Adoption: Theory, policy, and practice.* London: Cassell.

U.S. Department of Health and Human Services, Administration for Children and Families. (2002). *The AFCARS report.* Washington, DC: Author.

U.S. Department of Health and Human Services, Administration for Children and Families. (2008). *Trends in foster care and adoption: FY 2002–FY 2006.* Retrieved May 13, 2008, from http://www.acf.hhs.gov

U. S. Department of State, Office of Children's Issues. (2008). *Intercountry adoption.* Retrieved April 30, 2008, from http://travel/state.gov

Warren, S. B. (1992). Lower threshold for referral for psychiatric treatment for adopted adolescents. *Journal of the American Academy Child and Adolescent Psychiatry, 31,* 512–527.

Wegar, K. (1995). Adoption and mental health: A theoretical critique of the psychopathological model. *American Journal of Orthopsychiatry, 65,* 540–548.

Cassandra Simmel

ADOPTION

SEE Volume 1: *Adopted Children;* Volume 2: *Adoptive Parents.*

ADVERSITY, CHILDHOOD

SEE Volume 1: *Resilience, Childhood and Adolescence;* Volume 2: *Risk.*

AFFIRMATIVE ACTION, IMPLICATIONS FOR YOUTH

SEE Volume 1: *Policy, Education;* Volume 2: *Policy, Employment; Racism/Race Discrimination.*

AGE NORMS

The presence of elementary school girls in salons getting facials, men in their 60s starting families, junior high school boys drinking martinis—on some level, these scenarios give one pause. Although there is nothing remarkable about getting a facial, starting a family, or drinking a martini, the *age* of the actors doing these things is just not right. In life course parlance, these behaviors are off-time, either too early or too late. They violate age norms, or shared expectations about when life events, transitions, or behaviors *ought* to occur.

Age is a key factor that organizes social life (Riley, Foner, and Waring, 1988). And the sequence of socially defined, age-graded events and roles that an individual enacts over time is the cornerstone of the life course perspective (Elder, 1975). That is, most individuals in the United States can expect to begin formal education around age 6, complete high school around age 18, transition to marriage in their 20s and early 30s, maintain stable employment during adulthood, retire from the labor force in their 60s, and die thereafter. Although increased heterogeneity in the timing of transitions has occurred over time, this general sequencing is still widespread. Its stability arises from two related processes. First, social institutions (such as the educational system) and institutional transitions (such as retirement) that have age-related boundaries attached play a key role in ensuring the sequence of life events described above. Second, age norms then add onto this institutional role.

This entry focuses on this latter concept of age norms, describing what it means, outlining its appeal and weaknesses, and providing applications of it in research on adolescence. What should be noted up front is that age norms are more often assumed in research than tested directly despite theory that age norms matter during this stage of rapid physical, social, emotional, and cognitive development.

DEFINING THE CONCEPT

The concept of age norms gained prominence through the work of Bernice Neugarten and colleagues on the Kansas City Studies of Adult Life. Conducted during the late 1950s and early 1960s, the study helped make age research more prominent in sociology, emphasizing the social and psychological elements of age that guide human development. In their seminal article on age norms, Neugarten and colleagues (1965) posited that age norms are a pervasive system of expectations regarding age-appropriate behaviors that lets individuals know when they *should* or *should not* engage in a particular behavior or transition to a new status. These norms, they argued, are embedded into the cultural fabric of everyday life, illustrated in phrases such as "Act your age," "She's

too young to be having sex," or "He's too old to be unmarried." Operating at the group level, age norms create social expectations for behavior and provide a social clock or timetable for major life events. As such, individuals are aware both of the social clock and of their own timing relative to others, easily describing themselves as "early," "late," or "on-time" with regard to different transitions and events. Finally, age norms, like all norms, are supported and enforced through a variety of sanctions placed upon the transgressor (Neugarten, Moore, and Lowe, 1965; Settersten and Mayer, 1997).

Using survey responses from a sample of middle-class adults in two Midwestern cities, Neugarten and colleagues found that nearly all agreed that a "right" age for different life transitions exists (e.g., marriage, parenthood, to be settled into a career). For example, more than 85% of respondents thought that the right age for a woman to marry was between ages 19 and 24. From these results, they concluded that the normative pattern of adult development, especially in the 1950s and early 1960s, comes about, in part, as age norms and age expectations "operate as prods and brakes upon behavior, in some instances hastening an event, in others delaying it" (Neugarten et al. 1965, p. 711). This construct and its related ideas (social clock, being off-time, etc.) have been incorporated into a life course perspective, generating a host of studies focused on the significance of age in society and the timing of life events.

LIMITATIONS OF THE CONCEPT

Although intuitively appealing and a starting point for much life course scholarship, a great deal of conceptual ambiguity, weak measurement, and limited empirical evidence is associated with the concept of age norms (Hagestad, 1990; Elder, 1978; Marini, 1984; Modell, 1997). Using the marriage example from above, when looking at the timing of an event such as age at marriage, are researchers observing an actual norm or simply a statistical regularity? As Marini (1984) noted, an observed behavior that is very common may be a *custom* but not a *norm*, the former referring to a collective expectation of what an individual *will* do and the latter referring to a collective expectation of what an individual *ought* to do. Similarly, in relying on statistical regularities to infer the presence of a norm, one may overlook the possibility that no actual norm exists, but rather a range of behaviors are acceptable. Continuing with the marriage example, looking at union formation behavior among young adults at the start of the 21st century, most are single, neither married nor cohabiting with a partner. Does this statistical regularity mean that to be single is a norm or a behavior that one ought to engage in, whereas to be married or cohabiting is nonnormative? Or, are

Teens Smoking. *Adolescents smoke cigarettes in a café. The image of adolescents engaging in "problem behaviors" such as smoking goes against most age norms.* **AP IMAGES.**

being married or cohabiting in young adulthood simply other, less common options (Settersten and Mayer, 1997)? Together, these questions limit the scientific utility and application of this concept.

TWO EXAMPLES OF AGE NORMS IN RESEARCH ON ADOLESCENCE

Despite these limitations, age norms and the consequent social psychological implications of being off-time underlie much of the research on adolescent development. Although the salience of age in human development is waning over time (Neugarten, 1996), some argue that the first two decades of life remain a developmental period where age-grading is more or less consistent (Hagestad, 1990) and, thus, a period where age norms may matter most to development. To that end, three areas of research have been reviewed—moving from the general to the more specific—that build upon or test the role of age norms in adolescent development.

First, much of the research on "problem behaviors" in adolescence (e.g., smoking, drinking, drug use, sexual intercourse) is predicated, in part, on the general belief

that engaging in such behaviors in the second decade of life violates the shared expectations of about what an adolescent is—one who is priceless, innocent, and in need of protection—and what behaviors in which she *ought* to engage. That is, smoking is bad for one's health regardless of age, but the image of a smoking 12-year-old *feels* especially egregious. Similarly, the physical and, to some extent, emotional risks associated with sexual intercourse among two consenting 14-year-olds who love and respect one another and use contraception are minimal. Yet most adults, including many academics, and many young people themselves *feel* that such a transition is inappropriate. In both cases, the feelings of inappropriateness triggered by these off-time or early transitions are reflected in public policies (sexual education programs in schools) and law (age restrictions on purchasing cigarettes).

As another example, much of the research on pubertal timing in girls' lives draws on the notion of being off-time as a factor in explaining early-maturing girls' increased likelihood of engaging in problem behaviors (e.g., Ge, Conger, and Elder, 1996; Stattin and Magnusson, 1990). These studies assume that a departure from the normative developmental schedule is less socially desirable and even stressful for young people (Neugarten, 1979). Because early maturing girls depart from this timetable and are visibly different from their peers at a moment when being like everyone else matters a great deal and because they are perceived by others as older than they really are, early maturers enter the social world of adolescence sooner, doing so with neither the support of their larger peer group nor the development time needed to acquire and integrate the skills needed to confront the new tasks in adolescence.

The final example of the role of age norms in adolescent development comes from a study on race/ethnic differences in age, sequencing norms, and behaviors. Patricia East (1998) used survey data to chart the timing and sequencing of girls' sexual, marital, and birth expectations. She found that the normative timetable by which girls saw their lives unfolding was constructed differently by race and ethnicity. Latinas desired early and rapid transitions for marriage and birth. African-American girls perceived the youngest desired age for first sexual intercourse and the greatest likelihood of nonmarital fertility. Finally, whites perceived older ages for each.

SEE ALSO Volume 1: *Drinking, Adolescent; Drug Use, Adolescent; Elder, Glen H., Jr.; Puberty; Sexual Activity, Adolescence; Transition to Marriage; Transition to Parenthood;* Volume 2: *Roles.*

BIBLIOGRAPHY

East, P. L. (1998). Racial and ethnic differences in girls' sexual, marital, and birth expectations. *Journal of Marriage and the Family, 60,* 150–162.

Elder, G. H., Jr. (1975). Age differentiation and the life course. *Annual Review of Sociology, 1,* 165–190.

Elder, G. H., Jr. (1978). Approaches to social change and the family. *American Journal of Sociology, 84,* S1–34.

Ge, X. J., Conger, R. D., & Elder, G. H. (1996). Coming of age too early: Pubertal influences on girls' vulnerability to psychological distress. *Child Development, 67,* 3386–3400.

Hagestad, G. (1990). Social perspectives on the life course. In R. H. Binstock & L. K.George (Eds.), *Aging and the social sciences* (pp. 151–168). Academic Press: San Diego, CA.

Marini, M. M. (1984). Age and sequencing norms in the transition to adulthood. *Social Forces, 63,* 229–244.

Modell, J. (1997). What do life course norms mean? *Human Development, 40,* 282–286.

Neugarten, B. (1979). Time, age, and the life cycle. *American Journal of Psychiatry, 136,* 887–894.

Neugarten, B. L., Moore, J. W., & Lowe, J. (1965). Age norms, age constraints, and adult socialization. *The American Journal of Sociology, 70,* 710–717.

Neugarten, D. A., ed. (1996). The meanings of age: Selected papers of Bernice L. Neugarten. Chicago: University of Chicago Press.

Riley, M. W., Foner, A., & Waring, J. (1988). Sociology of age. In N. J. Smelser (Ed.), *Handbook of sociology* (pp. 243–290). Newbury Park, CA: Sage.

Settersten, R. A., Jr., & Mayer, K. U. (1997). The measurement of age, age structuring, and the life course. *Annual Review of Sociology, 23,* 33–61.

Stattin, H., & Magnusson, D. (1990). *Pubertal maturation in female development.* Hillsdale, NJ: Lawrence Erlbaum Associates.

Shannon E. Cavanagh

AGGRESSION, CHILDHOOD AND ADOLESCENCE

Aggressive behavior among children is recognized as a major risk factor for subsequent developmental maladjustment, both for the perpetrators and the victims. For decades, attempts to understand and prevent childhood aggression have focused on physical aggression. Physical aggression is usually defined as the use of physical force against another person either with an object (e.g., stick, rock, knife) or without (e.g., slap, push, punch, kick, bite). However, increasing press coverage of school-based incidences of peer victimization as well as films such as *Mean Girls* (2004) and books such as *Queen Bees and Wannabes* (2002) have drawn attention to the fact that children's aggressive behavior incorporates more than the infliction of bodily harm. Thus, children can hurt their peers through more subtle forms of aggression, for example through social exclusion or rumor spreading.

Different labels have been used to describe these more subtle forms of aggression, including *indirect aggression, relational aggression,* and *social aggression.* All three terms refer to aggressive behavior that is intended to damage another's self-esteem or social status, but indirect aggression is mainly covert in nature (e.g., spreading rumors, social exclusion from the group) whereas relational aggression can be both covert and overt (e.g., threatening to withdraw friendship, divulging secrets). Social aggression encompasses both overt and covert behaviors and also includes nonverbal aggressive behavior (e.g., ignoring someone, negative facial expressions or body movements). A review of the literature reveals, however, that the three labels essentially refer to the same overarching construct. In this entry, the term *social aggression* will be used. Unlike physical aggression, social aggression does not cause any immediate physical injuries. Many adults therefore view social aggression among children as less serious than physical aggression. However, social aggression has been found to activate the same areas of the brain that register physical pain and is considered by the victims to be as harmful as physical aggression.

SEX DIFFERENCES IN THE DEVELOPMENTAL COURSE OF AGGRESSION

With the exception of early infancy, when physical aggression first appears, males are considerably more physically aggressive than females throughout the life course. This sex gap in physical aggression gradually widens over the course of childhood and adolescence and reaches a peak in young adulthood. In contrast, girls are already more socially aggressive than boys during the preschool period, when this type of behavior first emerges. The sex gap in social aggression also continuously widens over the course of childhood, with a peak in adolescence, but decreases considerably thereafter. By young adulthood, males and females show relatively similar levels of social aggression (Côté, 2007).

What may explain these divergent trajectories for the two sex groups? One possible explanation refers to differential rates of *cognitive maturation* in infancy between boys and girls. Of particular importance in this context may be the fact that girls develop expressive language skills sooner than boys, which may enable girls to solve conflict situations verbally instead of with physical force and to use socially aggressive strategies that involve the manipulation of others (e.g., rumor spreading). Another explanation may lie in differential *socialization experiences* of the two genders. Several studies show that socializing agents such as parents, teachers, and peers tend to encourage gender-normative behaviors and discourage gender-non-normative behaviors, and this pattern is also found with respect to aggression. Parents and teachers are more likely to disapprove of girls' than of boys' expression of anger and are more likely to use firm directives and follow through on requests for good behavior in girls than in boys. Similarly, peers view physical aggression as less normative for girls than for boys. The differential structure of girls' and boys' peer contexts may also facilitate a greater use of social aggression in girls. Boys tend to engage in more rough-and-tumble play than girls, which may lead to more serious aggressive behavior when conflicts arise. In contrast, girls' play is more oriented toward intimacy and social inclusion within relatively small groups. As a result, girls' peer groups may offer more opportunities for the use of socially aggressive strategies such as rumor spreading or social exclusion than boys' peer groups.

ETIOLOGY OF PHYSICAL AGGRESSION

For several decades, the prevailing theory of the etiology (or origin) of aggression among scholars was that it is a learned behavior—occurring either as a response to a provocation or threat or as an instrumental means of goal attainment—that results from observing and imitating aggressive role models. Such aggressive role models may be observed either in the family, the peer group, the neighborhood, or in the mass media. Empirical data seemed to support the social learning theory of aggression. Aggression, especially physically aggressive behavior, is more prevalent in children and adolescents who are exposed to marital violence and child abuse, whose parents use harsh discipline and physical punishment, who have highly aggressive siblings or friends, who live in disadvantaged neighborhoods characterized by high crime, and who report frequent exposure to violence in films, music, and print media.

Experimental studies also provided empirical support for observation and imitation of aggressive role models as the primary explanatory mechanism of aggressive behavior. For example, a now classic study by Albert Bandura and colleagues (the "Bobo doll" study) showed that children who observed an adult physically abuse a puppet were more inclined to later use physical aggression themselves against the puppet than children who had not witnessed the adult's aggressive behavior (Bandura, Ross, and Ross, 1961). Empirical support for the idea that aggression is a learned behavior also came from robust criminological statistics showing that arrests for violent offenses (i.e., physical aggression) appear first in preadolescence, increase sharply during adolescence, and decrease slowly thereafter (Sampson and Laub, 1993).

More recently, researchers have challenged the idea that physical aggression is a learned behavior that peaks in adolescence (Tremblay and Nagin, 2005). They argue that, whereas the increase in violent crime during adolescence is in line with a social learning model of aggression, the decrease of violent crime in adulthood is not. Moreover, both observational and questionnaire-based studies suggest that physical aggression (i.e., kicking, biting, hitting, or pushing another person) is already prevalent in toddlerhood. Perhaps even more noteworthy, recent longitudinal studies indicate that physical aggression increases in the first years of life up to a peak at around 30 months of age and then decreases steadily thereafter. These data suggest that, rather than being a result of social learning, physical aggression appears to be part of most—if not all—young children's potential behavioral repertoire. Of course, considerable individual differences exist with respect to the frequency and intensity with which such behavior is enacted. More importantly, over the course of development, most children learn *not* to use physical aggression in interpersonal interaction.

Despite the general decrease in physical aggression after its peak in toddlerhood, recent studies show that up to 15% of children remain on a stable and high trajectory of physical aggression throughout the preschool years and beyond. What can explain these distinct developmental patterns? At least part of these interindividual differences may be due to genetic factors. A large body of research suggests that at least 50% of the variance in physical aggression during childhood is explained by genetic factors (DiLalla, 2002). There is also evidence, however, that genetic effects on physical aggression diminish with age. Together, these findings suggest that genetic factors may play a considerable role in explaining interindividual differences in children's initial propensity toward physical aggression, whereas socialization may influence to a large extent whether and how quickly children learn to replace physical aggression with socially more acceptable behavior.

ETIOLOGY OF SOCIAL AGGRESSION

Although most children seem to "unlearn" physical aggression in favor of more socially acceptable alternatives, these alternatives may not necessarily comprise only prosocial behavior. Instead, some children may revert to more covert, socially aggressive strategies to obtain their goals or to seek revenge against others. Unlike physical aggression, social aggression is usually not observed before the preschool period. Given its manipulative and often circuitous nature, social aggression requires a certain amount of verbal skills as well as an understanding of others' intentions and emotions, which only start to emerge at around 4 years of age (Sutton, Smith, and Swettenham, 1999). Once these skills develop, however, some children seem to use social aggression more and more frequently. Thus, whereas physical aggression appears to peak in toddlerhood and (for most children) gradually decline thereafter, social aggression increases with age.

By adolescence, social aggression is by far the predominant type of aggression in social interaction, and remains such thereafter. Indeed, research with adults suggests that psychological harassment, which is equivalent to social aggression, is a major problem in the workplace and one of the main reasons for absenteeism and sick leave. The gradual increase in social aggression over the course of middle childhood and into adolescence suggests that social aggression may be more of a learned behavior than physical aggression. Support for this notion also comes from genetically informative research, which showed that genetic effects only account for only around 20% of the variance of social aggression (Brendgen, Dionne, Girard, Boivin, et al, 2005).

If one type of aggression gradually replaces another, one would expect that it is essentially the same children who first display high levels of physical aggression in infancy and toddlerhood and then high levels of social aggression in middle childhood and adolescence. Findings from cross-sectional studies indeed reveal that many aggressive children use both forms of aggression. Moreover, there is evidence that most children whose social aggression increases from 2 to 8 years of age have displayed moderate to high levels of physical aggression in early childhood. Additional evidence for a common root of physical and social aggression comes from genetically informed research (such as studies of twins), which shows that the two types of aggression are in fact explained by the same genetic factors (Brendgen, et al, 2005).

By the same token, however, this research reveals that physical aggression and social aggression share relatively few environmental predictors after the common genetic factors are taken into account. Together, current research thus suggests that a generalized—and to a significant extent genetically driven—individual disposition for aggressive behavior may shift from physical to social aggression as children mature. Whether and when this shift occurs, however, seems to be determined by the extent to which the child is exposed to social environmental influences that discourage the use of physical aggression and tolerate, or even reward, the use of social aggression.

ENVIRONMENTAL PREDICTORS AND DEVELOPMENTAL OUTCOMES OF PHYSICAL AND SOCIAL AGGRESSION

The finding that physical aggression and social aggression share relatively few environmental predictors after the common genetic factors are taken into account raises the question what these environmental predictors may be. Empirical evidence suggests that the family context and particularly parental behaviors are among the main environmental predictors of child physical aggression (Loeber and Farrington, 2000). Thus, harsh disciplinary practices toward the child and a lack of warmth have been shown to foster physically aggressive behavior in the child, and these effects are exacerbated in families living in chaotic circumstances characterized by disorder and noise. Notably, these environmental effects are found even when controlling for genetic effects on physical aggression. Children who are exposed to such a stressful family environment at a young age are less likely to learn how to regulate their behavior and more likely to show continuously high levels of physical aggression.

Parental influences on social aggression have been less frequently examined so far. Evidence suggests, however, that one way children may learn socially aggressive behavior is by observing parents' use of manipulative tactics such as love withdrawal or guilt induction—either toward each other or toward the child (Casas, Weigel, Crick, Ostrov et al., 2006). Parents are not the only source of influence on child behavior, of course, and arguably one of the most important additional influences comes from the peer group. Thus, research shows that affiliation with highly physically aggressive friends may enhance the effect of an existing genetic liability for physical aggression in a child. In contrast, friends' social aggression seems to foster socially aggressive behavior even in children without a genetic predisposition to such behavior. In other words, having highly socially aggressive friends may foster socially aggressive behavior even in children who do not have any preexisting genetic liability for this type of behavior.

Socializing agents such as parents, peers, or teachers may also foster physical or social aggression in the child without even using these behaviors themselves. Social learning theory postulates that a lack of punishment will cause an individual to continue an undesirable behavior that may lead to potential rewards (e.g., obtaining an object or exacting revenge against someone). In line with this notion, a lack of parental monitoring has been related to physical aggression in the offspring. Lack of punishment by socializing agents may play an even greater role in children's use of social aggression. By preschool age, children already believe that socially aggressive responses to provocations are more acceptable than physically aggressive responses. Similarly, parents and teachers tend to view social aggression as less serious than verbal and physical aggression, and they are also less sympathetic to the victims of social aggression. Adults are also much less likely to intervene in instances of social aggression, and if they do intervene, they are less inclined to discipline the socially aggressive perpetrator.

The lack of negative response from the social environment toward the use of social aggression suggests that the developmental outcomes for this type of behavior may differ from those of physical aggression. A plethora of studies have documented the numerous negative consequences of physically aggressive behavior for the perpetrator; these range from problematic relationships with parents, teachers, and the peer group to academic difficulties (low grades, grade retention, high school dropout), as well as later sexual risk behavior (e.g., multiple partners, teen pregnancy) and delinquency (e.g., gang membership, drug use, theft, violence) (Loeber and Farrington, 2000). In addition, there is evidence for a link between physical aggression and internalizing problems (e.g., depression). Do socially aggressive children and adolescents suffer the same fate? Empirical evidence suggests that negative consequences may only occur for individuals displaying extreme levels of social aggression. Specifically, even when accounting for potentially co-occurring physical aggression, children displaying extremely high levels of social aggression are at risk for subsequent delinquency as well as anxiety, depression, and social withdrawal (Crick, Ostrov, and Werner, 2006).

These negative outcomes seem to be even more pronounced for boys than for girls. One possible explanation for this may be that gender non-normative behavior (such as social aggression in boys) may incur more negative reactions from the social environment, which in turn may result in greater adjustment problems in the children displaying gender-non-normative behavior. Social aggression does not seem to be related to negative developmental outcomes in children who display less extreme levels of this type of behavior, however. In fact, some scholars propose that social aggression entails certain benefits for the perpetrator (Werner and Crick, 2004). For example, because social aggression often involves shared contempt and the purposeful exclusion of a third party from a small friendship circle, it may promote cohesiveness and closeness among the perpetrators. Social aggression has also been linked to perceived popularity in the peer group and the receipt of prosocial attention, thus affording the perpetrator a measure of social power over others. The potential for such positive consequences may also explain why, over the course of development, social aggression gradually becomes the predominant type of aggressive behavior for most individuals.

FUTURE DIRECTIONS IN THE STUDY OF AGGRESSION

Although socialization likely plays a crucial role in explaining the development of aggressive behavior in children and adolescents, the links between a putative environmental variable such as the use of corporal punishment by parents and aggression in the child may reflect the genetic transmission of aggressive behavior. Genetically informative research designs such as twin studies make it possible to estimate the contribution of genetic and environmental factors to the development of aggressive behavior. In addition, the etiological mechanisms linking environmental and genetic influences can be explored. Two relevant etiological mechanisms in this context are a possible *gene-environment interaction* and a *gene-environment correlation* (Moffitt, 2005). A gene-environment interaction is indicated if, for example, the effect of an environmental risk factor such as physical maltreatment on aggression is stronger in children with a greater genetic risk for aggressive behavior. In contrast, a gene-environment correlation reflects a mechanism whereby individuals evoke or select their environment as a function of heritable traits.

These environmental features may then help maintain or exacerbate the child's aggressiveness. For example, a heritable characteristic such as aggression may in turn trigger aggressive responses from the environment, a process referred to as *evocative Gene-Environment correlation*.

Alternatively, an *active Gene-Environment correlation* would be indicated if aggressive children seek out specific social environments, such as peers with similar behavioral characteristics, as a function of their genetic disposition toward this behavior. Studies indeed provide support for gene-environment interaction and gene-environment correlation processes in predicting childhood aggression.

Even genetically informative studies, however, cannot provide proof of the causality of effects between aggressive behavior and putative predictors or outcomes of such behavior. A test of causality can only be achieved through experimental manipulation. In recent decades, however, ethical concerns have been voiced about research involving the direct manipulation of aggressive behavior or its putative antecedents. One innovative way to circumvent these pitfalls may be offered by experimental intervention studies that include both a treatment group and a control group. For example, the causal link between aggression and its putative antecedents can be tested by reducing the hypothesized antecedent risk factor through intervention and by examining whether aggressive behavior also decreases subsequently. With innovative methods and analytical tools such as these, future research will yield an ever better understanding of the etiology of aggression. By the same token, findings from this research will contribute to the optimization of preventive interventions targeting early risk factors (e.g., parenting behaviors) and to the development of school policies targeting context factors (e.g., teacher awareness and classroom management rules) to help stem aggressive behavior in all its forms (Zins, Elias, & Maher, 2007).

SEE ALSO Volume 1: *Bullying and Peer Victimization; Genetic Influences, Early Life; Media Effects; Mental Health, Childhood and Adolescence; Peer Groups and Crowds; Socialization; Socialization, Gender; Theories of Deviance.*

BIBLIOGRAPHY

Bandura, A., Ross, D., & Ross, S. A. (1961). Transmission of aggression through imitation of aggressive models. *Journal of Abnormal Social Psychology, 63,* 575–582.

Brendgen, M., Dionne, G., Girard, A., Boivin, M., Vitaro, F., & Pérusse, D. (2005). Examining genetic and environmental effects on social aggression: A study of 6-year-old twins. *Child Development, 76,* 930–946.

Casas, J. F., Weigel, S. M., Crick, N. R., Ostrov, J. M., Woods, K. E., Jansen Yeh, E. A., et al. (2006). Early parenting and children's relational and physical aggression in the preschool and home contexts. *Journal of Applied Developmental Psychology, 27,* 209–227.

Côté, S. M. (2007). Sex differences in physical and indirect aggression: A developmental perspective. *European Journal on Criminal Policy and Research, 13,* 183–200.

Crick, N. R., Ostrov, J. M., & Werner, N. E. (2006). A longitudinal study of relational aggression, physical aggression, and children's social–psychological adjustment. *Journal of Abnormal Child Psychology, 34,* 131–142.

DiLalla, L. F. (2002). Behavior genetics of aggression in children: Review and future directions. *Developmental Review, 22,* 593–622.

Loeber, R., & Farrington, D. (2000). Epidemiology of juvenile violence. *Child and Adolescent Psychiatric Clinics of North America, 9,* 733–748.

Moffitt, T. E. (2005). The new look of behavioral genetics in developmental psychopathology: Gene-environment interplay in antisocial behaviors. *Psychological Bulletin, 131,* 533–554.

Sampson, R. J., & Laub, J. H. (1993). Crime in the making: Pathways and turning points through life. *Crime Delinquency, 39,* 396.

Sutton, J., Smith, P. K., & Swettenham, J. (1999). Social cognition and bullying: Social inadequacy or skilled manipulation? *British Journal of Developmental Psychology, 17*(3), 435–450.

Tremblay, R. E., & Nagin, D. S. (2005). The developmental origins of physical aggression in humans. In R. E. Tremblay, W. W. Hartup, & J. Archer (Eds.), *Developmental origins of aggression* (pp. 83–106). New York: Guilford.

Werner, N. E., & Crick, N. R. (2004). Maladaptive peer relationships and the development of relational and physical aggression during middle childhood. *Social development, 13*(4): 495–514.

Zins, J. E., Elias, M. J., & Maher, C. A. (2007). *Bullying, victimization, and peer harassment: A handbook of prevention and intervention.* New York: Haworth Press.

R. Mara Brendgen

ALCOHOL USE, ADOLESCENT

SEE Volume 1: *Drinking, Adolescent.*

ANTISOCIAL BEHAVIOR

SEE Volume 1: *Aggression, Childhood and Adolescence; Bullying and Peer Victimization; Crime, Criminal Activity in Childhood and Adolescence; Theories of Deviance;* Volume 2: *Crime, Criminal Activity in Adulthood.*

ASSIMILATION

The question of immigrants' progress lies at the heart of the immigration debate in the early 21st century. Since the mid-1990s, the debate has centered on the question of whether the predominantly non-European immigrants are ever able to assimilate into mainstream American society. Consequently, the matter of immigrant incorporation generates the most uncertainty and controversy (Zhou & Lee, 2007). The assimilation perspective has dominated much sociological thinking on immigrant adaptation since the turn of the 20th century. Central to this perspective are the assumptions that there is a natural process by which diverse ethnic groups come to share a common culture and gain equal access to the opportunity structure of the host society, that this process entails the gradual abandonment of old-world cultural and behavioral patterns in favor of new ones, and that once, set in motion, this process moves inevitably and irreversibly toward assimilation.

THE CLASSICAL ASSIMILATION PERSPECTIVE

Classical assimilation scholars generally assume that the host society consists of a single mainstream dominated by a majority group (White Anglo-Saxon Protestants, or WASPs). Migration leads to a situation of the *marginal man* in which ethnic minority groups are pulled toward the host culture but are drawn back by the culture of their origin (Park, 1928; Stonequist, 1937). As time passes, however, diverse ethnic groups from underprivileged backgrounds go through the painful process through a natural race relations cycle of contact, competition, and accommodation as group members abandon their old ways of life to *melt* completely into the host society's mainstream (Park, 1928). These scholars also acknowledge the potency of institutional factors such as family socioeconomic status (SES), phenotypical ranking, and racial/ethnic subsystems in determining the rate of assimilation. In particular, the assimilation of some ethnic minorities is especially problematic because the subordination of those groups often is based on ascribed characteristics, or characteristics one is born with, such as skin color, language of origin, and religion. The process of assimilation of readily identifiable minority groups, especially African Americans, is likely to be confined within racial-caste boundaries, leading to intergroup differences in the pace of assimilation (Warner & Srole, 1945).

Milton Gordon (1964) devised a typology of assimilation to capture the complexity of the process, ranging from cultural, structural, marital, identificational, attitude-receptional, and behavior-receptional assimilation to civic assimilation. In Gordon's view, cultural assimilation, or acculturation, is a necessary first step and is considered the top priority on the agenda of immigrant adjustment but does not lead automatically to other forms of assimilation (e.g., large-scale entrance into the institutions of the host society or intermarriage). In certain circumstances acculturation may take place and continue indefinitely even when no other forms of assimilation occur. Ethnic groups may continue to be distinguished from one another because of spatial isolation and lack of contact, and their

full assimilation depends ultimately on the degree to which these groups gain the acceptance of the dominant group in the host society. Structural assimilation, in contrast, is the "keystone of the arch of assimilation" that inevitably leads to other stages of assimilation (Gordon, 1964, p. 81). Although vague about how groups advance from one stage to another and what causes the change, Gordon anticipates that most ethnic groups eventually will lose all their distinctive characteristics and cease to exist as ethnic groups as they pass through different stages of assimilation.

From the classical assimilation perspective, distinctive ethnic traits such as old-world cultures, native languages, and ethnic enclaves are sources of disadvantages. Those disadvantages affect assimilation negatively, but the negative effects are reduced greatly in each of the successive generations because native-born generations use English as the primary means of communication and become more and more similar to the mainstream American population in life skills, manner, and outlook. Although complete acculturation to the dominant culture may not ensure all ethnic groups full social participation in the host society, immigrants are expected to free themselves from their old cultures so that they can begin rising up from marginal positions. Between the 1920s and the 1950s the United States seemed to have absorbed the great waves of immigrants who arrived primarily from Europe. Past sociological studies indicated progressive trends of socioeconomic achievement across immigrant generations and increasing rates of intermarriage as determined by educational attainment, job skills, length of stay since immigration, English proficiency, and levels of exposure to American culture (Alba, 1985; Handlin, 1973; Lieberson & Waters, 1988).

In the 21st century Richard Alba and Victor Nee (2003) revamped the assimilation perspective in *Rethinking the American Mainstream*. They argued that contemporary institutional changes, from civil rights legislation to immigration law, combined with individual incentives and motivation, have reshaped the context of immigrant reception profoundly, making it more favorable for the assimilation of newcomers and their children despite persistent racial discrimination and economic restructuring. Instead of assuming a single, unilateral WASP mainstream into which immigrants are expected to assimilate, Alba and Nee reconceptualized the American mainstream as one that encompasses "a core set of interrelated institutional structures and organizations regulated by rules and practices that weaken, even undermine, the influence of ethnic origins per se" (p. 12). This mainstream may include members of formerly excluded ethnic or racial groups, and it may contain not just the middle class or affluent suburbanites but the working class or the central-city poor. Individual experiences of intergenerational mobility among immigrants are thus not dissimilar to those in the mainstream.

Upward, horizontal, or downward social mobility is possible for immigrants and their children as much as it is for those in mainstream society.

However, according to Alba and Nee (2003), the process of assimilation varies from individual to individual and from group to group, depending on two causal mechanisms. The first is a set of proximate causes that involve an individual's or group's purposive action and social networks (particularly exchange mechanisms of social rewards and punishments within a primary group and community) and the forms of capital (human, social, and financial) an individual or group possesses. The second is a set of distal causes that are embedded in larger social structures such as the state and the labor market. Alba and Nee suggest that all immigrants and their descendants eventually assimilate, but not necessarily in a single direction as predicted by the classical theory. They believe that "an expectation of universal upward mobility for any large group is unrealistic" (p. 163). This theoretical framework helps explain how immigrants, particularly those of non-European origin and working-class background, incorporate into the mainstream at different rates and by different measures. Despite their definition of the mainstream as inclusive, however, their notion of successful assimilation explicitly refers to incorporation into the middle class, not the working or lower class (Zhou & Lee, 2007).

ANOMALIES

Since the late 1960s the classical assimilation perspective and its application to more recently arrived non-European immigrant groups has been challenged. Instead of eventual convergence into the middle-class mainstream as predicted by assimilation theories, several anomalies appear to be significant. The first concerns persistent ethnic differences across generations. Classical theories predict assimilation as a function of the length of U.S. residence and the number of succeeding generations, but this is not how it seems to work. Prior research has revealed an opposite pattern: the longer the U.S. residence, the more maladaptive the outcomes, whether measured in terms of school performance, aspirations, or behavior (Portes & Rumbaut, 2001; Suárez-Orozco & Suárez-Orozco, 2001; Telles & Ortiz, 2007). Moreover, even small differences in parental educational and occupational status result in substantial differences in children's educational and occupational mobility. In a study of the Irish, Italian, Jewish, and African Americans, for example, Joel Perlmann (1988) showed that even with family background factors held constant, ethnic differences in levels of schooling and economic attainment persisted in the second generation and later generations and that schooling was not equally commensurate

with occupational advancement for African Americans compared with European Americans across generations.

Another anomaly is what Herbert Gans (1992) describes as the second generation decline. Gans notes three possible scenarios for the contemporary new second generation: education-driven mobility, succession-driven mobility, and niche improvement. He observes that immigrant children from less advantaged socioeconomic backgrounds had a much harder time than other middle-class children succeeding in school and that a significant number of the children of poor, especially dark-skinned poor, immigrants had multiple risks of being trapped in permanent poverty in an era of stagnant economic growth and in the process of Americanization because these immigrant children "will either not be asked, or will be reluctant, to work at immigrant wages and hours as their parents did but will lack job opportunities, skills and connections to do better" (pp. 173–174). Gans anticipated downward mobility for many immigrants, including some of those from middle-class backgrounds, and dismal prospects for children of the less fortunate who must confront high rates of unemployment, crime, alcoholism, drug use, and other pathologies associated with poverty and the frustration of rising expectations.

Still another anomaly relates to the counterintuitive phenomena associated with contemporary immigration. In the fastest-growing knowledge-intensive industries in the United States, foreign-born engineers and other highly skilled professionals disproportionately take key technical positions and even ownership positions, such as Chinese and Indian immigrants in Silicon Valley. Highly skilled and economically resourceful immigrants appear to skip several rungs on the mobility ladder and bypass the conventional enclave-to-suburbia route immediately upon arrival. In immigrant enclaves, ethnic commercial banks, corporate-owned expensive restaurants, and chain supermarkets stand side by side with traditional rotating credit associations, coffee and tea houses, and mom-and-pop stores, opening alternative paths to social mobility without the loss of ethnic distinctiveness (Zhou, 1992).

In urban public schools neither valedictorians nor delinquents are atypical among immigrant children regardless of timing and racial or socioeconomic backgrounds (Zhou & Bankston, 1998). Although immigrant children are overrepresented on lists of award winners and in academic fast tracks, many others are vulnerable to multiple high-risk behaviors, school failure, membership in street gangs, and youth crime. Even Asian Americans, the so-called model minority, have seen a steady rise in youth gang membership. Some Asian gang members are from suburban middle-class families, attend magnet schools, and are good students. Some of the notorious

Asian gangs include the Flying Dragons, the Fuk Ching, the Viet Ching, and the Korean Power. These anomalies indicate a significant gap between theory and reality.

SEGMENTED ASSIMILATION

Segmented assimilation as a middle-range theory emerged in the early 1990s with the 1993 publication of "The New Second Generation" by Alejandro Portes and Min Zhou in the *Annals of the American Academy of Political and Social Sciences.* The theory is built on the empirical observations that the host society is highly stratified by class and race/ethnicity, that immigrant social mobility is contingent on ethnic specificities and structural circumstances, and that immigrants arrive with different amounts and kinds of resources to cope with resettlement and socioeconomic incorporation.

Unlike classical assimilation theories that posit an irreversible and unidirectional path leading to eventual incorporation into an undifferentiated, unified, and white middle-class mainstream by all immigrants, the segmented assimilation theory conceives of the mainstream society as being shaped by systems of class and racial stratifications. It emphasizes the interaction between race/ethnicity and class and between group membership and larger social structures that intentionally or unintentionally exclude nonwhites. It attempts to delineate the multiple modes of incorporation that emerge among contemporary immigrants and their offspring, account for their different destinies of convergence (or divergence) in the new homeland, and addresses the ways in which particular contexts of exit and reception of national-origin groups affect outcomes.

From this perspective, the process of assimilation may take multiple pathways, sometimes with different turns leading to varied outcomes. Three main patterns are discernible. The first is the time-honored upward mobility pattern dictating acculturation and economic integration into the normative structures of mainstream middle-class America. This is the old-fashioned path of severing ethnic ties; unlearning old-world values, norms, and behavioral patterns; and adapting to the WASP core associated with the middle class. The second pathway is the downward mobility pattern that dictates acculturation

Pledge of Allegiance. *An integrated classroom recites the Pledge of Allegiance.* © **BETTMANN/CORBIS.**

and parallel integration into the margins of American society. This is the path of adapting to native subcultures in direct opposition to the WASP core culture or creating hybrid oppositional subcultures associated with native groups trapped in the margins of the host society or the bottom rungs of the mobility ladder. The third pathway is socioeconomic integration into mainstream American society with lagged and selective acculturation and deliberate preservation of the values and norms, social ties, and ethnic institutions of the immigrant community. This is the path of deliberately reaffirming ethnicity and rebuilding ethnic networks and structures for socioeconomic advancement into middle-class status.

The segment of society into which an immigrant or ethnic group assimilates is determined by the unique contexts of exit and reception. The context of exit entails a number of factors, including the premigration resources that immigrants bring with them (e.g., money, knowledge, and job skills), the social class status already attained by immigrants in their homelands, motivations, and the means of migration. The context of reception includes the positioning of the national-origin group in the system of racial stratification, government policies, labor market conditions, and public attitudes and the strength and viability of the ethnic community in the host society. Segmented assimilation theory focuses on the interaction of these two sets of factors, predicting that particular contexts of exit and reception can create distinctive cultural patterns and strategies of adaptation, social environments, and tangible resources for the group and give rise to opportunities or constraints for the individual independent of individual socioeconomic and demographic characteristics.

Whereas the unique contexts of exit and reception lead to distinct modes of incorporation for immigrant and refugee groups, different modes of incorporation explain variations in the contexts in which individuals strive to "make it" in their new homeland. For example, to explain why immigrant Chinese or Korean children generally do better in school than do immigrant Mexican or Central American children even when they come from families with similar income levels, live in the same neighborhood, and go to the same school, one must look to the unique contexts in which those children grow up. Among the various contextual factors that may influence academic outcomes, one stands out among Chinese and Koreans: an ethnic community with an extensive system of supplementary education, including nonprofit ethnic language schools and private institutions for academic tutoring, enrichment, standardized test drills, college preparation and counseling, and extracurricular activities aiming mainly at enhancing the competitiveness of children's prospects for higher education. The ethnic system of supplementary education is built not only on the strong human capital and financial resources that Chinese and Korean immigrants brought with them to the new country but also on their experience with a competitive educational system in the homeland (Zhou & Kim, 2006). Mexican and Central American communities lack similar ethnic social structures that generate resources conducive to education. Moreover, the children of Mexican and Central American immigrants who live in the same neighborhoods as Chinese and Korean immigrants largely are excluded from these ethnic resources.

Empirically, segmented assimilation is measured by a range of observable socioeconomic indicators, such as educational attainment, employment status, income, and home ownership. For the children of immigrants indicators of downward assimilation include dropping out of school, having children early, and being arrested or sentenced for a crime, for these variables are strong predictors of future low educational attainment, occupational status, income, and likelihood of home ownership. Numerous qualitative and quantitative works have produced evidence that supports segmented assimilation predictions that the second generation is likely to assimilate upwardly, downwardly, or horizontally into an American society that is highly segmented by class and race and to do so in different ways.

From the segmented assimilation perspective, downward assimilation is only one of several possible outcomes. Curiously, the segmented assimilation theory often is misinterpreted as suggesting and predicting a single outcome—downward assimilation—and therefore criticized for being pessimistic about the immigrant second generation. Nonetheless, to refute the segmented assimilation theory or state that the second generation will move into the mainstream middle class sooner or later, one must demonstrate that both of the following cases are false: the proportions of those falling into the major indicators of downward assimilation—high school dropouts, teenage pregnancies, and arrests for breaking the law—are insignificant for each national-origin or ethnic group and that the differences in outcomes are randomly distributed across different national-origin or ethnic groups regardless of the modes of incorporation of those groups. Despite the fact that the majority of the second generation in the early 21st century is likely to follow the path of upward social mobility taken by the children and grandchildren of earlier immigrant waves, those who are at high risk of falling through the cracks leading to downward assimilation would need external supports, such as quality schools, language assistance, after-school programs, and organized youth leadership activities from the local community and the state.

SEE ALSO Volume 1: *Bilingual Education; Immigration, Childhood and Adolescence; Oppositional Culture; Racial Inequality in Education; Socialization, Race.*

BIBLIOGRAPHY

Alba, R. D. (1985). *Italian Americans: Into the twilight of ethnicity.* Englewood Cliffs, NJ: Prentice-Hall.

Alba, R. D., & Nee, V. (2003). *Remaking the American mainstream: Assimilation and contemporary immigration.* Cambridge, MA: Harvard University Press.

Gans, H. J. (1992). Second-generation decline: Scenarios for the economic and ethnic futures of the post-1965 American immigrants. *Ethnic and Racial Studies, 15*(2), 173–192.

Gordon, M. M. (1964). *Assimilation in American life: The role of race, religion, and national origins.* New York: Oxford University Press.

Handlin, O. (1973). *The uprooted.* (2nd ed.). Boston: Little, Brown.

Lieberson, S., & Waters, M. (1988). *From many strands: Ethnic and racial groups in contemporary America.* New York: Russell Sage Foundation.

Park, R. E. (1928). Human migration and the marginal man. *American Journal of Sociology, 33*(6), 881–893.

Perlmann, J. (1988). *Ethnic differences: Schooling and social structure among the Irish, Italians, Jews, and blacks in an American city, 1880–1935.* New York: Cambridge University Press.

Portes, A., & Rumbaut, R. G. (2001). *Legacies: The story of the immigrant second generation.* Berkeley: University of California Press.

Portes, A., & Zhou, M. (1993). The new second generation: Segmented assimilation and its variants among post-1965 immigrant youth. *Annals of the American Academy of Political and Social Sciences, 530,* 74–96.

Stonequist, E. V. (1937). *The marginal man: A study in personality and culture conflict.* New York: C. Scribner's Sons.

Suárez-Orozco, C., & Suárez-Orozco, M. M. (2001). *Children of immigration.* Cambridge, MA: Harvard University Press.

Telles, E., & Ortiz, V. (2007). *Generations of exclusion: Mexican Americans, assimilation, and race.* New York: Russell Sage Foundation.

Warner, W. L., & Srole, L. (1945). *The social systems of American ethnic groups.* New Haven, CT: Yale University Press.

Zhou, M. (1992). *Chinatown: The socioeconomic potential of an urban enclave.* Philadelphia, PA: Temple University Press.

Zhou, M., & Bankston, C. L. III. (1998). *Growing up American: How Vietnamese children adapt to life in the United States.* New York: Russell Sage Foundation.

Zhou, M., & Kim, S. S. (2006). Community forces, social capital, and educational achievement: The case of supplementary education in the Chinese and Korean immigrant communities. *Harvard Educational Review, 76*(1), 1–29.

Zhou, M., & Lee, J. (2007). Becoming ethnic or becoming American? Tracing the mobility trajectories of the new second generation in the United States. *Du Bois Review, 4*(1), 1–17.

Min Zhou

ATTACHMENT THEORY

Close relationships are central to individuals' physical and emotional well-being, and attachment theory has been a central framework that researchers have used to study close relationships. Although most attachment research has attempted to elucidate the infant-mother relationship, attachment has been conceptualized as a life course phenomenon that "plays a vital role … from the cradle to the grave" (Bowlby, 1969, p. 208). Accordingly, attachment theory and research have expanded to reflect the life course focus and illuminate attachment-related processes and sequelae during the preschool years, middle childhood, adolescence, and adulthood. The publication of *Handbook of Attachment* in 1999 (Cassidy & Shaver, 1999) and the publication of a revised and updated second edition in 2008 are testimony to the growth and breadth of attachment theory and research since its emergence in the 1960s and 1970s.

BOWLBY'S THEORY OF ATTACHMENT

While working at a school for maladjusted children and training as a child psychiatrist at the British Psychoanalytic Institute, John Bowlby developed a conviction that a child's early experiences in the family, especially with the mother, are fundamental to psychological development and well-being. This view was in sharp contrast to the psychoanalytic theory of the time, which discounted experience and emphasized inner fantasy as the root of emotional disturbance.

In gathering evidence about the role of early experience, Bowlby focused on maternal separation and loss, and his first published paper, "Forty-Four Juvenile Thieves: Their Characters and Home Life" (1944), showed how the absence of a consistent caregiver was associated with later delinquency. In 1951, in response to an invitation from the World Health Organization (WHO) to report on the mental health of homeless children, Bowlby published *Maternal Care and Mental Health,* which underscored his major premise about the centrality of a continuous and warm relationship with the mother (or "mother substitute") for a child's psychological well-being and the negative consequences of maternal deprivation (e.g., prolonged separation). However, questions about how and why maternal deprivation is so disruptive remained unanswered in the WHO report. Bowlby developed his ideas further and explicated his theory of attachment in the three-volume work *Attachment and Loss* (Bowlby, 1969, 1973, 1980). Drawing on theory and research in psychoanalysis, evolutionary biology, ethology, cognitive development, and control systems theory, Bowlby posited that attachment serves the biological function of protection by maintaining contact between the caregiver and the infant. From an evolutionary perspective such maintenance of proximity ultimately aids in the survival of the species.

Bowlby (1969) delineated three hierarchically organized components of attachment. Attachment behaviors (e.g., crying, smiling, approaching) maintain proximity to the caregiver. Those behaviors are organized within the individual into an attachment behavioral system whose goal is maintenance of proximity to the caregiver; the goal of the attached individual is felt security. The concept of a behavioral system is rooted in control systems theory, and the system is analogous to a thermostat in that it regulates the proximity between the infant and the caregiver as circumstances vary in the environment (e.g., threats, location of caregiver) and the infant (e.g., tired, sick).

The attachment system is one of multiple behavioral systems and complements the child's exploratory system. That is, whereas the attachment figure provides comfort in times of stress, that figure also acts as a "secure base" from which the child may explore the environment in nonthreatening and relatively stress-free situations. The attachment bond is the tie that an infant has to his or her caregiver. In contrast to attachment behaviors, which are situational, an attachment bond exists over time and does not depend only on the presence of attachment behaviors.

Bowlby's theory of attachment has proved to be an effective perspective from which to examine children's socioemotional development, and his discussion of the internal working model provided a new way to conceptualize continuity in development. The internal working model is described as a mental representation of an infant's relationship with his or her attachment figure, which is constructed continuously as the infant interacts with the environment (Bowlby, 1969, 1973). This internal representation of the parent-child attachment relationship provides the mechanism of continuity between the quality of children's early relationships with parents and their later socioemotional adjustment. Attachment theory posits that a child carries the internal working model of the parent-infant relationship forward into his or her close relationships with others. Central to carrying forward the attachment relationship is the child's need to maintain a coherent sense of self, and in doing that the child may behave in a way that evokes the same treatment in later relationships that has occurred within his or her early attachment relationships (Sroufe & Fleeson, 1986).

PATTERNS OF INFANT-MOTHER ATTACHMENT

Bowlby's theory of attachment put forth groundbreaking notions regarding the mother-infant bond, but it was the pioneering work of Mary Salter Ainsworth that led to a proliferation of attachment research that continues in the early 21st century. Ainsworth joined Bowlby's research unit at the Tavistock Clinic in London from 1951 to 1954, a time when Bowlby was constructing his theory of

attachment. Influenced by Bowlby's thinking and the ethological methods employed in his laboratory, Ainsworth went on to conduct two studies—the first in Uganda and the second in Baltimore, Maryland—that employed detailed naturalistic observations of mothers and infants. Those studies provided some of the first empirical evidence of Bowlby's conceptualization of infant-mother attachment. Moreover, the development and validation of the Strange Situation among the 26 mother-infant pairs in the Baltimore study (Ainsworth & Wittig, 1969; Ainsworth, Blehar, Waters, & Wall, 1978) was a critical methodological contribution that gave researchers an empirical method to test the tenets of attachment theory.

The Strange Situation is a laboratory-based paradigm designed to heighten an infant's attachment behavioral system and typically is administered to infants between 12 and 18 months of age. During the eight episodes of the Strange Situation, which occur over the course of approximately 20 minutes, the infant is exposed to increasing, although moderate, levels of stress (e.g., interaction with a strange adult, separation from the mother). From observations of infant behavior during the Strange Situation, especially behavior during two reunion episodes with the mother, Aisnworth and her colleagues (1978) derived three classifications of infant attachment: secure (type B), insecure-avoidant (type A), and insecure-resistant (type C).

In normative U.S. samples approximately 65% of infants are classified as secure, 20% as avoidant, and 15% as resistant (van Ijzendoorn & Kroonenberg, 1988). Infants classified as secure tend to seek out the mother in the reunion episodes of the Strange Situation and appear happy to see her return. If distressed during a separation, the infant is able to be comforted by the mother during the reunion. Infants classified as avoidant tend to demonstrate lower levels of distress during separations and greater avoidance of the mother (e.g., turning away, ignoring) during reunions. Infants classified as insecure-resistant become highly distressed during separations from the mother, yet in the reunion episodes they have difficulty being comforted by the mother. Their behavior may alternate between seeking contact and angrily pulling away from or resisting contact with the mother.

Ainsworth and her colleagues (1978) showed that the attachment patterns that emerged in the Strange Situation at 12 months of age were related to patterns of mother-infant interaction assessed in extensive home observations (approximately 72 hours per family) during the first year of life. That is, an infant's behavior in the Strange Situation seemed to reflect his or her prior experience and current expectations of the mother's responsiveness and availability. Ainsworth and her colleagues (1978) showed that mothers of infants classi-fied as secure tended to be sensitive to the infant's needs during everyday interactions in the home and responded to the infant's distress in an appropriate, timely, and sensitive fashion. When confronted with the moderate stress of the Strange Situation, those infants' behavior (e.g., happy to see mother return, approach mother if distressed) seemed to reflect an expectation that the mother would be available and effective in responding to signals of distress and the need for proximity. In contrast, mothers of insecure-avoidant infants tended actively to reject the infants' signals of distress and attempts to maintain proximity, especially close physical contact, and those infants minimized displays of distress in the Strange Situation as a strategy to maintain prox-imity to the mother. Mothers of insecure-resistant infants tended to be inconsistently responsive to the infants' distress, and those infants heightened their attachment behavior in the Strange Situation as a way to maintain proximity. This correspondence between Ainsworth's intensive naturalistic observations in the home and infant behavior in the brief laboratory episodes makes the Strange Situation a powerful methodological tool.

Some infants demonstrate both avoidance and resistance or odd behaviors (e.g., repetitive rocking) in the Strange Situation and do not fall clearly into one of the three attachment classifications described above. In investigating these unclassifiable infants, Mary Main and colleagues (Main & Solomon, 1990; Main & Weston, 1981) developed a fourth attachment classifi-cation (disorganized/disoriented; type D) that reflects the lack of a coherent attachment strategy. Infants in high-risk populations (e.g., exposure to maltreatment, parental psychopathology) have been overrepresented in the disorganized classification.

ANTECEDENTS OF INFANT-MOTHER ATTACHMENT

Since the 1978 Baltimore study of Ainsworth and col-leagues hundreds of researchers have examined maternal behavior, particularly sensitivity, as an antecedent of infant-mother attachment security as it is assessed in the Strange Situation. In a meta-analysis of 66 studies, De Wolff and van IJzendoorn (1997) reported that the sen-sitivity-security association was significant yet modest (i.e., 0.24). The modest extent of that association may be due in part to the use of global ratings of maternal sensitivity in low-stress situations (e.g., play). In line with Bowlby's (1969) emphasis on attachment as a biobeha-vioral system of protection, some have argued that sensi-tivity to infant distress is paramount to the formation of a secure attachment (Goldberg, Grusec, & Jenkins, 1999; Thompson, 1998), and in one study sensitivity to infant

distress versus nondistress at age 6 months predicted infant-mother attachment security at age 15 months (McElwain & Booth-LaForce, 2006).

Complementing correlational evidence of sensitivity-security associations, intervention studies have provided support for sensitivity as a causal factor in the develop-ment of a secure infant-mother attachment (Bakermans-Kranenburg, van IJzendoorn, & Juffer, 2003). For instance, in a low-income sample of 100 Dutch infants who were high in irritability at birth, a brief intervention aimed at fostering maternal reading of infant cues and sensitive responsiveness was administered between 6 and 9 months to half the sample. The results indicated that infants in the intervention group were more likely to receive secure attachment classification at the end of the first year (68% versus 28%) (van den Boom, 1994).

Although maternal sensitivity is important, it is not the only potential antecedent of a secure attachment. Child characteristics also must be considered, and the role of child temperament in particular has elicited debate. Critics of attachment theory argue that security as assessed in the Strange Situation simply may reflect individual differences in child temperament. However, evidence of a direct temperament-security association has been mixed, and intervention studies such as the one by D. C. van den Boom (1994) provide strong counter-evidence of a direct effect of temperament. That is, even in infants high in irritability, intervention efforts to bol-ster maternal sensitivity increased the chances of a secure attachment. In contrast, a high rate of insecure attach-ment (72%) was found in the control group, and the combination of infant irritability and low family income that characterized this sample might have impeded sensi-tive responsiveness among the mothers in the control group. More complex models of the way child tempera-ment interacts with risk or protective factors to predict attachment security are needed.

The mother-infant dyad is situated in a larger con-text. From an ecological perspective, the quality of the marital relationship, levels of nonspousal social support, and maternal psychological well-being are factors that may affect the infant-mother attachment relationship directly or indirectly (Belsky, 1999). Attention also has been paid to the effect of maternal employment on infant-mother attachment. In contrast to Bowlby's focus on prolonged and chaotic separations, separations expe-rienced by children in day care tend to be brief and routine. Nonetheless, because of the large number of U.S. infants who experience nonmaternal care, investigat-ing its effect on attachment is essential.

The NICHD Study of Early Child Care launched in 1991 is the most comprehensive study to date of this question. The participants were 1,364 children and their families from 10 sites across the United States, and an array of assessments in home, laboratory, and day care settings were conducted beginning at one month of age. The findings indicated that maternal sensitivity and responsiveness predicted attachment security at 15 months, whereas child care variables (hours, onset, and quality) did not (NICHD Early Child Care Research Network, 1997). Child care variables, however, interacted with maternal sensitivity so that infants who experienced lower-quality child care, more hours per week in care, or more changes in care in combination with low maternal sensitivity had a higher likelihood of insecure attachment.

SEQUELAE OF INFANT-MOTHER ATTACHMENT

In line with the tenet of attachment theory that early attachment relationships influence children's subsequent functioning, a second major area of attachment research has focused on the sequelae of attachment. In the process infant-mother attachment has been related to a host of outcomes, including social competence, peer acceptance and status, friendship quality, and behavioral problems. However, there is a range of views about how much attachment theory should be expected to predict (Belsky & Cassidy, 1994; Thompson, 1998). A narrow view suggests that because attachment between the mother and the infant is a close, intimate relationship, early attachment should be related to the subsequent quality of the mother-child relationship as well as the child's mode of relating in other close, intimate relationships. A broader view posits that attachment should be related to children's interpersonal interactions with others more generally (e.g., sociability, empathy). An even broader view asserts that attachment provides a basic foundation for later development in multiple domains, including personality, cognition, and language.

Both quantitative and qualitative reviews seem to concur with a more narrow view of attachment (Belsky & Cassidy, 1994; Schneider, Atkinson, & Tardiff, 2001; Thompson, 1998; van IJzendoorn, Dijkstra, & Bus, 1995). Associations with early attachment seem to be most consistent when outcomes include indices of the mother-child relationship or the child's relationships with close others (e.g., friends), although even those associations tend to be modest. Less consistent and weaker associations emerge between early attachment and children's more general social competencies, interactions with unfamiliar others, and cognitive abilities.

In extending and refining predictions about the sequelae of attachment, the next generation of research should attempt to identify processes by which early attachment exerts an influence on later functioning. In light of Bowlby's notion of the internal working model and others' extensions of his theory, children's processing of social information (Main, Kaplan, & Cassidy, 1985), mentalistic understanding of others (Fonagy & Target, 1997), and regulation of emotion (Cassidy, 1994) appear to be prime candidates for potential intervening mechanisms. Furthermore, most research has examined secure-insecure differences on child outcomes because of the relatively low numbers of children in each of the insecure groups (avoidant, resistant, disorganized). However, as described above, the different insecure groups experience different patterns of caregiving, and when examined separately, children in the avoidant and resistant groups exhibit distinct behavioral and psychological outcomes. Thus, detection of attachment-related differences may be hindered by combining children into one "insecure" group. Future research should continue to illuminate the differential sequelae of the insecure groups.

BEYOND MOTHERS

Infants typically develop attachment bonds with multiple caregivers, although the primary caregiver is usually the preferred attachment figure during times of stress. Research on infant attachment to fathers, day care providers, and teachers has provided a needed addition to the study of the mother-infant dyad (Belsky, 1999). In light of the increasing time fathers are spending in the caregiving role in recent years, a better understanding of the antecedents and sequelae of infant-father attachment is an important direction for further inquiry. Relatedly, attachment figures during infancy and childhood are most often parents and other adult caregivers, yet during adolescence and adulthood close friends and romantic partners may become central (Hazan & Shaver, 1994). As attachment research and theory move beyond the mother-child dyad, important considerations arise regarding the integration and organization of working models across partners and time, concordance versus discordance of attachment security with multiple attachment figures, and the joint contributions of multiple attachments for interpersonal functioning across the life course.

BEYOND INFANCY

Although theory and research have focused on attachment during infancy as measured through the use of the Strange Situation, Bowlby conceptualized attachment as a lifelong phenomenon; accordingly, attachment research beyond infancy has burgeoned. Age-appropriate measures of attachment have been developed and include modified Strange Situation procedures and observations of secure base behavior in the home using a Q-sort methodology for preschool- and school-age children as well as self-report and narrative measures for children, adolescents, and adults. After Ainsworth's Strange Situation, the most extensively used and validated attachment measure is the Adult Attachment Interview (AAI) developed by Main and colleagues (Main et al., 1985). The AAI is a semistructured interview in which adult respondents are asked about childhood experiences. Labor-intensive coding of interview transcripts involves assessing attachment classifications in a way that is based not on the content of the responses (what happened in the individual's childhood) but on their coherence (responses are truthful, relevant, and succinct). In this regard, the AAI attempts to capture the adult's state of mind with respect to attachment. Paralleling the infant classificatory system, the AAI classifications include autonomous (secure), dismissing (avoidant), preoccupied or enmeshed (resistant), and unresolved (disorganized).

With the development of attachment assessments beyond infancy, it has been possible to examine the continuity of attachment patterns across time and, if discontinuity is detected, whether intervening events account for change (i.e., lawful discontinuity). Attachment theory posits that primary attachment relationships become consolidated during the first 5 years of life, and although children are open to new experience, they become resistant to change later in development unless there has been a major change in the environment (Bolwby, 1969). Evidence for continuity, however, is mixed (Thompson, 1998; Waters, Hamilton, & Weinfield, 2000). For instance, among long-term longitudinal studies, stability in attachment assessed in infancy by means of the Strange Situation and in late adolescence or young adulthood by means of the AAI has shown high stability in two studies (64–77%) but low stability in two others (39–51%). Importantly, however, in several of these long-term studies instability of attachment appears to be lawful in some cases in that intervening life events (e.g., parental mental illness, child maltreatment) predict change in attachment status from infancy to adulthood (Waters et al., 2000).

In keeping with infant research on attachment patterns, investigations of attachment beyond infancy have focused primarily on individual differences. Although this focus has been productive, explication of development beyond infancy is needed. According to Bowlby (1969), development of attachment occurs early in life, and the last stage—the "goal-corrected partnership"—emerges by age five. Some attachment theorists suggest, however, that the preschool and adolescent years, with their rapid advances in neurological and cognitive capacities, may be especially fruitful periods for considering qualitative change in the attachment behavioral system and related representations (Crittenden, 2000; Thompson, 1998). In this regard, delineating developmental

changes in and reorganization of the internal working model as individuals move through the life course will be essential. Although important in its own right, further theorizing about developmental change in attachment also will move methodological assessment of attachment forward in important ways.

Attachment theory and research have emerged as the central framework for understanding close relationships across the life course. Research spurred by attachment theory spans infancy to adulthood, children and families from a diverse array of cultures, and biological, cognitive, and emotional processes. Perhaps most important, Bowlby's theory of attachment originated from his observations of children struggling in real-world circumstances. Fittingly, the decades of research that followed Bowlby's and Ainsworth's work are being utilized to make recommendations and consider implications for clinical practice, intervention, and policy (Berlin, Ziv, Amaya-Jackson, & Greenberg, 2005; Oppenheim & Goldsmith, 2007).

SEE ALSO Volume 1: *Bowlby, John; Child Care and Early Education; Dating and Romantic Relationships, Childhood and Adolescence; Family and Household Structure, Childhood and Adolescence; Maternal Employment; Parent-Child Relationships, Childhood and Adolescence; Parenting Style; Poverty, Childhood and Adolescence;* Volume 2: *Dating and Romantic Relationships, Adulthood.*

BIBLIOGRAPHY

Ainsworth, M. D. S., Blehar, M. C., Waters, E., & Wall, S. (1978). *Patterns of attachment: A psychological study of the strange situation.* Hillsdale, NJ: Lawrence Erlbaum Associates.

Ainsworth, M. D. S., & Wittig, B. A. (1969). Attachment and exploratory behavior of one-year-olds in a strange situation. In B. M. Foss (Ed.), *Determinants of infant behavior IV.* London: Methuen.

Bakermans-Kranenburg, M. J., van Ijzendoorn, M. H., & Juffer, F. (2003). Less is more: Meta-analyses of sensitivity and attachment interventions in early childhood. *Psychological Bulletin, 129*(2), 195–215.

Belsky, J. (1999). Interactional and contextual determinants of attachment security. In J. Cassidy & P. R. Shaver (Eds.), *Handbook of attachment: Theory, research, and clinical applications* (pp. 249–264). New York: Guilford Press.

Belsky, J., & Cassidy, J. (1994). Attachment: Theory and evidence. In M. Rutter & D. F. Hay (Eds.), *Development through life* (pp. 373–402). Oxford and Boston: Blackwell Scientific Publishers.

Berlin, L. J., Ziv, Y., Amaya-Jackson, L., & Greenberg, M. T. (2005). *Enhancing early attachments: Theory, research, intervention, and policy.* New York: Guilford Press.

Bowlby, J. (1969). *Attachment and loss.* Vol. 1: *Attachment.* New York: Basic Books.

Bowlby, J. (1973). *Attachment and loss.* Vol. 2: *Separation, anxiety, and anger.* New York: Basic Books.

Bowlby, J. (1980). *Attachment and loss.* Vol. 3: *Loss, sadness, and depression.* New York: Basic Books.

Cassidy, J. (1994). Emotion regulation: Influences of attachment relationships. In N. A. Fox (Ed.), The development of emotion regulation: Biological and behavioral considerations. *Monographs of the Society for Research in Child Development, 59*(240), 228–249.

Cassidy, J., & Shaver, P. R. (Eds.). (1999). *Handbook of attachment: Theory, research, and clinical applications.* New York: Guilford Press.

Crittenden, P. M. (2000). A dynamic-maturational approach to continuity and change in pattern of attachment. In P. M. Crittenden & A. H. Claussen (Eds.), *The organization of attachment relationships: Maturation, culture, and context* (pp. 343–357). Cambridge, U.K., and New York: Cambridge University Press.

De Wolff, M. S., & van IJzendoorn, M. H. (1997). Sensitivity and attachment: A meta-analysis on parental antecedents of infant attachment. *Child Development, 68*(4), 571–591.

Fonagy, P., & Target, M. (1997). Attachment and reflective function: Their role in self-organization. *Development and Psychopathology, 9,* 679–700.

Goldberg, S., Grusec, J. E., & Jenkins, J. M. (1999). Confidence in protection: Arguments for a narrow definition of attachment. *Journal of Family Psychology, 13*(4), 475–483.

Hazan, C., & Shaver, P. R. (1994). Attachment as an organizational framework for research on close relationships. *Psychological Inquiry, 5*(1), 1–22.

Main, M., Kaplan, N., & Cassidy, J. (1985). Security in infancy, childhood, and adulthood: A move to the level of representation. In I. Bretherton & E. Waters (Eds.), Growing points of attachment theory and research. *Monographs of the Society for Research in Child Development, 50*(1–2), 66–104.

Main, M., & Solomon, J. (1990). Procedures for identifying infants as disorganized/disoriented during the Ainsworth Strange Situation. In M. T. Greenberg, D. Cicchetti, & E. M. Cummings (Eds.), *Attachment in the preschool years: Theory, research, and intervention* (pp. 121–160). Chicago: University of Chicago Press.

Main, M. & Weston, D. R. (1981). The quality of the toddler's relationship to mother and to father: Related to conflict behavior and the readiness to establish new relationships. *Child Development, 52*(3) 932–940.

McElwain, N. L., & Booth-LaForce, C. (2006). Maternal sensitivity to infant distress and nondistress as predictors of infant-mother attachment security. *Journal of Family Psychology, 20*(2), 247–255.

NICHD Early Child Care Research Network. (1997). The effects of infant child care on infant-mother attachment security: Results of the NICHD Study of Early Child Care. *Child Development, 68*(5), 860–879.

Oppenheim, D., & Goldsmith, D. F. (2007). *Attachment theory in clinical work with children: Bridging the gap between research and practice.* New York: Guilford Press.

Schneider, B. H., Atkinson, L., & Tardiff, C. (2001). Child-parent attachment and children's peer relations: A quantitative review. *Developmental Psychology, 37*(1), 86–100.

Sroufe, L. A., & Fleeson, J. (1986). Attachment and the construction of relationships. In W. W. Hartup & Z. Rubin (Eds.), *Relationships and development* (pp. 51–71). Hillsdale, NJ: L. Erlbaum Associates.

Thompson, R. A. (1998). Early sociopersonality development. In W. Damon (Series Ed.) and N. Eisenberg (Vol. Ed.), *Handbook of child psychology.* Vol. 3: *Social, emotional, and personality development* (5th ed.) (pp. 25–104). New York: Wiley.

Van den Boom, D. C. (1994). The influence of temperament and mothering on attachment and exploration: An experimental manipulation of sensitive responsiveness among lower-class mothers with irritable infants. *Child Development, 65*(5), 1457–1477.

Van IJzendoorn, M. H., Dijkstra, J., & Bus, A. G. (1995). Attachment, intelligence, and language: A meta-analysis. *Social Development, 4*(2), 115–128.

Van IJzendoorn, M. H., & Kroonenberg, P. M. (1988). Cross-cultural patterns of attachment: A meta-analysis of the strange situation. *Child Development, 59*(1), 147–156.

Waters, E., Hamilton, C. E., & Weinfield, N.S. (2000). The stability of attachment security from infancy to adolescence and early adulthood: General introduction. *Child Development, 71*(3), 678–683.

Nancy L. McElwain

ATTENTION DEFICIT/ HYPERACTIVITY DISORDER (ADHD)

According to the *Diagnostic and Statistical Manual* (fourth edition, text revision [*DSM–IV–TR*]; 2000), Attention deficit hyperactivity disorder (ADHD) is characterized by a pervasive and persistent lack of attention and/or heightened activity level. Children with ADHD are defined as quantifiably distinct from their peers as they have problems paying attention, staying on task, and remaining sedentary. These particular behavioral patterns are especially problematic in the school setting as children displaying ADHD traits produce work that is disorganized and incomplete and they are easily distracted by extraneous stimuli. Symptoms of hyperactivity include fidgeting, squirming in one's seat, talking excessively, and/or acting as though one is driven by a motor (American Psychiatric Association [APA], 2000).

ADHD diagnoses have risen steadily since the 1970s in the United States, yet many other countries report little, if any, ADHD among child and adolescent populations (Breggin, 2002). Historically speaking, ADHD is a relatively new phenomenon, and although millions of American children now carry the ADHD label, this was not always the case. In the 1950s, ADHD did not exist in the United States. In the 1970s, an estimated 2,000 American children (the vast majority of whom were boys) were diagnosed as "hyperactive" and the standard method of treatment was behavior-modification therapies. In 2003 the Centers for Disease Control and Prevention reported approximately 4.4 million American children had been diagnosed with ADHD (again, the majority are boys), and the accepted method of treatment was daily doses of Methylphenidate (MPH), often referred to by the brand name of Ritalin (Breggin, 2002; Stolzer, 2005).

Although it is accurate to report that ADHD diagnoses are increasing in many westernized countries, scholars have pointed out that 80 to 90% of MPH produced worldwide is prescribed for American children in order to control behaviors that have just recently been classified as pathological (Leo, 2000). Relatively recently, typical childhood behaviors such as not paying attention and being physically active in confined classrooms has been classified by the *DSM–IV–TR* as a verifiable mental disorder. According to the *DSM–IV–TR* (APA, 2000), symptoms of ADHD include "fidgeting," "running or climbing excessively," "often has difficulty playing quietly," and "often fails to give close attention to details or makes careless mistakes in schoolwork." Symptoms must be present in two or more settings (e.g., home, school, or various social settings), although it is unlikely that the child will display the same level of dysfunction in all settings as symptoms typically worsen in environments that require "monotonous" and "repetitive" tasks (p. 86). Conversely, ADHD symptoms are minimal (or absent) when the child is receiving positive reinforcement, is under close supervision, is in an interesting environment, and/or is engaged in an activity that they find enjoyable. Typically, symptoms of ADHD are present before the age of 7 (APA, 2000).

SCIENCE BEHIND THE FINDINGS

Scholars have suggested that the *DSM*'s diagnostic criteria have serious scientific flaws. Fred Baughman (2006) asked: At what point does hyperactivity become "persistent"? How does one tell the difference between "normal" childhood behavior and pathology? Baughman also asked: What is "typical"? typical for a particular classroom? a particular geographical location? or perhaps a particular culture? Also, the *DSM–IV–TR* does not in any way control for gender differences in behavior patterns—differences that can be quantified across cultures, across historical time, and across mammalian species (Bjorklund & Pellegrini, 2002; Stolzer, 2005).

As is the case with all psychiatric disorders, members of the APA vote on which disorders meet the criteria for inclusion in the latest *DSM*. "The Purpose of the *DSM* is to provide clear descriptions of diagnostic categories in order to enable clinicians and investigators to diagnose, communicate about, study, and treat people with various mental disorders" (APA, 2000, p. xxxvii). It is interesting to note that homosexuality was for many years defined as

a psychiatric disorder by the APA and, as such, was included in the *DSM* until 1978. Defining homosexuality as a mental disorder clearly illustrates the subjective nature of the *DSM* and clarifies that perceptions of what constitutes a legitimate mental disorder can change over time. Despite the combined efforts by the APA and the pharmaceutical industry, which both actively promote ADHD as a neurologically based brain disorder, there exists no scientific evidence to substantiate this claim (Baughman, 2006; Breggin, 2002). No neurological or metabolic tests are performed to confirm the existence of ADHD. Rather, diagnostic testing typically follows a prescribed pattern:

1. The child is having difficulty in school;

2. The parents are called in for a conference and are informed that in order to get the child the help he/she needs, a formal assessment must be conducted;

3. The formal ADHD assessment is done using a standardized checklist of behaviors for the specified child;

4. The child is then referred to a physician (most often to a general practitioner);

5. The physician relies on a standardized ADHD behavior checklist, and if the child exhibits six out of nine of the ADHD behavior patterns, he/she is formally diagnosed with ADHD;

6. Psychotropic medication is the prescribed treatment plan.

Many scientists have actively refuted the reliability and validity of current ADHD assessment procedures (Baughman, 2006; Carey, 2002; Stolzer, 2007). According to published research, ADHD assessment tests are highly subjective and vary tremendously from one rater to the next (Carey, 2002). The answers contained on the assessment tests are limited to *never, rarely, sometimes, often*, and *always*. Scientifically speaking, these are not operationally defined terms and clearly carry multiple meanings depending on the perceptions of a specific rater. At the present time, these terms are not universally quantified, and this fact most certainly decreases both the reliability and validity of the ADHD diagnostic process (Breggin, 2002; Carey, 2002). Other scholars have suggested that the status of the rater (e.g., the teacher, parent, or physician) is not considered in the course of assessment. Tolerance level, understanding of normative developmental processes, gender, age, personality type, education, and cultural background are variables that heavily influence rater perception, yet these variables are not controlled for in any quantifiable way (Carey, 2002; Stolzer, 2007).

ECONOMIC CONTEXT

Any discussion of ADHD must address the economic context surrounding this disorder. In 1975, Americans enacted legislation (often referred to as the Mainstreaming Act) that allowed children with physical disabilities access to the public school system. In 1991, children with behavioral and/or learning disabilities were included in this amendment, and since that time, ADHD diagnoses have skyrocketed across America (albeit this is not the situation in other countries). Under the 1991 American amendment, schools that enroll students with disorders such as ADHD receive federal funding. That is, the more children who are diagnosed with behavioral disorders, the more money the individual school receives. It is also interesting to note the disparities that exist with regard to ADHD rates among American students. Private American schools receive no federal money for children diagnosed with ADHD and typically have extremely low rates of ADHD among their student populations. Conversely, public schools receive federal money for each child diagnosed, and the rates of ADHD among public school students are as high as 60% in some American school districts (Baughman, 2006).

GENDERED AND DEVELOPMENTAL INFLUENCES

Throughout human history, males and females have followed very different developmental trajectories. According to Peter Jensen and colleagues (1997), males evolved in an environment that required elevated activity levels. As males perfected this "hyperactive" way of being, this distinct and valuable male trait was not only highly desirable but was in fact integral to the survival of the human species. As compulsory schooling became the norm in most societies, uniquely male traits were not at all adaptive in the newly constructed classroom setting. According to Bjorklund and Pellegrini (2002), the high activity levels currently in observed children can be directly linked to humanity's ancient and evolutionary past. School systems in the modern era require sedentary learning (e.g., sitting in desks for extended periods of time), and these relatively new expectations coupled with the proliferation of new childhood psychiatric disorders has, according to Jensen and colleagues, fueled the unprecedented rise of ADHD diagnoses across much of the United States.

From a developmental perspective, childhood has been altered dramatically over a relatively short time period. The unstructured, outdoor roaming of the past has been replaced by sedentary, adult-monitored play. Television, computers, and electronic video games now engulf children at every developmental stage. Children are continually immersed in artificial light and temperature, are surrounded by four walls with no access to the natural elements, and are expected to remain sedentary for hours at a time (Stolzer, 2005; Wilson, 1993). A review of the literature indicates that aggression, hyperactivity, and inattentiveness decrease when children are exposed to the outdoors, have freedom to engage in large motor activity, are interested in the subject matter, and are involved in one-on-one interaction with a caring and competent adult (Breggin, 2002). Although the *DSM–IV–TR* currently classifies ADHD-typed behaviors as pathological, Breggin insisted that inattentiveness, disorganization, high activity level, and getting bored easily with mundane tasks is not only developmentally appropriate but is in fact observable across cultures and across historical time with regard to child populations.

IMPORTANT PREDICTORS OF ADHD

ADHD is diagnosed by particular behavioral patterns that include fidgeting, excessive running or climbing, not paying attention to instructions, and difficulty playing quietly. These behaviors must be persistent and must be displayed more frequently and more severely than is typically observed in individuals who do not have ADHD (APA, 2000). However, according to the Surgeon General of the United States (1999), diagnosing an individual with a mental disorder is open to many different interpretations that are rooted in value judgments that may vary across cultures. The Surgeon General also stated that diagnosing a disorder such as ADHD is rather precarious, as there are no definitive markers (e.g., lesions, lab tests, or brain abnormality) that can positively identify a particular mental disorder. According to Peter Jensen and James Cooper (2000), the belief that ADHD is neurological in nature is not supported by scientific evidence as current assessment procedures ignore the complex and diverse range of variables associated with particular childhood behaviors. Furthermore, researchers have pointed out that drawing precise and accurate boundaries between typical child behavior and abnormal behavior patterns is difficult at best. The science of accurately diagnosing ADHD is especially problematic because of the ongoing processes of cognitive, emotional, and physical development. By their very nature, children are ever changing, thus making stable measurements and/or diagnoses extremely complex. Using adult criteria for mental illness in children and adolescents is also scientifically questionable as many of the symptoms of adult pathology are characteristics of normal development in child populations (Surgeon General of the United States, 1999).

KEY DISPARITIES IN THE CAUSES AND TREATMENT OF ADHD

Proponents of the "disordered brain" hypothesis insist that ADHD is the result of an atypical neurological system and that pharmaceutical drug intervention is necessary to correct a chemical imbalance within the brain. The pharmaceutical industry has a vested economic interest in promoting this disordered brain hypothesis and has been quite successful in using a multimedia advertising campaign. Parenting magazines, television, and physician offices routinely distribute materials that refer to ADHD as a brain disorder, although no scientific data supports this assertion (Baughman, 2006; Breggin, 2002; Jensen & Cooper, 2000; Surgeon General of the United States, 1999).

Jensen and Cooper (2000) postulated that ADHD-typed behavior is highly adaptive and served human beings well until the advent of compulsory schooling. Baughman (2006) hypothesized that ADHD assessment tests actually measure adults' frustrations with typical and historically documented child behaviors. According to Baughman, a respected pediatric neurologist, "In the overwhelming majority of cases, the underlying issue is either a clash between a normal child and the requirements of his adult controlled environment or the product of diagnostic zeal in a newly deputized teacher-turned-deputy brain diagnostician" (p. 215). Baughman clearly pointed out the controversial nature of ADHD: Is this disorder a brain malfunction requiring pharmaceutical intervention? Or is ADHD a remnant of an evolutionary past that does not fit in with the rigid structure of the American school system?

MEDICATION

The overwhelming majority of children diagnosed with ADHD are prescribed MPH in order to control undesirable behaviors (Baughman, 2006). Although it is well known that MPH can reduce disruptive behaviors and increase compliance and sustained attention, very seldom are the dangerous effects of this drug discussed openly (Stolzer, 2007). The Food and Drug Administration (FDA) has classified MPH as a Schedule II drug along with morphine, opium, and barbiturates as these types of drugs have been proven to be highly addictive and to cause a wide range of physiological atrophy (Breggin, 2002).

MPH has been found to produce severe withdrawal symptoms, irritability, suicidal feelings, headaches, and Tourette's syndrome, a condition that causes both physical and verbal "tics" (Breggin, 1999; Novartis Pharmaceutical Corporation, 2006). The drug also has been associated with weight loss, disorientation, personality changes, apathy, social isolation, depression, insomnia, increased blood pressure, cardiac arrhythmia, tremors, weakened immunity, growth suppression, agitation, fatigue, accelerated resting pulse rate, visual disturbances, drug dependency, anorexia, nervousness, aggression, liver dysfunction, hepatic coma, angina, and toxic psychosis (Breggin, 1999; Novartis, 2006).

According to the pharmaceutical firm Novartis (2006), MPH is a central nervous system stimulant; however, the mode of therapeutic action in ADHD is not known. Novartis openly states that the specific etiology or cause of ADHD is unknown and that no single diagnostic test can definitively diagnose ADHD. Novartis acknowledges that the effectiveness of MPH for long-term use (i.e., more than two weeks) has not been established in controlled trials and has stated unequivocally that the safety of long-term use of the drug in child populations has not yet been determined.

LABELING EFFECTS

For more than 40 years, social scientists have been aware of the deleterious effects of labeling children and adolescents. Once an official label is affixed, adults' perceptions of the individual child can actually bring about expected behavior via a process called the "self-fulfilling prophecy" (Feldman, 2007; Rosenthal & Jacobson, 1968). With regard to the ADHD label, the problem is assumed to be within the individual child, requiring no alteration of the familial, contextual, physical, or socioemotional variables that surround the child. The children are not taught that they themselves can control their behavior (e.g., develop an internal locus of control). Rather, the child is convinced by adults that the only way to control behavior is through pharmaceutical intervention. In this way, individual self-efficacy is compromised, and behavior is collectively defined as being outside of the child's control.

FUTURE DIRECTION
OF ADHD RESEARCH

The pharmaceutical industry currently monopolizes ADHD research by systematically promoting ADHD as a neurological disorder, funding major medical conferences relating to ADHD, funding ADHD research, providing financial incentives for physicians who prescribe specific ADHD drugs, and funding groups such as CHADD (Children and Adults with Attention Deficit Disorder) who openly promote psychotropic drug use in child populations (Breggin, 2002; Jureidini & Mansfield, 2001). In the future, it is imperative that unbiased, empirical research is conducted to increase our understanding of the highly varied nature of ADHD. Furthermore, laws must be enacted that guarantee that scientifically based, objective research is guiding conventional therapeutic practice (Stolzer, 2007). Certainly, it is easier to medicate children than to collectively address the wide-ranging factors that are affecting child and adolescent populations in the modern era. Perhaps future researchers will concentrate on the multitude of individual and society-level variables that affect developmental processes and, in doing so, will significantly increase our understanding of the vacillating complexities associated with the ever-developing human.

SEE ALSO Volume 1: *Disability, Childhood and Adolescence; Learning Disability.*

BIBLIOGRAPHY

American Psychiatric Association. (2000). *Diagnostic and statistical manual of mental disorders* (4th ed., rev.). Washington, DC: Author.

Baughman, F. (2006). *The ADHD fraud: How psychiatry makes "patients" of normal children.* Oxford, U.K.: Trafford.

Bjorklund, D. F., & Pellegrini, A. D. (2002). *The origins of human nature: Evolutionary developmental psychology.* Washington, DC: American Psychiatric Association.

Breggin, P. (1999). Psychostimulants in the treatment of children diagnosed with ADHD: Risks and mechanisms of action. *International Journal of Risk and Safety in Medicine, 12,* 3–35.

Breggin, P. R. (2002). *The Ritalin fact book.* Cambridge, MA: Perseus Books.

Carey, W. (2002). Is ADHD a valid disorder? In P. S. Jensen & J. R. Cooper (Eds.), *Attention deficit hyperactivity disorder: State of the science: Best practices.* Kingston, NJ: Civic Research Institute.

Feldman, R. S. (2007). *Child development.* (4th ed.). Upper Saddle River, NJ: Prentice Hall.

Jensen, P. S., & Cooper, J. R. (Eds.). (2000). *Attention deficit hyperactivity disorder: State of the science, best practices.* Kingston, NJ: Civic Research Institute.

Jensen, P. S., Mrazek, D., Knapp, P. K., Steinber, L., Pfeffer, C., & Schowalter, J. (1997). Evolution and revolution in child psychiatry: ADHD as a disorder of adaptation. *Journal of the American Academy of Child and Adolescent Psychiatry, 36*(12), 1572–1679.

Jureidini, J., & Mansfield, P. (2001). Does drug promotion adversely influence doctor's abilities to make the best decisions for patients? *Australasian Psychiatry, 9,* 95–100.

Leo, J. (2000). Attention deficit disorder: Good science or good marketing? *Skeptic, 8*(1), 63–69.

Novartis Pharmaceuticals Corporation. (2006). Ritalin LA (package insert). East Hanover, NJ: Elan Holdings.

Rosenthal, R., & Jacobson, L. (1968). *Pygmalion in the classroom: Teacher expectations and pupils' intellectual development.* New York: Holt, Rinehart, & Winston.

Stolzer, J. (2005). ADHD in America: A bioecological analysis. *Ethical Human Psychology and Psychiatry, 7*(1), 65–75.

Stolzer, J. M. (2007). The ADHD epidemic in America. *Ethical Human Psychology and Psychiatry, 9*(2), 37–50.

Surgeon General of the United States. (1999). *Mental health: A report of the Surgeon General.* Washington, DC: United States Department of Health and Human Services. Retrieved April 16, 2008, from http://www.surgeongeneral.gov/library

Wilson, E. O. (1993). Biophilia and the conservation ethic. In S. R. Kellert & E. O. Wilson (Eds.), *The biophilia hypothesis.* Washington, DC: Island Press.

J.M. Stolzer

AUTISM

Autism is one of five pervasive developmental disorders. These developmental disorders were identified as pervasive because they affect more than one domain of development (as opposed to a specific developmental disorder that affects only one domain of development, such as a reading disorder). The other four pervasive developmental disorders are Asperger's Disorder, Pervasive Developmental Disorder–Not Otherwise Specified (PDD–NOS), Rett's Disorder, and Childhood Disintegrative Disorder. These disorders vary in terms of timing, severity, and nature of symptoms. However, all pervasive developmental disorders involve deficits in social functioning and repetitive behaviors.

Current prevalence estimates suggest that 1 in every 150 children in the United States is affected by a pervasive developmental disorder (Centers for Disease Control, 2007). Additionally, autism occurs four to five times more often in boys than in girls (Volkmar, Szatmari, & Sparrow, 1993). Current estimates of prevalence in the United States are similar to those of other countries, though significantly greater than in earlier decades. In 1979 autism prevalence was 2 to 5 children per every 10,000 (Wing & Gould, 1979). The cause of this increase in autism prevalence is widely debated. Some argue that this number reflects a true increase in cases of autism. Others argue that the change in prevalence is because of increased awareness and more accurate diagnosis of autism (especially high-functioning autism). Researchers have not fully resolved this debate.

Autism is a characterized by deficits in three areas: (a) social interaction, (b) communication, and (c) repetitive behaviors or interests. Although mental retardation is more common in children with autism than in the general population, not all children with autism also have mental retardation. Furthermore, some experts have argued that standard IQ tests, which rely heavily on language, underestimate the intelligence of children with autism.

Children with autism comprise a very heterogeneous group, reflecting variability in the nature and severity of symptoms. For example, children with a severe presentation of autism (e.g., low-functioning autism) may be (a) entirely uninterested in social interaction; (b) have no verbal language skills; and (c) engage in repetitive behaviors such as rocking back and forth, flapping their hands, or banging their heads. By contrast, children with a milder presentation of autism (e.g., high-functioning autism) may (a) appear interested in social interaction but lack the social skills to maintain appropriate social relationships, (b) have age-appropriate language skills yet have difficulty using language socially (e.g., reading body language or knowing how to start or end a conversation appropriately), and (c) have an unusual and all-consuming interest (e.g., being intensely interested in one species of moth such that it consumes all of their free time and is the only thing they want to discuss). Researchers and theorists have speculated that the heterogeneity in autism may represent different subtypes or causes of autism. However, this has yet to be clearly determined by research.

Autism was first described by physician Leo Kanner (1894–1981) in 1943. Kanner described 11 children with what he termed *extreme autistic loneliness*. His original case studies described the children's physical and psychological functioning and proposed that autism was a disorder of biological origin. However, over time, he discarded discussion of biological causes and focused instead on family-related causes of autism. Specifically, he suggested that mothers who were consistently cold and rejecting of their children caused their children to turn inward and develop autism. Often times, this is referred to as the *refrigerator mother* theory of autism. This theory has been widely disproven by decades of research and is no longer recognized as a valid theory of autism.

CURRENT THEORIES OF AUTISM

Although Kanner's refrigerator mother theory has been widely debunked, current theorists continue to debate (a) what constitutes the core deficit of autism and (b) the causes of autism. Core deficit theories of autism seek to better understand the disorder by clearly identifying the defining features of the disorder. One prominent theory focuses on the role of joint attention in the development and maintenance of autism. Before children are able to use words to communicate, they use eye contact and nonverbal gestures for two communicative purposes: (a) joint attention (nonverbal sharing) and (b) requesting (Bates, Camaioni, & Volterra, 1975). For instance, imagine that an 18-month-old child is seated at a table with her mother when her mother accidentally knocks a plate onto the floor. The toddler watches the plate drop to the floor, smiles, and then makes eye contact with her mother as if to say, "Wow! Did you see that?" In other words, the toddler is making eye contact with her mother in order to share her interest and enjoyment of this surprising event. This is in contrast to the toddler engaging in eye contact to request something. To exemplify the distinction between joint attention and requesting eye contact, imagine that the same toddler is sitting at the table trying to turn on a musical toy without success. After a while, the toddler makes eye contact with her mother, while at the same time handing her the toy. In other words, the toddler is using eye contact and a giving gesture to request help.

This functional distinction appears to be particularly important for understanding autism. In typically developing children, joint attention emerges at 6 months of age and is well-developed by 18 months of age. However, children with autism uniformly show severe deficits in joint attention eye contact. They engage in joint attention eye contact far less frequently than typically developing children (Mundy & Sigman, 1989) and children with other developmental delays (Kasari, Freeman, Mundy, & Sigman, 1995). It is important to note that requesting eye contact may remain relatively intact in a subset of children with autism. Thus, children with autism have a specific deficit in eye contact for social sharing, not a global deficit in eye contact.

Theorists have proposed that an innate drive to engage in joint attention is necessary for providing social learning opportunities that promote normal brain development and social development. Because children with autism lack this innate drive to engage in joint attention, they miss out on necessary social experiences. This, in turn, results in atypical brain development and perpetuation of social difficulties (Mundy & Neal, 2001).

Other theories of core deficits of autism focus on the role of cognition, social or otherwise. One of the most prominent social cognitive theories of autism proposes that autism reflects *mindblindness*, which is reflected by a core deficit in Theory of Mind (ToM). ToM reflects a child's capacity to understand that other people have thoughts, feelings, and beliefs that are independent and, perhaps, different from their own. Understanding ToM allows one to understand and predict another person's behavior. Typically developing children usually develop ToM by age 4. By contrast, there is a large amount of research showing that children with autism lack a well-developed ToM (Baron-Cohen, Leslie, & Frith, 1985). It has been suggested that this deficit in ToM reflects deficits in information-processing abilities that are social in nature. It has also been suggested that joint attention is a precursor to ToM, although there is little evidence to support this argument. Nonetheless, like joint attention, ToM deficits do appear to be universally, and specifically, impaired in children with autism.

Current theories on the cause of autism focus on different mechanisms. The debate often focuses on the role of nature (heredity and neurobiology) versus nurture markers have been found in different and relatively small subsets of children. For instance, abnormalities on chromosome 16 were found in only 1% of cases. No genetic marker has been found to be present in every child with autism, further spurring the debate about different genetic pathways to and subtypes of autism.

Other biological factors that have been implicated in autism include differences in brain structure and function, as well as differences in autoimmune indicators. Research on neurobiological factors associated with autism has suggested higher rates of neurological disorders such as epilepsy. Research regarding structural and functional changes in the brain is mixed. However, accumulating evidence appears to point to some children with autism having larger brain volumes, as well as differences in the structure and function of areas of the brain important for attention, language, and emotions. Research also suggests a potential autoimmune role in autism. Studies of children with autism have found higher rates of autoantibodies and, in particular, antibrain autoantibodies.

Exploration of environmental causes of autism is one of the mostly hotly debated areas in autism. No one topic has been more controversial than the debate over the role of the measles, mumps, and rubella (MMR) vaccine. Some have proposed that thimerosal, a mercury-based preservative used in many childhood vaccines, causes autism. Others have proposed that it is the MMR vaccine itself that causes autism. Although many parents still contend that the MMR vaccine caused their child's autism, a large body of evidence suggests otherwise. Specifically, several large population studies have failed to find a link between the MMR vaccine and autism. One of the most compelling studies was a naturalistic study of the incidence of autism in the Kohoku district of Japan during a several-year period in which the district discontinued giving the MMR vaccine altogether. Despite there not being a single MMR vaccination given after 1993, incidence of autism increased at similar rates in the Kohoku district as seen elsewhere in the world where MMR vaccinations were routinely given (Honda, Shimizu, & Rutter, 2005). This study provides compelling evidence that the MMR vaccine is not likely to be a main cause of autism, nor can the administration of this vaccine explain the rise in autism prevalence in recent years. It is also notable that, as a precaution, thimerosal was removed from all childhood vaccinations. Despite this fact, autism prevalence rates have continued to rise. Thus, it also seems very unlikely that thimerosal is a main cause of autism.

Another proposed environmental mechanism involves severe food allergies and intolerance. Specifically, it has been suggested that severe intolerance to casein (milk protein) and gluten (wheat protein) may exacerbate symptoms

Autistic Child. *An autistic child plays with a toy. Children with autism comprise a very heterogeneous group reflected by variability in the nature and severity of symptoms.* CUSTOM MEDICAL STOCK PHOTO, INC. REPRODUCED BY PERMISSION.

(environment). Evidence for the role of nature comes from a large body of research showing that autism is largely heritable. Family and twin studies have revealed higher rates of autism and other pervasive developmental disorders among family members of children with autism. Indeed, having one child with autism increases the likelihood of having a subsequent child with autism 50- to 100-fold what would be expected in the general population (Folstein & Rutter, 1988). This inherited vulnerability, however, may not be specific to autism. Studies have also shown higher rates of other learning disabilities (e.g., dyslexia) in family members of children with autism than in the general population.

Research examining the role of specific genes in autism has yet to find one specific autism gene. Instead, researchers such as Dietrich Stephan (2008) have linked autism to several genes and to a specific deletion or duplication of material on chromosome 16 (Weiss, Shen, & Korn, in press). However, the identified genetic of autism. Currently, evidence for this theory is largely anecdotal. Furthermore, it may only be relevant to a subset of children with autism who also have gastrointestinal disorders. This is an area that would benefit greatly from well-controlled research studies.

One compelling area of research on environmental mechanisms involves the prenatal environment. Previous studies have suggested a link between prenatal and birth complications and autism. Similarly, studies have shown increased risk for autism following prenatal exposure to German measles, valproic acid, and thalidomide. Additionally, prenatal exposure to abnormally high levels of testosterone has also been implicated in autism. Finally, a recent study identified maternal antibodies that increase the risk of having a child with autism. These researchers suggest that these maternal antibodies may mistakenly attack fetal brain tissue, which later results in autism (Braunschweig et al., in press). However, not all mothers of children with autism have these antibodies. Thus, these antibodies likely play a role in only a subset of children with autism.

In conclusion, many different genetic, neurobiological, and environmental pathways to autism have been proposed. Research has not found a link between any one factor and the development of autism in all children. As such, it appears likely that there are multiple pathways to autism based on different combinations of genetic or biological risk and environmental exposure.

TREATMENT

Autism experts agree that early identification and intervention is crucial to reducing severity of symptoms and optimizing functioning in children with autism. Studies on the infant siblings of children with autism have identified several early warning signs for autism, including a lack of joint attention eye contact or not responding to one's name when called at age 1. Other possible warning signs include delayed language development and motor milestones (e.g., walking).

Parents of children with autism have many treatment options to choose from including, but not limited to, educational interventions, speech therapy, occupational therapy, sensory integration therapy, applied behavioral analysis, facilitated communication, social skills instruction, play-based treatments, and medical interventions. Parents of children with autism often use multiple types of treatments at the same time, on the order of 20 to 60 hours of treatment per week. For example, a report from the Interactive Autism Network (2008) suggests that children with autism are enrolled in an average of five different treatments at any one time.

To date, only three treatments have been shown to be effective in randomized, controlled studies: applied behavior analysis (ABA), joint attention and symbolic play intervention, and the structured teaching of the Treatment and Education of Autistic and Related Communication-Handicapped Children (TEACCH) organization. Research on ABA found that young children with autism who received 40 hours per week of therapy for 2 years showed significant increases in IQ and improvement in educational functioning when compared to children who received 10 hours or less of ABA per week (Lovaas, 1987). Young children with autism who participated in either a joint attention or social skills intervention for a length of 6 weeks at 30 minutes per day were found to more frequently engage in joint attention and demonstrate more sophisticated play skills than children who received no treatment (Kasari, Freeman, & Paparella, 2006). Finally, a study of the effectiveness of the TEACCH program found that children who received treatment for 4 months showed greater improvement in developmental test scores than children who received no treatment (Ozonoff & Cathcart, 1998).

Aside from these three instances, there is very little scientific evidence to support the effectiveness of the vast majority of treatments for autism. This does not imply that these treatments are necessarily ineffective but that their effectiveness has not yet been tested in well-controlled research studies. Several large comparative studies of autism treatments are underway with the expectation that the results of these studies will better inform treatment decisions in future decades.

SEE ALSO Volume 1: *Attention Deficit/Hyperactivity Disorder (ADHD); Disability, Childhood and Adolescence; Genetic Influences, Early Life; Learning Disability.*

BIBLIOGRAPHY

Baron-Cohen, S., Leslie, A. M., & Frith, U. (1985). Does the autistic child have a "theory of mind"? *Cognition, 21*(1), 37–46.

Bates, E., Camaioni, L., & Volterra, V. (1975). The acquisition of performatives prior to speech. *Merrill–Palmer Quarterly, 21*, 205–226.

Braunschweig, D., Ashwood, P., Krakowiak, P., Hertz-Picciotto, I., et al. (in press). Autism: Maternally derived antibodies specific for fetal brain proteins. *Neurotoxicology.*

Centers for Disease Control. (2007). *Prevalence of the autism spectrum disorders in multiple areas of the United States, surveillance years 2000 and 2002: A report from the Autism and Developmental Disabilities Monitoring (ADDM) network.* Retrieved May 21, 2008, from www.cdc.gov/od

Folstein, S. E., & Rutter, M. L. (1988). Autism: Familial aggregation and genetic implications. *Journal of Autism and Developmental Disorders, 18*, 3–30.

Honda, H., Shimizu, Y., & Rutter, M. L. (2005). No effect of MMR withdrawal on the incidence of autism: A total population study. *Journal of Child Psychology and Psychiatry, 46*(6), 572–579.

Interactive Autism Network. (2008). *IAN research findings: Treatment data.* Retrieved May 21, 2008, from www.autismspeaks.org/inthenews

Kanner, L. (1943). Autistic disturbances of affective contact. *Nervous Child, 2,* 217–250.

Kasari, C., Freeman, S., Mundy, P., & Sigman, M. D. (1995). Attention regulation by children with Down syndrome: Coordinated joint attention and social referencing looks. *American Journal on Mental Retardation, 100,* 128–136.

Kasari, C., Freeman, S., & Paparella, T. (2006). Joint attention and symbolic play in young children with autism: A randomized controlled intervention study. *Journal of Child Psychology and Psychiatry, 47*(6), 611–620.

Lovaas, O. I. (1987). Behavioral treatment and normal educational and intellectual functioning in young autistic children. *Journal of Consulting and Clinical Psychology, 55,* 3–9.

Mundy, P. M., & Neal, A. R. (2001). Neural plasticity, joint attention, and autistic developmental pathology. *International Review of Research in Mental Retardation, 23,* 139–168.

Mundy, P. M., & Sigman, M. (1989). Specifying the nature of social impairment in autism. In G. Dawson (Ed.), *Autism: New perspectives on diagnosis, nature, and treatment* (pp. 3–21). New York: Guilford.

Ozonoff, S., & Cathcart, K. (1998). Effectiveness of a home program intervention for young children with autism. *Journal of Autism and Developmental Disorders, 28*(1), 25–32.

Stephan, D. A. (2008). Unraveling autism. *The American Journal of Human Genetics, 82,* 7–9.

Volkmar, F. R., Szatmari, P., & Sparrow, S. S. (1993). Sex differences in pervasive developmental disorders. *Journal of Autism and Developmental Disorders, 23*(4), 579–591.

Weiss, L. A., Shen, Y., & Korn, J. M. (in press). Association between microdeletion and microduplication at 16p11.2 and autism. *New England Journal of Medicine.*

Wing, L., & Gould, J. (1979). Severe impairments in social interaction and associated abnormalities in children: Epidemiology and classification. *Journal of Autism and Developmental Disorders, 9*(1), 11–29.

A. Rebecca Neal

B

BANDURA, ALBERT

1925–

Albert Bandura, born on December 4 in the Canadian hamlet of Mudare, Alberta, is the founder of social cognitive theory, which has been applied across disciplines such as psychology, sociology, communication studies, and management to study human motivation, thought, and action. Social cognitive theory posits that people intentionally direct and orchestrate their own life course within ever-changing social environments that impose constraints and offer opportunities. Individuals are not passive recipients of environmental influences. Instead, they guide their own self-development, choosing and creating environments. Children learn to regulate their behavior, thoughts, affect, and motivation, thus influencing their life course. This capacity for self-directedness is a distinctive feature of the theory as it has been a distinctive feature of Bandura's life.

Bandura's mother migrated from the Ukraine, his father from Poland. Despite economic hardship, the family valued education and the finer aspects of life. As Bandura has stated, "My mother was a superb cook, and my father played a sprightly violin" (Pajares, 2004, p. 1). An appreciation of music and cuisine has characterized Bandura throughout his life. The only school in Mudare lacked resources, but rather than holding back the young Bandura, this promoted his self-directed learning. He received a bachelor's degree from the University of British Columbia in 1949 and earned a master's degree (1951) and a doctorate (1952) from the University of Iowa. In 1952 he married, his wife Ginny becoming his life companion and mother of their two daughters. Ban-

dura joined the psychology department at Stanford University in 1953 and has remained there. He has received many honors and awards, including 14 honorary doctorates and an Award for Outstanding Lifetime Contribution to Psychology from the American Psychological Association (2004). He has served on more than 30 editorial boards of scholarly journals and has written nine books and hundreds of journal articles. In 2002 Bandura was ranked as one of the most eminent psychologists of the 20th century and the most frequently cited living psychologist (Haggbloom, 2002).

Early in his career Bandura moved beyond the prevailing behaviorist and psychoanalytic approaches by providing a social learning analysis of social modeling that did not rely on associationistic conditioning and trial-and-error learning. From his classic Bobo doll experiments, in which children were shown models punching an inflatable doll, Bandura established the cognitive factors of learning through observation. Children learn novel physical and verbal responses by observing a model. Their reproduction of the modeled responses, however, is dependent on whether they expect punishment or reward for that behavior. Social modeling enables people to transcend their everyday lives by exposing them to new ideas. The principles of social modeling have been generalized to television dramas in developing countries to address problems such as literacy, family planning, the status of women, and the spread of HIV/AIDS (Bandura, 2002). Television modeling provides strategies and incentives to guide people toward managing problems they encounter in their daily lives and thus transform their lives for the better.

Albert Bandura. LINDA A. CICERO/STANFORD NEWS SERVICE.

bilities to produce specific attainments and goals, in an article published in 1977. Strong self-efficacy beliefs provide the staying power that allows children to master academic skills and develop and sustain a wide spectrum of social behaviors. Unless people believe they can attain their goals, they have little motivation to attempt to do so. Self-efficacy beliefs extend beyond individual beliefs to collective efficacy: an individual's belief about a group's capabilities.

Bandura later addressed the issue of moral agency. Early in their lives children develop personal standards that do not condone aggression and immoral conduct. However, they also develop the capacity to disengage their moral standards. The concept of moral disengagement explains how people who do not countenance violent and immoral conduct can commit atrocities against humanity. For example, soldiers killing "the enemy" is justified in the name of righteous ideologies and nationalistic imperatives.

Concern for the improvement of the human condition is pivotal to social cognitive theory. This theory not only offers an explanation of agentic human functioning based on social modeling and self-regulatory processes but also provides a basis for people to develop and strengthen their efficacy beliefs so that they can improve their own and others' life circumstances through personal and social change.

SEE ALSO Volume 1: *Aggression, Childhood and Adolescence; Moral Development and Education;* Volume 2: *Agency.*

BIBLIOGRAPHY

Bandura, A. (1977). Self-efficacy: Toward a unifying theory of behavioral change. *Psychological Review, 84,* 191–215.

Bandura, A. (1986). *Social foundations of thought and action: A social cognitive theory.* Englewood Cliffs, NJ: Prentice Hall.

Bandura, A. (2002). Health promotion by social cognitive means. *Health Education & Behavior, 31,* 143–164.

Bandura, A. (2005). The evolution of social cognitive theory. In K. G. Smith & M. A. Hitt (Eds.), *Great minds in management: The process of theory development* (pp. 9–35). Oxford and New York: Oxford University Press.

Bandura, A., Barbaranelli, C., Caprara, G. V., & Pastorelli, C. (1996). Mechanisms of moral disengagement in the exercise of moral agency. *Journal of Personality and Social Psychology, 71*(2), 364–374.

Haggbloom, S. J. (2002). The 100 most eminent psychologists of the 20th century. *Review of General Psychology, 6,* 139–152.

Pajares, F. (2004). *Albert Bandura: Biographical sketch.* Retrieved April 30, 2008, from http://www.des.emory.edu/mfp/bandurabio.html

In recognition of the broadening scope of his research and to distinguish his approach from the multiple versions of social learning theory, Bandura expounded social cognitive theory in the 1986 book *Social Foundations of Thought and Action: A Social Cognitive Theory.* Drawing on a large body of research, he combined the central role of cognitions, self-regulatory processes, and motivations into an agentic theory of self-development, adaptation, and change that emphasizes social influences. Individuals are viewed as self-reflecting, self-regulating, and capable of mapping their own life courses. Self-regulation develops from personal standards constructed from a diverse array of social influences and is not simply a mimicking of what children have been taught, have been evaluatively prescribed, or have seen modeled. As children age, they are better able to regulate their behavior by means of self-evaluative reactions that are based on personal standards; however, most behavior is regulated by the interplay of self- and social-evaluative reactions.

Self-efficacy is the cornerstone of the agentic social cognitive theory of human behavior. Bandura introduced the concept of self-efficacy, a judgment of personal capa-

Kay Bussey

BEHAVIORAL GENETICS

SEE Volume 1: *Genetic Influences, Early Life;* Volume 2: *Genetic Influences, Adulthood;* Volume 3: *Genetic Influences, Later Life.*

BILINGUAL EDUCATION

Bilingual education encompasses a range of instructional programs in which children are taught in two languages. In the U.S. context, bilingual programs provide instruction in English and in a second language. Bilingual education originated to serve language minority students; however, programs now serve both language-minority and native-English speakers. Numerous contextual factors—human and material resources, administrative support, parent education levels, neighborhood and school context—contribute to great variation in program offerings and outcomes. Complicating the matter, the end goal of bilingual education has been fraught with conflict since its current incarnation under the Bilingual Education Act of 1964 (BEA).

As a compensatory education model, the BEA legislated the use of language-minority students' primary language for instruction as a means to acquire English. Transitional bilingual programs (often referred to as early-exit or late-exit bilingual) maintain as their goal English language acquisition and a transition into English-only instruction as quickly as possible. Alternately, maintenance bilingual education programs promote biliteracy in both English and the second language; dual-immersion, maintenance, heritage language, and two-way bilingual programs fall in this category. Whether the end goal is biliteracy and biculturalism, or only English acquisition, often determines student outcomes and achievement.

HISTORY AND DEVELOPMENT OF BILINGUAL EDUCATION

Bilingual education is not new to this century; the United States has a long history of native language instruction. As early as 1694, German-speaking residents of the colony of Pennsylvania in Philadelphia established schools using their mother tongue. German-language and other bilingual schools operated through the end of World War I in an environment of linguistic neutrality, if not tolerance. After World War I, rampant nationalism and nativism nearly brought an end to bilingual education; by 1923, 34 states had designated English as the sole language of instruction permitted in public and private elementary schools. Such restrictions, however, were curtailed after World War II when U.S. military personnel returned, having been frustrated by an inability to communicate with their allies, who came from societies where linguistic pluralism was the norm. After the Soviet Union launched its Sputnik satellite, the U.S. federal government passed the National Defense Education Act in 1958; Titles VI and IX of that act focus on the retention and expansion of foreign language resources in the United States.

The revival of bilingual education in the second half of the 20th century began in 1963 with the opening of Coral Way Elementary School in southern Florida. Established by Cuban refugees anticipating a prompt return to Cuba, the bilingual program at Coral Way provided a means of maintaining their children's native tongue. As a result of the civil rights movement, the Elementary and Secondary Education Act (ESEA) of 1965 was amended to include the BEA of 1968, which became ESEA Title VII. The amendment included bilingual education programs in federal education policy and authorized the use of federal funds to support primary language education for speakers of languages other than English. Although the BEA permitted primary language instruction, it did not mandate it.

In 1974 *Lau v. Nichols* went before the U.S. Supreme Court on behalf of Chinese language-minority students in San Francisco challenging the inequality of access to education for nonnative speakers of English. The court found for the plaintiffs, noting that "students who do not understand English are effectively foreclosed from any meaningful education" (*Lau v. Nichols*, 1974). Effectively, the court decided, the provision of the same books, curricula, and teachers, all in English, is insufficient to ensure access to an equal education. Although *Lau* mandated linguistic support services, no mandate existed for any specific academic programs or pedagogic practices; bilingual education was listed as one of many possible curricular options.

In response to *Lau*, California was one of the first states in the nation to enact a comprehensive bilingual-education bill, the Chacón-Moscone Bilingual-Bicultural Education Act of 1976. This piece of legislation mandated that school districts must provide primary language instruction whenever ten non-English proficient (NEP) or fifteen limited English-proficient (LEP) students of the same language group were present in the same grade level. Chacón-Moscone provided schools with detailed instructions about the type of language support that should be provided and, notably, proclaimed bilingual education a right of limited English-proficient students. There was almost no enforcement of Chacón-Moscone, however. Even when most widely offered in California, bilingual education only served approximately 30% of students who would qualify (California Department of Education, 2008). In 1998, Californians passed Proposition 227 effectively banning bilingual education and instruction in languages other than English. Since then, Massachusetts and Arizona have also passed English-Only amendments in order to curtail bilingual education efforts.

Story Time. *Teacher reads a Spanish language book to students.* GALE, CENGAGE LEARNING.

Definition of terms Since the passage of the BEA and the *Lau* decision, federal, state, and local entities have developed many terms to describe the growing immigrant language-minority population in public schools from kindergarten through twelfth grade. Federal regulations refer to those language-minority students deemed limited English proficient as LEP; states, districts, and schools often prefer the terms *English language learner* (ELL) or *English learner* (EL) to designate such students. *Lau* requires that ELL students receive some sort of services, the most common of which is English as a Second Language (ESL) instruction. Bilingual classrooms provide instruction in the students' primary language, but offer ESL services as well. ESL classes, however, do not generally involve bilingual instruction, though at times the primary language is used for clarification.

As mentioned already, bilingual programs may be classified as transitional or maintenance; transitional programs generally provide some instruction in the student's primary language in the first few years of schooling, with a move to an English-only instructional setting by either second grade (early-exit) or fourth grade (late-exit).

Maintenance bilingual programs continue instruction in the primary language after the student has gained proficiency in English, with the goal of biliteracy; these programs are also geared only toward language-minority students. However, dual language and two-way immersion programs incorporate both native English speakers and language-minority students in the same classroom wherein both groups are taught academic content in English and Spanish, for example, while receiving ESL instruction, and Spanish as a second language (SSL) instruction. The goal in these classrooms is biliteracy for all students. Despite its prevalence in popular discourse, however, bilingual education remains a relative rarity; most ELLs receive little to no instruction in the primary language (Zehler et al., 2003).

Significance of research in bilingual education Bilingual researchers and educators currently point to the loss of a natural resource: bilingual proficiency among language-minority children in our society. Children of immigrants lose the home or primary language within the first generation without active intervention (Rumbaut & Portes, 2001). The study of bilingual education is fundamental for social, civic, cultural, and economic reasons. Bilingual education provides one avenue through which schools can engage immigrant language-minority students (and their parents) who might not otherwise be drawn into the democratic process. Bilingual education is one facet of the larger field of multicultural education, which posits the importance of the home culture, language, and identity; bringing the family and the school together as partners in the education of the next generation (Baker, 2006). Socially, bilingual education validates the culture of the home in the context of the host society, the school. Through bilingual education and primary language maintenance and development, schools work to develop students' identities in a positive reflection of the larger sociopolitical situation. Although the goals of bilingual education are ultimately pedagogic, focused on the academic achievement of language-minority students, the role of bilingual education in these students' social and cultural integration merits consideration.

Themes and theories underlying bilingual education
Pedagogically, bilingual education is based in part on the theory of linguistic transfer (Baker, 2006); academic and cognitive skills learned in the primary language will transfer after sufficient proficiency is mastered in the second language. In addition, bilingual education draws heavily on Second Language Acquisition (SLA) and sociolinguistic theory wherein achievement depends not only on the purpose for which the language is acquired, but also the context in which the language is acquired. Finally, SLA theory posits that two forms of language exist: social and academic. Social language, or basic interpersonal communicative skills (BICS), negotiates meaning with the use of contextual clues (Cummins, 1984). Many transitional bilingual education programs move students into English-Only instruction as soon as they demonstrate BICS proficiency. In contrast, a primary goal of maintenance bilingual education is to foster academic language, or the development of Cognitive Academic Language Proficiency (CALP) in both languages. CALP is more complex, entailing both cognitive skills and knowledge of academic content. In an academic setting it is imperative that both social and academic language (BICS and CALP) are used to ensure an additive form of bilingual education. Here, social language serves as a foundation for learning the more complex academic language required for academic achievement.

COGNITIVE BENEFITS OF BILINGUALISM AND BILINGUAL EDUCATION

Before the 1960s, bilingual students were considered disadvantaged; their performance on standardized intelligence tests suggested a deficit in comparison with the scores given to their monolingual counterparts. Many of these studies in fact measured English language proficiency rather than intelligence (Ovando, Collier, & Combs, 2003). Peal and Lambert (1962) conducted a landmark study that controlled for student background variables and found that the intellectual experience of the bilingual outpaced that of the monolingual in concept formation, mental flexibility, and intellectual capacity. Numerous studies since then have confirmed these findings (August & Hakuta, 1997) suggesting the long-term benefits for bilingual education.

LONG-TERM EFFECTS OF BILINGUAL EDUCATION

It is unlikely that transitional models of instruction will produce either high levels of academic achievement or bilingualism because they are based on assimilation, rather than acculturation. Bilingual education models that focus on biliteracy and cultural pluralism promote primary language maintenance while English is being learned, allowing for selective acculturation (Rumbaut & Portes, 2001) and ultimately, bilingualism. The long-term effects of bilingual education depend on the quantity and quality of the primary language instruction provided, as well as the context of the program itself.

FUTURE OF THE FIELD

After passage of *English-Only* legislation in several states, bilingual education underwent an increase in dual immersion and two-way programs. Native English speaking parents now advocate for the development of second language proficiency in their children. No longer relegated to the compensatory education models of the mid-1960s, researchers recognize the value of bilingual education as enrichment-added education. Research and practice indicate a shift to bilingual education as enrichment; the target audience has expanded beyond language-minority students to include their native English speaking classmates. However, although the field may shift from compensatory to enrichment, the federal government continues to operationalize bilingual education and primary language instruction as a means to an end. With the authorization of the No Child Left Behind Act in 2001, Title VII and its focus on bilingual education disappeared; in its place, Title III now deals with English language acquisition. Pedagogically and culturally, the future of bilingual education rests on its success in incorporating meaningful and challenging curriculum, theory-based best practice and educators who are well versed in language acquisition theory.

SEE ALSO Volume 1: *Assimilation; Immigration, Childhood and Adolescence; Policy, Education; Racial*

Inequality in Education; School Tracking; Vocational Training and Education.

BIBLIOGRAPHY

August, D., & Hakuta, K. (1997). *Improving schooling for language minority students: A research agenda.* Washington, DC: National Academies Press.

Baker, C. (2006). *Foundations of bilingual education and bilingualism.* (4th ed.). Clevedon, England: Multicultural Matters.

California Department of Education (2008). *Limited English Proficient Services and Enrollment: R-30 Language Census, LEPS by Type of Service 1989–1999.* Retrieved March 3, 2008, from http://www.cde.ca.gov

Crawford, J. (1999). *Bilingual education: History, politics, theory and practice.* (4th ed.). Los Angeles: Bilingual Educational Services.

Cummins, J. (1984). *Bilingualism and special education: Issues in assessment and pedagogy.* San Diego, CA: College-Hill.

Gándara, P. (2005). Learning English in California: Guideposts for the nation. In M. M. Suárez-Orozco, C. Suárez-Orozco, & D. B. Qin (Eds.), *The new immigration: An interdisciplinary reader* (pp. 219–232). New York: Routledge.

Lau v. Nichols, 414 U.S. 563 (1974).

Ovando, C. J., Collier, V. P., & Combs, M. C. (2003). *Bilingual and ESL classrooms: Teaching in multicultural contexts.* (3rd ed.). New York: McGraw-Hill.

Peale, E., & Lambert, W. E. (1962). The relation of bilingualism to intelligence. *Psychological Monographs, 76,* 1–23

Rumbaut, R. G., & Portes, A. (2001). *Ethnicities: Children of immigrants in America.* Berkeley: University of California Press.

Zehler, A., Fleischman, H. L., Hopstock, P. J., Stephenson, T. G., Pendzick, M., & Sapru, S. (2003). *Descriptive study of services to LEP students and LEP students with disabilities.* Arlington, VA: Development Associates, for the U.S. Department of Education. Retrieved March 3, 2008, from http://www.ncela.gwu.edu

Rebecca Callahan
Soria E. Colomer

BIRACIAL YOUTH/ MIXED RACE YOUTH

The 2000 U.S. Census brought multiracial identification to national attention through a policy change that allowed respondents to check multiple boxes in answering the question about race. As a result of that change multiracial people emerged as one of the fastest-growing populations in the United States between 1990 and 2000; 41% of the multiracial population was under age 18, compared with 25% of the monoracial population. In 2000 there were 2.9 million multiracial persons under age 18, accounting for 3.95% of the youth population in the United States. After the overturn in 1967 of laws banning interracial marriage, intermarriage in all racial groups increased dramatically (Lee & Bean, 2004). Multiracial births reached 5.3% of all births in 2000 (National Center for Health Statistics, 2002). Multiracial youth are a growing demographic whose experiences are relevant to the experiences of all young people growing up in a multiethnic society.

Historically, research on multiracial people was stymied by cultural norms against acknowledging multiracial ancestry, small samples, and a lack of theoretical grounding or hypothesis testing. The marginal man theory of Park (1928) and Stonequist (1935) argues that multiracial people are marginalized and isolated by all monoracial groups. That theory still informs the work of some researchers despite evidence that multiracial people are integrated and well-adjusted members of society.

THEORIES AND THEMES IN THE LITERATURE ON MULTIRACIAL YOUTH

Much of the literature on identity development among multiracial persons starts by challenging the notion of race as a single and fixed aspect of identity. Racial identity develops slowly for these youth, and their sense of racial boundaries is more fluid than fixed (Johnson, 1992). Theories of multiracial identity development focus on the phases and/or tasks that youth typically complete or accomplish in establishing their identities, with variations for different racial mixes. A second theme of the literature has been an examination of the developmental implications of having a flexible racial identity. In this entry literature on identity development is differentiated from literature on the impact of identity on youth outcomes.

After Park's and Stonequist's work on multiracial people there was a dearth of attention to the topic for 40 years. In the 1970s, when the biracial baby boom began, theories of biracial identity and development focused on adaptation. Research from that era suggests that multiracials and monoracials have equivalent, though slightly different, processes of racial identity development (Thornton & Wason, 1996). However, the paucity of data prevented empirical testing of those theories, and critics charged that they did not consider the issue of choosing between identities or asserting a multiracial identity in a world that was unaware of multiraciality or interested in sanctioning it rather than understanding it (Daniel, 2002).

Current theories focus on mulitracials as a unique group for whom identity can be fluid across time and contexts. These theories address the conflict and guilt associated with choosing one identity over others and the resolution of those conflicts as a person comes to accept,

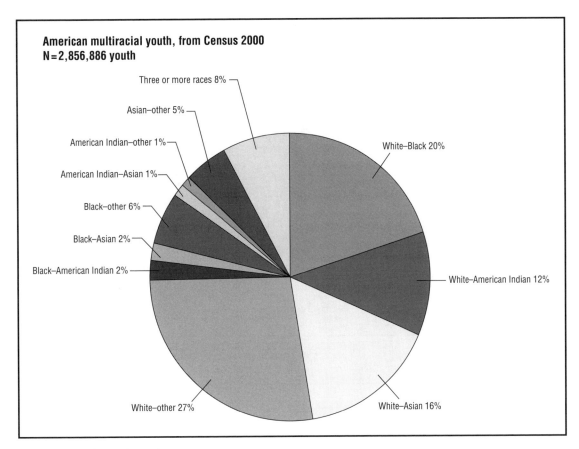

American multiracial youth, from Census 2000
N = 2,856,886 youth

Three or more races 8%

Asian–other 5%

American Indian–other 1%

American Indian–Asian 1%

Black–other 6%

Black–Asian 2%

Black–American Indian 2%

White–other 27%

White–Black 20%

White–American Indian 12%

White–Asian 16%

American Mutliracial Youth. Census 2000. CENGAGE LEARNING, GALE.

integrate, and assert all the parts of his or her identity. In this entry these theories are divided into three types: developmental phase theories loosely based on Piaget's stages of cognitive development, task theories based on Erikson's identity development model, and a third group that was designed specifically to explain multiracial identity.

Phase Theories The phase theories (Collins, 2000) typically begin with awareness, a stage describing young children's experiences of personal identity: becoming aware of skin tone and its connection with group membership. The second phase, choice, typically includes a growing awareness of cultural differences based on skin tone, along with an internal struggle to embrace internally and claim publicly a particular racial identity. Struggle, the third phase, involves confusion and guilt over having chosen a particular identity and in so doing rejecting those of some family members and peers. Multiracial youth may struggle in claiming an identity that is inconsistent with the norm by which children are categorized in accordance with the race of the lower-status parent (the norm of hypodescent). The fourth phase consists in strategizing ways to resolve the struggle stage

by convincing significant others that one's choice is legitimate and/or by broadening one's own conception of racial identity to include context or time-dependent racial identity. The final phase includes integrating all of one's identities and accepting oneself as a multiracial person whose identity is not compromised or determined by others. There may be recognition of multiracial as the appropriate reference group and/or a sense that all of one's different racial identities are valuable.

Task Theories The task theories (Gibbs, 1987), which are based on Erikson's (1968) general theory of youth identity development, focus on the particular challenges of racial identity development for multiracial people. Erikson argued for the formation of a stable (and monoracial) identity through the accomplishment of tasks such as gaining autonomy and independence from parents, developing positive peer relations, developing a sexual identity, and exploring career options. Multiracial task theorists argue for a flexible but integrated identity that changes as needed over time and across context. Although proponents of Eriksonian theory might consider such variability an unhealthy sign, multiracial theorists recognize it as a strategy for negotiating

Biracial Youth. *Because Jackson, Mississippi's schools are largely divided along racial lines, Kim Stamps, center, home schools her biracial children; son, Alkebu-lan, 12, left, and daughter, Abyssinia, 10.* AP IMAGES.

various social expectations and multiple truthful ways to identify. Thus, to the traditional Eriksonian developmental tasks (establish peer relations, sexual identity, and career choice), multiracial task theorists add the tasks of integrating racial identities and managing others' expectations for racial identification.

Multiracial Identity Theories Proponents of theories of multiracial identification argue that people's identities are fluid and are shaped by inherited influences, traits, and context. Root (1999) explained how all these influences are filtered through the lenses of generation, class, gender, and history of race relations to produce a variable but healthy racial and ethnic identity. Poston's (1990) model is similar to Root's, but it also explains how the relative status of the various ethnic groups in a person's background affect that person's choice of racial identity, along with physical appearance, language, age, and political involvement. Poston's and Root's models consider physiological factors as well as environmental factors at the micro (family, peer group), meso (school, neighborhood), and macro (societal, national) levels. These two theories were designed to capture the identity formation of people of all racial mixtures.

In contrast, the 2001 model of Rockquemore and Brunsma focuses exclusively on how the relatively small population of Black-White people self-identifies. Despite its size, this population merits special attention because of the social distance between Blacks and Whites in North American society and the resulting social pressures on part-

Black youth to identify as Black. Rocquemore and Brunsma's model features four identity types: singular (Black or White), border (biracial), protean (sometimes Black, sometimes White), and transcendent (no race). They found that the "one-drop rule" constrains part-Black people, even those with only one Black great-grandparent, to identify as Black. The one-drop rule is enforced by Whites and Blacks alike; to maintain political and social group strength, the Black community has developed an interest in maintaining this oppressive rule (Davis, 2001). In contrast to part-Blacks, Rockquemore and Brunsma argued that nonBlack multiracial people are not subjected to the one-drop rule. The theories of multiracial identity development were devised largely to fill gaps in identity theories that had focused on the particular racial and cultural issues of distinct groups.

However, most of the issues facing monoracial minority youth also are faced by most multiracial youth. Except for those who look and act White, multiracial young people face ethnic discrimination from Whites. All multiracial youth experience ethnic discrimination from members of ethnic groups who think they are not "ethnic enough" to be legitimate members of their group. In contrast, some part-White youth may benefit from the privileges and networks of the White parts of their families, whereas others are cut off from all or most of their White relatives. Like the varieties of monoracial minority youth, multiracial young people develop in extremely varied ethnic and cultural contexts.

STATE OF THE FIELD

Changes in social interaction norms and data-gathering norms increased the visibility of multiracial youth and amplified empirical research on that population. Most empirical research on multiracial youth has made use of samples identified by ancestry regardless of whether the respondents assert a multiracial identity (Rockequemore & Brunsma, 2001; Hitlin, Brown, & Elder, 2006; Herman, 2004; Harris & Sim, 2002; Brunsma, 2005). These studies consider the types and determinants of identity as well as the effects of identity on developmental outcomes. Herman (2004) found that physiognomy, ethnic identity, race of coresident parents, racial makeup of the neighborhood, and racial makeup of the school are associated with reported race among multiracial high school students. Rockquemore and Brunsma's (2001) study shows that among part-Black multiracials skin tone is not much of an influence on biracial identity; what matters is the multiracial person's assumption of how others perceive his or her appearance. Brunsma (2005) found that the parents of part-White multiracial preschoolers often listed their children's identities as mixed or White. Similarly, Xie and Goyette (1997) found that the parents of part-Asian biracial children used seemingly arbitrary criteria to classify their children's race. Among parents of part-Hispanic children, 27 to 30% categorized their children as multiracial, depending on the other options. Finally, Hitlin et al. (2006) found that multiracial youth change their racial identities over time.

The new survey data have necessitated methodological decisions and research on the management of race data. For example, to summarize information about a group of multiracial people, scholars must sort them into groups. There are many different "mixes" of multiracial youth. The early literature focused on Black-White mixes, but mixes with Asian, American Indian, Middle Eastern, and Hispanic are also part of the literature. For biracial subjects in relatively populous categories such as Asian-White, Hispanic-White, Black-White, American Indian-White, and Black-Hispanic, categorizing respondents is a straightforward task. However, for the small number of multiracial subjects, the researcher must group them in an uninformative "other" category or choose the biracial category that "best" describes them.

CONTEXT AND RACIAL IDENTIFICATION

Regardless of a multiracial youth's particular racial ancestry, the tasks of discovering and asserting a racial identity are complex. Although they often are forced to designate a single racial identification that ignores one or more of their racial ancestries, multiracial youth typically do not hold a single racial identity. Little or no research exists on

differences between racial identity and racial identification that allow young people to designate more than one racial or ethnic group for both identity and identification. However, several articles have used survey data to consider the factors that affect racial identification among multiracial youth. Herman (2004) and Hitlin et al. (2006) look at adolescents' self-identifications, and Brunsma (2005) looks at parents' identifications of their children. Although the respondents differ in these studies, the findings are largely similar: Parent and youth reports of racial identification are influenced by contextual factors such as neighborhood and school racial composition, regional history of racial categorization types, immigration status, language use at home, race of peers and coresident parent(s), skin tone, and racial ancestry. For example, multiracial youth who live in wealthier and whiter neighborhoods are more likely to identify as White, whereas those who attend predominantly White schools and are members of ethnic social crowds are more likely to identify as nonWhite (Herman, 2004). Some of these relationships may be endogenous: A youth's racial identity may influence that youth's parents' choice of a school or neighborhood with a particular racial composition, and it is likely that racial identity is associated with a youth's choice of peers.

For multiracial youth, contexts contain people and symbols of different racial and ethnic backgrounds, and this may contribute to value and cultural incongruence. Multiracial young people cope with this in part by having fluid racial identities over time and across contexts. This strategy allows different aspects of a multiracial youth's racial identity to become salient at different times and in different places, similarly to the way monoracial youth experience healthy inconsistency in cultural identity or peer crowd identity. Although theorists have argued that holding an inconsistent racial identity is detrimental, Root (1997) showed that it is a typical and healthy part of multiracial adolescent development.

Context influences much of an adolescent's exposure to stressors. Examples include being the only minority group member in a high-track math course or living in a neighborhood where most people are of a different race. Such incongruent racial contexts are challenging for all youth, and multiracial youth almost never have congruent racial contexts that consist of people who all share their particular racial mixes. Some scholars argue that multiracial adolescents lack a sense of racial belonging because no single race group embraces or can support them (Gibbs, 1998). Others argue that belonging everywhere and nowhere strengthens and diversifies their identities because they learn to span boundaries and switch codes appropriately in each context. Herman shows how racial congruence within a context acts as an intermediary between identity and outcomes (Herman, 2004). For example, belonging to an ethnic crowd and having a smaller percentage of White students in a youth's school are associated with minority racial identification, whereas residing with a White parent is associated with White identification.

Time, generation, and geographic location also are associated with racial identification among multiracial populations. Twine (1997) provided ethnographic evidence of multiracial young women's changing identities, particularly during adolescence, when dating begins. Part-Black girls who are raised in White communities are most likely to experience a change in identity from nonBlack to Black as they start dating. Hitlin et al. (2006) showed that multiracial youth's racial identifications on surveys typically change over time by adding a category, subtracting a category, or changing categories. Generation of immigration affects racial identity in the expected sense: The further in time they are from immigration, the less likely multiracial part-Asian youth are to identify as Asian (Herman, 2004). Some scholars argue that the political and social history of a geographic location can affect people's racial identity (Root, 1997). In Louisiana, which has a history of slavery and strict enforcement of the one-drop rule, most part-Black people identify as Black, whereas in Oklahoma, which has a history of tolerance for racial mixing among American Indians, Blacks, and Whites, many more part-Black people identify as mixed.

FUTURE DIRECTIONS

In 2000 the U.S. Census and other large, nationally representative data sets began to change their demographic race questions to allow a "check all that apply" option. This change has increased the visibility of the multiracial population and also may have encouraged multiracial people who formerly identified with only one race group to recognize the legitimacy of expressing their multiple ancestries. There is evidence that simply asking students to categorize photos with both a "check all that apply" item and a "check the one box that best describes you" item increases the number who report a multiracial heritage for themselves. Elsewhere research is being done on the impact of holding a particular identity on various developmental outcomes. For example, holding a minority identity is associated with higher self-esteem and fewer depressive symptoms but lower school achievement except in the case of Asian identifiers (Udry, Li, & Hendrickson-Smith, 2003). Increasing attention has been paid to multiracial youth in the clubs and activities aimed at their population (Renn, 2004), the magazines they can subscribe to, the organizations they can join, and their celebrity status. Ideally, these organizations and activities can help foster positive multiracial

identities, and they are sources of information and camaraderie.

SEE ALSO Volume 1: *Assimilation; Identity Development; Immigration, Childhood and Adolescence; Racial Inequality in Education.*

BIBLIOGRAPHY

Brunsma, D. (2005). Interracial families and the racial identification of mixed-race children: Evidence from the early childhood longitudinal study. *Social Forces, 84*(2), 1131–1157.

Collins, J. F. (2000). Biracial Japanese American identity: An evolving process. *Cultural Diversity and Ethnic Minority Psychology, 6*(2), 115–133.

Daniel, G. R. (2002). *More than black? Multiracial identity and the new racial order.* Philadelphia: Temple University Press.

Davis, F. J. (2001). *Who is black? One nation's definition.* University Park: Pennsylvania State University Press.

Erikson, E. (1968). *Race and the wider identity.* New York: Norton.

Gibbs, J. T. (1987). Identity and marginality: Issues in the treatment of biracial adolescents. *American Journal of Orthopsychiatry, 57,* 265–278.

Gibbs, J. T. (1998). Biracial adolescents. In J. T. Gibbs, L. N. Huang, and associates (Eds.), *Children of color: Psychological interventions with culturally diverse youth* (pp. 305–332). San Francisco: Jossey-Bass.

Harris, D., & Sim, J. (2002). Who is multiracial? Assessing the complexity of lived race. *American Sociological Review, 67*(4), 614–627.

Herman, M. (2004). Forced to choose: Some determinants of racial identification among multiracial adolescents. *Child Development, 75*(3), 730–748.

Hitlin, S. J., Brown, J. S., & Elder, G. H., Jr. (2006). Racial self-categorization in adolescence: Multiracial development and social pathways. *Child Development, 77*(5), 1298–1308.

Johnson, D. J. (1992). "Developmental pathways: Toward an ecological theoretical formulation of race identity in black-white biracial children." In M. P. Root (Ed.), *Racially mixed people in America* (pp. 37–49). Thousand Oaks, CA: Sage.

Lee, J., & Bean, F. D. (2004). America's changing color lines: Immigration, race/ethnicity, and multiracial identification. *Annual Review of Sociology, 30,* 221–242.

National Center for Health Statistics. (2002). *Births: Final data for 2001.* National Vital Statistics Reports, Vol. 51, 2. Washington, DC: Department of Health and Human Services. Retrieved March 1, 2008, from http://www./cdc.gov/nchs/data/nvsr/nvsr51/nvsr51_02.pdf

Park, R. (1928). Human migration and the marginal man. *American Journal of Sociology, 33*(6), 881–893.

Poston, W. S. (1990). The biracial identity development model: A needed addition. *Journal of Counseling and Development, 69*(2), 152–155.

Renn, K. A. (2004). *Mixed race students in college: The ecology of race, identity and community on campus.* Albany: State University of New York Press.

Rockquemore, K. A., & Brunsma, D. (2001). The new color complex: Phenotype, appearances, and (bi)racial identity. *Identity, 3*(1), 225–246.

Root, M. P. (1997). Biracial identity. In G. G. Bear, K. M Minke, & A. Thomas (Eds.), *Children's needs II: Development, problems, and alternatives* (pp. 751–759). Bethesda, MD: National Association of School Psychologists.

Root, M. P. (1999). The biracial baby boom: Understanding ecological constructions of racial identity in the 21st century. In R. Sheets & E. Hollins (Eds.), *Racial and ethnic identity in school practices: Aspects of human development* (pp. 67–90). Mahwah, NJ: Lawrence Erlbaum Associates.

Stonequist, E. (1935). The problem of the marginal man. *American Journal of Sociology, 41*(1), 1–12.

Thornton, M., & Wason, S. (1996). Hidden agendas, identity theories and multiracial people. In M. P. Root (Ed.), *The multiracial experience: Racial borders as the new frontier* (pp. 101–120). Thousand Oaks, CA: Sage.

Twine, F. W. (1997). Brown-skinned white girls: Class, culture, and the construction of white identity in suburban communities. In R. Frankenberg (Ed.), *Displacing whiteness: Essays in social and cultural criticism* (pp. 214–243). Durham, NC: Duke University Press.

Udry, J. R., Li, R. M., & Hendrickson-Smith, J. (2003). Health and behavior risks of adolescents with mixed-race identity. *American Journal of Public Health, 93*(11), 1865–1870.

Xie, Y., & Goyette, K. (1997). The racial identification of biracial children with one Asian parent: Evidence from the 1990 census. *Social Forces, 76*(2), 547–570.

Melissa R. Herman

BIRTH CONTROL, ADOLESCENT USE

SEE Volume 1: *Sex Education/Abstinence Education; Sexual Activity, Adolescent; Transition to Parenthood;* Volume 2: *Birth Control.*

BIRTH WEIGHT

Birth weight is the first weight of a fetus or newborn and is obtained shortly (ideally less than an hour) after birth. Babies born weighing more than 2,500 grams (5.5 pounds) typically are categorized as having normal birth weight, whereas babies weighing less than that are categorized as having low birth weight. Babies weighing less than 1,500 grams (3.3 pounds) frequently are placed in a separate category of very low birth weight. These conventional thresholds are based on extensive evidence showing that babies and children born in the lower weight ranges have much higher risks of infant mortality, poor health, disability, and developmental/cognitive delays.

A baby's weight at birth results from the rate at which the baby grew in utero and the duration of the pregnancy (the gestational age of the baby). Low birth weight therefore results from growth restriction (being small for gestational age) and/or preterm birth (birth before 37 weeks of gestation). Growth restriction and preterm birth are related to a range of factors that include: (a) the fetus/baby's characteristics (e.g., sex, congenital anomalies); (b) the mother's health and behavior throughout her life (e.g., her birth weight, childhood nutrition, history of smoking while pregnant, use of prenatal care services); and (c) the past and present social and physical environment (e.g., maternal socioeconomic conditions during childhood and adulthood, exposure to toxins).

Although data before the mid-20th century are sparse, it appears that birth weights increased with industrialization. For example, between 1900 and 1950 rates of low birth weight in Europe appear to have dropped from between 15% and 17% to between 7% and 9%. Since the 1960s rates of low birth weight in industrialized nations have remained relatively stable at approximately 8%. There has, however, been improvement in birth-weight-specific survival, particularly among the smallest infants. In 1960 a baby with very low birth weight in the United States had about a 50% chance of surviving until age one, whereas in 2000 it had about a 75% chance. Much of this improvement in survival can be credited to advances in neonatal intensive care. Although these mortality reductions represent a significant accomplishment, very low birth weight survivors face several physical health challenges, and higher survival rates in this group ultimately can increase adult disability rates.

BIRTH WEIGHT DIFFERENTIALS

Although low birth weight has been studied extensively in industrialized nations, the vast majority (95.6%) of babies with low birth weight worldwide are born in developing countries. In 2000 the rate of low birth weight in developing nations (16.5%) was more than double the rate in most industrialized nations and similar to the rates in Europe about 100 years earlier. These dramatically higher rates of low birth weight in developing nations reflect large international disparities in income, maternal health, food supply, shelter, and health care.

Disparities in birth weight in rich industrialized nations also can be stark. In the United States in 2005 the rate of low birth weight among non-Hispanic Black Americans was twice the rate among Whites (14% versus 7%, respectively). Rates of low birth weight in industrialized nations also tend to be significantly higher among younger mothers and mothers with incomes below the poverty line.

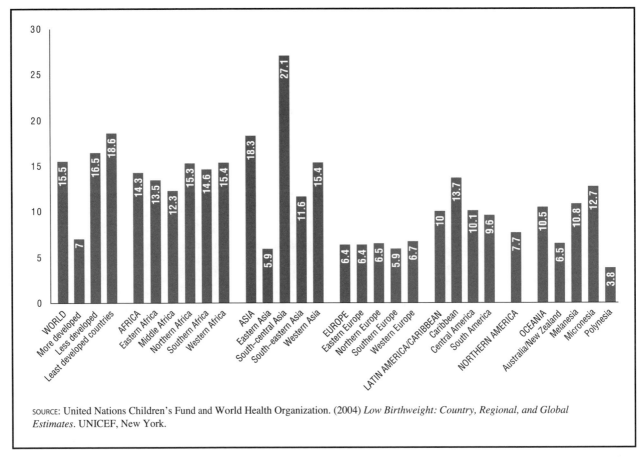

Figure 1. *Percentage of low birth weight infants by United Nations regions, 2000.* CENGAGE LEARNING, GALE.

CONSEQUENCES OF BIRTH WEIGHT

Birth weight is an important topic of study because size at birth is a key indicator of the future opportunities, or life chances, of a baby. Low birth weight has an effect on life chances by increasing the risk of infant mortality roughly 20-fold. Among those who survive the first year, low birth weight can limit life chances by hindering good health and cognitive development. Babies with very low birth weight are at increased risk for a range of serious disabilities, such as cerebral palsy, mental retardation, hearing/vision impairment, and chronic lung or gastro-intestinal problems. Slightly larger babies (those between 1,500 and 2,499 g [3.3 and 5.5 lb]) are significantly less likely to face these disabilities but still may encounter a range of more subtle disadvantages. Relative to those in the normal range of birth weight, infants with low birth weight on average require more familiarization time before they can recognize people and objects and typically are less adept at manipulating objects. Children with low birth weight also tend to score lower on IQ tests than their heavier counterparts. Such developmental

disadvantages may affect school progress. By age 18 years, individuals with low birth weight are less likely to have graduated from high school and more likely to have been left back a grade.

Low birth weight can continue to affect well-being throughout adulthood. Individuals with low birth weight earn about 8% less in the adult labor market than do their counterparts with normal birth weight. Also, adults with low birth weight have elevated risks of cardiovascular disease and type 2 diabetes. The consequences of low birth weight may even be transmitted across generations. Parents who were born at a low birth weight are significantly more likely to have an infant with low birth weight and may pass on to their children some of the disadvantages associated with low birth weight.

CHALLENGES FOR RESEARCH AND PRACTICE

An important concern for contemporary researchers and policy makers is identifying the relative importance of particular, distinct determinants of birth weight, such as parental income during pregnancy, maternal childhood conditions, prenatal care, and nutrition. Understanding the unique benefit generated by, say, an increase in income or access to prenatal care is important in designing interventions to increase birth weights and reduce the disparities discussed above. However, it can be difficult to estimate precisely the benefits of specific inputs because women who had low incomes during pregnancy also are likely to have had disadvantaged childhoods, to have forgone prenatal care, and to have had poorer nutrition. Thus, it is frequently unclear whether an association between poverty during pregnancy and low birth weight reflects a true effect of income or the effect of correlated, unmeasured factors, such as childhood conditions and prenatal care. A central goal for research is identifying strategies (e.g., trials, natural experiments) that will generate more accurate estimates of particular underlying causes.

The large gap in birth weight between Whites and African Americans also has been debated. Because low socioeconomic status is positively associated with low birth weight and African Americans are concentrated disproportionately in lower socioeconomic categories, many argue that higher rates of low birth weight among Blacks result primarily from their socioeconomic disadvantages. When Whites and Blacks of similar socioeconomic status are compared, disparities in birth weight are reduced significantly; however, even holding income constant cannot eliminate the White-Black gap entirely. Socioeconomic explanations of racial disparities also are challenged by the Hispanic health paradox. Hispanics in the United States face several socioeconomic disadvantages. However, their rates of low birth weight are frequently lower than White rates, challenging the argument that low socioeconomic status causes low birth weight. As Hispanic immigrants become more acculturated to the United States, however, this birth weight advantage tends to decline, leading many to argue that the Hispanic health paradox reflects healthier cultures and lifestyles among recent immigrants, such as a healthier diet and a stronger network of community support.

Other researchers have proposed biological/genetic explanations for racial differences in birth weight. Although Black infants are twice as likely as White infants to be born at a low weight, Black infants have better birth-weight-specific survival rates than Whites do. In other words, African-American infants with low birth weight are more likely to survive than are comparable White infants. This has led some researchers to suggest that racial disparities in birth weight reflect a type of biological adaptation. These arguments suggest that over many generations, populations exposed to deprived environments may experience natural selection for smaller babies because in an environment of limited nutrition, smaller babies may have survival advantages related to

BARKER HYPOTHESIS

According to the Barker hypothesis, environmental influences that impair growth and development early in life are important causes of chronic disease in adulthood. In the 1990s David Barker and his colleagues published several studies showing that individuals who were born at a low birth weight are significantly more likely to have heart disease and type 2 diabetes later in life. Those findings led Barker to hypothesize that when a fetus experiences nutritional deprivation during key stages of pregnancy, its development is altered in anticipation of postnatal scarcity (organ structure and function are "reprogrammed" to maintain metabolic thriftiness later in life). These in utero adaptations may have conferred significant survival benefits during historical periods when starvation was a serious concern. However, in the caloric abundance of contemporary industrialized nations metabolic thriftiness may increase the risk of adult diseases. The Barker hypothesis has received a great deal of attention in academia and the media and has helped popularize many life course concepts.

metabolic efficiency. Overall, there is little evidence for genotypic bases of racial disparities in birth weight or racial categorizations more generally, and research in this area has been controversial.

Several unanswered questions remain about associations between birth weight and adult outcomes, including low education attainment or wages, cardiovascular disease, and diabetes. One question is whether these associations really reflect a causal effect of birth weight. Low birth weight is associated with several environmental disadvantages, and so individuals who were born at a low weight are likely to have faced a variety of challenges throughout their lives—such as a low-quality education, poor diet, and a disadvantaged neighborhood—all of which may be alternative explanations for later outcomes. Much of the research in this area is concerned with developing strategies to isolate the effects of birth weight by, for instance, comparing siblings or twins who are likely to have had similar backgrounds.

Assuming that birth weight has a causal effect on adult well-being, several questions remain about the possible pathways through which these effects work. One of the most prominent hypotheses in this area is the Barker hypothesis, which argues that environmental stressors cause changes in fetal development that ultimately increase the risk of cardiovascular disease and type 2 diabetes. Although the Barker hypothesis provides a compelling explanation for associations between birth weight and chronic disease, it does not speak as effectively to associations between birth weight and later socioeconomic outcomes. An important ongoing research question is how different mechanisms, especially brain development, IQ, and physical health, may account for associations between birth weight and adult socioeconomic status.

SEE ALSO Volume 1: *Health Differentials/Disparities, Childhood and Adolescence; Illness and Disease, Childhood and Adolescence; Infant and Child Mortality.*

BIBLIOGRAPHY

Bennett, F. (1997). The LBW, premature infant. In R. T. Gross, D. Spiker, & C. Haynes (Eds.). *Helping low birth weight, premature babies: The infant health and development program* (pp. 3–16). Stanford, CA: Stanford University Press.

Conley, D., Strully, K. W., & Bennett, N. G. (2003). *The starting gate: Birth weight and life chances.* Berkeley: University of California Press.

Osmond, C., & Barker, D. J. (2000). Fetal, infant, and childhood growth are predictors of coronary health disease, diabetes, and hypertension in adult men and women. *Environmental Health Perspectives, 108* (Supplement 3), 545–553.

National Center for Health Statistics, Centers for Disease Control. (2005). *Health, United States, 2005.* Retrieved May 1, 2008, from http://www.cdc.gov/nchs/hus.htm

United Nations Children's Fund and World Health Organization. (2004). *Low birthweight: Country, regional, and global estimates.* Geneva: Worth Health Organization; New York: Author.

Ward, W. P. (1993). *Birth weight and economic growth: Women's living standards in the industrializing west.* Chicago: University of Chicago Press.

Williams, D., & Collins, C. (1995). U.S. socioeconomic and racial differences in health: Patterns and explanations. *Annual Review of Sociology, 21,* 349–386.

Wintour, E. M., & Owens, J. A. (Eds.). (2006). *Early life origins of health and disease.* New York: Springer Science+Business Media.

Kate W. Strully

BODY IMAGE, CHILDHOOD AND ADOLESCENCE

Body image is a complex multifaceted concept that generally encompasses how people perceive and feel about their physical appearance, including the weight, size, shape, or athletic ability of their bodies. Constructing one's body image typically involves the process of judging the body against socially constructed and socially learned ideals about attractiveness or beauty. These body ideals are learned over the life course (beginning sometimes as early as preschool) within the social and cultural contexts (e.g., families, countries) where daily life unfolds. Though macrotrends in body image ideals are often identifiable (e.g., the emphasis on thinness as a sign of beauty in women in most developed countries), these trends are not universal and do not influence all individuals uniformly. In particular, body concerns of boys and girls tend to emphasize different aspects of the body and tend to refer to different ideals. Boys often strive to increase their body size to conform to ideals that equate muscularity with masculinity, whereas girls often attempt to decrease body size to conform to ideals that equate being thin with being feminine.

WHY BODY IMAGE MATTERS

Though body ideals vary between boys and girls, research (particularly studies published since 1995) has clearly demonstrated that among children and adolescents, both boys and girls, body image can be a complicated and sometimes painful issue. This is also not a small social problem. Although reliable nationally representative prevalence rates of body image problems are difficult to construe, numerous studies using diverse sampling frames and methods consistently report that a significant number of youth experience some type of body dissatisfaction. This is of concern because body image problems can have serious consequences for the physical and mental health and well-being of girls and boys.

The most common response of girls to body dissatisfaction and a desire to be thin is to engage in calorie-restrictive dieting that can lead girls to develop eating problems and disorders and can put girls at risk of obesity, stunted growth, bone-density loss, anxiety, and depression. Among boys, the desire to increase muscle mass sometimes leads to the use of weight-gain supplements and anabolic steroids. Body dissatisfaction among boys can also lead to depression (Smolak, 2004). Finally, in a culture that emphasizes antifat attitudes, the prevalence of overweight and obesity among youth is on the rise (Ogden, Flegal, Carroll, & Johnson, 2002). Overweight children often experience painful social sanctions for being overweight. Teasing overweight boys and girls about their weight can lead to higher levels of emotional distress and a higher frequency of binge eating (Neumark-Sztainer, Falkner, Story, Perry, Hannan, & Mulert, 2002). To help youth make healthy choices about eating and weight-change behaviors, one must first understand how the fear of overweight and the desire for thinness or muscularity develop in the early life course.

Historical and Demographic Trends One only has to look to Hollywood to see how body ideals for girls and women have changed over time. In the 1950s, female body ideals were significantly heavier and curvier than ideals at the turn of the 21st century. The image of Marilyn Monroe starkly contrasts images of movie stars in 2008 (who are noticeably thinner and less curvaceous). This historical change in ideal female bodies is frequently commented on in the literature on body image; however, historical trends in body ideals for men are less often discussed. Some limited evidence suggests that body ideals for men, as displayed in magazines or action figures, have become increasingly muscular since the 1970s.

Though there are identifiable trends in body ideals in the United States, not all adolescents and children perceive these ideals in a uniform manner. For example, the prevalence of weight-change practices and body dissatisfaction varies significantly among adolescents of different racial and ethnic groups. In particular, African-American girls seem to have a more flexible definition of the ideal body than White girls (Nichter, 2000). Additionally, African-American boys and girls are more likely to choose a heavier ideal body type for either gender. This flexibility means that African-American girls are more likely to be satisfied with their bodies (despite being heavier on average) and, even when African-American girls do perceive themselves as overweight, they are less likely to experience negative emotional consequences.

African-American adolescents generally experience protection against negative body images, but this is not necessarily true for Asian-American or Latina girls (very little is known about Asian-American or Latino boys). A large study of adolescent weight control found that approximately 40% of Asian-American and White girls were currently dieting, compared with 33% of Latinas and 22% of African-American girls. Disordered eating was originally considered a White, middle-class problem; however, research in 2008 suggests that it is not an insignificant problem for girls of any race or ethnic group. Further, some evidence suggests that Latinas and Asian-American girls may engage in disordered-eating and experience body dissatisfaction at the same rate as white girls (Neumark-Sztainer, Story, Falkner, Teuhring, & Resnick, 1999).

Developmental Trends In addition to differences in body concerns among boys and girls among different ethnic and racial groups, research suggests that body concerns change as children age and as their bodies and cognitive and social abilities develop. Although research on children is limited by an inadequate ability to measure body image in early childhood, children develop the awareness that fat is *bad* sometimes as early as preschool. As children age, this knowledge is increasingly incorporated into how children evaluate their bodies and body dissatisfaction becomes more common (Ricciardelli & McCabe, 2001).

The middle-school years are a critical time in the development of body image. Both girls and boys tend to experience increased body dissatisfaction during these years (Rosenblum & Lewis, 1999). Several developmental changes may make the middle-school years more difficult for adolescents in terms of their body image. First, during middle school, adolescents' concern with how their peers view them and whether they are socially accepted generally increases. Second, physical appearance takes on new meaning and plays an important role in how adolescents experience social life in schools (Eder, Evans, & Parker, 1995). Third, puberty generally begins during the middle-school years. For girls, this can produce drastic physiologic changes that may move girls further away from normative body ideals. For boys, puberty can bring them closer to normative male ideals by causing them to grow taller and by allowing them to develop a more muscular physique. Middle-school boys, like girls, experience decreased body satisfaction, but boys tend to recover a positive body image more quickly than girls, possibly because of the different role puberty plays in bringing boys closer and girls further from gendered body ideals (Littleton & Ollendick, 2003). Though puberty usually begins in middle school, pubertal development can also vary substantially among adolescents; therefore, the timing of puberty also matters to adolescents' body images. Fitting in or standing out because of early or late maturation can exacerbate the difficulties that many boys and girls experience as their bodies change during this period (Martin, 1996).

Limitations In 2008 demographic and developmental research on body image provides clear evidence that body image is important to consider across groups and throughout child development, but several limitations of this research are present. First, although the focus is changing, girls have received more attention in the overall literature than boys. Second, research using rich psychological measures of body image tends to use relatively small and predominantly White samples of youth (though research is increasingly including youth from different socioeconomic, racial, or ethnic backgrounds). Currently, no nationally representative data are available that include the complex measures developed by psychologists that are the most reliable way to investigate the multidimensional aspects of body image. Third, existing measures of body image were developed for girls (primarily White) and may not have much to do with the body concerns of boys or nonWhite girls. Fourth, despite the growing awareness that body image construction

Weight-Loss Camp. *Overweight teen and pre-teen boys and girls swimming at Camp Kingsmont, a weight management camp for teenagers with weight issues.* © **KAREN KASMAUSKI/CORBIS.**

begins in early childhood, it is not yet known how to measure body image in elementary-school aged children reliably (Smolak, 2004). Therefore, before researchers can understand the true demographic and developmental trends in body image, they must identify more ways to measure body image in different populations, and these measures should be included in nationally representative studies of children and adolescents.

THE ROLE OF FAMILIES

Children first begin to learn body ideals in the primary context of their lives—their families. Within families, especially around mealtimes, children learn values related to food and fat. If parents express body image concerns or model behaviors, such as dieting, in front of their children, their children are more likely to experience increases in body concern. Moreover, when parents tease or express concern about their children's weights, this can harm children's emotional well-being. Parents may have good intentions, such as trying to help their child maintain a healthy body, but children's perceptions of pressure about weight can increase the likelihood that they may develop problematic negative body images.

Much of the literature on the role of families in adolescent body image has focused primarily on girls' relationships with their mothers. The research that has investigated daughters *and* sons suggests that adolescent daughters are more likely than sons to discuss weight with their mothers and to report observing mothers modeling weight-change behaviors. The book *Fat Talk* by Mimi Nichter (2000) provides a full qualitative view of what girls and their parents say about weight, bodies, and dieting and of the complex ways that parents influence their daughters. Sometimes the parents' intention is to help their daughters maintain a healthy body or to intervene when they observe unhealthy weight-loss strategies; in other instances, parents reinforce normative body ideals that equate beauty with being thin. Nichter also investigated the role of families in the construction of African-American girls' body images. She found that African-American girls receive more positive than negative comments from family members and that they experience more family and community support for constructing their own flexible definitions and portrayals of beauty than their White peers.

THE ROLE OF FRIENDS AND PEERS

As children age, extrafamilial relationships, particularly with friends, become an increasingly salient context where social interactions either reinforce or contradict body ideals learned in early childhood in families. Peers first begin to influence children's body image in elementary school where appearance-related teasing is often common. By adolescence, being accepted by peers takes on even greater meaning, and the opportunities and mechanisms whereby friends influence body image diversify. Fitting in to body ideals can affect how accepted or ostracized adolescents feel. The *peer appearance context* (Jones, 2004) of adolescents' daily social lives becomes an important source of information on acceptable weight and weight-control behaviors. In this context, adolescents learn weight-control behaviors by watching or talking with peers. Teasing reinforces the stigma attached to deviating from body ideals. Friends can either protect or exacerbate the drive to conform to body ideals of peers. For example, when girls have friends who practice weight control, they themselves are more likely to practice weight control and to feel pressured to lose weight.

FAT TALK, SOCIAL COMPARISON, AND THE MEDIA

In addition to friends exerting direct pressure to lose weight, *fat talk* among girls is a common way that girls construct their image of an ideal body among their friends (Nichter, 2000). Fat talk occurs when girls discuss (or overhear other girls discussing) how fat or dissatisfied with their bodies they are. In these informal discussions, girls are expected to chime in and express their own body dissatisfaction. Disparaging one's body is often the expected norm. The result of these interactions is that girls construct an image of what is *too fat* that frequently does not correspond to medical definitions of overweight. Though fat talk can result in increased feelings of body dissatisfaction, not all girls engage in fat talk for the same reasons and a girl's motivation for participating can moderate the impact fat talk has on her self-concept. In particular, tone must distinguish between girls who express body dissatisfaction to fit in, conform, or be friendly, and girls who have internalized thinness ideals and may be on the path to eating orders or depression.

In addition to fat talk, social comparison of one's body with those of friends, peers, and representatives of ideal bodies (such as fashion models in magazines) is another way that adolescents socially construct their feelings and perceptions of their bodies. The media provides ample examples of body ideals. Frequent exposure to these idealized bodies can negatively affect the body images of children and adolescents.

Because mass media are large diffuse social institutions, how these corporations create the body images of youth involves a complex multifaceted process. Even if adolescents see media images as unrealistic or express a preference for more realistic images, their own sense of body image may still be harmed if they assume that significant others will use media images to judge them (Milkie, 1999). Importantly, not all adolescents respond to exposure to the media in the same way. Some boys and girls have personal resources, such as self-esteem or social support, which allow them to limit the extent to which they internalize (or personally believe in) the body ideals portrayed in the media. One thing that the media and social comparison with media icons does highlight is that children and youth do not learn body ideals simply in direct interactions with parents or friends. Diffuse social messages about bodies are prevalent in developed societies and even in the absence of direct pressure from significant others, these messages can be internalized. For both boys and girls, internalization of normative body ideals is a key step toward unhealthy or negative body images; therefore, understanding the circumstances under which children and youth internalize these messages is fundamental.

SCHOOL CULTURE

Because the internalization of potentially problematic body ideals is most likely to develop within subcultures that emphasize being thin or muscular, it is crucial to investigate the primary social contexts of adolescents' lives. Family and friends are contexts that have received a lot of attention in the literature on body image; however, the role schools play in fostering adolescent cultures that affect body image has not been thoroughly investigated. As children age, schools become an increasingly salient context. By adolescence, schools serve as the primary location of adolescents' social and academic lives. Schools have long been recognized as an important venue for the formation of peer cultures with specific values and codes of behavior (Coleman, 1961; Eder, Evans, & Parker, 1995). School-based adolescent cultures are known to affect a range of adolescent values and behaviors (e.g., smoking). In fact, one of the only studies of school cultures and body image by Eisenberg, Neumark-Sztainer, Story, and Perry, (2005) found that adolescent girls who attend schools where a large proportion of the female student body practices unhealthy methods of weight control (e.g., diet pills, or laxative abuse) are more likely to engage in unhealthy behaviors themselves. Future work on body image should expand this area of research to look at more schools across the United States (and other countries) and to include boys.

CONCLUSION

Developing a healthy or positive body image is an important part of children's development into happy, healthy adults. Understanding how to encourage youth toward healthy lifestyles without damaging their emotional well-being is an important challenge for parents, researchers, and professionals who work with youth. Current research has clearly established that body image is not a problem that can be ignored for girls or boys, young children or adolescents. Continuing to investigate the multifaceted aspects of body image across the early stages of the life course and the complicated factors that influence how children's body images develop over this period are important steps toward creating policies and interventions that empower children to lead healthy, well-adjusted lives.

SEE ALSO Volume 1: *Eating Disorders; Media and Technology Use, Childhood and Adolescence; Obesity, Childhood and Adolescence; Peer Groups and Crowds; Self-Esteem; Socialization, Gender.*

BIBLIOGRAPHY

Coleman, J. S. (1961.) *The adolescent society: The social life of teenagers and its impact on education.* New York: Free Press of Glencoe.

Eisenberg, M. E., Neumark-Sztainer, D., Story, M., & Perry, C. (2005). The role of social norms and friends' influences on unhealthy weight-control behaviors among adolescent girls. *Social Science & Medicine, 60,* 1165–1173.

Eder, D., Colleen Evans, C. C., & Parker, S. (1995). *School talk: Gender and adolescent culture.* New Brunswick, NJ: Rutgers University Press.

Jones, D. C. (2001.) Social comparison and body image: Attractiveness comparisons to models and peers among adolescent girls and boys. *Sex Roles, 45,* 645–664.

Jones, D. C. (2004). Body image among adolescent girls and boys: A longitudinal study. *Developmental Psychology, 40,* 823–835.

Littleton, H. L., & Ollendick, T. (2003). Negative body image and disordered eating behavior in children and adolescents: What places youth at risk and how can these problems be prevented? *Clinical Child and Family Psychology Review, 6,* 51–66.

Martin, K. A. (1996). *Puberty, sexuality, and the self: Boys and girls at adolescence.* New York: Routledge.

Milkie, M. (1999). Social comparisons, reflected appraisals, and mass media: the impact of pervasive beauty images on black and white girls' self-concepts. *Social Psychology Quarterly, 62,* 190–210.

Neumark-Sztainer, D., Falkner, N., Story, M., Perry, C., Hannan, P. J., & Mulert, S. (2002.) Weight-teasing among adolescents: Correlations with weight status and disordered eating behaviors. *International Journal of Obesity, 26,* 123–131.

Neumark-Sztainer, D., Story, M., Falkner, N. H., Teuhring, T., & Resnick, M. D. (1999). Sociodemographic and personal characteristics of adolescents engaged in weight loss and weight/muscle gain behaviors: Who is doing what? *Preventive Medicine, 28,* 40–50.

Nichter, M. (2000.) *Fat talk: What girls and their parents say about dieting.* Cambridge, MA: Harvard University Press.

Ogden, C. L., Flegal, K. M., Carroll, M. D., and Johnson, C. (2002.) Prevalence and trends in overweight among US children and adolescents, 1999–2000. *Journal of the American Medical Association, 288,* 1728–1732.

Ricciardelli, L. A., & McCabe, M. P. (2001). Children's body image concerns and eating disturbance: A review of the literature. *Clinical Psychology Review, 21,* 325–344.

Rosenblum, G. D., & Lewis, M. (1999). The relations among body image, physical attractiveness, and body mass in adolescence. *Child Development, 70,* 50–64.

Smolak, L. (2004). Body image in children and adolescents: Where do we go from here? *Body Image, 1,* 15–28.

Anna Strassmann Mueller

BOWLBY, JOHN
1907–1990

In creating attachment theory, John Bowlby, who was born in London, was guided by a unique combination of family, personal, and educational influences. After rejecting a naval career, he studied preclinical sciences and psychology (including child psychology) at the University of Cambridge from 1925 to 1928 at the suggestion of his father, an eminent surgeon. Reluctant to pursue further with medical training, Bowlby spent the year after graduation as a volunteer teacher at two progressive schools for maladjusted children. That experience convinced him that children's emotional problems frequently are linked to disruptions in early family relationships and led to his decision to become a child psychiatrist and psychoanalyst. Bowlby's painful memories of losing his favorite nanny as a young child and later being sent to a boarding school also might have played a role in his choice of profession (Holmes, 1993).

CLINICAL WORK AND EARLY PUBLICATIONS

In 1936 Bowlby had completed his psychiatric training and began to work at the London Child Guidance Clinic, where he noticed that the symptoms for which children were referred often could be explained by probing their parents' childhood experiences. His concurrent training as a child analyst under the supervision of Melanie Klein was filled with tension because of his strongly held view that actual relationships, not internal drives, are

John Bowlby. SIR RICHARD BOWLBY.

the instigators of children's emotional disturbances. When Bowlby's work with children was interrupted during World War II, he gained expertise in experimental design and statistics as member of a group of army psychiatrists and psychologists charged with improving officer selection procedures. That training is evident in the quasi-experimental design of a seminal study linking children's stealing behavior to early family disruptions (Bowlby, 1944).

In 1945, as director of the department for parents and children at the Tavistock Clinic in London, Bowlby founded a research group to study the responses of young hospitalized and institutionalized children experiencing long-term separation from their parents. The results of that study, along with the documentary film *A Two-Year-Old Goes to Hospital* (1952) by Bowlby's collaborator James Robertson, contributed to the liberalization of hospital visiting rules for parents. Based on that work, the World Health Organization (WHO) commissioned Bowlby to write a research-based report on the well-being of homeless children in postwar Europe. The report concluded that to facilitate healthy development, "the infant and young child should experience a warm, inti-

mate, and continuous relationship with his mother (or permanent mother substitute) in which both find satisfaction and enjoyment" (Bowlby, 1952, p. 13).

LATER WORKS AND THEIR RECEPTION

Seeking theoretical grounding for his recommendations, Bowlby searched the literature on ethology, evolutionary biology, cybernetics/systems theory, information processing, memory, the emotions, and developmental psychology. His goal was to reinterpret psychoanalytic views of human relationships in light of contemporary science. However, in 1958, when Bowlby presented his first version of attachment theory to the British Psychoanalytical Society, the reception was hostile. Undeterred, he expanded his early theorizing into the trilogy *Attachment and Loss* (1969, 1973, 1980). In those volumes he showed an unusual ability to recognize cutting-edge ideas before they gained general acceptance.

In *Attachment* (1969) Bowlby reviewed research on social bonds in birds and primates to propose that fleeing to an attachment figure when frightened and using that figure as a secure base for exploration increases an infant's survival chances. He also contended that an attachment figure's prompt and appropriate responsiveness to a child's expression of fear or distress engenders a more secure, trusting, and happy relationship that fosters the growth of self-reliance. His argument relied on Mary Ainsworth's 1967 short-term longitudinal studies of infant–mother attachment in Uganda and her then-unpublished work in the United States. Their collaboration had begun in London in 1945.

In *Separation* (1973) Bowlby stressed that young children experience fear not only in the presence of unlearned and cultural "clues to danger" but also in the absence of an attachment figure. He regarded the systems controlling escape and attachment behaviors as distinct members of a larger family of stress-reducing and safety-promoting systems whose function is to maintain an organism within a defined (safer) relationship to its environment. Additionally, he proposed an epigenetic model of healthy and unhealthy developmental pathways from infancy to adulthood, drawing on the work of C. H. Waddington (1957). He also noted the link between secure attachment and warm but limit-setting parenting styles during early and later childhood.

In *Loss* (1980) Bowlby fleshed out ideas on the role of representation (internal working models) in relationship functioning and personality development, a topic he had introduced in the first two volumes. In the context of discussing clinical case studies of bereavement, he presented tentative ideas about the distortion of representation by defensive processes, building on emerging findings on cognition and memory.

The initial impact of attachment theory on developmental psychology was due largely to the empirical work of Mary Ainsworth and co-workers. However, as attachment research moved beyond infancy, Bowlby's trilogy began to exert a more direct influence. This is evident in longitudinal attachment studies from birth to young adulthood in the United States, Germany, and Israel. It is also evident in research on attachment representations in children, parents, and couples as well as in a growing number of clinical studies (see Cassidy & Shaver, 2008).

Partly as a result of changes in societal norms, Bowlby has been criticized for excessive emphasis on the maternal role, insufficient attention to fathers, and devaluation of women's work outside the home. Studies on child–father attachment and the effects of nonfamilial child care instigated by these criticisms have refined and expanded his theory. Promising new avenues for attachment research are also opening up in affective and social neuroscience.

SEE ALSO Volume 1: *Attachment Theory; Parent-Child Relationships, Childhood and Adolescence; Parenting Style.*

BIBLIOGRAPHY

Ainsworth, M. D. S. (1967). *Infancy in Uganda: Infant care and the growth of love.* Baltimore: Johns Hopkins Press.

Ainsworth, M. D. S., Blehar, M. C., Waters, E., & Wall, S. (1978). *Patterns of attachment: A psychological study of the strange situation.* Hillsdale, NJ: Lawrence Erlbaum.

Bowlby, J. (1940). The influence of early environment in the development of neurosis and neurotic character. *International Journal of Psychoanalysis, 21,* 154–178.

Bowlby, J. (1944). Forty-four juvenile thieves: Their characters and home life. *International Journal of Psychoanalysis, 25,* 19–53, 107–128.

Bowlby, J. (1952). *Maternal health and mental health.* Geneva: World Health Organization.

Bowlby, J. (1958). The child's tie to his mother. *International Journal of Psychoanalysis, 39,* 350–377.

Bowlby, J. (1960a). Grief and mourning in infancy. *Psychoanalytic Study of the Child, 16,* 3–39.

Bowlby, J. (1960b). Separation anxiety. *International Journal of Psychoanalysis, 41,* 89–113.

Bowlby, J. (1969). *Attachment and loss: Vol. 1. Attachment.* New York: Basic Books.

Bowlby, J. (1973). *Attachment and loss: Vol. 2. Separation.* New York: Basic Books.

Bowlby, J. (1980). *Attachment and loss: Vol. 3. Loss, sadness, and depression.* New York: Basic Books.

Bretherton, I. (1992). The origins of attachment theory: John Bowlby and Mary Ainsworth. *Developmental Psychology, 28*(5), 759–775.

Cassidy, J., & Shaver, P. R. (Eds.). (2008). *Handbook of attachment: Theory, research, and clinical applications* (2nd ed.). New York: Guilford Press.

Grossmann, K. E., Grossmann, K., & Waters, E. (2005). *Attachment from infancy to adulthood: The major longitudinal studies.* New York: Guilford Press.

Holmes, J. (1993). *John Bowlby and attachment theory.* New York: Routledge.

Robertson, J. (1952). *Film: A two-year-old goes to hospital* [Motion picture]. London: Child Development Research Unit; New York: New York University Film Library.

Waddington, C. H. (1957). *The strategy of the genes: A discussion of some aspects of theoretical biology.* London: Allen & Unwin.

Inge Bretherton

BREASTFEEDING

Infancy is the first and most vulnerable stage of the life course of the child and a significant passage in the life course of the mother. Because they occur at this critical juncture, decisions about infant feeding reflect concerns about optimizing infant and child health but also about dividing familial roles and resources and negotiating the cultural norms for what constitutes a good mother. Although current medical opinion unanimously endorses breastfeeding over bottle feeding, social researchers, who view breastfeeding as a sociocultural as well as biological process, disagree about the importance of breastfeeding in contemporary society for both mothers and children.

HISTORICAL BACKGROUND

Historically, breastfeeding often symbolized a mother's moral duty to her child and to the larger society. In the eighteenth and nineteenth centuries, for example, maternal breastfeeding was part of new democratic values and was thought to instill strength and virtue in the nation's future citizenry. Nonetheless, alternative practices such as wet nursing (hiring a woman other than the baby's biological mother to breastfeed the infant) and bottle feeding with animal milk or other foods continued because circumstances could make maternal breastfeeding difficult or impossible but also because some groups, particularly in the affluent classes, retained strong cultural preferences against it (Fildes, 1986). By the early 20th century, however, cultural prescriptions for the good mother shifted and breastfeeding rates declined as public sanitation, hygienic water and supplies of cow-milk, and the prestige of science made physician-prescribed infant formulas the more appealing, modern practice (Apple, 1987).

The late 20th century marked the revival of breastfeeding prescriptions, a turnaround in rates, and, somewhat paradoxically, the extension of medical authority over what had been the less medicalized, less modern practice. Authority over other aspects of childbearing and child rearing had shifted a century earlier, particularly for the middle class, from kin and religious leaders to medical professionals for the final word on what was best for child and mother. Christian groups aiming to strengthen the traditional family against scientific authority first embraced natural birth and breastfeeding in the 1940s and 1950s, but resistance to medicalization grew with the 1960s back-to-nature ethos and feminist demands for woman-centered health care (Blum, 1999). With increased scientific understanding of the immunological properties of human milk, the partial incorporation of feminist demands into hospital birthing, and the involvement of the U.S. government in breastfeeding promotion, medical authority became predominant. Thus, social research on breastfeeding is an outgrowth of the critical feminist response. However, although there is a shared sociocultural perspective, the recent surge in social research on breastfeeding is divided on the validity of medical claims, the role of government, and whether breastfeeding among all mothers should be prioritized by feminists and sociologists.

A SIDE NOTE ON GLOBAL ISSUES

The movement toward breastfeeding in the United States, Britain, and other developed nations spurred controversy over the use and rampant promotion of formula in developing nations. The severe risks to infants and children in forgoing breastfeeding amid poverty, contaminated water, and unsanitary conditions have led to repeated international protests against formula producers as well as international agreements under the auspices of the World Health Organization and the United Nations Children's Fund (UNICEF) to promote breastfeeding and restrict the marketing of substitutes (Dykes, 2007). However, social scientists point out that breastfeeding is not free or without consequences (Blum, 1999; Carter, 1995; Smith, 2004). In conditions of scarcity lactation can compromise mothers' nutritional status and ability to perform household and paid work, and this can put the entire family and the nursing baby at risk. Breastfeeding is always more than its product, human milk; it is a social practice involving trade-offs of women's time and activities.

MEDICAL RESEARCH: IS BREAST BEST?

The health risks to children of advanced nations in forgoing human milk are less clear, leaving social research divided. The current American Academy of Pediatrics recommendation, which is supported by other medical professionals, is that mothers should breastfeed for a minimum of 1 year (i.e., using no formula or other milks), the first 6 months exclusively, as the only source of infant nutrition (AAP 2005). The U.S. government promotes this standard, as it has set steadily increasing national goals. The goals for 2010 are for a 75% initiation rate, 50% at 6 months, and 25% at 1 year (U.S. Department of Health and Human Services, 2003). Rates over time have responded to such campaigns but continue to fall below targets. Initiation rates in the first decade of the 21st century were about 70%, with just over a third of mothers still breastfeeding at 6 months, and less than 20% at 1 year. Moreover, many mothers rely on formula; by 3 months 30 to 40% of breastfeeding mothers are supplementing (in addition to the 30% who never breast-fed), and at 6 months less than 15% report relying on breast milk alone (Centers for Disease Control and Prevention, 2007). This rapid turn to formula supplementation indicates that most mothers attempt to breastfeed, but for a number of reasons, the standard of using breast milk exclusively is unrealistic. Rates for the United Kingdom and Australia are similar, with high drop-off and supplementation rates (Lee, 2007; Murphy, 1999; Bartlett, 2005).

The most well established benefits of human milk include decreased risk for common ear, respiratory, and gastric-diarrheal infections among infants and young children (Centers for Disease Control and Prevention, 2007; U.S. Department of Health and Human Services, 2003). The evidence for more dramatic claims—reduced risk of childhood cancers, sudden infant death syndrome, diabetes, and lifetime obesity, along with a higher IQ—is equivocal. Although many social researchers appear to accept medical claims without skepticism, several have raised important criticisms.

The sociologist Linda Blum, as part of a larger ethnographic study, examined 30 years of late-20th-century infant-feeding advice, finding that it oversimplifies complex results and thus exaggerates the benefits of breastfeeding. She also noted that because breastfeeding rates are correlated strongly with income, education, and race, the distinct contribution of breast milk to infant and child health is difficult to determine even with advanced statistical techniques because researchers do not yet understand the myriad ways in which social advantage is health-enhancing (Blum, 1999).

Jules Law (2000) found that much medical research is flawed because unexamined normative assumptions favoring the gendered division of care giving are conflated with breastfeeding. Some research on breast versus formula feeding has used infant hospitalization as an outcome variable, ignoring the tendency of physicians to assess breast-fed babies with full-time mothers as better cared for and thus better off at home compared with formula-fed babies with other caregivers. Thus, formula feeding may be associated with higher rates of hospitalization but not higher rates of illness. Such circular studies also have claimed that breastfeeding leads to mother-child bonding when it is used as both an outcome and a presumptive indicator of bonding. Though retained by breastfeeding advocates, this notion of bonding—of a critical early stage in which separation from the mother causes children's pathology—was based on questionable analogies from studies of war orphans and monkeys and was discredited by subsequent research (Blum, 1999; Eyer, 1992).

PUBLIC HEALTH: SHOULD MOTHERS BE MANIPULATED?

The sociologist Joan Wolf (2007) has shown that medical journals are riddled with conflicting findings on breast-feeding; she demonstrated how this equivocal research may be misused by breastfeeding promotion campaigns under government sponsorship. The American Public Health Association Code of Ethics states that public health, in its tapping of public resources and intent to change behavior, should address only fundamental causes of disease. However, formula feeding in advanced nations has not been shown to be a primary cause for any known disease; therefore, according to Wolf, it should not be interpreted as a danger to children.

Other social researchers also take issue with the public health approach, citing its role in creating unnecessary guilt on the part of bottle-feeding mothers for the implied lack of care for their children (Dykes, 2007). In a longitudinal qualitative study the British sociologist Elizabeth Murphy found that the obligation to follow expert advice is profoundly moral, heightened by the maternal imperative to place children's interests first; thus, bottle-feeding mothers experience stigmatized, deviant identities (Murphy, 1999). In a related survey, Ellie Lee found a large minority of mothers reporting guilt, failure, uncertainty, and worry about the harm done by formula feeding; mothers who had planned to breastfeed but found it difficult after complicated births reported the most distress (Lee, 2007). Murphy, Lee, and Wolf built on the sociology of risk, which has shown that that even well-informed publics misunderstand risk calculation and probabilistic statements of likelihood, translating imperfect options such as bottle versus breast into absolutes of "hazard or safety" (Wolf, 2007, p. 613). Moreover, "bombarded with [expert] advice about how to reduce their risk of everything," most take personal, individual responsibility for problems with persistent social roots (Wolf, 2007, p. 612).

Social researchers largely agree that most obstacles to breastfeeding are social and that such obstacles should be changed to avoid blaming individual mothers. Changing workplace policies that limit mothers' ability to combine labor force participation and infant care are a clear priority, particularly in the United States with its

Breastfeeding Protest. *Lisa Pierce Bonifaz, of Boston, breastfeeds her 11–week old baby Marisol during a breastfeeding protest near the Delta Airlines ticket counter in Boston's Logan International Airport. Mothers were there to protest the removal of a nursing mother from a Delta commuter flight in Burlington, VT.* AP IMAGES.

lack of nationally guaranteed paid family leave. For some, such reforms should aim to enlarge the range of positive care giving choices for mothers and families rather than increase breastfeeding rates; but others argue for policy change primarily so that more mothers will breastfeed.

SHOULD BREASTFEEDING BE THE PRIORITY?

Two Australian political economists have contributed significantly to policy arguments. Judith Galtry argues for policies based on the International Labor Organization's Maternity Protection Convention; in addition to paid leave, this includes flexible work programs and paid breastfeeding breaks during the workday. Comparing the high, moderate, and low breastfeeding nations of Sweden, the United States, and Ireland through a review of quantitative studies, Galtry discovered that only Sweden's generous paid leaves, along with its many gender equity measures, raise both breastfeeding initiation and duration rates (Galtry, 2003). Julie Smith argued somewhat differently for changed incentive structures, calculating that if human milk were given a market value that included mothers' labor and consumption costs, mothers no longer would be compelled to choose the ostensibly cheaper formula (Smith, 2004). However, Galtry ignored larger factors that contribute to "Sweden's enviable child health statistics" (Galtry, 2003, p. 174), notably generous anti-poverty and universal health care measures. Smith echoed manipulative, fear-inciting rhetoric with her contention that "artificial [i.e., formula] feeding is the tobacco of the 21st century" (Smith, 2004, p. 377).

Contemporary cultural studies scholarship on breastfeeding similarly argues for prioritizing breastfeeding, exaggerating its contribution to infant and child health and the risks posed by bottle feeding. Such studies analyzing popular culture, however, add that priority be given to breastfeeding for mothers to increase their sensual pleasure, empowerment, and resistance to having their bodies thought of as sex objects (Bartlett, 2005). Following earlier sociological work (Blum, 1999; Carter, 1995), cultural studies confirm that norms for women's sexual bodies pose another major obstacle to breastfeeding. That is, contemporary norms of heterosexuality define breasts as objects of display for the male gaze. Though infant-feeding advice trivializes mothers' embarrassment and counsels learning to nurse discreetly, numerous instances of mothers being sanctioned for public breastfeeding demonstrate that the conflict between maternal breasts and sexualized breasts is an obstacle with social, not individual, roots. Laws in several nations protect public breastfeeding, removing it from categories of lewd and lascivious or indecent conduct, yet the sight of maternal breasts continues to threaten strongly held cultural norms (Bartlett, 2005; Blum, 1999, 2005). Thus, cultural scholars argue for breastfeeding as a "creative corporeal model" for women's empowerment (Bartlett, 2005, p. 178).

RACE AND CLASS DIFFERENCES?

Cultural scholars who contend that breastfeeding can be empowering and pleasurable for all mothers often write about their own experience, noting but discounting the fact that they mother within privileged social locations (Bartlett, 2005; Hausman, 2004). Many sociologists, in contrast, question whether breastfeeding is objectively the best choice for all mothers and their children (Blum, 1999; Carter, 1995; Law, 2000; Lee, 2007; Murphy, 1999). Qualitative studies consistently find that institutional and cultural obstacles loom larger in the lives of lower-income mothers who confront more stressful lives, competing health and family needs, rigid workplaces, and scarce privacy, with single mothers being particularly vulnerable. Middle-class women also find breastfeeding difficult, chaotic, and autonomy-compromising (Blum, 1999; Carter, 1995; Dykes, 2007; Lee, 2007; Murphy, 1999).

Nonetheless, public health officials repeatedly assume that less-privileged mothers must be ill informed in light of their lower propensity to breastfeed (Centers for Disease Control and Prevention, 2007; U.S. Department of Health and Human Services, 2003). Ethnographic studies strongly suggest that this assumption is false; Blum (1999) and Carter (1995) demonstrated that working-class mothers are just as informed and concerned with what is best for their children as middle-class mothers. In addition to these obstacles, Blum found that African-American low-income mothers often chose not to breastfeed to resist racist legacies that cast them as close to nature, ostensibly oversexed, and in need of monitoring (Blum, 1999). Bartlett has argued similarly for Australian aboriginal women, who, particularly when migrating to urban areas, confront racialized stereotypes and scrutiny of their mothering practices, leading to lower rates of breastfeeding (Bartlett, 2005).

FUTURE RESEARCH: AT THE BREAST OR AT THE PUMP?

For many mothers in contemporary society, infant feeding at the breast has been melded with experience with the pump, a manual or electric device to speed the expression of breast milk. Breast pumps are ubiquitous in advice and public health literature, portrayed as a handy tool for mothers returning to the workplace and other everyday activities while providing the very best for their babies. The limited research on their use suggests that most women find breast pumping unpleasant, time-consuming, physically draining, and professionally compromising (Blum, 1999; Dykes, 2007). Blum noted the irony of public discussion increasingly collapsing the *natural* practice of feeding at the breast with mothers at the pump. Future research should focus more centrally on this phenomenon, which challenges depictions of breastfeeding as a creative corporeal model of womanly empowerment as well as of intimacy and attachment between mother and child. Breast pumping may offer a positive option for partners who share parenting (Dykes, 2007); however, Blum found that African-American mothers who relied on kin networks tended to reject breastfeeding because of reliance on this unpleasant practice (Blum, 1999).

Future research also might expand on Galtry's cross-national comparisons of infant-feeding norms and practices. Although it is important to compare developing nations to ameliorate high infant mortality, comparing advanced nations may shed greater light on the efficacy of varied forms of policy support for diverse families and care giving arrangements. Social movement researchers also might compare breastfeeding activism within advanced nations. Although voluntary organizations dedicated to breastfeeding support have been scrutinized (Blum, 1999), little attention has been paid to recent breastfeeding demonstrations, the public "nurse-ins" of U.S. "lactivists" (Blum, 2005) and Australian "breast-fests" (Bartlett, 2005). Equally compelling are questions of why breastfeeding activism in other advanced nations focuses more centrally on antiglobal, anticorporate efforts.

Social research on breastfeeding reflects continuing concerns for children's health and well-being. However, it also reflects concerns about changing gendered divisions of caregiving and questions about which women can be good mothers who contribute to the nation's future.

SEE ALSO Volume 1: *Attachment Theory; Illness and Disease, Childhood and Adolescence; Infant and Child Mortality; Parent-Child Relationships, Childhood and Adolescence.*

BIBLIOGRAPHY

American Academy of Pediatrics. (2005). *Breastfeeding and the use of human milk.* Retrieved May 2, 2008, from http://www.aappolicy.aappublications.org

Apple, R. D. (1987). *Mothers and medicine: A social history of infant feeding, 1890–1950.* Madison: University of Wisconsin Press.

Barlett, A. (2005). *Breastwork: Rethinking breastfeeding.* Sydney, Australia: UNSW Press.

Blum, L. M. (1999). *At the breast: Ideologies of breastfeeding and motherhood in the contemporary United States.* Boston: Beacon Press.

Blum, L. M. (2005). Breast versus bottle in the "real world" or what I did last summer. *SWS Network News, 22*(4), 5–6. Retrieved May 5, 2008, from http://www.socwomen.org/newsletter/QuarkDec05.pdf

Carter, P. (1995). *Feminism, breasts, and breastfeeding.* New York: St. Martin's Press.

Centers for Disease Control and Prevention. (2007). *Breastfeeding trends and updated national health objectives for exclusive breastfeeding—United States, birth years 2000–2004.* Retrieved May 29, 2008, from http://www.cdc.gov/mmwr/preview/mmwrhtml/mm5630a2.htm

Dykes, F. (2007). *Breastfeeding in the hospital: Mothers, midwives, and the production line.* New York: Routledge.

Eyer, D. 1992. *Mother-infant bonding: A scientific fiction.* New Haven CT: Yale University Press.

Fildes, V. A. (1986). *Breasts, bottles, and babies: A history of infant feeding.* Edinburgh, UK: Edinburgh University Press.

Galtry, Judith. 2003. "The Impact on Breastfeeding of Labour Market Policy and Practice in Ireland, Sweden, and the USA." *Social Science & Medicine* 57: 167-177.

Hausman, B. L. 2004. "The feminist politics of breastfeeding." *Australian Feminist Studies* 19: 273–285.

Law, J. (2000). The politics of breastfeeding: Assessing risk, dividing labor. *Signs, 25*(2), 407–450.

Lee, E. (2007). Health, morality, and infant feeding: British mothers' experiences of formula use in the early weeks. *Sociology of Health and Illness, 29*(7), 1075–1090.

Murphy, E. (1999). "Breast is best": Infant feeding decisions and maternal deviance. *Sociology of Health and Illness, 21,* 187–208.

Smith, J. (2004). Mothers' milk and markets. *Australian Feminist Studies, 19*(45), 369–379.

U.S. Department of Health and Human Services. (2003). *HHS blueprint to boost breast-feeding.* Retrieved May 29, 2008, from http://www.fda.gov/fdac/features/2003/303_baby.html

Wolf, J. B. (2007). Is breast really best? Risk and total motherhood in the national breastfeeding awareness campaign. *Journal of Health Politics, Policy and Law, 32*(4), 595–636.

Linda M. Blum
Jennifer J. Esala

BRONFENBRENNER, URIE
1917–2005

Urie Bronfenbrenner, a developmental psychologist, changed the way in which scientists theorize about and conduct research in human development. In his formulation of the bioecological theory of human development, he developed a theoretical paradigm that emphasized the dynamic relations between the individual and a multi-leveled ecological context. His emphasis on theory, multidisciplinary scholarship, research designs that both test as well as formulate hypotheses, and the interplay between science and policy has powerfully transformed research in human development and its application in programs serving children and families.

Bronfenbrenner was born in Moscow and came to the United States at the age of 6. He settled with his parents in a small town in upstate New York, where his father, a neuropathologist, worked at the New York Institution for the Mentally Retarded. In 1934 he attended Cornell University and majored in psychology. At the time, the field of psychology was dominated by lab experimentation and defined the individual in terms of a set of unconnected, maturationally based systems such as sensation, perception, emotion, and motivation. It was not until he met Frank Freeman (1898–1986) at Cornell that Brofenbrenner's training supported a developmental model integrating genetically based characteristics and the environment. Under Freeman's

Urie Bronfenbrenner. AP IMAGES.

individual is conceptualized as developing within "a set of nested structures, each inside the next, like a set of Russian dolls" (Bronfenbrenner, 1979, p. 3). Bronfenbrenner's theories moved the field from the study of "the science of the strange behavior of children in strange situations with strange adults for the briefest possible periods of time" (p. 19) to the study of individuals in the natural contexts in which development occurs.

In his groundbreaking 1979 book, *The Ecology of Human Development*, Bronfenbrenner described four interrelated ecological levels: (a) the micro-system, which contains the developing individual, such as the family unit; (b) the meso-system, the interrelationships between micro-systems, such as the link between the home and school context; (c) the exo-system, those contexts that do not directly involve the developing person but have an influence on the micro-system, such as parents' place of employment; and (d) macro-systems, the highest level of the ecological model, involving culture, policy, and other macro-institutions that create consistency in the underlying systems. In later versions of the model (Bronfenbrenner & Morris, 2006), he directed attention to characteristics of the individual person—biology, psychology, and behavior—that underlie individual development, as, in his words, there came to be "too much [research on] context without development" (Bronfenbrenner, 1986, p. 288). He incorporated the contexts that were typically the domain of other social science disciplines and, in linking them to the individual, invited interdisciplinary scholarship on human development across the life course.

Bronfenbrenner highlighted four defining properties of the bioecological paradigm: process, person, context, and time (PPCT). Proximal processes, or the interactions between individuals and persons, objects, or symbols in their environment, were conceptualized as the "engines of development." In influencing development, they are thought to vary as a function of characteristics of the person, the immediate or distal context, and across time. This model is dynamic in nature. It incorporates the notion of time at the micro-, meso-, and macro-levels and has the individual as both a producer of change in developmental processes (through effects on proximal processes) and a product of the process of development. In short, the model nests the individual in an ecological framework but considers change in development as a function of the exchanges between the individual and an actively changing context.

With a framework that highlights the plasticity of development in the context of risk, Bronfenbrenner was a long supporter of integrating research with practice to improve the human condition. He was one of the

mentorship, Bronfenbrenner went on to receive his M.A. from Harvard under Walter Fenno Dearborn (1878–1955), followed by a Ph.D. in developmental psychology from the University of Michigan. Literally 24 hours after completing his Ph.D., he was inducted into the U.S. Army. As part of a unit of the Office of Strategic Services (now the Central Intelligence Agency), he met psychologist Kurt Lewin (1890–1947) and several other leading scientists. With Lewin, Bronfenbrenner received what he believed was his true graduate training by spending the evening hours singing songs and thinking about human behavior and development. Following World War II (1939–1945), Bronfenbrenner was briefly at the University of Michigan before taking a joint professorship in psychology at the then-named College of Home Economics at Cornell University. At his death, Bronfenbrenner was the Jacob Gould Schurman Professor Emeritus of Human Development and of Psychology in the College of Human Ecology at Cornell University.

Bronfenbrenner is probably best known for the bioecological model of human development, in which the founders of Head Start, the nation's largest federally sponsored early childhood development program providing comprehensive programming to low-income preschool children. As he has described quite eloquently, science should inform policy and vice versa. To Bronfenbrenner, his theoretical and policy focus were as mutually reinforcing as the dimensions of his ecological model. His call to action pervades his writings: "The responsibilities of the researcher extend beyond pure investigation, especially in a time of national crisis. Scientists in our field must be willing to draw on their knowledge and imagination in order to contribute to the design of social interventions: policies and strategies that can help sustain and enhance our most precious human resources—the nation's children" (1988, p. 159).

BIBLIOGRAPHY

Bronfenbrenner, U. (1979). *The ecology of human development: Experiments by nature and design.* Cambridge, MA: Harvard University Press.

Bronfenbrenner, U. (1986). Recent advances in research on human development. In R. K. Silbereisen, K. Eyferth, & G. Rudinger (Eds.), *Development as action in context: Problem behavior and normal youth development* (pp. 287–309). New York: Springer-Verlag.

Bronfenbrenner, U. (1988). Strengthening family systems. In E. F. Zigler & M. Frank (Eds.), *The parental leave crisis: Toward a national policy* (pp. 143–160). New Haven, CT: Yale University Press.

Bronfenbrenner, U. (2005). *Making human beings human: Bioecological perspectives on human development.* Thousand Oaks, CA: Sage.

Bronfenbrenner, U., & Morris, P. (2006). The bioecological model of human development. In R. M. Lerner & W. Damon (Eds.), *Handbook of child psychology: Vol. 1. Theoretical models of human development* (6th ed.). New York: John Wiley.

Pamela Morris

BULLYING AND PEER VICTIMIZATION

In the decade since the highly publicized school shootings of the late 1990s, school bullying has received increased attention from scholars, journalists, and school administrators. However, significant research had already accumulated well before the shootings at Columbine High School in Colorado in 1999. Dan Olweus (1978) is widely considered to have founded the subfield in the late 1970s, partly in response to the highly publicized suicide of a victim of bullying. Unfortunately, there are many tragic bullying stories, and their consequences are a significant part of why it is important to study. Even so,

bullying stands out as a research topic not only because of its horrific consequences but because of its paradoxical nature. It is ubiquitous, yet mysterious: ubiquitous in that it is unlikely that a primary or secondary school student makes it through the day without at least witnessing a bullying event, but mysterious in that the motives for bullying are difficult to discern—often for the bullies themselves—and the benefits obtained unclear.

With contributions from psychology, sociology, education, public health, and other social sciences from across the industrialized world, research on bullying is diverse with respect to both discipline and nationality. The language barriers of the latter are trivial when compared with the former, so one might expect chaos. Instead, a relatively high degree of consensus exists in bullying research, particularly with regard to its consequences and the profiles of victims. However, some areas of confusion remain, including definitions, the role of race, the social status and mental well-being of bullies, and the extent to which it is instrumental (i.e., done in order to achieve some desired, immediate outcome) or pathological. This entry discusses both consensus and confusion, beginning with definitions and measures of bullying, followed by a theoretical overview and concluding with promising new directions.

DEFINITIONS OF BULLYING

Espelage and Swearer (2003), in their review of the literature, point out that "perhaps, the most challenging aspect of bullying prevention programming is reaching a consensus on a definition of bullying" (p. 368). There are nearly as many definitions as there are researchers, but many are variations of the one proposed by Olweus (1978, 1993), who defined bullying as "engaging in negative actions against a less powerful person repeatedly and over time." However influential this definition, in practice, measuring relative power is difficult and often circular: Researchers may only know if one student is less powerful than another if the former has already been victimized by the latter. An additional limitation concerns the requirement that the abuse be both repeated and enduring: A student who pours milk on a classmate in the cafeteria need not repeat the act for it to have lasting consequences.

Kenneth Rigby (2002), another leader among researchers on bullying, has combined elements from several definitions: "Bullying involves a desire to hurt + hurtful action + a power imbalance + typically repetition + an unjust use of power + evident enjoyment by the aggressor and generally a sense of being oppressed on the part of the victim" (p. 51). If all these elements must be present, this definition becomes even more restrictive. Tattum and Tattum (1992) conceived of bullying as a "willful, conscious desire to hurt another and put him/her under stress," although perhaps this definition goes too far in the other direction because even thinking about harming another person would qualify. What virtually all definitions of bullying have in common is that it represents intentional harm and is one sided. Additionally, the consensus is that it can involve a wide range of behaviors, generally categorized as physical aggression, direct verbal abuse, and indirect or relational aggression (which includes spreading rumors and ostracism).

MEASURING BULLYING

Researchers use diverse methods to measure bullying, coding behaviors they observe directly, asking subjects to report their own involvement, or asking teachers, parents, or peers to report on the behaviors of subjects. Naturally, the choice of method is likely to influence results inasmuch as they address, or fail to address, two key obstacles to accurate and unbiased measurement of bullying: social desirability bias and stereotype. Concerning the former, many roles are unappealing to either parents or teens, but the bully is unusual for the low regard in which he or she is held by both. Powerful or popular though they may be, bullies are invariably portrayed by the media as cowardly, insecure, and disliked. This is evident not just in popular culture but also in academic studies describing clusters of aggressive, popular pupils who are also widely disliked (e.g., Farmer, Estell, Bishop, O'Neal, & Cairns, 2003). Desirability bias can lead to underreporting, and studies that compare bully and victim reports have found results consistent with this scenario (Veenstra et al., 2007).

The second obstacle to accurate measurement is a stereotypic view of bullying as direct aggression. One study of pupils' definitions found that they were often limited to overt aggression (Naylor, Cowie, Cossin, de Bettencourt, & Lemme, 2006) and another found that less than 20% of pupils identified indirect or psychological abuse as bullying (Boulton, 1997). For schoolchildren, the term *bully* is strongly associated with direct aggression and substantially less so with indirect aggression (Smith, Cowie, Olafsson, & Liefooghe, 2002). For these reasons, approaches relying exclusively on self-reports or those that use the term *bully* are less than ideal, although they have the advantages of ease of administration and comparability across studies.

Surveys with multiple informants are probably the most accurate, however. Some ask teachers to rate students, but teacher accuracy is often low (Leff, Kupersmidt, Patterson, & Power, 1999) and varies according to the teachers' attention to the problem and opportunities to observe it. Accuracy may improve when students themselves are asked about their classmates' behaviors.

Some studies ask each student to nominate a (generally fixed) number of bullies, whereas others ask them to rate *all* their classmates on their aggressive behaviors, both of which produce consistent and accurate measures.

CAUSES

Despite the diversity of contexts and focal variables, much of the literature on bullying can fit into what might be labeled an individual-pathological framework, which suggests that bullying is a consequence of mental health problems such as depression, anger, anxiety, low self-esteem, or low empathy. Researchers have found that bullies have significantly lower self-esteem than bystanders (O'Moore & Kirkham, 2001), are more depressed (Roland, 2002), and have lower global self-worth, scholastic competence, social acceptance, and greater behavioral conduct problems (Austin & Joseph, 1996).

However, other studies have found no significant differences in self-esteem between bullies and bystanders (Olweus, 1993) and still others have found that bullies have *higher* self-esteem or that bullying enhances self-regard (Kaukiainen et al., 2002; Rigby & Slee, 1992). These findings pose a significant challenge to the individual-pathological model, but other weaknesses play a part as well. The social psychological processes by which some depressives, for example, become bullies and others victims are generally not specified, nor are broader contextual factors taken into consideration. These contextual factors are important, particularly to the extent that they influence both bullying and mental health.

To that end, some researchers have focused on factors that operate at the family, peer, or higher levels of aggregation. The theoretical frameworks applied at these levels are diverse. Some can be labeled transitive models, which suggest that adolescents who are abused abuse others in turn. This argument is most common at the family level, where it is found that parents who are aggressive or neglectful, use corporal punishment, or engage in serious conflicts are more likely to have children who bully (Smith & Myron-Wilson, 1998). They also operate at the peer level, as evidenced in some accounts of *bully-victims*, whose aggression is a response to victimization.

Other studies approximate classic criminological theories such as social control or social learning theory. Social influence, the key component of social learning theory, is seen as an explanation for patterns of homophily, by which bullies are friends with other bullies, which are observed in adolescent friendship networks (Mouttapa, Valente, Gallaher, Rohrbach, & Unger, 2004). In fact, a longitudinal study of sixth through eighth graders found even stronger homophilic effects for bullying than for other aggressive behaviors (Espelage

Victim of Bullying. *The motives for bullying are difficult to discern.* GALE, CENGAGE LEARNING.

& Holt, 2001). Other studies adopt, at least implicitly, variants of social control theory, suggesting that weak attachments to school, for example, may lead to aggressive behavior (DeWit et al., 2000). Despite the value of social ecological approaches, few studies have included factors operating in multiple contexts.

Although these theoretical frames have expanded the horizons of bullying research, they share with their psychological counterparts an implicit view of bullying as pathological. They generally ignore the possibility that bullying is instrumental in enhancing and maintaining status and are difficult to reconcile with other findings that bullies are often popular among their peers. In contrast, some researchers (Hawley, Little, & Card, 2007; Pelligrini & Long, 2002) argue that bullying is, at least in part, designed to accomplish social goals. One study found that target characteristics influenced the status gains obtained from bullying, as boys who bullied other boys won more prestige than those who tended to bully girls (Rodkin & Berger, in press).

THE DEMOGRAPHICS OF BULLYING

As already indicated, theories of bullying are often only implied, and many studies have produced findings that are unattached to any particular perspective. Gender is perhaps the only variable universally included in studies of bullying. Generally, boys are more likely than girls to bully their peers (Nansel, Overpeck, Haynie, Ruan, & Scheidt, 2001; Olweus, 1993), although it is possible that some of this effect is due to de-emphasis of relational aggression, in which girls are as likely, or more likely, to engage (Crick, 1996). Relational aggression is a form of bullying whereby rumors, ostracism, and similar tactics are used to destroy or damage the relationships, reputations, and social status of peers.

In comparison with gender, research examining the role of race in bullying is relatively slim and offers mixed results. Studies from outside the United States often find no racial significant differences in bullying or victimization (Boulton, 1995; Smith et al., 2002), though some have found that minorities are more likely to be victimized (Rigby, 2002; Wolke, Woods, Stanford, & Shulz, 2001). One national study in the United States found that African American students were less likely to be victimized (Nansel et al., 2001). It seems likely that the broader racial context is more important for bullying than the racial or ethnic background of particular individuals.

Most research has focused on children and adolescents in elementary or middle school rather than high school. The studies that have incorporated subjects from all levels of education have generally found that bullying increases over the course of elementary school, peaks in middle school, and then slowly declines in later adolescence (Smith et al., 2002).

CONSEQUENCES

Although the causes of bullying remain unclear and contested, its consequences for victims are unfortunately all too clear. Victimization by bullies has been linked to most mass-casualty school shootings over the past two decades (Vossekuil, Fein, Reddy, Borum, & Modzeleski, 2002). Victimization is also significantly related to suicidal ideation (Kaltiala-Heino, Rimpelä, Marttunen, Rimpelä, & Rantanen, 1999), social isolation (Hodges & Perry, 1999), anxiety and depression (Baldry, 2004), low self-esteem (O'Moore & Kirkham, 2001), physical health problems (Ghandour, Overpeck, Huang, Kogan, & Scheidt, 2004), and diminished academic performance and school attachment (Woods & Wolke, 2004). Many of these effects can last well into adulthood.

It is not only the victims who suffer negative consequences from bullying. Most studies distinguish among victims, pure bullies, and bully-victims. For most outcomes, bully-victims are at least as likely to suffer negative consequences as pure victims and are often worse off. However, bullies also experience difficulties. Aside from their increased likelihood of mental health problems (which are generally viewed as precursors but rarely tested longitudinally), they face increased risk of criminal convictions later in young adulthood (Olweus, 1993) and are also more likely to have difficulty maintaining positive relationships as adults (Rigby, 2001).

FUTURE DIRECTIONS

Several promising new directions for bullying research have opened up. First, scholars are increasingly committed to the consideration of multiple factors operating in multiple contexts. This trend is likely to culminate in the application of Bronfenbrenner's (1979) social ecological framework, which emphasizes the interaction between contexts as much as the contexts themselves. However, data for such efforts remain scarce.

Second, social network methods, by which students nominate those whom they bully and those who bully them, have recently been applied to bullying (Rodkin & Berger, in press; Veenstra et al., 2007). Social network analysis offers several advantages over prior approaches. First, it easily incorporates multiple perspectives on a relationship, thus mitigating reporting bias. Second, it allows for sophisticated analyses at both the individual and aggregate levels. Blockmodeling, for example, categorizes individuals based on *whom* they bully, not just *whether* they bully. Compared with simple aggregate prevalence rates, network data provide more nuanced information such as centralization and structure. Most important, network data raise a host of new questions about who bullies whom. The prevalence of intergender and interracial bullying, for example, should be of great theoretical and practical interest.

Finally, the increasing commitment of schools toward bullying prevention provides unique opportunities for experimental intervention designs. The Olweus Bullying Prevention Program (Olweus, 2005) is among the most widely used and has been shown to significantly reduce bullying. However, these results are based on a small number of experimental tests, and a meta-analysis of school interventions (including the Olweus program) found that very few had any significant effects, those that were found did not last long, and effects generally did not extend to all groups (Vreeman & Carroll, 2007). Another randomized experimental design found significantly greater declines in victimization in the treatment schools, although the prevalence of bullies was not significantly different (Jenson & Dieterich, 2007). Substantial work remains to be done in the area of prevention. Thus, although network data and social ecological approaches may be pushing the field toward significant new insights, they will be incomplete, if not hollow, if they are not used to improve the lives of children and adolescents.

SEE ALSO Volume 1: *Aggression, Childhood and Adolescence; Friendship, Childhood and Adolescence; Peer Groups and Crowds; School Violence.*

BIBLIOGRAPHY

Austin, S., & Joseph, S. (1996). Assessment of bully/victim problems in 8- to 11-year-olds. *British Journal of Educational Psychology, 66,* 447–456.

Baldry, A. (2004). The impact of direct and indirect bullying on the mental and physical health of Italian youngsters. *Aggressive Behavior, 30,* 343–355.

Boulton, M. (1995). Patterns of bully/victim problems in mixed race groups of children. *Social Development, 4,* 277–293.

Boulton, M. (1997). Teachers' views on bullying definitions, attitudes, and abilities to cope. *British Journal of Educational Psychology, 67,* 223–233.

Bronfenbrenner, U. (1979). *The ecology of human development: Experiments by nature and design.* Cambridge, MA: Harvard University Press.

Crick, N. (1996). The role of overt aggression, relational aggression, and prosocial behavior in the prediction of children's future social adjustment. *Child Development, 67,* 2317–2327.

DeWit, D., Offord, D., Sanford, M., Rye, B., Shain, M., & Wright, R. (2000). The effect of school culture on adolescent behavioural problems, self-esteem, attachment to learning, and peer approval of deviance as mediating mechanisms. *Canadian Journal of School Psychology, 16,* 15–38.

Espelage, D., & Holt, M. (2001). Bullying and victimization during early adolescence: Peer influences and psychosocial correlates. In M. Geffner, & C. Young (Eds.), *Bullying behavior: Current issues, research and interventions.* New York: Hayworth.

Espelage, D., & Swearer, S. (2003). Research on school bullying and victimization: What have we learned and where do we go from here? *School Psychology Review, 32,* 365–383.

Farmer, T., Estell, D., Bishop, J., O'Neal, K., & Cairns, B. (2003). Rejected bullies or popular leaders? The social relations of aggressive subtypes of rural African-American early adolescents. *Developmental Psychology, 39,* 992–1004.

Ghandour, R., Overpeck, M., Huang, Z., Kogan, M., & Scheidt, P. (2004). Headache, stomachache, backache and morning fatigue among adolescent girls in the United States. *Archives of Pediatric Adolescent Medicine, 158,* 797–803.

Hawley, P., Little, T., & Card, N. (2007). The allure of the mean friend: Relationship quality and processes of aggressive adolescents with prosocial skills. *International Journal of Behavioral Development, 31,* 170–180.

Hodges, E., & Perry, D. (1999). Personal and interpersonal antecedents and consequences of victimization by peers. *Journal of Personality and Social Psychology, 76,* 677–685.

Jenson, J., & Dieterich, W. (2007). Effects of a skills-based prevention program on bullying and bully victimization among elementary school children. *Prevention Science, 8,* 285–296.

Kaltiala-Heino, R., Rimpelä, M., Marttunen, M., Rimpelä, A., & Rantanen, P. (1999). Bullying, depression, and suicidal ideation in Finnish adolescents: A school survey. *British Medical Journal, 319,* 348–351.

Kaukiainen, A., Salmivalli, C., Lagerspetz, K., Tamminen, M., Vauras, M., Maki, H., et al. (2002). Learning difficulties, social intelligence and self concept, connections to bully–victim problems. *Scandinavian Journal of Psychology, 43,* 269–278.

Leff, S., Kupersmidt, J., Patterson, C., & Power, T. (1999). Factors influencing teacher identification of peer bullies and victims. *School Psychology Review, 28,* 505–517.

Mouttapa, M., Valente, T., Gallaher, P., Rohrbach, L. A., & Unger, J. (2004). Social network predictors of bullying and victimization. *Adolescence, 39,* 315–335.

Nansel, T., Overpeck, M., Haynie, D., Ruan, J., & Scheidt, P. (2001). Bullying behaviors among U.S. youth: Prevalence and association with psychosocial adjustment. *Journal of the American Medical Association, 285,* 2094–2100.

Naylor, P., Cowie, H., Cossin, F., de Bettencourt, R., & Lemme, F. (2006). Teachers' and pupils' definitions of bullying. *British Journal of Educational Psychology, 76,* 553–576.

Olweus, D. (1978). *Aggression in the schools: Bullies and whipping boys.* Washington, DC: Hemisphere Press.

Olweus, D. (1993). *Bullying at school: What we know and what we can do.* Cambridge, MA: Blackwell.

Olweus, D. (1999). Sweden. In P. K. Smith et al. (Eds.), *The nature of school bullying: A cross national perspective* (pp. 224–249). London: Routledge.

Olweus, D. (2005). A useful evaluation design, and effects of the Olweus Bullying Prevention Program. *Psychology, Crime and Law, 11,* 389–402.

O'Moore, M., & Kirkham, C. (2001). Self-esteem and its relationship to bullying behaviour. *Aggressive Behavior, 27,* 269–283.

Pelligrini, A., & Long, J. (2002). A longitudinal study of bullying, dominance, and victimization during the transition from primary school through secondary school. *British Journal of Developmental Psychology, 20,* 259–280.

Rigby, K. (2001). Health consequences of bullying and its prevention in schools. In J. Juvonen, & S. Graham (Eds.), *Peer harassment in school: The plight of the vulnerable and victimized* (pp. 310–323). New York: Guilford Press.

Rigby, K. (2002). *New perspectives on bullying.* London: Jessica Kingsley Publishers.

Rigby, K., & Slee, P. (1992). Dimensions of interpersonal relations among Australian school children, reported behavior attitudes towards victims. *Journal of Social Psychology, 133,* 33–42.

Rodkin, P., & Berger, C. (in press). Who bullies whom? Social status asymmetries by victim gender. *International Journal of Behavioral Development.*

Roland, E. (2002). Aggression, depression, and bullying others. *Aggressive Behavior, 28,* 198–206.

Smith, P., & Myron-Wilson, R. (1998). Parenting and school bullying. *Clinical Child Psychology and Psychiatry, 3,* 405–417.

Smith, P., Cowie, H., Olafsson, R., & Liefooghe, A. (2002). Definitions of bullying: A comparison of terms used, and age and gender differences, in a fourteen-country international comparison. *Child Development, 73,* 1119–1133.

Tattum, D., & Tattum, E. (1992). Bullying: A whole-school response. In N. Jones, & E. Jones (Eds.), *Learning to behave: Curriculum and whole school management approaches to discipline* (pp. 60–72). London: Kogan Page.

Veenstra, R., Lindenberg, S., Zijlstra, B., Oldehinkel, A., De Winter, A., Verhulst, F., et al. (2007). The dyadic nature of bullying and victimization: Testing a dual perspective theory. *Child Development, 78,* 1843–1854.

Vossekuil, B., Fein, R. A., Reddy, M., Borum, R., & Modzeleski, W. (2002). The final report and findings of the Safe School Initiative: Implications for the prevention of school attacks in the United States. Washington, DC: U.S. Secret Service and U.S. Department of Education. Available at http://secretservice.tpaq.treasury.gov/ntac/

Vreeman, R., & Carroll, A. (2007). A systematic review of school-based interventions to prevent bullying. *Archives of Pediatric Adolescent Medicine, 161,* 78–88.

Wolke, D., Woods, S., Stanford, K., & Shulz, H. (2001). Bullying and victimization of primary school children in England and Germany: Prevalence and school factors. *British Journal of Psychology, 92,* 673–696.

Woods, S., & Wolke, D. (2004). Direct and relational bullying among primary school children and academic achievement. *Journal of School Psychology, 42,* 135–155.

Robert Faris

C

CANCER, CHILDHOOD AND ADOLESCENCE

SEE Volume 1: *Illness and Disease, Childhood and Adolescence;* Volume 3: *Cancer, Adulthood and Later Life.*

CHILD ABUSE

All people experience some negative events in their childhood. At what point do unpleasant experiences cross the line into abuse? Child abuse can be divided into three major areas: emotional, physical, and sexual. All forms of abuse can have negative consequences on child development, with the impact lasting into adulthood. Although there are no clear-cut criteria for what constitutes abuse, the behaviors that meet legal and psychological definitions are discussed for all three areas.

EMOTIONAL ABUSE

Emotional abuse can include both active forms of insulting and terrorizing as well as acts of omission such as neglect. These descriptions capture two separate types of behavior: The first is the active behavior of insulting, ridiculing, terrorizing, or threatening a child through words; the second is the failing to provide attention to children through neglect, ignoring their physical needs (food, shelter, or medical attention), or failing to support them emotionally (not attending school events, ignoring a teacher's requests, or not spending time interacting with them). Other terms used to describe emotional abuse include *psychological maltreatment, psychological abuse, verbal abuse, verbal aggression, psychological damage,* or *mental injury.*

Emotional Maltreatment Emotional maltreatment is important because it is inherent in all types of child abuse and some researchers argue that it connects the cognitive, affective, and interpersonal problems related to sexual abuse, physical abuse, and neglect (Brassard, Hart, & Hardy, 1993). There is consensus that emotional abuse aggravates other forms of abuse; however, research on emotional abuse is not as plentiful as in other areas of child abuse.

It is a challenge to identify or categorize nonphysical injuries. Without visible physical injuries, it is very difficult to determine the line between insensitive or poor parenting and emotional abuse. For example, bragging about one's own children offends the Chinese-honored virtue of humility. Historically, the Chinese parenting strategy of shaming a child, even in public, has been considered a necessary part of discipline. Clinicians must balance sensitivity toward ethnic differences in parenting practices with assessment of potential harm to the children.

One example of terrorizing-like emotional abuse might be threatening a child with physical violence. Among adults, threatening bodily harm is the crime of assault. For children, the use of physical threats is widespread in North America. Verbal aggression is often considered a subtype of emotional abuse or psychological maltreatment. Whether this should be considered abusive is under debate. North American society tolerates parents using threats of violence or intimidation, as long as that

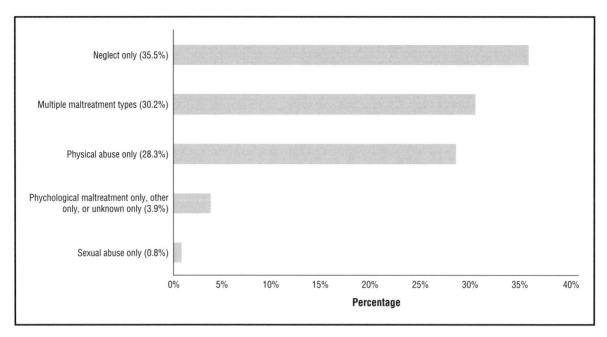

Child Abuse and Neglect Fatalities by Maltreatment Type in 2004. CENGAGE LEARNING, GALE.

violence is couched in the form of physical discipline. One naturalistic study observing parents in public reported the use of threats to spank, hit, punch, hurt, pop, or beat children (Davis, 1996). Author Phillip Davis argued that the use of a culturally approved label such as *spanking* legitimized the behavior. This observational study also found that 50% of the adults who threatened a child went on to hit the child. Parents might lose the social acceptance of their verbal aggression or physical intimidation if they threatened to physically attack or injure children rather than relying on euphemisms such as *spank* (Straus, 2000).

Neglect Neglect is sometimes considered a passive form of maltreatment, in that it often consists of failing to parent appropriately. Extreme failures of neglect might include failing to provide the necessities of life (food, medical care, or clean housing). Emotional neglect may be as subtle as denying a child attention, reassurance, or acceptance. Both emotional and physical neglect have been identified as causes of delayed development or "failure-to-thrive" diagnoses among children (Iwaniec, 1997). Symptoms of failure to thrive include slow weight gain and height growth and psychosocial development that is significantly below norms and not caused by organic illness.

The definition of parental neglect may vary because it is often a parental failure to meet community standards, and standards vary from one community to the next. One example of neglect could be leaving a child at home alone. Whereas the community standards may

be clear on whether leaving a 4-year-old child home alone is neglectful, there may be less agreement on whether leaving a 10-year-old child home alone constitutes neglect.

Prevalence rates of emotional abuse are estimated in several different ways. The U.S. Department of Health and Human Services (DHHS) compiles the child abuse cases reported to child protective services (CPS) in each state. In the 2005 report on child maltreatment by DHSS (2007), 7.1% of children were identified as having suffered emotional abuse. In comparison, 62.8% of victims experienced neglect. Another estimate of prevalence rates are the National Incidence Studies (NIS) mandated by the U.S. Congress. The NIS included both CPS investigations and estimates from professionals who may encounter child abuse. The NIS-3 results released in 1996 indicated that nearly 22% of the more than 1.5 million children who were abused in 1993 in the United States were physically neglected (Sedlak & Broadhurst, 1996). It was estimated that 13.7% of the children identified were emotionally neglected.

Consequences The impact of emotional abuse is varied and widespread. Identifying the exact consequences of emotional abuse can be difficult because negative consequences may appear slowly at different developmental stages as impaired emotional, cognitive, or social abilities. Psychological abuse has been linked to low self-esteem; hostility and higher aggression; anxiety, depression, interpersonal sensitivity, and dissociation; and shame and anger. Various studies have attempted to isolate the negative correlation between emotional abuse and the impact of physical abuse. For example, Collete Hoglund and Karen Nicholas (1995) argued that emotional abuse, but not physical abuse, correlates with higher levels of shame. Emotional abuse inflicted by parents may relate to hostility in the victims specifically because victims develop a sense that other people do not value them or are unlikely to be kind (Nicholas & Bieber, 1994). In that research, no such relationship with hostility was found for physical abuse alone. In another study comparing the impact of different types of child abuse, emotional abuse was revealed to double the risk for feeling suicidal and psychopathy (Mullen, Martin, Anderson, Romans, & Herbison, 1996).

PHYSICAL ABUSE

Physical abuse consists of any assault on a child by a caregiver. What constitutes an assault is controversial. Some researchers would argue that any form of physical discipline can be considered a physical assault (Straus, 2001). Physical abuse can include slapping, scalding, burning with a cigarette, hitting with an object, punching, or kicking a child. David Kolko (2002) pointed out that the definition of physical abuse is a social judgment as to what constitutes an excessive form of parent-to-child discipline. There is no clear line where physical discipline crosses from nonabusive to abusive behavior. That line will vary depending on state statutory definitions, the injuries sustained by the child, prior history of child welfare interactions, judgments made by the caseworkers, and other social characteristics.

The prevalence rates of physical abuse differ somewhat among estimates. The 2005 DHHS (2007) report on child maltreatment estimated that 16.6% of the CPS investigations involved cases of physical abuse. This percentage follows the pattern of decreasing reports of physical abuse. Between 1992 and 2003, a 36% decline in physical abuse cases was reported (Jones, Finkelhor, & Halter, 2006). The older NIS-3 estimated that nearly 25% of the children identified as harmed by abuse in 1993 were physically abused. Community samples report low rates of childhood physical abuse around or below 10% (Mullen et al., 1996).

One example of physical discipline that can escalate into physical abuse is spanking. Disciplining children by slapping or hitting them, although falling under some research definitions as a form of physical abuse, is a contentious and potentially problematic practice. One study found that only 55% of parents support the idea of spanking, although more than 94% have spanked their children at one time or another (Straus, 2001). Some countries in Europe have banned corporal punishment entirely, even in homes. Sociologist Murray Straus, developer of the

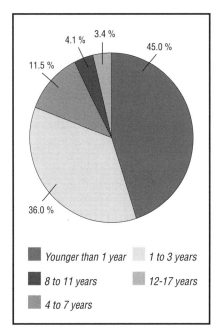

Child Abuse and Neglect Fatalities Victims by Age, 2004. *This chart shows that majority of child abuse victims are under one year old.* CENGAGE LEARNING, GALE.

Conflict Tactics Scale, has criticized the fact that researchers hoping to discover ways to end physical abuse routinely ignore the risk factor of spanking. Even if the risk factor for corporal punishment is moderate, the prevalence of this risk factor (94%) means it can have a greater impact on public health than a high risk factor with low prevalence (Straus, 2000). The problem is that although corporal punishment is usually the precursor to physical abuse, most spankings do not lead to physical abuse.

Consequences Children who are physically abused can experience lasting damage that goes beyond the immediate injuries and pain inflicted. Even corporal punishment that does not exceed accepted cultural norms in North America (e.g., moderate use of spanking) now appears to have a negative psychological impact on children. The negative psychological effects can be social or interpersonal, behavioral, intellectual, or cognitive. For example, physical abuse appears to have a unique relationship with anger in victims—a relationship that does not appear for sexual or emotional abuse. A history of physical abuse has also been linked to increased mental health issues, eating disorders, depression, low self-esteem, and marital problems. Experiencing childhood physical abuse relates to increased rates of feeling suicidal, whereas battering experienced as an adult does not. Cynthia Perez and Cathy Widom (1994) reported that physical abuse was a predictor of lower IQ and lower academic performance.

Higher rates of physical illness, anxiety, anger, and depression were reported by adults almost 40 years after their physical abuse occurred (Springer, Sheridan, Kuo, & Carnes, 2007).

SEXUAL ABUSE

Sexual abuse of a child includes any sexual activity to which a child cannot or does not consent. Nearly every U.S. state has legislation that defines sexual contact with a child under 14 years old as illegal and abusive. Additionally, sexual activity with children ages 14 to 18 may be considered illegal if the perpetrator is significantly older than the child, if the perpetrator is a caregiver or in a position of authority over the child, or if the activity is against the child's will.

Sexual abuse occurred in less than 10% of the CPS determinations in the 2005 child maltreatment report (DHHS, 2007). The NIS-3 estimated that 14% (more than 300,000) of the children identified as abused in 1993 were sexually abused (Sedlak & Broadhurst, 1996). In the 12 years following 1993, however, there was a significant drop in the number of sexual abuse cases reported in the annual child maltreatment reports (47%) and in self-report measures among youths (Jones et al., 2006).

Retrospective estimates of childhood sexual abuse present higher prevalence rates perhaps because they capture victimization that occurred over a number of years. Rates of childhood sexual victimization reported by college students remain consistently around 20 to 25% of women and 5 to 15% of men. Adult rates of disclosure among the general public are similar to college populations: 27% of women and 16% of men (Finkelhor, Hotaling, Lewis, & Smith, 1990). The rates are even higher in other populations—46% in a psychiatric clinic (Wurr & Partridge, 1996) and 66% in a clinic that treats sexually transmitted diseases (Senn, Carey, Vanable, Coury-Doniger, & Urban, 2007).

Consequences Similar to other types of abuse, sexual abuse has been linked with a variety of negative psychological, emotional, and interpersonal difficulties. The first area is trauma-specific consequences. A large number of sexually abused children meet the diagnostic criteria for posttraumatic stress disorder. Cognitive distortions are the second problem area. Victims of sexual abuse may develop feelings of self-blame, lowered self-esteem, or feelings of helplessness. Emotional distress is a third problem area for some victims. Anxiety, depression, and anger have all been found to correlate with sexual abuse, and these types of distress can continue at rates higher than seen in nonabuse victims into adulthood.

Another area of impact is seen in problematic externalizing behaviors such as aggression and high-risk sexual activity. Finally, the fifth area of concern is that there is evidence that sexual abuse can be associated with poorer educational achievement; however, it may be the limitations in the social and family contexts within which child maltreatment occurs that is the true interference.

CO-OCCURRENCE

Much of the research on child abuse has focused on one type of abuse without assessing or holding constant the other types of abuse. Current research is focusing on the cumulative effects of experiencing multiple types of trauma. Physical abuse and emotional abuse are often associated or co-occurring. Experiencing both types of abuse appears to compound their potential negative impact. Emotional abuse appears to also co-occur with sexual abuse and neglect. Researchers are also beginning to look at the abuse occurring within the entire family unit, as child abuse and domestic abuse co-occur in a majority of instances (Edleson, 1999).

Children's reactions to abuse vary widely. The severity and length of negative symptoms experienced may depend on the severity and length of the abuse, characteristics of the abuser, family support, or mental health, disposition, age, or gender of the child. It is also important to recognize that not all victims of child abuse will experience all or even any of the negative consequences. Jean McGloin and Widom (2001) found that 22% of abused children showed no long-term symptoms but demonstrated strong psychosocial resilience.

THE ABUSERS

Despite the popular fear that children are being preyed on by the stranger lurking in the bushes, it is parents or family members who commit the majority of abuse against children. According to the 2005 child maltreatment report, 79.4% of the abuse investigated through CPS was perpetrated by parents. Another 6.8% of the abuse was committed by other relatives (DHHS, 2007). Specifically, 76.5% of physical abuse was committed by parents and 7.3% by another relative. For neglect, 86.6% was perpetrated by a parent and 4.5% by another relative. For sexual abuse, 26.3% of the perpetrators were parents and 28.7% were another relative. In addition, another 9.2% were people in caregiver positions (legal guardian, day-care provider, or the romantic partner of the parent). The remaining perpetrators were either neighbors or friends (4.9%), other individuals (23.3%), or unknown (7.7%).

CHILD ABUSE PREVENTION

Ideally, distinguishing abusive behavior from acceptable childrearing practices should be simple and consistent across the diverse groups in North America. In reality, however, definitions of appropriate parenting practices are culturally bound. For example, the Vietnamese have a tradition that is still in use of hitting a disobedient child with a bamboo stick. Clinicians regularly have to make decisions based on behaviors that could be labeled either as a culturally specific parenting practice or a mild instance of abuse. Caseworkers must be culturally competent in order to assess true threats and communicate with at-risk families.

Child abuse remains a serious social problem in the United States. Despite the decline in rates of physical and sexual abuse reported in the annual child maltreatment reports, at least 1,460 children died in 2005 because of child abuse or neglect (DHHS, 2007). Child abuse prevention requires strengthening the parenting skills of families at risk. New resources for parents are still not reaching all of the families in need. Future child abuse research should examine intervention programs and practices of child protection investigators to ensure that at-risk populations are being served in the best manner possible. Further information and support can be found through the National Center for Missing and Exploited Children (www.missingkids.com) or the National Children's Advocacy Center (www.nationalcac.org).

SEE ALSO Volume 1: *Foster Care; Mental Health, Childhood and Adolescence; Parent-Child Relationships, Childhood and Adolescence; Poverty, Childhood and Adolescence; Resilience, Childhood and Adolescence;* Volume 2: *Domestic Violence.*

BIBLIOGRAPHY

Brassard, M. R., Hart, S. N., & Hardy, D. B. (1993). The psychological maltreatment rating scales. *Child Abuse & Neglect, 17*(6), 715–729.

Briere, J., & Runtz, M. (1990). Differential adult symptomatology associated with three types of child abuse histories. *Child Abuse & Neglect, 14*(3), 357–364.

Childhelp. *Prevention and treatment of child abuse.* Retrieved April 4, 2008, from http://www.childhelp.org

Davis, P. W. (1996). Threats of corporal punishment as verbal aggression: A naturalistic study. *Child Abuse & Neglect, 20*(4), 289–304.

Department of Health and Human Services. *Child maltreatment 2005.* Updated March 30, 2007. Retrieved June 11, 2008, from http://www.acf.hhs.gov/programs

Edleson, J. L. (1999). The overlap between child maltreatment and woman battering. *Violence against Women, 5*(2), 134–154.

Finkelhor, D., Hotaling, G., Lewis, I. A., & Smith, C. (1990). Sexual abuse in a national survey of adult men and women: Prevalence, characteristics, and risk factors. *Child Abuse & Neglect, 14*(1), 19–28.

Hoglund, C. L., & Nicholas, K. B. (1995). Shame, guilt, and anger in college students exposed to abusive family environments. *Journal of Family Violence, 10*(2), 141–157.

Iwaniec, D. (1997). An overview of emotional maltreatment and failure-to-thrive. *Child Abuse Review, 6*(5), 370–388.

Jones, L. M., Finkelhor, D., & Halter, S. (2006). Child maltreatment trends in the 1990s: Why does neglect differ from sexual and physical abuse? *Child Maltreatment, 11*(2), 107–120.

Kolko, D. J. (2002). Child physical abuse. In J. E. B. Myers, L. Berliner, J. Briere, C. T. Hendrix, C. Jenny, & T. A. Reid (Eds.), *The APSAC handbook on child maltreatment* (2nd ed., pp. 21–54). Thousand Oaks, CA: Sage.

McGloin, J. M., & Widom, C. S. (2001). Resilience among abused and neglected children grown up. *Development and Psychopathology, 13*(4), 1021–1038.

Mullen, P. E., Martin, J. L., Anderson, S. E., Romans, S. E., & Herbison, G. P. (1996). The long-term impact of the physical, emotional, and sexual abuse of children: A community study. *Child Abuse & Neglect, 20*(1), 7–21.

Nicholas, K. B., & Bieber, S. L. (1994). Perceptions of mothers' and fathers' abusive and supportive behaviors. *Child Abuse & Neglect, 18*(2), 167–178.

Perez, C. M., & Widom, C. S. (1994). Childhood victimization and long-term intellectual and academic outcomes. *Child Abuse & Neglect, 18*(8), 617–633.

Sedlak, A. J., & Broadhurst, D. D. (1996). *Executive summary of the third National Incidence Study of Child Abuse and Neglect.* Washington, DC: U.S. Department of Health and Human Services. Retrieved June 11, 2008, from http://www.childwelfare.gov/pubs

Senn, T. E., Carey, M. P., Vanable, P. A., Coury-Doniger, P., & Urban, M. (2007). Characteristics of sexual abuse in childhood and adolescence influence sexual risk behavior in adulthood. *Archives of Sexual Behavior, 36*(5), 637–645.

Springer, K. W., Sheridan, J., Kuo, D., & Carnes, M. (2007). Long-term physical and mental health consequences of childhood physical abuse: Results from a large population-based sample of men and women. *Child Abuse & Neglect, 31*(5), 517–530.

Straus, M. A. (2000). Corporal punishment and primary prevention of physical abuse. *Child Abuse & Neglect, 24*(9), 1109–1114.

Straus, M. A. (2001). *Beating the devil out of them: Corporal punishment in American families and its effects on children* (2nd ed.). New Brunswick, NJ: Transaction

Wurr, C. J., & Partridge, I. M. (1996). The prevalence of a history of childhood sexual abuse in an acute adult inpatient population. *Child Abuse & Neglect, 20*(9), 867–872.

M. Alexis Kennedy

CHILD CARE AND EARLY EDUCATION

With more than 12 million children in the United States under the age of 6 in non-parental care every day, participation in early care and education settings has become relatively commonplace (Children's Defense Fund [CDF], 2005). In fact, most children have experienced some level of non-parental care by the time they enter kindergarten.

Because children's early care and education experiences can provide them with the foundational skills necessary for later school and life success, understanding what programs produce short- and long-term benefits is critical.

DEFINING EARLY CARE AND EDUCATION

A variety of terms are used to describe children's early experiences in non-parental care. The term *early care and education* (ECE) encompasses all arrangements that provide care, supervision, and education to infants, toddlers, and preschoolers prior to formal school entry. These experiences can occur in a home or center setting as well as in a public school system and are often thought of as either informal or formal arrangements. Informal care includes home-based settings such as *family child care*, which is most frequently provided in a private residence other than the child's home, and *nanny* or *relative care*, which is typically provided in the child's home by someone other than a parent or in a relative's home. Formal care experiences include *center-based care* and *pre-kindergarten programs*. *Center based care* is typically provided in a non-residential facility for children ages 0 to 6 and can have different sponsors, including religious organizations, independent owners, universities or social service agencies. *Pre-kindergarten programs* are generally associated with a K–12 school system and include a year or two of learning in which children develop skills necessary for the kindergarten classroom (CDF, 2005; Magnuson & Waldfogel, 2005).

Informal arrangements are generally funded by parent fees and tend to be considered of lower quality than formal arrangements. In contrast, formal care arrangements can be either privately (e.g., most infant and toddler programs) or publicly (e.g., Early Head Start and Head Start programs or pre-kindergarten programs) funded and typically vary in the populations they serve. For example, publicly funded programs such as Head Start are intended to serve children from families with fewer economic resources and are not available to middle and upper income families. Most early care and education settings offer full or part day programs that can last from as little as 1 year to as many as 5 years.

BASIC TRENDS IN EARLY CARE AND EDUCATION RESEARCH

Changes in the goals of early care and education, as well as shifts in the economic needs of families, have contributed to an overall increase in enrollment in these programs and thus to an urgent need to understand the impacts of these experiences on children's development. In the United States, early care and education began primarily as a support for employed parents and helped

Head Start. *Pre-school teacher Keiwana Jones, 23, adjusts the headphone for three-year-old Kacia Gay as she prepares to play an educational computer game at Fundamentals Academy, in Cuyahoga County, OH, as James Wallace, 3, waits patiently for his turn. Fundamentals Academy is one of many Head Start Centers in Cuyahoga County that is struggling to cope with the changes in eligibility rules.* **AP IMAGES.**

encourage economic self-sufficiency, especially among single and low-income women (Lamb & Ahnert, 2006). Early care and education also served to acculturate children, especially immigrants, and their parents, and to enrich children's lives in a variety of domains. Over time, the purposes of early care and education have shifted to focus on the academic and social preparation of children for entry into formal schooling. Indeed, kindergarten is no longer reserved for pre-academic foundational learning, but instead is used to teach skills once taught in first grade. As such, early care and education settings must now provide the foundational skills children need to succeed in kindergarten and beyond (National Association for the Education of Young Children [NAEYC], 1996). In a recent review of the early childhood education literature, Hyson and colleagues outlined two primary goals of early care and education: a) to develop cognitive skills necessary for academic success (e.g., representational thinking, self-regulation, and planning) and b) to develop emotional competencies (e.g., emotional security and emotional regulation) (Hyson, Copple, and Jones, 2006).

Research suggests that early care and education programs can provide promising avenues for preparing children for formal schooling and for deflecting negative academic and behavioral trajectories. These experiences also offer long-term economic benefits—both personal and societal (Barnett, 1998). The positive effects of ECE appear to be particularly strong for children from less-advantaged backgrounds (Votruba-Drzal, Coley, and Chase-Lansdale, 2004). Yet little is known about which specific features of these experiences support children from different backgrounds and with different developmental needs. Despite the fact that there is ample evidence suggesting that high-quality ECE experiences promote positive outcomes, child care participation, especially extensive hours in care, has also been found to predict higher levels of aggression in young children (NICHD ECCRN, 2005). It is important to note that although children's aggressive behaviors were elevated, they were typically still within the normal range of problems.

Just as the purposes of early care and education have shifted, so, too, has the demand. According to the U.S. Department of Labor, participation in early care and education programs has increased by approximately 26% between 1998 and 2008, although these rates differ considerably by child and family demographics. For example, older children are more likely to be placed in formal care arrangements such as preschool or pre-kindergarten than are younger children. Similarly, families with the highest and lowest incomes are most likely to use center-based care and preschool settings (Magnuson and Waldfogel, 2005). These settings are more accessible to low-income families than middle-income families because of government subsidies and publicly funded programs such as Head Start. Even with the rising participation in ECE, these programs are not offered universally. In fact, the universality of early care and education settings differs considerably by state, with some states providing public preschool for all children (e.g., Oklahoma or Georgia) and others leaving the care and education of young children to individual families (e.g., Mississippi). Thus, as the National Research Council (2003) has suggested, researchers must continue to investigate systematically which programs and practices work, for whom, and under what conditions in order to generate effective early care and education policies and practices and to determine whether universal versus targeted programs are necessary.

THEMES AND THEORIES IN EARLY CARE AND EDUCATION

A variety of topics drive research and practice in the field of early care and education. Of particular relevance to this entry are quality, developmentally appropriate practice, and short- and long-term impacts. Policymakers are just beginning to recognize the importance of quality early care and education in building the foundational skills necessary for later learning and life success. However, the definition of quality is quite heterogeneous, in part because quality is often described more in terms of desired outcomes than necessary inputs. Indeed, the majority of research on ECE describes outcomes in terms of "what happens to children" as a result of their educa-

tional experiences rather than what practices are necessary to produce positive outcomes (Bennett, 2000). Research trends in ECE, however, have led to increased efforts to understand *how* providers and teachers behave in the classroom and *what* is being taught. As such, a growing number of studies examine which quality indicators, including developmentally appropriate practices, are necessary to produce short- and long-term outcomes.

Quality In the United States, quality has typically been measured and monitored using instruments that assess process and structural characteristics of programs. Process quality includes caregiver sensitivity and responsiveness as well as cognitive stimulation and is typically measured via observations of activities and interactions in the childcare setting, including interactions with caregivers and peers and language stimulation. Some measures focus primarily on the experiences of individual children (e.g., Observational Rating of the Caregiving Environment [ORCE]) whereas other measures focus primarily on the experiences of the group (e.g., Infant/Toddler Environmental Rating Scale [ITERS], Early Childhood Environmental Rating Scale [ECERS]).

In general, process quality assessments offer an attempt to quantify the care and education children receive. Structural quality includes child-adult ratio, group size, and the formal education and training of teachers and is typically measured via observations or reports of structural features of the classroom. Structural quality is more easily quantified than process quality and as a result, these features tend to appeal more to policymakers because they may be formally regulated. Studies suggest that both process and structural quality across settings in the United States is fair to minimal (Cost, Quality and Outcomes Study Team, 1995; NICHD ECCRN, 2005). It is worth noting that high structural quality does not guarantee high process quality, although many believe structural quality is necessary for providers to offer sensitive caregiving and age-appropriate activities (Phillips, Howes, & Whitebook, 1992).

There are several other important indicators of quality to consider when examining ECE policy and practice, including federal and state level structural indicators, such as licensing and government subsidies for programs, as well as funding mechanisms. As with process and structural quality at the local level, federal and state structure also tends to be relatively poor. For example, despite the fact that licensed care providers typically offer more sensitive and responsive care than do unlicensed providers, only about half of U.S. states require that providers be licensed. With respect to funding, parent fees rather than government subsidies typically fund early care and education programs. As a result, not all children are guaranteed a preschool education. For example, in the United States the average state expenditure on children enrolled in early childhood education programs in 2007 was approximately $3,642, compared with $11,286 in K-12 (National Institute for Early Education Research, 2007).

Best Practice Another topic of particular relevance to researchers and practitioners in the field of early care and education has to do with developmentally appropriate practice for young children. In 1987 the National Association for the Education of Young Children released a set of best practice guidelines that were based on five interrelated dimensions of early childhood practice:

1. creating a caring community of learners,

2. teaching to enhance development and learning,

3. constructing appropriate curriculum,

4. assessing children's development and learning,

5. establishing reciprocal relationships with families (NAEYC, 1996).

These guidelines are meant to support practitioners in their daily practice and are based on extensive knowledge about how children develop and learn. Importantly, they are intended to be both age appropriate and individually appropriate and contribute to a child's positive development in a variety of domains. Indicators of developmentally appropriate practice include the presence of active learning experiences and varied instructional strategies, as well as a balance between teacher-directed and child-directed activities (NAEYC, 1996). Programs that use an integrated curriculum as well as learning centers to engage children in a variety of topics are also considered developmentally appropriate. It is important to note that although these guidelines are generally accepted as a strong indicator of high-quality early care and education programs, not all practitioners agree that they reflect best practice for young children. Among the criticisms are that the guidelines are too prescriptive, that they discourage self-reflective practice among teachers, and that they are not culturally sensitive (Novick, 1996).

Impacts One of the most robust findings in the early care and education literature is that children who attend high-quality preschool programs are better at following directions, joining in activities, waiting and/or taking turns, problem solving, and relating to teachers and parents than children who do not attend preschool programs (for reviews, see Clarke-Stewart & Allhusen, 2005; Lamb & Ahnert, 2006; NICHD ECCRN, 2005). Further, children in settings with higher process-quality have more secure relationships with their mothers and caregivers and perform better on standardized tests of

cognitive and language ability. Children also exhibit fewer behavior problems when they are in settings in which their teachers are more sensitive and responsive to their needs. Similarly, children in settings with high structural quality, indexed by low child-staff ratios, for example, are better able to initiate and participate in conversations, are typically more cooperative, and show less hostility during interactions with others (Clarke-Stewart & Allhusen, 2005; Howes, 1997; Lamb & Ahnert, 2006; NICHD ECCRN, 2005;). It is worth noting that a limited body of evidence suggests that the effect of early childhood care and education on kindergarten reading and math skills accrues to children who start their care between ages two and three; starting before two or after three results in fewer gains (Loeb, Bridges, Bassok, Fuller, & Rumberger, 2007).

Evidence of the long-term effects of early childhood education programs is mixed. Experimental studies such as the Perry Preschool Project and the North Carolina Abecedarian Project suggest that there are long-term benefits of high-quality early education programs, including higher levels of educational attainment, lower levels of juvenile crime and arrests, and lower rates of public assistance (Barnett, 1998). A limited number of nonexperimental studies have identified modest effects of high-quality programs on children's development through second grade, including greater receptive language ability, math ability, cognitive and attention skills, and social skills, as well as fewer behavior problems (NICHD ECCRN, 2005; Cost, Quality, & Child Outcomes Study Team, 1995). Moreover, children who experience high-quality stable early care and education settings also have more secure attachments with adults and peers, engage in more complex play, and are less likely to be retained a grade or be assigned to special education (Clarke-Stewart & Allhusen, 2005; Lamb & Ahnert, 2006). Other studies have found no evidence of long-term benefits.

There is also a growing body of research examining the specific impacts of developmentally appropriate practice on children's social, behavioral, and academic outcomes. Results suggest that on average, children who attend preschool and kindergarten classrooms where developmentally appropriate practices occur exhibit greater academic success in the early grades than do other children. Children who attend developmentally appropriate classrooms also exhibit fewer stress-related outcomes and greater motivation levels than other children (Charlesworth, Hart, Burts, & DeWolf, 1993). Developmentally appropriate practices appear to be even more important for children at risk for academic failure. Among a sample of predominantly African-American and Hispanic children who participated in the Head Start/Public School Transition Project, children who attended more developmentally appropriate classrooms exhibited significantly higher achievement than their peers who did not attend such classrooms (Huffman & Speer, 2000). There is also some evidence of long-term effects of developmentally appropriate practices (DAP) on children's outcomes. For example, children who attended developmentally appropriate classrooms had higher graduation rates and higher monthly incomes, as well as fewer arrests, through early adulthood (Barnett, 1998).

STUDYING EARLY CARE AND EDUCATION: RESEARCH CHALLENGES AND FUTURE DIRECTIONS

Although the field of early care and education has made tremendous strides in identifying features of high-quality programs and best practices for children, additional systematic studies of which particular programs and practices work, for whom, and under what conditions are needed (NRC, 2003). Indeed, the vast majority of research in this field has been based on non-experimental or observational studies. Because parents' choices about early care and education programs are often constrained by their background characteristics, including socioeconomic status, isolating the effects of ECE programs on children's subsequent behavior and achievement from the effects of family factors is difficult. That is, a child may exhibit better skills in a given domain not because of her preschool experiences per se but because her family values a particular set of skills and has sought out, and provided, many other opportunities for the child to build these skills. Moreover, the variability in type, quality, and duration of early care and education settings makes it difficult to understand the full impact of these experiences on children's development. Nevertheless, non-experimental studies offer critical information about beneficial features of ECE settings.

In the coming years the field of early care and education would benefit from rigorous experimental studies in which one group of children is randomly assigned to classrooms or programs with a specific set of features, while another group of children who are eligible to attend these classrooms and programs are assigned to classrooms and programs that do not offer these features. By randomly assigning children to these different groups, the researcher is able to determine the impacts of the classroom or program features on children's outcomes because any initial differences between children who experience the program and children who do not are considered to be due solely to chance.

For both practical and ethical reasons, however, random assignment is not always possible. In this case, researchers should make every effort to take advantage of natural experiments (also referred to as quasi-experimental designs) in which a change in a policy or practice occurs. Because participants have no control over these policy or practice changes, they offer a non-biased estimate of program effects. Not only will experimental or quasi-experimental studies provide the estimates of causal impact that are necessary for creating policies and for changing practices in early care and education, but they will also be important for informing ongoing debates about universal versus targeted programs and interventions. Finally, methodological advances in the study of development also provide researchers with a valuable set of tools for understanding the effects of early care and education experiences on children's outcomes (for a review, see McCartney, Burchinal & Bub, 2006). Employing these methods whenever possible will continue to advance the study of early care and education.

SEE ALSO Volume 1: *Cognitive Ability; Maternal Employment; Racial Inequality in Education; School Readiness; Socioeconomic Inequality in Education; Stages of Schooling.*

BIBLIOGRAPHY

Barnett, W. S. (1998). Long-term effects on cognitive development and school success. In W. S. Barnett & S. S. Boocock (Eds.), *Early care and education for children in poverty: Promises, programs, and long-term results* (pp. 11–44). Albany: State University of New York.

Bennett, J. (2000). *Goals, curricula, and quality monitoring in early childhood systems.* Paper presented at the Consultative Meeting on International Developments in ECEC. New York: The Institute for Child and Family Policy, Columbia University.

Charlesworth, R., Hart, C. H., Burts, D. C., & DeWolf, M. (1993). The LSU Studies: Building a research base for developmentally appropriate practice. In S. Reifel (Ed.), *Perspectives in developmentally appropriate practice: Advances in early education and day care*, Vol. 5 (pp. 3–28). Greenwich, CT: JAI.

Children's Defense Fund (CDF). (April 2005). *Child care basics.* Retrieved June 11, 2008, from http://www.childrensdefense.org

Clarke-Stewart, K. A., & Allhusen, V. D. (2005). *What we know about childcare.* Boston: Harvard University Press.

Cost, Quality, & Child Outcomes Study Team. (1995). *Cost, quality, and child outcomes in child care centers, technical report.* Denver: University of Colorado at Denver, Department of Economics, Center for Research in Economic and Social Policy.

Howes, C. (1997). Children's experiences in center-based child care as a function of teacher background and adult: Child ratio. *Merrill-Palmer Quarterly, 43*(3), 405–425.

Huffman, L. R., & Speer, P. W. (2000). Academic performance among at-risk children: The role of developmentally appropriate practices. *Early Childhood Research Quarterly, 15*, 167–184.

Hyson, M., Copple, C., & Jones, J. (2006). Early childhood development and education. In W. Damon, L. M. Lerner, K. A. Renninger, I. E. Sigel (Eds.), *Handbook of child psychology*, Vol. 4, *Child psychology in practice*, 6th ed. New York: Wiley.

Lamb, M. E., & Ahnert, L. (2006). Nonparental child care: Contexts, concepts, correlates, and consequences. In W. Damon, L. M. Lerner, K. A. Renninger, I. E. Sigel (Eds.), *Handbook of child psychology*, Vol. 4, *Child psychology in practice*, 6th ed. New York: Wiley.

Loeb, S., Bridges, M., Bassok, D., Fuller, B., & Rumberger, R. W. (2007). How much is too much? The influence of preschool centers on children's social and cognitive development. *Economics of Education Review, 26*, 52–66.

Magnuson, K. A., & Waldfogel, J. (2005). Early childhood care and education: Effects on ethnic and racial gaps in school readiness. *The Future of Children, 15*, 169–196.

McCartney, K., Burchinal, M., & Bub, K. L. (Eds.). (2006). Best practices in quantitative methods for developmentalists. *Monographs of the Society for Research in Child Development, 71(3)*. Boston: Blackwell Publishing.

National Association for the Education of Young Children (NAEYC). (1996). *Developmentally appropriate practice in early childhood programs serving children from birth through age 8*. Retrieved June 11, 2008, from http://www.naeyc.org

National Institute for Early Education Research. (2007). *The state of preschool–2007: State preschool yearbook*. Newark, NJ: Rutgers Graduate School of Education.

National Research Council. (2003). *Strategic education research partnership*. Washington, DC: The National Academies Press.

NICHD Early Child Care Research Network (ECCRN) (2005). *Child care and child development*. New York: Guilford Press.

Novick, R. (1996). *Developmentally appropriate and culturally responsive education: Theory in practice*. Retrieved June 11, 2008, from http://www.nwrel.org

Phillips, D. A., Howes, C., & Whitebook, M. (1992). The social policy context of child care: Effects on quality. *American Journal of Community Psychology, 20*, 25–51.

Votruba-Drzal, E., Coley, R. L., & Chase-Lansdale, P. L. (2004). Child care and low-income children's development: Direct and moderated effects. *Child Development, 75*, 296–312.

Kristen L. Bub

CHILD CUSTODY AND SUPPORT

One out of every three children in the United States is born to an unmarried parent, and at least half of all American children will spend some time living apart from one of their parents by the age of 15 (Andersson, 2002). In 2004, 14 million parents had custody of 21.6 million children under the age of 21, and 83% of these custodial parents were mothers (Grall, 2006). Put another way, 27% of all children under the age of 21

resided with only one parent. These trends suggest that there may be a large portion of the population eligible for child support from a nonresidential parent at some time before they reach the age of majority, either as a result of parental divorce or separation, or being born to an unmarried parent. Understanding the implications of child custody for child well-being are important above and beyond the effect of divorce and being born to an unmarried parent on child well-being. The parent with whom the child lives provides daily interaction that is critical for child development and it is largely through this primary parent that the child has access to economic resources (Fox & Kelly, 1995).

CHILD CUSTODY

The early American judicial system relied heavily on English common law where children were property and fathers had absolute power and legal obligations to protect, support, and educate their children. As a result, until the middle of the 19th century, the legal system was patriarchal and fathers were granted custody in divorce cases with judges having great control in custody decisions. By the late 19th century, states adopted the Tender Years Doctrine—a guiding principle that existed in most states until the late 20th century—which held that the best interest of the child, especially very young or female children, in divorce cases was maternal custody. At the same time, courts favored placing older children with the same-sex parent, which often meant separating siblings. Although there are few published court decisions on this matter, judges recognized that maternal care was paramount for infants but in one case a judge declared that a boy of 3 years was at a tender age, but the same is not true once that boy is 5 years. The women's rights movement of the 1920s further increased maternal preference in custody decisions as the legal status of women increased over this time. The assumption of maternal preference was also supported by Freudian psychoanalytic theory (which emphasized the mother's roles as the primary nurturer) and psychological theories on the development of infant attachment. As a result of these changes, maternal custody remained the norm for many decades.

Demographic changes in the middle of the 20th century, marked by dramatic increases in the divorce rate, once again challenged presumption of custody. Fathers' claims of discrimination, constitutional challenges of equal protection, and the entry of women into the workforce led most states to reject the model of maternal custody in favor of more gender-neutral policies. The Uniform Marriage and Divorce Act in 1970 provided for a best interest of the child standard of practice in custody cases, a practice adopted by most states. Judges determined custody based on the individual needs and interests of the child rather than the gender or rights of the parents. Despite this, most custody determinations continued to be made in the mother's favor. Contemporary advocates have claimed that children should spend time with both parents, resulting in policies of shared physical custody, which has become the preferred option in many states.

Decisions concerning custody arrangements following a divorce can be very informal reached privately between parents, or alternatively decided more formally through the judicial system. National estimates of custody arrangements and requests are not available, however a study in California suggests that at least 50% of parents make private decisions between themselves about custody and visitation, and an additional 30% settle these issues after further negation (Maccoby & Mnookin, 1992). When parents are not able to reach negotiated agreements the formal legal process including judicial hearings, pretrial settlement conferences, and custody trials is used. These custody disputes are expensive and can require up to 3 years for settlement (Kelly, 1994).

Custodial outcomes reveal how parents and the legal system assess the well-being of children as well as the state of gender relations among parents (Seltzer, 1990). Examination of contemporary custody decisions has found that the likelihood of father custody is increased in cases where children are older, and particularly when the oldest child is a boy (Fox & Kelly, 1995). Greer Fox and Robert Kelly (1995) also find that the likelihood of father custody following divorce is decreased when mothers are highly educated or when fathers are unemployed. The influence of fathers' income on custody decisions is ambiguous. Fox and Kelly found that fathers with higher incomes are less likely to receive custody of children, whereas other studies found that the likelihood was higher (Christensen, Dahl, & Rettig, 1990). Maria Cancian and Daniel Meyer (1998) found that father custody is less likely when total income is high, but the mother has a higher share of the total income.

Empirical research has examined the effects of custody arrangements on child outcomes and finds that several important child outcomes may vary based on the gender of the single parent. Children residing in father-custody families have a significantly higher risk of drug use, more problems at school, take part in risky health behaviors more frequently, and exhibit more problematic behavior and are slightly disadvantaged in terms of cognitive skills. Adults who grew up in a single-father household obtain approximately one-half year less of education than their counterparts who grew up in single-mother households (Downey, Ainsworth-Darnell, & Dufur, 1998), and do worse even once socioeconomic status has been taken into account (Biblarz & Raferty,

1999). Children living in single-father families generally do just as well or better on indicators of mental and physical health than children residing with two biological parents, but exhibit worse access to health care compared to children residing in other family structures (Leininger & Ziol-Guest, 2008).

These types of studies are unable to assert causality due to the observational nature of the study design and, since relatively little is known about single-father families, it is difficult to identify the potential causal mechanisms driving the results. These findings could indicate selection effects, namely that fathers are less likely to maintain custody of children who are in poor health or that fathers are more likely to take custody of children who are exhibiting problem behaviors. Very little research exists that illustrates the direction of these selection effects. Previous literature on women who relinquish custody of their children suggests that they do so because of economic difficulties, emotional problems, fears about lengthy custody hearings, and abuse (Herrerias, 1995). However, changes in the labor market experiences of women and fathers' increasing interest in being the sole caretaker than before also influence the types of fathers who retain custody (Greif, 1995).

CHILD SUPPORT

In 1950 the federal government implemented a policy designed for child support collections at a time when the prevalence of single-parent families (primarily mother-headed) was low. Congress required state welfare agencies to notify law enforcement agencies when welfare benefits were being distributed to families where the child had been abandoned by one of the parents. Law enforcement then would attempt to locate the nonresident parent and collect child support. Between 1950 and 1975, child support policies at the federal level were primarily focused on these children.

Since the late 1970s, Congress has enacted policy to strengthen the private child support system, particularly for low-income children. An important goal has been to shift the cost of raising children away from the government and onto parents (especially absent parents). While many of the changes have been targeted at increased financial responsibility by absent parents, there has also been much more focus, recently, on increasing the rights of the absent biological parent (usually the father). Prior to 1975, child support orders were the outcome of private negotiations between parents and the courts. The outcome of these negotiations varied, as did enforcement, with much of the enforcement done by mothers following a divorce, often requiring lawyers and court action. The private court nature of these negotiations also resulted in different outcomes for low-income and higher-income parents who were able to afford good lawyers and avoid support orders altogether. By contrast, low-income fathers often could not afford lawyers and had unrealistic orders imposed. Further, orders were often based on the minimum amount of money required to raise a child, differentially affecting low-income fathers relative to higher ones.

The child support enforcement system was established when Congress enacted Title IV-D of the Social Security Act of 1975. The federal Office of Child Support Enforcement (OCSE) was established to oversee the federal-state child support program and lets states operate their own child support agency (the IV-D program) in accordance with the federal laws. The goal of this policy was to ensure that public support was not going to support children who could be supported by the nonresident parent. Since 1975 Congress has acted on several occasions to increase the purview of child support enforcement. For example, in 1980 and 1984 the law was changed to include all children regardless of household income and welfare status and provide services universally. In 1993, changes required states to develop in-hospital paternity programs, while the 1996 welfare reform instituted changes in child support payment penalties and paternity establishment.

Most children receive little or no child support from the noncustodial parent due to the lack of a formal child support order or because either no payments or only partial payments were made on an existing child support order. In 2004 40% of custodial parents did not have a child support agreement (36% of custody mothers and 60% of custodial fathers). Among families due child support, 45% received all that was due, 31% received some of what was due, and the balance received none of what was due. Research suggests that children whose families received child support received, on average, 16% of their total family income from child support (Grall, 2006). Child support is an especially important source of income for poor children who receive it, representing between 26–50% of total annual income (Grall, 2006). Further, empirical research also indicates that child support income may be more beneficial to children than income from other sources and that child support itself is associated with positive outcomes for children.

Custodial fathers are less likely to have child support orders, compared to custodial mothers. According to findings from the four-state Child Custody and Child Support Project (CCCSP), child support orders exist in about one-third of father custody cases, whereas orders are in place for more than 80% of the mothers who retain physical custody (Christensen, Dahl, & Rettig, 1990). Further, Susan Stewart (1999) found that 36% of noncustodial mothers contributed child support when

their children resided with other relatives, whereas only 27% of noncustodial mothers contributed child support when their children lived with the father.

Few studies examine the characteristics of fathers who secure child support orders following divorce. J. Thomas Oldham (1994) suggests that fathers are less likely to receive orders compared to mothers because of judge opinions on the following: They may be hesitant to require women to pay support, custodial fathers have higher incomes than non-custodial mothers so they may suggest that fathers are less in need of child support, or mothers may give up custody of their children in exchange for no payment of support required. Findings suggest that fathers with low incomes and earnings, especially lower than their wives, were more likely to have child support orders (Greif & DeMaris, 1991).

Child support enforcement and child support receipt may influence the behavior of both the custodial and noncustodial parent, and also may affect various child educational, developmental, and cognitive outcomes. A small but growing empirical literature focuses on the noneconomic impacts of child support receipt on children. Child support has been estimated to play an important role in children's cognitive test scores, perceived scholastic competence, and reading and math scores on standardized tests above and beyond its influence on total income, effects that endure even after controlling for unobserved characteristics of fathers and families. Studies also find that these associations extend to educational attainment—specifically that higher child support, independent of the effect on total income, is associated with greater school completion and that income from child support may be more important than income from other sources. Researchers also have investigated whether child support receipt is associated with behavioral and developmental outcomes. The evidence is mixed, suggesting weak associations between receipt and reduction of behavior problems, but no significant relationship between receipt and cognitive stimulation in the home or behavioral development.

FUTURE DIRECTIONS FOR CHILD CUSTODY AND CHILD SUPPORT RESEARCH

Child custody and child support populations, practices, and policies are ever-changing and the next wave of child custody and support research needs to contend with and address such issues. Specifically, an increase in the number of father-custody households highlights the importance of examining this family structure as it may differ from mother-custody households. Further, important population changes require focusing on child support within the context of multiple-partner fertility and incarcerated noncustodial parents. Additionally, joint and shared custody practices have shifted the dynamic both to how custody is examined and how child support payments are relevant in these cases. Finally, evaluating policy changes as a result of welfare reform reauthorization is going to be important.

Future research needs to concentrate on the changing dynamics of family structure and custody and how it relates to child well-being. By 2006 single-father families represented 14% of all single-parent families with children. While this represents only 5% of all families with children, single-father families are one of the fastest growing family types, increasing at a rate faster than single-mother families (Meyer & Garasky, 1993). Some research explores differences in child well-being associated with residing in a single-father family; however, more research needs to be conducted pertaining to the health of children, as much of the existing research regarding the effects of family structure on child outcomes often compares the experiences of children living with single mothers to those living with married parents.

Perhaps one of the most important changes in contemporary family structure is the prevalence of multiple-partner fertility, that is, families in which at least one partner has a child by someone else. Figures from Wisconsin suggest that more than half of mothers on welfare had children with more than one partner. Among the entire child support enforcement caseload, for 9% of mothers, both the mother and at least one father had children with more than one partner; 16% of the mothers had children with only one father but, also, the father had children with more than one mother; and 6% of mothers had children with more than one father who only had children with her (Cancian, Cook, & Meyer, 2003). Future research should focus on how these families manage child support obligations.

Incarcerated noncustodial parents are an important topic of study because incarceration poses a real barrier to child support payment and employment, and because it is an important child support enforcement tool. In one study, 29% of noncustodial fathers were institutionalized primarily in prisons (Sorensen & Zibman, 2001), estimates similar to those from the Fragile Families study where 27% of fathers were known to have been incarcerated (Western, 2006). Additionally, 55% of male inmates in state facilities and 63% of male inmates in federal facilities are parents of children under the age of 18 (Mumola, 2000), and between 22–26% of state inmates are part of the child support caseload (Griswold & Pearson, 2003). Incarceration is an important contributor to growing child support arrears as well. Future research needs to investigate whether policies and practices related to child support orders for incarcerated non-

custodial parents should change, and what the outcomes are as a result of the modification of orders due to this changing circumstance.

As noted above, joint and shared custody arrangements have become much more common in contemporary society. Previous research suggests a positive association between joint legal custody and child support and interactions (Seltzer, 1998). However, child support guidelines in these cases are less clear, and different states have developed different approaches to how child support is calculated. Future research should examine what the differential approaches used by states has on child support compliance and child well-being, as well as whether guidelines influence parental behavior.

Prior to welfare reform in 1996, federal law mandated that states *pass through* the first $50 of collected child support each month to families receiving cash assistance. Welfare reform changed the pass-through policy by granting states the option to determine their own pass-through policies; all but 16 states eliminated the pass-through policy. Welfare reform reauthorization in 2006 changed the pass-through policies once again, making federal participation more generous. States can pass through $100 per month for one child or $200 per month to a family with two or more children receiving cash welfare and the federal government will resume the cost share of the pass-through. Previous studies indicate that more generous pass-through policies are associated with higher probability of paternity establishment and child support receipt. Given these latest policy changes, research should determine how these policies affect transfers to children and families. Preliminary work suggests that if all states adopted this pass-through policy, the average amount of child support received while on cash welfare would more than double (Wheaton & Sorensen, 2007).

CONCLUSIONS

Given the changing nature of family structure in the United States, the number of children affected by custody decisions and eligible for child support has grown significantly. Public policy has historically responded to these changes by increasing involvement in what previously has been considered the private domain. Moving forward, as more children interact with the judicial and policy system, researchers will need to continue to understand the relationships between these systems and child well-being.

SEE ALSO Volume 1: *Child Abuse; Family and Household Structure, Childhood and Adolescence; Parent-Child Relationships, Childhood and Adolescence; Policy, Child Well-Being; Poverty, Childhood and Adolescence;* Volume 2: *Fatherhood; Motherhood; Noncustodial Parents.*

BIBLIOGRAPHY

Andersson, G. (2002). Children's experience of family disruption and family formation: Evidence from 16 FFS countries. *Demographic Research, 7*(7), 343–364.

Biblarz, T.J., & Raferty, A.E. (1999). Family structure, educational attainment, and socioeconomic success: Rethinking the "pathology of matriarchy." *American Journal of Sociology, 105,* 321–365.

Cancian, M., Cook, S., & Meyer, D.R. (2003). *Child support in complicated TANF families.* Report to the Wisconsin Department of Workforce Development.

Cancian, M., & Meyer, D.R. (1998). Who gets custody? *Demography, 35*(2), 147–157.

Christensen, D.H., Dahl, C.M., & Rettig, K.D. (1990). Noncustodial mothers and child support: Examining the larger context. *Family Relations, 39*(4), 388–394.

Downey, D.B., Ainsworth-Darnell, J.W., & Dufur, M.J. (1998). Sex of parent and children's well-being in single-parent households. *Journal of Marriage and the Family, 60*(4), 878–893.

Fox, G.L., & Kelly, R.F. (1995). Determinants of child custody arrangements at divorce. *Journal of Marriage and the Family, 57*(3), 693–708.

Grall, T.S. (2006). *Custodial mothers and fathers and their child support: 2003.* Washington, DC: U.S. Government Printing Office. Retrieved May 25, 2008, from http://www.census.gov/prod/2006pubs/p60-230.pdf

Greif, G. L. (1995). Single fathers with custody following separation and divorce. *Marriage & Family Review, 20,* 213–231.

Greif, G.L., & DeMaris, A. (1991). When a single custodial father receives child support. *American Journal of Family Therapy, 19*(2), 167–176.

Griswold, E., & Pearson, J. (2003). Twelve reasons for collaboration between departments of correction and child support enforcement agencies. *Corrections Today, 65*(3), 88–90 and 104.

Herrerias, C. (1995). Noncustodial mothers following divorce. *Marriage and Family Review, 20,* 233–255.

Leininger, L.J., & Ziol-Guest, K.M. (2008). Reexamining the effects of family structure on children's access to care: The single-father family. *Health Services Research, 43,* 117–133.

Maccoby, E. E., & Mnookin, R. H. (1992). *Dividing the child: Social and legal dilemmas of custody.* Cambridge, MA: Harvard University Press.

Meyer, D.R., & Garasky, S. (1993). Custodial fathers: Myths, realities, and child support policy. *Journal of Marriage and the Family, 55*(1), 73–89.

Mumola, C.J. (2000). *Incarcerated parents and their children.* Washington, DC: U.S. Department of Justice. Retrieved May 25, 2008, from http://www.ojp.usdoj.gov/bjs/pub/pdf/iptc.pdf

Oldham, J. T. (1994). The appropriate child support award when the non-custodial parent earns less than the custodial parent. *Houston Law Review, 31,* 585–616.

Seltzer, J. A. (1990). Legal and physical custody arrangements in recent divorces. *Social Science Quarterly, 71,* 25–266.

Seltzer, J.A. (1998). Father by law: Effects of joint legal custody on nonresident fathers' involvement with children. *Demography, 35*(2), 135–146.

Sorensen, E., & Zibman, C. (2001). Getting to know poor fathers who do not pay child support. *Social Service Review, 75*(3), 420–434.

Stewart, S.D. (1999). Nonresident mothers' and fathers' social contact with children. *Journal of Marriage and the Family, 61*(4), 894–907.

Western, B. (2006). *Punishment and inequality in America.* New York: Russell Sage Foundation.

Wheaton, L., & Sorensen, E. (2007). *The potential impact of increasing child support payments to TANF families.* Retrieved May 25, 2008, from www.urban.org/UploadedPDF/411595_child_support.pdf

Kathleen M. Ziol-Guest

CHRONIC ILLNESS, CHILDHOOD

SEE Volume 1: *Illness and Disease, Childhood and Adolescence.*

CIVIC ENGAGEMENT, CHILDHOOD AND ADOLESCENCE

The topic of civic engagement on the part of U.S. citizens is routinely investigated in and outside of the social sciences. Questions pertaining to political participation and civil society target all age groups, but in recent years, scholars have become interested in the question of civic engagement among youth. Normally, the definition of youth centers on those who are eligible to vote, between the ages of 18 and 24, or sometimes referred to as *generation next*; however, researchers have also begun to focus their lens on childhood and adolescence, primarily because of the impact previous civil rights struggles have had on this group. Overall, it is believed that this generation is more liberal on a number of topics in relationship to their older counterparts, especially with regard to gay marriage and women's rights. Social scientists have focused on civic engagement among children and adolescents in relationship to three areas: voluntary organizations, youth leadership and development, and youth activism. Focusing on youth, particularly adolescents, allows scholars to expand the definition of activism to include a range of political and social activities.

VOLUNTEER AND AFTER-SCHOOL ACTIVITIES

Because civic engagement and political participation is often linked to young adults over 18, studies of civic participation among adolescents often focuses on volunteer activities. Teenagers are often encouraged to participate in after-school and volunteer activities, for instance, as a way to encourage and mold future political participation. Volunteer organizations vary in their definition and can begin in childhood and often include organizations such as the Girls Scouts, Boy Scouts, and Little League sports. Some of these organizations originate and are affiliated with the schools that youths attend, or are already included as part of the school infrastructure. Religious institutions, such as synagogues and churches, also offer youth a way to become involved in civic activities. Research questions typically focus on the relationship between the particular activity, age of the youth, and future civic engagement. Through these organizations, youth establish social networks, peer relationships, and may even deepen relationships with their parents, who may be involved in similar civic activities.

The importance of after-school activities in regard to youth involvement in voluntary associations and civic engagement is made especially clear in Daniel McFar-

land's and Reuben Thomas's (2006) study. Using quantitative data to study school-related activities, parental engagement, and other social networks, McFarland and Thomas found that those organizations where youths were encouraged and trained in the areas of public speaking, vocal, dance, and instrument training enhance and encourage future civic engagement and political participation. Therefore, as the authors suggest, cuts in performing arts programs such as drama and music will have deleterious effects on the breadth and depth of American political and civic involvement. As this and other recent studies suggest, voluntary and after-school activities have a significant impact on the leadership potential of youth. Sociologist Amy Best (2000) even found that innocuous events, such as the senior prom, are important sites for communicating civic responsibilities, democracy, and political ideology among youths.

Similar in importance to childhood voluntary activities, parental civic engagement is a significant influence on childhood and adolescent involvement in political activities. Overall, there is a concern that the decline in social capital among youths and adults alike will contribute to the lack of political participation on the part of those in their youth; therefore, the political participation of parents, peers, and the neighborhood a youth grows

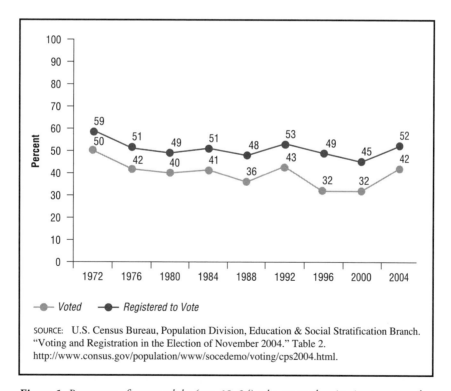

SOURCE: U.S. Census Bureau, Population Division, Education & Social Stratification Branch. "Voting and Registration in the Election of November 2004." Table 2. http://www.consus.gov/population/www/socedemo/voting/cps2004.html.

Figure 1. Percentage of young adults (ages 18–24) who reported registering to vote and voting in Presidential elections, 1972–2004. **CENGAGE LEARNING, GALE.**

up in also affects their own participation. While volunteer and extracurricular organizations may remedy this problem somewhat, the social capital in one's own neighborhood and social circle may have a larger impact on future civic involvement. For instance, in their research on young citizens' knowledge of political issues, Scott Wells and Elizabeth Dudash (2007) found that youths gleaned most of their political information from the Internet, their family, and their friends. They use this knowledge to determine their individual political strategies and future voting patterns. Additionally, and perhaps more importantly, their family's voting patterns and political beliefs were held in higher regard than media sources. Overall, most research on the political knowledge of youths (adolescents in particular) is related to their parent's political involvement, knowledge, and willingness to share with their children.

YOUTH LEADERSHIP AND DEVELOPMENT

Among teenagers, scholars have also begun to study more traditional forms of youth and leadership development as a way to understand youth civic engagement. Moving away from volunteer and after-school activities, youth leadership organizations set out to train the youth to assume leadership roles in their communities, often on par with adults. These organizations vary from being youth-led organizations to organizations led by adult allies who train the youth as organizers. The overall assumption in these organizations is that youths are as deeply and equally invested in the well-being of their communities as adults, who are often positioned as leaders. The topics of the organizations vary, ranging from educational change, art, speech and debate, and spoken word poetry. In each case, the overall goal is to engage and support the youth as leaders in their community, able to make decisions for the good of all. There are several goals in these organizations: (a) to have youths and adults work together as allies; (b) to provide youths with leadership capabilities that will be sustained over time; and (c) to create opportunities for the youth to understand and incorporate their own experience into their organization's strategies.

Teenagers as a group are overwhelmingly disenfranchised, and are not allowed to participate in traditional forms of civic engagement, such as voting. Therefore, their relationships with adults are imperative in achieving their civic goals. Over time, working with adults allows the youth to learn how to take on leadership roles in peer situations with people who are older than them and outside of their family. For instance in Wendy Wheeler's

and Carolyn Edlebeck's (2006) work on youth civic engagement, it was concluded that youth-adult relationships should be intentional, elucidating the ways that youth and adults work together, envision change with one another, and produce strategies to that effect. A central aspect of this relationship is self-reflective activities in which youths build self-esteem; build alliances across race, class, and gender lines; and begin to connect what sociologist C. Wright Mills (1916–1962) described as personal troubles to public issues.

YOUTH ACTIVISM

Similar to the research on youth leadership and development, scholars have begun to focus on youth activism as part of their analysis of youth civic engagement. Questions surrounding what inspires youths to participate in social change and politics dominate these discussions. Recognizing that contemporary forms of activism may not fit into previous models of social change, sociologists have begun to ask questions about how youth activism in the early 21st century might compare to activism of the 1960s, 1970s, and 1980s. For instance, sociologist Todd Gitlin's (2003) book, *Letters to a Young Activist*, tackles this issue directly. In particular, Gitlin questions how the youth can organize in what many refer to as the post–civil rights era when the discussions of 1960s activism mark cultural and political understandings of social change. Perhaps because of this, social scientists have turned their eye toward popular culture as a potential mechanism for civic engagement.

Many scholars have focused on popular culture as an important mechanism for engaging youth in political consciousness and political debate. For instance, popular cultures such as hip-hop have received considerable attention because of their ability to draw youth to social change events. Similarly, the relationship between hip-hop culture, politics, and post–civil rights youth is the focus of both academic and popular discourses. Many approaches focus, importantly, on the relationship between the hip-hop generation and mainstream democratic processes. During the 2004 and 2000 presidential elections, Russell Simmons (b. 1957), cofounder of Def Jam records, was central in organizing the *hip-hop vote*. For youth of color, in particular, popular culture is an important site for political mobilization.

SEE ALSO Voume 1: *Activity Participation, Childhood and Adolescence; Political Socialization; Religion and Spirituality, Childhood and Adolescence; Sports and Athletics.*

Campaign Volunteers. *Barack Obama smiles for a group photo with young volunteers.* AP IMAGES.

BIBLIOGRAPHY

Best, A. (2000). *Prom night: Youth, schools, and popular culture.* New York: Routledge.

Gitlin, T. (2003). *Letters to a young activist.* New York: Basic Books.

Kitwana, B. (2002). *The hip-hop generation: Young Blacks and the crisis in African-American culture.* New York: Basic Civitas Books.

McFarland, D. A., & Thomas, R. J. (2006). Bowling young: How youth voluntary associations influence adult political participation. *American Sociological Review, 71*(3), 401–426.

Wells, S. D., & Dudash, E. A. (2007). Wha'd'ya know?: Examining young voters' political information and efficacy in the 2004 election. *American Behavioral Scientist, 50*(9), 1280–1289.

Wheeler, W. (with C. Edlebeck). (2006). Leading, learning, and unleashing potential: Youth leadership and civic engagement. *New Directions for Youth Development, 2006*(109), 89–97.

Andreana Clay

CLAUSEN, JOHN
1914–1996

John Adam Clausen, an American sociologist who was born on December 20, in New York City, is considered the father of the sociology of mental health. He was a primary figure in the sociology of the life course whose academic and organizational contributions include the conceptualization of planful competence, pioneering work in the social psychological study of mental illness, leadership of the Socioenvironmental Laboratory at the National Institute of Mental Health at its inception, and overseeing one of the most important data archives in life course studies. Clausen died on February 15 in California.

Clausen's most influential life course concept was planful competence, an umbrella term that links three dimensions developed early in life that influence adult outcomes: intellectual investment, dependability, and self-confidence. Planful competence involves awareness of one's intellectual abilities, social skills, and emotional responses to others. It includes the ability to recognize and develop one's interests and options and to think about ways to maximize those options. The theory suggests that individual rationality and possibilities for choice have replaced traditional determinants of life outcomes. Adolescents with greater planful competence evidence higher self-knowledge and self-confidence and realistic goal setting. They are more likely to achieve later-life goals because these capacities inhibit unwise choices. In *American Lives* (1993) Clausen used life

John Clausen. PHOTO COURTESY OF UNIVERSITY OF
CALIFORNIA, BERKELEY, DEPARTMENT OF SOCIOLOGY.

histories and quantitative analysis to show that planfully
competent adolescents had more stable careers and marriages and found more fulfillment in later life.

Clausen's early life trajectory illustrates the concept
of planful competence. Born in New York City, Clausen
recognized his strengths in mathematics and science at an
early age. He began undergraduate studies in engineering
at Cornell but after two years switched to economics, a
field more closely aligned with his broadening interests.
After graduation Clausen's ability to follow his own
interests and recognize the impact of those around him,
even in the face of adversity, cost him a job. He was fired
from a bank position for wearing a button supporting the
1936 presidential candidate Franklin D. Roosevelt in
defiance of the organization's requirement to wear but-
tons supporting Alf Landon, the Republican candidate.

Clausen returned to Cornell, earning a master's
degree in sociology in one year while working full-time
as a caseworker for the welfare department. What Clausen
learned in that work experience helped develop his focus
on the social psychological aspects of sociology. In 1938 he
entered the University of Chicago, where his training was
shaped by Samuel A. Stouffer, Harold Blumer, and Ernest
Burgess. During those years Clausen spent time as a
research assistant at the Institute for Juvenile Research,
studying ex-convicts and delinquents, and developed "an
interest in what you can learn by looking very intensely at
an individual life history" (Clausen, 1991). During that

time he married Suzanne Ravage, with whom he had four
sons during their 54 years together.

Clausen worked at the Virginia State Planning Board
until World War II, when he joined the War Department, where he contributed to the *American Soldier*
series. After the war he briefly took a position as an
assistant professor of sociology at Cornell. In 1948 he
was appointed to the newly established National Institute
of Mental Health (NIMH) and within three years
became chief of the Laboratory for Socio-Environmental
Studies. In that capacity he brought together talented
young scholars, including Erving Goffman, Morris
Rosenberg, Leonard Pearlin, William Caudill, Marion
Radke-Yarrow, Carmi Schooler, and Melvin Kohn. The
laboratory played a major role in establishing the importance of social science at NIMH.

In 1960 Clausen was named director of the Institute
for Human Development (IHD) at the University of
California at Berkeley and became a professor in the
Department of Sociology, where he remained for the rest
of his career. The IHD center conducted three well-
known longitudinal studies that followed individuals
from their early years in the 1920s and 1930s into late
life: the Oakland Growth Study, the Guidance Study,
and the Berkeley Growth Study. Clausen recruited Glen
Elder to work on the Oakland Study, resulting in the
pioneering study *Children of the Great Depression* (1974).
During his years as director Clausen oversaw the merging
of the three studies, leading to Jack Block and Norma
Haan's *Lives through Time* (1971) and the anthology
Present and Past in Middle Life (Eichorn, Clausen, Haan,
& Honzik, 1981).

Clausen contributed numerous chapters and papers
to the field of life course research and wrote influential
books such as *The Life Course: A Sociological Perspective*
(1986) and *American Lives* (1993). This body of work
reflected his lifelong concern with individual development, linking childhood and adulthood, including the
ways people exert agency over their lives. Clausen was
praised for his logic and rigor in both theory and methods and won many scholarly honors.

SEE ALSO Volume 1: *Data Sources, Childhood and
        Adolescence; Elder, Glen H., Jr.;* Volume 2: *Agency.*

BIBLIOGRAPHY
Clausen, J. A. (1986). *The life course: A sociological perspective.*
    Englewood Cliffs, NJ: Prentice-Hall.

Clausen, J. A. (1991). Adolescent competence and the shaping of
    the life course. *American Journal of Sociology, 96*(4), 805–842.

Clausen, J. A. (1991). "Interview with John Clausen." *Berkeley
    Faculty Live Interview Series.* Interviewed by Jon Stiles, Arona
    Ragins, Michelle Motoyshi, University of California,
    Berkeley, Department of Sociology. Retrieved April 9, 2008,

from http://sociology.berkeley.edu/index.php?page=faculty
live#clausen

Clausen, J. A. (1993). *American lives: Looking back at children of the great depression*. New York: Free Press.

Eichorn, D. H., Clausen, J. A., Haan, N., Honzik, M. P., & Mussen, P. H. (Eds.). (1981). *Present and past in middle life*. New York: Academic Press.

Elder, G. H., Jr. (1974). *Children of the great depression: Social change in life experience*. Chicago: University of Chicago Press.

Christopher P. Kelley
Steven Hitlin

COGNITIVE ABILITY

A precise definition of the term *human cognitive ability* is elusive even though it is a subject widely studied by psychologists, sociologists, behavioral geneticists, and other scientists. In the most general sense, cognitive ability can be defined as the capacity to perform a set of cognitive tasks, or "task[s] in which appropriate or correct mental processing of information is necessary for successful performance" (Carroll, 1993, p. 10). In practice, however, concepts and measurements of cognitive ability are more narrowly defined to capture individual capacity to perform specific tasks that predict educational and labor market success, such as verbal and mathematical skills.

The literature also makes a distinction between *cognitive ability* and *cognitive achievement* (e.g., educational attainment). Ability is more likely to be genetically determined whereas achievement is more acquired through individual effort, schools, and peers, for example. Also, in contrast to achievement, which is more heavily influenced by motivations and opportunities in later life, ability is most malleable in childhood and tends to stabilize in adolescence (Carroll, 1993).

Attempts to measure ability are limited by both conceptual and practical problems. Some argue that assessments of children under age 6 cannot be accurately obtained. Others argue that ability tests are more likely to capture test-taking skills than true underlying cognitive traits. However, several cognitive assessments exist that are reliable (i.e., produce the same results when individuals are retested or tested in alternative ways) and can be efficiently administered to large representative samples (Rock & Stenner, 2005). Two of the most commonly used are the Peabody Picture Vocabulary Test–Revised and the Woodcock-Johnson Psycho-Educational Battery–Revised tests. The former tests vocabulary size among children ages 3 to 5 whereas the latter can assess individuals age 3 to 95 and is a more comprehensive test of four broad abilities: reading, mathematics, written language, and general

knowledge. Other noteworthy tests include the Early Childhood Longitudinal Study–Kindergarten Battery, the Wechsler Preschool and Primary Scale of Intelligence–Revised, and the Stanford-Binet Intelligence Scale.

MAIN CHARACTERISTICS OF COGNITIVE DEVELOPMENT

Two main conclusions can be drawn from research on cognitive ability. First, there is consensus that neither nature nor nurture fully explains the development of human cognition; rather, interactions between the two produce cognitive outcomes (Plomin, 1994). These studies compare the cognitive outcomes of siblings who were raised together and siblings who were raised apart, or compare monozygotic twins (100% similarity in genetic makeup) and dizygotic twins (50% similarity in genetic makeup). The findings demonstrate that whereas cognitive functioning is a heritable trait, environment goes a long way in shaping cognitive development. For example, meta-analyses of adoption and twin studies involving more than 10,000 twin pairs found that at least half of the variation in cognitive ability across individuals could be attributed to environmental influences (Chipeur, Rovine, & Plomin, 1990).

Second, early childhood is the most critical stage in cognitive development. From a biological perspective, this period is most crucial because brain development occurs most rapidly during early childhood (Shore, 1997). Additionally, social science research shows that most of the genetic and environment effects are realized by the time children reach adolescence (Guo, 1998; Keane & Wolpin, 1997).

PREDICTORS OF COGNITIVE OUTCOMES

The environmental predictors of cognitive ability include both prenatal and postnatal factors. Maternal behavior can influence cognitive outcomes even before birth. For example, alcohol and drug-use during pregnancy can impair healthy fetal development.

Family Income and Poverty Perhaps the most widely studied and best understood environmental factors are family income and poverty. First, the effect of family income is most prominent when economic hardship is experienced in early childhood relative to any other period of development (Duncan & Brooks-Gunn, 1997). Guo (1998) found that the effect of cumulative poverty over childhood (from birth to age 8) is more detrimental for childhood ability than the effect of cumulated poverty from birth to adolescence on adolescent ability. Duration of poverty also matters. Studies that have contrasted multiyear to single-year measures of income and demonstrate that single-year measures understate the influence of poverty because much of child poverty is persistent rather than transitory (Duncan, Brooks-Gunn, & Klebanov, 1994; Korenman, Miller, & Sjaastad, 1995). Possible mechanisms through which income may influence cognitive function include (but are not limited to) childrearing practices and quality of the home learning environment, parental employment and childcare provisions, family structure and change, and neighborhoods. Each of these factors can also have an independent effect on cognitive outcomes.

Parenting Behaviors and Home Learning Environments
Studies based on naturalistic and laboratory settings document important socioeconomic variation in childrearing practices that may influence child cognitive development. For example, socioeconomically advantaged parents are more likely to talk to their children, use complex vocabulary, engage in educationally oriented activities, and encourage child-initiated conversation (Hoff, 2003; Lareau, 2003). Additionally, parents facing economic stress tend to be more volatile and are more likely to use harsh physical punishment (Bradley, Corwyn, McAdoo, & Coll, 2001; Elder, 1999). What is less clear is to what extent these differences go on to affect children's cognitive development.

Studies show that the characteristics of children's home environment relate to child cognitive outcomes and also mediate the relationship between income and cognitive outcomes. Quality of home environment is often measured using the Home Observation and Measurement of the Environment (HOME) scale, which is a composite measure of the material resources at home (e.g., toys and books), childrearing practices (e.g., discipline, maternal warmth), and physical organization of the home (e.g., disorganized and cluttered). Home measures account for up to half of the statistical relationship between poverty and cognitive outcomes (Korenman et al., 1995; Smith, Brooks-Gunn, & Klebanov, 1997). However, because Home scores aggregate such a wide assortment of characteristics, it is difficult to identify the specific aspect of home environment (i.e., material vs. childrearing practices) that matter.

Parental Employment and Childcare Maternal employment dramatically increased in the late 20th and early 21st century, and parents looked outside of the family for alternative childcare options. As a result, the type and quality of nonparental childcare plays an increasingly important role in influencing children's early learning environment.

Both randomized trials (see Currie, 2001, and Karoly et al., 1998, for a detailed review) and examinations of large-scale programs such as Head Start (Currie & Thomas, 1995) show that developmentally appropriate preschool programs have the potential to enhance verbal development and math reasoning, especially among disadvantaged children who may not be receiving appropriate cognitive stimulation at home. Low quality, informal childcare (e.g., care by untrained relatives), on the other hand, may have negative effects. However, high quality preschools alone are not enough. Gains in cognitive outcomes are lost if children are subsequently placed into low quality schools (Garces, Thomas, & Current, 2002).

Studies have found that high quality childcare may partially mediate the potentially negative effects of early maternal employment (Han, 2005; National Institute of Child Health and Human Development Early Child Care Research Network, 1997, 1998). Access to high quality childcare, however, is varied in the United States.

Family Structure and Change Numerous studies show that family structure (e.g., single parenthood and nonmarital birth) and changes (i.e., divorce, separation, cohabitation, and remarriage) have the potential to cause instability in children's lives (Amato, 2000; Demo & Acock, 1988). However, it may not be family structure or change, per se, that causes lower cognitive outcomes; rather, preexisting conditions that jointly determine family formation and cognitive outcomes (e.g., poverty, mother's age and education) may partially explain why children born into nonmarital unions have worse reading scores than children of mothers who are continuously married (Cooksey, 1997). Additionally, many studies find that the association between family structure and change disappears once maternal characteristics such as mothers' education, IQ, age at first birth, and smoking behavior during pregnancy are accounted for (Carlson & Corcoran, 2001; Fomby & Cherlin, 2007).

OTHER SOCIODEMOGRAPHIC DISPARITIES: RACE/ETHNICITY AND SEX

White–Black and White–Hispanic gaps in cognitive outcomes are well documented; limited data has often precluded the study of other ethnic groups such as Asian Americans. Studies show that White advantage in test scores begins before children enter school and lasts into adulthood (e.g., the White–Black gap in verbal tests among preschool children ranges from one fourth of a standard deviation to more than one standard deviation; Rock & Stenner, 2005).

Theories arguing a genetic basis for these gaps persist, in spite of the fact that no genetic evidence has substantiated these claims. The weight of the evidence shows that a complex and interrelated set of environmental factors produces these gaps. This argument is based on the following findings: (a) gaps in test scores have declined during the 20th century; (b) test scores of Black or mixed-raced children who were raised by adoptive White parents are substantially higher than their nonadoptive counterparts; and (c) environmental changes can have a substantial impact on test scores (Jencks & Phillips, 1998).

Racial differences in family backgrounds can be stark. For example, one study estimated that 10% of White children, 37% of Hispanic children, and 42% of Black children live in poverty and account for 50 to 80% of the gap in verbal and math scores (Duncan & Magnuson, 2006). However, because family background is related to so many characteristics that also affect cognitive development (e.g., parenting practice, maternal health, neighborhood experience), teasing out the specific family effects is problematic.

Relative to the White–Black test gap, sex differences are small, ranging from one fifth to one third of a standard deviation. Gender gaps in test scores only begin to emerge in high school with math scores slightly favoring boys and larger gaps in verbal scores favoring girls (Hedges & Nowell, 1995). Environmental explanations, rather than genetic ones, are most compelling given that these gaps only emerge later in adolescence and have been declining over the past several decades (Friedman, 1998).

The most prominent explanations point to gendered socialization, which differentially promotes verbal and mathematical competency among boys and girls. For example, some scholars argue that schools discourage female success in the math and sciences (see American Association of University Women, 1995). Additionally, sex differences in out-of-school experiences may also matter (Downey & Yuan, 2005; Entwisle, Alexander, & Olson, 1994). For example, boys are more likely to perform activities that enhance quantitative skills (e.g., using computers, participation in math/science clubs) whereas girls are more involved in activities that tend to promote verbal/reading skills (e.g., reading, art classes).

Long-Term Effects The literature suggests two important longer-term trends: (a) Child cognitive development has lasting effects on adult outcomes and (b) both racial and sex gaps widen over the life course. First, cognitive development in early childhood is predictive of a variety of adult outcomes ranging from educational achievement, criminality, health outcomes, and labor market performance (Rouse, Brooks-Gunn, & McLanahan, 2005). Additionally, comparing the effects of education and cognitive ability on occupational standing, John Warren, Jennifer Sheridan, and Robert Hauser (2002) showed that whereas the effect of education declines over time, ability has a positive and persistent effect over the life course. Second,

both racial and sex gaps in test scores widen when children enter schools, suggesting that school quality may contribute to producing disparities. These findings, coupled with studies emphasizing the importance of early childhood environment, have led prominent scholars to argue that early childhood is fundamental in setting individuals on trajectories that become increasingly difficult to alter as individuals reach adulthood (Heckman, 2006).

FUTURE DIRECTIONS

Although the literature shows that cognitive ability is influenced by both genetic and environmental factors, it also suggests several important avenues for future research. First, future research should better identify the relative importance of genetic versus environmental factors and provide a better understanding of how this relationship may change over the life course. For example, attempts to quantify genetic determination of cognitive ability provide mixed results, and estimates vary dramatically by age. For example, from infancy to childhood, 20% to 40% of cognitive outcomes can be attributed to genetic factors (Kovas, Harlaar, Petrill, & Plomin, 2005; Plomin, 1994). In late adulthood, nearly 60% can be accounted for by shared genetic traits.

Second, methodological challenges remain in identifying the causal effect of environmental characteristics, such as family background. For example, family income is highly correlated with other characteristics of the family that can also affect ability, so it is difficult to identify causality. Although sibling and longitudinal data can help tease out unobserved family and individual factors that bias estimates, perhaps the best solution lies in the implementation of large-scale experiments.

Third, there is considerable racial variation in how environment affects cognitive outcomes. For example, studies have shown that neighborhood characteristics, maternal employment, and preschool quality have a greater effect on the cognitive development of White versus Black children. Are these differences due to racial differentials in family organization, access to goods and services, or demographic composition (e.g., income, education)? Future research can provide insight into the culturally specific pathways through which cognitive outcomes can be affected.

SEE ALSO Volume 1: *Academic Achievement; Child Care and Early Education; Family Process Model; Genetic Influences, Early Life; Learning Disability; Socialization, Gender.*

BIBLIOGRAPHY

Amato, P. R. (2000). The consequences of divorce for adults and children. *Journal of Marriage and the Family, 62*(4), 1269–1287.

American Association of University Women. (1995). *How schools shortchange girls: The AAUW report: A study of major findings on girls and education.* New York: Marlowe.

Bradley, R. H., Corwyn, R. F., McAdoo, H. P., & Coll, C. G. (2001). The home environments of children in the United States, Part I: Variations by age, ethnicity, and poverty status. *Child Development, 72*(6), 1844–1867.

Carlson, M. J., & Corcoran, M. E. (2001). Family structure and children's behavioral and cognitive outcomes. *Journal of Marriage and the Family, 63,* 779–792.

Carroll, J. B. (1993). *Human cognitive abilities: A survey of factor-analytic studies.* Cambridge, U.K.: Cambridge University Press.

Chipeur, H. M., Rovine, M., & Plomin R. (1990). LISREL modelling: Genetic and environmental influences on IQ revisited. *Intelligence, 14,* 11–29.

College Board. (1991). *Profile of SAT and achievement test takers.* Princeton, NJ: Educational Testing Service.

Cooksey, E. C. (1997). Consequences of young mothers' marital histories for children's cognitive development. *Journal of Marriage and the Family 59*(2), 245–261.

Currie, J. (2001). Early childhood education programs. *The Journal of Economic Perspectives, 15*(2), 213–238.

Currie, J., & Thomas, D. (1995). Does Head Start make a difference? *American Economic Review, 85*(3), 341–364.

Demo, D. H., & Acock, A. C. (1998). The impact of divorce on children. *Journal of Marriage and the Family, 50*(3), 619–648.

Downey, D. B., & Vogt Yuan, A. S. (2005). Sex differences in school performance during higher school: Puzzling patterns and possible explanations. *The Sociological Quarterly, 46*(2), 299–321.

Duncan, G. J., & Brooks-Gunn, J. (1997). Income effects across the life span: Integration and interpretation. In G. J. Duncan & J. Brooks-Gunn (Eds.), *Consequences of growing up poor* (pp. 596–610). New York: Russell Sage Foundation.

Duncan, G. J., & Magnuson, K. A. (2006). Can family socioeconomic resources account for racial and ethnic test score gaps? *The Future of Children, 15*(1), 15–34.

Duncan, G. J., Brooks-Gunn, J., & Klebanov, P. K. (1994). Economic deprivation and early-childhood development. *Child Development, 65*(2), 296–318.

Elder, G. H., Jr. (1999). *Children of the great depression: Social change in life experience.* (25th anniversary ed.). Boulder, CO: Westview Press.

Entwisle, D. R., Alexander, K. L., & Olson, L. A. (1994). The gender gap in math: Its possible origins in neighborhood effects. *American Sociological Review, 59,* 822–838.

Fomby, P., & Cherlin, A. (2007). Family instability and child well-being. *American Sociological Review, 72,* 181–204.

Friedman, L. (1989). Mathematics and the gender gap: A meta-analysis of recent studies on sex differences in mathematical tasks. *Review of Educational Research, 59*(2), 185–213.

Garces, E., Thomas, D., & Currie, J. (2002). Longer-term effects of Head Start. *The American Economic Review, 92*(4), 999–1012.

Guo, G. (1998). The timing of the influences of cumulative poverty on children's cognitive ability and achievement. *Social Forces, 77*(1), 257–287.

Han, W.-J. (2005). Maternal nonstandard work schedules and child cognitive outcomes. *Child Development, 76,* 137–154.

Heckman, J. J. (2006). Skill formation and the economics of investing in disadvantaged children. *Science, 312*(5782), 1900.

Hedges, L. V., & Nowell, A. (1995). Sex differences in mental test scores, variability, and numbers of high-scoring individuals. *Science, 269,* 41–45.

Hoff, E. (2003). The specificity of environmental influence: Socioeconomic status affects early vocabulary development via maternal speech. *Child Development, 75*(5), 1368–1378.

Jencks, C., & M. Phillips. (1998). The Black–White test score gap: An introduction. In C. Jencks & M. Phillips (Eds.), *The Black–White test score gap* (pp. 1–51). Washington, DC: Brookings Institution Press.

Karoly, L. A., Greenwood, P. W., Everingham, S. S., Hoube, J., Kilburn, M. R., Rydell, C. P., et al. (1998). *Investing in our children: What we know and don't know about the costs and benefits of early childhood interventions.* Santa Monica, CA: RAND Corporation.

Keane, M. P., & Wolpin, K. I. (1997). The career decisions of young men. *Journal of Political Economy, 105*(3), 473–522.

Korenman, S., Miller, J. E., & Sjaastad, J. E. (1995). Long-term poverty and child development in the United States: Results from the NLSY. *Children and Youth Services Review, 17,* 127–155.

Kovas, Y., Harlaar, N., Petrill, S. A., & Plomin, R. (2005). "Generalist genes" and mathematics in 7-year-old twins. *Intelligence, 33,* 473–489.

Lareau, A. (2003). *Unequal childhoods: Class, race, and family life.* Berkeley: University of California Press.

National Institute of Child Health and Human Development Early Child Care Research Network. (1997). The effects of infant child care on infant–mother attachment security: Results of the NICHD Study of Early Child Care. *Child Development, 68,* 860–879.

National Institute of Child Health and Human Development Early Child Care Research Network. (1998). Early child care and self-control, compliance, and problem behavior at 24 and 36 months. *Child Development, 69,* 1145–1170.

Plomin, R. (1994). *Genetics and experience.* Newbury Park, CA: Sage.

Rock, D., & Stenner, A. J. (2005). Assessment issues in the testing of children at school entry. *The Future of Children, 15*(1), 15–34.

Rouse, C., Brooks-Gunn, J., & McLanahan, S. (2005). Introducing the issue. *The Future of Children, 15*(1), 5–13.

Shore, R. (1997). Rethinking the brain: New insights into early development. New York: Families and Work Institute.

Smith, J., Brooks-Gunn, J., & Klebanov, P. (1997). Consequences of growing up poor for young children. In G. J. Duncan & J. Brooks-Gunn (Eds.), *Consequences of growing up poor.* New York: Russell Sage Foundation.

Weinberg, N. (1999). Cognitive and behavioral deficits associated with parental alcohol use. In M. E. Hertzig & E. A. Farber (Eds.), *Annual progress in child psychiatry and child development* (pp. 315–331). Philadelphia, PA: Psychology Press.

Warren, J. R., Sheridan, J., & Hauser, R. M. (2002). Occupational stratification across the life course: Evidence from the Wisconsin Longitudinal Study. *American Sociological Review, 67*(3), 432–455.

Amy Hsin

COLEMAN, JAMES S.
1914–1996

James S. Coleman (1926–1995) was one of the preeminent sociologists of the second half of the 20th century. His writings span many subjects including the sociology of education, mathematical sociology, political organization, diffusion of innovation, rational choice theory, and numerous other topics. He was born in Bedford, Indiana. He graduated from DuPont Manual High School in Louisville, Kentucky, in 1944, received a Bachelor of Science degree from Purdue University in 1949, and earned a Ph.D. in sociology from Columbia University in 1955.

During his studies at Columbia he was a research associate at the Bureau of Applied Social Research. His career of research and teaching was spent at the University of Chicago (1956–1959 and 1973–1995) and Johns Hopkins University in Baltimore (1959–1973). His contributions to understanding the life course and human development focus primarily on youth and adolescence, although his body of work is ultimately much broader and encompasses organizations and informal groups in addition to individual trajectories.

Along with many journal articles, Coleman's major publications on youth, adolescence, and education include *The Adolescent Society* (1961), *Equality of Educational Opportunity* (*EEO*, with Campbell, Hobson, McPartland, Mood, Weinfeld, & York, 1966), *High School Achievement: Public, Catholic, and Private Schools Compared* (with Hoffer & Kilgore, 1982), *Public and Private High Schools: The Impact of Communities* (with Hoffer, 1987), and *Parents, Their Children, and Schools* (edited with Schneider, 1993). Two important policy reports were *Youth: Transition to Adulthood* (1973) and *Becoming Adult in a Changing Society* (with Húsen, 1985). The publication of *EEO* was a monumental event that brought Coleman squarely into the media and policy spotlights. The report was one of the earliest examples of a large-scale social survey being used to shape public policy, and it challenged widely held assumptions about the dominance of school funding and teacher traits in determining student achievement. *EEO* emphasized the influence of family background and the characteristics of one's classmates on learning. The full body of Coleman's research, however, must be considered to appreciate his influence.

Without attempting to summarize all of Coleman's theoretical contributions and empirical findings on youth, adolescence, and education, a couple of dominant themes can be highlighted. First, Coleman was a keen observer of broad social changes occurring in the United States (and elsewhere) during the 20th century. He was fascinated by the dual advents (social inventions, really)

James Samuel Coleman. PHOTO COURTESY OF THE ASA.

of nearly universal participation in secondary schooling *and* adolescence as a distinct stage of life during which youth were segregated in formal institutions set apart from the rest of society. The broad lesson Coleman took away was that informal social systems would emerge within schools fostering values and norms among youth that would often be at odds with the formal goals of the adult-designed institutions.

Second, Coleman assumed social actors were goal directed and purposive. When his writings offered policy prescriptions, he urged designers of social systems to take into account the preferences and choices of actors. Wise designs of schools, workplaces, and other institutions of socialization would structure incentives in ways that would motivate people to work toward organizational goals.

An insightful essay by Heckman and Neal (1996) juxtaposes Coleman the rational choice theorist with Coleman the empiricist. According to these authors, Coleman the theorist consistently described goal-directed actors and the primacy of people's choices made in response to incentives. Coleman the empiricist was a true inductive scientist, bringing his own hypotheses to a project but also learning from the data and revising his vision in light of the evidence.

Coleman was often criticized for neglecting to articulate explicit quantitative models suggested by theory before beginning data analysis. He was also criticized for not explicitly modeling the consequences of choices and self-selection into social settings; he left himself open to criticisms of not having properly separated spurious correlations from genuine causation. In response to the first criticism, both Coleman and Heckman defend splendidly Coleman's refusal to follow rigid econometric conventions (see Postscript 2 in Heckman and Neal, 1996). The second criticism is harder to dismiss.

In several essays late in his career, Coleman reflected on the themes and tensions in his collective works on youth and education. He saw great analytic power in how *The Adolescent Society* sought to understand youths' behaviors and outcomes only after conceptualizing and measuring the goals and interests of the youth themselves. Such work stressed the norms and bases of popularity characterizing social systems, and the social location of any particular actor within a social system. He saw considerable limitations to studies (including *Equality of Educational Opportunity*) that sought to explain youths' behaviors and outcomes as a function of, or measured against, administrative goals. Coleman fundamentally wanted to use schools as microcosms that could reveal more general principles about society at large. Given this goal, Coleman was aware insights are severely limited if the primary questions are, "What are the official goals of a school, and how well are these met? How effective are schools in facilitating or generating academic achievement? What inequalities exist in educational resources, and how do these translate into inequalities in the outcomes of schooling?" He was much more interested in discovering the emergent norms, values, bases of status, and social cleavages of adolescents' worlds—and how these shaped behaviors, attainments, and the community's functioning.

SEE ALSO Volume 1: *Cultural Capital; Intergenerational Closure; Parental Involvement in Education; Racial Inequality in Education; School Culture; Social Capital; Socioeconomic Inequality in Education.*

BIBLIOGRAPHY

Coleman, J. S. (1961). *The adolescent society.* Glencoe, IL: The Free Press.

Coleman, J. S. (1973). *Youth: Transition to adulthood.* Report of the Panel on Youth of the President's Science Advisory Committee. Washington, DC: U.S. Government Printing Office.

Coleman, J. S. (1991). Reflections on schools and adolescents. In D. L. Burleson (Ed.), *Reflections: Personal essays by thirty-three distinguished educators.* Bloomington, IN: Phi Delta Kappa Educational Foundation.

Coleman, J. S. (1994). A vision for sociology. *Society, 32,* 29–34.

Coleman, J. S., Campbell, E. Q., Hobson, C. J., McPartland, J., Mood, A. M., Weinfeld, F. D., et al. (1966). *Equality of Educational Opportunity.* Washington, DC: U.S. Government Printing Office.

Coleman, J. S., & Hoffer, T. (1987). *Public and private high schools: The impact of communities.* New York: Basic Books.

Coleman, J. S., Hoffer, T., & Kilgore, S. (1982). *High school achievement: Public, catholic, and private schools compared.* New York: Basic Books.

Coleman, J. S., & Húsen, T. (1985). *Becoming adult in a changing society.* Paris: U.N. Organization for Economic Cooperation and Development.

Coleman, J. S., & Schneider, B. (Eds.). (1993). *Parents, their children, and schools.* Boulder, CO: Westview Press.

Heckman, J. J., & Neal, D. (1996). Coleman's contributions to education: Theory, research styles, and empirical research. In J. Clark (Ed.), *James S. Coleman* (pp. 81–102). London: Falmer Press.

Stephen B. Plank

COLLEGE CULTURE

College culture refers to patterns of student life within a college or university campus. It encompasses *collective* behaviors and values—as distinguished from the mere compilation of *aggregate* data about numerous individual students. Furthermore, the concept emphasizes continuity of these social patterns, in contrast to transient or haphazard episodes. In sum, a college culture is a complex social network that endures over time through elaborate rituals and symbols. It includes, for example, inviting a student to be the member of a prestigious honor society or extending a pledge of initiation as part of a sorority or fraternity. It also includes the scars one acquires when turned down for inclusion in such groups. As such it has become an integral rite of passage for late adolescents in American society. Thanks to the seminal work of sociologist Burton Clark (1971), characterizations of college culture emphasize institutional belief and loyalty as central to affiliation. These help to create and to transmit a college's culture to newcomers over generations, contributing to a legendary memory about campus events, often called the "institutional saga."

An excellent source for literature review of college culture as explored from a variety of disciplines is the monograph by George Kuh and Elizabeth Whitt, *The Invisible Tapestry* (1988). Placing this concept within research on college students is covered in two classic works: Nevitt Sanford's anthology *The American College:*

A Psychological and Social Interpretation of the Higher Learning (1961) and, more recently, Ernest Pascarella and Patrick Terenzini's *How College Affects Students: A Third Generation of Research* (2005).

HISTORICAL TRENDS

Observation of college culture has a rich tradition first associated with memoirs about clustering tendencies of students at medieval universities. At the University of Paris and elsewhere in Europe, migrating students formed associations reflecting their national roots. Banding together, they acquired distinctive colors and codes, giving rise to the characterization of "Town versus Gown." This distinguished local citizens from university students, who were required to wear academic gowns. Student clashes with landlords and merchants around charges of price gouging on rents, food, and wine, led to higher authorities such as popes, bishops, and kings bestowing permanent legal protections on university students. Within this framework of academic rights and responsibilities, student cultures flourished.

In contrast to the urban universities on the European continent and in Scotland, English universities around the 16th century underwent a structural change, and the college cultures within these universities veered in a new direction. At Oxford and Cambridge, the crucial unit became the residential "colleges" that formed a honeycomb within the university framework. So, although the university conferred degrees and administered examinations, undergraduate life and learning took place within a self-contained residential college. Each college had its own charter, endowment, faculty, curriculum, admissions plans, and traditions of student life. Construction of a spacious university campus and its college quadrangles gave rise by the late 19th century to an elaborate array of teams and clubs that were run by students, for students. A student could join with fellow debaters, athletes, or musicians whose shared experience often surpassed the formal course of studies as a source of a student's primary identity.

This "collegiate way" was successfully transplanted to colleges of the American colonies and, later, the United States (Rudolph, 1962). Groups formed within the college flourished as extracurricular activities and were a major source of socialization into American life. Student clubs at American colleges pioneered serious interest in reading and writing fiction, establishing book collections and libraries, projects in scientific field work, sophistication in debate, forming athletic teams, and discussing politics. There was a discernible life cycle to these student-initiated activities. After creation by students, the success of an activity led to attempts by administrators and faculty to prohibit the activities because they were seen as diverting student attention away from formal studies and even undermining the strict, formal mission of collegiate education at the time. Such interventions usually failed, as students persisted in forming renegade groups for literary societies, athletic teams, and publications. After presidents recognized the futility of suppression, they tried to salvage some official control by incorporating student activities into the formal university structure.

Faculty accounts of student life in the late 19th century invoked colorful metaphors. Undergraduates at Yale brought to mind thoroughbred ponies at play, cantering and frolicking (Santayana, 1892) in the campus meadow. Shifting from turf to surf, another depiction was that of fellow passengers in a boat—a common fate while rowing from freshmen to senior status. Students who had to drop out due to poor grades were mourned as sailors who had fallen overboard and missed all the good times (Santayana, 1892).

COLLEGE CULTURE IN THE LIVES OF STUDENTS AND IN AMERICAN SOCIETY

A college culture socializes assorted students into a coherent group. In the United States this usually has focused on undergraduates. However, the concept is sufficiently elastic to include other constituencies, including graduate students and professional school students. Indeed, one of the classic studies of college culture is *Boys in White*, based on an ethnographic study of medical students at the University of Kansas (Becker, Geer, Hughes, & Strauss, 1961).

Sociologists Burton Clark and Martin Trow (1967) identified the concept of "student subcultures." According to their model, a campus was a fluid configuration of subgroups, many of which provided a student with a primary affiliation within the context of the general campus culture. This conceptualization allowed an analyst to impose a loose template on any particular college or university—and then refine the relative strength of each subculture both to confirm the universal model and simultaneously to capture the precise variations for the case study. Also, Clark and Trow's typology was neutral. It did not say that the persistence of a strong subculture of fraternities and sororities was good or bad; rather, the researcher had to identify and interpret on a case-by-case basis. It even allowed for subtle, important distinctions. One might find that a campus had a distinctive subculture—for example, being a place where performing arts enjoyed a strong tradition of prestige—yet there was no imperative that this same subculture would be—or should be—found elsewhere.

Although one can speak generally about a college culture, historians and sociologists have parsed this broad concept into components. Helen Horowitz (1987) characterized campus life, whether in the 17th century or 20th century, as having three layers: insiders, outsiders, and rebels. Groupings were related to prestige, power, wealth, race/ethnicity, and gender. She identified what she called the "College Men" as the group with inordinate power to set the tone and reward system for the entire institution. Later, with the emergence of women's colleges as well as coeducation, Horowitz noted that the "College Women" could be added to her original concept. "Rebels" referred to relatively small groups—ranging from reform-minded student editors to innovative clubs—who did not seek the affirmation of the dominant "inside" culture. "Outsiders" typically were those students denied full acceptance into the prestigious groups, suggested in the early 21st century by the social exclusion of "nerds" or racial minorities from fraternities and sororities.

MAIN THEMES AND THEORIES IN RESEARCH

Central to research and theories on college cultures is that students have the power to create a world of their own within the institutional regulations and campus environment. One editor of *The Saturday Review* described college cultures to be no less than a "city-state" established by students within the institution (Canby, 1936). For students in the late 19th century the collective motto was, "Don't Let Your Studies Interfere with Your Education." Implicit in this good-natured banner was the serious fact that college culture was not the same as those of the faculty or administration. College culture demonstrated the ability to defer when necessary to official rules. Yet the enduring college was in the quite separate arena by and for students themselves. The college culture was sufficiently strong that it even led behavioral scientists to write about a "hidden curriculum" of what was taught and learned, as distinguished from the formal curriculum (Snyder, 1971).

If a college culture had the capacity to resist control by adult groups, it also had the leverage to split the student body into factions. Although admission to a college was supposed to provide entrée to campus citizenship, this usually worked only so long as a college was small and homogeneous. However, when enrollments expanded rapidly and included increasing student diversity in religion, ethnicity, family income, and geography, a major function (often a dysfunction) of college cultures was to break down shared experiences of students and to fragment them by exclusion within the campus. This relates closely to historian Horowitz's typology of insiders and outsiders within a student body. This was most controversial when membership in an internal subculture was based less on merit and more on ascribed characteristics.

The power of exclusion was a formidable detriment to integration of diversity for women and for racial and ethnic minorities. One partial response was for outsider groups to form their own subcultures. Hence, one finds exclusion from established fraternities may have led to creation of Jewish fraternities or African American fraternities as a counterbalance to the dominant Greek letter system. Cultural patterns reinforced by the Greek letter societies were complex. To one extreme, they were hailed for promoting loyalty to fellow members and teamwork. To another extreme, fraternities were often associated with encouraging ritualized student drinking and extended immaturity. In sum, the crucial research question was whether a college culture was inclusive or exclusive in campus life.

NEW DIRECTIONS IN RESEARCH

The bulk of systematic research on higher education since the 1950s has drawn heavily from psychology and the behavioral sciences. Yet for the serious study of college culture, an appropriate and sorely needed discipline has been anthropology. Perhaps one of the most original and provocative studies of college cultures was the book by anthropologist Michael Moffat, *Coming of Age in New Jersey* (1989). Relying on strategies of fieldwork usually associated with the analysis of distant cultures, anthropologist Moffat chose to undertake an ethnographic study of a dormitory at a large state university—his own campus, Rutgers University in New Jersey. An important feature of the campus residence hall was that it was coeducational. Moffat focused on how young women and men cooperated to create an environment with norms and values in the relatively new format where members of both sexes shared living spaces. The findings were counter to fears of concerned parents and clergy. Coeducational dormitories did not promote licentious behavior or rampant promiscuity. To the contrary, students emphasized respect for privacy and mutual concern for the welfare of members of the dormitory community. Instead of amorous relations, men and women tended more toward sibling interactions of brothers and sisters. A promising sign of continued reliance on anthropology to study college culture is the recent book, *My Freshman Year* (Nathan, 2005).

The hegemony of such student organizations as Greek letter fraternities and sororities to dominate American college culture for centuries now faces a new perspective. Whereas selection into a fraternity or sorority was long seen by insiders and outsiders as advantageous

College Students. PHOTO BY LEITA ETHERIDEGE-SIMS. COURTESY OF LANCE LOGAN SIMS, MPH.

to adult affiliations and success, signs of change have surfaced. At the University of California, Santa Barbara, for example, data indicate that membership in a fraternity tended to be counterproductive as a base from which to gain campus leadership. At Yale there were surprising signs that seniors who had been "tapped" for membership in elite senior societies were making decisions that would have been unthinkable to earlier generations of undergraduates: Invitees were rejecting membership invitations. One explanation was that exclusive student groups not based on merit were seen as detrimental to adult life, especially public office, where discrimination by gender, race, religion, or ethnicity have become a liability. How this groundswell of refusal develops in the 21st century will be important.

A fertile area of reconsideration involves attention to students' family affiliations as part of college life, based on criticisms of an earlier model in which a new student who entered college was seen as poised to discard family

affiliation to become a citizen in the campus (Tinto, 1987). Many sociologists saw this transitional split as necessary if students were to internalize new cognitive skills, values, curriculum, and networks. It was captured in the expression of "breaking home ties" as a feature of "going away to college." A revised perspective is that a fulfilling college experience need not require this break. For many students, maintaining family ties has been recast as helpful to navigating the new demands of campus life. This is an important deliberation given concern about undergraduate attrition and alienation (Braxton, 2000).

The diversity of American higher education also has meant that college culture requires recognition of new constituencies. Commuter students often have been ignored as marginal or disengaged. This neglect makes little sense if commuters dominate an institution's student profile. One novel study focused on "The Commuter's Alma Mater" to identify how nontraditional students worked among themselves and with campus officials to reconfigure services and facilities to fit their social patterns (Mason, 1993).

Worth final note is that college culture is a topic long central to memoirs and fiction (Kramer, 2004). A challenge for social and behavioral scientists is how to give adequate consideration to this genre of American literature as an important source of data as part of their own systematic scholarly research. It could mean reliance on Owen Johnson's 1912 novel *Stover at Yale* to describe the round of life and values at Yale and other historic East Coast campuses prior to World War I. In the early 21st century, one might consider how well or how poorly a popular Hollywood movie such as *Animal House* provides a reasonably accurate portrayal of fraternities and sororities as strong influences on the college culture of a state university.

SEE ALSO Volume 1: *College Enrollment; Drinking, Adolescent; School Culture; Sexual Activity, Adolescent.*

BIBLIOGRAPHY

Becker, H. S., Geer, B., Hughes, E. C., Strauss, A. L. (1961). *Boys in white: Student culture in medical school.* Chicago: University of Chicago.

Braxton, J. M. (Ed.). (2000). *Reworking the student departure puzzle.* Nashville, TN: Vanderbilt University Press.

Canby, H.S. (1936). *Alma mater: The gothic age of the American college.* New York: Farrar & Rinehart.

Clark, B. R. (1971). Belief and loyalty in college organization. *Journal of Higher Education, 42*(6), 499–515.

Clark, B. R., & Trow, M. (1967). *Determinants of college student subcultures.* Berkeley, CA: Center for Research and Development in Higher Education.

Haskins, C. H. (1957). *The rise of the universities.* Ithaca, NY: Cornell University.

Horowitz, H. L. (1987). *Campus life: Undergraduate cultures from the end of the eighteenth century to the present.* New York: Knopf.

Kramer, J. E. (2004). *The American college novel: An annotated bibliography* (2nd ed.). Lanham, MD: Scarecrow.

Kuh, G. D., & Whitt, E. J. (1988). *The invisible tapestry: Culture in American colleges and universities.* Washington, DC: ASHE-ERIC.

Mason, Tisa (1993). *The Commuter's Alma Mater.* Williamsburg: College of William & Mary doctoral dissertation.

Moffat, M. (1989). *Coming of age in New Jersey: College and American culture.* New Brunswick, NJ: Rutgers University.

Nathan, R. (2005). *My freshman year.* Ithaca, NY: Cornell University.

Pascarella, E., & Terenzini, P. (2005). *How college affects students: A third generation of research.* San Francisco, CA: Jossey-Bass.

Rudolph, F. (1962). *The American college and university: A history.* New York: Knopf.

Sanford, N. (Ed.). (1962). *The American college: A psychological and social interpretation of the higher learning.* New York: Wiley.

Santayana, G. (1892). A glimpse of Yale. *Harvard Monthly, 15,* 89–96.

Snyder, B. R. (1971). *The hidden curriculum.* New York: Knopf.

Tinto, V. (1987). *Leaving college: Rethinking causes and cures of student attrition.* Chicago: University of Chicago Press.

John R. Thelin

COLLEGE ENROLLMENT

College enrollment entails taking classes at a college or university to fulfill the requirements for a bachelor's degree or associate's degree. In most cases enrollment for a bachelor's degree is done at a 4-year college or university that provides a general liberal arts education with a specific focus or major in preparation for a professional career. Enrollment for an associate's degree, in contrast, typically takes place at a 2-year community college where the focus is on preparing for midlevel and semiprofessional occupations in the local labor market. The boundaries between the two educational missions have eroded as the demand has increased for 2-year community colleges to provide remedial instruction to academically unprepared students who can transfer later to a 4-year college or university (Rosenbaum, 2001). College enrollment generally occurs in young adulthood in the years immediately after high school graduation, and students take an average of 4.6 years to complete a bachelor's degree and 3.6 years to complete an associate's degree.

In the first decade of the 21st century, college enrollment rose in the United States: Almost 15 million students were enrolled in 2005, up from 7.4 million in 1970. This was due in part to the economic and social benefits that accrue to those who earn postsecondary credentials. On average, recipients of a bachelor's degree have higher lifetime earnings and lead healthier, longer lives than their peers who have some or no college experience. Although the majority of college graduates do not go to graduate school, a bachelor's degree is a prerequisite for a more advanced degree and, on average, a more lucrative career in law, business, medicine, or science.

HISTORICAL TRENDS

College enrollment, like school enrollment in general, is fueled largely by the demand for labor. Before the Civil War, when the domestic economy was driven primarily by manual labor and agriculture, the need for education beyond the primary years was low, and youths left school at an early age. After the war and into the 20th century, the industrial revolution brought a growing need for a more educated workforce as industry and large-scale manufacturing came to dominate the domestic economy. In response, school enrollment surged: In the Northeast, the center of industrial production at that time, high school enrollment grew from 4.9% of youth in 1890 to 52% in 1940. Postsecondary education, however, remained the province of affluent young people who were preparing for white-collar professional jobs in business or medicine.

In the second half of the 20th century, the economy became increasingly reliant on technological, quantitative, and communication skills, and this created a demand for a college-educated workforce. Consequently, school enrollment among college-age youth rose dramatically: In 1950, 9% of those 20 to 21 years old, the prime college-going age, were enrolled in school. By 2000 that figure had more than quintupled to 48.7%. By 2004 the majority of high school seniors were enrolled in college (60.3%), making postsecondary enrollment the modal experience for young people after leaving high school in the United States. In light of those changes in the economy and the demand for labor, college enrollment as a distinct component of young adulthood and the life course more generally is, historically, a relatively new phenomenon.

COLLEGE ENROLLMENT IN THE EDUCATIONAL LIFE COURSE

The transition to college involves a unique set of changes that extend beyond the gradually increasing rigor of coursework that typically accompanies the passage from grade to grade. In high school most students work among a common set of teachers and classmates and take classes in one or two buildings. Academic progress is tracked by a guidance counselor or school administrator with whom contact is frequent and personal. Postsecondary institutions, in contrast, tend to be larger than high schools: The average public high school has approximately 760 students whereas the average 4-year public college or university has about 8,800. Because of their size, colleges have complex and often confusing bureaucratic mechanisms to organize and process students. For example, administrative necessities such as tuition and payments, course scheduling, and financial aid are housed in different offices or buildings, and a professor who may or may not teach students often serves as an academic adviser. The volume of students enrolled in a college often limits access to those resources and makes the interactions less personal.

The curricular organization of college, particularly with respect to time requirements and coursework, is less regulated than it is in high school; consequently, students have greater autonomy and more choices. For example, high school students generally are expected to be present on school grounds Monday through Friday, approximately between 8 A.M. and 3 P.M. (25 hours per week), whereas full-time college students spend less time in the classroom (typically 15 to 20 hours per week). In high school the curriculum is designed to meet state graduation requirements, and high school students thus have

JUNIOR/COMMUNITY COLLEGE

■

One of the most notable recent changes in the postsecondary educational system in the United States has been the gradual growth in 2-year community college enrollment, which accounted for 43% of all postsecondary enrollment in 2005, up from 31% in 1970. These schools, which are funded by state and local government agencies, typically have open admissions policies by which all individuals regardless of academic proficiency can enroll in courses. Because of their affordability and accessibility, 2-year community colleges often serve economically and academically disadvantaged youths. In terms of their educational mission these schools serve a dual function: providing youth with basic skills and training for midlevel and semiprofessional occupations and giving less prepared youth an opportunity to acquire course credits and remedial instruction before transferring to a 4-year college or university. Two-year colleges thus serve both as labor market intermediaries and as part of the pipeline to a bachelor's degree. However, only 13% of youths who start off in these schools complete an associate's degree (Berkner, He, & Cataldi, 2002) and only 26% transfer to a 4-year college or university.

limited options in the courses they can take. In college the curriculum is organized to give students a core competency in a specific discipline, which typically is decided on during the first or second year of enrollment. In contrast to the more general subject matter in high school, the major-driven college curriculum gives students an opportunity to focus on their specific interests and develop skills unique to their planned occupations. Moreover, college students have more course options available to them and therefore have more flexibility to schedule classes so that they can meet obligations to their classes, families and friends, and, in many cases, jobs.

In addition to the structural and organizational changes, the transition to college marks the first time in the educational life course that the majority of enrolled students live outside their parents' or guardian's home while taking classes, mostly in dormitories and residence halls. Considered in the social demography literature a semiautonomous living arrangement, dorm life lacks the responsibilities associated with renting an apartment or owning a home but involves physical and emotional separation from the family, an experience that is often daunting and stressful for those just out of high school. Historically, in the United States the move out of the parental home accompanied marriage, and together those experiences served as key markers in the transition to adulthood. As the age at first marriage has increased along with rates of college enrollment immediately after high school, the first forays out of the parental home have become linked more closely with college enrollment than with marriage.

PERSPECTIVES ON COLLEGE PERSISTENCE

These structural, curricular, and organizational changes, which are most salient during the transition to college, often are overwhelming, particularly for disadvantaged and academically unprepared students. Thus, rates of college dropout are highest during the first year: Among first-time college freshmen enrolling immediately out of high school, 34.7% either dropped out at some point in the first year or did not return the next fall. Currently, only 13% of youths who attend a 2-year community college earn an associate's degree and about 51% of those who attend a 4-year college or university earn a bachelor's degree.

In light of the importance of a highly educated labor force and the socioeconomic benefits that accrue to those with a bachelor's degree, much theoretical and empirical attention has been paid to identifying the determinants of sustained enrollment, or persistence. Most research on postsecondary persistence is approached from one of three perspectives: the role of social structure, individual ability and preparation, and the relationship between the student and the university.

A social structural approach, which is informed by the conflict perspective in sociology, implicates college enrollment and persistence in larger processes of educational stratification. With limited slots available in the higher education system, inequalities in social status, particularly race/ethnicity and socioeconomic background, lead to an unequal allocation of students to institutions of varying stature and quality; this creates very different enrollment experiences. For example, racial/ethnic minorities and less affluent students are more likely to enroll in public schools, 2-year community colleges, and academically less selective schools (Karen, 2002), which tend to have fewer support systems and resources than do private and more academically competitive colleges. These institutional differences exacerbate the difficulties associated with the transition to college and thus contribute to different rates of persistence. Among students entering college in 2003–2004, 43.4% of Blacks and 37.2% of Hispanics left college without a degree within the first 3 years compared with 31.2% of Whites. Similarly, 43.7% of students whose

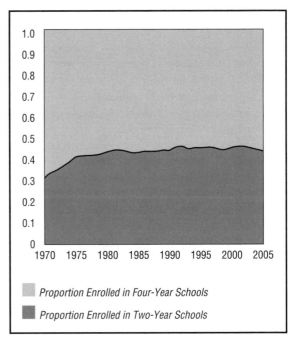

Figure 1. *Proportion of total postsecondary enrollment, 1970–2005.* CENGAGE LEARNING, GALE.

parents did not attend college left college without a degree within 3 years compared with 17.7% of students whose parents earned bachelor's degrees.

Whereas structural explanations emphasize allocation processes that are determined by ascribed characteristics (characteristics one is born with), individual-level perspectives focus on achieved characteristics such as academic preparation and orientation toward the future. Most of this research is informed by William Sewell and colleagues' Wisconsin Model of Status Attainment, which was one of the first empirical studies that used longitudinal data to identify the individual-level mechanisms that link family background with educational attainment (Sewell, Haller, & Portes, 1969). Their analyses showed that strong academic preparation in school (as measured by grades and/or standardized test scores) and ambitious plans for one's future education (as measured by how far students expected to go in school) were strong predictors of how far individuals would progress in the educational system. Those relationships have been replicated in a large body of research within the sociology of education: Students who have a solid foundation in academic subjects and who express commitment to higher education before they enter are best able to navigate the transition to college and persist through degree completion (Deil-Amen & Turley, 2007).

In contrast with the structural and individual perspectives, which emphasize characteristics of students before they enter college, more recent research in the higher education community has explored the emerging relationships students have with the university, particularly the process through which students integrate into the academic and social communities of a school (Tinto, 1993). Upon entry into college, students are detached physically and/or academically from the communities of their past: their families, high schools, and peer groups, as well as their local areas of residence in many cases. This transitional period tends to be isolating, stressful, and disorienting as entering students encounter the structural, organizational, and housing changes described above. Consequently, it is at this time that students are most likely to drop out. However, once students establish and maintain relationships with faculty members, school personnel, and classmates, they feel a stronger bond with and commitment to the university and are more likely to remain enrolled.

Although they have different emphases, these three perspectives complement one another in that the identified characteristics and processes often work in tandem. For example, racial/ethnic minorities and young people with limited economic resources tend to receive the poorest academic training in the elementary and secondary years and as a result are less ready for the rigor and demands of the college curriculum. If these students enroll in college, they usually attend less competitive schools such as 2-year community colleges, which provide fewer opportunities and resources to facilitate integration. With limited preparation before enrollment and less contact with classmates and faculty once they are enrolled, disadvantaged youths are the most likely to leave before finishing a degree. In this way college enrollment intensifies educational stratification processes rooted in a person's social origins and serves to reproduce existing status differences in adulthood.

Once students weather the transitional obstacles associated with the first year, the odds of persisting until degree completion increase substantially. For example, 74% of students who enroll continuously without a break in enrollment in a 4-year school complete a bachelor's degree compared with 20% of those who take a break in enrollment of 4 months or longer.

DEVELOPMENTAL GROWTH

Although persistence is one of the most frequently studied outcomes because of its bearing on degree completion and socioeconomic well-being in adulthood, other important developmental changes accompany the college enrollment experience. With respect to academics, as students work their way through their degree requirements, they increasingly focus on courses within their majors and as a result develop more specialized skills

and proficiencies. Many students participate in internships and co-op programs that provide real-world experience that sometimes leads to full-time work after graduation. These curricular and noncurricular experiences are associated with growth in abstract reasoning and communication skills and an intellectual orientation toward problem solving (Pascarella & Terenzini, 1991).

College students also form health and lifestyle patterns that are shaped by their environment and position in the life course. With less parental supervision and, for most on-time traditional students, without the full responsibilities of a family or career, many college students experience a prolonged adolescence in which they have greater freedom to make choices about their behaviors and activities. In some cases this autonomy can compromise the health of enrollees. Between 1989 and 2001 the number of students on campus seeking help for depression, stress, anxiety, and eating disorders rose substantially. Moreover, compared with their nonenrolled peers, college students are more likely to binge drink and gamble.

THE CHANGING POSTSECONDARY LANDSCAPE

As enrollment has grown and the role of colleges as a central institution in young adulthood has emerged, a host of changes have continued to alter the postsecondary landscape. For example, although they historically enrolled at lower rates, women reached parity with men by the late 1970s and gradually began to surpass them. In the first decade of the 21st century, women had a sizable enrollment advantage: Among graduating seniors in 2004, 74.8% of women enrolled in college compared with 65.4% of men. Despite this progress, women remain less likely to major in science and engineering, majors that lead to highly valued and compensated jobs in a global, technology-driven economy.

At the same time that college is becoming a prerequisite for socioeconomic well-being in adulthood, it is becoming more expensive. In the 2006–2007 school year, tuition at 4-year public schools accounted for 12.1% of median family income, up from 4.9% in the 1976–1977 school year. If this trend persists without options for financial aid that keep pace with it, the role of college in reproducing inequality will expand rather than narrow.

Two-year community colleges provide a more affordable alternative to 4-year colleges and universities, an option that is particularly attractive for those with limited economic resources (Dougherty, 1994). Enrollment in these schools has risen, accounting for 56.6% of all postsecondary enrollees in 2005, up from 31.4% in 1970. In addition, there are increasing opportunities to

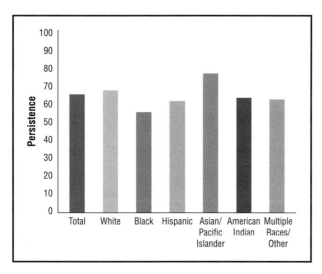

Figure 2. *Rates of Postsecondary Persistence for 2003–2004 Beginning Postsecondary Students, as of June 2006.* CENGAGE LEARNING, GALE.

BIBLIOGRAPHY

Adelman, C. (1999). *Answers in the tool box: Academic intensity, attendance patterns, and bachelor's degree attainment.* Washington, DC: U.S. Department of Education, Office of Education Research and Improvement.

Berkner, L, He, S., & Cataldi, E. F. (2002). *Descriptive summary of 1995–96 beginning postsecondary students: Six years later* (NCES 2003–151). Washington, DC: National Center for Education Statistics, U.S. Department of Education.

Deil-Amen, R., & Rosenbaum, J. E. (2003). The social prerequisites of success: Can college structure reduce the need for social know-how? *Annals of the American Academy of Political and Social Science, 586,* 120–143.

Deil-Amen, R., & Turley, R. L. (2007). A review of the transition to college literature in sociology. *Teachers College Record, 109*(Spec. Issue).

Dougherty, K. D. (1994). *The contradictory college: The conflicting origins, impacts, and futures of the community college.* Albany: State University of New York Press.

Karen, D. (2002). Changes in access to higher education in the United States: 1980–1992. *Sociology of Education, 75,* 191–210.

Pascarella, E. T., & Terenzini, P. T. (1991). *How college affects students.* San Francisco, CA: Jossey-Bass.

Phipps, R. A. (2004). *How does technology affect access in postsecondary education? What do we really know? Report of the National Postsecondary Educational Cooperative Working Group on access-technology* (NPEC 2004–831). Jessup, MD: U.S. Department of Education.

Rosenbaum, J. E. (2001). *Beyond college for all: Career paths for the forgotten half.* New York: Russell Sage Foundation.

Sewell, W. H., Haller, A. O., & Portes, A. (1969). The educational and early occupational attainment process. *American Sociological Review, 34,* 82–92.

Slutske, W. S. (2005). Alcohol use disorders among U.S. college students and their non-college-attending peers. *Archives of General Psychiatry, 62,* 321–327.

Tinto, V. 1993. *Leaving college: Rethinking the causes and cures of student attrition.* Chicago: University of Chicago Press.

Robert Bozick

enroll in shorter-term programs that award specialized certifications in fields such as information technologies and paralegal services (Deil-Amen & Rosenbaum, 2003).

As new technology and innovative methods of production continue to dominate the economy, the need for new and adaptable skills has implications for older workers, who often lack those skills. Colleges have responded to this change by reaching out to adults in need of skill development and/or retraining. In 2005 those age 25 or older accounted for 17.7% of all undergraduate enrollees. In addition to affecting labor demand, technology has altered the way postsecondary education is administered. Many courses use the Internet for lectures and assignments, and nearly all university libraries are linked to an expansive network of cyber-repositories. Additionally, online programs and distance learning are readily available; 89% of public 4-year colleges and universities offer some form of distance learning.

As these and other trends continue to alter the postsecondary landscape, college enrollment as a distinct component of young adulthood—an economically feasible experience immediately after high school that is centered on a physical campus—may give way to new forms of training and skill development and consequently may have different implications for the relationship between higher education and the life course.

SEE ALSO Volume 1: *Academic Achievement; Cognitive Ability; College Culture; Gender and Education; High-Stakes Testing; Policy, Education; Racial Inequality in Education; School Tracking; Socioeconomic Inequality in Education;* Volume 3: *Lifelong Learning.*

CONSUMPTION, CHILDHOOD AND ADOLESCENCE

The lives and experiences of children and adolescents cannot be understood with any depth without accounting for and addressing the world of commercial goods and media in which they are situated. Historically speaking, the world of consumer goods increasingly has come to define the contours and content of childhood and adolescence. Over the course of the 20th century and into the 21st, commercial products, retail spaces, advertising, marketing, and media not only have provided the markers of various age-based identities but also have helped delineate and create some of the widely recognized stages of the early life course—such as the toddler, "tween," and teenager. Children and adolescents, as well, have taken an active part in the historical shaping of the life course in their changing roles and activities as consumers and audiences.

HISTORY OF YOUTHFUL CONSUMPTION

At the opening of the 20th century, commercial interest in children was generally sparse and unremarkable. In the United States, some goods for children could be found advertised in a few magazines for parents, on select shelves in department stores, and described on a few pages in mail-order catalogs for companies such as Sears, Roebuck or Montgomery Ward. Products for children at this time consisted of some ready-made clothing; breakfast foods such as cold cereal or hot oats; simple toys such as dolls, hobby horses, and gun-and-target sets; books; and nursery ware (Cook, 2004; Cross, 1997; Jacobson, 2004; Kline, 1993; Leach, 1993). Besides the candy counter in drug and general stores, few, if any, commercial spaces were designed with children in mind or intended for their patronage.

By 1920, creating markets for children's goods had become an increasingly collective, organized, and purposeful endeavor. Merchants and manufacturers instituted trade organizations and published trade journals as a way of coordinating efforts to capture the child and mother trade. By this time, a number of industries had come to recognize children and mothers as key patrons for their products and had organized themselves to cater to these new clienteles. The toy industry led the way with *Toys and Novelties Playthings* in 1903 and *Toys and Novelties* in 1909. The first trade journal devoted exclusively to children's clothing, *The Infants' Department*, began publication in 1917. In 1924, the *Horn Book* was founded to promote books for children, although as early as 1874 *Publishers Weekly* listed children's books.

Consumption, goods, and commercial contexts prove to be significant factors in shaping the understandings of childhood and youth and in fact the very definitions, boundaries, and transitions between stages or phases of the early life course. This effect was felt in the 1920s and 1930s as the United States toy industry positioned itself and its products as integral to the "healthy" and "appropriate" development of children. According to Ellen Seiter (1993), the growing toy industry was buttressed by a new ethos of hands-on parenting promulgated by *Parents* magazine (est. 1926). This new parenting brought the relatively new theories of developmental psychology into the marketplace by popularizing the "idea that childhood is

divided in to discrete, observable stages" that can be supported and even hastened by giving the right toy to a child at the right time in her or his growth (p. 65).

The children's clothing industry played a similar but distinct role in helping to reshape childhood, particularly girlhood. In 1920 only three size ranges in girls' clothing existed—newborn to 1.5 years, 1.5 years to 6 years, and 6 to 14 years. By 1950 there were seven size ranges: infant, toddler, children's, girls', subteen, junior miss, and miss, which covered girls from birth through about age 16. The increasing differentiation of the life course for young females over this time stemmed in part from age grading in the educational system—for example, nursery schools, kindergartens, and middle schools—and their ensuing effects on the social lives of girls. The finer distinctions in clothing sizes and styles both reinforced and helped produce differentiations in age and social status in American girlhood. Merchants in the 1930s, 1940s, and 1950s reported that girls often sought to "jump" to "more sophisticated" styles in order to appear older than they were, actions that prompted merchants, and eventually manufacturers, to add new styles or change age-style designations to accommodate the girls' behavior and preferences.

The cases of the toy and clothing industries point to a general feature of child and youth consumption—namely, that advertising and marketing appeals often include parents as well as children. The historical inclination and impetus, however, has steadily and strongly favored addressing and appealing to children directly, instead of or in addition the parent. The middle-class home had become a site of sentimental domesticity—an emotional haven set apart from the cold, calculating world of work and money—and thereby increasingly oriented toward children's wants and desires. This attitude paradoxically also helped shape the emerging world of consumer goods and spaces for children (Cook, 2004; Zelizer, 1985).

Throughout the 1920s and 1930s, children's goods in department stores steadily acquired their own spaces, child-oriented iconography, specialized personnel, and selling techniques as well as their own separate departments. Some stores dedicated entire floors to juvenile merchandise. More than simply housing children's goods, the overall layout, the height of the fixtures and mirrors, and the color schemes of these departments and floors were designed intentionally to appeal to the child's point of view, rather than that of the mother, with the express purpose of inducing children to request goods from their mothers. A similar impetus to appeal to the child directly could be found with the rise of radio in the 1920s and the concomitant child-specific programming that ensued (Cross, 1997). In addition, comic books, films made for children's viewing, and later television programs arose and configured themselves with a child audience and child market in mind.

Appealing to and addressing children and youth as direct and primary consumers had the cumulative effect of identifying and defining youth and childhood as distinct and internally differentiated phases of life identified by increasingly nuanced, age-graded goods, media, and commercial spaces. The designation of the "the toddler," for instance—which is now a general, named stage of the life course—arose initially almost entirely as a merchandising and manufacturing category of children's clothes (Cook, 2004).

The rise of teenagers and their accompanying youth culture in the 1950s and 1960s represents perhaps the most prominent example of the convergence between consumer culture and the formation and expression of age-based identities. The teenager came into prominence beginning in the mid-1940s. A new national media for teen girls cultivated an audience of peers with the publication of *Seventeen* magazine in 1944, followed by *Miss America* and *Junior Bazaar* in 1945. Grace Palladino (1996) discusses some of the various versions of teenagerhood in the 1950s, including swingers, bobby soxers, and hepcats, whose subcultural lingo, comparatively open sexuality, and freedom worried many parents and schoolteachers. Teenagers and youth in the 1960s and early 1970s took their place on the national stage as visible proponents and symbols of counterculture rebellion.

A cultural phenomenon, "the teenager" has come to represent a new kind of adolescence realized through the medium of consumption of such things as fast food, rock music, distinctive dress, and forms of speech that differentiate their world from that of adults. High school—the institutional "home" of adolescent culture—provides for peer contact among teenagers and the regular opportunity to gauge oneself vis-à-vis others. In ensuing decades, the experiences one has in high school are heralded by psychologists, educators, and in the popular imagination as some of the most important of one's youth and perhaps of one's life. The dramas and rites surrounding prom night, with its focus on heterosexual dating, conspicuous display, and expectations of romance, become stories relived and retold, for better or worse, throughout one's adult life (Best, 2000). The social divisions and social labeling that occurs during adolescence often take place in the context of the high school social relations and interactions. As Murray Milner, Jr. (2004) discussed, "geeks," "freaks," "cool kids," and "brains"—social types typically found in American high schools—acquire social definition in contrast to one another by displays of taste through popular culture goods.

The dominant trend of the relationships among consumption, childhood, and adolescence over the first seven decades of the 20th century was one in which commercial goods and contexts, in part, shaped and

Toy Debut. *Two children look at a new toy.* AP IMAGES.

created new categories and understandings of the early life course. Many of these age designations—as found with and through various means of consumption—helped define childhood or youth in contradistinction to adults and, in so doing, delineated various, more or less bounded identities of adolescence and childhood.

By the 1980s, observers and commentators expressed concern about the apparent "loss" of childhood. Neil Postman (1982) famously argued that childhood was disappearing due to the predominance of electronic media—particularly television. Television, unlike print media, does not require language literacy to be consumed and so does not effectively pose a barrier to participation. As a "total disclosure medium," contended Postman, television opens the "back stage" of adult society to children—that is, depictions of and discussions about sexuality, money, parenting strategies—that used to be gradually and deliberately revealed to children by adults. Hence, childhood is losing its distinctiveness. Joshua Meyrowitz (1985) offered an alternative view that, instead of a one-way movement of childhood disappearing into adulthood, adults are also behaving and dressing in ways traditionally associated with children; hence, he argued, it is more useful to pay attention to the blurring boundaries between childhood and adulthood.

YOUTHFUL CONSUMPTION TODAY

Concerns about the loss or blurring of childhood in the late 20th and early 21st century continue to be voiced by academics, educators, politicians, and parents with increasing alarm. The rise and ubiquity of digital media—in particular, the Internet—shatters barriers to various kinds of information in ways that make broadcast television's break from print media seem minor, because the Internet, unlike television, does not determine the direction of the flow of information. Twenty-first century moral panics about children's ability to access materials deemed unfit for them, such as pornography, sit side-by-side with concerns about how children can be accessed and tracked by sexual predators as well as by marketers, requiring children to be knowledgeable about and prepared for such eventualities.

The phenomenon of "kids getting older younger" arose in the 1990s in light of concerns about the apparent "adultification" of children and youth. Much of this concern has, as in the past, focused on girls and sexuality. Observers point to girls as young as 9 years old discarding Barbie dolls as "babyish" in favor of Bratz dolls, whose style and "attitude" are said to be flamboyant, "street," and sexual, as an indication of age blurring. The term *tweens*—boys and girls approximately 7 to 12 or 13 years old—was coined initially by marketers in the early 1990s but has since become an accepted designation by adults as well as children for this age range (Cook & Kaiser, 2004). The clothing and personal styles of tween girls remains a constant topic of concern for mothers and in public discussion at stores such as Claire's, an international chain, that offer a variety cosmetics, jewelry, and lace underwear for preteen girls. In the 21st century, some parents take their prepubescent children to spas for makeovers, pedicures, and other body treatments once thought the reserve of adult women. For boys, much of the concern centers on exposure to violence in digital video games and to sexualized images of girls and women.

The cultural landscape of childhood and youth remains in a state of flux in large part because the various stages and designations of the early life course are now thoroughly enmeshed with the goods and images of the commercial marketplace. Public concern about the over-commercialization of childhood and adolescence has spurred public discussion about the role of advertising and marketing in public life. Social critics such as Juliet Schor (2004) argue that the overall materialistic values promoted by advertisers and marketers have a negative effect on children's self-image, their physical health, and

their relationship to peers and family. Beyond a few safeguards such as rating systems for television, films, and video games and limitations on what kind information can be collected from children on the Internet, few public policy initiatives are in place that address children and commercial life to any significant degree.

SEE ALSO Volume 1: *Activity Participation, Childhood and Adolescence; Body Image, Childhood and Adolescence; College Culture; School Culture; Media Effects; Peer Groups and Crowds.*

BIBLIOGRAPHY

Best, A. (2000). *Prom night: Youth, schools, and popular culture.* New York: Routledge.

Cook, D. T. (2004). *The commodification of childhood.* Durham, NC: Duke University Press.

Cook, D. T., & Kaiser, S. B. (2004). Betwixt and be tween: Age ambiguity and the sexualization of the female consuming subject. *Journal of Consumer Culture, 4*(2), 203–227.

Cross, G. (1997). *Kids' stuff.* Cambridge, MA: Harvard University Press.

Jacobson, L. (2004). *Raising consumers.* New York: Columbia University Press.

Kline, S. (1993). *Out of the garden: Toys and children's culture in the age of TV marketing.* London: Verso.

Leach, W. (1993). *Land of desire.* New York: Pantheon.

Milner, M. Jr. (2004). *Freaks, geeks, and cool kids: American teenagers, schools and the culture of consumption.* New York: Routledge.

Meyrowitz, J. (1985). The adultlike child and the childlike adult: Socialization in the electronic age. In H. Graff (Ed.), *Growing up in America: Historical experiences* (pp. 612–631). Detroit, MI: Wayne State University Press.

Palladino, G. (1996). *Teenagers: An American history.* New York: Basic Books.

Postman, N. (1982). *The disappearance of childhood.* London: W. H. Allen.

Schor, J. B. (2004). *Born to buy.* New York: Scribners.

Seiter, E. (1993). *Sold separately: Children and parents in consumer culture.* New Brunswick, NJ: Rutgers University Press.

Zelizer, V. (1985). *Pricing the priceless child.* Princeton, NJ: Princeton University Press.

Daniel Thomas Cook

CRIME, CRIMINAL ACTIVITY IN CHILDHOOD AND ADOLESCENCE

The brute fact of age always complicates efforts to understand the nature of crime and victimization. Study upon study shows criminal behavior to increase steadily through the adolescent years, peak in late adolescence or early adulthood, and then decline rapidly and steadily through the remainder of the life span. Although the generality and stability of this trend has provoked controversy, the general conclusion that crime is strongly clustered in adolescence is not seriously in dispute. That offending typically occurs so early in the life span makes it particularly significant for life course scholars, both for understanding what factors influence crimes among teens and for its implications for the unfolding life course.

Although seldom explicitly acknowledged, the age distribution of crime has organized much criminological research. Most importantly, it has meant that the study of crime, at least for much of the 20th century, has been the study of juvenile delinquency. Here, researchers traditionally focused on teenagers and sought to understand what differentiates offending among them. Early studies in the social ecology of crime used arrest statistics of juveniles from police to develop a general theory of social disorder and crime (Shaw & McKay, 1969). Yet in the 1950s and 1960s, arrest data fell into disrepute as social scientists drew attention to the organizational factors that influenced their collection and suggested, with powerful consequences, that such data told much more about the activities of the police than about the criminal activities of individuals. In light of this, researchers increasingly turned to self-report surveys to study crime, but again focused their attention almost exclusively on the activities of adolescents. Regardless of the overall pattern of offending over the life span, the main focus of researchers was crime among adolescents, and explanations of such crime were used, for better or worse, to explain offending in general.

The traditional focus on juveniles has shaped both how crime is understood and how it is studied. In the latter respect, researchers have focused on measuring juvenile delinquency. The concept of juvenile delinquency is interesting. In one respect, it encompasses many of the behaviors that both the public and social scientists regard as crime. This includes actions such as vandalism, theft, violence, and drug use. At the same time, it also includes behaviors that are often seen as outside the domain of official crime statistics, such as skipping school, cheating on tests, breaking curfew, drinking, and even smoking. To make matters more complicated, serious efforts to compare the types of acts captured by self-report survey studies with acts that routinely appear in police and court data conclude that there are some significant differences in both the range of acts under study (typically excluding white-collar crimes such as fraud and embezzlement) and the severity of incidents (typically excluding serious violence such as aggravated

rape and murder). Still, the most thorough assessments also conclude that there is nothing inherent in self-report approaches or the study of juveniles that precludes investigation of the severity or range of youthful deviant acts. Instead, it is more a question of what types of issues the researcher wishes to study and what types of populations (i.e., school, street, or incarcerated) of adolescents are included.

TRENDS IN JUVENILE DELINQUENCY

Studying delinquency over time is complicated by the fact that definitions of offending, as well as social policy regarding adolescent offenders, have been quite variable. In some time periods, societies were quite resistant to the idea of criminalizing adolescent behavior and would often go to quite elaborate lengths to ignore or downplay juvenile delinquency. The "decriminalization" of delinquency in the 1970s and early 1980s, for example, saw the widespread adoption of policies designed to divert delinquents away from the courts and prisons. In contrast, more recent years, including much of the 1990s and early 21st century, have seen a hardening of attitudes and policies such that adolescents are increasingly subject to the criminal justice system.

In light of such historical variability, criminologists interested in trends over time typically look to homicide data, given that homicides are particularly visible to law enforcement (i.e., very high rates of detection and reporting), police have little to no discretion for such offenses, and definitions are very consistent over time and across jurisdictions. Drawing upon such data, most criminologists agree that while the overall pattern of offending across the age span and the general concentration among adolescents has not changed appreciably, there have been some notable historical shifts in the overall volume of delinquency. Contrary to much theory and expectation, the decades following World War II (1939–1945), a period of considerable prosperity, saw significant increases in rates of delinquency. After this, rates stabilized through the early 1980s, but then dramatically increased through the early 1990s.

Drawing upon data from the Federal Bureau of Investigation's supplemental homicide report (2003), the number of juvenile homicides rose from approximately 1,000 in 1984 to more than 3,600 10 years later. Equally interesting, as quickly as juvenile homicides increased, they decreased. After peaking in 1994, homicides showed steady and consistent declines through the late 1990s and early 21st century, such that homicides, as well as rates of juvenile violence more generally, are as low at the start of the 21st century as they have been at any point since the early 1960s. While a variety of explanations have been offered, there is considerable consensus that the homicide boom and bust was a simple function of the rapid spread of crack cocaine and the resulting "arms-race" among inner city males and the equally rapid decline in the popularity of crack, stabilization of the drug trade matured, and large scale incarceration.

THE DEMOGRAPHY OF DELINQUENCY

Any understanding of the nature of delinquency begins with accounting how much offending actually takes place and how specific groups are differentially involved. In the first respect, minor delinquency is quite common in that almost everyone commits *some* delinquent act at some point in their adolescent years. Here, most surveys show that upward of 90% of teenagers have hit someone, stole something, damaged or destroyed property, or used drugs and alcohol at some point during their adolescent years. At the same time, serious delinquency is very rare given that very few adolescents engage in serious, frequent offending and even fewer engage in the most serious forms of violence. In light of this, it is useful to consider the general demography of offending, and focus attention on both the more mundane and more serious types of delinquency.

One important aspect of delinquency is the pervasive and enduring sex differences in offending. Considering homicide, males accounted for 92% of all offenders, outnumbering females by more than 9 to 1. Sex differences are also prevalent among more common types of delinquency. For example, Howard Snyder and Melissa Sickmund (2006) report that almost twice as many males (11%) as females (6%) belong to gangs, have vandalized property (47% compared to 27%), or sold drugs (19% compared to 12%). Differences are even smaller with respect to theft and assault. Overall, males are much more likely to be involved in delinquency than females and such differences are evident across a range of contexts and time periods.

A second demographic feature of delinquency, and arguably more contentious, is racial differences in delinquency. On one hand, researchers have amassed considerable data showing significant race differences in offending. On the other hand, critics argue that data showing race differences comes from agencies, notably police and courts, that have historically shown significant biases against racial minorities. Starting with homicide, African-American adolescents account for one-half of all juvenile homicides, even though they account for only 14–15% of the adolescent population. White adolescents account for the vast majority of other teen homicides

(47%), whereas Asian Pacific Islanders and American Indians account for 1% and 2%, respectively. For other offenses and delinquency measured through self-report, race differences are less extreme and considerably more variable. For example, while twice as many African-American adolescents have been suspended from school when compared with Whites (56% compared to 28%) and African-American adolescents have higher rates of assault (36% compared to 25%), differences for vandalism, theft, and drug dealing are more similar across each race (Snyder & Sickmund, 2006). In general, there is a consensus that African Americans are slightly overrepresented among adolescent offenders.

A final feature of the demography of delinquency is the age patterning of offending within adolescence. The distribution of delinquency through adolescence is important to consider, especially because certain crimes (e.g., drinking and *soft* drug use) may be gateways to more serious offending (e.g., *hard* drug use or robbery). For homicide, it is clear that offending increases with advancing age through adolescence. For example, those 13 years of age have rates of offending of only 0.32 per 100,000. This increases to 1.16 for 14-year-olds, 2.29 for 15-year-olds, and 4.08 for 16-year-olds. By age 17, the rate of homicide offending is 28 times greater (9.00) than that of 13-year-olds. Other arrest data such as that from the Federal Bureau of Investigation (2003) suggest a similar pattern, whereas self-report data, again, is more varied. That offending increases through adolescence features prominently in life course and developmental theories, as well as criminal justice policy, although its exact implications are still being debated.

DELINQUENCY AS BEHAVIOR: UNDERSTANDING THE FACTS

Although there are a number of important theories and a long tradition of testing theories of crime, several scholars have noted that all theories seem to have reasonable empirical support. When such claims are subjected to scrutiny, it quickly becomes apparent that most researchers typically attribute particular variables to particular theories, find some statistically significant association, and conclude empirical support for the theory. Unfortunately, such conclusions lack credence when one recognizes that variables have various meanings and interpretations, and that the broad expectations of any given theory require more than just the presence of a particular association.

Consider, for example, the almost universally observed finding that delinquents have delinquent peers. Many researchers find this association, attribute it to the social learning camp (i.e., that young people learn deviant behavior from their deviant peers), and then conclude strong support for social learning perspectives (Akers, 1998).

Juvenile Offenders. © BILL GENTILE/CORBIS.

The skeptic, however, might suggest that (a) having delinquent peers implies nothing about their capacity or activity in teaching criminal values and behavior; (b) friendship patterns are not randomly occurring, and individuals select and are selected into peer groups based on characteristics such as shared values and shared interests (e.g., delinquency); and (c) reporting delinquent peers may itself be reporting delinquency given that much delinquency occurs in groups (i.e., co-offending). Acknowledging such considerations casts doubt on delinquent peers as evidence of social learning and thus challenges its contribution to the overall support of social learning perspectives.

An alternative way of thinking about empirical support is to consider the key aspects of criminal behavior and ask what theory (or theories) best accounts for such facts. Several features are noteworthy. First, crime as an activity requires very little time commitment. Even among the most extensive and hard-core delinquents, the vast majority of their time is spent on noncriminal endeavors. Second, there appears to be virtually no specialization in offending. The teenage robber is also stealing property from houses and cars, illegally selling guns,

skipping school, getting into fights, vandalizing property, and using drugs and alcohol. Third, most juvenile delinquency involves co-offending; that is, it tends to be a group activity. Fourth, the social networks and peer relationships of delinquents are quite porous and ephemeral. The peer group of any adolescent, let alone a delinquent one, includes a mix of delinquent and nondelinquents; and adolescents, including delinquent ones, change peer groups with considerable frequency.

Fifth, efforts to actually show a well-defined learning process among delinquents have generally failed. Delinquents clearly have delinquent associations, but the role of delinquent attitudes and beliefs is much more equivocal (i.e., delinquents show the same conventional values as nondelinquents) and little evidence exists of tutelage and reinforcement. (In contrast, evidence abounds about the intensive and extensive efforts of parents and teachers to promote prosocial, antideviant behavior.) Sixth, delinquents clearly have weaker connections to conventional institutions. Their popularity among peers is questionable, they are less connected to and involved with parents and family, they do worse in school, and they are less involved in a range of school-like activities. The latter point may be particularly important because it suggests that delinquents may be poor learners when compared to nondelinquents. This raises an interesting dilemma for theories that emphasize positive socialization—given that successful socialization of any sort requires that one is a capable learner. Finally, delinquents clearly have a wide array of social strain (e.g., poor peer relationships, greater family conflict, and poor grades). Yet the link between strain and delinquency is heavily contingent upon propensity to engage in delinquency. In other words, not all people react equally to strain and those who already show delinquent tendencies are most likely to adopt delinquent "solutions" to their problems.

The patterns described above are not in serious dispute and, in the aggregate, highlight three important features of criminal behavior. First, it is difficult to see evidence of significant and successful socialization among deviant youth. Clearly, deviance can be and is a group activity among adolescents. But this does not imply that deviance is a product of successful socialization of any meaningful sort. The social networks of delinquent adolescents are not necessarily conducive to strong and effective socialization, particularly socialization that would (presumably) counteract the influence of all prior and all peripheral socialization. The evidence of extensive or profound value commitments among delinquents is sparse at best and connotes the idea of a failure of socialization rather than a success.

Second, crime and delinquency appear as remarkably simplistic activities with nothing of substance to learn. Any random individual already possesses most, if not all, the necessary skill for virtually any delinquent endeavor. Moreover, even if someone did need to learn some type of crime, it would not be difficult or require extensive tutelage. The (fragile) networks of adolescents are characterized by a generality of deviance that is a mixed bag of petty violence, pecuniary offenses, destructive acts, and substance abuse.

Third, there are important differences in the social relationships of delinquents and nondelinquents, with respect to families, peers, friends, and authorities. Adolescents are not really delinquents as much as they drift between delinquent and conforming behavior, between delinquent and conventional social groups. Indeed, the fact that this encyclopedia entry repeatedly refers to *delinquents* throughout the discussion should be seen as nothing more than a shorthand convention for the more accurate designation of "those more likely to engage in delinquency."

So what perspective best accounts for the key features of crime in the early stages of the life course? For several decades, some criminologists argued that traditional theories of crime have asked the wrong question in focusing on why people commit crimes. Instead, control perspectives begin with the question of why people *do not* commit crimes. Here, the starting point is that the intrinsic benefits of crime are self-evident and widely applicable. Violence is a response to social disputes that does not require patience or understanding. Theft provides immediate rewards to the offender without the requirement of hard work, significant effort, or long-term investment. Drugs and alcohol provide quick highs without concern for the future or its effects on others. Indeed, a wide variety of crime is thrilling and exciting (and the prospect of getting caught or besting police and law enforcement merely add to the thrill) for some youths. Control theories emphasize the idea that crime is *not* like all other behaviors in that it is intrinsically a self-interested activity, requires no real skill or planning, and provides immediate benefits to the offender.

In offering such a view of crime, control perspectives focus on the idea of restraint. If crime is intrinsically rewarding and is so for most young people, what is it that restrains people from engaging in it? Travis Hirschi's (1969) seminal contribution, *Causes of Delinquency*, answered this question by emphasizing the power of the social bond. Specifically, Hirschi argued that if crime is really just self-interested behavior, individuals are free(er) to act in such a way when their bond to society is weak or broken. Social bonds, according to Hirschi, involve the following:

- Attachment to others in society that makes one sensitive to the opinions and values of others;

- Commitment to conventional activities that involve investments of time, energy, and self that ultimately are *costs* in the face of engagement of crime;

- Involvement in conventional activities that make a person simply too busy to engage in deviant behavior;

- Belief in the values of the dominant society, values that are common within the society or common within the social group.

The key distinguishing feature of control perspectives is that they assume social values are inherently against crime, that no social group could be organized by criminal values (and thus values such as the Hollywood stereotype of organized crime as hierarchical, structured, and efficient is quite literally fiction), and that socialization, particularly effective socialization, will be inherently against crime. Thus, individuals who violate the values of society do so because they are not committed to others and their values and are thus more immune from social influence. It is this that frees them to act in their own self-interest. From this perspective, crime is neither forced nor required and thus it is really a question of more or less likelihood of deviant behavior. For the contemporary adolescent, the issue is how attached, committed, and involved they are with school, peers, and families and how these determine their likely involvement in crime.

Future work needs to recognize the behavioral reality of adolescent delinquency and consider the ways in which it is connected to early life experiences and later life outcomes. While such work is clearly on the table, the vitality of future research will depend heavily on the degree to which theory and research recognize the capacity of individuals as actors at any given age and how they select and accept or avoid and challenge the socialization efforts of those around them. Ironically, scholars have for more than a century viewed adolescence as a period of "storm and stress" (Hall, 1904), but have been remarkably reticent to consider the full implications of this. From a control theory perspective, any understanding of offending in adolescence and over the life course would benefit greatly from a questioning of the power and practice of socialization and the recognition of individuals as constructors of both their life circumstances and their activities.

SEE ALSO Volume 1: *Aggression, Childhood and Adolescence; Juvenile Justice System; School Violence; Theories of Deviance;* Volume 2: *Crime, Criminal Activity in Adulthood; Incarceration, Adulthood.*

BIBLIOGRAPHY

Agnew, R., Brezina, T., Wright, J., & Cullen, F. (2002) Strain, personality traits, and delinquency: Extending general strain theory. *Criminology, 40*(1), 43–72.

Akers, R. (1964). Socioeconomic status and delinquent behavior: A retest. *Journal of Research in Crime and Delinquency, 1*(1), 38–46.

Akers, R. (1998). *Social learning and social structure: A general theory of crime and deviance.* Boston: Northeastern University Press.

Bernard, T. (1990). Twenty years of testing theories: What have we learned and why? *Journal of Research in Crime and Delinquency, 27*(4), 325–347.

Elliott, D., Huizinga, D., & Ageton, S. (1985). *Explaining delinquency and drug use.* Beverly Hills, CA: Sage.

Federal Bureau of Investigation. (2003). *Crime in the United States, 2002: Uniform crime reports.* Washington, DC: U.S. Government Printing Office.

Greenberg, D. (1985). Age, crime, and social explanation. *American Journal of Sociology, 91*(1), 1–21.

Hall, G. (1904). *Adolescence.* New York: Appleton.

Hindelang, M., Hirschi, T., & Weis, J. (1981). *Measuring delinquency.* Beverly Hills, CA: Sage.

Hirschi, T. (1969). *Causes of delinquency.* Berkeley: University of California Press.

Hirschi, T., & Gottfredson, M. (1983). Age and the explanation of crime. *American Journal of Sociology, 89*(3), 552–584.

Sampson, R., & Laub, J. (1993). *Crime in the making: Pathways and turning points through life.* Cambridge, MA: Harvard University Press.

Shaw, C., & McKay, H. (1969). *Juvenile delinquency and urban areas.* Chicago: University of Chicago Press.

Snyder, H., & Sickmund, M. (2006). *Juvenile offenders and victims: 2006 national report.* Washington, DC: National Center for Juvenile Justice. Retrieved May 25, 2008, from http://ojjdp.ncjrs.org/ojstatbb/nr2006/downloads/NR2006.pdf

Steffensmeier, D., Allan, E., Harer, M., & Streifel, C. (1989). Age and the distribution of crime. *American Journal of Sociology, 94*(4), 803–831.

U.S. Department of Justice, Federal Bureau of Investigation. (2006, September). *Crime in the United States, 2005,* Table 38. Retrieved June 12, 2008, from http://www.fbi.gov/ucr/05cius

Williams, J. R., & Gold, M. (1972). From delinquent behavior to official delinquency. *Social Problems, 20*(2), 209–229.

Wolfgang, M., Figlio, R., & Sellin, T. (1972). *Delinquency in a birth cohort.* Chicago: University of Chicago Press.

Ross Macmillan

CRIME AND VICTIMIZATION, CHILDHOOD AND ADOLESCENCE

SEE Volume 1: *Child Abuse;* Volume 2: *Crime and Victimization, Adulthood; Domestic Violence.*

CULTURAL CAPITAL

The noted French sociologist Pierre Bourdieu first developed the sociological concept of cultural capital (Bourdieu, 1977, 1984, 1986, 1996; Bourdieu & Passeron, 1990; Swartz, 1997). The concept, in its most simplistic form, encompasses the distinctive sets of social class-based cultural and behavioral attributes (e.g., disposition, style, worldview, knowledge), material possessions (e.g., music, art work), and institutional representations (e.g., educational degrees, honors, other credentials), which tend to distinguish individuals of dominant class (or middle-to-upper class) background from those of dominated (or lower class) backgrounds. At the heart of the concept of cultural capital is the more widely familiar notion of economic capital (i.e., money, property, business ownership, other assets). In industrialist and post-industrialist capitalist societies, economic capital is routinely exchanged for material possessions, services, and property. In addition, individuals, and indeed, groups, have different levels of economic capital at their disposal in any given historical period. Differing levels of economic capital at the group level are often used to define distinct social classes. Bourdieu's concept of cultural capital, along with other forms of noneconomic capital (social and symbolic), extends the more fundamental notion of *capital* and *exchange* to other less readily recognizable areas of social relations and everyday life.

Bourdieu (1986) classified cultural capital into three broad types: (a) embodied forms—such as dispositions, tastes, aspirations, and knowledge; (b) objectified forms—such as the material objects representative of *high* culture; and (c) institutionalized forms—including formalized symbols recognizing competence or distinction (e.g., degrees, academic honors, memberships). Although cultural and economic capital are highly associated, cultural capital is a much more effective form of capital in legitimating existing social inequality, given that its possession is widely viewed in terms of competence rather than economic wealth. The possession of cultural capital is of great advantage to its holder because it facilitates continued accumulation—those with high levels tend to do well both educationally and professionally—and because increasing levels of cultural capital, exemplified by educational degrees and other formal credentials, can eventually be converted back into economic capital through occupational placement.

However, in Bourdieu's view, the particular sets of dispositions and symbolic possessions that, statistically speaking, distinguish middle and upper class people from those of lower and working class backgrounds are not inherently superior. Rather, their relative value (as capital) is arbitrary, and purely based on their association with social groups of high status and power within society. As such, the concept of cultural capital provides an explanation of the ways in which more subtle, symbolic *possessions*, less visible and recognized as economic capital, function as a form of capital, which can be exchanged for (or converted into) other forms of capital to maintain or elevate one's position in the social hierarchy.

Bourdieu's concept is contextually grounded within a larger theory of social reproduction; that is, a comprehensive model of how existing social inequalities are transmitted from one generation to the next. Integral to understanding the concept of cultural capital are Bourdieu's related concepts of habitus and field, which, respectively, help to explain the formation of cultural capital, and to contextualize its relative power. Habitus represents the internalization of the conditions associated with one's socioeconomic position in society; that is, it constitutes the mental constructs and behavioral inclinations that have developed across one's formative life experiences, shaped by one's position in the social hierarchy. As such, habitus represents the master patterns of dispositions, tastes, and perspectives, which have developed across time, shaped by one's cumulative experiences.

Field, for Bourdieu, represents the context in which individuals compete (although not necessarily consciously) for resources, power, and position within the social hierarchy. Features of habitus, depending on one's set of life experiences, can become cultural capital—in that they place the individual at a relative advantage—within specific fields of action. For Bourdieu, however, those in higher social positions are much more likely to be relatively advantaged across many different fields of competition, because they have been socialized into the dominant culture and therefore have a better understanding of the *rules of the game,* which have been shaped historically by dominant class interests.

CULTURAL CAPITAL AND SOCIAL REPRODUCTION IN MODERN SOCIETY

In modern, developed societies, Bourdieu argues, cultural capital has come to play a pivotal role in reproducing social inequities, whereas in feudal societies, or even nondemocratic forms of capitalist society found in past centuries, economic capital, possessed by the few, was sufficient to control most, if not all, sectors of societal operations and social relations. Status ascription, the overt transmission of both economic capital and social status from one generation to the next, was an accepted fact, just as royalty is passed on from a king to his progeny. In modern meritocratic societies, however, although economic capital is still passed from one generation to the next, it is much less acceptable for children to simply inherit social status (e.g., occupational positions, political posts) from their parents. Because cultural/behavioral attributes tend to closely correspond to one's socio-economic position in society, individuals originating from higher social positions tend to value and recognize these same attributes—attributes more commonly exhibited within their higher status social surroundings—in others. As such, the possession of these symbolic attributes, when recognized by others in similarly high social positions, functions as a cue, of sorts, to signify one's *appropriateness* for group inclusion.

This *cultural* recognition, however, often occurs at a subconscious level and is usually interpreted (by the individuals, themselves) as competence, intelligence, language facility, and natural ability. This subconscious process of mutual recognition provides relatively continuous access to other forms of capital readily available within higher social strata. Moreover, the *culturally specific* knowledge of the field of competition (i.e., the rules of the game), and associated level of comfort acting within this field, enables higher status individuals, who possess these attributes, to successfully navigate the field and access other forms of capital, whether cultural (e.g., a degree from a prestigious university), social (inclusion into powerful social networks), economic (high paying employment), or symbolic (general recognition of talent or competence). It is the possession and activation (Lareau & Horvat, 1999) of such attributes and field-specific knowledge that together constitute cultural capital.

With the expansion of the education sector across the past century in modern industrialist societies, schooling and educational credentialing have developed into an increasingly important and, theoretically, merit-based route to social mobility. Bourdieu argued that schools and educational institutions, more generally, now constitute a key field of play in which individuals, and groups, must vie to secure or elevate their social status. In Bourdieu's view, educational institutions, from preschool to postgraduate study, have been structured historically in close correspondence with existing social inequalities. Similarly, the orientations and expectations of teachers and administrators have been shaped by experiences tied to their own social class backgrounds—generally middle and upper middle class. As such, the subtle process of class-based cultural recognition (and lack of recognition) continuously occurs within schools and classrooms between teachers and students, administrators and parents.

Children from middle- and upper-class backgrounds, whose cultural dispositions and experience-based attributes tend to align closely with those of their teachers (i.e., children who possess cultural capital), are much more likely to experience success in school. Their levels of *requisite* knowledge and their behavioral patterns/styles, possessed by virtue of their out-of-school experiences (which are in turn shaped by their position in the social class hierarchy) are rewarded within schools in the form of good grades, positive praise, and heightened academic opportunities (e.g., college-level tracking, admission to more selective schools). Moreover, these cultural/behavioral attributes are legitimated by schools, because their educational success is *christened* as the logical consequence of hard work and/or natural ability.

In contrast, children from lower socioeconomic backgrounds, who have little access to *cultural capital*—because they have developed significantly different habitus across their lives than have their more middle- and upper class peers—are therefore *disadvantaged* in school. They, unlike their more advantaged peers, do not share the class-based cultural dispositions of teachers and administrators, and therefore, are less likely to succeed in school. As such, they are compelled to operate within a *foreign* system, one that expects them to have already mastered certain linguistic and behavioral patterns and associated social and cognitive skills, which are, in turn, directly implicated in academic success. These students experience the *cultural* misalignment between home (social class) and school in the forms of academic failure, disciplinary action, and frustration. Schools, therefore, may inadvertently work to reproduce existing social inequities. This represents one possible explanation of why middle- and upper-class children are much more likely to successfully complete each stage of schooling (Gambetta, 1997), enter postsecondary institutions of learning and obtain degrees (Horvat, 2001), and secure better paid professional employment, thus maintaining their relatively advantaged position in society.

CULTURAL CAPITAL'S TREATMENT IN RESEARCH

The concept of cultural capital offers a more nuanced understanding of the ways in which class-based inequalities are reproduced, at the level of the individual, from generation to generation in modern differentiated societies. Embedded within a class-based theoretical framework, it provides an alternative perspective on the subtle processes by which individuals, shaped by their class-specific environments, differentially interact with societal institutions in statistically predictable ways to reproduce their objective conditions of existence. However, since its initial conception and development, the concept of cultural capital has been interpreted and *operationalized* in manifold ways in social science research.

Unfortunately, the concept has often been defined for the purpose of measurement and hypothesis testing in ways that have not been faithful to Bourdieu's larger theoretical project (Lareau & Weininger, 2003). Quantitative studies, which have primarily used secondary survey–based data sets not designed specifically to measure the concept of cultural capital, have been able to only partially capture the concept's full meaning (for a complete review of this literature, see Lareau & Weininger, 2003). Most often, these studies, which primarily examine the impact of cultural capital on educational outcomes, define the concept in terms of leisure activities associated with *high brow* culture, such as interest in and/or visits to theaters and museums and knowledge about the classical arts (De Graaf, De Graaf, & Kraaykamp, 2000; DiMaggio, 1982; DiMaggio & Mohr, 1985; Kalmijn & Kraaykamp, 1996; Katsillis & Robinson, 1990). Some studies have extended the concept to more mundane activities, such as library visits, parent/child reading behaviors, and educational resources in the home (De Graaf, 1986; Kalmijn & Kraaykamp, 1996; Roscigno & Ainsworth-

Darnell, 1999; Sullivan, 2001). Some researchers have defined the concept in opposition to academic ability or technical skills (DiMaggio, 1982; DiMaggio & Mohr, 1985), whereas others have, to varying extents, incorporated such skill levels into the concept (Farkas, Grobe, Sheehan, & Shuan, 1990; Sullivan, 2001).

The results of such studies have tended to support the relationship between cultural capital and educational achievement and attainment, although the size (or meaningfulness) of these relationships has been questioned (Kingston, 2002). However, the possession of cultural capital, so defined, has not always been as tightly linked with social class as Bourdieu's theory of social reproduction would predict. In fact, some researchers have found that within the educational context, the possession of cultural capital tends to yield greater benefits to individuals from lower socioeconomic backgrounds than to those in higher social class positions (Aschaffenburg & Maas, 1997; DiMaggio, 1982; Kalmijn & Kraaykamp, 1996). That is, the possession of cultural capital may provide people from lower class backgrounds with the means to attain *social mobility* through schooling (DiMaggio, 1982). Those richer in economic capital may have less need of cultural capital (and the academic success that it brings) to retain their social position into adulthood.

Qualitative studies of cultural capital have proved a bit more enlightening than survey-based research, because they have focused on discovering the processes by which parents interact with children and schools, charting the variety of ways in which class-based differences in knowledge, styles, and comfort level influence both family-to-school relationships and child development (Lareau, 2000, 2003; Lareau & Horvat, 1999). These studies have contributed to the further development of the concept by providing more concrete examples of actual class-based differences in practices and dispositions (i.e., manifestations of habitus) at the ground level within specific contexts. However, qualitative work, arguably, has not been able to increase the operational clarity of the concept, nor has it improved its measurability. As a result, the concept remains largely abstract and subject to different interpretations.

Limitations of the Concept Most research on cultural capital that has been produced to date, including that of Bourdieu himself, has neglected to examine the ways in which cultural capital is acquired by some proportion of the lower class (Carter, 2003). The concept, therefore, has not been able to explain how some people from socially disadvantaged backgrounds are able to succeed in school and take advantage of educational opportunities to improve their social status as adults. Bourdieu was specific in defining cultural capital, and the habitus that

it engenders, in terms of statistical occurrences, a clear acknowledgement that some variation in the possession of cultural capital does exist between individuals within, as well as between, social class positions. However, no clear explanation exists for how this occurs, especially with regard to variation in cultural capital evident within lower and working class strata. If cultural capital accrues in relation to one's objective, class-based living conditions, then how do some people from socially disadvantaged positions develop the requisite cultural/behavioral attributes to succeed in school and to achieve social mobility? Have they simply been able to supersede the effects of the objective conditions in which they were socialized, and if so, how? Have they learned the cultural dispositions in school, which according to Bourdieu's theory, expects but does not explicitly teach these dispositions? These theoretical problems continue to challenge the conceptualization and use of cultural capital in the research.

FUTURE DIRECTIONS

The concept of cultural capital continues to provide a rich foundation for research into the widely documented statistical association between social class background and educational attainment. Future research will need to further explore the ways in which families advantage their children in educational and other contexts; the degree to which differences in practices, expectations, and aspirations are based in social class positioning; and how institutions respond to such differences. Moreover, as Bourdieu's overall theoretical project features the school so prominently in the intergenerational transmission of social status, it is essential that more research be conducted within schools and classrooms to examine the extent to which teachers implicitly recognize and reward existing stores of cultural capital as academic competence in students from more advantaged backgrounds. Studies will need to carefully examine the practices of schools and teachers to distinguish between the explicit instruction of skills and implicit expectations of personal styles and habits. Finally, more research will need to focus on differences in educational and occupational choices made by youth at key institutional junctures to determine the degree to which such choices are made with conscious and strategic intent (as in a rational choice model) or are reflections of more subtle and generalized class-based dispositions derived from the subconscious schema of what constitutes one's appropriate place in the social structure.

SEE ALSO Volume 1: *Coleman, James; Human Capital; Intergenerational Closure; Racial Inequality in Education; School Culture; Social Capital; Socioeconomic Inequality in Education.*

BIBLIOGRAPHY

Aschaffenburg, K., & Maas, I. (1997). Cultural and educational careers: The dynamic of social reproduction. *American Sociological Review, 62*, 573–587.

Bourdieu, P. (1977). Cultural reproduction and social reproduction. In J. Karabel & A. H. Halsey (Eds.), *Power and ideology in education* (pp. 487-511). New York: Oxford University Press.

Bourdieu, P. (1984). *Distinction: A social critique of the judgment of taste.* London: Routledge & Kegan Paul.

Bourdieu, P. (1986).The forms of capital. In J. G. Richardson (Ed.), *Handbook of theory and research for the sociology of education* (pp. 241-258). New York: Greenwood Press.

Bourdieu, P. (1996). *The state nobility: Elite schools in the field of power.* Stanford, CA: Stanford University Press.

Bourdieu, P., & Passeron, J. (1990). *Reproduction in education, society and culture* (2nd ed.). London: Sage Publications.

Carter, P. L. (2003). "Black" cultural capital, status positioning, and schooling conflicts for low-income African American youth. *Social Problems, 50*, 136–155.

De Graaf, P. M. (1986). The impact of financial and cultural resources on educational attainment in the Netherlands. *Sociology of Education, 59*, 237–246.

De Graaf, N. D., De Graaf, P. M., & Kraaykamp, G. (2000). Parental cultural capital and educational attainment in the Netherlands: A refinement of the cultural capital perspective. *Sociology of Education, 73*, 92–111.

DiMaggio, P. (1982). Cultural capital and school success: The impact of status culture participation on the grades of U.S. high school students. *American Sociological Review, 47*, 189–201.

DiMaggio, P., & Mohr, J. (1985). Cultural capital, educational attainment, and marital selection. *American Journal of Sociology, 90*, 1231–1261.

Farkas, G., Grobe, R., Sheehan, D., & Shuan, Y. (1990). Cultural resources and school success: Gender, ethnicity, and poverty groups within an urban district. *American Sociological Review, 55*, 127–142.

Gambetta, D. (1987). *Were they pushed or did they jump? Individual decision mechanisms in education.* New York: Cambridge University Press.

Horvat, E. M. (2001). Understanding equity and access in higher education: The potential contribution of Pierre Bourdieu. In J. C. Smart (Ed.), *Higher education: Handbook of theory and research* (pp. 195-238). New York: Agathon Press.

Kalmijn, M., & Kraaykamp, G. (1996). Race, cultural capital, and schooling: An analysis of trends in the United States. *Sociology of Education, 69*, 22–34.

Katsillis, J., & Robinson, R. (1990). Cultural capital, student achievement, and educational reproduction: The case of Greece. *American Sociological Review, 55*, 270–279.

Kingston, P. W. (2001). The unfulfilled promise of cultural capital theory. *Sociology of Education, 74*, 88–99.

Lareau, A. (2000). *Home advantage: Social class and parental intervention in elementary education* (updated ed.) Lanham, MD: Rowman & Littlefield.

Lareau, A. (2003). *Unequal childhoods: Class, race, and family life.* Berkeley: University of California Press.

Lareau, A., & Horvat, E. M. (1999). Moments of social inclusion and exclusion: Race, class, and cultural capital in family-school relationships. *Sociology of Education, 72*, 37–53.

Lareau, A., & Weininger, E. B. (2003). Cultural capital in educational research: A critical assessment. *Theory and Society, 32*, 567–606.

Roscigno, V. J., & Ainsworth-Darnell, J. (1999). Race, cultural capital, and educational resources: Persistent inequalities and achievement returns. *Sociology of Education, 72*, 158–178.

Sullivan, A. (2001). Cultural capital and educational attainment. *Sociology, 35*, 893–912.

Swartz, D. (1997). *Culture and power.* Chicago: University of Chicago Press.

Dan H. Weinles
Erin McNamara Horvat

CULTURAL IMAGES, CHILDHOOD AND ADOLESCENCE

Childhood is a key stage in the life course, and cultural and historical approaches to childhood have moved in two somewhat contradictory directions. One approach has focused on children's lived experience—on children's voices, perceptions, behavior, and experiences. This line of investigation has focused on the highly specific circumstances in which children have grown up, emphasizing differences rooted in class, ethnicity, gender, geographical region, and historical era. This approach treats children as agents who play an active role in their own social, cognitive, physical, and moral development, constructing their own cultural and social identities, and reshaping cultural sensibilities.

The other approach, derived largely from cultural studies, focuses on childhood as a cultural category that reflects adult nostalgia, anxieties, expectations, and desires, and that is imposed on children. This approach, which is less concerned with the lived experience of individual children than with cultural symbolism and representation, examines the symbolic and psychological meanings that adults attach to the child. This cultural approach explores the shifting divide between adults and children, the representation of childhood in literary and visual culture and social and political discourse, and how artists, educators, psychologists, physicians, and poets have played a crucial role in defining childhood, in essentialist terms, as a sacred, symbolic category, defined in opposition to adulthood yet embodying adult preoccupations with asexual innocence, vulnerability, and spontaneity.

Ambivalence has characterized cultural images of childhood. This life stage has often been viewed positively, as a time of innocence, freedom, creativity, and imagination; as such, it has offered a critical vantage point for apprising the pretensions, blindnesses, and weaknesses of adults. But images of childhood also contained explicitly or implicitly negative elements. Children have been regarded as naïve, incompetent, irrational, emotional, dependent, and vulnerable. Children, from this perspective, required protection and sheltering from the adult world, and growing up could be seen in negative terms, as corruption. Bad children could easily be regarded not simply as naughty or mischievous but as precociously or inherently wicked, or even demonic.

Superficially, it would seem that these two approaches could scarcely be more divergent, with one approach dealing with "real" children and the other with adult representations and conceptions of childhood. However, the two approaches are only superficially contradictory. Cultural conceptions of childhood, age, and gender inevitably color observations of children's behavior and shape the institutions and practices that structure children's lives.

In the 17th-century American colonies, childhood was generally regarded as a life stage that should be passed through as quickly as possible. The New England Puritans regarded children as sinful, even bestial creatures, reflecting a religious belief in original sin and humanity's innate depravity. As the Reverend Cotton Mather put it: "Are they Young? Yet the Devil has been with them already. . . . They go astray as soon as they are born. They no sooner step than they stray, they no sooner lisp than they ly." The Puritans regarded crawling as animal-like, toys as devilish, and play as lacking in value. In fact, however, the Puritans were obsessed with children, regarding the young as a trust from God and the key to creating a godly society. The Puritan emphasis on education reflected this preoccupation with childhood.

By the time of the American Revolution in the late 18th century, a new conception of childhood was arising. Urban, middle-class parents began to regard their offspring as innocent, malleable, and fragile creatures who needed to be sheltered from the adult world. These parents kept their children at home much longer than in the past. Instead of putting them out to work, they put them in school instead. Childhood, according to this view, was the opposite of adulthood. At a time when the preindustrial social order was breaking down, and increasingly commercial and urban patterns of life were appearing, children were an important symbol; they were regarded as asexual and pure and were associated with nature and organic wholeness.

It is important to note, however, that at the same time that urban, middle-class parents romanticized their children as little angels, working-class and farm children became more valuable economically than ever before. Although reformers (later known as "child-savers") romanticized the bootblacks, newsboys, match girls, and "street Arabs" found on urban streets, there was a tendency to sharply differentiate between the innocence of middle-class children and the precocity and incipient corruption of these street children (who were part of what was called "the dangerous classes").

Toward the end of the 19th century, new images of childhood appeared. The emerging popular culture industry—literature, photography, and the movies—played a particularly important role in disseminating images of heartwarming infants, wide-eyed waifs hungering for a home, curly-haired cherubs, and savvy street urchins. Among images of girlhood, there was an assortment of Cinderellas, Pollyannas, princesses, tomboys, bobby-soxers, and, later in time, prepubescent Lolitas. Among images of boyhood were mischievous scamps, rambunctious ragamuffins, little rascals, angry and alienated adolescents, and, more recently, a parade of pranksters, burnouts, stoners, and homeboys.

These images, in turn, reflected shifts in public anxieties, aspirations, fears, and fantasies. At the beginning of the 20th century, the Victorian image of the child as a little angel gave way to a new ideal. The new standard was spunky, sassy, naughty, and cute, a response to public worries that boyhood, in particular, was becoming too constrained and female-dominated. The earliest comics promote the image of the adorably willful child. First came the Yellow Kid, among the first American comic strip characters and the prototype for Dennis the Menace, Bart Simpson, and other gap-toothed rascals and troublemakers. Then came Buster Brown, the little rich kid with a blond pageboy haircut who was always getting into mischief, but a milder brand than the Yellow Kid. Mary Pickford, the movies' first influential screen child (despite the fact that she was an adult), embodied the new ideal of childhood. She was naughty and coquettish but also innocent and sweet. She was followed by the first true child star, 6-year-old Jackie Coogan, who reinforced the image of the ideal child as rambunctious, excitable, and energetic.

These positive images of childhood contributed to making the early 20th century a period of reform. Child-savers embarked on a crusade to universalize a middle-class notion of childhood, by launching "pure milk" campaigns, enacting compulsory school attendance laws, establishing "mothers' pensions" that allowed single mothers to keep their children at home (rather than institutionalizing them), and restricting and ultimately outlawing child labor.

The Great Depression brought many new images of childhood. The Depression sparked fears of a lost generation of children, who might fall into crime and be susceptible to demagogues. One popular Depression-era image of childhood was of "angels with dirty faces." These included the Little Rascals and the Dead End Kids, the urban offspring of Tom Sawyer and Huck Finn.

It was during the Depression that Walt Disney became synonymous with children's movies and his films developed their trademark traits. The Disney studio self-consciously reworked fairy tales, myths, and classic children's stories, erasing elements that it considered inappropriate for kids and making the stories more didactic and moralistic. Thus, for Pinocchio (1940) to become a real boy, he must prove himself "brave, truthful, and unselfish." *Snow White and the Seven Dwarfs* (1937) emphasized proper gender behavior. Foreshadowing later Disney films, the heroine finds fulfillment in housework and makes marriage her life's ultimate goal.

The most popular child star of the 1930s was Shirley Temple, who topped the box office every year from 1935 to 1938. She was America's little darling, tap-dancing and singing through the Depression in 50 short films and features by the time she was 18. Part of her attraction was her cuteness, charm, dimpled cheeks, and bouncing curls. During a time when the nation's self-confidence had been battered, she provided reassurance. She was adults' ideal girl—independent, even-tempered, flirtatious, and infectiously optimistic. She was undeniably talented: She could sing, dance, act, and melt the heart of the grouchiest sourpuss. Escapist fantasy, too, was part of her appeal. Lacking a mother in almost all of her movies, she was free from domestic constraints. But her appeal went beyond escapism. In many films, she served as a "spiritual healer" who resolved family disputes, bridged class differences, and restored adults' confidence in themselves. Oblivious to class and racial differences, she moved easily between poor and wealthy homes without ever being greedy or envious.

At the end of the decade, a new cinematic stereotype appeared, supplanting even Shirley Temple in popularity. This was the all-American teen, personified by Mickey Rooney and Judy Garland in the Andy Hardy movies, which focused on middle-class teenagers' crushes, infatuations, and humorous and embarrassing mishaps. Such "Kleen Teens" as Deanna Durbin, Roddy McDowell, Dickie Moore, Lana Turner, and Jane Withers provided the caricature that the troubled, misunderstood, and alienated teen characters of 1950s films rebelled against.

During World War II, a highly sentimental view of childhood appeared on the screen, one that bore little resemblance to children's actual wartime experiences. A deeply romanticized view of childhood was apparent in movies such as *National Velvet* (1944), *A Tree Grows in Brooklyn* (1945), and, shortly after the war, *Miracle on 34th Street* (1947).

The 1950s marked the beginning of the end of innocence. Before World War II, the mystery and otherness of childhood had been rarely depicted in popular culture, but the mounting influence of popularized versions of Freudianism resulted in new images of children. Beginning in the 1950s, movie-goers saw depraved children, anticipated in *Mildred Pierce* in 1945, then realized in *The Bad Seed* (1956) and such later films as *Children of the Damned* (1963) and *The Exorcist* (1973). In the 1970s, children were depicted as precocious, miniature adults, as in *Harold and Maude* (1971), and as emotional footballs, such as in *Kramer v. Kramer* (1979). Another key theme was the death of childhood innocence, seen in Martin Scorsese's *Taxi Driver* (1976) and Louis Malle's *Pretty Baby* (1978). Few American films before the 1960s explored children's psychological life or tried to see the world through children's eyes; but many, like the Our Gang comedies and Walt Disney cartoons, tried to depict the world of a child's imagination.

During the 1950s, amused condescension gave way to concern and bewilderment. No longer were portraits of children exclusively images of wholesome naughtiness, mooning boys, and puppy love. Kids increasingly became a vehicle for exploring the confusions of modern society. The cute child was replaced by the evil child, such as Rhoda Penmark, the 8-year-old pigtailed murderer in *The Bad Seed* (1956). The movies also brought to the screen rebellious and alienated adolescents, as well as the world of leather-clad juvenile delinquents, switchblades, and drag racing.

During the 1960s, there were attempts to recapture an image of childhood innocence, evident in such movies as *Mary Poppins* (1964), *Chitty Chitty Bang Bang* (1968), *The Sound of Music* (1965), *40 Pounds of Trouble* (1962), and *Oliver!* (1968). But there were also more psychologically nuanced portraits of childhood. *To Kill a Mockingbird* (1962) viewed racism through the eyes of a child. *The Effects of Gamma Rays on Man-in-the-Moon Marigolds* (1972) portrayed the psychological and emotional abuse of a child.

During the 1950s and 1960s, specific genres of movies were, for the first time, marketed directly to the young, including science fiction films, motorcycle and juvenile delinquent movies, and beach blanket and surfer films, reflecting the large cohort of Baby Boomers who were born between 1945 and 1964. After 1970 the targeting of children and adolescents became much more intensive and self-conscious. One recurrent formula involved a teenage outcast, mocked by her popular, style-setting classmates, who has a makeover and ends up going to the high-school prom with the handsomest boy on the football squad. Yet especially striking were deeply disturbing images of youthful depravity. The movies portrayed kids as demons in such films as *Carrie* (1976) and *The Exorcist* (1973), as prostitutes in *Pretty Baby* (1978) and *Taxi Driver* (1976), and as incipient murderers in *Basketball Diaries* (1995). Portraits of indifferent, uninvolved, unobservant, and uncomprehending teachers and clueless, disconnected, self-deceived, and self-absorbed parents became much more common. The impact of family breakdown and disconnection was a particularly popular theme, apparent in movies as diverse as *WarGames* (1983), *ET: The Extra Terrestrial* (1982), and the *Home Alone* films.

A number of the most memorable recent American films dealing with childhood paint particularly unsettling portraits of the psyche and culture of the young. There was *River's Edge* (1986), based on the true story, which looks at how a group of working-class northern California teens responds after one of the boys murders his girlfriend. It paints a picture of emotionally numbed kids disconnected from the adults around them. Others were *Thirteen* (2003), which shows an adolescent world of body piercing, self-mutilation, tattoos, sexually provocative clothing, underage sex, and casual drug use, and the Columbine-inspired *Elephant* (2003), which portrays high schools as a brutal Darwinian world of cliques, taunting, and tormenting, culminating in violence.

These highly negative representations of childhood reflected, and perhaps reinforced, the tendency at the end of the 20th century to treat children, especially delin-quent children, in more controlling ways. The trend toward trying juvenile offenders in adult courts, the increase in testing in school, and the adoption of school dress codes and zero-tolerance discipline policies reflected a view of the young as a "tribe apart" and a declining tolerance for certain kinds of "childish" behavior that had been celebrated in popular culture a century earlier.

SEE ALSO Volume 1: *Consumption, Childhood and Adolescence; Media and Technology Use, Childhood and Adolescence;* Volume 2: *Cultural Images, Adulthood;* Volume 3: *Cultural Images, Later Life.*

BIBLIOGRAPHY

Cross, G. (2004). *The cute and the cool: Wondrous innocence and modern American children's culture.* New York: Oxford University Press.

Holland, P. (2004). *Picturing childhood: The myth of the child in popular imagery.* London: Tauris.

Jenkins, H. (1998). *The children's culture reader.* New York: New York University Press.

Levander, C. F., & Singley, C. J. (Eds.). (2003). *The American child: A cultural studies reader.* New Brunswick, NJ: Rutgers University Press.

Mintz, S. (2004). *Huck's raft: A history of American childhood.* Cambridge, MA: Belknap Press of Harvard University Press.

Steven Mintz

D

DATA SOURCES, CHILDHOOD AND ADOLESCENCE

This entry contains the following:

I. GENERAL ISSUES

Longitudinal studies follow the same group of people across multiple points in time. Over the last century, social scientists have conducted thousands of longitudinal studies. These studies, which serve as the lifeblood of life course research, are quite diverse in design, execution, focus, and purpose, as evidenced in the list below:

- Some studies are short term, lasting weeks or months, whereas others span years or even decades.

- Some studies are sponsored by the government for public use, whereas others are private enterprises.

- Some studies rely on surveys and require various kinds of statistical analysis, whereas others rely on ethnography and require qualitative analysis.

- Some studies focus on the individual's own characteristics, whereas others focus on the individual's relationships and social settings.

- Some studies are representative of the nation or some other well-defined population, whereas others were created with less systematic sampling frames.

Together, these diverse kinds of longitudinal studies have been used to construct the broad literature on the life course that is reviewed in this encyclopedia. This section focuses on longitudinal studies examining the early stages of the life course: childhood, adolescence, and young adulthood. A complete review of such studies would fill volumes, so a small sample of important, influential studies was selected because, first and foremost, they have been critical resources for life course researchers but also because they represent the great diversity in methods, approaches, and organization in longitudinal research described above.

These studies do more than investigate young people over time. Of equal importance, they all target the main concepts of life course research, including trajectories, transitions, timing, and context.

BIBLIOGRAPHY

Elder, G. H., Jr., & Giele, J. Z. (1998). *Methods of life course research: Qualitative and quantitative approaches.* Thousand Oaks, CA: Sage.

Phelps, E., Furstenberg, F. F., & Colby, A. (Eds.). (2002). *Looking at lives: American longitudinal studies of the 20th century.* New York: Russell Sage.

Robert Crosnoe

II. BALTIMORE STUDY

Most major longitudinal studies of the life course have been either nationally representative or focused on middle-class, white samples. The Baltimore Study, however, offered a window into how life unfolds for young people growing up under disadvantaged circumstances. This distinct focus reflects the study's origins as an evaluation of a prenatal service program for pregnant teenagers.

Funded by a consortium of private foundations as well as by the U.S. Department of Health and Human Services, the Baltimore Study began in 1966 with a sample of 399 pregnant adolescents in obstetric care at a hospital with a high rate of teen births in Baltimore. The majority of these young women entered a comprehensive prenatal and postpartum program aimed at improving the health of children and promoting the life prospects of mothers, with the larger goal of breaking the intergenerational cycle of poverty. A minority of the young women were randomly assigned to the regular hospital program.

Study personnel reviewed medical records and interviewed the mothers and their parents at various intervals. In 1972, the young mothers were systematically compared to classmates who had delayed becoming mothers themselves. The mothers and their children were reinterviewed in 1983. At this point, life history calendars were also completed dating back to the time of the focal child's birth. Mothers, children, and then grandchildren were followed up again in 1987 and 1995 (when the children of the original study were in their mid-20s), with life history calendars again completed.

The sociologist Frank Furstenberg (b. 1940) and his research collaborators have produced a series of books and articles documenting the experiences of the young mothers in the Baltimore Study. His 1976 book, for example, explained how these young women had become mothers at such an early age—largely through accidental rather than motivational means—and how their transi-

tion into motherhood had disrupted their lives, although some were able to return to school after having their children and, consequently, proved to be more resilient in the long run. A 1987 book, written with Jeanne Brooks-Gunn and Philip Morgan, took advantage of the two-decade follow-ups to examine how young mothers' lives and careers had shaped their children's development. For the most part, the children of the young Baltimore mothers had many problems as they grew up. These poor outcomes resulted from a combination of adverse circumstances (e.g., economic stress and family instability) with the timing of such events in children's lives. In other words, not only did they experience hardship, they experienced hardship at particularly vulnerable times in their lives.

The Baltimore Study provides a glimpse into the mechanics of the life course and the links between the life courses of family members (e.g., the linked lives concept in life course theory). Because of its focus on a portion of the American population that was historically overlooked by sociologists, the study's findings are particularly relevant to research in the social sciences.

More information about access to the Baltimore Study can be found on the Web site supported by the Department of Sociology Population Studies Center at the University of Pennsylvania. The original sample included 399 women, and the timeframe of the study is from 1966 through 1995.

BIBLIOGRAPHY

Furstenberg, F F. (1976). *Unplanned parenthood: The social consequences of teenage childbearing.* New York: Free Press.

Furstenberg, F. F., Brooks-Gunn, J., & Morgan, S. P. (1987). *Adolescent mothers in later life.* New York: Cambridge University Press.

University of Pennsylvania, Department of Sociology Population Studies Center. The Baltimore Study. Retrieved April 10, 2008, from http://www.pop.upenn.edu/baltimore/index.html

Robert Crosnoe

III. BENNINGTON WOMEN'S STUDY

From 1935 to 1939, social psychologist Theodore Newcomb (1903–1984) examined the political attitudes and experiences of the student body of Bennington, a private women's college in New Hampshire. The study is widely regarded as a classic.

Although Bennington was a historically liberal environment, its expensive tuition meant that, during the Great Depression (1929–1939), most of its students came from economically advantaged families that were also politically conservative. This mismatch between the sending families and the receiving college made for an interesting dynamic that has great relevance to life course research. In short, young women tended to enter Bennington with politically conservative views, but would often graduate from Bennington 4 years later with liberal views. For example, each class became progressively more liberal in voting patterns than the one below it (Newcomb, 1943). Moreover, follow-up interviews conducted decades later revealed that this liberalism had endured, with the vast majority of the Bennington alumni voting Democratic and participating in progressive political activities in the 1980s (Alwin, Cohen, & Newcomb, 1991).

Why and how did this happen? At Bennington, the faculty members tended to be liberal, as did many popular students. Liberalism, therefore, had high status. In a new, potentially frightening environment far from home, the young women starting college tended to adopt the prevailing norms and attitudes of the campus culture. This strategy, conscious or not, was self-protective. It helped them adjust and fit in. Yet as they became more immersed in campus life, they began to internalize the Bennington worldview, becoming true believers so to speak. Furthermore, after they left the Bennington campus, the views that they had developed there selected them into marriages, communities, and careers that reinforced their liberalism. In other words, they moved in liberal circles and married liberal men, which further supported their political worldviews. This phenomenon illustrates a key idea of life course research: Life experiences can accumulate in a self-propagating way.

The basic phenomenon documented by the Bennington study still resonates with views about the college experience in the early 21st century. More generally, the lives of the Bennington women demonstrate how experiences at one stage of life set the stage for experiences at subsequent stages of life in self-fulfilling ways.

The Bennington Women's Study is not publicly available at this time. The original sample included 527 women. The time frame of the original study was 1935 to 1939 with follow-up studies in 1960 and 1984.

BIBLIOGRAPHY

Alwin, D., Cohen, R. L., & Newcomb, T. (1991). *Political attitudes over the life span: The Bennington women after 50 years.* Madison: University of Wisconsin Press.

Newcomb, T. (1943). *Personality and social change: Attitude formation in a student community.* New York: Dryden Press.

Robert Crosnoe

IV. BERKELEY GUIDANCE STUDY AND THE OAKLAND GROWTH STUDY

The Berkeley Guidance Study and the Oakland Growth Study—following cohorts of young people born between 1920 and 1921 and 1928 and 1929 respectively—were conducted separately but are often grouped together. Both began around the Great Depression (1929–1939) era at the Institute of Human Development at the University of California at Berkeley, and, together, they have been seminal to the development of the life course paradigm.

The Oakland Growth Study was launched in 1932 under the direction of Harold Jones (1894–1960) and Herbert Stolz (1886–1971). In its original incarnation, it ran until 1939. Designed to study patterns of normal physical, emotional, and social development, it followed a group of 167 fifth and sixth graders living in Oakland, California, as they grew into teenagers. Multiple follow-ups were conducted decades later, spanning five intervals through 1981. Perhaps the most famous work associated with the Oakland Growth Study is the book *Children of the Great Depression* by Glen Elder Jr., which was first published in 1974. Focusing on the historical era in which the data were collected, Elder demonstrated how children adapted to the dramatic economic changes of their lives. As expected, he found that they had been deeply affected by their experiences during this period, with many of their troubles resulting from disruption to relationships within their families. Yet the overarching conclusion of this book stressed the resilience of the Oakland children.

The Berkeley Guidance Study began in 1928 under the direction of Jean MacFarlane (1894–1989), who enrolled 248 infants born in Berkeley, California, during 1928 and 1929. The study ran through World War II (1939–1945), with two additional data collections in 1959 and 1969. Again, Elder exploited the longitudinal and historical nature of the data to craft the basic principles of his life course approach. A series of articles he wrote with Avshalom Caspi and Daryl Bem (1988) on the long-term consequences of being shy or explosive during childhood is a good example. The team reported that shy boys faced delayed transitions into marriage and parenthood when they became adults and that, in the long run, they had lower occupational attainment than other boys. Another finding from this research was that explosive boys faced downward socioeconomic mobility relative to their peers and had higher rates of divorce when they grew up. In both cases, a personality characteristic selected young people into settings that reinforced their original dispositions. As demonstrated by the Bennington Women's Study, which was conducted between 1935 and 1939, life pathways took on a highly cumulative form.

A 1993 book by the sociologist John Clausen (1914–1996), *American Lives*, detailed how the original Oakland and Berkeley study members turned out and articulated the concept of planful competence (e.g., human agency) to explain who met with success in work and family and who did not. The theme of all of these works emanating from the Berkeley and Oakland studies is that lives are lived the long way and need to be studied as such.

These two studies are not publicly available at this time. The original Oakland sample included 167 school-aged children, with a timeframe of 1939 through 1981. The original Berkeley sample included 248 school-aged children, with a timeframe of 1928 through 1969.

BIBLIOGRAPHY

Caspi, A., Elder, G. H., Jr., & Bem, D. (1988). Moving away from the world: Life course patterns of shy children. *Developmental Psychology, 24*(6), 824–831.

Clausen, J. A. (1993). *American lives: Looking back at the children of the Great Depression*. New York: Free Press.

Elder, G. H., Jr. (1974). *Children of the Great Depression: Social change in life experience*. Chicago: University of Chicago Press.

Robert Crosnoe

V. EARLY CHILDHOOD LONGITUDINAL STUDIES

The adolescent-focused data sets collected by the National Center for Education Statistics (NCES) reveal just how much inequality was set before secondary school began. The Early Childhood Longitudinal Studies (ECLS) were designed to identify the origins of such inequalities. These studies follow two cohorts.

The Kindergarten Cohort (ECLS-K) is a nationally representative sample of 22,782 American kindergartners in more than 1,000 programs. The first data collection occurred in the fall of 1998, when the children were just beginning kindergarten. They were given diagnostic tests in several arenas (e.g., oral language skills, math knowledge, and so forth), and their parents, teachers, and school administrators were interviewed. This protocol was repeated in the spring of their kindergarten year. A 25% random subsample was followed up in the fall of first grade, primarily to better understand the role of summer break in achievement disparities, and then the full sample was followed up in the spring of 2000 (the end of first grade for most of the original sample children), 2002 (end of the third grade), and 2004 (end of the fifth grade).

The Birth Cohort (ECLS-B) is a nationally representative sample of 14,000 infants born in 2001. Data collection began when they were 9 months old. The primary form of data collection was a parent interview, but children also were assessed by study personnel at home (e.g., evaluation of cognitive skills and physical measurements). This protocol was repeated when the children were 2 years old and then when they were preschool age in 2005. At the 2005 data collection, the children's child-care providers and preschool teachers were interviewed. In the fall of 2006, when the majority of children entered kindergarten, they were again followed up, with their teachers also interviewed. The remainder of the sample was followed up when they entered kindergarten the following year (2007).

ECLS-K allows the study of school readiness—who enters the formal educational system with the most developed academic and academic-related skills—and its consequences for early learning and achievement. ECLS-B came after ECLS-K, but it allows for inquiries into why the differences in school readiness documented in ECLS-K came to be. Because of their nationally representative samples and large numbers of low-income, racial minority, and immigrant children, the ECLS studies are well-positioned to study inequality.

The ECLS data sets are relatively new and, as such, have not produced as many studies as the other NCES data sets. What has come out so far, however, has been informative. One of the first monographs based on ECLS-K, by Valerie Lee and David Burkham in 2002, documented just how much inequality existed among different race and ethnic groups and different social classes at the very beginning of formal schooling. This "inequality at the starting gate" is a bellwether for future inequalities across the life course, primarily because the educational system is so cumulative and because subgroup differences in educational attainment powerfully predicts subgroup differences in occupational attainment, earnings, family formation, and health throughout adulthood. Thus, ECLS-K targets one of the first major life course transitions—the transition from the private world of the family to the public world of the school—that helps to establish life trajectories for individual children as well as differences in these trajectories across various segments of the American population. Although still in their beginning stages, the ECLS data sets have great potential to influence life course research because they provide a national picture of the stages of life that precede the stages captured by most national studies, such as the National Educational Longitudinal Study (NELS), the National Longitudinal Study of Adolescent Health (Add Health), and the National Longitudinal Survey of Youth (NLSY).

The ECLS data sets are publicly available through the National Center for Education Statistics. More information about access can be found on the Web site supported by the Institute of Education Sciences. The original ECLS-K sample included 22,782 kindergartners, and the timeframe of the study is currently from 1998 through 2004 (with future data collections planned). The original ECLS-B sample included 14,000 infants, and the timeframe of the study is currently from 2001 through 2007 (with future data collections planned).

BIBLIOGRAPHY

Institute of Education Sciences. *Early Childhood Longitudinal Program* (ECLS). Retrieved March 18, 2008, from http://nces.ed.gov/ecls

Lee, V.E., & Burkham, D.T. (2002). *Inequality at the starting gate: Social background differences in achievement as children begin school.* Washington, DC: Economic Policy Institute.

Robert Crosnoe

VI. NATIONAL EDUCATIONAL LONGITUDINAL STUDY

Like the pioneering High School and Beyond study (HS&B), the National Educational Longitudinal Study (NELS) was designed by the National Center for Education Statistics (NCES) for research on the link between school contexts and educational pathways. NELS had the added mission of providing better information on the link between home and school. NELS is valuable to life course research because it allows the examination of schooling experiences that set the stage for the rest of life in connection to the family and school contexts that are major settings of early life.

NELS began in 1988 with a nationally representative sample of 24,599 eighth graders enrolled in 1,052 schools across the United States. After the base year data collection, which included interviews with students, parents, teachers, and school administrators, follow-up interviews were conducted in 1990, 1992, 1994, and 2000. Parents were reinterviewed in 1992, and teachers and school administrators were interviewed at all time points through 1992. High school and college transcripts were also collected during the 1990s. It is important to note that efforts were made to follow dropouts as well as those who remained in school, and that some young people were added to the sample after 1988 in order to maintain the representativeness of the sample as a whole. The student is the unit of the analysis, but data were collected from multiple actors in the students' lives as they transitioned through, and then out of (by graduating or dropping out), the educational system and higher education.

As already mentioned, NELS is especially rich in relation to home-school connections and other forms of social capital. James Coleman and Barbara Schneider (1993) provide examples, using NELS, of how parents can work the system for their children and of how schools can thwart or facilitate the educational participation of parents. Such information has shed valuable light on the persistence of socioeconomic and racial disparities in academic achievement and educational attainment. Other major topics relevant to the life course that are often studied with NELS include pathways to dropping out and the disruptiveness of school transitions.

For example, Robert Croninger and Valerie Lee (2001) reported that teachers—another important source of social capital—were protective factors in the lives of young people most at risk for dropping out because of their family or academic circumstances. This form of linked lives—a student connected to a teacher by virtue of classroom assignment—proved to be a turning point for many young people on a clear trajectory toward dropping out from school. As another example, Kathryn Schiller's 1999 study documented that transitioning from middle school to high school often leads to declining rates of academic achievement because of disrupted social relations. Yet young people who performed poorly academically in middle school tended to increase their achievement when only a small portion of their middle school peers made the transition to high school with them. In other words, some students are better able to start fresh when they attend a different school than most of their middle school peers. A transition that is often seen as problematic, therefore, could be positive depending on the context in which it occurred.

NELS provides data that can be used to construct academic histories across different family and institutional contexts. Moreover, its time frame allows researchers to understand how these histories forecast adult experiences. In this way, NELS allows researchers to take a life course perspective on education.

The NELS data set is publicly available through the National Center for Education Statistics. More information about access can be found on the Web site supported by the Institute of Education Sciences. The original sample included 24,599 eighth graders, and the timeframe of the study is from 1988 through 2000.

BIBLIOGRAPHY

Croninger, R. G., & Lee, V. E. (2001). Social capital and dropping out of high school: Benefits to at-risk students of teachers' support and guidance. *Teachers College Record, 103*(4), 548–581.

Institute of Education Sciences. *National Education Longitudinal Study of 1988 (NELS 88).* Retrieved May 29, 2008, from http://nces.ed.gov/surveys/nels88

Schiller, K. S. (1999). Effects of feeder patterns on students' transition to high school. *Sociology of Education, 72*(4), 216–233.

Schneider, B., & Coleman, J. S. (1993). *Parents, their children, and schools.* Boulder, CO: Westview Press.

Robert Crosnoe

VII. NATIONAL LONGITUDINAL STUDY OF ADOLESCENT HEALTH

Add Health is the moniker for this large, ongoing study of adolescents in the United States. Sponsored by the National Institute of Child Health and Human Development and other funding agencies and operated out of the Carolina Population Center in Chapel Hill, North Carolina, Add Health is a nationally representative sample of seventh to twelfth graders during the 1994 and 1995 school year. Its original purpose was to provide information on the social contexts of the health and health behaviors of the American youth.

Add Health began with a census-like survey of more than 90,000 students in its 132 focal middle schools and high schools. From there, a representative sample of 20,745 students was selected across schools for the primary data collection. Wave I, in which adolescents were interviewed at home, occurred in 1995. The adolescent interviews were supplemented with parent interviews, school administrator interviews, and census information on neighborhoods. Wave II followed up the Wave I youth, except seniors, in 1996. Wave III followed up the Wave I youth, including seniors, in 2001. Wave IV is currently under way, with most sample members being in their late 20s. Add Health has several special features, including a sibling subsample and biomarker data that allow for genetic inquiries, a network component in which friendship patterns within schools are mapped out, a partner module in which romantic partners were identified and in some cases interviewed, and an education component that includes extensive academic information from high school transcripts.

Add Health has provided insight into major adolescent health issues, including obesity, depression, suicide, substance use, and, in particular, sexual activity. Many studies have linked together these various aspects of health, with Stephen Russell's and Kara Joyner's (2001) studies of the mental health and health behaviors of gay and lesbian youth being a widely cited example. Yet the influence of Add Health extends beyond health. The peer networks data have been especially informative. For example, Dana Haynie (2001) demonstrated that the well-documented influences of friends on adolescents' engagement in risky behavior was in fact largely conditional on the position of the adolescents and their friends in larger networks of social relations in their schools. As another example, James Moody (2001) demonstrated that many racially integrated schools in the Add Health sample were in fact largely segregated by race in terms of everyday social interaction.

Add Health is a resource for understanding many aspects of adolescence. Because the sample has been followed over time, these adolescent experiences can be linked to young adult outcomes, facilitating understanding of continuity and change across the life course. Moreover, its ability to link biological and social data puts it at the vanguard of new movements to understand the interplay of genes and the environment in the life course.

The Add Health data set is publicly available through the Carolina Population Center. More information about access can be found at the Carolina Population Center Web site supporting the study. The original core sample included 20,745 adolescents, and the timeframe of the study is currently from 1994 through 2002 (with future data collections planned).

BIBLIOGRAPHY

Guo, G., & Stearns, E. (2002). The social influences on the realization of genetic potential for intellectual development. *Social Forces, 80*(3), 881–910.

Haynie, D.L. (2001). Delinquent peers revisited: A network approach for understanding adolescent delinquency. *American Journal of Sociology, 106,* 1013–1057.

Moody, J. (2001). Race, school integration, and friendship segregation in America. *American Journal of Sociology, 107*(3), 679–716.

National Longitudinal Study of Adolescent Health (*Add Health*). Retrieved May 2, 2008, from http://www.cpc.unc.edu/addhealth

Resnick, M. D., Bearman, P. S., Blum, R. W., Bauman, K. E., Harris, K. M., Jones, J., et al. (1997). Protecting adolescents from harm: Findings from the National Longitudinal Study of Adolescent Health. *Journal of the American Medical Association, 278*(10), 823–832.

Russell, S. T., & Joyner, K. (2001). Adolescent sexual orientation and suicide risk: Evidence from a national study. *American Journal of Public Health, 91*(8), 1276–1281.

Robert Crosnoe

VIII. NATIONAL LONGITUDINAL SURVEY OF YOUTH

The National Longitudinal Survey of Youth (NLSY) refers to a collection of studies overseen by the Bureau of Labor Statistics as part of an umbrella program of national longitudinal studies on the life events of Americans. Two iterations of NLSY have been particularly useful to researchers interested in adolescence as the foundation for later stages of the life course.

NLSY79 began in 1979 with a sample of 12,686 young people who had been born in the late 1950s and early 1960s. This sample included an over-sample of African-American, Hispanic, and economically disadvantaged youths as well as a subsample of youth who had enlisted in the armed forces by 1978. A mixture of personal interviews and telephone interviews, NLSY79 ran annually through 1994 and then biennially up through 2008. It also included some additional data collection components, including surveys of schools and the collection of high school transcripts. Primary subject areas include labor market experiences, education, military experience, health, behavior, and family life. Beginning in 1986, the children of the NLSY79 sample members also were included in the data collection.

NLSY97 was similar in form and function to its predecessor, but focuses on young people in a different era and birth cohort. Its purpose was to gauge the transition between adolescence and adulthood, specifically how young people move from school into work and family life. It began in 1997 with a sample of 8,984 youths born between 1980 and 1984. Like NLSY79, it over-sampled the African-American and Hispanic youth and runs annually, with a parent interview in 1997 and a school supplement including surveys of school administrators and the collection of high school transcripts. NLSY97 covers many of the main subject areas as NLSY79, including labor market experiences, education, and family formation, and it also catalogs major event histories.

Several widely cited studies demonstrate the value of the various NLSY data sets to life course research. First, Daniel Lichter and colleagues (1992) used the NLSY79 survey data to construct relationship histories for young women, and then linked the data with the U.S. census to identify sex ratios in their local areas. They reported that socioeconomic attainment improved women's chances of marrying. Moreover, they were able to conclude that the quantity of *marriageable* (i.e., employed and financially stable) men in the area predicted their odds of getting married and trumped other important factors (e.g., welfare history, family background) in explaining why African-American women were less likely to marry than young white women. In other words, a key relationship pathway in the life course was dependent on larger social and economic contexts that shaped interactions among people.

Second, Richard Strauss's and Harold Pollack's 2001 study used the children of the NLSY79 respondents to track trends in obesity over the 1980s and 1990s. Their results detailed the dramatic increase in obesity among young people during this period, providing sound documentation of the obesity epidemic that had generated so much coverage in the national media. Yet, more than just establishing a basic population health trend, these data provide a historical and cultural context for understanding the role of obesity in the early life course. Third, the NLSY has been used extensively by criminologists. Its long-term frame and focus on the transition between adolescence and adulthood has been especially useful for establishing patterns of entry into and desistance from crime behavior across different stages of life and how these patterns are related to life course transitions and social relationships.

The guiding purpose of the NLSY, in all of its incarnations, has been to understand the kinds of adults that young people turn out to be. This purpose positions the NLSY as a pre-eminent tool for studying, on a national scale, the early stages of life within a life course framework.

The NLSY data set is publicly available through the Bureau of Labor Statistics. More information about access can be found at the Web site supported by the U.S. Department of Labor, Bureau of Labor Statistics. The original NLSY79 sample included 12,686 youth. It began in 1979 and is ongoing. The original NLSY97 sample included of 8,984 youth. It began in 1997 and is ongoing.

BIBLIOGRAPHY

Lichter, D. T., Kephart, G., McLaughlin, D. K., & Landy, D. J. (1992). Race and the retreat from marriage: A shortage of marriageable men? *American Sociological Review, 57*(6), 781–799.

Piquero, A. R., Brezina, T., & Turner, M. G. (2005). Testing Moffitt's account for delinquency abstention. *Journal of Research in Crime and Delinquency, 42*(1), 27–55.

Strauss, R. S., & Pollack, H. (2001). Epidemic increase in childhood overweight, 1986–1998. *Journal of the American Medical Association, 286*(22), 2845–2848.

U.S. Department of Labor (Bureau of Labor Statistics). *National longitudinal studies survey.* Retrieved May 29, 2008, from http://www.bls.gov/nls

Robert Crosnoe

IX. STANFORD-TERMAN STUDY

Begun in 1922, the Stanford-Terman Study is the longest-running longitudinal study of Americans. The study was created by Lewis Terman (1877–1956), a psychologist at Stanford University, to determine the kinds of adults that bright children became. A sample of 857 male and 671 female children in California schools who scored in the gifted range on the Stanford-Binet IQ test was assembled. The sample members (referred to as *Termites*) ranged from young children to teenagers and generally came from White middle-class homes. They took diagnostic tests, and they and their parents were interviewed.

After the initial data collection, the original sample members were followed up in fairly regular intervals through 1992. Because of budget constraints, the primary means of data collection was a mail-in questionnaire, but other forms of data were collected in less systematic fashion over the years, including press clippings, written responses to open-ended questions about family and work, and notes and letters from the Termites to Terman and his successors. In the 1980s, the sociologist Glen Elder Jr., and his research assistants (1993) developed codebooks for military service and health for this archived supplemental material. The resulting codes were coupled with the questionnaire data to create life record files for the Stanford-Terman subjects.

As expected, based on their family backgrounds and IQ scores, the Stanford-Terman children obtained high levels of education. Many had extraordinary career successes, reaching high posts in the government, making significant scientific breakthroughs, and gaining renown in television and movies. Yet others struggled, and the rates of divorce, suicide, and depression were higher than expected. One recurring theme in the data is that membership in the study, which was quite famous in its day, provided a sense of identity, purpose, and pride.

As the Stanford-Terman Study continued from decade to decade, it was used for many other purposes beyond its initial scope of studying the long-term consequences of intelligence. It has been especially useful for investigating trajectories of individual behavior over time, the role of transitions and turning points in peoples' lives, and how individuals respond to historical events. George Vaillant's 1983 study of the lives of alcoholics is a fascinating example. Also influential has been the work of Elder and others on the timing of life events, which has shown how one's year of birth determined whether Termites hit major events (e.g., the Great Depression [1929–1939], World War II [1939–1945], and so forth) at "good" or "bad" times in their lives and ultimately influenced how their lives turned out in the long run.

Started for one specific purpose, the Stanford-Terman Study has been used more often for other purposes. Although a highly specialized sample, its sheer longevity makes it incredibly valuable to life course research, especially on the long-term consequences of early experiences.

The Stanford-Terman Study is archived at the Henry A. Murray Research Archive at Harvard University. More information about access can be found at the Web site supported by the archive. The original sample included 1,528 boys and girls, and the timeframe of the study is from 1927 through 1986.

BIBLIOGRAPHY

Elder, G. H., Jr., Pavalko, E., & Clipp, C. (1993). *Working with archival data: Studying lives.* Newbury Park, CA: Sage.

Henry A. Murray Research Archive. The Stanford-Terman Study. Retrieved April 10, 2008, from http://www.murray.harvard.edu

Shanahan, M. J., Elder, G. H., Jr., & Miech, R. A. (1997). History and agency in men's lives: Pathways to achievement in cohort perspective. *Sociology of Education, 70*(1), 54–67.

Terman, L. M., & Oden, M. H. (1959). *Genetic studies of genius,* Vol. 5: *The gifted group at midlife; Thirty-five years of follow-up on the superior child.* Stanford, CA: Stanford University Press.

Vaillant, G.E. (1983). *The natural history of alcoholism: Causes, patterns, and paths to recovery.* Cambridge, MA: Harvard University Press.

Robert Crosnoe

DATING AND ROMANTIC RELATIONSHIPS, CHILDHOOD AND ADOLESCENCE

From the plays of William Shakespeare to contemporary Hollywood films, popular culture has highlighted the importance of adolescent dating and romance. Social science research has emerged to confirm that these romantic experiences are an important part of the life course and human development. These experiences contribute to the development of identity and interpersonal skills, as well as provide a forum for sexual activity. Adolescent romance may also act as a gateway to romantic and sexual relationships later in the life course. This entry provides an overview of adolescent dating and romantic relationships in the United States, including key terms and concepts, theoretic perspectives, some basic trends about the prevalence of these experiences and the consequences for adolescents, and the direction of current and future research in this field. Although it may be easy to dismiss romance as fleeting and inconsequential material for teen comedies, the prevalence and centrality of these relationships in adolescent life coupled with empiric research documenting far-reaching consequences suggest the importance of critically examining adolescent romantic relationships.

TERMS AND CONCEPTS

Although the terms *dating* and *romantic relationship* are often used interchangeably, it is useful to differentiate the two, particularly when discussing adolescents. The term *romantic relationship* refers broadly to a range of behaviors and emotions that occur between two persons. Although *dating* also refers to these experiences, it is difficult to separate from the concept of a "date," which is a specific behavior that may or may not happen within an adolescent romantic relationship. Other terms are used by adolescents themselves to describe these romantic experiences, and these vary temporally and geographically (e.g., going steady). Despite the wide array of conceptualizations of romantic activity, recent academic publications primarily use romantic relationship to describe these experiences.

Researchers have also grappled with conceptualizing romantic experiences and measuring the incidence of these experiences. For instance, two major survey instruments used in the study of adolescent relationships use different questions. The National Longitudinal Study of Adolescent Health (Add Health) asks respondents to list any same or opposite sex *special romantic relationships*, as well as asks about nonromantic sexual relationships. After extensive pretesting, the Toledo Adolescent Relationship Study (TARS) employed a section that asks about opposite sex *dating*, using the definition that dating means "when you like a guy [girl], and he [she] likes you back," but that it does not require going out on a formal date (Giordano, Longmore, & Manning, 2006).

The concept of romantic relationships contains multiple dimensions, including emotional and physical elements. The emotional dimension captures feelings such as physical attraction, a sense of intimacy, love, and personal closeness. The physical dimension includes sexual activity, but also nonsexual contact, such as holding hands and kissing. However, relationships may also be characterized by behaviors (such as going out alone on a formal date) or by the social recognition of the relationship by peers.

Although the activities and emotions just given form a foundation for defining romantic relationships, romantic experiences in adolescence take many different forms. Some relationships, particularly those in later adolescence, may rival that of adult relationships in terms of emotional intensity and physical activity. However, other relationships may be low on the emotional dimension, may contain no physical contact, or in some cases are characterized by neither dimension. Relationships most commonly form with partners of the opposite sex, but same-sex romantic relationships also occur in this period.

Romantic encounters have long been an important element in adolescence, but the implications of these relationships have shifted with cultural changes regarding gender, sexuality, and the increasing age of marriage. Relationships in adolescence were formerly tied closely to marriage, but they are less so connected now with the increasing age at first marriage (Fields, 2003). Relationships also were once characterized by formal rules of behavior, particularly around gender norms and appropriate levels of physical intimacy. Although this traditional model of romance may have largely disappeared for contemporary adolescents, romantic relationships continue to be central to the teenage experience and are still prevalent, with most adolescents reporting some type of romantic experience by age 18 (Carver, Joyner, & Udry, 2003).

PERSPECTIVES ON ROMANTIC RELATIONSHIPS

Early theories of adolescent romance focused on the developmental implications, highlighting these experiences as a crucial component of the life stage (Sullivan, 1953). Erikson's classic description of identity formation (1968) cites romantic relationships as crucial to the development of emotional intimacy. Interactions in childhood revolve around the family and same-sex peer groups, but adolescence reflects a time for mixed-sex peer groups, and the emergence of romantic pairs.

Developmental theories have paid specific attention to how the form and function of romantic relationships change over time and across adolescence. Shulman and Seiffge-Krenke (2001) summarize developmental research and theory identifying four sequences: initiation, affiliation, intimate, and committed. The initiation stage includes limited contact, with the primary benefit of giving the adolescent confidence that he or she can find a romantic partner. Affiliation involves going out in peer groups, and these relationships are generally characterized by partner companionship. In intimate relationships, there is a more clearly identified *couple* orientation, and these relationships may involve sexual activity. The final phase is a committed relationship, which is long term and involves deep intimacy and caring, and may resemble adult relationships.

Consistent with this perspective of the changing nature of relationships over the adolescent period, research finds that the duration of relationships varies with the age of the adolescent. Although Carver, Joyner, and Udry (2003) found an average relationship duration of 14 months, adolescents age 14 and younger reported an average duration of 5 months, and those 16 and older reported an average of 21 months. One element in extending this development perspective is situating these romantic experiences on a trajectory into adult relationships (including romantic, cohabiting, and marital unions). Research finds that adolescents who form romantic relationships report higher expectations to marry after they reach adulthood (Crissey, 2005) and are more likely to form relationships, cohabit, and marry in early adulthood (Meier & Allan, 2007; Raley, Crissey, & Muller, 2007). This research supports Meier and Allan's (2007) depiction of adolescent romantic relationships as the *social scaffolding* for romantic experiences in early adulthood.

Gender is particularly important in the study of heterosexual adolescent romance. Although both boys and girls have romantic relationships, these experiences may not be the same. Cultural expectations of traditional gender roles are frequently most salient in romantic relationships, and Feiring (1999) argues that the development of romantic relationships coincides with and encourages the development of gender identity with girls and boys experiencing greater conformity to their gender roles when they are engaged in romantic relationships. In addition, the centrality of dating is also likely to vary by gender, with romance traditionally having greater importance for girls compared to boys (Hudson, 1984). Interest in heterosexual relationships is ubiquitous in female peer groups, and being attractive to potential mates consumes a considerable amount of the time and energy, but research has not found this in male peer groups (Holland & Eisenhart, 1990). However, boys are certainly not oblivious to romance, as research finds that adolescent boys are also emotionally invested in romantic relationships (Giordano et al., 2006).

The romantic experiences of adolescents also may be influenced by characteristics such as race, socioeconomic status, and family background. For instance, research has found that Black adolescents are less likely to form romantic relationships compared with their White peers (Crissey, 2005; Meier & Allen, 2007). Some evidence suggests that relationship characteristics vary by race, including dimensions such as duration and physical activity (Carver et al., 2003; Crissey, 2005; Meier & Allen, 2007). The development of the TARS romantic relationship section specifically addressed this by recognizing that the concept of a formal date is "strongly class-linked and would tend to exclude lower socio-economic status (SES) youth" (Giordano et al. 2006, p. 268).

A crucial component of these theoretical perspectives on adolescent romance is the recognition that these relationships emerge in a critical time for physical development. The pubertal period includes the emergence of sexual attraction, as well as the desire to engage in physical and sexual contact, and to form romantic relationships. The sexual component of adolescent experiences has received considerable attention, both from researchers and within public discourse. A large body of research on romantic relationships is rooted in the idea that these relationships serve as a main venue for adolescent sexual activity, and therefore provide the potential for pregnancy and contracting sexually transmitted infections. Research has found that adolescents who form romantic relationships are more likely to have sex (Kaestle, Morisky, & Wiley, 2002). Early sexual debut is particularly a concern for younger adolescents. However, compared to sexual partners who just met, adolescents in relationships are more likely to use contraception (Manning, Longmore, & Giordano, 2000).

Media attention has recently pointed to the sexual component of adolescent experiences with reports on the rise of nonromantic sexual encounters, termed *hooking up* and *friends with benefits*. These reports suggest that the conventional scripts of dating are extinct and that romance is no longer a feature of adolescent life. Published research suggests that this is not true. Although the norms of relationships have changed over the past few decades, teenagers still overwhelmingly participate in romantic relationships and report that these emotional experiences are desirable (Carver et al., 2003; Giordano et al., 2006).

Although the deviance perspective on romantic relationships focuses on sexual activity, researchers have also noted the potential for other negative consequences from romantic relationships. As recognized in the developmental perspective, romance is important for emotional development. Because these relationships are frequently emotionally charged, negative experiences may be particularly detrimental. Larson, Clore, and Wood (1999) note that romantic relationships are "the single largest source of stress for adolescents" (p. 35). Relationships may include arguments and infidelity, and they frequently end in a break-up. These experiences may lead to negative emotional consequences including depression, anger, and jealousy (Joyner & Udry, 2000; Larson et al., 1999).

Romantic relationships may also be problematic when considering differences between the partners. The developmental benefits of romantic relationships depend on age and emotional maturity. However, romantic relationships routinely occur between partners of different ages. This age gap can have various consequences for the younger participant, such as increased relationship intensity and duration, exposure to an older peer group, and involvement in activities such as going out alone or having sex. For example, young adolescent females with older partners have an increased risk of sexual activity and of not using contraception (Kaestle, et al., 2002; Manning, et al., 2000).

GAY AND LESBIAN RELATIONSHIPS

The bulk of romantic experiences in adolescence are heterosexual, and consequently the vast majority of the research on romance considers opposite sex interactions. However, more recent research has paid specific attention to the romantic experiences of sexual minority youth (including gay, lesbian, bisexual, and transgendered adolescents). Research from the TARS study found that 8% of respondents report being bisexual and 1% report being homosexual, whereas about 6% of respondents in Add Health report same-sex attraction. In the Add Health study, 2.2% of boys and 3.5% of girls who had a romantic relationship in the 18 months before the interview occurred had reported a partner of the same sex (Carver et al., 2003). Although the literature documents higher incidence of such negative outcomes as depression, school problems, and substance use for sexual minority youth, researchers suggest moving out of this *at risk* model to understand the experiences of these adolescents completely (Russell, Driscoll, & Truong, 2002).

RELATIONSHIPS AND ABUSE

Romantic relationships can also be a venue for negative interactions such as coercion, violence, and abuse. The Centers for Disease Control and Prevention (CDC) estimates that about 1.5 million high school students, which is nearly 9% of the population, have been hit, slapped, or otherwise physically hurt by a romantic partner (CDC, 2006). According to a 2006 survey conducted by Teenage Research Unlimited, more than 60% of adolescents who have been in a romantic relationship report that their partner has made them feel bad or embarrassed about themselves. Furthermore, this study found that adolescents in romantic relationships experience pressure to have sex with their partners, physical and emotional abuse, and controlling behaviors such as repeated calls and text messages.

CURRENT AND FUTURE DIRECTIONS

Despite the prominent role of romantic relationships in popular culture, the topic is only now amassing a large body of scholarly research with increasing recognition of the potential risks and rewards of these relationships. Researchers continue to explore the developmental benefits by connecting romantic experiences to the formation of adult relationships and marriage. By locating adolescent romantic relationships as a midpoint between the family of origin and adult relationships, research is uncovering how adolescent romance factors into development across the life course. The focus on negative consequences including those resulting from sexual activity, and more recently on emotional well-being, is beginning to widen to include implications for outcomes such as substance use and educational performance. Research dedicated to understanding the heterogeneity of romantic experiences, including differences by race, class, family background, and for sexual minority youth, continues to emerge.

Research can also continue to explore how the meanings and consequences of romantic relationships are changing with cultural and technologic changes. Gone are the days of the distant pen pal. Relationships can now be easily maintained in real time between partners on the opposite side of the globe using cell phones, e-mail, text messages, webcams, and other technologic modalities. With the expansion of virtual social networks, adolescents may form deep emotional bonds with people they will never meet in person. In fact, they may even form bonds with a figment of someone else's imagination. Media are filled with reports of the hazards of this type of interaction, but researchers are just beginning to explore how these technological changes are influencing adolescent development.

Romantic relationships continue to be a fixture on the modern adolescent landscape, even if these experiences bear little resemblance to the conventional notion of teenage romance. Romance continues to be an important component in the development of interpersonal skills and identity formation, as well as providing a training ground for romantic experiences in adulthood. Relationships also contain the potential for risk, including physical outcomes such as contracting a sexually transmitted infection, unintended pregnancy, and physical abuse, as well as a range of negative emotional outcomes. Researchers need to continue to study these romantic experiences to help parents, teachers, and teens themselves understand and navigate the adolescent period.

SEE ALSO Volume 1: *Health Behaviors, Childhood and Adolescence; Peer Groups and Crowds; Puberty; Sex Education/Abstinence Education; Sexual Activity, Adolescent; Transition to Marriage; Transition to Parenthood.*

BIBLIOGRAPHY

Carver, K. P., Joyner, K., & Udry, J.R. (2003). National estimates of adolescent romantic relationships. In P. Florsheim, *Adolescent romantic relations and sexual behavior: Theory, research, and practical implications* (pp. 23–56). Mahwah, NJ: Lawrence Erlbaum and Associates.

Centers for Disease Control and Prevention. (2006). Physical dating violence among high school students—United States, 2003. *MMWR Morbidity and Mortality Weekly Report, 55,* 532–535

Crissey, S. R. (2005). Race/ethnic differences in marital expectations of adolescents: The role of romantic relationships. *Journal of Marriage and Family, 67,* 697–709.

Erikson, E. H. (1968). *Youth: Identity and crisis.* New York: Norton.

Feiring, C. (1999). Gender identity and the development of romantic relationships in adolescence. In W. Furman, B. B. Brown, & C. Feiring (Eds.), *The development of romantic relationships in adolescence* (pp. 211–234). New York: Cambridge University Press.

Fields, J. (2003). *America's family and living arrangements.* Washington, DC: U.S. Census Bureau.

Giordano, P. C., Longmore, M. A., & Manning, W. D. (2006). Gender and the meanings of adolescent romantic relationships: A focus on boys. *American Sociological Review, 71,* 260–287.

Holland, D. C., & Eisenhart, M. A. (1990). *Educated in romance: Women, achievement, and college culture.* Chicago: University of Chicago Press.

Hudson, B. (1984). Femininity and adolescence. In A. McRobbie & M. Nava, *Gender and generation* (pp. 31–53). London: Macmillan.

Joyner, K., & Udry, J. R. (2000). You don't bring me anything but down: Adolescent romance and depression. *Journal of Health and Social Behavior, 41,* 369–391.

Kaestle, C. E., Moriskey, D. E., & Wiley, D. J. (2002). Sexual intercourse and the age difference between adolescent females and their romantic partners. *Perspectives on Sexual and Reproductive Health, 34,* 304–309.

Larson, R. W., Clore, G. L., & Wood, G. A. (1999). The emotions of romantic relationships: Do they wreak havoc on adolescents? In W. Furman, B. B. Brown, & C. Feiring (Eds.), *The development of romantic relationships in adolescence* (pp. 19–49). New York: Cambridge University Press.

Manning, W. D., Longmore M. A., & Giordano, P. C. (2000). The relationship context of contraceptive use at first intercourse. *Family Planning Perspectives, 32,* 104–110.

Meier, A., & Allen, G. (2007). Romantic relationships from adolescence to young adulthood: Evidence from the National Longitudinal Study of Adolescent Health. Minnesota Population Center Working Paper Series No. 2007-03. University of Minnesota, Minneapolis, MN. Retrieved July 2, 2007 from http://www.pop.umn.edu/research/mpc-working-papers-series/2007-working-papers-1/2007-03-romantic-relationships-from-adolescence-to-young-adulthood-evidence-from-the-national-longitudinal-study-of-adolescent-health

Raley, R., Kelly, S. C., & Muller, C. (2007). Of sex and romance: Late adolescent relationships and young adult union formation. *Journal of Marriage and Family, 69,* 1210–1226.

Russell, S. T., Driscoll, A. K., and Truong, N. (2002). Adolescent same-sex romantic attractions and relationships: Implications for substance use and abuse. *American Journal of Public Health, 92,* 198–202.

Shulman, S., & Seiffge-Krenke, I. (2001). Adolescent romance: Between experience and relationships. *Journal of Adolescence, 24,* 417–428.

Sullivan, H. S. 1953. *The Interpersonal Theory of Psychiatry.* New York: Norton.

Sarah R. Crissey

DAY CARE

SEE Volume 1: *Child Care and Early Education.*

DELINQUENCY

SEE Volume 1: *Crime, Criminal Activity in Childhood and Adolescence; Theories of Deviance.*

DEVELOPMENTAL SYSTEMS THEORY

The contemporary study of human development is focused on concepts and models associated with developmental systems theories (Cairns & Cairns, 2006; Gottlieb, Wahlsten, & Lickliter, 2006; Lerner, 2002, 2006) and on their use in understanding behavior across the life span (Baltes, Lindenberger, & Staudinger, 2006). The roots of these theories are linked to ideas in developmental science that were presented at least as early as the 1930s and 1940s (e.g., Maier & Schneirla, 1935; Novikoff, 1945a, 1945b; Von Bertalanffy, 1933), if not even significantly earlier—for example, in the concepts used by late 19th century and early 20th century founders of the study of child development (e.g., Cairns & Cairns, 2006). Developmental systems theory provides an innovative and important frame for the study of the human life span.

In the early 21st century, developmental systems theories are understood as a family of conceptual models that promote a holistic, or integrated, view of human development. Development is seen as a change process involving mutually (bidirectionally) influential relations among all parts of the individual (e.g., genes, hormones, brain, emotions, thoughts, and behaviors) and all levels of the ecology, or contexts, of life (such as families, peer groups, schools, after-school programs, businesses, faith institutions, neighborhoods, or the physical setting), all of which vary across time and therefore history. As such, in this theoretical approach, the whole individual is seen as "greater than the sum" of his or her parts (he or she is the "multiplication" of the parts, not the addition of them).

Accordingly, within a developmental systems perspective, the study of development does not attempt to isolate for analysis individual components of the overall system (e.g., genes, the person, the family). Because each part of the system is related to all other parts, the function and meaning of any part of the system is derived from this relation. Features of the individual (his or her genes or personality) and features of the context (his or her peer group, school, or culture) must be studied together, as they each influence each other across time. In other words, given the integration of all these levels, from genes to the physical environment and history (temporality), a developmental systems perspective emphasizes that complex and changing relationships exist between individuals and their

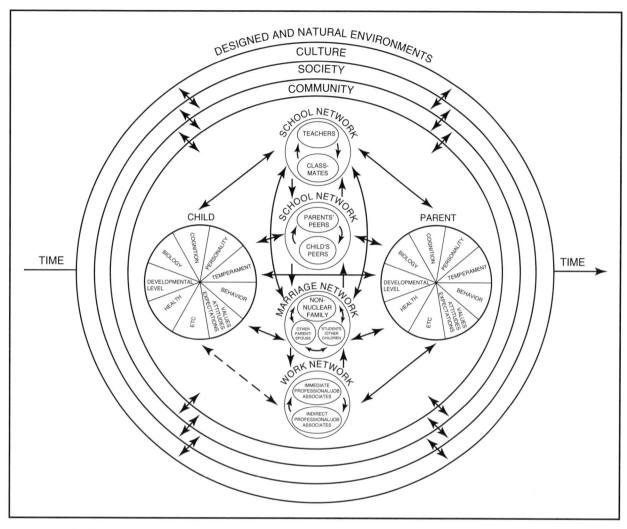

Figure 1. *A developmental systems view of human development: Parent–child relations and interpersonal and institutional networks are embedded in and influenced by particular community, societal, cultural, and designed and natural environments, all changing across time (across history).* **CENGAGE LEARNING, GALE.**

ecology. This systematic integration of systems is known as *relationism,* and it stands in contrast to theories that attempt to dichotomize development, such as in the now passé nature versus nurture controversy. Figure 1 illustrates the integrated relations within the developmental system, here in regard to the links between a child, a parent, and the other levels within their ecology of human ecology.

Because all levels of the system are interrelated, they are also mutually influential. The characteristics and actions of a person affect and are affected by the features of all the other levels of organization within his or her social and physical ecology. Moreover, changes in the person (such as cognitive, emotional, and physical development across childhood and adolescence) mean that the individual will differentially affect the context. Individuals with different intellectual abilities and interests, with

varying identities and purposes, or with different physical and health attributes and needs require different interpersonal and intellectual resources or educational or health programs to grow and prosper in healthy ways. In turn, environmental changes can elicit or require the development of new behaviors in individuals if healthy or positive behavior (i.e., adaptive functioning) is to exist.

Therefore, by influencing the contexts that influence them, individuals contribute to their own development. There is a bidirectional arrow (⟷) between individuals and contexts. This bidirectional arrow signifies the mutually influential relationships between the person and all the other levels of organization in his or her biological, psychological, social, cultural, and physical world.

As discussed by Paul Baltes and colleagues (2006), these contextual changes can occur normatively across age, such as in a young person's transition from elementary school to middle school or from high school to college or to the world of work. In addition, these changes can take place normatively across history (e.g., the invention of new electronic devices may make methods of interpersonal communication different for a new generation of youth in comparison to their parents' generation). Moreover, there can be nonnormative life events (such as accidents, diseases, or death of a loved one) and nonnormative historical events (wars, hurricanes, or economic depressions) that can require behavior and development to change if healthy functioning is to occur.

In short, to understand the character of human development across life, relations among all the levels of organization within the ecology of human development must be viewed holistically. In addition, the mutually influential relations between people and their real-life contexts (i.e., individual $\longleftrightarrow$ context relations) need to be studied across the life span, because changes in person and/or in the context at any point within life or history can alter significantly the relations a person has with his or her world. Furthermore, because of the sensitivity of human development to changes in the ecology, developmental systems theory stresses that development happens in the real world, not in contrived laboratory situations. Changes in families, schools, businesses, or the physical environment can change the course of life.

IMPLICATIONS FOR RESEARCH AND PRACTICE

This developmental systems theoretical approach has important implications for research and practice. For instance, the evaluation of an educational program must be framed by trying to understand the relation between an individual's unique qualities and the specific facets of a program. Do children's levels of cognitive development, interests, motivations, or behavioral skills match the content of a curriculum or its method of presentation (such as lectures or online delivery)? Does this fit between children and context apply also to adolescents, adults, and the aged, who vary in sex, race, ethnicity, religion, socioeconomic circumstances, or culture? A developmental systems perspective would provide an important framework for researchers and practitioners who seek to advance or promote for diverse individuals knowledge about, involvement in, or skills associated with such programs.

From these examples, it is evident that the unit of analysis within this theoretical approach is the relationship between an individual and the multiple levels that comprise his or her ecology, rather than the individual or the ecology alone. A person's development is determined by fused (i.e., inseparable and mutually influential) links among the multiple levels of the ecology of human development, including variables at the levels of inner biology (such as genes, the brain), the individual (such as personality, temperament, values, purposes, or cognitive style), social relationships (with peers, teachers, and parents), sociocultural institutions (educational policies and programs), and history (normative and nonnormative events, such as elections and wars, respectively).

The dynamic (i.e., mutually influential) changes that exist across the developmental system create openness and flexibility in development. The presence of such potential for change in development implies that there is a potential for plasticity (systematic change) across life. In turn, the plasticity of development means that one may be optimistic about the ability to promote positive changes in human life by altering the course of individual $\longleftrightarrow$ context relations. In comparison to developmental perspectives that regard people as passive recipients of environmental stimulation (e.g., Bijou & Baer, 1961) or as automatons ("lumbering robots"; Dawkins, 1976) controlled by the set of genes acquired at conception, viewing development as a matter of at least relatively plastic individual $\longleftrightarrow$ context relations suggests that each person is an important producer of his or her own development.

People can play an important role in determining the nature of their relationships with their contexts through characteristics of physical, mental, and behavioral individuality, including the setting of goals or purposes (Damon, Menon, & Bronk, 2003) and the actions they take to pursue their objectives (Baltes et al., 2006; Freund & Baltes, 2002). Through the purposes they set—for instance, their selection of goals (S), their skills and strategies for optimizing their paths towards their goals (O), and their abilities to compensate (C) when goals are blocked or purposes are not reached—individuals influence their own developmental trajectories (e.g., Baltes et al., 2006; Freund & Baltes, 2002).

Interaction between a child and a teacher may illustrate this active agency of individuals. A child with an "easy" temperament (e.g., an ability to rapidly adjust to new events and stimuli, a positive mood, and a long attention span) is likely to elicit positive, attentive responses from his or her teacher. Such responses may, in turn, promote further positive behaviors from the child. Ultimately, a healthy, adaptive teacher $\longleftrightarrow$ child relationship is supported by such relations (Chess & Thomas, 1999). In turn, a child with a "difficult" temperament (e.g., slow to adjust to new stimuli, a negative mood, and a short attention span) may elicit negative reactions from a teacher; this temperament might contribute to problematic teacher $\longleftrightarrow$ child relations. In both cases, the respective behaviors of both the child and his or her teacher have influenced the behaviors of the other person in the relationship, and the child is therefore co-shaping the course of his or her own development. By underscoring the active contribution that each individual has on his or her developmental trajectory, the developmental systems perspective brings the importance of individual differences to the fore: As each individual interacts in a unique way with his or her context, he or she may develop differently from other individuals.

Therefore, from the developmental systems perspective, development is not seen as a simple, linear, cause-and-effect process but as a complex, flexible process whereby the actions and purposes of the individual play a causal role (Brandtstädter, 2006). Moreover, the reason that developmental systems theories place a strong emphasis on ecological validity (i.e., the importance of understanding people in settings representative of their real-world settings), as opposed to ecologically unrepresentative laboratory settings, is because this contribution of the person to his or her own development occurs within the actual ecology of human development—in the homes, schools, faith institutions, after-school programs, businesses, and physical settings of a community. Thus, a strength of developmental systems theories is that, rather than concentrating on a limited aspect of a person's functioning or focusing on people in contrived situations, it focuses on the diversity and complexity of human development, as it takes place in the contexts within which individuals actually spend their lives across the breadth of the entire life span.

In sum, the ideas of plasticity and optimism within the developmental systems perspective provide a theoretical foundation for applying developmental science to promote positive change across the life span. By devising programs and policies that have the flexibility to maximize the fit between diverse individuals and the settings of human development, the probability of positive development may be optimized.

CONCLUSIONS

The interrelated features of contemporary developmental systems theories involve concepts such as relationism, the integration of levels of organization, historical embeddedness (temporality), relative plasticity, and diversity. These concepts lead to themes ranging from the importance of context for understanding human development to the idea that one may be optimistic that the application of developmental science can result in the promotion of positive development for diverse individuals across the breadth of the human life course. Developmental systems theories provide rich and varied conceptual tools for describing, explaining, and enriching the course of human development.

SEE ALSO Volume 1: *Cognitive Ability; Genetic Influences, Early Life; Social Development.*

BIBLIOGRAPHY

Baltes, P. B., Lindenberger, U., & Staudinger, U. M. (2006). Life span theory in developmental psychology. In W. Damon (Series Ed.) & R. M. Lerner (Volume Ed.), *Handbook of child psychology: Vol. 1. Theoretical models of human development* (6th ed., pp. 569–664). Hoboken, NJ: Wiley.

Bijou, S. W., & Baer, D. M. (1961). *Child development: A systematic and empirical theory* (Vol. 1). New York: Appleton-Century-Crofts.

Brandtstädter, J. (2006). Action perspectives on human development. In W. Damon (Series Ed.) & R. M. Lerner (Volume Ed.), *Handbook of child psychology: Vol. 1. Theoretical models of human development* (6th ed., pp. 516–568). Hoboken, NJ: Wiley.

Cairns, R. B., & Cairns, B. D. (2006). The making of developmental psychology. In W. Damon (Series Ed.) & R. M. Lerner (Volume Ed.), *Handbook of child psychology: Vol. 1. Theoretical models of human development* (6th ed., pp. 89–165). Hoboken, NJ: Wiley.

Chess, S., & Thomas, A. (1999). *Goodness of fit: Clinical applications from infancy through adult life.* Philadelphia, PA: Brunner/Mazel.

Damon, W., Menon, J., & Bronk, K. C. (2003). The development of purpose during adolescence. *Applied Developmental Sciences, 7*(3), 119–128.

Dawkins, R. (1976). *The selfish gene.* New York: Oxford University Press.

Freund, A. M., & Baltes, P. B. (2002). Life-management strategies of selection, optimization, and compensation: Measurement by self-report and construct validity. *Journal of Personality and Social Psychology, 82,* 642–662.

Gottlieb, G., Wahlsten, D., & Lickliter, R. (2006). The significance of biology for human development: A developmental psychobiological systems perspective. In W. Damon (Series Ed.) & R. M. Lerner (Volume Ed.), *Handbook of child psychology: Vol. 1. Theoretical models of human development* (6th ed., pp. 210–257). Hoboken, NJ: Wiley.

Lerner, R. M. (2002). *Concepts and theories of human development.* (3rd ed.). Mahwah, NJ: Lawrence Erlbaum.

Lerner, R. M. (2006). Developmental science, developmental systems, and contemporary theories of human development. In W. Damon (Series Ed.) & R. M. Lerner (Volume Ed.), *Handbook of child psychology: Vol. 1. Theoretical models of human development* (6th ed., pp. 1–17). Hoboken, NJ: Wiley.

Maier, N. R. F., & Schneirla, T. C. (1935). *Principles of animal psychology.* New York: McGraw-Hill.

Novikoff, A. B. (1945a). Continuity and discontinuity in evolution. *Science, 101,* 405–406.

Novikoff, A. B. (1945b). The concept of integrative levels and biology. *Science, 101,* 209–215.

Von Betralanffy, L. (1933). *Modern theories of development.* London: Oxford University Press.

Richard M. Lerner
Alicia Doyle Lynch
Michelle Boyd

DIABETES, CHILDHOOD

SEE Volume 1: *Illness and Disease, Childhood and Adolescence.*

DISABILITY, CHILDHOOD AND ADOLESCENCE

According to the *International Classification of Functioning, Disability and Health*, Second Edition (ICIDH-2) (World Health Organization, 1999), not all impairments are disabling. Most scholars working in the field of disability studies in the early 21st century would define an *impairment* is an anatomic or physiologic trait or condition, the effects of which sometimes may be ameliorated by appropriate professional intervention. The term *disability* is used to describe conditions with social consequences. Because society stigmatizes people with disabilities and creates physical and social barriers to their full participation in society, they are at a disadvantage in relation to more typical individuals. This disadvantage is shared by families of children with disabilities (Darling, 1979). From the time they know or suspect that their children may have impairments, parents and other family members must adapt to this knowledge and to the reactions of others in society. The impact on family roles continues throughout childhood, adolescence, and beyond. The nonstandard life course trajectories of people with disabilities (and their families) reflect on how disability is socially constructed (Irwin, 2001).

RESEARCH ON PARENTS

Most social science research on children with disabilities has focused on parents and families rather than on the children themselves. Many early studies concentrated exclusively on mothers; more recently, fathers and other family members have been included as well. Much early research dwelled on pathologic family reactions and suggested that childhood disability had a negative effect on family integration (Mandelbaum & Wheeler, 1960). More recent studies have noted some positive effects and have suggested that negative effects are more the result of societal barriers than of family pathology (Seligman & Darling, 2007).

Because most families have little experience with disability before the birth of an affected child, they are usually poorly prepared for the diagnosis and its consequences. This lack of preparation is sometimes complicated by medical professionals who withhold the truth about the child's condition, resulting in parental anomie (Darling, 1994). Interactional difficulties faced by parents during the early months postnatally involve negative reactions from family members, friends, and strangers. Usually by the end of infancy, most parents have resolved their anomie and have developed some strategies for coping with the reactions of others.

The goal of most families of children with disabilities is to achieve a lifestyle that is as close as possible to the norm for families with nondisabled children. The achievement of a normalized lifestyle may be related less to the degree of a child's impairment or parents' coping abilities than to the opportunity structure within which the family resides (Seligman & Darling, 2007). Barriers to normalization include lack of access to appropriate medical care, educational opportunities, child care, and other resources. Consequences include restrictions on parents' employment or career paths and social opportunities. Families who achieve normalization during the school years may again encounter difficulties when their children reach adolescence or adulthood, because of issues relating to limited opportunities for independent living.

The literature on childhood disability contains numerous accounts of difficult interactions between families and the professionals who work with them, including physicians, teachers, therapists, and others. Some of this difficulty derives from the conflicting roles of parents and professionals (Seligman & Darling, 2007) and from the continued presence of professional dominance (Leiter, 2004). Some newer training programs for medical professionals have been incorporating the family's point of view, resulting in some decrease in the power imbalance in the professional-family relationship (Darling & Peter, 1994).

STUDIES OF CHILDREN

As already noted, most of the literature in this field has been centered on the effects on the family rather than on the reactions of the children themselves. However, a small social science literature on children does exist. Some early studies suggested that children with disabilities had lower self-esteem than other children. Like early studies of parents, this body of research had methodologic flaws, including reliance on clinical samples and lack of comparison groups (Darling, 1979). More recent research on children reflects the findings of studies of parents—that difficulties stem more from social disadvantage than from limitations inherent in the children's impairments themselves (Middleton, 1999).

In a discussion of children with disabilities in South Africa and other countries, Philpott and Sait (2001) argue that this population has been excluded from both children's programs and disability programs. They take the view that in a context of poverty, disabled children are especially vulnerable to neglect and exclusion.

CURRENT TRENDS
AND POLICY IMPLICATIONS

Whereas earlier research tended to employ a medical model and to focus on parents' and children's coping strategies, more recent studies have tended to be based in a "social" model (Oliver, 1996) and to focus on social change. Federal legislation, like the Americans with Disabilities Act (ADA) and the Individuals with Disabilities Education Act (IDEA) in the United States and comparable legislation in other countries, has reflected a shift from a charity perspective to a rights perspective (Scotch, 2001). In the United States especially, children with disabilities are increasingly being educated in inclusive settings with nondisabled children. Future research must address the effects on children of greater integration into the societal mainstream.

SEE ALSO Volume 1: *Attention Deficit/Hyperactivity Disorder (ADHD); Autism; Cognitive Ability; Illness and Disease, Childhood and Adolescence; Learning Disability;* Volume 2: *Disability, Adulthood;* Volume 3: *Disability and Functional Limitation, Later Life.*

BIBLIOGRAPHY

Darling, R. B. (1979). *Families against society: A study of reactions to children with birth defects.* Beverly Hills, CA: Sage.

Darling, R. B. (1994). Overcoming obstacles to early intervention referral: The development of a video-based training model for community physicians. In R. B. Darling & M. I. Peter (Eds.), *Families, physicians, and children with special health needs: Collaborative medical education models* (pp. 135–148). Westport, CT: Greenwood.

Darling, R. B., & Peter, M. I. (Eds). (1994). *Families, physicians, and children with special health needs: Collaborative medical education models.* Westport, CT: Greenwood.

Irwin, S. (2001). Repositioning disability and the life course: A social claiming perspective. In M. Priestley (Ed.), *Disability and the life course: Global perspectives* (pp. 15–25). New York, Cambridge University Press.

Leiter, V. (2004). Parental activism, professional dominance, and early childhood disability. *Disability Studies Quarterly, 24,* 1–16.

Mandelbaum, A., & Wheeler, M. E. (1960). The meaning of a defective child to parents. *Social Casework, 41,* 360–367.

Middleton, L. (1999). *Disabled children: Challenging social exclusion.* Oxford, England: Blackwell Science.

Oliver, M. (1996). *Understanding disability: From theory to practice.* New York: St. Martin's Press.

Philpott, S., & Sait, W. (2001). Disabled children: An emergency submerged. In Priestley, M. (Ed.), *Disability and the life course: Global perspectives* (pp. 151–166). New York, Cambridge University Press.

Scotch, R. (2001). *From goodwill to civil rights: Transforming federal disability policy.* Philadelphia: Temple University Press.

Seligman, M., & Darling, R. B. (2007). *Ordinary families, special children: A systems approach to childhood disability.* New York: Guilford.

World Health Organization (1999). *International classification of functioning, disability and health.* Geneva: Author.

Rosalyn Benjamin Darling

DIVORCE, EFFECTS ON CHILDREN

SEE Volume 1: *Child Custody and Support; Family and Household Structure, Childhood and Adolescence;* Volume 2: *Divorce and Separation.*

DRINKING, ADOLESCENT

Adolescence is a period of the life course often characterized by an increase in engagement in behaviors that pose a risk of harm, including alcohol use. During adolescence many individuals begin to experiment with a variety of substances; alcohol is the most widely used. Research has examined the ways in which alcohol use progresses throughout this developmental period as well as across the life course. Scientists have gained a more complete understanding of adolescent drinking by identifying risk and protective factors that predict alcohol use and developing theories to explain these behaviors. Innovative research to address unanswered questions is ongoing, due to the importance of understanding adolescent alcohol use and its implications for the health of growing individuals.

DEFINING AND STUDYING
ALCOHOL USE IN ADOLESCENCE

Research attempting to indentify antecedents and consequences of adolescent alcohol use typically examines four distinct constructs: age of initiation (or onset), frequency of alcohol use, intensity of alcohol use, and heavy episodic (or binge) drinking. Taken together, measures of initiation, frequency, and intensity illustrate the overall pattern of an individual's consumption of alcohol. The definition of the *age of initiation*, also called onset, differs slightly based on what researchers are interested in examining. Age of initiation has been operationalized as the age at one's first drink (more than a few sips), first regular use (i.e., weekly use), or first episode of drunkenness. Early

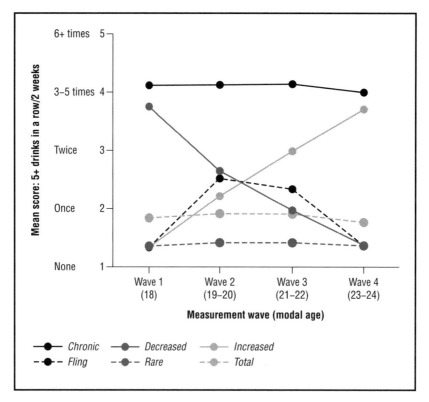

Figure 1. *Mean scores for five or more drinks in a row in the past two weeks by binge drinking trajectory.* **CENGAGE LEARNING, GALE.**

alcohol use is considered detrimental largely because of its association with continued use, association with deviant peer group selection, contribution to injuries and accidents, and negative effects on the developing brain during childhood and adolescence.

Frequency of alcohol use describes how regularly or how often adolescents consume alcohol. Some adolescents use alcohol sporadically whereas others are habitual users. Often individuals are asked to mentally aggregate their behavior. For example, they may be asked to report *how often*, on average, over the past 12 months they consumed alcoholic beverages, from never, to once a week, to every day.

Measures of *intensity*, or *quantity*, generally assess how *much* an individual drinks on an average drinking occasion or on a peak drinking occasion. Combined with knowledge of an individual's gender, weight, and the amount of time over which the drinks were consumed, researchers use total quantity of drinks to calculate a person's peak *blood alcohol concentration* (BAC). BAC is an estimate of the level of intoxication an individual experienced, with high levels associated with specific physiologic effects (e.g., impaired judgment, blacking out). Therefore, intensity of alcohol use is associated with acute health risks that may be relatively minor, such as injury, or severe, such as death due to alcohol poisoning.

One specific way to describe the intensity of alcohol use is *heavy episodic drinking*, also called *binge drinking*. Heavy episodic drinking describes an occasion when a man consumed five or more drinks in a row or a woman consumed four or more drinks in a row. The discrepancy in the number of drinks by gender takes into account the relative weight differences and differences in ability to metabolize alcohol. Therefore, the gender difference in number of drinks reflects the amount of alcohol necessary to reach a similar BAC (i.e., become intoxicated). This measure is intended to describe an individual's drinking behavior over a relatively short amount of time (e.g., an evening), rather than a "binge" over several days.

Adolescent alcohol use is widespread. Data from the 2006 Monitoring the Future Study indicated that alcohol use had been tried by 41% of eighth graders, 62% of tenth graders, and 73% of twelfth graders in the United States and that 20%, 41%, and 56% in these three grades had been drunk at least once in their lifetime. Further, heavy episodic drinking at least once in the prior 2-week period was reported by 11% of eighth graders, 22% of tenth graders, and 25% of twelfth graders (Johnston, O'Malley, Bachman, & Schulenberg, 2007). Although alcohol use is common, it has many risks and negative consequences. Immediate consequences include fighting, injury, risky sexual behavior, victimization, alcohol poisoning, and drunk-

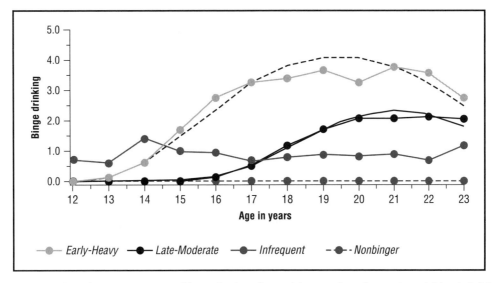

Figure 2. *Growth curve trajectories of binge drinking from adolescence through emerging adulthood. Solid lines represent estimated growth trajectories. Dashed lines represent observed means of binge drinking at each age.* CENGAGE LEARNING, GALE.

driving fatalities (Hingson, Heeren, Zakocs, Kopstein, & Wechsler, 2002). Long-term consequences include academic failure, alcoholism, and deficits caused by alcohol's effects on the developing adolescent brain. Each year in the United States, about 5,000 people under age 21 die as a result of motor vehicle crashes, unintentional injuries, homicides, and suicides that involve underage drinking (NIAAA, 2004/2005).

DEVELOPMENTAL TRENDS IN ALCOHOL USE

In general, people's use of alcohol tends to increase during adolescence before peaking and then decreasing during the transition to adulthood (see reviews by Maggs & Schulenberg, 2004/2005; Schulenberg & Maggs, 2002). Some social role changes accompany these average drinking increases (e.g., college entrance) and decreases (e.g., spousal and parenting roles). Within these broad normative trends are specific pathways (or trajectories) followed by individuals or small groups of people with more similar origins, developmental course, and outcomes of their alcohol use (see Figure 2). The most common, and lowest-risk, trajectory reflects drinking patterns of abstainers, light drinkers, very rare heavy drinkers, or individuals who rarely drink at high levels regardless of age (i.e., a low, flat line for amount of heavy drinking over time). A second pattern is stable-moderate drinking, described by some heavy drinking during adolescence and early adulthood but without dramatic escalation to severely problematic levels.

Groups of more problematic alcohol users have also been identified. Chronic heavy drinkers typically begin

using alcohol at relatively young ages (by middle adolescence) and continue to use at high rates into their twenties. A second more problematic group, late-onset heavy drinkers, tends to initiate drinking slightly later—for example late in high school—but to increase steeply in drinking and continue heavy use into early adulthood. A third more problematic trajectory, for individuals often called "fling" drinkers, exhibits heavy drinking within a developmentally limited time period and desists by late adolescence or early adulthood. In the realm of clinical psychology, Zucker (1995) has identified four types of alcoholism that are distinguished by their antecedent causes, courses, and outcomes. One of these types, *developmentally limited alcoholism*, shows a similar pattern to the fling drinkers, that is, time-limited, peer-focused heavy drinking that reduces spontaneously with the successful assumption of adult family and career roles.

ALCOHOL-RELATED RISK AND PROTECTIVE FACTORS DURING ADOLESCENCE

Certain personal and social-environmental characteristics have been consistently associated with heavier adolescent alcohol use (Hawkins, Catalano, & Miller, 1992). The antecedents of adolescent alcohol use can be conceptualized into two categories: risk factors and protective factors. *Risk factors* are variables (e.g., personal, family, environmental characteristics) that predict a higher likelihood of a negative outcome, in this case of using or abusing alcohol (Maggs & Schulenberg, 2005). Some commonly researched risk factors for adolescent alcohol use include genetic vulnerability, childhood impulsivity, sensation seeking proclivity, psychiatric diagnoses, family history of alcoholism, heavy drinking peers, positive alcohol related expectancies, and childhood trauma (Griffin, Scheier, Botvin, & Diaz, 2000; NIAAA, 1997). In addition, early initiation of alcohol use in adolescence (prior to age 14–16) is a noted risk factor for later problems including heavy alcohol use, drug use, and driving after drinking (Hawkins et al., 1997).

Gender is also a risk factor: Men tend to drink more than women. In addition, individuals who do not live with two biologic parents and who have parents who use alcohol more heavily and have more symptoms of alcoholism are more likely to use heavily themselves. Male heavy drinkers in late adolescence are also especially likely to have exhibited more externalizing symptoms, such as delinquency and aggressivity (Maggs & Schulenberg, 2004/2005). In regard to ethnicity, White and Hispanic high school students are more likely to report using alcohol than their Black peers (Johnston, O'Malley, Bachman, & Schulenberg, 2007). Disparities in risk are also found among varying levels of socioeconomic status, with higher rates of alcoholism found among families of lower socioeconomic status (Ellis, Zucker, & Fitzgerald, 1997).

In contrast, *protective factors* are variables that predict a higher likelihood of a positive outcome. In this context, protective factors would be characteristics that are likely to result in abstaining from alcohol use, later initiation or lower levels of alcohol use, or fewer alcohol-related negative consequences. Examples of such protective factors include school commitment, academic achievement, religious involvement, prosocial peer involvement, peer acceptance, self-esteem, parental attachment, parental involvement, and structured free time. In many studies, the lack of an established risk (e.g., not having parents who drank heavily) might also be considered as protective (Hawkins et al., 1992).

When accounting for possible risk or protective factors of alcohol use during adolescence, the concepts of equifinality and multifinality must be considered (Cicchetti & Rogosch, 1996). That is, the same problem may have different causes, and not all people exposed to a given risk factor will develop the problem. These terms, therefore, highlight the varying nature of risk and protective factors and help to explain why some people do not develop problems despite exposure to significant risk factors, and why some individuals do develop problems despite little exposure to risk factors. Equifinality involves the idea that different patterns of risk and protective factors may lead to the same outcome. For example, heavy alcohol use in adolescence may be caused, in part, by a family history of alcoholism for some youth and by peers who use alcohol heavily for others. Multifinality denotes the idea that a given pattern of risk and protective factors may lead to many different outcomes. For

example, parental alcoholism has been shown to increase the likelihood of alcohol misuse and dependence in some people, whereas it causes others to abstain from alcohol completely (Sher, 1991).

Much is known of the vast array of potential risk and protective factors for adolescent alcohol use. However, due to the notions of equifinality and multifinality, our understanding of exactly how risk and protective factors work together to predict an individual's behavior or how they are more or less influential at different phases of an individual's life remains somewhat limited.

THEORIES OF ADOLESCENT ALCOHOL USE

Several important theories have been applied to alcohol use in adolescence and early adulthood (see review by Chassin et al., 2004). The overarching *developmental-contextual perspective* (Baltes, 1987; Lerner, 1982) asserts that development occurs across the lifespan in multiple domains of functioning (e.g., cognitive, interpersonal, emotional) as individuals select and accommodate to multiple contexts. These processes are especially evident during transitions, and the adolescent transition to alcohol use is no exception. Several factors influence the timing and course of alcohol use for individuals, as already discussed. Three main theoretical perspectives on adolescent alcohol use are reviewed here.

Social control theory is a dominant perspective in sociologic delinquency research that argues that adolescents who are not connected to institutions (e.g., schools, religious organizations) and role models (e.g., parents, teachers) in society are more likely to engage in risk behaviors, including alcohol use (Elliott, Ageton, & Canter, 1979; Hirschi, 1999). Several versions of the theory exist, but the unifying concept is that delinquent behavior is the result of weakened ties to the values and norms of conventional society. As a result of weak bonds to people and organizations that would provide norms for prosocial behavior, people may use alcohol or other drugs. Delinquency is learned through imitation and differential reinforcement. Social control theory is a general deviance hypothesis that describes the quality of informal social controls as shapers of risk behavior initiation, behavior maintenance, and behavior change.

Second, *problem behavior theory* is a well-known and oft-cited perspective to explain substance use (Jessor & Jessor, 1977; Donovan, Jessor, & Costa, 1988). Problem behavior is defined as behavior that (a) is considered inappropriate for adolescents, (b) departs from the social and legal norms of society at large, and (c) tends to elicit social control responses from authoritative institutions. This definition is broad, and therefore problem behavior theory contends that adolescents with these characteristics

Keg Party. *Alcohol is the most widely used substance among adolescents.* AP IMAGES.

tend to engage in multiple problem behaviors, including alcohol use, illicit drug use, delinquency, and precocious sexual behavior. The prediction of both problem and conventional behaviors is hypothesized to be possible through the influences of demographic characteristics, and, primarily, through the influences of the *personality system* (i.e., motivations, beliefs, and attitudes) and *perceived-environment system* (i.e., peer and parental supports and approval of behaviors).

Third, from the tradition of experimental psychology, the *theory of reasoned action* is a widely acknowledged model for understanding behavioral intentions in many domains, including substance use. This theory asserts that attitudes and expected consequences from drinking, as well as the perceived social norms of alcohol use, explain the behavior (Fishbein & Ajzen, 1975). *Attitudes* are defined as an individual's belief that a behavior will lead to a given set of consequences (e.g., having more fun, having a hangover), weighted by the value attached to those outcomes (e.g., very important to me to experience or avoid). Subjective *social norms* are defined as the combination of normative beliefs (i.e., perceived approval of drinking by others) and motivation to comply with these beliefs. The theory of reasoned action focuses on attitudes about drinking and beliefs about social norms for use and resulting consequences. It has also been extended into the theory of planned behavior by Ajzen (2001) to include perceptions of control over the behavior.

TRENDS IN RESEARCH ON ALCOHOL USE

Trends in research on alcohol use and abuse among adolescents are focused largely on improving the quality of

data. Innovative approaches include assessing adolescent drinking expectancies, behaviors, and consequences on a series of specific days (or drinking occasions) to understand better why and how adolescents use alcohol and their experiences when they do. For example, new developments in statistics such as multilevel models allow researchers to ask whether people drink more on days they are in a better mood compared with occasions when they are in a worse mood, instead of just asking whether people who are more or less happy drink more than others. Such data will allow researchers to understand more fully the reciprocal relations of alcohol use and constructs such as affect, sexual behavior, other drug use, and sleep patterns of adolescents.

To improve understanding of normative developmental changes in alcohol use, person-centered analysis of substance use over time is becoming increasingly popular. Statistical strategies include latent transition analysis, which models how people move in and out of discrete categories (e.g., current alcohol user or not) over time, and trajectory analysis, which fits a curve to the patterns of alcohol use (see Figure 2), are increasingly popular. These statistical approaches enable researchers to identify developmental patterns of alcohol use across early, middle, and late adolescence that differentiate potentially problematic alcohol users. In addition, these approaches may help prevention efforts to identify adolescents who are particularly at risk for future alcohol-related harm and addiction based on early indicators.

In addition, genetically informed designs—for example twin and adoption studies—are able to control for biologic differences in sensitivity to alcohol and for environmental factors such as patterns of parental use. These designs provide important information about the causes and effects of alcohol use initiation and maintenance (Fowler et al., 2007). Furthermore, because individuals who initiate alcohol use earlier are also likely to continue using it, the true influences of early initiation on later alcohol problems and dependence is not firmly established.

Approaches for the prevention of alcohol use by adolescents and interventions to reduce harmful use are being developed. Efficacious programs include approaches that teach refusal skills (Botvin & Griffin, 2002), focus on changing perceived social norms for use (Walters & Neighbors, 2005), correct expectancies of alcohol's positive effects (Baer, Kivlihan, Blume, McKnight, & Marlatt, 2001), or use environmental strategies (e.g., checking identification, reducing alcohol outlet density) (Imm et al., 2007). Randomized trials of prevention programs that are designed to reduce risk factors (e.g., delay initiation of use) or increase protective factors (e.g., increase positive leisure time use) to produce positive changes in outcomes of interest (e.g., fewer alcohol-related negative consequences) can also inform an understanding of the development and course of alcohol use in the lives of individuals.

SEE ALSO Volume 1: *College Culture; Drug Use, Adolescent; Health Behaviors, Childhood and Adolescence; Mental Health, Childhood and Adolescence; Peer Groups and Crowds; School Culture; Theories of Deviance;* Volume 2: *Health Behaviors, Adulthood;* Volume 3: *Health Behaviors, Later Life.*

BIBLIOGRAPHY

Ajzen, I. (2001). Nature and operation of attitudes. *Annual Review of Psychology, 52,* 27–58.

Baer, J. S., Kivlihan, D. R., Blume, A. W., McKnight, P., & Marlatt, G. A. (2001). Brief intervention for heavy-drinking college students: 4-year follow-up and natural history. *American Journal of Public Health, 91,* 1310–1316.

Baltes, P. B. (1987). Theoretical propositions of life-span developmental psychology: On the dynamics between growth and decline. *Developmental Psychology, 23,* 611–626.

Botvin, G. J., & Griffin, K. W. (2002). Life skills training as a primary prevention approach for adolescent drug abuse and other problem behaviors. *International Journal of Emergency Mental Health, 4,* 4147.

Chassin, L., Hussong, A., Barrera, M., Molina, B. S. G., Trim, R., & Ritter, J. (2004). Adolescent substance use. In R. M. Lerner & L. Steinberg (Eds.), *Handbook of adolescent psychology* (2nd ed., pp. 665–696). New York: Wiley.

Cicchetti, D., & Rogosch, F.A. (1996). Equifinality and multifinality in developmental psychopathology. *Development and Psychopathology, 8,* 579–600.

Donovan, J. E., Jessor, R., & Costa, F. M. (1988). Syndrome of problem behavior in adolescence: A replication. *Journal of Consulting and Clinical Psychology, 56,* 762–765.

Elliott, D. S., Ageton, S. S., & Canter, R. J. (1979). An integrated theoretical perspective on delinquent behavior. *Journal of Research in Crime and Delinquency, 16,* 3–27.

Ellis, D. A., Zucker, R. A., & Fitzgerald, H. E. (1997). The role of family influences in development and risk. *Alcohol Health and Research World, 21,* 218–226.

Fishbein, M. & Ajzen, I. (1975). *Belief, attitude, intention and behavior: An introduction to theory and research.* Reading, MA: Addison-Wesley.

Fowler, T., et al. (2007). Exploring the relationship between genetic and environmental influences on initiation and progression of substance use. *Addiction, 101,* 413–422.

Griffin, K. W., Scheier, L. M., Botvin, G. J., & Diaz, T. (2000). Ethnic and gender differences in psychosocial risk, protection, and adolescent alcohol use. *Prevention Science, 1,* 199–212.

Hawkins, J. D., Catalano, R. F., & Miller, J. Y. (1992). Risk and protective factors for alcohol and other drug problems in adolescence and early adulthood: Implications for substance abuse prevention. *Psychological Bulletin, 112,* 64–105.

Hawkins, J. D., Graham, J. W., Maguin, E., Abbot, R., Hill, K. G., & Catalano, R. F. (1997). Exploring the effects of age of alcohol use initiation and psychosocial risk factors on subsequent alcohol misuse. *Journal of Studies on Alcohol, 58,* 280–290.

Hingson, R. W., Heeren, T., Zakocs, R. C., Kopstein, A., & Wechsler, H. (2002). Magnitude of alcohol-related mortality and morbidity among US college students ages 18–24. *Journal of Studies on Alcohol, 63,* 136–144.

Hirschi, T. (1999). Social bond theory. In F. T. Cullen & R. Agnew (Eds.), *Criminological theory: Past to present (Essential readings)* (pp. 167–174). Los Angeles: Roxbury.

Imm, P., Chinman, M., Wandersman, A., Rosenbloom, D., Guckenburg, S., & Leis, R. (2007). *Preventing underage drinking: Using Getting to Outcomes with SAMHSA Strategic Prevention Framework to achieve results.* Newport Beach, CA: RAND Corporation. Available at: www.rand.org

Jessor, R. & Jessor, S. L. (1977). *Problem behavior and psychosocial development: A longitudinal study of youth.* New York: Academic Press.

Johnston, L. D., O'Malley, P. M., Bachman, J. G., & Schulenberg, J.E. (2007). *Monitoring the future: National survey results on drug use, 1975-2006: Vol. 1. Secondary school students* (NIH publication No. 07-6205), Bethesda, MD: National Institute on Drug Abuse.

Lerner, R. M. (1982). Children and adolescents as producers of their own development. *Developmental Review, 2,* 342–370.

Maggs, J. L., & Schulenberg, J. (2004/2005). Trajectories of alcohol use during the transition to adulthood. *Alcohol Research and Health, 28,* 195–201.

Maggs, J. L., & Schulenberg, J. (2005). Initiation and course of alcohol use among adolescents and young adults. In M. Galanter (Ed.), *Recent developments in alcoholism* (pp. 29–47). New York: Kluwer Academic/Plenum.

National Institute on Alcohol Abuse and Alcoholism (NIAAA) (1997). Youth drinking: Risk factors and consequences. *Alcohol Alert,* No. 37. Bethesda, MD: Author.

National Institute on Alcohol Abuse and Alcoholism (NIAAA) (2004/2005). The scope of the problem. *Alcohol Research and Health, 28,* 111–120.

Ouellette, J. A., Gerrard, M., Gibbons, F. X., & Reis-Bergan, M. (1999). Parent, peers and prototypes: Antecedents of adolescent alcohol expectancies, alcohol consumption, and alcohol-related life problems in rural youth. *Psychology of Addictive Behaviors, 13,* 187–197.

Schulenberg, J. E., & Maggs, J. L. (2002). A developmental perspective on alcohol use and heavy drinking during adolescence and the transition to young adulthood. *Journal of Studies on Alcohol, Supplement No. 14,* 54–70.

Sher, K. J. (1991). *Children of alcoholics: A critical appraisal of theory and research.* Chicago: University of Chicago Press.

Walters, S. T., & Neighbors, C. (2005). Feedback interventions for college alcohol misuse: What, why and for whom? *Addictive Behaviors, 30,* 1168–1182.

Zucker, R. A. (1995). Pathways to alcohol problems and alcoholism: A developmental account of the evidence for multiple alcoholisms and for contextual contributions to risk. In R. A. Zucker, G. M. Boyd, and J. Howard (Eds.), *The development of alcohol problems: Exploring the biopsychosocial matrix of risk* (pp. 255–289). NIAAA Research monograph 26. Rockville, MD: U.S. Department of Health and Human Services, National Institute on Alcohol Abuse and Alcoholism.

Megan E. Patrick
Caitlin Abar
Jennifer L. Maggs

DRUG USE, ADOLESCENT

Use of illicit drugs by adolescents has been a concern for many years. Because most developed societies in Europe, Asia, and North America prohibit the use of tobacco and alcohol by those under the age of 16 or 18, and laws are in place proscribing the use of marijuana, cocaine, and other psychoactive substances, adolescent drug use falls under the purview of the juvenile justice system. Furthermore, it is a relatively common behavior that is studied by scholars from several academic disciplines, including medicine, public health, psychology, sociology, economics, and anthropology. Concerns about drug use have also led to well-funded federal, state, and local government programs designed to detect, prevent, and treat adolescent drug use and users. The study of adolescent drug use includes examining the reasons for use, patterns and consequences of use, prevention programs designed to identify and curb use, and treatment for those who develop problems associated with use. Important research has also focused on two distinct developmental issues: (a) development of use from one type of drug to another; and (b) consequences of adolescent drug use for social and psychological development and the achievement of life course milestones.

WHY ADOLESCENTS USE DRUGS

Several biological and social scientific theories attempt to explain why adolescents use drugs; however, they rarely distinguish among types of drugs. Biological theories tend to address risk factors such as low impulse control or impaired neurochemical functioning. For instance, some genetic-based theories argue that adolescents with impaired dopamine, serotonin, or monamine oxidase (MAO) functioning—which are associated with impulsivity, aggression, and sensational-seeking personality traits—are at greater risk of drug use and problems related to use.

Psychological theories have increasingly adopted cognitive-affective or social learning approaches. Cognitive-affective theories suggest that adolescents who know where to find and successfully use drugs are more likely to use them. A similar approach focuses on refusal self-efficacy to point out that some youth are incapable of refusing to use drugs when they are offered, whereas others have stronger refusal skills, perhaps because of personality traits, such as emotional stability or conscientiousness.

Social learning theory focuses on relations with others, in particular by studying the interplay between an adolescent's definition of drug use and the definitions—or what some call *beliefs* or *cognitions*—of the adolescent's family members, friends, important adult figures, and role models. According to this perspective, three stages culminate in the development of regular drug use: initiation based on observing and imitating significant others; continuation based on social reinforcement by significant others; and the adoption of beliefs that positive outcomes are associated with drug use, but few costs (e.g., more satisfactory social relationships, enjoying the effects of a drug, low risk of being caught and punished). Cognitively focused versions of this theory further point out that self-efficacy about the use of drugs can develop through observing the actions or listening to the beliefs of significant others. For example, if close friends discuss the merits of use and how to use drugs without getting caught, an adolescent's self-efficacy concerning use is magnified.

Other social psychological and sociologically based theories emphasize weak attachments to conventional sources of socialization (e.g., parents, schools) and stronger attachments to peers who may encourage drug use. Adolescents who experience weak attachments to parents or schools, stronger attachments to peers, and values that are conducive to drug use, such as alienation or lack of conformity, are at high risk of use. Furthermore, adolescents who have stressful personal or social environments may find that drug use either alleviates the stress or offers a way to cope with the stress that is unavailable through conventional means. The Social Development Model (SDM) expands this attention to social factors by pointing out that some influences are more salient during certain developmental periods. In childhood, family influences are more important and thus when they are fractured or disrupted, the stage is set for risky behaviors such as drug use. During adolescence, peer influences emerge and become more important than family or school influences. According to SDM, drug use is usually adopted if adolescents become involved with peers who use drugs. But this is more likely if, during earlier developmental periods, they had relatively few positive social interactions at home or in school, they were not taught adequate interpersonal skills by their parents and siblings, and they received little positive reinforcement at home or at school.

Other theories argue that adolescents who use drugs tend to be oriented toward short-term objectives at the expense of long-term goals; have low self-esteem that is bolstered by drug use, perhaps because of peer-acceptance; or are poorly supervised or supported by parents. Some researchers have attempted to combine various aspects of these theories into one model, such as Jessor's Problem Behavior Theory (PBT) (Jessor & Jessor, 1997) or Oetting and Beauvais's (1987) Peer Cluster theory. These theories tend to distinguish among proximal influences (e.g., peers) and distal influences (e.g., poor family relations, temperament) on drug use.

Another popular theory does not try to explain directly why adolescents use drugs, but rather focuses on the sequencing of drug use. Known as the *gateway hypothesis*, it proposes that a well-established pattern of drug use exists that begins with inhalants or cigarettes and alcohol, shifts to marijuana, and then, for some adolescents, ends up with cocaine and other illegal drugs (e.g., LSD, heroin). Somewhere along this progressive pathway, some adolescents develop a substance-use disorder, such as drug dependence or drug abuse, which includes unsuccessful attempts to decrease use, tolerance (requiring more of a drug to get the same effect), or withdrawal symptoms (physiological discomfort manifest when the person stops using a particular drug). The gateway sequence is often described as a funnel, where few users move on to the use of other drugs. Moreover, although the sequencing has been fairly well-established, debate continues over whether the use of one drug causes the use of another, or whether certain drugs act as gateways to other drugs merely for social, psychological or cultural reasons. One promising notion is that adolescents who move farther into the sequence, especially those who become dependent, tend to have mental health problems. Moreover, research suggests that those who move on tend to be more frequent users or use a greater variety of drugs (e.g., regular use of cigarettes, alcohol, *and* marijuana prior to cocaine use).

CORRELATES OF ADOLESCENT DRUG USE

In addition to theories that try to explain the reasons for drug use, substantial information has been reported on the most common correlates or risk factors associated with use. These correlates may be divided into intrapersonal, interpersonal, institutional, and community factors. The intrapersonal factors that are positively associated with adolescent drug use include physiological and psychological conditions such as impaired dopamine, MAO, or serotonin functioning that may have a genetic basis; attention deficit/hyperactive disorder (ADHD); oppositional defiant disorder/conduct disorder (ODD/CD), which is often manifest by aggression, delinquency, and nonconformity; depressive disorders, particularly in girls (but not in boys); impulsivity; early pubertal onset; poor cognitive skills or school performance; and a sensation-seeking personality. Attitudes and norms that favor drug use, or failure to see the risks of use, are, not surprisingly, positively associated with use. Evidence is inconsistent regarding the association between demographic characteristics (e.g., sex, ethnicity, socioeconomic status) and adolescent drug use, although African Americans report a lower prevalence of some types of drug use (e.g., cigarettes) than Whites in many surveys.

Teenage Boys Smoking Marijuana. *During adolescence, peer influences emerge and become more important than family or school influences.* **INGRAM PUBLISHING/GETTY IMAGES.**

The most frequently studied interpersonal influences are relations with family members and friends. Many studies indicate that drug use or dependence/abuse among parents, poor parenting skills, or lack of parent-child relations are associated with a higher risk of drug use among adolescents. The dynamics of these relationships are not fully understood but seem to stem from early childhood experiences in which parents rely on authoritarian or permissive methods with children, exhibit overly aggressive or avoidant parenting strategies, fail to supervise children, or themselves provide models of drug use to children. A combination of these factors may lead to rejection by conventional peers and acceptance by other adolescents who have experienced the same difficult family lives. These types of peer relations, which often mirror coercive family relations, increase the risk of misbehaviors, including the use of drugs such as cigarettes, marijuana, and cocaine. If poor family relations, lack of parental supervision, and generally lower involvement in family activities continue during adolescence, a spiral of delinquency and drug use is increasingly likely. For example, studies showing that adolescents who regularly share meals with their families or talk to their parents frequently are at lower risk of drug use may reflect the consequence of this developmental pathway away from or toward drug use. Finally, other studies indicate that having an older sibling who uses drugs increases the likelihood of drug use among adolescents.

Evidence is clear and consistent that the most powerful risk factor involves friends who use drugs. Solitary use, at least initially, is rare. Rather, adolescent drug users usually have friends who also use, and they tend to use together. The complication for research on this issue concerns the causal sequence: Does associating with certain peers lead to drug use, or does drug use lead to associating with others who use drugs? Studies have frequently relied on adolescents' reports of their friends' behaviors to estimate the association with peer use. However, this has led to an overestimation of peer effects because adolescents tend to estimate their friends' use as corresponding to their own level of use. More rigorous studies that query adolescents and their friends about drug use suggest that peer influences are not as strong as previous studies indicated. Nevertheless, it is likely that peer influences are the consequence of other intrapersonal and family-based factors that increase the risk of associating with, as well as influencing, friends' behaviors.

Institutional factors that are associated with adolescent drug use involve families, but also include schools, religious organizations, and community influences. Although the most frequently studied family influences involve relations with parents or siblings, there is also evidence that family structure is associated with drug use. Adolescents who live with two biological parents appear to be at lowest risk of drug use, whereas those living with a single father or with no biological parent are at highest risk. The reasons for these associations are not well understood nor are they accounted for fully by other factors (e.g., family relations, socioeconomic status). A promising hypothesis, however, is that they may reflect stressful living conditions or custody decisions wherein the most difficult adolescents are placed with fathers or tend to be placed in alternative care arrangements.

Some evidence shows that adolescents who attend poorly organized schools, schools with indifferent teachers and administrators, or schools that include a high concentration of drug users are at heightened risk of drug use. However, school policies and programs designed to discourage drug use have been shown to have little effect. Evaluations of several large, school-based drug prevention programs, such as Drug Abuse Resistance Education (DARE), indicate that they are not effective in deterring adolescent drug use over the long term. Moreover, drug testing programs, which many secondary schools have adopted, appear to have little effect on rates of use among students. The most promising school-based prevention programs are those that focus on commitments, norms, or intentions not to use drugs, have *booster* sessions led by peers about a year after initial program delivery, use interactive delivery methods (e.g., frank discussions between adolescents and program leaders), and have parallel community programs designed to prevent adolescent drug use.

Attention has been growing about the influence of religious beliefs, practices, and organizations on adolescent drug use. Religious influences have been addressed at both individual and aggregate levels. At the individual level, studies have focused on whether personal religious beliefs (e.g., belief in God; seeing religion as particularly important in one's life), religious practices (e.g., personal prayer, attendance at religious services), or religious peers and family members affect adolescent drug use. Some evidence suggests that all these factors are negatively associated with cigarette, marijuana, and cocaine use, although these factors are not as influential as the family and peer factors discussed earlier.

At the aggregate level, studies suggest that attending schools with a higher concentration of religious adherents decreases the likelihood of drug use. In addition, belonging to a religious denomination that specifically prohibits certain forms of drug use, such as cigarette smoking, diminishes the probability of use.

Another way of viewing the potential influence of religion is to consider whether individual-level effects are amplified or buffered by aggregate level effects. The *moral communities* hypothesis, for example, proposes that the influence of personal religiousness is efficacious only when it is supported by community- or school-level religious norms or beliefs. In particular, proponents of this view argue that religious beliefs and practices decrease the risk of adolescent drug use primarily in communities that have a high concentration of members who share religious beliefs and practices but are less influential in more secular communities. Research has found inconsistent results, although at least two studies suggest that religious adolescents who attend high schools with many religious students are at particularly low risk of drug use.

Finally, community influences have a modest effect on adolescent drug use. Communities experiencing high unemployment, more transience, or a lack of communal trust in neighbors may provide poorer social environments for adolescents, thus increasing the risk of drug use.

TRENDS IN ADOLESCENT DRUG USE

Two large national surveys in the United States serve as the basis for studying trends in adolescent drug use. The *Monitoring the Future* (MTF) program, conducted by researchers at the University of Michigan, is a series of national surveys of high school and junior high students that have been completed annually since 1975 (see, for example, Johnston, O'Malley, Bachman, & Schulenberg, 2007). The *National Survey of Drug Use and Health* (NSDUH) (formerly known as the *National Household Survey on Drug Abuse* or NHSDA), conducted by the Substance Abuse and Mental Health Services Administration (SAMHSA), has collected national surveys of U.S. residents, ages 12 and older, periodically since the early 1970s (see, for example, Office of Applied Studies, 2007). Identifying historical trends in adolescent drug use before the 1970s is difficult because consistent and rigorous surveys were not conducted; however, most experts agree that the 1960s was a period of increasing

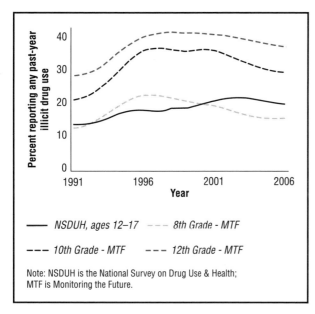

Figure 1. *Trends in any past-year adolescent drug use.* CENGAGE LEARNING, GALE.

experimentation with illicit drugs such as marijuana and hallucinogens.

The MTF and NSDUH surveys show a similar ebb and flow of drug use since the 1970s, with similar trends across drug types. However, certain forms of drug use have become more or less popular across the years. The highest level of adolescent drug use occurred in the late 1970s and has since decreased. The most dramatic decrease has occurred for cigarettes; the prevalence of use has decreased continually since the late 1970s. However, there has also been an increase in the use of prescription-type drugs such as the painkiller hydrocodone, many of which are diverted and sold in the so-called *gray market*.

Since the early 1990s, adolescent drug use increased and then decreased for some groups. For example, Figure 1 shows the percentage of adolescents who reported any illicit drug use in the past year from 1991 through 2006 based on three groups of students from the MTF surveys and those ages 12 to 17 years from the NSDUH. In the MTF, levels of use increased through the late 1990s and have since decreased slightly. However, the NSDUH shows a leveling off in the mid-2000s. (Note: The lines in the graphs represent smoothed running averages to minimize the influence of random fluctuations from year to year.)

Trend analyses, such as those that involve emergency department admissions and drug users seeking treatment, suggest that some types of drugs have increased in both popularity and in presenting problems for adolescents.

These include the so-called *club drugs* such as methylenedioxymethamphetamine (MDMA, or ecstasy), gamma hydroxybutyrate (GHB), and flunitrazepam ("roofies"). However, the prevalence of use is low enough in general population surveys to make conclusions about long-term trends highly unstable.

CONSEQUENCES OF ADOLESCENT DRUG USE

Several studies have examined the short- and long-term consequences of adolescent drug use. Although most adolescents who use drugs quit using by young adulthood and do not suffer any negative consequences, a minority either continues to use or are at heightened risk of developmental problems or disrupted life-course milestones. The causal patterns or effects are not clear, yet strong correlational evidence suggests that heavier forms of drug use and abuse negatively affect normal development in adolescents.

Some specific findings from longitudinal research are that: (a) heavy drug use impairs educational milestones, with an increased risk of school drop-out, truancy, and poor school performance; (b) drug use is associated with precocious sexual experimentation, teenage pregnancy, and a higher risk of sexually transmitted diseases; (c) illicit drug use that continues from adolescence into adulthood is associated with higher levels of occupational turnover, including a heightened risk of being fired or of multiple job quits; and (d) heavier users of marijuana and other illicit drugs tend to be more involved in delinquency and criminal behavior and experience more arrests and mental health problems in early adulthood.

FUTURE RESEARCH ON ADOLESCENT DRUG USE

Adolescent drug use remains an important topic for research because it continues to be a common behavior among young people. For many adolescents, drug use is a temporary and quite benign behavior that has few long-term consequences. Nevertheless, identifying who is at highest risk of experiencing problems remains an important issue. Moreover, there is a need to expand research attention to the comorbidity of drug use problems and mental health problems, the genetic bases of dependence and abuse, and the long-term developmental consequences of drug use for young, middle, and older adults.

SEE ALSO Volume 1: *College Culture; Drinking, Adolescent; Health Behaviors, Childhood and Adolescence; Peer Groups and Crowds; School Culture; Theories of Deviance;* Volume 2: *Health Behaviors, Adulthood;* Volume 3: *Health Behaviors, Later Life.*

BIBLIOGRAPHY

Banken, J. A. (2004). Drug abuse trends among youth in the United States. *Annals of the New York Academy of Sciences, 1025*, 465–471.

Cuijpers, P. (2002). Effective ingredients of school-based drug prevention programs: A systematic review. *Addictive Behaviors, 27*, 1009–1023.

Dishion, T. J., Nelson, S. E., & Bullock, B. M. (2004). Premature adolescent autonomy: Parental disengagement and deviant peer process in the amplification of problem behavior. *Journal of Adolescence, 27*, 515–530.

Elkins, I. J., McGue, M., & Iacono, W. G. (2007). Prospective effects of attention deficit/hyperactivity disorder, conduct disorder, and sex on adolescent substance use and abuse. *Archives of General Psychiatry, 64*, 1145–1152.

Flory, K., Lynam, D., Milich, R., Leukefeld, C., & Clayton, R. (2004). Early adolescent through young adult alcohol and marijuana trajectories: Early predictors, young adult outcomes, and predictive utility. *Development and Psychopathology, 16*, 193–213.

Hoffmann, J. P., & Johnson, R. A. (1998). A national portrait of family structure and adolescent drug use. *Journal of Marriage and the Family, 60*, 633–645.

Jessor, R., & Jessor, S. L. 1997. *Problem behavior and psychosocial development: A longitudinal study of youth.* New York: Academic Press.

Johnston, L. D., O'Malley, P. M., Bachman, J. G., & Schulenberg, J. E. (2007). *Monitoring the future: National survey results on drug use, 1975-2006. Volume I: Secondary school students.* Bethesda, MD: National Institute on Drug Abuse.

Kandel, D. (Ed.) (2002). *Stages and pathways of drug involvement: Examining the gateway hypothesis.* New York: Cambridge University Press.

Macleod, J., Oakes, R., Copello, A., Crome, I., Egger, M., Hickman, M., et al., (2004). Psychological and social sequelae of cannabis and other illicit drug use by young people: A systematic review of longitudinal, general population studies. *The Lancet, 363*, 1579–1588.

Oetting E. R., & Beauvais, F. (1987, April). Peer cluster theory: Socialization characteristics and adolescent drug use. *Journal of Counseling Psychology, 34*, 205-213.

Office of Applied Studies. (2007). NSDUH: Results from the 2006 National Survey of Drug Use and Health: National findings. Rockville, MD: Substance Abuse and Mental Health Services Administration. Retrieved March 6, 2008, from http://www.oas.samhsa.gov/nsduhLatest.htm

Petraitis, J., Flay, B. R., & Miller, T. Q. (1995). Reviewing theories of adolescent substance use: Organizing pieces in the puzzle. *Psychological Bulletin, 177*, 67–86.

Silberg, J., Rutter, M., D'Onofrio, B., & Eaves, L. (2003). Genetic and environmental risk factors in adolescent substance use. *Journal of Child Psychology and Psychiatry, 44*, 664–676.

Wallace, J. M., Yamaguchi, R., Bachman, J. G., O'Malley, P. M., Schulenberg, J. E., & Johnston, L. D. (2007). Religiosity and adolescent substance use: The role of individual and contextual influence. *Social Problems, 54*, 308–327.

John P. Hoffmann

E

EATING DISORDERS

To say that the pervasiveness of eating disorders in the United States is distressing may be an understatement. The incidence rates for eating disorders range from no fewer than 8 per 100,000 persons in the general population per year for anorexia nervosa to 12 or more incidences for bulimia nervosa. Incidence rates for those most at risk are even higher. One study found an incidence rate for anorexia nervosa to be in excess of 70 per 100,000 females ages 15 to 19 (Lucas, Crowson, O'Fallon, & Melton, 1999). Even more startling is the mortality rate. Persons with anorexia nervosa are more likely to die than persons with any other psychiatric disorder. However, studies on comorbidity show that persons suffering from anorexia nervosa and other eating disorders often suffer from anxiety disorders as well, including obsessive–compulsive disorder and social phobia, the onset of which often occurs prior to an eating disorder. Thus it may be inaccurate to say that anorexia nervosa is the only causal factor in the mortality rates of these individuals. Having said that, the cause of death most commonly associated with persons with anorexia nervosa is suicide (Birmingham, Su, Hlynsky, Goldner, & Gao, 2005). This, coupled with the fact that those most at risk for anorexia nervosa are girls ranging in age from 14 to 18 years old, makes understanding and studying eating disorders even more critical (Keel, 2005). The National Institute of Mental Health agrees and in 2005 provided more than $20 million for research on eating disorders (Chavez & Insel, 2007a).

DEFINITIONS

Eating disorders are psychiatric diagnoses. Therefore official definitions found in the American Psychiatric Association's *Diagnostic and Statistical Manual for Mental Disorders* (Fourth Edition, text revision [*DSM–IV–TR*]; American Psychiatric Association [APA], 2000) must be understood.

To be clinically diagnosed with anorexia nervosa, a patient must: be underweight; be fearful of being fat; view his or her body from a distorted perspective; and, for women, experience amenorrhea, or the absence or cessation of one's menstrual periods (APA, 2000). To be underweight is to weigh 85% or less than the minimally defined acceptable weight for one's height and age. For example, if guidelines suggest that a person should weigh at the minimum 100 pounds, then weighing 85 pounds or less would constitute an underweight individual. Being afraid of becoming fat even if one is underweight and seeing one's body from a distorted perspective characterizes the second and third criteria for anorexia nervosa. Finally, women who should be menstruating but have not done so for three menstrual cycles are experiencing amenorrhea (APA, 2000). There are two subtypes of anorexia nervosa: restricting type and binge-eating/purging type. These subtypes allow clinicians to specify whether an individual also engages in binging and purging behaviors in addition to anorexia nervosa symptoms.

Bulimia nervosa is characterized by eating an excessive amount of food, lacking self-control during a binge, and then responding to the guilt associated with the overeating by engaging in compensatory behavior such as vomiting, fasting, or overexercising (APA, 2000). A binge is defined as eating more food than most ordinary people would consume within a specific time period under comparable circumstances while at the same time feeling at a loss of control over one's behavior. The *DSM–IV–TR* criteria states that for a diagnosis of bulimia nervosa, the binge eating and purging or non-purging compensatory behavior must occur at least twice over the course of each week for a period of 3 months. The *DSM–IV–TR* also identifies poor body image influencing feelings of self-worth as a diagnostic criterion.

As with anorexia nervosa, the *DSM–IV–TR* also distinguishes two subtypes of bulimia nervosa: purging type and nonpurging type. These subtypes discern between patients who vomit or use laxatives (purge) and those who employ nonpurging behaviors such as over-exercising or fasting to compensate for overeating.

Anorexia nervosa and bulimia nervosa are probably the most widely recognized eating disorders, but they are not the most pervasive. Eating disorder not otherwise specified (EDNOS) is the most frequently diagnosed eating disorder (Machado, Machado, Gonçalves, & Hoek, 2007). The prevalence rate of EDNOS diagnosis among adults in treatment for eating disorders is estimated at 60% (Fairburn & Bohn, 2004). This diagnosis is a catchall term for those who do not meet the complete clinical definition for either anorexia nervosa or bulimia nervosa. For example, if a woman meets all of the criteria of anorexia nervosa but by definition is not underweight or is still menstruating, she will be diagnosed as EDNOS. In the same manner, a man who eats an extraordinary amount of food, similar to a person with bulimia nervosa, but does not engage in inappropriate behavior to rid his body of the calories will be diagnosed as EDNOS. In this latter case, the man will be classified with binge-eating disorder, a condition that falls under EDNOS in the *DSM–IV–TR* (APA, 2000).

Prior to clinical diagnosis, friends and family members may look for warning signs if they suspect a loved one has an eating disorder. These may include but are not limited to being preoccupied with food or weight, being secretive about eating, avoiding social situations with food, and engaging in out-of-the-ordinary rituals with their food. In addition, a person with an eating disorder may express mood shifts or may dress in layers to hide weight loss (Ciotola, 1999). The medical consequences associated with eating disorders can be life threatening. Complications from eating disorders can range from constipation and tooth decay to infertility and cardiac problems (Rome & Ammerman, 2003).

HISTORICAL PERSPECTIVE

Evidence of self-starvation can be traced back to ancient civilizations in Egypt and Greece. Similarly, many historical religious accounts mention fasting rituals as a way to enhance prayer. Bell (1985) recounted the lives of saints throughout history such as Saint Catherine of Siena (1347–1380), who refused to eat, and he termed this behavior *holy anorexia*. The "discovery" of anorexia nervosa, however, is usually credited to William Gull, a prominent British physician, who first published a paper on anorexia nervosa in 1874 (Keel, 2005). Some authors credit not only Gull but also a lesser known French psychiatrist, Ernest-Charles Lasègue, who, working independently of Gull, published his own writings on anorexia in the year 1873.

In contrast to anorexia nervosa's long documented history, bulimia nervosa has a much more recent discovery. Research on this disorder first appeared in a 1979 article in the journal *Psychological Medicine* by Gerald Russell. In this paper Russell introduced the reader to patients who, like those with anorexia nervosa, are fearful of being fat but instead of starving themselves overeat and then purge. Following Russell's article, research on bulimia nervosa flourished, as did media attention to this disorder. The 1980s also saw a marked increase in the number of persons seeking treatment for bulimia nervosa (Theander, 2002).

THE STUDY OF EATING DISORDERS

The spectrum of research on eating disorders is impressive. Some studies explore genetic and neurobiological factors associated with eating disorders, whereas others focus on the outcomes of various treatments options, including pharmacological and psychosocial therapies (Chavez & Insel, 2007a). Other studies explore how family environments influence eating disorders (Laliberté, Boland, & Leichner, 1999), whereas still others observe eating disorders as a social problem. For example, Hesse-Biber, Leavy, Quinn, and Zoino (2006) argued that eating disorders are caused in part by economic and social causes: Media, diet, fitness, plastic surgery, and other industries both promote and profit from society's culture of thinness ideal.

Eating disorders research is as pertinent as ever, as scholarly journals such as *Eating Disorders: The Journal of Treatment and Prevention*, *International Journal of Eating Disorders*, and *European Eating Disorders Review* are dedicated entirely to the publication of studies on eating disorders. Discipline-specific journals, such as *American Psychologist*, have also recognized the importance of adding to the state of knowledge of anorexia nervosa, bulimia nervosa, and EDNOS and have dedicated entire issues to the special topic of eating disorders.

Limitations nevertheless exist within the research on eating disorders. One of the most disconcerting disparities has been the lack of diversity in the study populations. Despite the popular notion that eating disorders affect only White middle-class women, evidence suggests that eating disorders do not discriminate and are found to exist across racial and ethnic groups (Franko, Becker,

Thomas, & Herzog, 2007). In spite of this finding, very little research on eating disorders has focused on girls and women of color (Smolak & Striegel-Moore, 2001). This is problematic in both a theoretical and an applied sense. As a result of the lack of research on diverse populations, clinicians, health care professionals, and others on the front lines of detecting eating disorders are dismissing or misinterpreting eating disorder symptoms in patients of color. Studies show that racial and ethnic minorities are less likely than White patients to be asked by their health care providers about eating disorders and are not being referred for treatment to the same degree (Becker, Franko, Speck, & Herzog, 2003).

Once in treatment, racial and ethnic minority patients may face additional biases because many treatment and prevention programs do not take into account the diversity of experiences and are instead targeted primarily at White women's experiences (Smolak & Striegel-Moore, 2001). This is important to consider, as evidence suggests that how eating disorders manifest themselves may vary among different racial and ethnic populations. One study found significant variation in eating disorders symptoms among different ethnic groups, revealing the greatest frequency of laxative use among Native Americans and the least usage of diuretics among Asians. (Franko et al., 2007). Another study examined differences among Black and White women diagnosed with binge-eating disorder and found variation in frequency of binging and concern for body image and weight (Pike, Dohm, Striegel-Moore, Wilfley, & Fairburn, 2001).

FUTURE RESEARCH AREAS ON EATING DISORDERS

Future research on eating disorders will continue to explore diverse populations including not only racial and ethnic minorities but also men and gay, lesbian, bisexual, and transgender populations. The study of eating disorders will also expand to include more research on women who experience eating disorders later in life.

Research on eating disorders in men indicates a relationship between sexual identity and eating disorder behaviors: Gay and bisexual men are more likely than heterosexual men to develop an eating disorder (Feldman & Meyer, 2007). Early discussions on lesbian women and eating disorders suggests that lesbians may be immune to the unrealistic beauty ideals of a patriarchal society because they are not interested in attracting male partners. More recent data indicates, however, that lesbian and bisexual women suffer from eating disorders much like heterosexual women (Feldman & Meyer, 2007). Future studies will continue to explore the differences in lesbian, bisexual, and heterosexual women's experiences with eating disorders.

Some researchers (e.g., Forman & Davis, 2005) are responding to the need to study eating disorders within an older population of women who may experience similar symptoms but have different causes than younger women. Research suggests that reasons for middle-age women's dissatisfaction with their body image are likely related to the body changes associated with aging. Diagnosis and treatment may also pose different issues for middle-age women than younger women. For example, family counseling for younger women may include parents, whereas support for middle-age women is more likely to come from a partner (Forman & Davis, 2005).

Finally, the future of research on eating disorders is likely to address the influence of the Internet. In addition to numerous other concerns with the Internet, researchers are turning their attention to proanorexia (pro-ana) and probulimia (pro-mia) web sites. These web sites provide a forum for visitors to share tips and tricks, "thinspirations," photos of thin celebrities, and other secrets in an effort to promote an eating disorder lifestyle. Anna Bardone-Cone and Kamila Cass (2007) conducted one of the first experiments on the impact of viewing a pro-anorexia web site. College women who visited the pro-anorexia web site were more likely to exhibit body dissatisfaction and lower self-esteem than their counterparts who visited a fashion or a home decor web site. They were also more likely to exercise after their experience with the pro-anorexia web site. Further research on the influence of these web sites on eating disorders is sure to follow.

DSM–V

In 2007 the *International Journal of Eating Disorders* dedicated a supplemental issue of their journal to a discussion of the future of eating disorders in preparation for the forthcoming (2012) publication of the *DSM–V*. The contributors to this issue recognized the importance of diagnostic criteria for treatment and other outcomes and raised important questions and concerns. Chavez and Insel (2007b), for example, asked if amenorrhea is a necessary criterion for diagnosing anorexia nervosa. They also called attention to how a binge in bulimia nervosa is operationalized. The classification of EDNOS was also noted as worthy of discussion given the prevalence of its diagnosis. Within this discussion of EDNOS, attention to binge-eating disorder, the most commonly assigned category of this diagnosis, is relevant as well (Chavez & Insel, 2007b). In any case, the discussion and study of eating disorders promises to continue and develop over time.

SEE ALSO Volume 1: *Body Image, Childhood and Adolescence; Mental Health, Childhood and Adolescence; Obesity, Childhood and Adolescence; Puberty; Socialization, Gender;* Volume 2: *Obesity, Adulthood.*

BIBLIOGRAPHY

American Psychiatric Association. (2000). *Diagnostic and statistical manual of mental disorders* (4th ed., text rev.). Washington, DC: Author.

Bardone-Cone, A. M., & Cass, K. M. (2007). What does viewing a pro-anorexia website do? An experimental examination of website exposure and moderating effects. *International Journal of Eating Disorders, 40*, 537–548.

Becker, A. E., Franko, D. L., Speck, A., & Herzog, D. B. (2003). Ethnicity and differential access to care for eating disorder symptoms. *International Journal of Eating Disorders, 33*, 205–212.

Bell, R. M. (1985). *Holy anorexia*. Chicago: University of Chicago Press.

Birmingham, C. L., Su, J., Hlynsky, J. A., Goldner, E. M., & Gao, M. (2005). The mortality rate from anorexia nervosa. *International Journal of Eating Disorders, 38*, 143–146.

Chavez, M., & Insel, T. R. (2007a). Eating disorders: National Institute of Mental Health's perspective. *American Psychologist, 62*, 159–166.

Chavez, M., & Insel, T. R. (2007b). Special issue on diagnosis and classification [Foreword]. *International Journal of Eating Disorders, 40*(S3), S2.

Ciotola, L. (1999). Eating disorders: An overview. In M. H. Immell (Ed.), *Eating disorders* (pp. 14–17). San Diego, CA: Greenhaven Press.

Fairburn, C. G., & Bohn, K (2005). Eating disorder NOS (EDNOS): An example of the troublesome "not otherwise specified" (NOS) category in *DSM–IV. Behavior and Research Therapy, 43*, 691–701.

Feldman, M. B., & Meyer, I. H. (2007). Eating disorders in diverse lesbian, gay, and bisexual populations. *International Journal of Eating Disorders, 40*, 218–226.

Forman, M., & Davis, W. N. (2005). Characteristics of middle-aged women in inpatient treatment for eating disorders. *Eating Disorders: The Journal of Treatment and Prevention, 13*, 231–243.

Franko, D. L., Becker, A. E., Thomas, J. J., & Herzog, D. B. (2007). Cross-ethnic differences in eating disorder symptoms and related distress. *International Journal of Eating Disorders, 40*, 156–164.

Hesse-Biber, S., Leavy, P., Quinn, C. E., & Zoino, J. (2006). The mass marketing of disordered eating and eating disorders: The social psychology of women, thinness, and culture. *Women's Studies International Forum, 29*, 208–224.

Keel, P. K. (2005). *Eating disorders*. Upper Saddle River, NJ: Pearson Prentice Hall.

Laliberté, M., Boland, F. J., & Leichner, P. (1999). Family climates: Family factors specific to disturbed eating and bulimia nervosa. *Journal of Clinical Psychology, 55*, 1021–1040.

Lucas, A. R., Crowson, C. S., O'Fallon, W. M., & Melton, L. J. (1999). The ups and downs of anorexia nervosa. *The International Journal of Eating Disorders, 26*, 397–405.

Machado, P. P. P., Machado, B. C., Gonçalves, S., & Hoek, H. W. (2007). The prevalence of eating disorders not otherwise specified. *International Journal of Eating Disorders, 40*, 212–217.

Pike, K. M., Dohm, F. A., Striegel-Moore, R. H., Wilfley, D. E., & Fairburn, C. G. (2001). A comparison of Black and White women with binge eating disorders. *The American Journal of Psychiatry, 158*, 1455–1460.

Rome, E. S., & Ammerman, S. (2003). Medical complications of eating disorders: An update. *Journal of Adolescent Health, 33*, 418–426.

Russell, G. (1979). Bulimia nervosa: An ominous variant of anorexia nervosa. *Psychological Medicine, 9*, 429–448.

Smolak, L., & Striegel-Moore, R. H. (2001). Challenging the myth of the golden girl: Ethnicity and eating disorders. In R. H. Striegel-Moore & L. Smolak (Eds.), *Eating disorders: Innovative directions in practice and research* (pp. 111–132). Washington, DC: American Psychological Association.

Striegel-Moore, R. H., & Bulik, C. M. (2007). Risk factors for eating disorders. *American Psychologist, 62*, 181–198.

Theander, S. S. (2002). Literature on eating disorders during 40 years: Increasing number of papers, emergence of bulimia nervosa. *European Eating Disorders Review, 10*, 386–398.

Michelle Napierski-Prancl

ELDER, GLEN H., JR.
1934–

Glen H. Elder Jr. (born in Cleveland, Ohio), a U.S. sociologist, is the primary architect of the life course perspective on human development, which has had an immense influence on scientific research in sociology, psychology, and other disciplines. The basic parameters of this theoretical perspective emerged from Elder's own investigations of the ways in which children, adults, and the elderly chart out their lives over time within the constraints imposed by their environments, social positions, and historical circumstances.

Biography within context is one of the core themes of life course research. Fittingly, Elder's own personal history sheds light on his scientific accomplishments. He was raised in Cleveland and its suburbs by his parents, who were high school teachers and athletic coaches. As a teenager, he moved with his family to a dairy farm in Pennsylvania so that his father could pursue a lifelong dream of working the land. From there, Elder moved on to Pennsylvania State University, where he received his bachelor's degree (1957) and also met his wife. He then attended the University of North Carolina–Chapel Hill, where he earned a Ph.D. in sociology (1961). His first faculty position was at the University of California, Berkeley, where he realized the value of studying how people were affected by the rapid, dramatic changes of modern society and where he began his pioneering work on archived longitudinal samples of young people born in the early 20th century. After moving with his wife and three sons several more times to take on new professional appointments, most notably at

Glen H. Elder. PHOTO COURTESY OF GLEN H. ELDER.

Elder was born. By pure happenstance, these studies spanned the Great Depression. Elder retrieved, organized, and coded these historical data and then analyzed the resulting data set with a special emphasis on the changing economic circumstances of the children's families. The enduring, hopeful message of this study is that children are amazingly resilient in the face of early economic adversity.

What is notable about this work was that it situated children's development within larger social contexts, not just in their families or their communities, but also in the structure of U.S. society (e.g., the class system) as well as particular historical eras. That doing so now seems self-evident is a testament to the influence of this study. At the time, this approach was groundbreaking. This book served as the genesis of life course theory, which Elder has been refining ever since. This theory orients researchers toward asking specific kinds of questions when they design studies dealing with human lives and then provides tools for helping them interpret the findings of these studies.

Life course theory has five basic principles:

1. Life-Span Development: Human development and aging are lifelong processes.

2. Agency: Individuals construct their lives through the choices and actions they take within the opportunities and constraints of history and social circumstance.

3. Time and Place: The life course of individuals is embedded and shaped by the historical times and places they experience over their lifetime.

4. Timing: The developmental antecedents and consequences of life transitions, events, and behaviors vary according to their timing in a person's life.

5. Linked Lives: Lives are lived interdependently and sociohistorical influences are expressed through this network of shared relationships.

These principles are evident in Elder's later works, including more than a dozen books and hundreds of journal articles. As one example, he performed a similar reconfiguration of another historical, longitudinal data source—the Stanford-Terman study, which has followed a group of intellectually gifted children for more than 80 years—to assess the long-term impact on men's lives of having served during World War II (in the United States, 1941–1945). More recently, he joined with sociologist Rand Conger (Iowa State University) to conduct a decade-long study of farm families living through the collapse of the agricultural economy in Iowa in the 1980s. In both cases, Elder has driven home the simple but all important idea that human lives are lived in time and place.

Cornell University in Ithaca, New York, and then to Boys Town in Nebraska, Elder returned to the University of North Carolina in 1984 as the Howard W. Odum Distinguished Professor of Sociology, a position that he held until his official retirement in 2007. Along the way, Elder served as the president of the Society for Research in Child Development and vice president of the American Sociological Association, and as a fellow in the Sociological Research Association, American Psychological Association, and the Gerontological Society of America. He was elected to the American Academy of Arts and Sciences in 1988.

Elder's work at the Institute of Human Development at Berkeley resulted in what is arguably his greatest scholarly achievement: *The Children of the Great Depression.* This book, first published in 1974, reported on the adjustment of young Californians who came of age during the trying economic times of the Depression era (1929–1939). These youth had participated in two longitudinal studies of psychological development before

SEE ALSO Volume 1: *Clausen, John; Data Sources, Childhood and Adolescence; Parent-Child Relationships, Childhood and Adolescence; Poverty, Childhood and Adolescence;* Volume 2: *Agency;* Volume 3: *Cohort.*

BIBLIOGRAPHY

Elder, G. H., Jr. (1974). *Children of the great depression: Social change in life experience.* Chicago: University of Chicago Press.

Elder, G. H., Jr. (1998). The life course as developmental theory. *Child Development, 69,* 1–12.

Elder, G. H., Jr., & Conger, R. D. (2000). *Children of the land: Adversity and success in rural America.* Chicago: University of Chicago Press.

Robert Crosnoe

ELEMENTARY SCHOOL

SEE Volume 1: *Stages of Schooling.*

EMPLOYMENT, YOUTH

Employment is increasingly recognized as an important developmental context of adolescence, one that may operate alongside or in conjunction with families, schools, and friendship groups. Adolescent employment is also a topic of great interest to educators and policy makers, who, along with parents, want to ensure that adolescents are adequately prepared for adult roles. Of particular concern is how investments in employment stand in relation to education in facilitating a successful transition into meaningful and financially rewarding work in adulthood. Government policy regulates adolescent employment as to the hours and times of day teens may work, as well as prohibiting work deemed too dangerous for youth. Policy efforts have also been aimed at helping economically disadvantaged youth gain work experience that facilitates their transition from school to work.

PREVALENCE AND HISTORICAL TRENDS

Nearly all high school students are employed at some point during the school year (U.S. Dept. of Labor, 2000; Entwisle, Alexander, & Olson, 2000). Summer employment is more common than school-year employment (Perreira, Harris, & Lee, 2007), though nearly all research on adolescent employment is concerned with school-year employment.

Employment of teens enrolled in school became more common in the period between the 1940s and the 1980s,

though it has been largely stable since then (Greenberger & Stenberg, 1986; Warren & Cataldi, 2006; for a discussion of earlier situations, see Mortimer, 2003). The hours adolescents typically spend in paid work have been stable historically (Warren & Cataldi, 2006). A key to understanding employment trends among students, however, is to remember the trend of rising school enrollment. Warren and Cataldi (2006) make an important distinction between adolescent and student employment, clarifying the source of this apparent growth in student employment before 1980. In their analysis of Whites and Blacks between 16 and 17, they found that rising employment among boys stemmed from a decline in the proportion of adolescent boys who worked but who were not enrolled in school. They conclude that the trend among boys "should be seen as a trend toward greater rates of school enrollment among employed men" (p. 119). Among girls, the rise in employment through 1970 largely stemmed from a decline in the proportion of girls neither working nor enrolled in school (e.g., young homemakers). After 1970, however, rising student employment among girls involved more students working.

PATTERNS OF EMPLOYMENT ACROSS ADOLESCENTS

Employment becomes more common with age. Estimates from national surveys vary some, in part based on whether adolescents are asked about current employment or typical school-year employment. In the Monitoring the Future surveys, 36% of girls and 41% of boys in eighth grade, 38% of girls and 45% of boys in tenth grade, and 68% of both boys and girls in twelfth grade report school year employment (Safron, Schulenberg, & Bachman, 2001). Estimates by age from the National Longitudinal Survey of Adolescent Health show a similar pattern (Perreira et al., 2007). Although most studies involve high school students, the adolescent work career often starts earlier (Entwisle et al., 2000; Mortimer, 2003).

With advancing age, adolescents tend to work more hours per week, and the type of work they do changes (Steinberg & Cauffman, 1995; Mortimer, 2003). Estimates from the Monitoring the Future study indicate that of those working for pay, 15.6% of girls and 24.4% of boys in the tenth grade, and 37.7% of girls and 44.3% of boys in twelfth grade work more than 20 hours per week (Safron, Schulenberg, & Bachman, 2001). As they mature, teens move from largely informal work (e.g., babysitting and yard work) into formal employment, and the tasks they do become more complex (Steinberg & Cauffman, 1995; Mortimer, 2003; Entwisle, Alexander, & Olson, 2005). Older teens most commonly work as restaurant staff and retail store clerks

Food Service. *A manager and an employee at Boston Market discuss how many chickens should be roasted.* © **ERIK FREELAND/CORBIS.**

(Steinberg & Cauffman, 1995), although they are also represented in a range of other jobs (Mortimer, 2003).

Since the 1980s, girls have been equally likely to be employed, though among workers boys tend to work more hours per week than girls (Schoenhals, Tienda, & Schneider, 1998; Warren & Cataldi, 2006; Perreira et al., 2007). Studies consistently find that Black and Hispanic adolescents are less likely to be employed than non-Hispanic white adolescents, though among workers they tend to work the same number or slightly more hours per week (D'Amico, 1984; Bachman & Schulenberg, 1993; Steinberg & Cauffman, 1995; Schoenhals et al., 1998; Warren & Cataldi, 2006). First- and second-generation immigrant adolescents are also considerably less likely to work than third-generation adolescents, owing in part to differences in work participation by race/ethnicity (Perreira et al., 2007). Finally, although many studies find no major differences in paid work participation by socio-economic background, some evidence suggests that employment is lower among adolescents from very poor and from very wealthy families (Steinberg & Cauffman, 1995; Schoenhals et al., 1998). In contrast, the average number of hours worked among employed students is consistently related to socioeconomic status. Students from families with higher socioeconomic status work fewer hours per week (Bachman & Schulenberg, 1993; Schoenhals et al., 1998).

POTENTIAL BENEFITS AND RISKS

Adolescent employment holds the potential for both beneficial and detrimental consequences in both the short and longer terms. From one point of view, employment provides important socializing experiences in preparation for adulthood. Mortimer (2003), for example, noted that paid work introduces adolescents to the world of work, allowing them to gain experience and develop foundational skills. Based on a similar rationale, a series of national commissions have extolled the benefits of working during adolescence (e.g., National Commission on Youth, 1980; Panel on Youth, 1974). Through paid work, adolescents would learn responsibility, the value of money, and practical skills that would be useful in their adult jobs. Mortimer further notes that even jobs that adolescents do not expect to hold as adults may stimulate thinking about one's occupational future and help young people learn how to balance school and work in a way that facilitates later attainment.

Conversely, scholars and educators have worried that paid work detracts from important educational pursuits, including studying, participation in extracurricular activities, and regular attendance at school. This argument usually assumes a zero-sum model in which an hour adolescents spend at work is an hour less to spend on other important tasks. Concerns have been raised about whether time in paid work also competes with time allotted to developmental processes of importance during adolescence including self-discovery and the development of autonomy and social responsibility. Greenberger and Steinberg (1986) have argued this position, for example, and suggested that employment during adolescence promotes *pseudomaturity*—that is, social maturity gained through the worker role, but without psychological maturity. They have also suggested that the poor quality of contemporary teens' work experience fosters negative attitudes toward work, encourages workplace deviance, and heightens the use of alcohol and drugs to cope with workplace stress. Additional concerns have been raised about the safety of adolescent workers (e.g., NRC, 1998) and whether employment weakens parental authority and exposes young workers to somewhat older peers who are more likely to drink and use other drugs (e.g., McMorris & Uggen, 2000).

Scientific evaluation of these arguments began in earnest in the 1980s. Research over the past several decades offers evidence with which parents, educators, and policymakers can weigh the costs and benefits of paid work in adolescence. Most of that research has focused on employment status and the number of hours adolescents work rather than the nature of their work.

Consistent with the view that paid work is developmentally beneficial, studies have shown that hours worked in high school are linked to higher rates of labor force participation, lower unemployment, and higher earnings in young adulthood (Marsh, 1991; Carr, Wright, & Brody, 1996; Mortimer, 2003). For girls, employment is also associated with gains in self-reliance (Greenberger & Steinberg, 1986) and greater knowledge of the work world (D'Amico, 1984).

Fueling concerns over the risks of employment, however, studies also find that work hours are associated with cigarette, alcohol, and other drug use (Steinberg & Dornbusch, 1991; Bachman & Schulenberg, 1993; McMorris & Uggen, 2000; Mortimer, 2003; Pasternoster, Bushway, Brame, & Apel, 2003), delinquency and related problem behaviors (Marsh, 1991; Steinberg & Dornbusch, 1991; Bachman & Schulenberg, 1993; Pasternoster et al., 2003), poorer health behaviors such as missing sleep, skipping breakfast, and less frequent exercising (Bachman & Schulenberg, 1993), more time in

unstructured social activities such as recreation, dating, and riding around for fun (Safron, Schulenberg, & Bachman, 2003), and poor academic performance, including lower grades and educational aspirations, less time doing homework, and higher rates of absenteeism and school drop out (D'Amico, 1984; Marsh, 1991; Steinberg & Dornbusch, 1991; Bachman & Schulenberg, 1993; Warren & Lee, 2003).

Some of these relationships are nonlinear, however, and indicate that moderate employment offers benefits over, or is at least equivalent to, not working. Those who work only a few hours (i.e., 1–5 per week) get more sleep, and are more likely to eat breakfast and to exercise more often than those not working, and those working between 6 and 10 hours behave in much the same way on these dimensions as those not working (Bachman & Schulenberg, 1993). Moderate work hours have also been linked to positive academic outcomes including higher grades (Steinberg & Dornbusch, 1991; Mortimer, 2003), more time doing homework (Steinberg & Dornbusch, 1991), more time spent in school activities (Safron et al., 2003) and equivalent or lower rates of dropping out of high school (D'Amico, 1984; Warren & Cataldi, 2006). As a result, much of the attention has shifted to *intensive* employment, usually defined as working more than 20 hours per week.

As research in this area has matured, two observations have shaped the way in which adolescent employment is understood: (a) that it might not have the same effects for all groups of adolescents, and (b) that students who do work, and do so at varying intensities, differ beforehand in many ways, which could explain why working is correlated with the various benefits and risks already noted in this entry.

Varying Effects Evidence is growing that the effects of working depend on various adolescent characteristics, including whether earnings are being saved for college (Marsh, 1991) or are being used to support oneself or one's family (Newman, 1996), racial or ethnic group membership (Johnson, 2004), and whether a student is likely to go onto higher education (Entwisle et al., 2005). For example, important tangible benefits of paid work may exist, and there may be less to lose, for youth who find little appeal in higher education or do not have the money to pursue it. The effects of working may also be conditional on adolescents' developmental histories. Apel and colleagues (2007) found that the effects of moving into intensive work hours for the first time at age 16 on substance use and criminal activity depends on adolescents' earlier trajectories with respect to these behaviors. In particular, they find that heavy work involvement may be of benefit to some at-risk youth.

Selection Processes Importantly, although the research discussed in this entry shows that employment, and more often work intensity, in adolescence is associated with a great many aspects of adolescent behavior and well-being, little agreement exists on whether working has a causal effect on these outcomes. Most critical is a concern about whether these relationships may be spurious, owing to preexisting differences among students in socioeconomic background, academic ability, motivation, work orientations, and other characteristics. Similarly, the causal order may be reversed, such that students with lower grades, who use alcohol and other drugs, and who get into trouble seek out employment at higher intensities as an alternative arena in which to succeed or as a source of income to support their desired lifestyles. It has also been suggested that both working longer hours and behaviors such as substance use are manifestations of an underlying syndrome—that one does not necessarily cause the other (Bachman & Schulenberg, 1993).

In response to concerns about whether employment or work hours has causal effects on behavior, well-being, and later attainment, studies have taken youths' selection into employment of varying intensities more seriously over time. Most studies find that adjusting for covariates greatly reduces or eliminates the association between work hours and the outcome of interest. For example, by adjusting for pre-existing differences among students, scholars have shown that the relationships between work hours and both grades and homework time is spurious (Schoenhals et al., 1998; Warren, LePore, & Mare, 2000).

Even adjusting for covariates, however, studies do find a robust association between work hours and higher substance use (McMorris & Uggen, 2000; Pasternoster et al., 2003), school dropout (Warren & Lee, 2003), and absenteeism (Schoenhals et al., 1998), but also that steady, moderate, work hours are related to higher educational attainment (Mortimer & Johnson, 1998), especially among young people with low educational promise (Staff & Mortimer, 2007).

Other approaches to addressing selection processes are emerging. Using propensity score matching, Lee and Staff (2007) find the effect of intensive employment and school drop out is not totally spurious. However, among those with a high propensity to work intensively, who tended to be from families with lower socioeconomic status and to have weaker school performance and lower chances for postsecondary education, work hours had no effect on the likelihood of drop out. Using fixed and random effects models to adjust for unobserved heterogeneity among students, Pasternoster and colleagues (2003) find the relationship of work intensity with delinquency, substance use, and problem behaviors is spurious.

FUTURE DIRECTIONS

Future research on adolescent employment will inevitably continue to grapple with the question of causality. Continued attention to methods for addressing observed and unobserved differences prior to employment is needed.

Signs are accumulating that more attention is being paid to the nature of the work adolescents do, and not only to the hours they put in. Despite repeated calls to consider this issue, until very recently, only a few studies considered what work adolescents were performing. Mortimer (2003) provides one of the most comprehensive analyses of the precursors and consequences of the quality of work experiences. Moreover, she demonstrates that questions about work hours and the nature of the work experience are inextricably linked. For example, the temporal pattern of investment in work across the high school years is related to adolescents' assessments of their work, with those that worked longer hours experiencing greater learning opportunities, but also having more demanding and stressful jobs.

With the same panel, Staff and Uggen (2003) find that characteristics of *good* adult jobs, like autonomy, status, and pay, are not necessarily good for adolescents. They found adolescent autonomy was associated with increased school deviance, alcohol use, and probability of arrest. Jobs that promote status with peers were also associated with increased school deviance and alcohol use. Higher wage jobs were associated with an increased probability of arrest. In contrast, jobs that were perceived to be compatible with school were linked to decreased alcohol use and probability of arrest, and opportunities to learn on the job were associated with decreased alcohol use. Entwisle and colleagues (2005) also point to a gradual and orderly movement into formal work, levels of job stress, and using earnings for family support as important variants in adolescents' jobs when it comes to school completion. Based on these and related studies, our understanding of adolescent employment has become considerably more nuanced in the last few decades.

SEE ALSO Volume 1: *Academic Achievement; Drinking, Adolescent; Drug Use, Adolescent; Human Capital; Peer Groups and Crowds; Self-Esteem; Vocational Training and Education; School to Work Transition;* Volume 2: *Careers; Employment.*

BIBLIOGRAPHY

Apel, R., Bushway, S., Brame, R., Haviland, A. M., Nagin, D. S., & Paternoster, R. (2007). Unpacking the relationship between adolescent employment and antisocial behavior: A matched samples comparison. *Criminology, 45*(1), 67–97.

Bachman, J. G., & Schulenberg, J. (1993). How part-time work intensity relates to drug use, problem behavior, time use, and satisfaction among high school seniors: Are these consequences or merely correlates? *Developmental Psychology, 29*, 220–235.

Carr, R. V., Wright, J. D., & Brody, C. J. (1996). Effects of high school work experience a decade later: Evidence from the national longitudinal survey. *Sociology of Education, 69*, 66–81.

D'Amico, R. (1984). Does employment during high school impair academic progress? *Sociology of Education, 57*, 152–164.

Entwisle, D. R., Alexander, K. L., & Olson, L. S. (2000). Early work histories of urban youth. *American Sociological Review, 65*, 279–297.

Entwisle, D. R., Alexander, K. L., & Olson, L. S. (2005). Urban teenagers: Work and dropout. *Youth and Society, 37*, 3–32.

Greenberger, E., & Steinberg, L. (1986). *When teenagers work: The psychological and social costs of adolescent employment.* New York: Basic Books.

Johnson, M. K. (2004). Further evidence on adolescent employment and substance use: Differences by race and ethnicity. *Journal of Health and Social Behavior, 45*, 187–197.

Lee, J. C., & Staff, J. (2007). When work matters: The varying impact of work intensity on high school dropout. *Sociology of Education, 80*, 158–178.

Marsh, H. W. (1991). Employment during high school: Character building or a subversion of academic goals? *Sociology of Education, 64*, 172–189.

McMorris, B. J., & Uggen, C. (2000). Alcohol and employment in the transition to adulthood. *Journal of Health and Social Behavior, 41*, 276–294.

Mortimer, J. T. (2003). *Working and growing up in America.* Cambridge, MA: Harvard University Press.

Mortimer, J. T., & Johnson, M. K. (1998). Adolescent part-time work and educational achievement. In K. Bormann & B. Schneider (Eds.), *The adolescent years: Social influences and educational challenges* (pp. 183–206). 97th Yearbook of the National Society for the Study of Education. Chicago: University of Chicago Press.

National Commission on Youth (1980). *The Transition of Youth to Adulthood: A Bridge Too Long.* Boulder, CO: Westview Press.

National Research Council. (1998). *Protecting youth at work: Health, safety and development of working children and adolescents in the United States.* Washington, DC: National Academy Press.

Newman, K. S. (1996). Working poor: Low-wage employment in the lives of Harlem youth. In J. A. Graber, J. Brooks-Gunn, & A. C. Petersen (Eds.), *Transitions through adolescence: Interpersonal domains and context* (pp. 323–343). Mahwah, NJ: Lawrence Erlbaum.

Panel on Youth of the President's Science Advisory Committee (1974). *Youth: Transition to adulthood.* Chicago: University of Chicago Press.

Paternoster, R., Bushway, S., Brame, R., & Apel, R. (2003). The effect of teenage employment on delinquency and problem behaviors. *Social Forces, 82*, 297–335.

Perreira, K. M., Harris, K. M., & Lee, D. (2007). Immigrant youth in the labor market. *Work and Occupations, 34*, 5–34.

Safron, D. J., Schulenberg, J. E., & Bachman, J. G. (2001). Part-time work and hurried adolescence: The links among work intensity, social activities, health behaviors, and substance use. *Journal of Health and Social Behavior, 42*, 425–429.

Schoenhals, M., Tienda, M., & Schneider, B. (1998). The educational and personal consequences of adolescent employment. *Social Forces, 77*, 723–762.

Staff, J., & Mortimer, J. T. (2007). Educational and work strategies from adolescence to early adulthood: Consequences for educational attainment. *Social Forces, 85*, 1169–1194.

Staff, J., & Uggen, C. (2003). The fruits of good work: Early work experiences and adolescent deviance. *Journal of Research in Crime and Delinquency, 40*, 263–290.

Steinberg, L., & Cauffman, E. (1995). The impact of employment on adolescent development. *Annals of Child Development, 11*, 131–166.

Steinberg, L., & Dornbusch, S. M. (1991) Negative correlates of part-time employment during adolescence: Replication and elaboration. *Developmental Psychology, 27*, 304–313.

Steinberg, L., Fegley, S., & Dornbusch, S. M. (1993). Negative impact of part-time work on adolescent adjustment: Evidence from a longitudinal study. *Developmental Psychology, 29*, 171–180.

U.S. Department of Labor. (2000). *Report on the Youth Labor Force.*

Warren, J. R., & Cataldi, E. F. (2006). A historical perspective on high school students' paid employment and its association with high school dropout. *Sociological Forum, 21*, 113–143.

Warren, J. R., & Lee, J. C. (2002). The impact of adolescent employment on high school dropout: Differences by individual and labor-market characteristics. *Social Science Research, 32*, 98–128.

Warren, J. R., LePore, P. C., & Mare, R. D. (2000). Employment during high school: Consequences for students' grades in academic courses. *American Educational Research Journal, 37*, 943–969.

Monica Kirkpatrick Johnson

ENGLISH AS A SECOND LANGUAGE

SEE Volume 1: *Assimilation; Bilingual Education.*

ERIKSON, ERIK
1902–1994

Psychoanalyst Erik Homburger Erikson's two most important contributions are a theory of the life cycle showing the integration of personal development and social context from earliest childhood to oldest age and a theory of the development of identity in adolescence based on earlier childhood experiences. Born in Frankfurt, Germany, on June 15, the illegitimate son of a brief affair by his affluent Danish mother, Erikson was adopted by his mother's second husband, pediatrician Theodor Homburger. Following his mediocre career in a classical secondary school

Joan and Erik Erikson. TED STRESHINSKY/TIME LIFE
PICTURES/GETTY IMAGES.

(a school emphasizing study of Greek and Latin, together
with European history) and late adolescence wandering
about Germany, periodically studying art and searching
for his own identity, he was invited by a childhood friend
to join him teaching in a school in Vienna that was
primarily for children who were in analysis with pioneering child analyst Anna Freud (1895–1982). After seeking a
personal analysis with Miss Freud and completing formal
psychoanalytic education at the newly created Vienna
Psychoanalytic Institute, Erikson and his family left
Vienna for the United States following Adolf Hitler's rise
to power in 1933.

With his impressive Viennese psychoanalytic credentials, and a gift for child psychoanalysis, even though he
had no formal university education, Erikson had little
difficulty securing research appointments in a number of
developmental and psychological studies, first at Harvard
University, then at Yale University and later at the University of California, Berkeley. During this time he took
part in two ethnographic projects, studying the Native

American Sioux tribe of the Plains while at Yale and then
the Yurok of the Northwest Coast while at Berkeley.
After consulting with his family, Erikson changed his
name from Homburger to Erikson when he became a
U.S. citizen in 1939. Erikson was uncomfortable with
empirical American developmental psychological study
and the American research university as a setting for his
own work. He was diffident in relations with his colleagues, and resentful of the time away from his own
writing, including papers that eventually became the
chapters of *Childhood and Society* (1950) and several
significant psychoanalytic papers.

Erikson accepted an appointment to the Austen
Riggs Center of Stockbridge, Massachusetts, which was
devoted to the study and treatment of psychological
disorders, and somewhat later served as a professor at
Harvard, where at middle age he taught his first university courses including a very popular undergraduate survey of his work. Retiring in 1970, Erikson earned a
Pulitzer Prize and a National Book Award for his psycho-
biographical study of the South Asian charismatic leader
Mohandas Gandhi. Erikson died on May 12 in the Cape
Cod, Massachusetts, community where he had spent his
last years as his health declined, surrounded by his wife,
Canadian-born Joan, and his two sons and a daughter.

Much of Erikson's work built on and expanded earlier work of Sigmund Freud, especially Freud's argument
that biologically based drives across the first years of life
are a powerful determinant of personality and the foundation for a child's effort to resolve conflicts regarding
desire and rivalry with the parents during the preschool
years. Erikson accepted Freud's emphasis on a psychological development as cumulative or epigenetic. Freud
had maintained that the relative degree of satisfaction or
frustration across the first years of life determined the
manner in which subsequent developmental phases were
experienced. Erikson maintained, however, that Freud
had failed to realize the potential of this scheme for
normal psychological development and had also failed
to emphasize the social context of early psychological
development. Further, Erikson held that Freud's perspective on personality development was important for developmental study from infancy through oldest age.

Erikson's portrayal of cumulative personality development emphasizes both the adaptive and the problematic outcomes for further psychological development
related to each developmental stage. Maturational forces
provide the impetus for the first three stages of trust
versus mistrust (infancy), autonomy versus shame and
doubt (toddlerhood), and initiative versus guilt (preschool years). Social forces become the impetus for later
developmental stages including industry versus inferiority
(elementary school years), identity versus role confusion

(adolescence), intimacy versus isolation (youth), generativity versus stagnation (the settled years of adulthood), and integrity versus despair (later life). Whereas Vaillant and Milofsky (1980) suggested that the adult life cycle is best portrayed as developmental tasks rather than stages, Joan Erikson maintained that a ninth stage dealing with very late life should be added to this account of the life cycle (Erikson & Erikson, 1997). Vaillant and Milofsky suggested that phases of the life cycle posed for adolescence and youth should include a separate subphase of career consolidation versus self-absorption and that of the adult years should include the generativity-related subphase concern of keeper of the meaning versus rigidity.

Erikson's other major achievement was a detailed discussion of adolescence and the issue of identity or the search for a sense of personal sameness or continuity in one's own life (Erikson, 1959/1980). Biographer Friedman (1999) has highlighted these themes in Erikson's life: his identity as an illegitimate Danish-born son who did not know his biological father and who as an adolescent wandered throughout Germany searching for meaning and coherence, and his changing his name when granted citizenship to reflect his Danish parentage. Founded on his clinical work and his biography of Martin Luther (1958), Erikson maintained that much of the psychological distress of his young patients had been misunderstood as major psychopathology when it was more likely the effort to resolve the identity conflict of adolescence and youth. Erikson (1959/1980) wrote a number of essays on youth and identity.

Erikson's portrayal of the life cycle has been criticized as culturally specific. Further, Erikson's emphasis on cumulative development does not permit recognition of the interplay of person and society over both time and place. Dannefer (1984) argued that both ontogenetic and life-span models of development should be replaced with a life course model that considers sociohistorical context, particularly one's birth cohort.

SEE ALSO Volume 1: *Cognitive Ability; Identity Development; Moral Development and Education.*

BIBLIOGRAPHY

Dannefer, D. (1984). Adult development and social theory: A paradigmatic reappraisal. *American Sociological Review, 49,* 100–116.

Erikson, E. H. (1958). *Young man Luther: A study in psychoanalysis and history.* New York: Norton.

Erikson, E. H. (1980). *Identity and the life cycle.* New York: Norton. (Originally published 1959)

Erikson, E. H. (1985). *Childhood and society* (35th anniversary ed.). New York: Norton. (Originally published 1950)

Erikson, E. H., & Erikson, J. M. (1997). *The life cycle completed* (extended ed. with new chapters on the ninth stage of development by J. M. Erikson). New York: Norton.

Friedman, L. J. (1999). *Identity's architect: A biography of Erik H. Erikson.* New York: Scribner.

Vaillant, G. E., & Milofsky, E. (1980). Natural history of male psychological health: IX. Empirical evidence for Erikson's model of the life cycle. *The American Journal of Psychiatry, 137,* 1348–1359.

Bertram J. Cohler

EXTRACURRICULAR ACTIVITIES

SEE Volume 1: *Activity Participation, Childhood and Adolescence; Civic Engagement, Childhood and Adolescence; Sports and Athletics.*

F

FAMILY AND HOUSEHOLD STRUCTURE, CHILDHOOD AND ADOLESCENCE

The structure and composition of American families has changed dramatically since the mid-20th century. Around 1950, most American children were born into marital unions and about three-quarters remained in "traditional" nuclear families—defined as families with two biological parents married to each other, full siblings only, and no other household members—through childhood and adolescence. In the early 21st century, the family structure histories of American children are far more complex. Increases in nonmarital childbearing, divorce, and cohabitation, combined with declines in marriage and remarriage, have translated into more dynamic relationship histories for adults and more complex living arrangements for their children. Although snapshot estimates indicate that the majority of children live with both biological parents, life course estimates suggest that more than half of all children will spend some time living outside of a "traditional" nuclear family (Bumpass & Lu, 2000).

These changes are dramatic and have generated a large, multidisciplinary literature that has both documented family change and explored the implications of these changes for adults and, especially, children. These changes have also spilled into the national policy domain. The Personal Responsibility and Work Opportunity Reconciliation Act of 1996 was the first federal law to explicitly promote marriage and encourage the formation of two-parent families. This federal commitment continued into the 21st century. As part of the 2005 reauthorization of the Temporary Assistance to Needy Families program, the administration of George W. Bush expanded this legislation by providing additional money for programs aimed at increasing healthy marriages and two-parent families.

At the heart of both the research and policy initiatives is the question of whether the basic functions of the family—ensuring children's social and emotional adjustment and economic well-being in childhood and beyond—are compromised by changes in the structure and composition of the family. In other words, do deviations from the "traditional" family *cause* children to engage in more problem behaviors, do less well in or drop out of school, be more depressed, and ultimately fail to successfully transition to adulthood? This entry provides a general review of the literature that tries to address this question.

Before describing trends in family structure and their implications for children, it is important to note that the changes in the family have not occurred in a vacuum. Instead, they are a part of a larger set of changes in the nature of work, the economy, marriage, and gender roles (Bumpass, 1990). Between 1950 and the early 21st century, the U.S. labor market transitioned from a growing, largely manufacturing-based economy that offered a family wage to nearly all men and actively discriminated against women to a service-based economy that is less secure, is highly credentialed, and relies on the labor of women. At the same time, norms about gender roles and the importance of marriage (but not children) as well as expectations about personal happiness were realigned.

163

These changes, in turn, have transformed the structure and organization of the adult stage of the life course for contemporary American women and men. Moreover, although these changes have permeated all of American society, significant racial and ethnic differences in opportunities and constraints in the labor market have translated into important racial and ethnic differences in the structure of families. Together, these changes shape the normative context in which adults make decisions about romantic unions and in which children are raised (Casper & Bianchi, 2002).

AN OVERVIEW OF CHILDREN'S FAMILY STRUCTURE STATUS

Providing an overview of children's family structure status must begin with a definition of the term *family*. Although this seems like a relatively straightforward task, as the current debate about gay marriage suggests, deciding who is in the family is a hotly contested issue in the United States. Like all social institutions, the family is socially constructed. But more than other institutions, the American family is imbued with rich social and cultural meanings. Any discussion of the family, especially as it relates to children, is often one about values—as they relate to kids but also to ideas about gender, marriage, economic stability, and individual well-being. Thus, coming up with a definition that incorporates cultural meanings and reflects the lived experience of children remains a challenging task.

The U.S. Bureau of the Census definition of the family provides a good starting point. The Census Bureau defines the family as "A group of two or more people who reside together and who are related by birth, marriage, or adoption."

A household, by contrast, includes one or more people who occupy a residential unit. By these definitions, a cohabiting couple—gay or heterosexual—is *not* a family, simply a household. Meanwhile, all families maintain a household. One indication of the many changes in family behavior can be indexed by the declining proportion of households that are made up of families and the increasing proportion composed of nonfamilies.

Based on this definition of family, a snapshot of U.S. families with children under 18 from 2005 indicated that about 71% lived in two-parent families, 23% lived in single-mother families, and 3% lived in single-father families. No parents were in the home of about 4% of all children. (Kreider & Fields, 2005). As a point of comparison, in 1970 about 85% of all children lived in two-parent families, 11% lived in single-mother families, 1% lived in single-father families, and 3% lived with neither parent. As dramatic as these changes are, these statistics obscure the heterogeneity within these catego-

ries. For instance, two-parent families include both two-biological parent families and stepparent families, and, as discussed below, the adjustment of young people in these family forms differs in important ways. Similarly, single-mother and single-father families can also include a cohabiting partner (same or opposite sex), one not recognized within the aforementioned Census Bureau definition of family.

An emerging literature that relies on retrospective reports of family structure history and/or applies estimation procedures to conventional point estimates of family structure provide a more accurate picture of children's family experiences. These measures highlight the heterogeneity and fluidity that underlie these single point-in-time estimates (Bumpass & Lu, 2000; Teachman, 2003; Wu & Martinson, 1993). Dramatic increases in divorce rates in the 1970s fueled much of the literature regarding family structure and child well-being. Yet divorce is not the only significant change in the family. More than ever, increases in nonmarital fertility and changes in marriage, cohabitation, and remarriage have shaped the family trajectories of American youth. Particularly noteworthy are trends that define the two major "alternative" family structures: single-parent families and stepparent families.

Single-Parent Families The modal "alternative" family structure category is single-parent families (Bumpass & Lu, 2000). This family type has always been a part of the family structure regime in the United States. What has changed are the demographic factors driving this status. At the start of the 20th century, high mortality rates were largely responsible for the incidents of single-parent families, leaving widows to raise their children. By the 1960s and 1970s, most single-parent families were created through divorce or separation. In the early 21st century, nonmarital fertility is contributing to the growth of single-mother families. About a quarter of children lived in single-parent families in 2001, up from about 12% in 1970. In all, life course estimates suggest that about half of all U.S. children will spend some time in a single-parent family (Bumpass & Sweet, 1989).

Parental divorce is a major contributor to single-parent families. Divorce rates increased dramatically across the 20th century, peaking around 1980, and thereafter remained stable, even receding for some social groups (Raley & Bumpass, 2003). Still, divorce remains a common family structure transition for American youth. At the beginning of the 21st century, about a half of all marriages ended in divorce, with half of these involving children (Amato, 2000). In all, about 40% of all children will experience parental divorce by age 18. The likelihood of divorce varies by race and social class, with more disadvantaged families experiencing a greater likelihood of divorce or separation than others (Raley & Bumpass, 2003).

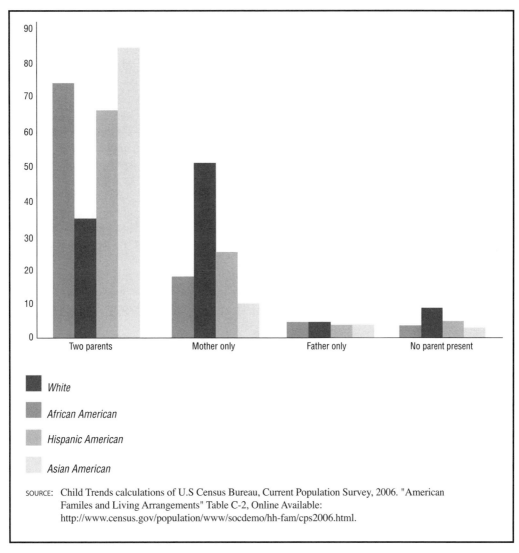

Figure 1. *Living arrangements of American children, by race and ethnicity, 2006.* CENGAGE LEARNING, GALE.

Another factor contributing to the high proportion of single-parent families is nonmarital fertility. In the early 21st century, the majority of all babies in the United States are born into married-parent families. Rates of nonmarital fertility, however, have risen markedly since the mid-20th century. In 1970 about 10% of all births occurred outside of marriage, which contrasted with a 40% figure for the early 21st century (Hamilton, Martin, & Ventura, 2006). Pronounced race/ethnic differences exist in the percentage of births to unmarried women. In 2005, about 70% of all births to non-Hispanic Black women, 63% of births to American Indian or Alaskan native woman, and 48% of births to Hispanic women occurred outside of marriage, compared with about 25% for non-Hispanic White women and 16% for Asian or Pacific Islander women (Hamilton et al., 2006). Although these births occurred outside a marital union, about 40% of them did occur in a cohabiting union.

This, too, varies by race/ethnicity, with about half of White and Latinos births and a quarter of African-American births occurring in such unions (Wildsmith & Raley, 2006).

The majority of young people in single-parent families—especially very young children—reside with their mother. Father-only families, although statistically rare, are becoming more common. Between 1960 and 2001, the proportion of father-only families grew from about 1.4% to 3.1%, representing an increase of more than 100% (Kreider & Fields, 2005).

Stepparent Families Residing in stepparent families is also a common experience for American children (Coleman, Ganong, & Fine, 2000). Like single-parent families, stepparent families have always been a part of the American family portrait. Historically, parental death was the leading precursor to this family structure. Moreover, most of these families were formed through remarriage among widowed mothers. In the early 21st century, divorce and, to a lesser extent, nonmarital fertility often precede this status. Similarly, contemporary stepparent families are initiated through either marriage or cohabitation. Overall, about 7% of children live in a stepparent family (Kreider & Fields, 2005). Yet almost one-third of U.S. children born in the early 1980s are expected to spend some time in a stepparent family (Bumpass, Raley, & Sweet, 1995).

A nontrivial number of cohabiting stepparent families are headed by gay and lesbian parents. Estimates based on the 2000 census indicate that about 1% of all U.S. households include same-sex partners, with these families equally divided among female and male couples. Among female same-sex partner households, about a third include children. Among male same-sex partner households, about 22% include children (Simmons & O'Connell, 2003).

WHAT ARE THE IMPLICATIONS OF THESE CHANGES?

What does this diversity in family structure mean for contemporary American youth? Although the magnitude and long-term implications of changes in family structure continue to be debated (Cherlin, 1999), scholars generally agree that children raised by two continuously married parents are, on average, better off on a host of indicators than are children in other family forms. The number of studies that explore these associations is vast, and these studies measure family structure and child adjustment in many ways. Nevertheless, some general associations can be outlined between the two most common alternative family structures—single-parent and stepparent families—and child well-being, measured in terms of emotional, social, and cognitive adjustment.

Single-Parent Families As mentioned above, divorce is a common factor leading to this family status and, for a long time, was the primary focus of family structure research. Compared to those in stable, married-parent families, children with divorced parents were, on average, more likely to be depressed, engage in problem behaviors such as minor delinquent acts, smoking, and underage drinking, drop out of high school, score lower on standardized tests, transition to first sex earlier, become pregnant as a teenager, and report poorer grades than others during childhood and adolescence (Amato, 2000; McLanahan & Sandefur, 1994). Given that the overwhelming majority of these young people reside with their mother, nearly all of the empirical work on single-parent families is based on mother-only families. Yet, as noted above, young people do reside in father-only families, and the few studies that have explicitly studied them indicate that these families are different from mother-only families in

important ways (e.g., higher income, less stable, different parenting styles). Overall, young people in these families often look about the same or worse than do those in mother-only families. Much of this effect is explained by differences in parenting practices and family instability (Harris, Cavanagh, & Elder, 2000).

The research on children born outside of marriage is less extensive, but the findings are similar. Compared to those raised in stable, married-parent families, children born outside of marriage report, on average, lower levels of emotional, social, and cognitive adjustment (Amato, 2005). Research underway includes the Fragile Families and Child Wellbeing Study, which focuses on nearly 5,000 children born between 1998 and 2000 who were disproportionately born to unmarried parents. This study, designed to better understand the ways nonmarital fertility affects development, promises to provide additional insights into the implications of nonmarital fertility for child and adolescent development.

Stepparent Families Researchers also have investigated the consequences for children residing in a stepparent family. Although the presence of a stepparent usually improves children's standard of living and means that two adults are available to monitor and supervise children's behavior, researchers consistently show that children in stepfamilies exhibit more problems than do children with continuously married parents and look about the same as do children who live with single mothers (Coleman et al., 2000; McLanahan & Sandefur, 1994).

Researchers have generally assumed that the pathway to this family form is universal (e.g., marriage → single parenthood → remarriage). Work in the early 21st century, however, suggests that the pathway to this family status can moderate its association with adolescent adjustment. For instance, Sweeney (2007) found that young people in stepfather families formed after divorce reported better mental well-being than did those in stepparent families formed after a nonmarital birth. Similarly, research by Manning and Lamb (2003) indicates that adolescents living with cohabiting stepparents fared worse than those living in married stepfamilies. In both cases, most of these differences were explained by differences in the socioeconomic circumstances of these families.

A small but growing area of family scholarship focuses on adolescent well-being in gay or lesbian families. Although the number of young people who report living in a gay or lesbian family remains small even in large, nationally representative studies such as Add Health (where parent's sexual orientation is not asked directly but is inferred by the gender of the parent's partner), the existing evidence suggests that adolescents living with same-sex parents do about the same on a host of emotional, behavior, and cognitive indicators as those living with opposite-sex parents (Patterson, 2006).

WHAT FACTORS EXPLAIN THESE LINKAGES?

Some combination of economic hardship, compromised parenting practices, and increased emotional stress have been used to explain the link between family structure and child adjustment. Research by McLanahan and Sandefur (1994) indicates that about half of the "divorce" effect observed in young people in single-parent families is explained by changes in the resident parent's financial status. Mothers typically gain custody following divorce and often experience a substantial decline in family income and an increase in economic instability. This economic uncertainty, in turn, can further exacerbate stress in the home environment, often prompting residential and school changes and altering the mother's work schedule (Amato, 2000). Single parents also may employ less effective parenting strategies than others (Amato, 2000; McLanahan & Sandefur, 1994). This comes about both because of increased economic stress and because mothers are less available to their children on a day-to-day basis. Heavier responsibilities plus diminished economic resources often leads to harsh and inconsistent parenting, less supervision, and weakened parental authority that undermines the parent–child relationship (McLanahan & Sandefur, 1994).

Parenting practices and family stress also play a role in explaining young people's adjustment in stepparent families. Compared with biological parents, stepparents often lack the legitimacy of biological parents and/or have less incentive to invest time in the children living in their homes. As a result, what they can and do offer young people is not always equivalent to what biological parents can offer. Children and adolescents in these families also report lower levels of parental support and closeness with both biological parents and stepparents (Goldscheider & Goldscheider, 1998; Sweeney, 2007). Living in stepparent families means that children and adolescents must adapt to new people in the household—not only a parent's partner but also stepsiblings. These young people must also balance their relationships with their parent's partner along with their own evolving relationships with their resident and nonresident parents. Doing so can be stressful for young people, who often balance feelings of guilt, jealousy, and friendship with the key adults in their lives (Amato, 2000).

WHAT ABOUT THE ROLE OF FAMILY INSTABILITY?

As compelling as the associations between family structure and child and adolescent adjustment are, the living arrangements of American children are far more dynamic than these static measures of family structure imply. From the perspective of a child, family structure can often include some combination of parental marriage, divorce, single parenthood, cohabitation, and remarriage (Bumpass & Lu, 2000; Teachman, 2003). Given this, what is missed when family structure is thought about and measured in this way? An emerging literature attempts to address this question through the instability and change perspective.

Building on stress theory, this perspective posits that changes in a parent's marital or romantic histories constitute a major stressor in a child's life. Beyond highlighting that family structure change is stressful, this perspective emphasizes the potentially cumulative nature of family structure change. Although many children never experience a family structure change, those who experience one family transition are at a greater risk of experiencing subsequent transitions and the concomitant stresses that they involve (Wu & Martinson, 1993). Thus, young people who experience multiple family transitions are expected to experience more compromised well-being than those who experience no such transitions or only one (Fomby & Cherlin, 2007; Teachman, 2003).

Research backs up this assertion. Beginning with early childhood, family instability in early childhood was associated with increases in behavioral problems at age 3 (Osborne & McLanahan, 2007) and at the transition to elementary school (Cavanagh & Huston, 2006). Family instability also was negatively associated with white children's problem behavior during middle childhood (Fomby & Cherlin, 2007). Finally, family instability was associated with the nature of young people's romantic relationships (Cavanagh, Crissey, & Raley, in press) and the likelihood of a premarital birth for all women during adolescence (Wu & Martinson, 1993).

WHAT ROLE DOES SELECTION PLAY IN THESE ASSOCIATIONS?

The empirical evidence linking family structure and child adjustment is quite impressive, but does this mean the link is causal? That is, does residing in a particular family structure *cause* compromised child well-being, or is the observed link the result of maternal and paternal characteristics that affect both the likelihood that parents' experience unstable romantic histories and that their children experience compromised well-being in childhood? Although no social group is immune from the family changes described above, there are important racial, ethnic, and social class differences in the likelihood of experiencing a stable two-parent family environment, on the one hand, and a family marked by change, on the other. These differences, in turn, are also related to how well children do on a host of indicators.

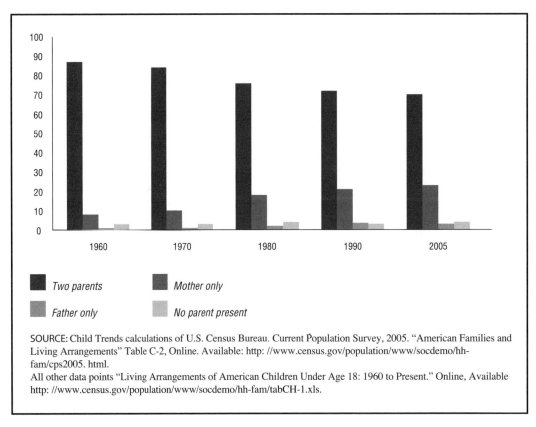

SOURCE: Child Trends calculations of U.S. Census Bureau. Current Population Survey, 2005. "American Families and Living Arrangements" Table C-2, Online. Available: http: //www.census.gov/population/www/socdemo/hh-fam/cps2005. html.
All other data points "Living Arrangements of American Children Under Age 18: 1960 to Present." Online, Available http: //www.census.gov/population/www/socdemo/hh-fam/tabCH-1.xls.

Figure 2. *Living arrangements of American children, 1960–2005.* CENGAGE LEARNING, GALE.

The only way of determining causality is by randomly assigning families to different family structure statuses—a design that is implausible for practical and ethical reasons. Short of that, researchers have used different statistical methods and included different indicators that tap selection processes. A 2007 study by Fomby and Cherlin highlights the important of selection in understanding the link between family structure and child adjustment. With indicators of family instability as well as a comprehensive list of maternal characteristics associated with selection, they found that, for African-American children, maternal characteristics explained all of the observed family instability effect on children's cognitive functioning and problem behavior. For Whites, selection processes also mattered, but an instability effect on problem behavior remained. In all, scholars agree that selection plays a significant role in the link between family structure and child adjustment, but that the experience of family structure change remains a factor in child well-being (Amato, 2000).

FUTURE DIRECTIONS

The first wave of family structure research focused on the impact of divorce and remarriage on child well-being. Much of this work relied on a fairly simple model of

family change where all children were assumed to be born in a marital union that eventually dissolved, with some experiencing a subsequent parental remarriage. This model held for most children (especially White, middle-class children) and provided compelling and consistent information about the association between divorce and child well-being. This model, however, has become outdated and does not reflect the realities of the lives of American children in the early 21st century. Despite pronounced racial and ethnic differences in family structure histories, the implications of family structure for children of color, especially African-American children, remains largely unclear.

Informed by changing fertility and marriage patterns along with ethnographic research that highlights the fluidity of family experiences, the second wave of family structure research may continue to look beyond static measures of family structure and attempt to incorporate the whole of children's family structure experiences. Divorce rates have stabilized, but nonmarital fertility continues to increase. Moreover, children born outside of marriage are at a greater risk of experiencing more family transitions, including the formation and dissolution of (multiple) parental marriages and cohabitations throughout their early life course. Together, these changes are shaping the lives of American children in the early 21st century.

These changes are also shaping the way some family scholars think about the family and its impact on children. The aforementioned Fragile Families and Child Wellbeing Study represents both an example of how family scholars are thinking about family structure and an exciting resource for studying the implications of nonmarital fertility and family instability for child and adolescent adjustment.

SEE ALSO Volume 1: *Child Custody and Support; Grandchildren; Parent-Child Relationships, Childhood and Adolescence; Policy, Child Well-Being; Poverty, Childhood and Adolescence;* Volume 3: *Grandparenthood.*

BIBLIOGRAPHY

Amato, P. R. (2000). The consequences of divorce for adults and children. *Journal of Marriage and the Family, 62,* 1269–1287.

Amato, P. R. (2005). The impact of family formation change on the cognitive, social, and emotional well-being of the next generation. *The Future of Children, 15,* 75–96.

Bumpass, L. L. (1990). What's happening to the family? Interactions between demographic and institutional change. *Demography, 27,* 483–498.

Bumpass, L. L., & Lu, H. H. (2000). Trends in cohabitation and implications for children's family contexts in the United States. *Population Studies, 54,* 29–41.

Bumpass, L. L., Raley, R. K., & Sweet, J. A. (1995). The changing character of stepfamilies: Implications of cohabitation and nonmarital childbearing. *Demography, 32,* 425–436.

Bumpass, L. L., & Sweet, J. A. (1989). National estimates of cohabitation. *Demography, 26,* 615–625.

Casper, L. M., & Bianchi, S. M. (2002). *Continuity and change in the American family.* Thousand Oaks, CA: Sage Publications.

Cavanagh, S. E., Crissey, S. R., & Raley, R. K. (in press). Family structure, parenting, and adolescent romantic relationships. *Journal of Marriage and Family.*

Cavanagh, S. E., & Huston, A. C. (2006). Family instability and children's early problem behavior. *Social Forces, 85,* 551–581.

Cherlin, A. J. (1999). Going to extremes: Family structure, children's well-being, and social science. *Demography, 36,* 421–428.

Coleman, M., Ganong, L., & Fine, M. (2000). Reinvestigating remarriage: Another decade of progress. *Journal of Marriage and the Family, 62,* 1288–1307.

Fomby, P., & Cherlin, A. J. (2007). Family instability and child well-being. *American Sociological Review, 72,* 181–204.

The Fragile Families and Child Wellbeing Study. Retrieved June 3, 2008, from Princeton University, Bendheim-Thoman Center for Research on Child Wellbeing Web site: http://www.fragilefamilies.princeton.edu

Goldscheider, F. K., & Goldscheider, C. (1998). The effects of family structure on leaving and returning home. *Journal of Marriage and the Family, 60,* 745–756.

Hamilton, B. E., Martin, J. A., & Ventura, S. J. (2006, November 21). *Births: Preliminary data for 2005* (Health E-Stats). Retrieved May 28, 2008, from Centers for Disease Control and Prevention Web site: http://www.cdc.gov/nchs/products/pubs/pubd/hestats/prelimbirths05/prelim births05.htm

Harris, K. M., Cavanagh, S. E., & Elder, G. H., Jr. (2000, March). *The well-being of adolescents in single-father context.* Paper presented at the annual meeting of the Population Association of America, Los Angeles, CA.

Kreider, R. M., & Fields, J. (2005). *Living arrangements of children: 2001* (Current Population Report P70-104). Washington, DC: U.S. Bureau of the Census.

Manning, W. D., & Lamb, K. A. (2003). Adolescent well-being in cohabiting, married, and single-parent families. *Journal of Marriage and Family, 65,* 876–893.

McLanahan, S., & Sandefur, G. (1994). *Growing up with a single parent: What hurts, what helps.* Cambridge, MA: Harvard University Press.

Osborne, C., & McLanahan, S. (2007). Partnership instability and child well-being. *Journal of Marriage and Family, 69,* 1065–1083.

Patterson, C. J. (2006). Children of lesbian and gay parents. *Current Directions in Psychological Science, 15,* 241–244.

Raley, R. K., & Bumpass, L. L. (2003). The topography of the divorce plateau: Levels and trends in union stability in the United States after 1980. *Demographic Research, 8,* 245–260.

Simmons, T., & O'Connell, M. (2003). *Married-couple and unmarried-partner households: 2000* (Census 2000 Special Report). Washington, DC: U.S. Bureau of the Census.

Sweeney, M. M. (2007). Stepfather families and the emotional well-being of adolescents. *Journal of Health and Social Behavior, 48,* 33–49.

Teachman, J. (2003). Childhood living arrangements and the formation of coresidential unions. *Journal of Marriage and Family, 65,* 507–524.

Thompson, E., McLanahan, S. S., & Curtin, R. B. (1992). Family structure, gender, and parental socialization. *Journal of Marriage and the Family, 54,* 368–378.

Wildsmith, E., & Raley, R. K. (2006). Race-ethnic differences in nonmarital fertility: A focus on Mexican American women. *Journal of Marriage and Family, 68,* 491–508.

Wu, L. L., & Martinson, B. C. (1993). Family structure and the risk of a premarital birth. *American Sociological Review, 58,* 210–232.

Shannon E. Cavanagh

FAMILY PROCESS MODEL

The family process model posits that the effects of poverty on child well-being go beyond the material resources afforded by higher incomes. According to the model (see Figure 1), poverty impacts children's development indirectly through its negative effect on family processes. The reasons are simple. Poverty is a highly disorienting and upsetting experience that, perniciously, can make parents doubt themselves and lose hope in the future. In this way, it introduces a level of stress and discord into the family that ultimately affects the social, emotional, and academic lives of children. From the family process literature, two general aspects of the home environment—parents' marital or romantic relationships and parenting behaviors—have been identified as primary avenues through which family processes link poverty to child development.

A wealth of empirical evidence has documented that growing up in poverty places children at risk for a wide range of physical, cognitive, and socioemotional problems (Seccombe, 2000). For example, parents and teachers report that low-income children are more likely to be aggressive, to experience symptoms of depression, and to receive lower scores on measures of academic achievement compared to their more affluent peers. Understanding the ways in which poverty affects children's well-being, therefore, is an important goal for social science researchers. In the large and growing body of literature on the development of economically disadvantaged children, a powerful explanation for the association between poverty and poor development has emerged: the family process model, which is the focus of this entry.

CHILDHOOD POVERTY
IN THE UNITED STATES

In 2006 about 17% of American children were raised in families with annual incomes that fell below the government poverty level of $20,614 for a family of four (U.S. Bureau of the Census, 2007). More American children live in poverty than was the case in the late 1970s and than children from any other industrialized nation. American children are also more likely to experience poverty than adolescents or adults. Although these statistics highlight the pervasiveness of childhood poverty in the United States, they do not provide a complete picture of poor American youth. Millions more families, with annual incomes just above the poverty level, also struggle to earn enough money for food and rent. Furthermore, the number of economically disadvantaged children, measured by multiple factors including family income, family structure, and educational attainment, is far greater than these basic statistics suggest.

Children from every racial/ethnic background live in poverty, but the likelihood of growing up in an impoverished family is much lower for White and Asian children than for African-American and Hispanic children. During the 10 years beginning with the mid-1990s, approximately 30% of African-American and Hispanic children lived in families with incomes below the poverty level, compared to about 10% of Asian and White children (Children's Defense Fund, 2004). Not only are African-American and Hispanic children more likely to live in poverty, but they are also more likely to live in high-poverty communities and to live in poverty over longer periods of time compared to poor White children.

Poverty and Child Development Not surprising, growing up in poverty places children at risk for a wide range of negative developmental outcomes. As mentioned, numerous studies have documented the association between poverty and poor physical, cognitive, social, and emotional development. For example, infants born to poor families are more likely to experience malnutrition, failure to thrive syndrome, and sudden infant death syndrome. During childhood, poverty is associated with poor performance on measures of cognitive functioning, as well as internalizing problems such as depression and anxiety and externalizing problems such as aggressive and antisocial behavior. Among adolescents, poverty is related to obesity and overall health. Poor adolescents are also more likely to become pregnant, associate with deviant peers, and experiment with illegal drugs than more affluent youth.

Research demonstrates that poverty also has a negative impact on academic achievement, and differences between poor and more affluent children can be seen at the very start of formal schooling. Economically disadvantaged children score significantly lower than both middle- and upper-class children on measures of reading and math achievement at the beginning of kindergarten and this problem is especially pronounced for poor racial/ethnic minority children. The substantial gap in academic competencies between poor and more affluent children persists throughout their educational careers. Poor children and adolescents earn lower grades and lower scores on achievement tests, they are more likely to be placed in lower curricular tracks and special education programs, and they are less likely to graduate from high school or enter into higher education than nonpoor youth (McLoyd, 1998).

Explanations for the Association Between Poverty and Poor Child Development Given that poverty affects child development within and across racial/ethnic groups, the next step is to understand how this occurs. Explanations for the association between poverty and children's well-being often center on the lack of material resources available to poor children and their families. For example, children raised in poverty often live in unsafe neighborhoods, attend ineffective schools, have poor diets, and receive little health care. According to financial capital models, poverty affects children directly by limiting material resources that are beneficial to children's development and well-being. Although some studies provide support for these models, the effects of poverty vary greatly from one outcome to another, and there is little consensus among researchers regarding the size of the

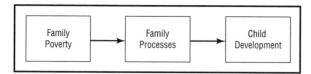

Figure 1. *A basic family process model.* **CENGAGE LEARNING, GALE.**

effects (Duncan, Yeung, Brooks-Gunn, & Smith, 1998; Haveman & Wolfe, 1995). Moreover, financial capital models overlook the possibility that one of the greatest influences of poverty may be related to nonmaterial family resources.

THE FAMILY PROCESS MODEL

One great advantage of the family process model is that it integrates two core developmental paradigms—ecological and life course theories—and thus bridges psychological and sociological approaches to the study of poverty and child development. From an ecological perspective (Bronfenbrenner, 1979), children develop within multiple, overlapping contexts and in increasingly complex interactions with their environments. The family process model draws on ecological theory by examining how processes in one context (family) influence the lives of children in other contexts (e.g., peer group). Developed most fully in the work of Elder (1998, 1999), life course theory views lives as interdependent trajectories embedded in social and historical contexts. The family process model takes a life course approach to studying child development by viewing children's lives as linked to their parents (i.e., parents' poverty disrupts children's development through its influence on parenting).

SUPPORT FOR THE FAMILY PROCESS MODEL

The general framework of the family process model draws heavily on studies of White families of the Great Depression (Elder, 1999). In several studies, Elder and colleagues examined the effects of economic loss during the depression on children's behavioral and socioemotional development. The results of this research indicated that economic loss had few direct effects on children's well-being. Instead, negative child outcomes occurred indirectly through the fathers' poor psychological functioning and negative parenting behaviors. Fathers who experienced severe financial loss were more likely to use punitive, rejecting, and inconsistent disciplinary practices, and these parenting behaviors were significantly related to children's socioemotional problems.

Following the pioneering work of Elder and his colleagues, McLoyd (1990) proposed a model to examine how poverty and economic loss affect African-American children's socioemotional development. According to this model, impoverished families often experience an excess of adverse life events, and the resulting psychological distress diminishes parents' capacity for supportive, consistent, and involved parenting, which, in turn, disrupts children's socioemotional functioning. Subsequently, family processes have linked poverty to a wide range of negative socioemotional outcomes during childhood and adolescence, including anxiety, depression, and poor social competence, as well as behavior problems related to compliance, impulse control, aggression, and drug use in youth of all races. Importantly, the family process model has also been applied to the educational experiences of children and adolescents. Numerous studies have provided evidence that a wide range of parenting behaviors, including emotional support and warmth, discipline strategies, education-related practices, and the presence of household rules and routines, explain at least some of the well-documented association between family income and academic outcomes (Burchinal, Roberts, Zeisel, Hennon, & Hooper, 2006; Conger et al., 1992, 1993; Gutman & Eccles, 1999; Raver et al., 2007; Yeung, Linver, & Brooks-Gunn, 2002; Mistry, Vandewater, Huston, & McLoyd, 2002).

MOVING BEYOND THE BASIC MODEL

Research has begun to extend the basic family process model in ways that have increased knowledge about child development in the context of economic disadvantage. For example, several studies have investigated the ways in which income (or the lack thereof) influences family processes. From this research, two primary pathways have emerged: parents' mental health and aspects of financial strain. In studies that vary widely with regard to family processes and child outcomes, poor parent psychological well-being (typically maternal depression) explains at least part of the association between income and family processes (Conger et al., 2002; Mistry et al., 2002; Parke et al., 2004; Vandewater & Lansford, 2005; Yeung et al., 2002). At the same time, a growing number of studies suggest that objective measures of economic hardship, such as low income, negatively affect parents' psychological well-being and parenting behavior through their impact on financial strain and stress (Gutman, McLoyd, & Tokoyawa, 2005; Mistry et al., 2002; Mistry, Biesanz, Taylor, Burchinal, & Cox, 2004).

Researchers have also sought to gain a better understanding of the extent to which the family process model applies to families across various racial, ethnic, and socioeconomic backgrounds. Although less is known about Asian and Hispanic families, studies of African-American families suggest that, in general, the family process model functions well for this racial group (Conger et al., 2002; Jackson, Brooks-Gunn, Huang, & Glassman, 2000). The model also appears to hold not just for families who live in extreme poverty but also for working- and middle-class families who experience economic loss (Conger et al., 1992, 1993). Furthermore, studies that draw on samples from rural, suburban, and urban areas suggest that the family process model can be applied to families living in a variety of geographic contexts (Conger et al., 2002; Gutman et al., 2005; Jackson, Brooks-Gunn, Huang, & Glassman, 2000).

FUTURE RESEARCH

Despite strong empirical and theoretical grounding for the family process model, gaps in the literature remain. For example, few researchers test the applicability of their proposed models for diverse racial/ethnic groups, making it difficult to determine the robustness of their findings for different races/ethnicities. Through early 2008, only three known studies have examined the equivalence of their conceptual models across various racial/ethnic groups. In an investigation of adolescents' academic achievement, Gutman and Eccles (1999) found that negative parent–adolescent relationships and school-based parental involvement explained the association between financial strain and achievement for both African-American and White families. No significant differences in the models between the two racial/ethnic groups were found. Parke and colleagues (2004), however, reported that paternal hostile parenting was related to adjustment problems for fifth-grade White children, whereas marital problems predicted poor adjustment for Mexican-American children.

In the most comprehensive investigation of model equivalence across race/ethnicity, Raver, Gershoff, and Aber (2007) examined the importance of a wide range of family processes in explaining the association between family income and measures of school readiness for African-American, Hispanic, and White families. Results of this study suggest that important differences may exist across the three racial/ethnic groups. For example, income was a stronger predictor of children's kindergarten achievement for African-American children than for Hispanic or White children. Material hardship was also more strongly related to parents' stress in African-American families. The positive association between parenting behavior and children's social competence, however, was stronger for White families than for racial/ethnic minority families. Taken together these studies suggest that racial/ethnic variation in family process models likely depends on a number of factors, including the family processes and developmental outcomes of interest, the stage of development, the gender of the parent, and the definition of economic disadvantage.

Research also needs to gain a better understanding of resilience in the context of poverty by incorporating protective factors into the family process model. To do so, researchers can focus on the association between poverty and family processes, the first piece of the model. Although poverty typically disrupts marital and parent–child relationships, differences exist in the home environments of poor families and are likely related to individual characteristics of family members as well as factors in work, school, and neighborhood settings. As one example, research indicates that economically disadvantaged parents are less optimistic about their children's educational chances than more affluent parents. For a variety of reasons, however, some poor parents are able to maintain positive beliefs about their children, despite their economic situation. In these families, optimistic beliefs about their children's educational careers may increase education-related parenting and thus represent a parent characteristic that protects against the impact of poverty on parenting.

Researchers can also identify protective factors by focusing on the second piece of the family process model: the association between family processes and child development. Investigating characteristics of children and their environments, especially those that are amenable to policy, is important for improving the well-being of economically disadvantaged children. For example, aspects of the school environment, such as characteristics of teachers and administrators and services and resources for families, may help to reduce the negative effect of disruptive family processes on low-income children's academic achievement.

In summary, the family process model has provided an excellent framework for understanding the importance of the family context in explaining the negative impact of poverty on children and adolescents' well-being. A large body of research has provided empirical evidence that the family process model can be applied to a wide range of developmental outcomes, to various developmental stages, to both boys and girls, and to families from diverse backgrounds. Investigating whether these models hold for different races/ethnicities and identifying protective factors, however, are important areas of future research, especially given their implications for policies aimed at improving the well-being of poor children.

SEE ALSO Volume 1: *Attachment Theory; Developmental Systems Theory; Elder, Glen H., Jr.; Parent-Child Relationships, Childhood and Adolescence; Policy, Child Well-Being; Poverty, Childhood and Adolescence; Resilience, Childhood and Adolescence.*

BIBLIOGRAPHY

Arnold, D. H., & Doctoroff, G. L. (2003). The early education of socioeconomically disadvantaged children. *Annual Review of Psychology, 54,* 517–545.

Black, M., & Dubowitz, H. (1991). Failure-to-thrive: Lessons from animal models and developing countries. *Journal of Developmental and Behavioral Pediatrics, 12*, 259–267.

Brody, G. H., Stoneman, Z., Flor, D., McCrary, C., Hastings, L., & Conyers, O. (1994). Financial resources, parent psychological functioning, parent co-caregiving, and early adolescent competence in rural two-parent African-American families. *Child Development, 65*, 590–605.

Bronfenbrenner, U. (1979). *The ecology of human development: Experiments by nature and design*. Cambridge, MA: Harvard University Press.

Brown, J. L., & Pollitt, E. (1996, February). Malnutrition, poverty, and intellectual development. *Scientific American, 274*, 38–43.

Burchinal, M., Roberts, J. E., Zeisel, S. A., Hennon, E. A., & Hooper, S. (2006). Social risk and protective child, parenting, and child care factors in early elementary school years. *Parenting: Science and Practice, 6*, 79–113.

Children's Defense Fund. (2004). *The state of America's children*. Washington, DC: Author.

Coleman, J. S. (1990). *Foundations of social theory*. Cambridge, MA: Belknap Press of Harvard University Press.

Conger, R. D., Conger, K. J., Elder, G. H., Jr., Lorenz, F. O., Simons, R. L., & Whitbeck, L. B. (1992). The family process model of economic hardship and adjustment of early adolescent boys. *Child Development, 63*, 526–541.

Conger, R. D., Conger, K. J., Elder, G. H., Jr., Lorenz, F. O., Simons, R. L., & Whitbeck, L. B. (1993). Family economic stress and adjustment of early adolescent girls. *Developmental Psychology, 29*, 206–219.

Conger, R. D., Wallace, L. E., Sun, Y., Simons, R. L., McLoyd, V. C., & Brody, G. H. (2002). Economic pressure in African American families: A replication and extension of the Family Stress Model. *Developmental Psychology, 38*, 179–193.

Crosnoe, R., Mistry, R. S., & Elder, G. H., Jr. (2002). Economic disadvantage, family dynamics, and adolescent enrollment in higher education. *Journal of Marriage and Family, 64*, 690–702.

Duncan, G. J., & Rodgers, W. L. (1988). Longitudinal aspects of childhood poverty. *Journal of Marriage and the Family, 50*, 1007–1021.

Duncan, G. J., Yeung, W. J., Brooks-Gunn, J., & Smith, J. R. (1998). How much does childhood poverty affect the life chances of children? *American Sociological Review, 63*, 406–423.

Eamon, M. K. (2001). Poverty, parenting, peer, and neighborhood influences on young adolescent antisocial behavior. *Journal of Social Service Research, 28*(1), 1–23.

Eccles, J. S., & Harold, R. D. (1993). Parent–school involvement during the early adolescent years. *Teachers College Record, 94*, 568–587.

Elder, G. H., Jr. (1974). *Children of the Great Depression: Social change in life experience*. Chicago: University of Chicago Press.

Elder, G. H., Jr. (1979). Historical change in life patterns and personality. In P. B. Baltes & O. G. Brim (Eds.), *Life-span development and behavior* (Vol. 2, pp. 117–159). New York: Academic Press.

Elder, G. H., Jr. (1998). The life course and human development. In W. Damon (Series Ed.) & R. M. Lerner (Vol. Ed.), *Handbook of child psychology, Vol. 1: Theoretical models of human development* (5th ed., pp. 939–991). New York: Wiley.

Elder, G. H., Jr. (1999). *Children of the Great Depression: Social change in life experience* (25th anniversary ed.). Boulder, CO: Westview Press.

Elder, G. H., Jr., Liker, J. K., & Cross, C. E. (1984). Parent–child behavior in the Great Depression: Life course and intergenerational influences. In P. Baltes & O. Brim (Eds.), *Life-span development and behavior* (Vol. 6, pp. 109–158). Orlando, FL: Academic Press.

Elder, G. H., Jr., Van Nguyen, T., & Caspi, A. (1985). Linking family hardship to children's lives. *Child Development, 56*, 361–375.

Furstenberg, F. F., Jr., Cook, T. D., Eccles, J., Elder, G. H., Jr., & Sameroff, A. (1999). *Managing to make it: Urban families and adolescent success*. Chicago: University of Chicago Press.

Goodman, E. (1999). The role of socioeconomic status gradients in explaining differences in U.S. adolescents' health. *American Journal of Public Health, 89*, 1522–1528.

Gutman, L. M., & Eccles, J. S. (1999). Financial strain, parenting behaviors, and adolescents' achievement: Testing model equivalence between African American and European American single- and two-parent families. *Child Development, 70*, 1464–1476.

Gutman, L. M., McLoyd, V. C., & Tokoyawa, T. (2005). Financial strain, neighborhood stress, parenting behaviors, and adolescent adjustment in urban African American families. *Journal of Research on Adolescence, 15*, 425–449.

Haveman, R., & Wolfe, B. (1995). The determinants of children's attainments: A review of methods and findings. *Journal of Economic Literature, 33*, 1829–1878.

Huston, A. C. (1999). Effects of poverty on children. In L. Balter & C. S. Tamis-LeMonda (Eds.), *Child psychology: A handbook of contemporary issues* (pp. 391–411). Philadelphia: Psychology Press.

Jackson, A. P., Brooks-Gunn, J., Huang, C. C., & Glassman, M. (2000). Single mothers in low-wage jobs: Financial strain, parenting, and preschoolers' outcomes. *Child Development, 71*, 1409–1423.

Jencks, C., & Phillips, M. (1998). The black–white test score gap: An introduction. In C. Jencks & M. Phillips (Eds.), *The black–white test score gap* (pp. 1–51). Washington, DC: Brookings Institution Press.

Lee, V. E., & Burkam, D. T. (2002). *Inequality at the starting gate: Social background differences in achievement as children begin school*. Washington, DC: Economic Policy Institute.

Lempers, J. D., Clark-Lempers, D., & Simons, R. L. (1989). Economic hardship, parenting, and distress in adolescence. *Child Development, 60*, 25–39.

Mayer, S. E. (1997). *What money can't buy: Family income and children's life chances*. Cambridge, MA: Harvard University Press.

McLanahan, S., & Sandefur, G. (1994). *Growing up with a single parent: What hurts, what helps*. Cambridge, MA: Harvard University Press.

McLoyd, V. C. (1990). The impact of economic hardship on black families and children: Psychological distress, parenting, and socioemotional development. *Child Development, 61*, 311–346.

McLoyd, V. C. (1998). Socioeconomic disadvantage and child development. *American Psychologist, 53*(2), 185–204.

McLoyd, V. C., & Wilson, L. (1991). The strain of living poor: Parenting, social support, and child mental health. In A. C. Huston (Ed.), *Children in poverty: Child development and public policy* (pp. 105–135). Cambridge, U.K.: Cambridge University Press.

Mistry, R. S., Biesanz, J. C., Taylor, L. C., Burchinal, M., & Cox, M. J. (2004). Family income and its relation to preschool children's adjustment for families in the NICHD Study of Early Child Care. *Developmental Psychology, 40,* 727–745.

Mistry, R. S., Vandewater, E. A., Huston, A. C., & McLoyd, V. C. (2002). Economic well-being and children's social adjustment: The role of family process in an ethnically diverse low-income sample. *Child Development, 73,* 935–951.

Parke, R. D., Coltrane, S., Duffy, S., Buriel, R., Dennis, J., Powers, J., et al. (2004). Economic stress, parenting, and child adjustment in Mexican American and European American families. *Child Development, 75,* 1632–1656.

Raver, C. C., Gershoff, E. T., & Aber, J. L. (2007). Testing equivalence of mediating models of income, parenting, and school readiness for white, black, and Hispanic children in a national sample. *Child Development, 78,* 96–115.

Samaan, R. A. (2000). The influences of race, ethnicity, and poverty on the mental health of children. *Journal of Health Care for the Poor and Underserved, 11,* 100–110.

Seccombe, K. (2000). Families in poverty in the 1990s: Trends, causes, consequences, and lessons learned. *Journal of Marriage and the Family, 62,* 1094–1113.

Sherman, A. (1994). *Wasting America's future: The Children's Defense Fund report on the costs of child poverty.* Boston: Beacon Press.

Smith, J. R., Brooks-Gunn, J., & Klebanov, P. K. (1997). Consequences of living in poverty for young children's cognitive and verbal ability and early school achievement. In G. J. Duncan & J. Brooks-Gunn (Eds.), *Consequences of growing up poor* (pp. 132–189). New York: Russell Sage Foundation.

Snow, C. E., Burns, M. S., & Griffin, P. (Eds.). (1998). *Preventing reading difficulties in young children.* Washington, DC: National Academy Press.

Sucoff, C. A., & Upchurch, D. M. (1998). Neighborhood context and the risk of childbearing among metropolitan-area black adolescents. *American Sociological Review, 63,* 571–585.

Taylor, R. D. (1994). Risk and resilience: Contextual influences on the development of African-American adolescents. In M. C. Wang & E. W. Gordon (Eds.), *Educational resilience in inner-city America: Challenges and prospects* (pp. 119–130). Hillsdale, NJ: Erlbaum.

U.S. Bureau of the Census. (2007). *Poverty.* Retrieved March 24, 2008, from http://www.census.gov/hhes/www/poverty/poverty.html

Vandewater, E. A., & Lansford, J. E. (2005). A family process model of problem behaviors in adolescents. *Journal of Marriage and Family, 67,* 100–109.

Wilson, W. J. (1987). *The truly disadvantaged: The inner city, the underclass, and public policy.* Chicago: University of Chicago Press.

Yeung, W. J., Linver, M. R., & Brooks-Gunn, J. (2002). How money matters for young children's development: Parental investment and family processes. *Child Development, 73,* 1861–1879.

Carey E. Cooper

FOSTER CARE

Foster care is a system in which licensed foster parents provide substitute care for children whose parents are unable to provide them with a safe environment. Between 1985 and 2005 the number of children in foster care increased from 276,000 to 513,000. Understanding who is in foster care, how the foster-care system operates, and why caseloads have grown is important for several reasons. First, foster children have higher rates of emotional, behavioral, developmental, and physical health problems than children outside of foster care. Second, the growth in foster-care caseloads has resulted in larger workloads for individual caseworkers, which is problematic because mistreatment of foster children is often blamed on the large number of cases caseworkers must manage. Third, government expenditures on foster care increase as the number of children in the system grows.

Before discussing the formal foster-care system, it is important to distinguish between formal foster care and other types of out-of-home care. Children in the formal foster-care system represent only a fraction of the children being cared for by someone other than their parents. Researchers from the Urban Institute estimate that approximately 2 million children lived in the care of someone other than their parents in 1997 (Ehrle & Geen, 2002; Ehrle, Geen, & Clark, 2001).

The type of out-of-home care is typically defined by whether the child is living with relatives and whether child-welfare services staff are involved (Ehrle et al., 2001). The most common type of out-of-home care is *private kinship care* (an estimated 1.3 million children in 1997), in which the caregivers are related to the child and there is no involvement by child-welfare services. The next most common arrangement is *public kinship foster care* (an estimated 500,000 children in 1997), in which child-welfare services are involved and relatives care for the child. In this case, child-welfare services may formally remove the child from the custody of his or her parents in which case the child is said to be in *formal kinship foster care* (an estimated 200,000 children in 1997). If child-welfare services are involved but do not obtain custody, the child is said to be in *voluntary kinship foster care* (an estimated 300,000 children in 1997). Finally, a smaller number of children are in *public foster care* (an estimated 200,000 children in 1997), whereby child-welfare services staff obtains custody of the child and places him or her with unrelated caregivers.

Understanding the different caregiving arrangements is important because most foster-care data pertains only to children in the custody of the state (children in public foster care and formal kinship foster care). Figure 1 shows the total number of children in foster care at the end of each fiscal year over the period from 1962 to

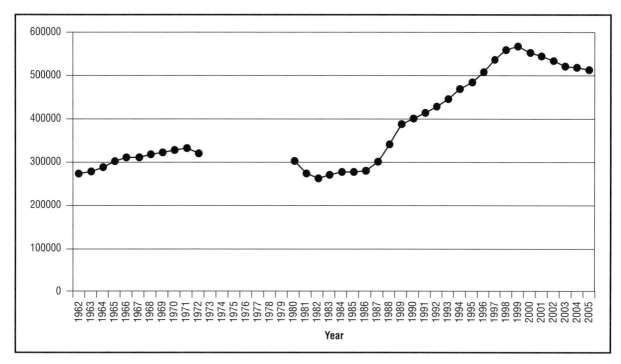

Figure 1. *Children in foster care.* CENGAGE LEARNING, GALE.

2005. Although data are not available for the period from 1973 to 1979, it appears that the number of children in care remained relatively constant from the early 1960s to the mid-1980s. Beginning in the mid-1980s, the number of children in foster care started to rise dramatically. The caseload peaked in 1999 before declining.

Researchers have suggested a number of factors to explain the increase in the number of children in care. These include changing economic conditions, social norms (e.g., the increase in the number of single-parent families), rates of parental substance abuse, rates of HIV/AIDS infection, rates of incarceration (particularly of women), and welfare policy. Early research highlights the role of the "crack cocaine epidemic" (U.S. Department of Health and Human Services, 1999; U.S. General Accounting Office, 1997). More recently, Swann and Sheran Sylvester (2006) studied the relationship between foster-care caseloads and a number of these factors and found a positive relationship between the rate of incarceration of women and the number of children in foster care and a negative relationship between welfare benefits and the number of children in foster care.

ENTRY INTO FOSTER CARE

In most cases, children come to the attention of child-welfare workers through reports of abuse or neglect. The complaint may come from a relative, teacher, doctor, law enforcement official, or someone else. It may involve physical or sexual abuse, neglect, or parental absence. Neglect results from parents who are present but are not providing adequate food, shelter, clothing, or appropriate medical care. Parental absence refers to parents who are either physically absent (e.g., because they are incarcerated) or present but incapacitated (e.g., because of drug abuse). Children may also come to the attention of child-welfare workers through their own behavior (e.g., through the juvenile justice system).

The exact path from the initial report to foster care depends on the circumstances and jurisdiction, but it will involve an investigation and the determination that removal from the home is necessary to ensure the safety and well-being of the child (Pecora, Whittaker, Maluccio, & Barth, 2000; Schene, 1998). A number of different factors bear on the ultimate decision to remove a child. These include the substantiation of the alleged abuse/neglect and the determination that the parent is unable or unwilling to protect the child. The specific type of placement will depend on the availability of relative caregivers or non-kin foster caregivers as well as the specific needs of the child (e.g., a child may have special needs that cannot be met by relatives).

In order to study the pathways into foster care, it would be ideal to have information about the children, their families, and the decision made at each step in the process. Unfortunately, such complete information does not exist. Consequently, researchers take two approaches. One is to study the characteristics of children in care. This approach focuses on the big picture rather than on individual steps in the process. For example, in a study of foster children in California in 1991, Lewit (1993) found that 20% of entrants were subject to physical or sexual abuse, whereas 68% were victims of neglect. In other findings, 26% of entrants were neglected because of parental absence or incapacitation, and 42% were victims of neglect because of inadequate care. Additionally, administrative data from the Adoption and Foster Care Analysis and Reporting System (AFCARS) shows that infants make up the largest percentage of entrants into foster care (U.S. Department of Health and Human Services, 2006). This report also shows that almost one-third of the entrants are African-American.

In the second approach, researchers explore the relationship between case (child and family) characteristics and a single decision made by child-welfare officials such as the decision to terminate parental rights. Lindsey (1992), for example, used data on children who had come "to the attention of public child welfare agencies throughout the United States" (p. 30), whereas Zuravin and DePanfilis (1997) focused on children already receiving child-welfare services in a large mid-Atlantic city. In both studies, the families had been reported to child protective services and been investigated. Given this, the research sought to understand what factors were related to the decision to remove the child from the home. Zuravin and DePanfilis found that children whose mothers abused drugs, had developmental problems, or suffered from mental health problems were more likely to be placed in foster care, whereas Lindsey found that children in families with uncertain income prospects were more likely to be placed in foster care.

CHARACTERISTICS OF FOSTER CHILDREN

Consistent with their high rate of entry into the foster-care system, studies find that Black children are overrepresented in the foster-care system (e.g., McRoy, 2005; U.S. Department of Health and Human Services, 2006). The overrepresentation of minorities in the foster-care system is an important topic that is surveyed by McRoy (2005).

Additionally, many researchers find that children in foster care are more likely than other children to have poor health, behavioral problems, difficulties in school, and other developmental issues (e.g., Committee on Early Childhood, Adoption, and Dependent Care, 2002; Fine, 1985; Jones Harden, 2004). Hochstadt, Jaudes, Zimo, and Schachter (1987) focused specifically on the health of children entering foster care and found high rates of developmental delays as well as emotional and behavioral problems.

Research suggests that stability during childhood is an important determinant of child development (Jones Harden, 2004), and a goal of child-welfare workers is to provide a stable environment for foster children. Unfortunately the data on foster children indicate that placement instability is common. For example, in 2001, 22% of children had experienced three or four placements during their current episode of care, 8% had experienced five or six placements, and 9% of children had experienced seven or more placements (U.S. House of Representatives, 2004). Webster et al. (2000) studied the relationship between case characteristics and placement instability for children in long-term foster care in California. Similar to other authors (e.g., Goerge, 1990), they found that being placed with kin is associated with greater stability.

EXITS FROM FOSTER CARE

Children may exit from foster care through a number of routes. These include reunification with parents, adoption, emancipation ("aging out"), running away, and death. The U.S. Department of Health and Human Services report discussed above includes descriptive information about the characteristics of children who exited foster care in fiscal year 2005. More than 10% of the children leaving foster care in 2005 were 17 years or older. Most of these children had "aged out" of the foster-care system, and they are of particular concern because studies suggest that many of these children are unprepared to live independently (Nixon, 2005). More than half of the children who left foster care were reunified with family, whereas 18% were adopted.

A number of studies (e.g., Benedict & White, 1991; Courtney, 1994) go beyond a simple description of the characteristics of children leaving care to explore the child, family, and placement characteristics associated with the probability of exiting foster care and the length of time in care when an exit occurs. Because many children experience more than one episode of foster care, additional research (e.g., Courtney, 1995; Festinger, 1996; Goerge, 1990) has explored both the probability of leaving care and the probability of reentering care.

It is important to remember, however, that although reunification and adoption are frequently cited goals of the foster-care system, all types of exits are not necessarily equally good. Furthermore, a particular child characteristic might increase the probability of some types of exits while decreasing the probability of other types of exits, and a different characteristic might reduce the probability of any type of exit.

McMurtry and Lie (1992) and Courtney and Wong (1996) studied how different child, family, and placement characteristics are related to different types of exits. Courtney and Wong used data on children who entered foster care in California in the first half of 1988 to look at exits in which children were reunified with family, were adopted, or ran away. They found that reunification typically happens very early in a foster-care spell, whereas adoption typically happens after several years in care. The likelihood of running away is highest early in the episode of care.

In terms of individual characteristics, Courtney and Wong (1996) found that being younger is associated with higher probabilities of being reunified, lower probabilities of being adopted, and lower probabilities of running away. Being African-American is associated with lower probabilities of reunification and adoption. This implies that African-American children will have, on average, longer stays in the foster-care system than children in other ethnic groups, a result also found by McMurtry and Lie (1992). Similarly, both studies found that significant health problems or disabilities reduce the probability of any of the three types of exits, suggesting that children with health problems may remain in the child-welfare system longer than other children.

There is also great concern about the amount of time children spend in foster care regardless of the type of exit. In fact, two key pieces of legislation, the Adoption Assistance and Child Welfare Act of 1980 and the Adoption and Safe Families Act of 1997, are focused in part on reducing the length of time children spend in foster care (Glisson, Bailey, & Post, 2000). From 1999 to 2001, the average time in foster care for children leaving care was 22.4 months (U.S. House of Representatives, 2004); however, almost 15,000 children exiting care in 2005 left after spending 5 or more years in care (U.S. Department of Health and Human Services, 2006).

LIFE AFTER FOSTER CARE

Relatively few studies explore the long-term consequences of foster care. An important reason for this is the difficulty of finding and interviewing former foster children (Barth, 1990). The studies that have been conducted have found mixed results. Maluccio and Fein (1985) reviewed a number of studies and concluded that foster care does not negatively affect children. In a separate review, Barth (1986) concludes that former foster children face significant educational and employment challenges but are otherwise similar to the general population. Barth (1990) studied 55 former foster children in the San Francisco area and found that they fare worse than the general population. More than one-third of the former foster children had been arrested, been convicted of a crime, or spent time in jail or prison since leaving foster care. Almost half had health problems, and more than one-third had significant uncertainty about their living arrangements.

CURRENT ISSUES IN FOSTER CARE

The role of kinship care continues to evolve and grow (Geen, 2004; Ingram, 1996). There are several reasons for this, including a shortage of non-kin foster parents and the Adoption and Safe Families Act's recognition of placement with kin as an acceptable long-term placement outcome. The growing role for kin caregivers raises a number of issues such as the services that should be made available to kin and appropriate licensing requirements. Prior to the widespread use of formalized kinship care, family members provided care without any involvement with or support from child-welfare workers. In other words, no licensing was required, and no services were provided. Studies have since shown that kin caregivers may have greater needs than non-kin caregivers but receive fewer services (Geen, 2004). Clarifying the rights and responsibilities of kinship caregivers will be an important continuing development in foster care.

Coordination among a number of different government systems and programs is becoming increasingly important because of legislation passed in the late 1990s. The Personal Responsibility and Work Opportunity Reconciliation Act of 1996 made dramatic changes in the operation of cash welfare programs (e.g., by implementing time limits), and it combined federal funding for many social service programs, including child-welfare programs, into a single grant. States are allowed to allocate money from this grant to programs in any way they see fit, which may affect budgets for child-welfare services. Additionally, the dramatic increase of incarcerated mothers has created special concerns in light of the Adoption and Safe Families Act's focus on expediting permanency decisions and moving children more quickly to adoption.

SEE ALSO Volume 1: *Adopted Children; Child Abuse; Policy, Child Well-Being; Poverty, Childhood and Adolescence;* Volume 2: *Adoptive Parents.*

BIBLIOGRAPHY
Barth, R. P. (1986). Emancipation services for adolescents in foster care. *Social Work, 31,* 165–171.

Barth, R. P. (1990). On their own: The experiences of youth after foster care. *Child and Adolescent Social Work Journal, 7,* 419–440.

Benedict, M. I., & White, R. B. (1991). Factors associated with foster care length of stay. *Child Welfare, 70,* 45–58.

Committee on Early Childhood, Adoption, and Dependent Care. (2002). Health care of young children in foster care. *Pediatrics, 109,* 536–541.

Courtney, M. E. (1994). Factors associated with the reunification of foster children with their families. *Social Service Review, 68,* 81–108.

Courtney, M. E. (1995). Reentry to foster care of children returned to their families. *Social Service Review, 69,* 226–241.

Courtney, M. E., & Wong, Y. I. (1996). Comparing the timing of exits from substitute care. *Children and Youth Services Review, 18,* 307–334.

Ehrle, J., & Geen, R. (2002). Kin and non-kin foster care: Findings from a national survey. *Children and Youth Services Review, 24,* 15–35.

Ehrle, J., Geen, R., & Clark, R. (2001). *Children cared for by relatives: Who are they and how are they faring?* Washington, DC: Urban Institute.

Festinger, T. (1996). Going home and returning to foster care. *Children and Youth Services Review, 18,* 383–402.

Fine, P. (1985). Clinical aspects of foster care. In M. J. Cox & R. D. Cox (Eds.), *Foster care: Current issues, policies, and practices* (pp. 206–233). Norwood, NJ: Ablex Publishing.

Geen, R. (2004). The evolution of kinship care policy and practice. *The Future of Children, 14,* 131–149.

Glisson, C., Bailey, J. W., & Post, J. A. (2000). Predicting the time children spend in state custody. *Social Service Review, 74,* 253–280.

Goerge, R. M. (1990). The reunification process in substitute care. *Social Service Review, 64,* 422–457.

Hochstadt, N. J., Jaudes, P. K., Zimo, D. A., & Schachter, J. (1987). The medical and psychosocial needs of children entering foster care. *Child Abuse and Neglect, 11,* 53–62.

Ingram, C. (1996). Kinship care: From last resort to first choice. *Child Welfare, 75,* 550–566.

Jones Harden, B. (2004). Safety and stability for foster children: A developmental perspective. *The Future of Children, 14,* 31–47.

Lewit, E. M. (1993). Children in foster care. *The Future of Children, 3,* 192–200.

Lindsey, D. (1992). Adequacy of income and the foster care placement decision: Using an odds ratio approach to examine client variables. *Social Work Research and Abstracts, 28,* 29–36.

Maluccio, A. N., & Fein, E. (1985). Growing up in foster care. *Children and Youth Services Review, 7,* 123–134.

McMurtry, S. L., & Lie, G. (1992). Differential exit rates of minority children in foster care. *Social Work Research and Abstracts, 28,* 42–48.

McRoy, R. G. (2005). Overrepresentation of children and youth of color in foster care. In G. P. Mallon & P. M. Hess (Eds.), *Child welfare for the twenty-first century: A handbook of practices, policies, and programs* (pp. 623–634). New York: Columbia University Press.

Nixon, R. (2005). Promoting youth development and independent living services for youth in foster care. In G. P. Mallon & P. M. Hess (Eds.), *Child welfare for the twenty-first century: A handbook of practices, policies, and programs* (pp. 573–582). New York: Columbia University Press.

Pecora, P. J., Whittaker, J. K., Maluccio, A. N., & Barth, R. P. (with Plotnick, R. D.). (2000). *The child welfare challenge: Policy, practice, and research* (2nd ed.). New York: Aldine de Gruyter.

Schene, P. A. (1998). Past, present, and future roles of child protective services. *The Future of Children, 8,* 23–38.

Sheran, M., & Swann, C. A. (2007). The take-up of cash assistance among private kinship care families. *Children and Youth Services Review, 29,* 973–987.

Swann, C. A., & Sheran Sylvester, M. (2006). The foster care crisis: What caused caseloads to grow? *Demography, 43,* 309–335.

U.S. Department of Health and Human Services, Children's Bureau. (2007). *Trends in Foster Care and Adoption—FY2000–FY2005.* Retrieved June 2, 2008, from http://www.acf.hhs.gov/programs/cb/stats_research/afcars/trends.htm

U.S. Department of Health and Human Services, Children's Bureau. (2006). *The AFCARS report: Preliminary FY 2005 estimates as of September 2006* (Report No. 13). Retrieved June 5, 2008, from http://www.acf.hhs.gov/programs

U.S. Department of Health and Human Services, Substance Abuse and Mental Health Services Administration. (1999, April). *Blending perspectives and building common ground: A report to Congress on substance abuse and child protection.* Washington, DC: U.S. Government Printing Office.

U.S. General Accounting Office. (1997). *Parental substance abuse: Implications for children, the child welfare system, and foster care outcomes* (GAO/T-HEHS-98-40). Washington, DC: U.S. Government Printing Office.

U.S. House of Representatives. (2004). *Background material and data on the programs within the jurisdiction of the House Committee on Ways and Means* (*The green book*). Washington, DC: U.S. Government Printing Office. Retrieved May 26, 2008, from http://www.gpoaccess.gov/wmprints

Webster, D., Barth, R. P., & Needell, B. (2000). Placement stability for children in out-of-home care: A longitudinal analysis. *Child Welfare, 79,* 614–632.

Zuravin, S. J., & DePanfilis, D. (1997). Factors affecting foster care placement of children receiving child protective services. *Social Work Research, 21,* 34–42.

Christopher A. Swann

FREUD, SIGMUND
1856–1939

Sigmund Freud lived for most of his life in Vienna, Austria. At the age of 29, he opened a private medical practice specializing in nervous disorders. Many of his first patients were women suffering from hysteria, a mental disturbance with disabling symptoms such as hallucinations and paralysis. Freud came to believe that these symptoms were the indirect expressions of early actual or imagined experiences of trauma that the patient had banished from consciousness, and he developed therapeutic methods (in which hypnosis was soon replaced by talk therapy) to bring these experiences back into consciousness in ways that alleviated the hysterical symptoms.

Founding a radically new psychology, which he named *psychoanalysis,* Freud began to develop a set of theories to assist his work with patients. As cornerstones to this effort, he formulated two related concepts: repression (a psychic process that works to protect the stability of the person's selfhood by banishing disturbing impulses and experiences from consciousness) and the unconscious (a mental structure that harbors the memories of these repressed experiences in such ways that often, years later, they reemerge in disguised and disabling ways to agitate the conscious self).

Bringing to light his patients' repressed memories led Freud to a series of findings that outraged many of his contemporaries. He discovered that, in many cases, these memories referred back to childhood sexual experiences, often involving seductions by an adult (usually the patient's father). The evolution of Freud's thinking about these findings proved pivotal to his theoretical developments. At first he believed the patients' accounts of sexual seduction to be literally true but then came to the conclusion that, in most cases, they were rather the product of early childhood wishful fantasies. This determination solidified what was to become a fundamental postulate of psychoanalytic psychology: *Psychical reality* (not factual reality) functions as the final arbiter in the shaping of the human personality, not only in childhood but throughout the life course. In the specific cases of his patients' repressed memories, moreover, Freud became convinced that what lie at the heart of their sexual content was a particular type of psychical reality: fantasies involving specific stages of *infantile sexuality*.

Although it was not until the 1920s that Freud put together a complete theory of the stages of psychological development, his writings on infantile sexuality contained most of its foundational elements. In this form, Freud's theory is organized as a sequential narrative of the shifts in the young child's preoccupation with different bodily organs, which serve for a time as the primary source of pleasurable sensations, defined broadly as sexual. In the child's first year (the *oral stage*), desires associated with the mouth become independent of the hunger drive and the focus is on pleasurable sensations. In the second year (the *anal stage*), desires associated with excretion give rise to a form of sexual pleasure connected to aggression. In the third to fifth year (the *phallic stage*), the focus of sexual pleasure shifts to the penis or clitoris, which, stimulated by masturbation and linked to fantasies involving the parent of the opposite sex, leads to a psychic drama that Freud called the Oedipus complex. In the throes of these fantasies, the child (in the case of the young boy) desires to take the place of his father in his relation to his mother, a goal that he is eventually forced to abandon. The child then banishes all desires of a sexual nature to the unconscious, where he or she remains repressed for 7 or 8 years (the *latency stage*). Only with

Sigmund Freud. GALE, CENGAGE LEARNING.

the onset of puberty (the *genital stage*) do these earlier sexual desires reemerge, reorganized in ways that complement the now dominant reproductive genital drive.

In its later elaborated form, Freud's human development theory overlays on this narrative of psychosexual stages an additional theory of psychological growth based on a tripartite division of psyche: id, ego, and superego. At birth, the infant is governed by the id, a bundle of psychic instinctual energies seeking pleasurable discharge. During the child's first year, in conjunction with the oral stage, a second psychic structure, the ego, emerges as an agency representing the psyche's interactions with the outside world and, in particular, as a direct reflection of the child's identification with his or her primary caretaker (usually the mother). Later, usually in the fifth year at the conclusion of the phallic stage, a third psychic structure, the superego, is established. As the source of ideals that inspire action and of self-accusatory judgments that inflict guilt, the superego is modeled on the parents' standards of right and wrong.

Since Freud's death, psychoanalytic practitioners have offered major modifications to his development theory. One focus has aimed at establishing a more complete understanding of the complex processes of psychological growth in the years preceding the phallic stage.

SEE ALSO Volume 1: *Parent-Child Relationships, Childhood and Adolescence;* Volume 3: *Self.*

BIBLIOGRAPHY

Freud, S. (1989). An autobiographical study. In P. Gay (Ed.), *The Freud reader.* New York: Norton. (Original work published 1925)

Freud, S. (1989). *An outline of psychoanalysis.* New York: Norton. (Original work published 1940)

Freud, S. (2000). *Three essays on the theory of sexuality.* New York: Basic Books. (Original work published 1905)

Gay, P. (1988). *Freud: A life for our time.* New York: Norton.

Wollheim, R. (1990). *Sigmund Freud.* Cambridge, U.K.: Cambridge University Press.

George Cavalletto

FRIENDSHIP, CHILDHOOD AND ADOLESCENCE

Friendship can generally be defined as a relationship characterized by mutual positive regard or liking. Other specific qualities such as intimate self-disclosure, loyalty, and social support often feature prominently in childhood and adolescent friendships, but these dynamics are shaped by and thus vary significantly according to such factors as historical era, age, gender, and social status of the individuals involved. To illustrate, girls more than their male counterparts are frequently socialized to value and engage in intimate self-disclosure with friends; thus it is difficult to argue that sharing intimacies is itself a defining feature of friendship. The amount of time young people spend together, the stability of these relationships, and what they do when they are together are also shaped by these broader social forces. But while the contours of friendship vary significantly across individuals and social groups, there is widespread scholarly agreement that friendships play a uniquely important role in the lives of children and adolescents.

THE IMPORTANCE OF FRIENDSHIP DURING CHILDHOOD AND ADOLESCENCE

Parents are key influences throughout the life course, as they appear early on the scene, structure multiple aspects of children's and adolescents' lives, and communicate intensively and extensively with them as they develop. Nevertheless, friendship experiences also are critically important because they influence the child's evolving sense of identity, provide many opportunities for relationship skill-building (that is, practice in how to relate to others), and shape the child's cultural understandings (the sense of what is right or wrong, cool, or to be avoided at any cost). Various aspects of friendship have been shown to influence important child and adolescent outcomes ranging from emotional well-being to teen pregnancy and involvement in violent behavior. But while scholarly interest often has focused on these consequential outcomes, it is important to consider the degree to which playing with and having fun with friends is often at the heart of the childhood and adolescent experience. Accordingly, forging a friendship can be considered an important developmental accomplishment in its own right, as well as an influence on behavior, well-being, and later relationship experiences.

HISTORICAL TRENDS

The advent of the Industrial Revolution is associated with a decline in the need for child labor, and the rise in compulsory education. In turn, the extensive periods of time children began to spend in public schools greatly increased exposure to similar-aged peers. Related to these developments, it could no longer be assumed that children would take up the occupations held by their parents, as had been commonplace in earlier eras. Children were increasingly required to forge their own identities, began to spend many hours outside the watchful eyes of their parents, and not surprisingly developed a heightened sense of the importance of relationships beyond the family context. More recent developments such as women's increased labor force participation have resulted in even younger children's exposure to same-age peers in child-care and nursery school settings.

Although these larger trends have been associated with a basic shift in the overall importance of friendships during childhood and adolescence, cohort changes likely have also affected the character of the friendships that are formed. For example, throughout history traditional gender norms and parental control of dating and courtship fostered much segregation along gender lines. More contemporary patterns, however, include a greater likelihood of experiencing cross-sex friendships, and many instances in which romantic relationships have evolved from what began as friendship relations. Technological changes, including the use of the Internet, cell phones, and text messaging, have also changed the landscape of friendship, both in terms of how friends communicate and in the possibilities for forging connections outside the child's immediate social environment. Finally, one consequence of recent shifts to a later average age at marriage is that friendships often continue as important sources of social support, reference, and socializing as adolescents navigate the transition to adulthood.

THEORETICAL PERSPECTIVES ON CHILD/ADOLESCENT FRIENDSHIPS

Developmental psychologists and sociologists alike have studied friendships during childhood and adolescence, but they have frequently emphasized different dynamics. Psychologists have often viewed friendship from the perspective of attachment theory, which focuses on the importance of early bonding with parents, especially the mother, as a pivotal phase of human development. According to this perspective, strong infant attachment results in feelings of comfort and security that are associated with greater willingness to explore and success within later relationships, including friendships. Researchers have thus studied the association between early family experiences and children's success in relationships with same-age peers, and these studies generally document carryover effects (that is, individuals characterized by secure early attachment are more likely to be well liked by peers and exhibit more skill in forging and maintaining friendships).

This perspective demonstrates how one phase of the life course may deeply influence the next one; theorists such as Sullivan (1953), however, developed the notion that while this is an important process, each phase of life also brings with it new opportunities and challenges. Accordingly, Sullivan argued that childhood friendships (what he called "chum" relationships) serve distinct functions for development, and often serve as an important corrective to what has occurred within the family unit.

Youniss and Smollar (1985) developed this perspective in more detail, focusing on the distinctive nature of children's and adolescents' relationships with friends as a contrast with their relationships with parents. These researchers argued that while the parent–child relationship is fundamentally hierarchical (parents have much more life experience to draw upon and more power in the relationship), friendships are generally much more egalitarian. In addition, unlike family bonds, friendships involve the element of choice. These basic features deeply influence the character of the two types of relations. The authors' contrasting depictions of parent–child and friendship relations are useful as they (a) suggest some limits of the idea of carryover effects from the world of family, (b) highlight specific ways in which friendships contribute uniquely to the child's development, and (c) provide a detailed portrait of how children actually experience these early relationships. For example, because the relationship is voluntary, children must work to maintain the friendship, must learn to take the other's perspective, and may be indulged less than is the case within the family. Because the relationship is forged between equals, children and adolescents often "cooperatively co-construct" plans of action, rather than receiving directives, as is more often the case within the context of parent–child interactions. Finally, because parents typically have such high levels of interest and investment in the child's future, communications often have a judgmental quality that contrasts with the more accepting stance of one's same-age peers. These factors influence other dynamics that are often found within friendships, such as feelings of intimacy and identification, and willingness to disclose one's fears and concerns.

Sociologists have also studied friendships extensively, with much of this research focused on the adolescent period. Although some research has examined basic dynamics within friendship, more attention has been given to problematic outcomes such as delinquency that may be subject to peer influences. Particularly in early investigations, excessive reliance on peers was viewed as problematic for youth, as these liaisons were seen as replacing the influence of parents. Although research did document a modest relationship between time spent with peers (particularly unstructured time) and problem outcomes such as delinquency, many studies also established that parents continue to be influential in the lives of children and adolescents. Contemporary theorizing has moved away from the tendency to position these two social relationships in fundamental opposition to one another, and also increasingly highlights that friends can be a positive or a negative influence on the developing child.

The interpretive perspective on childhood and adolescence is another influential theoretical perspective that presents an integrated perspective on parent and peer influences. This line of theorizing, elaborated by Corsaro (2003) in connection with research on preschool and grade school children, and Eder (1995) in investigations of middle school youths, focuses on the ways in which interaction and communication with one's same-age peers creates the immediate cultural world that children inhabit. Through talk and play with friends and other age-mates, children develop unique understandings about what will be esteemed, valued, or subject to ridicule, and also learn about social rules and obligations. As young people interact with one another, they necessarily draw from the larger culture (e.g., parent's attempts to socialize them in a particular direction) but inevitably do so in a creative, selective manner. As a result, children's worlds reflect but never duplicate that of their parents and other adults. Research in this tradition is important because it focuses attention not only on the general importance of interacting and communicating with friends but also on the content of this communication.

SOCIODEMOGRAPHIC CHARACTERISTICS OF CHILDREN AS INFLUENCES ON FRIENDSHIP PATTERNS

Developmental psychologists in particular have frequently explored the influence of age on friendship patterns, but there is also increased scholarly interest in the ways in which the child's other social statuses and locations, such as gender, social class, and race/ethnicity, influence the nature of friendship processes.

It is often stressed that the adolescent period is the phase of life when peers assume increased importance, but researchers have found that youngsters as early as preschool age show much interest in forging connections with their peers. Corsaro (2003), in observations of preschool-age children, frequently heard comments such as "We're friends, right?" a question that hints at both the importance and not-to-be-taken-for-granted nature of this type of relationship. Certainly researchers have documented that friendships become more complex and intimate as children mature. Early relationships often have a strong activity focus, whereas communication and intimacy loom larger in many adolescent relationships.

Nevertheless, research also shows that even children's early forays into the world of friendship are not only about the game that is being played but also about affection, perceived obligations, and a concern with maintaining these important connections. For example, Corsaro noted that while preschool children may sometimes refuse to include another child in their ongoing activities, what seems to be a selfish act often stems from their intense focus on play routines that involve sharing and reciprocity. Research on the adolescent period is

much more voluminous, however, and clearly shows that friendships become an especially important part of life during this phase of development. Opportunities (greater freedom of movement) and challenges (concerns about appearance, the opposite sex, being popular) associated with adolescence make close friends particularly valued as a safe haven or what Call and Mortimer (2001) describe as "an arena of comfort" during this time.

A significant body of research also has documented that gender influences the character of friendships as well as what is communicated within friendship circles. Even in childhood, young girls more often than boys engage in dyadic (or one-on-one) play, and talk and sharing of secrets is considered more central to developing relationships. Researchers have contrasted this with boys' stronger activity orientation, often focusing on games such as baseball or basketball that require a somewhat larger number of participants. Research has shown that during adolescence boys are less likely to engage in intimate self-disclosure with friends, gendered processes that potentially influence dynamics of later relationships (e.g., marriage relationships). Ethnographic studies by Eder (1995), Fine (1987), and Adler and Adler (1998) add to this portrait, emphasizing that boys' communications typically reward toughness and punish displays of vulnerability or weakness. Youniss and Smollar (1985), however, also highlighted that the lower scores of boys on intimate self-disclosure to friends was primarily attributable to the lack of intimacy reported by about 30% of their sample of adolescent boys. This finding suggests the importance of examining variations within samples of boys and girls, rather than focusing exclusively on aggregate gender differences.

Fewer studies have examined the influence of such factors as socioeconomic status and race/ethnicity on children's and adolescents' friendships, but research suggests more attention to these factors is warranted. Two alternative hypotheses about the likely influence of social class position have developed in the scholarly literature. Eckert's 1989 ethnography of a U.S. high school, for example, focused on ways in which socioeconomic factors play into adolescent social hierarchies. She observed that middle-class "jocks" typically have higher status than youths with lower status backgrounds, who are more likely to be considered "burnouts." Eckert hypothesized that friendship relations within these larger status groups differed as well. She suggested that middle-class youths were relatively more instrumental and willing to shift friendships as they became involved in different extra-curricular activities or moved on to college.

Focusing on burnouts, who often lacked involvement in structured activities and were less likely to succeed along traditional lines, Eckert observed that their

Play Time. *Through talk and play with friends, children develop unique understandings about what will be esteemed, valued, or subject to ridicule.* **AP IMAGES.**

friendships were longer lasting and assumed greater importance. A contrasting view is that experiences associated with higher socioeconomic status may provide resources and stability that allow young people to develop and maintain more intimate ties with friends. For example, lower-class youths are more likely to make frequent residential and school moves, and thus may have difficulty sustaining particular friendships. Living in unsafe neighborhoods has also been associated with a general wariness and lack of trust that, while an effective survival strategy, may also inhibit the development of highly intimate relationships.

Much of the research on race/ethnicity effects has focused on the degree to which children and adolescents form interracial ties. Research has demonstrated strong preferences for same-race and same-ethnic friendships, with recent studies documenting that Hispanic and Asian youth are more likely than others to exhibit heterogeneity in friendship choices. Researchers have shown that the racial composition of schools is a significant influence on friendship experiences. Moody (2001), for example, found the highest levels of friendship segregation in

moderately heterogeneous schools, but that very heterogeneous schools were associated with more integration of friendship ties. In addition, in those schools in which extracurricular activities were integrated, friendship segregation was also less pronounced.

Because this research shows that a majority of friendships are intraracial (or within one's own racial group), however, more research is needed on the everyday friendship experiences of youths who vary in their race and ethnic characteristics. As in the case of social class effects, contradictory hypotheses have been suggested, and the research evidence that bears on these ideas is less than definitive. Early on some researchers focusing on African-American youth developed a compensation hypothesis, suggesting that family structure differences and barriers to traditional success may foster an excessive dependence on peers. Some quantitative research, however, documents that African-American youth compared to White youth maintain a stronger family orientation, report spending less time with friends, and perceive less peer pressure. Some effects of race may ultimately be traced to socioeconomic status differences, but other differences in the character of friendship ties may stem from family or neighborhood effects and other cultural preferences. More research is also needed on friendship patterns of Hispanic and Asian-American youth, where forging links to these broader patterns is similarly important to pursue.

FRIENDSHIP EFFECTS ON DEVELOPMENTAL OUTCOMES

Researchers have shown that simply having friends is associated positively with children's emotional health and self-esteem, and that more socially competent youth also do better in these respects. Sociologists in particular, however, have focused much attention on the specific characteristics of friends as influences on a range of attitudes and behaviors. One of the challenges in this line of research is that while studies have repeatedly demonstrated that children and adolescents share similarities along a number of dimensions, what is less certain is whether this reflects an instance of "birds of a feather" flocking together or an actual influence process. Studies that follow young people over a period of time are best positioned to answer these questions, as they can distinguish between the initial similarity of friends and whether they tend to become more similar over time (findings that support the idea of mutual influence).

Research has shown that even when relying on longitudinal or multiwave studies, the delinquency of friends emerges as a significant predictor of a child's own delinquency and academic attainment is influenced by friends' orientations toward school and school performance. And while heterosexual relationships involve unique dynamics that transcend the peer context, research shows that affiliation with sexually permissive peers is a strong correlate of an adolescent's own sexual behavior choices. Even what appear to be individualistic behaviors such as suicide may be influenced by peer factors. For example, Bearman and Moody (2004) showed that having a friend who committed suicide was associated with a youth's own likelihood of having suicidal thoughts.

Although the research on peer effects generally documents effects of friends' attitudes and behaviors on a range of outcomes for both male and female adolescents, some research has noted gender differences. In a 2006 article, for example, Riegle-Crumb, Farkas, and Muller demonstrated that the characteristics of high school girls' friends influenced their math and science attainment, but this friendship effect was not found for boys. In addition, some research has found that involvement in mixed-gender groups decreases boys' involvement in delinquency, but tends to amplify the risk for girls.

FUTURE DIRECTIONS

The above review of current knowledge about child and adolescent friendships suggests some areas that warrant additional investigation. For example, it will be useful to explore changes in the way young people "do" friendship, from the increased reliance on e-mail and text messaging, to the less rigid lines between friendship and dating relationships. Because many of the foundational studies in this field have relied heavily on convenience samples of White youths, increased attention to the friendship experiences of youths varying in race/ethnicity and social class are especially needed. More research on friendships of sexual minority youth is also warranted, as the focus on sex, identity, and problem behaviors such as drug use has not provided a comprehensive understanding of the lives of gay, lesbian, and bisexual adolescents.

Another critical area to pursue involves more explicit study of linkages between the child's friendships and other social groups and organizations. For example, while the study of popularity and cliques has developed as a separate area of inquiry, researchers have noted that the child's social niche (i.e., whether popular, "alternative," or "druggie") channels friendship opportunities and potentially also influences the dynamics within these relationships. And while adolescence is characterized by freedom of movement relative to the greater supervision typical of childhood, additional research is needed on ways in which parents continue to influence adolescents' friendship choices. To illustrate, in a 2006 article Knoester, Haynie, and Stephens reported that the quality of the parent–child relationship, whether the parent chose a neighborhood because of its schools, and parental supervision all influenced whether the child was involved in a delinquent friendship network.

Because prior research has amply demonstrated that friendship experiences influence adolescent involvement in various prosocial and problem behaviors, a logical next step is to explore in more detail the interpersonal dynamics that foster these observed similarities between young people and their friends. Are there developmental shifts in the mechanisms underlying peer influence? How do male and female friendship groups differ in the ways in which friends influence one another? Warr (2002), focusing specifically on delinquency, outlined some potentially important mechanisms, such as the fear of ridicule and expectations of loyalty, but these need to be more systematically investigated. In addition, the dynamics involved in encouraging academic attainment or other outcomes might be distinctive in a number of important respects.

Finally, research on friendships, including studies of these influence processes, could benefit from theory and research that takes into account the element of human agency and capacities for change that are widely recognized but more difficult to study empirically. Many treatments of peer pressure depict the young person as essentially a passive recipient of these influence attempts.

A more comprehensive assessment would incorporate a view of adolescents as actively involved in their own attempts to influence, reacting selectively and creatively to communications from their friends, and sometimes changing networks to line up with new perspectives or life course changes.

SEE ALSO Volume 1: *Bowlby, John; Dating and Romantic Relationships, Childhood and Adolescence; Interpretive Theory; Peer Groups and Crowds; School Culture; Social Capital; Social Development; Youth Culture;* Volume 2: *Friendship, Adulthood; Social Support, Adulthood;* Volume 3: *Friendship, Later Life.*

BIBLIOGRAPHY

Adler, P. A., & Adler, P. (1998). *Peer power: Preadolescent culture and identity.* New Brunswick, NJ: Rutgers University Press.

Bearman, P. S., & Moody, J. (2004). Suicide and friendships among American adolescents. *American Journal of Public Health, 94,* 89–95.

Bukowski, W. M., & Sippola, L. K. (2005). Friendship and development: Putting the most human relationship in its place. *New Directions for Child and Adolescent Development, 2005*(109), 91–98.

Call, K. T., & Mortimer, J. T. (2001). *Arenas of comfort in adolescence: A study of adjustment in context.* Mahwah, NJ: Erlbaum.

Corsaro, W. A. (2003). *"We're friends, right?": Inside kids' culture.* Washington, DC: Joseph Henry Press.

Crosnoe, R. (2000). Friendships in childhood and adolescence: The life course and new directions. *Social Psychology Quarterly, 63,* 377–391.

Dunn, J. (2004). *Children's friendships: The beginnings of intimacy.* Malden, MA: Blackwell.

Eckert, P. (1989). *Jocks and burnouts: Social categories and identity in the high school.* New York: Teachers College Press.

Eder, D. (with Evans, C. C., & Parker, S.). (1995). *School talk: Gender and adolescent culture.* New Brunswick, NJ: Rutgers University Press.

Erwin, P. (1998). *Friendship in childhood and adolescence.* London: Routledge.

Fine, G. A. (1987). *With the boys: Little League baseball and preadolescent culture.* Chicago: University of Chicago Press.

Giordano, P. C. (2003). Relationships in adolescence. *Annual Review of Sociology, 29,* 257–281.

Knoester, C., Haynie, D. L., & Stephens, C. M. (2006). Parenting practices and adolescents' friendship networks. *Journal of Marriage and Family, 68,* 1247–1260.

Moody, J. (2001). Race, school integration, and friendship segregation in America. *American Journal of Sociology, 107,* 679–716.

Riegle-Crumb, C., Farkas, G., & Muller, C. (2006). The role of gender and friendship in advanced course taking. *Sociology of Education, 79,* 206–228.

Sullivan, H. S. (1953). *The interpersonal theory of psychiatry* (H. S. Perry & M. L. Gawel, Eds.). New York: Norton.

Warr, M. (2002). *Companions in crime: The social aspects of criminal conduct.* Cambridge, U.K.: Cambridge University Press.

Youniss, J., & Smollar, J. (1985). *Adolescent relations with mothers, fathers, and friends.* Chicago: University of Chicago Press.

Peggy C. Giordano

G

GAMBLING, CHILDHOOD AND ADOLESCENCE

SEE Volume 2: *Gambling.*

GANGS

The study of gangs as unsupervised peer groups has long centered on problems of adolescence. As most youth in gangs age, research has found, they "mature out" and move on to a conventional life, putting violent and criminal behavior behind. Some recent research, however, is reexamining this bedrock idea. The global era has brought a new urgency to understanding the life course of gang members.

A BRIEF HISTORY OF GANG RESEARCH AND THEORY

The study of gangs formally began in Chicago in the 1920s with the classic studies of Frederic Thrasher and the Chicago School of Sociology. For Thrasher, the gang was an adolescent experience, *interstitial* in both a spatial and temporal sense. Gangs formed in immigrant slums from second-generation youth, the children of immigrants, who were "in between" the controls of the old world and the freedoms of the new. For Thrasher and most subsequent gang researchers, the study of gangs has been mainly concerned with unsupervised, adolescent, male peer groups: how they form, behave, and eventually end.

The social scientists of the Chicago School discovered in the ecology of the industrial city patterns of behavior by youth in "disorganized," immigrant communities. Gangs formed from second-generation boys who rebelled against old world cultural traditions and controls, had loose ties to school and other conventional institutions, and were seduced by American culture. Early research in Los Angeles followed this model in studying Mexican immigrants None of the early studies had much to say about girls, because the gang experience was defined fundamentally as male adolescent alienation.

In the 1950s gangs started to draw more public attention. To counter periodic moral panics, the criminologist Walter Miller argued that whereas the numbers of gangs were relatively constant over time, it was *media attention* that varied. But he also sounded the alarm in several law enforcement surveys, indicating that the problem of gangs and law-violating groups had become national in scope, at least in large cities. The turbulent 1960s saw an ongoing theoretical dispute over the relative role of "lower-class culture" versus variations in neighborhood "opportunity structures" that mirrored liberal/conservative policy debates. An empirical test of theories by James F. Short and Fred L. Strodtbeck (1965) reaffirmed the salience of Thrasher's prior *group-process* perspective of adolescent behavior. Unresolved, however, both theoretically and in studies based on direct observation, was whether this process in a gang necessarily included criminality.

Although Albert Cohen's *reaction formation* thesis; i.e. his notion of *corner boy* behavior as a repudiation of *college-boy* norms, was not universally accepted, his focus on the adolescent, lower-class male epitomized the world

Gangs. *Three members of the Crips brandishing rifles.* GALE, CENGAGE LEARNING.

of gang research at the time. Minimized in nearly all this research was gender and ethnicity. Most studies concluded that gangs, as a male adolescent problem, would ultimately fade away once youth transitioned to adulthood in an improving economy. The optimism of modernity informed this era of gang research.

However, gang violence sharply increased in the 1960s and 1970s, raising questions about this confidence. Miller noted in 1975 that most gangs in the United States were White, but by the 1980s and 1990s research focused on the growing number of Black and Latino gangs. White gangs dwindled as their ethnic groups assimilated and moved to the suburbs away from their old, urban neighborhoods. The gang problem became intertwined with issues of race, crime, social protest, and the city.

Joan W. Moore (1978), almost singlehandedly, paid special attention to females, and her work framed gangs within the history of East Los Angeles barrios and the persisting conditions of poverty and inequality. She examined an issue of great importance for later research: Several generations of gangs in East LA, dating back to the 1920s, had not disappeared by the 1960s, but had become, in her words, "quasi-institutionalized." Jacobs'

study of Stateville, a prison outside Chicago, also noted adult influence in gangs and a protracted, close relationship between the prisons and the streets. A few years later Campbell did a seminal study of female gangs in New York, though the problems of girls and gangs remained marginal to criminology.

William Julius Wilson's (1987) analysis of deindustrialization and changes in black ghetto behavior formed a new context for the study of gangs. John Hagedorn's 1988 study of Milwaukee gang formation, along with Carl Taylor's study of male and female gangs in Detroit, Mercer Sullivan's comparative neighborhood study in New York, and Diego Vigil's study of Los Angeles "Barrio Gangs" looked at the impact of economic restructuring on gangs, but also at the history of local African American and Latino communities. These studies found that the gang experience, particularly for males, was no longer limited to adolescence. The lack of good jobs had been met by a parallel growth in the underground economy and the subsequent organization of drug sales by urban minority gangs. Drug selling, to cite Sullivan's book title, was just another way of "Gettin' Paid." Padilla summed up this approach with his book, "The Gang as an American Enterprise."

The crack wars of the late 1980s and early 1990s brought another surge of interest in studying gangs and attention to gang prevention and control programs. The National Institute of Justice (NIJ) funded a wave of gang research, much of it in St. Louis, and research began to look at gangs in many cities and of many different ethnic groups. The National Youth Gang Center (NYGC) followed up on Miller's earlier surveys by estimating the extent of the gang problem in the United States. Their surveys, mainly of law enforcement personnel, found nearly a million gang members in the mid-1990s, followed by gradual declines. In 2007, the NYGC estimated that there were about 25,000 gangs with a total of three-quarters of a million gang members, with more Latino gangs and gang members than other ethnicities. Their estimates, it is widely recognized, are unverifiable. What appears to be certain is that U.S. cities, large and small, since the late 1980s have had an identifiable gang problem that varies in scope and intensity but shows few signs of going away.

The rising rates of violence and increasing numbers of gangs in the 1990s led to a more intense focus on gang intervention and suppression programs. Malcolm Klein (1995) argued against a narrow law enforcement response, finding that repression strengthened, not weakened, gang cohesion. Similarly Irving Spergel argued for comprehensive models of gang intervention that slanted away from a one-sided reliance on law enforcement. Both Spergel and Klein, leading scholars on the topic, defined the gang traditionally as a basically male adolescent phenomenon and sharply distinguished it from "drug posses," prison gangs, and other, more adult-involved, street groups.

COMPARATIVE STUDIES OF GANGS

Girls' gang experiences continued to be neglected, though some efforts to raise the profile of female gangs were made (e.g. Chesney-Lind & Hagedorn, 1999). While law enforcement estimates of female gang participation are typically low, careful studies of neighborhoods in Los Angeles and Milwaukee found more than a third of all adolescent gang members were female. Significantly, however, one of the major differences between female and male gang involvement was the growing prevalence of adult roles in gangs for males. As girls aged, their role of mother pushed the role of gang member to the side in ways that for males, the role of father did not.

Most gang research continued to be based on the theory of the lack of controls on boys in poor, "disorganized" communities. For example, a typical definition drawn from a National Youth Gang Center online description of the parameters of the gang problem states:

"The terms 'youth gang' and 'street gang' are commonly used interchangeably and refer to neighborhood or street-based youth groups that are substantially made up of individuals under the age of 24" (Institute for Intergovernmental Research, 2008).

While the vast majority of gang members have always been juveniles, researchers in several cities have been exploring a more institutionalized gang that persists for decades and has considerable adult membership. Gangs in Chicago, as Conquergood, Venkatesh, Hagedorn (2008), and others described, were long standing presences in nearly every African-American, Mexican, and Puerto-Rican neighborhood in the city. Similarly, African-American and Mexican gangs in Los Angeles have persisted since the 1960s. Multi-generational gangs have become part of the landscape in many U.S. cities and present new challenges for neighborhood youth.

Such institutionalized gangs adapt to changing conditions and offer more to youth than just "gettin' paid." Gang rituals and symbols provide meaning, and members spin folklore or "rationalized myths" about gang history. Such gangs give solidarity and security for youth and offer a pathway to a future that includes continuity within their neighborhood spaces. Even prison has become an extension of the neighborhood, with homeboys or allied gang members in nearly every prison expected to help out any new inmate from the "hood." Spaces for some gangs, such as MS-13 or the Latin Kings, are transnational; their gangs now exist simultaneously in multiple cities, from Los Angeles or Chicago to San Salvador or Madrid. Gangs in both Chicago and Los Angeles have also occasionally dipped into politics and demonstrate what urban scholar Saskia Sassen calls a "presence," or a potential for social or political action.

This extension of adolescent gang life into adulthood is a global trend within urban spaces of *social exclusion*, a *Fourth World* to use Castells' (2000) term. The majority of the world is now urban, with nearly half of its population under the age of 25. More than 500 million youth live on less than $2 per day, the standard for poverty. More than 8 in 10 of the world's youth live in the developing world, and Africa alone may have 300 million slum dwellers. Mike Davis' (2006) "planet of slums" implicitly means a world of gangs, as groups of armed young men occupy ghettoes, favelas, and townships across the globe and represent a very real, alternative future for many youth.

Literature on "children in organized armed violence" or *coav* has studied comparatively the problems of youth from gang members to child soldiers growing up in cities across the world. Luke Dowdney's (2003, 2005) studies of Rio de Janiero drug factions highlight the problems children face growing up in a world of gangs. Diverse studies have found that many poor youth are turning to a variety of oppositional groups, including gangs, ethnic militias, drug cartels, and religious police. The lines between these groups have become harder to draw. Comparative studies of *coav* from Chicago to Cape Town to Mindanao to Rio de Janiero have established the global dimensions of this issue.

U.S. criminologists' approach to gangs has also taken an international comparative turn, but only as far as Europe. The Euro-gang research agenda of Mac Klein et al. has transplanted American concepts and concerns about adolescent gangs across the Atlantic to include Europe's "troublesome youth groups." Whereas European cities currently have few institutionalized gangs, conflict in the banlieues of Paris, Turkish gangs in Berlin, Afro-Caribbean gangs in London, and the emergence of Russian and Eastern European crime groups portend what may be a different, bleaker future. For example, Gloria La Cava and Rafaella Y. Nanetti report that in Albania the criminal economy includes up to 25% of the young men aged 18–25 and has become a "structural feature in Albanian life" (La Cava & Nanetti, 2000, p. 39). This startling finding exemplifies the new problems for the life course of youth gang members in many areas of the world.

FUTURE DIRECTIONS

Future directions for research include a more complete understanding of how institutionalized gangs and other armed groups have an impact on the life chances of children and youth. While the conventional problems of alienation, rebellion, and gender roles are present for adolescents everywhere, globalization has left youth in some cities with even more pressing problems. Although not all cities have institutionalized gangs, many others do, and the indication is that such groups are likely to increase. Young women may play a more involved role in gangs, though there is still very little research on girls' and women's gang lives. The political involvement of the Latin King and Queen Nation in New York City in the 1990s and their integration of women into their leadership has received little attention outside the important study by David Brotherton and Luis Barrios (2003). In some Colombian militias, women play a significant, and armed, role, although this does not appear to be a general pattern.

This focus on institutionalized gangs also highlights race and ethnicity as well as the importance of youth culture. While traditional gang research has deemphasized the salience of race, the world has been wracked by ethnic conflicts that include gangs and other groups of armed young men. Gang youth make meaning from

their ethnicity and religion, and their music, hip hop, is one of the strongest cultural forces in the world in the early 21st century. The nihilistic lure of gangsta rapper Snoop Dogg tells youth to "keep your mind on your money and your money on your mind." But youth, and their gangs, have more on their mind than money. One focus of research in the coming years will be to better understand the cultural meaning of various "resistance identities" of youth and their gangs as they attempt to make meaning out of a dangerous and forbidding world. As youth proceed along the life course in a "planet of slums," research needs to focus on the gang and similar armed groups as more than a transitional form.

SEE ALSO Volume 1: *Aggression, Childhood and Adolescence; Crime, Criminal Activity in Childhood and Adolescence; Theories of Deviance.*

BIBLIOGRAPHY

Brotherton, D., & Barrios, L. (2003). *Between black and gold: The street politics of the almighty Latin King and Queen Nation.* New York: Columbia University Press.

Castells, M. (2000). *The information age: Economy, society, and culture: End of millennium.* (Vol. III). Malden, MA: Blackwell.

Chesney-Lind, M., & Hagedorn, J. M. (Eds.). (1999). *Female gangs in America: Essays on girls, gangs, and gender.* Chicago: Lakeview Press.

Davis, M. (2006). *Planet of slums.* New York: Verso.

Dowdney, L. (2003). *Children of the drug trade: A case study of children in organised armed violence in Rio de Janiero.* Rio de Janiero: 7Letras.

Dowdney, L. (Ed.). (2005). *Neither war nor peace: International comparisons of children and youth in organised armed violence.* Rio de Janiero: 7Letras.

Hagedorn, J. M. (1998). *People and folks: Gangs, crime, and the underclass in a rustbelt city.* (2nd ed.). Chicago: Lakeview Press. (Original work published in 1988)

Hagedorn, J. M. (2008). *A world of gangs: Armed young men and gangsta culture.* Minneapolis: University of Minnesota Press.

Institute for Intergovernmental Research. (2008). *National Youth Gang Center (NYGC).* Retrieved May 22, 2008, from http://www.iir.com/nygc

Klein, M. W. (1995). *The American street gang: Its nature, prevalence, and control.* New York: Oxford University Press.

Klein, M., Kerner, H.-J., Maxsen, C. L., & Weitekamp, E. G. M. (Eds.). (2001). *The Eurogang paradox: Street gangs and youth groups in the U. S. and Europe.* Dordrecht, The Netherlands: Kluwer.

La Cava, G., and Rafaella Y. Nanetti. (2000). *Albania: Fight the vulnerability gap.* Washington, DC: The World Bank.

Miller, W. (1958). Lower class culture as a generating milieu of gang delinquency. *Journal of Social Issues, 14,* 5–19.

Moore, J. W. (1978). *Homeboys: Gangs, drugs, and prison in the barrios of Los Angeles.* Philadelphia: Temple University Press.

Moore, J. W. (1991). *Going down to the barrio: Homeboys and homegirls in change.* Philadelphia: Temple University Press.

Short, J. F., & F. L. Strodtbeck. (1965). *Group process and gang delinquency*. Chicago: University of Chicago.

Wilson, William Julius. (1987). *The truly disadvantaged*. Chicago: University of Chicago.

John M. Hagedorn

GAYS AND LESBIANS, YOUTH AND ADOLESCENCE

Gay and lesbian youth of the early 21st century are able to "come out" (i.e., to disclose their sexual identity to others) earlier than previous generations of sexual minority youth, find more supportive environments at schools through clubs (gay-straight alliances) and inclusive school policies, and have experienced a societal shift toward greater acceptance of sexual diversity. This picture is optimistic for future generations of gay and lesbian youth; at the same time, risks unique to sexual minorities are a part of daily life for many gay and lesbian youth. Same-sex attracted youth are more likely to attempt suicide (Russell, 2003) and experience increased rates of victimization at school due to sexual orientation (Morrow, 2004). These risks are beginning to be addressed through various media outlets (e.g., Kate Bornstein's book, *Hello Cruel World: 101 Alternatives to Suicide for Teens, Freaks, and Other Outlaws* (2006), or informative, supportive Internet resources for gay and lesbian youth, such as *The Safe Schools Coalition* Web site.

The research on gay and lesbian youth has mirrored society: Gays and lesbians have been a stigmatized population among youth, and until recently have been invisible in the daily lives of their families, schools, and communities, and also in the scientific literature. The first scientific attention to gay and lesbian youth came in the early 1970s; it took another 15 years for there to be more than a handful of published medical studies, each of which documented significant health risk among gay and lesbian adolescents. Then, in the mid-1980s, coinciding with the beginning of the HIV pandemic, attention to gay and lesbian youth began to grow.

Early studies focused almost exclusively on medical health risks and were based on what are termed "convenience" samples (even though such studies were hardly "convenient" for those pathbreaking researchers): Youth were recruited to participate in studies through programs that served the needs of gays and lesbians. The result was a decade or more of studies based on the experiences of youth that self-identified as gay and lesbian, many of whom were seeking services for the emotional and behav-

ioral health challenges that the researchers were attempting to understand. Thus, although these early studies almost certainly overestimated the magnitude of problems in the "general" gay and lesbian youth population (if indeed there exists a "general" population of gay and lesbian youth), they were important for establishing the existence of health disparities for gay and lesbian youth, thereby focusing attention on this hitherto ignored population. Since the mid-1980s, attention to this population has grown, and has been characterized by several themes: the definition and study of the population, the development of sexual orientation and identity, and the tension between risk and resilience in studies of gay and lesbian youth.

DEFINING AND STUDYING THE GAY AND LESBIAN YOUTH POPULATION

The early studies were defined by adolescent *sexual identity*: conception and labeling of oneself as gay, lesbian, bisexual, or heterosexual (Diamond, 2003). Yet most individuals come to understand same-sex sexuality before self-labeling or disclosing sexual identity. *Sexual orientation* is defined as one's sexual or emotional attractions to other persons. These attractions may be to the same sex (homosexual orientation), the other sex (heterosexual orientation), or to both the same and other sex (bisexual orientation). Realization of this self-concept is outwardly expressed as a sexual identity; alternatively, sexual orientation may be privately acknowledged but not publicly expressed, or the individual may be unaware of it consciously.

Sexual behavior refers to actual behavior between people. Sexual behavior may or may not be consistent with a person's sexual orientation or identity. It is important to note that diversity in *gender identity* includes youth who identify as transgender. Transgender youth may identify with a gender that is different from their biological sex; alternatively, they may not identify as either male or female, but rather with a combination of femaleness and maleness (Ryan & Futterman, 1998). Awareness about transgender identity has grown, along with increasing numbers of youth who self-identify as transgender. Since the early 1990s, researchers' attention has often focused on a broadly defined group of youth—sometimes termed "sexual minorities" (Russell, 2005)—including those who are gay, lesbian, bisexual, and transgender (GLBT).

Identifying and studying dimensions of sexuality (orientation, identity, and behavior) is challenging, particularly if the focus is on the adolescent years, during which time these dimensions are in development. The first studies relied on self-reports of sexual identity, and this approach continues to be important. It continues to

Gay–Straight Alliance Members. *The Gay–Straight Alliance, a group duplicated in high schools and colleges around the U.S., provides support and education for gay and straight students.* © **ED QUINN/CORBIS.**

be a challenge, however, to define sexual and gender identity because of historically changing cultural meanings of sexual identity labels: In the early 21st century, same-sex identified youth often use labels such as *queer* or *questioning*, or they refuse to use a label at all (Cohler & Hammack, 2007; Diamond, 2003). Sexual orientation has been studied through questions about sexual attractions, romantic attractions, emotional preferences, and multiple other indicators; there has not been strong measurement consensus for studying sexual orientation, although one review has provided guidance (Saewyc et al., 2004). Finally, some studies have relied on self-reports of same-sex sexual behavior to categorize the study population (Saewyc et al., 2004). Regardless of the measurement approach, several decades of research on sexual minority youth has been accumulated.

Given the challenges associated with identifying the population to be studied, and the historical changes in visibility and acceptability of GLBT people and identities, there is no definitive estimate of the proportion of the general adolescent population that are sexual minorities. Across multiple studies, however, the size of the adolescent sexual minority population has been found to range between 1% and 8% of the general population (Saewyc et al., 2004). At the same time, awareness of same-sex sexuality in adolescence has undoubtedly grown since the mid-1980s, and it is clear that there are now possibilities during adolescence that did not exist for older generations for sexual minority people to come out. Scholars have suggested a trend of earlier ages for coming out among cohorts in the beginning of the 21st century (Cohler & Hammack, 2007; Floyd & Bakeman, 2006; Ryan & Futterman, 1998). One study found that individuals from a cohort

who self-identified as gay or lesbian as adolescents during or after 1988 did so at younger ages compared to an earlier cohort that came out prior to 1988. The more recent cohort reported disclosure at an average age of 18 to a nonparent and 19 to 20 to a parent, whereas for the older cohort the average age of disclosure was 20 to nonparents and approximately 23 to parents (Floyd & Bakeman). The difference in coming-out ages for these cohorts is approximately the same as those reported in earlier work (Ryan & Futterman).

Explanations for these cohort differences rely on historical changes in social attitudes toward gays and lesbians, including the increased visibility of lesbian and gay people and issues in the media, and the public attention to multiple GLBT issues, including "gays in the military," marriage for same-sex couples, and efforts to legislate employment non-discrimination protections. Ryan and Futterman (1998) note that contemporary youth have access to GLBT role models (e.g., popular television programs such as *Will & Grace*, movies such as *Brokeback Mountain*, and sports figures such as Martina Navratilova) along with greater access to information regarding sexual orientation; these factors may allow sexual minority youth to feel more comfortable with their sexual identities, and therefore be more likely to come out than in the past.

SEXUAL ORIENTATION AND IDENTITY DEVELOPMENT

The development of gay and lesbian identity was originally conceptualized in stage models (Beaty, 1999). Several stage models were proposed in the 1970s and 1980s, each of which included the progression from feeling different, to coming out, to accepting, and finally integrating one's identity as gay, lesbian, or bisexual. These models established coming out as the primary task of adolescent identity development (Morrow, 2004). However, stage models do not accurately describe the daily experiences of GLBT youth (Diamond, 2003, 2005). For example, Diamond (2005) identifies important gender differences and acknowledges prior studies that show that the timing and sequencing of these stages may not be consistent for all youth. Thus, research should examine diversity among sexual minority youth in identity development, as well as the role of identity development experiences over the life course. It is generally regarded that a stages approach to the study of sexual identities is not the best conceptual framework. Nevertheless, the elements of these models—feelings of difference and awareness of sexual orientation, disclosure of sexual identities, and self-acceptance—continue to have meaning for understanding the experiences of GLBT adolescents (D'Augelli, 2005).

RISK OR RESILIENCE?

Without question, the emphasis in research has been on health and behavioral risk in the lives of sexual minority adolescents. Specifically, multiple studies using different measures and methods and in multiple countries document compromised health and well-being for adolescents based on same-sex sexual orientation, identity, or behavior, and adolescent transgender identity. In one sense, the focus on risk is consistent with the larger field of adolescent research, which historically has been concerned with identifying and preventing negative outcomes (although attention to positive development has grown since the late 1990s). Further, while the earliest studies focused on mental and physical health challenges of sexual minority youth, this emphasis was only reinforced by the imperatives of the HIV/AIDS crisis of the late 1980s, which prompted much of the research attention that began during that time. The inertia of this field of studies gained momentum and continues to focus on negative health and problem development. Yet, in spite of what many agree is an overfocus on risk, research continues to indicate that sexual minority youth remain a group that is at high risk. Clearly there is great need for studies to understand risk behaviors or statuses among adolescents in order to inform prevention and intervention; at the same time, additional research is needed on protective factors and on the development of positive outcomes (Russell, 2005).

Little noticed is that the majority of sexual minority adolescents grow up to be healthy and contributing members of society, despite societal prejudice and discrimination. The healthy development of the majority of GLBT youth suggests that, like many adolescents, they are resilient (Russell, 2005). What promotes healthy development and resilience? Promising advances have contributed to a shift from an exclusive focus on risk to consider the success and empowerment of GLBT youth (Cohler & Hammack, 2007).

Nevertheless, replacing negative outcomes with positive ones is not the full solution. Additional research is needed on the mechanisms that operate in the lives of sexual minority adolescents, prompting optimal or problematic adjustment. Consistent with the "minority stress" model (Meyer, 2003), a framework of normative versus unique risk and protective factors offers the possibility to clarify research questions and practical goals (Russell, 2003, 2005). Many of the risk factors and outcomes that have been described for sexual minority adolescents are normative—they are risks for all youth, such as compromised family or peer relationships, mental health problems, or substance abuse. New research is needed to identify the dimensions of risk or protective factors that are unique to sexual minority youth. What are the specific attitudes, behaviors, comments, or interactions that make family and peer relationships difficult for sexual minority youth? Are there factors unique to sexual minority adolescents that trigger depression or substance abuse? Alternatively, what specific interactions or behaviors might parents, siblings, and friends engage in that could promote healthy development? What factors unique among sexual minority adolescents might protect them from mental and behavioral health risk?

FUTURE DIRECTIONS

Questions of resilience are central in the newest generation of studies of sexual minority youth, and future studies will continue to identify distinctive experiences that protect or make them vulnerable, and that may be efficacious for intervening to prevent compromised health and to promote positive development. One trend in the research that will aid in accomplishing this goal is a shift from thinking about risk exclusively at the individual level to identifying sources of risk and protective factors in the broader environment in which adolescents grow and develop. Influenced by *ecological systems theories* of human development, studies in the early 21st century of sexual minority adolescents have focused on the important contexts that shape adolescents' lives: their families, schools, faith, and peers. Through studying these developmental settings, researchers are moving beyond documenting risk to identifying and contextualizing the sources of risk. Specifically, GLBT youth are not depressed because they are GLBT, but because they experience societal prejudice and discrimination at home, in school, in their faith communities, and among peers.

Much of the focus remains on risk, but a contextual resilience approach is promising. For example, it is simply assumed that GLBT youth will have compromised relationships with their parents; challenges in parent–adolescent relationships have been shown to predict maladjustment among GLBT youth (Morrow, 2004). Given the changes in public visibility and attitudes about GLBT people and issues over the course of past decades, some families may be accepting of GLBT adolescents, yet the role of accepting families in promoting positive development has not been explored.

Several other areas are deserving of attention in the research on sexual minority youth. Based on several decades of research, a consensus has emerged regarding sexual minority health disparities in adolescence and in adulthood. Very little is known, however, about the transition from adolescence into adulthood by sexual minorities. The study of trajectories of development will be crucial for a full understanding of the role of risk in the lives of sexual minority adolescents. For example, it is presumed that GLBT adolescents who experience compromised mental and behavioral health begin a trajectory that follows them into adulthood. Yet, given that contemporary self-identified GLBT youth are among the first cohorts to navigate adolescence with GLBT identities, is it possible that their adolescent experiences might provide the basis for learning to cope with GLBT stigma? Could contemporary cohorts of GLBT youth be at lower risk in 10 to 20 years for the health disparities seen among GLBT adults at the beginning of the 21st century? Longitudinal studies that follow GLBT youth into adulthood are needed to better understand the lasting influence of adolescent experiences on adult well-being.

New research is also needed on methods for identifying and studying sexual minority youth. The use of the Internet by youth and by researchers has exploded. Once viewed as suspect, online approaches to identifying and including sexual minority youth as research participants are now more common. Online research methods provide remarkable opportunities for the study of a marginalized, often invisible, and small population. At the same time, many questions about methodological strengths and weaknesses remain. An additional methodological challenge in studying all adolescents is the protection of research participants; this issue is particularly important for "at-risk" populations. The issue is compounded in the study of sexual minority youth because of the risks of disclosure: Seeking parental consent for research participation typically poses more risk for GLBT youth than does participation in social or behavioral studies. Innovations in methods for seeking adolescent consent for research participation that assure subject safety are needed, and the efficacy of new approaches needs to be empirically tested.

There is much still to learn about the health and development of sexual minority youth. In ending, it is important to shift perspective and acknowledge a final reason that research on sexual minority youth has been and will continue to be important: What is learned contributes to a better understanding of the health and development of all young people. This is not a trivial point because focusing on marginal or previously invisible populations allows the possibility to understand developmental process in the general population of adolescence in a new light, at times pointing to areas or topics that had been unexamined because they were thought to be "natural" or "normal." Research on GLBT identity development has, for example, made it possible to conceptualize heterosexual identity development (Striepe & Tolman, 2003). Thus, research on sexual minority adolescents and other marginalized or understudied populations is important not only for serving the needs of their group but also for understanding normative development in adolescence.

SEE ALSO Volume 1: *Dating and Romantic Relationships, Childhood and Adolescence; Identity Development; Sexual Activity, Adolescent;* Volume 2: *Gays and Lesbians, Adulthood; Socialization, Gender.*

BIBLIOGRAPHY

Beaty, L. A. (1999). Identity development of homosexual youth and parental and familial influences on the coming out process. *Adolescence, 34,* 597–601.

Bornstein, K. (2006). *Hello, cruel world: 101 alternatives to suicide for teens, freaks, and other outlaws.* New York: Seven Stories Press.

Cohler, B. J., & Hammack, P. L. (2007). The psychological world of the gay teenager: Social change, narrative, and "normality." *Journal of Youth and Adolescence, 36,* 47–59.

D'Augelli, A. R. (2005). Developmental and contextual factors and mental health among lesbian, gay, and bisexual youths. In A. M. Omoto & H. S. Kurtzman (Eds.), *Sexual orientation and mental health: Examining identity and development in lesbian, gay, and bisexual people* (pp. 37–53). Washington, DC: American Psychological Association.

Diamond, L. M. (2003). New paradigms for research on heterosexual and sexual-minority development. *Journal of Clinical Child and Adolescent Psychology, 32,* 490–498.

Diamond, L. M. (2005). What we got wrong about sexual identity development: Unexpected findings from a longitudinal study of young women. In A. M. Omoto & H. S. Kurtzman (Eds.), *Sexual orientation and mental health: Examining identity and development in lesbian, gay, and bisexual people* (pp. 73–94). Washington, DC: American Psychological Association.

Floyd, F. J., & Bakeman, R. (2006). Coming-out across the life course: Implications of age and historical context. *Archives of Sexual Behavior, 35,* 287–296.

Meyer, I. H. (2003). Prejudice, social stress, and mental health in lesbian, gay, and bisexual populations: Conceptual issues and research evidence. *Psychological Bulletin, 129,* 674–697.

Morrow, D. F. (2004). Social work practice with gay, lesbian, bisexual, and transgender adolescents. *Families in Society: The Journal of Contemporary Social Services, 85,* 91–99.

Russell, S. T. (2003). Sexual minority youth and suicide risk. *American Behavioral Scientist, 46,* 1241–1257.

Russell, S. T. (2005). Beyond risk: Resilience in the lives of sexual minority youth. *Journal of Gay and Lesbian Issues in Education, 2*(3), 5–18.

Ryan, C., & Futterman, D. (1998). *Lesbian and gay youth: Care and counseling.* New York: Columbia University Press.

Saewyc, E. M, Bauer, G. R., Skay, C. L., Bearinger, L. H., Resnick, M. D., Reis, E., et al. (2004). Measuring sexual orientation in adolescent health surveys: Evaluation of eight school-based surveys. *Journal of Adolescent Health, 35*(4), 345.e1–345.e15. Retrieved June 6, 2008, from http:// download.journals.elsevierhealth.com/pdfs/journals/1054-139X/PIIS1054139X04001612.pdf

The Safe Schools Coalition Web site. Available from http://www. safeschoolscoalition.org

Striepe, M. I., & Tolman, D. L. (2003). Mom, Dad, I'm straight: The coming out of gender ideologies in adolescent sexual-identity development. *Journal of Clinical Child and Adolescent Psychology, 32,* 523–530.

Stephen T. Russell
Russell Toomey

GENDER AND EDUCATION

Discussions and research concerning gender and education have traditionally focused on women's disadvantages throughout the educational system. In 1992 the American Association of University Women (AAUW) published their influential report, *How Schools Shortchange Girls.* The AAUW report spurred a great deal of research by social scientists and discussion in popular media about the status of women and girls in education. Since its publication, however, the tide has been changing. Social scientists are now focusing on the gender gap in education in which girls are advantaged, and the media is concerned about the plight of boys, as seen with the 2006 *Newsweek* magazine cover story titled "The Problem with Boys." Regardless of the current concern about which gender is more or less disadvantaged, the question of gender differences in education remains interesting to the public and of primary concern to researchers and policy makers.

TEST SCORES, GRADES, AND COURSE-TAKING

Gender differences in academic performance and educational trajectories begin as early as elementary school. In earlier grades, boys and girls have relatively similar scores on standardized tests of math and reading, but as children move through schooling, differences emerge (Willingham & Cole, 1997). By high school, boys on average score higher on mathematics tests, whereas girls on average score higher on reading tests. There is disagreement, however, on whether gender differences in test scores are declining over time. Some researchers argue that gender differences have declined in recent decades, whereas others argue that differences have remained stable since the 1960s.

Despite boys' consistent advantage on standardized math tests, girls receive better grades in school than boys. As early as the 1950s and 1960s, girls earned higher grades than boys in high school. This trend continues throughout all grades in school, and even in college; girls earn higher grades than boys in all subjects, including

math and science (Perkins, Kleiner, Roey, & Brown, 2004). In addition to earning higher grades than boys, girls show better reading skills than boys as early as kindergarten. Boys have more problems with reading throughout elementary school and are more likely to be diagnosed with reading disabilities, mental retardation, attention disorders, dyslexia, and speech problems than girls (Trzesniewski, Moffitt, Caspi, Taylor, & Maughan, 2006). Girls also demonstrate better social skills and are rated by their teachers as having better classroom behavior than boys (Downey & Vogt Yuan, 2005).

Historically, significant differences existed in the types of course work boys and girls took in high school, with boys taking more rigorous math and science courses than girls. These differences, however, have disappeared. In the early 21st century, girls and boys take the same number of rigorous math courses, and girls are more likely than boys to take biology and chemistry courses by high school graduation (Gallagher & Kaufman, 2005;). Girls also take more Advanced Placement (AP) and college preparatory courses and are more active in extracurricular activities than boys, with the exception of sports participation, where girls lag behind boys considerably (Freeman, 2004). Additionally, girls now have higher educational expectations than boys. In the 1950s and 1960s, boys had higher educational expectations than girls (Marini & Greenberger, 1978). Since the 1980s, however, girls report higher educational expectations than boys and are more likely than boys to expect to complete college in the United States and many other countries (Shu & Marini, 1998).

EDUCATIONAL ATTAINMENT

Beginning in kindergarten, girls and boys have differing educational trajectories. Parents are more likely to delay boys' entry into kindergarten. Boys comprise 60% of students who enter kindergarten one year after they are eligible, and boys are more likely than girls to repeat kindergarten. Throughout elementary school, boys are more likely to repeat a grade or more than girls. By high school, boys are more likely to drop out of school than girls. In 2005 almost 11% of males ages 16 to 24 were high school dropouts, compared to 8% of females (Snyder, Dillow, & Hoffman, 2007).

One of the most striking changes in the U.S. education system is the change in college completion rates of men and women, with women outpacing men for the first time. In 1960, 65% of all bachelor degrees were awarded to men. By 1982 women reached parity with men in college completion, and, since then, women's college completion has increased to the point where in 2005 women received 58% of all bachelor's degrees (Snyder et al., 2007). This trend is also occurring in the

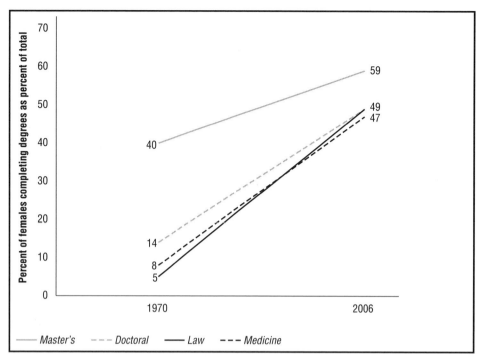

Figure 1. *Percentage of females completing graduate and professional degrees, 1970–2006.* CENGAGE LEARNING, GALE.

majority of industrialized countries with women on average representing 61% of students enrolled in university education (Organisation for Economic Co-operation and Development, 2007). In the United States, women's advantage in college-degree completion is more prominent in some racial groups than others. Women earn 66% of all bachelor's degrees earned by African Americans, 61% of degrees earned by Hispanics, 60% of degrees earned by Native Americans, 57% of degrees earned by Whites, and 55% of degrees earned by Asians (Snyder et al., 2007).

Although the female-favorable trend in college completion is striking, college fields of study remain highly sex-segregated. Although gender segregation in college declined substantially during the 1970s and 1980s, the decline stalled after this period (England & Li, 2006). Women earn only 20% of engineering, 28% of computer science, and 41% of physical science degrees. Women dominate the education and health professions fields, completing 77% of education degrees and 84% of health professions degrees (Freeman, 2004). Business, mathematics, social sciences, and history are less sex-segregated, with relatively equal numbers of men and women majoring in these subjects. The types of degrees earned by men and women differ, as do the types of colleges and universities they attend. Even though women are more likely to be enrolled in and complete college, the colleges they enroll in are less likely to be prestigious, selective schools.

This group largely includes prominent engineering schools. Women are more likely than men to enroll in two-year institutions, such as community colleges.

Regardless of the remaining sex segregation in college majors, which has important implications for women's and men's outcomes after school, women have made considerable progress in the attainment of graduate and professional degrees. Figure 1 shows the change in the percentage of women completing graduate and professional degrees from 1970 to 2006 (Snynder et al., 2007).

CAUSES AND EXPLANATIONS FOR GENDER DIFFERENCES IN EDUCATION

Gender differences in education have changed dramatically the since the 1960s. Traditionally, research focused only on how education advantaged men, but with the recent advancements of women in higher education, an increasing body of research has demonstrated a female advantage in many realms of education, especially in degree completion, grades, and classroom behaviors. Two main theoretical perspectives attempt to explain gender differences in education: biological perspectives and sociological perspectives. It should be noted, however, that these two perspectives are not always mutually exclusive.

Biological Perspectives Psychological and biological studies focus on differences in male and female brains, which cause males and females to excel at different subjects in school. Differences in the structure and organization of the brain lead to differing cognitive skills, which in turn, researchers argue, are the basis for gender differences in school performance and the predisposition to certain subjects. Females have advantages in verbal fluency, spelling, speech production, mathematical computation, and fine motor skills, which lead to advantages in standardized tests of reading and writing. Males have advantages in verbal analogies, mechanical reasoning, math word problems, memory for geometric configuration, spatial ability, and gross motor skills, which lead to advantages on standardized tests of math and science ability (Kimura, 1999). Despite these differences, research on cognitive sex differences tends to conclude that males and females are more similar than different and that cognitive abilities for the most part overlap. Although males and females have slightly different cognitive profiles, it is also agreed that males and females have equal cognitive abilities.

Sociological Perspectives. Sociological research on gender differences in schooling tends to ignore biological differences between the sexes, focusing instead on social factors that determine gender differences in the educational process. Gender role socialization, or the idea that social institutions socialize boys and girls according to traditional gender roles, influences children's and adults' expectations about their abilities, preferences, and opportunities. Parents' and teachers' perceptions of appropriate gender roles and skills may influence young boys' and girls' expectations and interests early in the life course. Some research finds that parents are equally involved in sons' and daughters' schooling, but other research suggests that parents may be more involved in sons' school activities and daughters' home activities and that parents have lower reading expectations of boys. Gender norms and socialization within the home can also affect boys' and girls' expectations about the future through parental role modeling. Girls tend to look to their mothers whereas boys look to their fathers when developing expectations about their future educational and occupational opportunities, which could reproduce traditional gender inequalities.

Teachers and schools also shape gender socialization and may treat males and females differently. Previously, girls were seen as disadvantaged in schools, with teachers calling on and praising boys more often in the classroom than girls (AAUW, 1992); some more recent arguments suggest, however, that boys are disadvantaged in schools. Teachers expect girls to be more studious and excel in the classroom, especially in reading and writing, whereas teachers expect more of boys in math and science. Teachers also rate girls as better students than boys, noting that girls are more cooperative and better communicators than boys (Downey & Vogt Yuan, 2005). It could be that girls are better students than boys, but it is also possible that schools are designed to reward behaviors that girls display naturally or that girls have been socialized to display, which could affect both teachers' and students' expectations (Mickelson, 1989).

Gender socialization also influences boys' and girls' school performance, and gender stereotypes may be of particular importance. Stereotype threat theory suggests that males and females are afraid of conforming to the traditional stereotypes of their gender, which negatively affects their performance in school and on standardized tests. Stereotype threat has been found to affect women when taking math tests, as there are widespread ideas that men outperform women on standardized tests, which causes additional stress and anxiety for women during test taking (Steele, 1997). This theory, however, has not been tested in the classroom, where women outperform men, but it is plausible that women are known to be better, more conscientious students, which may cause either anxiety or lowered expectations for men in the classroom.

IMPLICATIONS FOR GENDER DIFFERENCES IN EDUCATION

Gender differences in school experiences and educational outcomes have important long-term implications for economic and family outcomes for men and women. Women's increasing share of higher education has affected the labor market and men's and women's experiences in the labor market. In 2006 DiPrete and Buchmann found that over the previous 39 years, overall returns to higher education (or the amount by which income increased per year of schooling received) increased for both men and women but increased more rapidly for women. Between the 1970s and 1990s the gender wage gap, or the ratio of women's average earnings to men's average earnings, declined. Women in all segments of the earnings distribution saw increases in their wages, whereas women with high levels of human capital (in terms of education and labor force experience) saw the greatest increase in their wages.

Occupational sex segregation also fell between 1970 and 1990, although the rate of decline slowed in the second decade. This meant that more women entered prestigious and often better-paying positions in occupational sectors such as law, business, and the sciences. Nevertheless, women have not surpassed men in the labor market, as a gender wage gap still exists, and it seems that women's experiences in the labor market remain dampened to some degree. Given the continued gender segregation of college majors (Charles & Bradley, 2002), an important link exists between major choice and earnings, as women tend to choose majors that have lower earnings potential than men. Even for college-educated men and women with similar education credentials, standardized test scores, and majors, the gender gap in wages is reduced but remains significant

Gender differences in the educational process and educational attainment also have important long-term implications for an individual's family life. A negative relationship exists between a woman's education and divorce; the risk of divorce drops 6% for each additional year of schooling a woman receives. This is due, in part, to the fact that more educated individuals marry at later ages, but it is also due to marital homogamy (the marriage of like individuals) college-educated women are more likely to marry college-educated men, who have substantially lower rates of divorce than high school educated men. Moreover, after the mid-1970s, divorce rates fell among college-educated women whereas they continued to rise for less educated women. Higher educational attainment also is linked to fertility rates; college-educated women tend to have fewer children than women with only a high school education or less. Furthermore, women with less education are much more likely to have children outside of marriage than college-educated women. Education is the key determinant of fertility preferences, as rising levels of educational attainment substantially decreases fertility overall, which leads to societal changes in population size, density, and women's health.

THE FUTURE OF GENDER DIFFERENCES IN EDUCATION

Gender differences in education clearly have an impact on the future experiences and opportunities of students, and researchers need to continue to elucidate the different paths males and females experience in schools. Future research on this subject is likely to focus on the growing female-favorable gap in high school and college completion to determine the causes of the reversal of males' advantages in the classroom. Much research also needs to be conducted exploring the possible psychological and biological determinants of gender differences in cognitive ability and behaviors. One avenue that researchers may follow is to study very young children, at the beginning of or before elementary school. Studying the extent of and causes of gender differences in academic ability and experiences early in the life course may help tease out social, biological, and school factors that affect gender differences.

Finally, the vast majority of research on gender differences in education uses large-scale, nationally representative survey data to gather results, such as the National Education Longitudinal Survey, which is produced by the U.S. Department of Education and follows students from eighth grade through early adulthood, or the Early Childhood Longitudinal Study–Kindergarten Cohort, which follows students from entry to kindergarten through fifth grade. Very little research on gender differences in education explores these differences outside of the United States or uses qualitative data to answer questions.

Examining gender differences across educational systems and cultures may lead to new and interesting conclusions about the nature of gender differences in educational performance and may help researchers determine the differing effects that culture and educational systems have on gender differentials in schools. Also, qualitative data, using in-depth interviews and ethnographic research, could help answer questions about the daily experiences of boys and girls in school. For example, Lopez's (2003) study of low-income, second-generation Dominican, West Indian, and Haitian adolescents found that different socialization and gendered norms within families cause boys and girls from the same family to have very different educational outcomes. Also, Thorne's (1993) study of gender roles in public elementary schools provided insights into how boys and girls segregate themselves and adhere to traditional gender roles early in schooling. These studies are informative, albeit not common.

GENDER, EDUCATION, AND THE LIFE COURSE

As early as kindergarten, girls and boys have different experiences and outcomes in school, some of which advantage girls and others that advantage boys. These early experiences seem to set the stage for continued gender differences in the educational process, which in turn lead to differences in life outcomes such as marriage, family formation, and work. Consensus has yet to be reached on what causes gender differences in schooling, but it could be argued that both biological and social influences impact gender and education. One thing that is clear is that this topic will continue to capture the attention of researchers, policy makers, and the public.

SEE ALSO Volume 1: *College Enrollment; High School Dropout; Racial Inequality in Education; Socialization, Gender; Socioeconomic Inequality in Education;* Volume 2: *Gender in the Workplace.*

BIBLIOGRAPHY

American Association of University Women, Educational Foundation. (1992). *How schools shortchange girls: The AAUW report: A study of major findings on girls and education.* Washington, DC: Author.

Charles, M., & Bradley, K. (2002). Equal but separate? A cross-national study of sex segregation in higher education. *American Sociological Review, 67,* 573–599.

DiPrete, T. A., & Buchmann, C. (2006). Gender-specific trends in the value of education and the emerging gender gap in college completion. *Demography, 43,* 1–24.

Downey, D. B., & Vogt Yuan, A. S. (2005). Sex differences in school performance during high school: Puzzling patterns and possible explanations. *The Sociological Quarterly, 46,* 299–321.

England, P., & Li, S. (2006). Desegregation stalled: The changing gender composition of college majors, 1971–2002. *Gender and Society, 20,* 657–677.

Freeman, C. E. (2004, November). *Trends in educational equity of girls and women: 2004* (NCES Publication No. 2005-016). Washington, DC: National Center for Education Statistics. Retrieved June 16, 2008, from http://nces.ed.gov/pubs2005

Gallagher, A. M., & Kaufman, J. C. (Eds.). (2005). *Gender differences in mathematics: An integrative psychological approach.* Cambridge, U.K.: Cambridge University Press.

Kimura, D. (1999). *Sex and cognition.* Cambridge, MA: MIT Press.

Lopez, N. (2003). *Hopeful girls, troubled boys: Race and gender disparity in urban education.* New York: Routledge.

Marini, M. M., & Greenberger, E. (1978). Sex differences in educational aspirations and expectations. *American Educational Research Journal, 15,* 67–79.

Mickelson, R. A. (1989). Why does Jane read and write so well? The anomaly of women's achievement. *Sociology of Education, 62,* 47–63.

Organisation for Economic Co-operation and Development. (2007). *Education at a glance: OECD indicators, 2007.* Paris: Author.

Perkins, R., Kleiner, B., Roey, S., & Brown, J. (2004). *The High School Transcript Study: A decade of change in curricula and achievement, 1990–2000* (NCES Publication No. 2004-455). Washington, DC: National Center for Education Statistics. Retrieved June 16, 2008, from http://nces.ed.gov/pubs2004

Shu, X., & Marini, M. M. 1998. "Gender-Related Change in Occupational Aspirations." *Sociology of Education* 71 (1): 43–67.

Snyder, T. D., Dillow, S. A., & Hoffman, C. M. (2007, July). *Digest of educational statistics, 2006* (NCES Publication No. 2007-017). Washington, DC: National Center for Education Statistics. Retrieved June 16, 2008, from http://nces.ed.gov/pubs2007

Steele, C. M. (1997). A threat in the air: How stereotypes shape intellectual identity and performance. *American Psychologist, 52,* 613–629.

Thorne, B. (1993). *Gender play: Girls and boys in school.* New Brunswick, NJ: Rutgers University Press.

Trzesniewski, K. H., Moffitt, T. E., Caspi, A., Taylor, A., & Maughan, B. (2006). Revisiting the association between reading achievement and antisocial behavior: New evidence of an environmental explanation from a twin study. *Child Development, 77,* 72–88.

Willingham, W. W., & Cole, N. S. (1997). *Gender and fair assessment.* Mahwah, NJ: Erlbaum.

Anne E. McDaniel

GENETIC INFLUENCES, EARLY LIFE

The nature-versus-nurture debate is an important conceptual framework because it simplifies the very complex reality of genetic and environmental influences on development. The cost of this simplification is that individual differences are described as a function of *either* environmental *or* genetic characteristics rather than the simultaneous influence of genetic *and* environmental factors. This entry describes the ways in which researchers examine genetic influences among children and adolescents and then makes a case for simultaneously considering genetic and environmental causes. Although several relevant outcomes are linked to genetic factors such as personality, mental and physical health, and health-related behaviors, this entry focuses on cognitive functioning because of the reliability of the measure, the consistency of the findings, and the importance of the topic for social and behavioral researchers. In doing so, this entry reviews two important concepts that characterize the interplay between genes and environments. These concepts describe a situation in which a person's genes causes their environment (gene-environment correlation) or environmental settings that change the influence of someone's genes (gene-environment interaction). Following a discussion of these concepts, this section concludes with some general comments about this field.

STUDYING GENETIC INFLUENCES AMONG CHILDREN AND ADOLESCENTS

The scientific world of human genetics is rapidly changing, and hundreds of methods are currently available to researchers that describe the ways in which individuals are influenced by their genes. The complexity of this research, however, can be summarized by considering two broadly defined bodies of work: studies of twins and siblings in which genetic material is not available and studies in which study participants provide a physical specimen (e.g., saliva, blood, or tissue scraped from the inside of the cheek) that can be used to identify an individual's genes. The purpose of this entry is to introduce these two methods and summarize some of the major findings with respect to the influence of genetic factors among infants, children, and young adolescents.

Quantitative Genetic Studies Behavioral genetics is an interdisciplinary field composed of behavioral, social, and biologic scientists who are interested in describing the extent to which differences in individuals are explained by differences in their genes, differences in their environments, or some combination. The bulk of the research in this area is based on the straightforward observation that two siblings are more likely than two unrelated people to resemble one another in terms of their physical appearance, behavior, personality, and well-being, because full siblings are often raised in very similar settings (i.e., a similar environment) but they also inherit half their genes from the mother and half from the father; thus, they share about half the same genes. This same logic is extended to the comparison of twin pairs in which identical twins share all their genes and fraternal twins, similar to the manner in which full siblings only share (on average) half their genes. If one assumes that same-sex fraternal twins and identical twins are raised in relatively similar environments, then the excess similarity of identical twin pairs compared with fraternal twin pairs is believed to be due to the excess genetic similarity among these pairs.

Based on the results of these studies in siblings, the most consistent evidence for pronounced genetic influences is in the area of cognitive development even among very young infants. In one of the earliest studies (1972) to make this point, Ronald S. Wilson compared the statistical association between the test scores of two twins for the Bayley Mental Development Score among twin pairs from the ages of 3 months to 2 years. He demonstrated a greater similarity among identical twins ($r = .84$) compared with fraternal twins ($r = .67$) even as young as 3 months. Sandra Scarr (1993) made a similar point when she compared the similarity of IQ scores among identical twins ($r = .86$) and showed that this association is nearly the exact same as for the same person tested twice ($r = .87$)! Even more striking is that identical twins who are raised in different families ($r = .76$) still report a stronger correlation than fraternal twins who are raised in the same families ($r = .55$).

Cognitive functioning also emerges as one of the most heritable characteristics among children and early adolescents. Using data from the Colorado Adoption Project, Stacey Cherny and Lon Cardon (1994) demonstrated that genetics may account for 39% of reading skills among 7-year-old children and 36% among 12-year-old children. In one study David Reiss and colleagues (Reiss, Neiderheiser, Hetherington, & Plumin, 2000) assessed more than 700 twin and sibling pairs

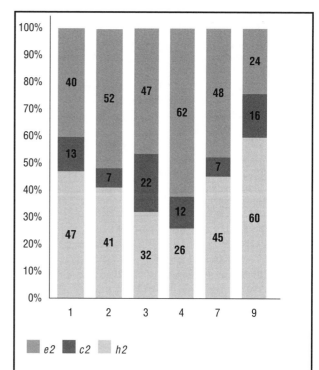

Note: Estimates for heritability (*h2*), shared environment (*c2*), and non-shared environment (*e2*) for each age are as follows: one (0.47, 0.13, 0.40), two (0.41, 0.07, 0.52), three (0.32, 0.22, 0.47), four (0.26, 0.12, 0.62), seven (0.45, 0.07, 0.48), nine (0.60, 0.16, 0.24). Estimates derived from Cherny, S. S., and Cardon, L. R. (1994). General cognitive ability. In J. C. DeFries, R. Plomin, and D. W. Fulker, (Eds.), *Nature and nurture during middle childhood* (pp. 46–56). Cambridge, MA: Blackwell Publishers. Data come from the Colorado Adoption Project using 87 adoptive sibling pairs, 102 non-adoptive pairs, and 300 singletons. Tests included the Bayley Mental Development Index (Bayley, 1969) for ages one and two, the Stanford-Binet Intelligence Scale (Terman and Merrill, 1973) for ages three and four, the Wechsler Intelligence Scale (Weschler, 1974) at age seven, and the SCAPTPC (Kent & Plomin, 1987) for age nine.

Figure 1. *Genetic influences among very young children: heritability estimates for 14 and 24 month old children for 10 behavioral and physical outcomes.* **CENGAGE LEARNING, GALE.**

from the ages of 10 to 18 years. They collected information on psychopathology and competence and then compared the similarity of these characteristic among pairs of identical twins, fraternal twins, full siblings, half-siblings, and unrelated siblings. They found that genetic factors increased in salience over time and suggested that genes may account for two thirds of individual differences in cognitive agency. Similar results were reported by François Nielsen (2006) using data from high school age adolescents for grade point average and a somewhat lower estimate for verbal IQ.

Quantitative geneticists often use these designs to calculate a value called the *heritability estimate*. These estimates describe the degree to which differences among people are due to genetic differences; they range from 0 (in which genes are not relevant) to 1 (in which genetic

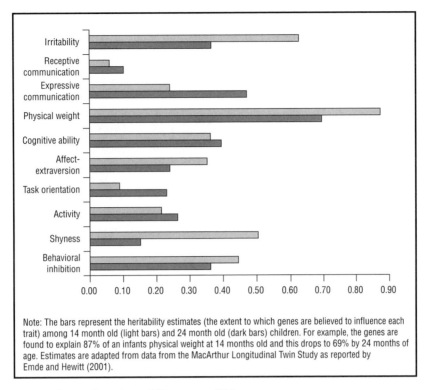

Note: The bars represent the heritability estimates (the extent to which genes are believed to influence each trait) among 14 month old (light bars) and 24 month old (dark bars) children. For example, the genes are found to explain 87% of an infants physical weight at 14 months old and this drops to 69% by 24 months of age. Estimates are adapted from data from the MacArthur Longitudinal Twin Study as reported by Emde and Hewitt (2001).

Figure 2. *Nature, nurture, and general cognitive ability among children.* **CENGAGE LEARNING, GALE.**

factors are fully responsible for a trait). Figure 1 provides a summary of heritability estimates for a number of different outcomes among very young children from the MacArthur Longitudinal Twin Study.

Studies with Measured Genetic Information Since the early 1990s, the field of behavioral genetics has changed dramatically. The availability of molecular information describing an individual's genetic makeup is now readily (and affordably) available to researchers. Carefully designed studies can identify broad regions of the human genome (the entire set of all chromosomes), specific genes (smaller sections of the human chromosome that carry genetic information), and even very small pieces of the human genome called single nucleotide polymorphisms (SNPs) that may be associated with the trait of interest.

It is beyond the limits of this entry to detail the complexity of the molecular studies that are currently underway; however, there are two main study designs currently employed by researchers. When specific genes are believed to be causally linked to some outcome, then it is relatively easy to measure an individual's genetic composition at a specific location on his or her entire genome. After an individual's genes are measured at that location, it is simply a matter of comparing the genes of

people who exhibit a particular behavior with those of a carefully constructed control group who do not exhibit the behavior. These studies are called *candidate gene* studies.

There are hundreds of candidate genes that have been linked to cognitive functioning and related outcomes, but efforts have primarily focused on the dopaminergic and serotonergic systems. Dopamine is one of a large number of neurotransmitters that is involved in the nervous system; this system is believed to mediate the reward pathway in the brain. Serotonin is also a neurotransmitter that is involved in brain development. The serotonergic system is linked to behavioral disinhibition, which has important implications for aggressiveness and impulsivity. Results from these studies focus on genes that influence the function of the receptors and the transporters for these two systems. Thus, although there are two main systems, multiple domains exist within each system that may operate independently from one another at times, but mostly they act in concert with one another. These systems are believed to influence cognitive development through more proximate behaviors such as temperament and hyperactivity.

The second type of molecular analysis involves studies that use information from the entire human genome to identify genetic risk (or protective) factors. Two distinct types of studies are conducted in which large amounts of genetic information are used: linkage analysis and genome wide association studies. *Linkage analysis* is used to identify the location of broad chromosomal regions that may contain genes (called quantitative trait loci or QTLs) that are in some way responsible for the emergence or stability of some trait. Danielle Posthuma and Eco J.C. de Geus (2006) reviewed findings from five whole-genome linkage scans designed to identify chromosomal locations that are associated with cognition. These studies found that areas on chromosomes 2 and 6 consistently predict both cognition and academic achievement, but results by Plomin et al. (2005) were far less conclusive among children.

Although linkage analyses are designed to identify the general location on the chromosome, *genome-wide association* studies are designed to identify specific alleles that may influence complex behaviors. The human chromosome is composed of nearly 8 billion bits of information, and people differ from one another when a protein at a specific location is different in the two people; each of these minute proteins is an SNP. The goal of genome wide association studies is to identify SNPs that are causally linked to the increased risk of a disease or disorder. To date, a very limited number of studies have identified SNPs that significantly influence cognitive functioning, and the effects are very small, accounting for less than 1% of the variance (Butcher, Davis, Craig, & Plomin, in press).

SOCIAL INFLUENCES ON GENETIC EFFECTS

The preceding discussion described ways to conceptualize genetic influences on children's development, and thus far the results have been structured around the nature-versus-nurture dichotomy. As described previously, this conceptual model serves an important purpose but is problematic because it does not consider several ways in which genes and environments interact. Genes do not exist outside of environments; environmental determinants have to *go through* people to affect some outcome. Therefore, to consider the effects to be independent or to be additive is to misinterpret the relative and combined influence of both sets of effects. Indeed, it is only the most basic behavioral genetics model in which characteristics are decomposed into only genetic and environmental components. Two important cautions exist with respect to gene-environment interplay that have important implications for social scientists. The first is called *social mediation;* it describes a situation in which specific genetic factors are regularly associated with a particular type of environment. The second is called *social moderation,* which is a situation in which specific genetic fac-

tors react differently in different environments. Also called *gene-environment correlation* and *gene-environment interaction,* respectively, these social influences bear heavily on the theoretical mechanisms that are behind genetic influences in early life.

Social Mediation Robert Plomin, John DeFries, and John Loehlin (1977) developed a typology of *social mediation* (also called *gene-environment correlation*) that has withstood the test of time largely intact. The first form of social mediation is called *evocative correlation,* which describes a situation in which people with given genetic characteristics tend to evoke similar reactions from other people. These reactions, in turn, shape that person's social context (or his or her environment). The most widely cited example of evocative correlations comes from work in which children who are genetically predisposed to have relatively irritable dispositions may be more likely than more peaceful children to *evoke* hostility and impatience from their parents, siblings, peers, or teachers.

The second form is called *passive correlation,* and it emphasizes the fact that children inherit genetic and environmental factors their parents. As described earlier in this entry, cognitive ability may be a highly heritable trait, and children raised by relatively intelligent parents may be more likely to be raised and socialized in an intellectually stimulating environment. Thus, the children passively inherit *both* the genetic characteristics related to cognitive development and the genetically influenced enriching environment. Finally, *active correlation* describes a situation in which someone's genes influence the type of social environment in which he or she chooses to interact. For example, children with genetically oriented cognitive skills may be attracted to more complicated games that both require and build problem-solving skills, which in turn influences positive cognitive development.

One of the central tenets of the sociologic perspective on the life course is the role of human agency; people actively construct their lives by means of those behaviors that they exhibit at different stages. Therefore, children are highly subject to the passive processes in their family, but over time they acquire more latitude in choosing and shaping their environments. Because passive and evocative gene-environment correlation (rGE) denote less agency and because the extent to which everyone's behaviors are limited by the relevant institutions in which they reside, people at either extreme of the life course are less likely to select into social groups as a function of their genes. In contrast, adults choose their settings—that is, with whom to associate, organizational memberships, and extent and quality of involvement—and, after those choices are made, they tend to evoke reactions within those chosen contexts.

NATURE VS. NURTURE DEBATE

The nature versus nurture debate reduces the cause of individual differences to *either* genetic (nature) *or* environmental (nurture) differences. Some scientists can answer this question by breeding animals with known characteristics across controlled environments to obtain a very precise measure of genetic and environmental influence. The task of determining the influence of genes is more difficult among humans because it is not possible (or moral) to experiment with environmental factors such as school quality, neighborhood safety, family stability, or health. Similarly, one cannot manipulate an individual's genes and then measure changes in his or her behavior. Therefore, a common way to describe genetic and environmental influences is to study twins, siblings, cousins, and adopted siblings. Because identical twins share all of their genes and fraternal twins share (on average) one-half of their genes, excess similarity of identical twins is believed to be due to their

genetic similarity. These studies estimate a value called "heritability," which ranges from a score of zero (where genes have no influence) to one (where genes are completely responsible). This heritability measure is a rough indicator of the "nature" effects, and the remaining amount is due to "nurture." According to these estimates, genetic factors account for 47% of the variation in cognitive ability at age 1 year.

Because the nature versus nurture debate focuses on either nature or nurture, it overlooks the more realistic observation that both nature and nurture matter; for example, sometimes nature causes people to be nurtured differently and sometimes the effects of nature depend on the nurture that one receives. Therefore, it is important to recognize that the interplay between genes and environments is far more complicated than the nature versus nurture paradigm.

One important example is the formation and maintenance of intimate relationships that involve assortative mating driven by active rGE, which strengthens over time. Most (if not all) behaviors known to drive assortative mating are also highly heritable, including, for example, mental health problems, various addictive behaviors, criminality and other forms of antisocial behavior, and education and intelligence. Thus, a spouse may contribute to the creation of an environment (particularly a household) that reflects his or her genetic factors and is shared by his or her partner. Such a pattern would be a variant of the classic passive correlation, which involves living in an environment created by a person from whom one is transmitted genetic material.

Social Moderation The gene-environment interaction perspective poses two related models: situations in which genetic effects depend on the environment and situations in which established environmental effects vary in their influence as a function of individual's genes. These models can be thought of as social moderation and genetic moderation, respectively. This section deals with the first orientation. The most relevant orientation for sociologists is that the environment serves as a trigger for the expression of a particular gene; a gene related to a particular outcome may only manifest as a cause in the presence of a triggering agent (*strong triggering*) or is

expressed markedly more so in the presence of the agent (*weak triggering*).

In an example of weak triggering, Guang Guo and Elizabeth Stearns (2002) showed the realization of genetically oriented verbal IQ is higher among children from families with greater access to social and economic resources. Specifically, they calculated a higher heritability of verbal IQ among adolescents for whom both parents were employed compared to those with at least one unemployed parent. They showed a similar association by race in which heritability estimates are higher among White children compared to Black children. Their argument is that family stability, educational resources in the home, school-level differences, and parental educational status denote critical resources to enable genetic factors to do their work. This perspective is central to a sociologic interpretation of gene by environment interaction effects because, as Bruce Link and Jo Phelan (1995) argued, the environment should be characterized as a "fundamental cause." That is, although genetic factors are critical to the etiology of antisocial behavior, these genetic factors depend on the social environment to initiate the cascade of events called *genetic expression*.

Rather than enabling genetic tendencies, the *social control* model refers to norms and structural constraints placed on people that limit their behavior; control stems from social structures or processes that maintain the social order (whether for the moral good or not). These controls might stem from strict legal enforcement, clear behavioral limitations associated with religion, highly organized and controlled educational settings, or broad macrolevel systems of stratification that limit particular individuals' mobility. Clear evidence for the broad social control model was presented in a study by Heath et al. (1985) that reviewed educational attainment among three birth cohorts in Norway. According to their estimates, the heritability of educational attainment was roughly 40% for men and women born before 1940. For men, changes in the traditional educational hierarchy provided greater access to the education system: The heritability increased to roughly 70%. For women, however, the heritability remained at 40%, reflecting the continuation of social norms and opportunities that controlled the educational opportunities and behaviors of women. The results of the study argued that the degree to which educational attainment was heritable was controlled among women but not men.

Genetic Moderation Genetic factors unique to individuals cause them to react differently to the same environmental stimuli. This gene-environment interaction perspective is shared by several genetic epidemiologists and is summarized in the following comment by Moffitt, Caspi, and Rutter (2006):

> Thus, findings of [gene-environment interaction studies] reframe the scientific question for environmental researchers. The question is not "Is there any environmental risk?" or "How big is the average effect of an environmental pathogen across all people exposed to it?" but rather "Who is at the greatest risk from an environmental pathogen?"

In two of the most widely cited examples of this perspective, Caspi and colleagues (2003) showed that well-established environmental risk factors operate differently as a function of individual's genes. In one example (Caspi et al., 2003), an adult's risk of major depression increases with the number of stressful life events that he or she may have experienced. However, among those people with two long alleles of a gene responsible for serotonin transmission (5-HTT), no clear association exists between chronic exposure to stressful life events and poor mental health. That is, people who are homozygous for this allele are particularly resilient to stressors that may otherwise lead to a major depressive episode. In a related study (Caspi et al., 2002), the researchers showed that childhood maltreatment does not appear to predict later forms of antisocial behavior among adults who have

a genotype that is linked to the level of monoamine oxidases (MAO).

CONCLUSION

Because social scientists are primarily interested in describing broad group-level relationships and interdependencies, very little emphasis has been placed on the role of genes. Emphasis is almost exclusively place on micro-, meso-, and macro-level environmental factors, and the debates revolve around the ways in which ascriptive characteristics place people within risky or supportive environments. Recently, however, interest has been renewed among sociologists in exploring the possibility that biologic characteristics are related to complex behaviors such as smoking, sexual behavior, and academic success (Guo an Tong, 2006; Nielsen, 2006). Because of the significance of the social environment in understanding the ways in which genes operate (e.g., moderation, mediation, identification), the long sociologic tradition of measuring and monitoring social environmental factors denotes an important contribution to behavioral genetic inquiry. Although heritability estimates have historically represented fixed parameters, sociologists are pushing for the understanding that they represent average values with a great deal of variation. Genes are important for understanding individual differences, but these influences can only be understood when they are situated in a particular context.

SEE ALSO Volume 1: *Academic Achievement; Cognitive Ability; Health Behaviors, Childhood and Adolescence; Health Differentials/Disparities, Childhood and Adolescence;* Volume 2: *Genetic Influences, Adulthood;* Volume 3: *Genetic Influences, Later Life.*

BIBLIOGRAPHY

Butcher, L. M., Davis, O. S. P., Craig, I. W., & Plomin, R. (in press). Genome-wide quantitative trait locus association scan of general cognitive ability using pooled DNA and 500K single nucleotide polymorphism microarrays. *Genes, Brain, and Behavior.*

Caspi, A., McClain, J., Moffitt, T. E., Mill, J., Martin, J., Craig, I. W., et al. (2002). Role of genotype in the cycle of violence in maltreated children. *Science, 297,* 851–854.

Caspi, A., Sugden, K., Moffitt T. E., Taylor, A., Craig, I. W., Harrington, H. L., et al. (2003). Influence of life stress on depression: Moderation by a polymorphism in the 5-HTT gene. *Science, 301,* 386–389.

Cherny, S. S., & Cardon, L. R. (1994). General cognitive ability. In J. C. DeFries, R. Plomin, & D. W. Fuller (Eds.), *Nature and nurture during middle childhood* (pp. 46–56). Cambridge, MA: Blackwell.

Emde, R. N., & Hewitt, J. K. (2001). *Infancy to early childhood: Genetic and environmental influences on developmental change.* New York: Oxford University Press.

Guo, G., & Stearns, E. (2002). The social influences on the realization of genetic potential for intellectual development. *Social Forces, 80*(3), 881–910.

Guo, G., & Tong, Y. (2006). Age at first sexual intercourse, genes, and social context: Evidence from twins and the dopamine D4 receptor gene. *Demography, 43*(4), 747–769.

Heath, A. C., Berg, K., Eaves, L. J., Scales, M. H., Corey, L. A., Sunder, J., et al. (1985). Education policy and the heritability of educational attainment. *Nature, 314*(6013), 734–736.

Link, B. G., & Phelan, J. (1995). Social conditions as fundamental causes of disease. *Journal of Health and Social Behavior, 35*, 80–94.

Moffitt, T. E., Caspi, A., & Rutter, M. (2006). Measured gene-environment interactions in psychopathology: Concepts, research strategies, and implications for research, intervention, and public understanding of genetics. *Perspectives on Psychological Science, 1*, 5–27.

Nielsen, F. (2006). Achievement and ascription in educational attainment: Genetic and environmental influences on adolescent schooling. *Social Forces, 85*, 193–216.

Plomin, R., DeFries, J. C., & Loehlin, J. C. (1977). Genotype-environment interaction and correlation in the analysis of human behavior. *Psychological Bulletin, 84*, 309–322.

Plomin, R., Turkic, D. M., Hill, L., Turkic, D. E., Stephens, M., Williams, J., et al. (2005). A functional polymorphism in the succinate-semialdehyde dehydrogenase (aldehyde dehydrogenase 5 family, member A1) gene is associated with cognitive ability. *Molecular Psychiatry, 9*, 582–586.

Posthuma, D., & de Geus, E. J. C. 2006. Progress in the molecular-genetic study of intelligence. *Current Directions in Psychological Science, 15*, 151–155.

Reiss, D., Neiderhiser, J. M., Hetherington, E. M., & Plomin, R. (2000). *The relationship code: Deciphering genetic and social influences on adolescent development.* Cambridge, MA: Harvard University Press.

Rutter, M. (2006). *Genes and behavior: Nature-nurture interplay explained.* Malden, MA: Blackwell.

Scarr, S. (1993). Developmental theories for the 1990s: Development and individual differences. *Child Development, 63*, 1–19.

Wilson, R. S. (1972). Twins: Early mental development. *Science, 175*, 914–917.

Jason D. Boardman
Michael J. Shanahan

GIFTED AND TALENTED PROGRAMS

SEE Volume 1: *Cognative Ability; School Tracking.*

GRAMMAR SCHOOL

SEE Volume 1: *Stages of Schooling.*

GRANDCHILDREN

This entry reviews the scholarly research about grandchildren and grandparent-grandchild relations within the context of multigenerational families. Because being a grandchild is the joint product of two family transitions, the nature of this family role cannot be separated from the roles that grandparents and parents perform, as well as the relationship between them. Where *grandparenting* and *parenting* are action verbs for how grandparents and parents enact their respective roles, being a grandchild has no comparable term. Thus, the roles played by grandchildren are more passively constructed and necessarily dependent on experiences within the wider family system.

GRANDPARENT ROLES IN RELATION TO GRANDCHILDREN

The study of grandparenting as an independent branch of inquiry emerged in the 1940s and 1950s as highlighted by several historical events—World War II (1939–1945) and the postwar baby boom. On the one hand, observers noted that grandparents helped their children's families adapt to the dislocations and hardships caused by war, and subsequently provided child care assistance to the growing number of households with young children. On the other hand, some argued that grandparents had become less relevant during the postwar economic expansion as a result of their children's geographic mobility and increasing reliance on professional sources for childrearing advice and help (Szinovacz, 1998).

Inspired by pioneering research on grandparenting styles by Bernice Neugarten and Karol Weinstein (1964), social scientific investigations into grandparenting surged in the 1980s when several important volumes were published on the topic by Vern Bengtson and Joan Robertson (1985) and Andrew Cherlin and Frank Furstenberg (1986)—*Grandparenthood* and *The New American Grandparent*, respectively. These works were instrumental in directing the attention of scholars to what had still been a somewhat marginal area in family research. Much of the literature of that period focused on characterizing grandparent roles themselves, using such descriptive labels as *fun-loving, companionate, formal,* and *remote* to describe how grandparents stylized their relationships with grandchildren and their families. Issues of race, class, gender, and timing (whether the role occurred at the *typical* stage in the life course, versus significantly earlier or later) were examined in relation to grandparenting styles.

In Cherlin and Furstenberg's national study of grandparents, no particular style emerged as a majority, underscoring the diverse and normatively ambiguous nature of grandparenting. Much discussion centered on how grandparents could resolve the tension between their wish to be fully involved in the lives of grandchildren and their need to adhere to the implicit "norm of noninterference" often expected of them. Research during this period also focused on the instrumental role of grandparents as guardians of family culture and continuity, lauding their contributions as the watchdogs of the family.

More contemporary evidence continues to show that grandparents are actively involved with their grandchildren, serving as mentors, confidants, and companions to them. The typical grandchild-grandparent relationship is one based on respect, unconditional love, mutual support, and friendship. A good deal of scholarly attention has been devoted to describing variation in the involvement of grandparents with their grandchildren based on structural, biological, and demographic characteristics of all three generations. Grandchildren tend to feel emotionally closer to their maternal than their paternal grandparents, feel closer to their grandmothers than their grandfathers, and are more intimate with grandparents who have better relationships with the lineal parent and in-law (Fingerman, 2004). Among all grandparents, maternal grandmothers are the most active contributors to their grandfamilies and maintain the closest relationships with grandchildren (Chan & Elder, 2000). Grandchildren also have more distant relationships with grandparents with fewer resources, specifically those who are older, less healthy, and widowed (Silverstein & Marenco, 2001).

Parental divorce has weakened some grandparent-grandchild relationships but has strengthened others (Kennedy & Kennedy, 1993). Because of custody decisions that favor mothers and higher rates of stepfamily formation among fathers, family disruption tends to strengthen ties to grandparents on the maternal side and weaken them on the paternal side (Uhlenberg & Hammill, 1998). Further, growth in the prevalence of step-grandparents has created new ambiguities in a relationship that already has uncertain guidelines.

INSTRUMENTAL FAMILY ROLES OF GRANDPARENTS AND GRANDCHILDREN

Scholarship on grandparents and their grandchildren achieved a sort of renaissance in the 1980s and 1990s when the number of grandparents raising grandchildren began to surge. The percentage of children living in grandparent-headed households almost doubled in the last 30 years of the 20th century, rising from 3% of children in 1970 to 5.5% in 1997 (Bryson & Casper, 1999). As of 2000, more than 2.4 million grandparents claimed primary responsibility for at least one coresident grandchild (Simmons & Lawler Dye, 2003). The number of grandchildren being raised by grandparents increased in all socioeconomic and ethnic groups, but rose most dramatically among African-American families in which the parental generation was incapacitated as a result of crack cocaine addiction, HIV or AIDS, and incarceration.

Research suggests that children being cared for full time by their grandparent(s) are at elevated risk for behavioral and emotional problems (Billing, Ehrle, & Kortenkamp, 2002), often assumed to be the consequence of the dire home environments that created the need for the grandparents' care in the first place. However, compared to day care, after-school programs, babysitters, nannies, and other formal sources of help, grandparents tend to have a less casual interest in the well-being of their grandchildren. One study, for instance, found that the health and school adjustment of children raised solely by grandparents was nearly equivalent to children raised by one biological parent (Solomon & Marx, 1995).

Families in which grandparents are raising grandchildren provide a dramatic, but only one, example of how the fortunes of three generations are mutually interdependent. Women's labor force participation provides grandparents additional opportunities to take care of grandchildren raised in dual earner and single-mother households. Mary Elizabeth Hughes and colleagues (2007) found that caring for grandchildren is quite common in the United States, with 40% of grandparents providing at least 50 hours of care per year for the children of working parents. In the United Kingdom one in five children less than 16 years old is looked after by their grandparents during the daytime (Clarke & Cairns, 2001), and a multinational European study found that 40–60% of grandparents reported taking care of grandchildren over a 1-year period (Attias-Donfut, Ogg, & Wolff, 2005).

How families benefit economically from grandparent-provided care has been the subject of several investigations. Emanuela Cardia and Serena Ng (2003) found that parents who received child-care services from grandparents improved their economic status by increasing their paid labor force participation and avoiding formal child-care costs. However, direct money transfers from grandparents had little economic effect on household resources because such assistance reduced the amount of time that parents worked in the paid labor force. Similar conclusions were reached in a study of immigrant grandparents in France, who tended to provide grandchild care to daughters with greater labor market potential than to those with greater economic need (Ralitza & Wolff, 2007). Together, these results support the notion that grandparents *strategically* invest their child care labor to optimize the overall well-being of their grandfamilies.

A key, but largely unanswered question that arises for researchers is whether grandchildren who receive social, emotional, and material resources from their grandparents later reciprocate by, for example, being caregivers for their grandparents. Grandchildren who received more early care from grandparents do tend to have closer relations with them in adulthood, arguably a precondition to being a caregiver. However, there is only limited evidence that grandchildren are prolific providers of support to their grandparents. A novel perspective developed by Debra Friedman and colleagues (2008) argues that grandparents have an incentive to invest in those grandchildren whose *parents* are most likely to be a caregiver to them, bringing into consideration the possibility that reciprocity with grandchildren may be indirect.

Among the most altruistic grandparents are those who care for their disabled grandchildren. A review of literature suggests that these heavily invested grandparents are greatly valued by their families, but found little systematic research concerning their capacity for meeting the demands of this challenging role, and the effectiveness of their intervention with respect to the well-being of the children under their care (Mitchell, 2007). Research by Jennifer Park and colleagues (2005) reveals that although grandparents are prolific providers of care for developmentally disabled, impaired, and special-needs grandchildren, they sometimes face difficulties in bonding with these grandchildren, particularly those exhibiting communication and behavioral difficulties.

PSYCHOSOCIAL INFLUENCE OF GRANDPARENTS ON GRANDCHILDREN

Family systems theory provides an overarching psychosocial paradigm that stresses the importance of looking beyond the parent-child relationship to more fully account for the familial forces that shape children's development and adaptation. Literature suggests that grandparents can be important contributors to the successful psychosocial development and well-being of their grandchildren in infancy and beyond. Suitable care and stimulation provided by grandmothers and grandfathers benefits the social, cognitive, and motor skills of infant grandchildren. The role of grandparents in mitigating adjustment difficulties of young grandchildren raised in nontraditional families has also been documented. However, in such circumstances, the involvement of grandmothers may also produce negative effects such as when it causes a reduction in maternal involvement (Chase-

Lansdale, Brooks-Gunn, & Zamsky, 1994) or when grandmotherhood occurs at such a young age that the role is not fully engaged (Burton, 1995).

There has been little systematic study of the continued importance of grandparents beyond childhood, in part because few studies of adolescents and young adults have inquired specifically about the role that grandparents play (or have played) in their lives. However, using national data, research by Sarah Ruiz and Merril Silverstein (2007) found that among children raised by single mothers or depressed mothers, those whose grandparents served as secure attachment figures tended to have better mental health in adulthood.

Grandparents play an important symbolic role with respect to their grandchildren by conveying core values and providing a cultural window into family history and traditions. Religion represents one of several intergenerational threads that bring grandparents and grandchildren closer together around a shared value system. Grandparents have been found to directly influence the religious orientation of their grandchildren without parental influences, suggesting a fundamental contribution to their worldview (Copen & Silverstein, 2007). Contact with grandparents also conditions grandchildren to think and act differently about their own and their family's aging. Those adults who had more exposure to their grandparents earlier in life tend to hold more positive attitudes about elderly people, express greater support for entitlement programs that benefit older adults, and are more likely to live with their aging parents.

SOCIOCULTURAL AND ECOLOGICAL PERSPECTIVES ON GRANDPARENT-GRANDCHILD RELATIONS

The type and level of grandparent involvement has a basis in cultural norms and in the economic organization of families, localities, and nations that emphasize or downplay the role of grandparents. Ethnic differences in grandparenting have long been observed, particularly with respect to African-American grandparents who tend to be more involved with their grandchildren, and more apt than other grandparents to discipline and guide their grandchildren (Hunter, 1997). African-American grandmothers closely identify with their role as grandparents and derive a great sense of meaning and accomplishment from contributing to the development of their grandchildren. The stronger, more authoritative role taken on by Black grandmothers has long cultural roots, reflecting a tradition of surrogate parenting and extended familism going back to the time of slavery.

Grandparent-grandchild relations in Hispanic families are typically viewed as stronger and more durable than those in non-Hispanic White families. However, Silverstein and Xuan Chen (1999), in a study of Mexican-American immigrant families, found that young adults who acculturated away from their native language and traditional customs tended to feel emotionally and socially detached from their more culturally traditional grandparents. The culture gap between grandchildren and their grandparents did not predict grandparents' assessments of their relationships with grandchildren. This pattern suggests that in immigrant families, grandchildren but not grandparents attribute the socioemotional distance between them to acculturation differences.

Local conditions also influence relations between grandparents and grandchildren. For instance, Valarie King and Glen Elder (1995), in their study of rural Iowa farm families, found that grandchildren tended to have stronger relations with their paternal grandparents than with their maternal grandparents, a reversal of the pattern found in urban families of Los Angeles (King et al., 2003) and the majority of other studies. The strength of paternal ties observed in rural Iowa is attributed to traditional patterns of inheritance in agricultural communities where land ownership is passed from father to son.

There are few cross-national studies of grandparenting and grandparent-grandchild relations. However, in a study of ten European nations, Karsten Hank (2007) found that older parents tended to live closer to children who had children of their own. That this relationship held in the more familistic nations of the Mediterranean as well as the *welfare-state* nations of Scandinavia provides evidence of the universal attraction of grandchildren. Further, it has become increasingly clear that the nature of grandparent-grandchild relations cannot be reduced to a simple dichotomy based on welfare-state characteristics. For instance, Gunhild Hagestad and Janneke Oppelaar (2004) report that grandparents in Norway care for their grandchildren at nearly twice the rate of grandparents in other European nations (including Spain). Paradoxically, grandparent-provided care was most common in the nation with the most generous public child care benefits, but also in the nation with the highest rate of women's labor force participation.

Evidence is strong that grandparents play a crucial role in maintaining the economic viability of families in the developing nations of Asia. It is not uncommon for grandparents in rural China to act as surrogate parents to their grandchildren when adult children migrate to find work in urban areas. In China and other Asian nations, child care labor provided by grandparents allows adult children to seek out more promising labor markets and send back remittances that benefit the grandparents and

young children left behind (Agree, Biddlecom, Chang, & Perez, 2002).

FUTURE DIRECTIONS

Discussed below are several areas for future research that will further existing knowledge about grandchildren and their relationships with grandparents.

Life-stage Considerations The timing of grandparents' contributions to their grandchildren has been little considered. In middle childhood, children are faced with the tasks of developing a sense of competence and developing their identities. Incomplete or inadequate resolution of these challenges can produce loneliness, anxiety, and unhappiness. Adolescents are faced with establishing autonomy and forming a consistent sense of self, and may be at odds with their parents. Young adults making key life decisions can benefit from the advice and support that grandparents may be best positioned to impart. Future research should address when and how grandparents are effective in helping their grandchildren cope with these different developmental challenges. The social implications of co-surviving grandparents and grandchildren have yet to be parsed, yet it is tempting to speculate that longer periods of joint survivorship between generations will increase opportunities for intergenerational support and exchange between them.

In addition, benefits provided by grandparents may persist long after their grandchildren leave childhood. There may be *sleeper* effects of early grandparent involvement that do not come to fruition until grandchildren begin managing adult roles, such as entering romantic relationships, beginning careers, and forming families of their own. Long-term longitudinal models will be required to trace the developmental outcomes of grandchildren as a function of their earlier intergenerational experiences to periods that postdate their grandparents' survival.

Demographic Change and Kin Supply Increased longevity has radically increased the amount of time that grandchildren will know their grandparents. Peter Uhlenberg (1996) notes that 80% of children born in the mid-1990s had at least one surviving grandparent compared to only 20% at the turn of the 20th century. The simple numerical truth is that because of longer life expectancies and declining fertility rates, there are, historically speaking, more grandparents per grandchild than ever before. Further, compositional changes in families because of divorce and remarriage have increased the number of step-grandparents in the population. Has this inversion in traditional family structures increased the competition for attending to grandchildren? If so, will grandchildren benefit from the increased attention they receive from grandparents?

Sociological, Biological, and Ecological Perspectives Bio-evolutionary theories ask whether grandparents are genetically predisposed to promote the success of their grandchildren—a phenomenon known as the *grandmother effect*. The principle of paternity-certainty posits that maternal grandmothers will have the strongest genetic incentive to invest in grandchildren and paternal grandfathers the weakest. Research has consistently shown evidence supporting this hypothesis in Western nations. However, in nations that have paternalistic cultures, the opposite pattern is realized, with more investment in grandchildren by paternal grandparents. These contradictory observations call into question the essentialist view of grandparent-grandchild relations, and, more generally, call for comparative analyses of grandchild investment strategies that take into account the implications of cultural and economic context across different regions of the world.

SEE ALSO Volume 1: *Cultural Capital; Family and Household Structure, Childhood and Adolescence; Parent-Child Relationships, Childhood; Sibling Relationships, Childhood and Adolescence; Social Capital;* Volume 3: *Grandparenthood.*

BIBLIOGRAPHY

Agree, E. M., Biddlecom, A. E., Chang M-.C., & Perez A. E. (2002). Transfers from older parents to their children in Taiwan and the Philippines. *Journal of Cross-cultural Gerontology, 17*(4), 269–294.

Attias-Donfut, C., Ogg, J., & Wolff, F. C. (2005). European patterns of intergenerational financial and time transfers. *European Journal of Aging, 2*(3), 161–173.

Bengtson, V. L., & Robertson, J. F. (Eds.). (1985). *Grandparenthood.* Beverly Hills, CA: Sage.

Billing, A., Ehrle, J., & Kortenkamp, K. (2002). *Children cared for by relatives: What do we know about their well-being.* Washington, DC: Urban Institute. Retrieved May 25, 2008, from http://www.urban.org/UploadedPDF/310486.pdf

Bryson, K., & Casper, L. M. (1999). Coresident grandparents and grandchildren. Washington, DC: U.S. Census Bureau. Retrieved May 25, 2008, from www.census.gov/prod/99pubs/p23-198.pdf

Burton, L. (1995). Intergenerational patterns of providing care in African-American families with teenage childbearers: Emergent patterns in an ethnographic study. In K.W. Schaie, V. Bengtson, & L. Burton (Eds.), *Adult intergenerational relations: Effects of societal change* (pp. 79–96). New York: Springer.

Cardia, E., & Ng, S. (2003). Intergenerational time transfers and child care. *Review of Economic Dynamics, 6*(2), 431–454.

Chan, C. G., & Elder, G. H., Jr. (2000). Matrilineal advantage in grandchild-grandparent relations. *The Gerontologist, 40*(2), 179–190.

Chase-Lansdale, P. L., Brooks-Gunn, J., & Zamsky, E. S. (1994). Young African-American multigenerational families in poverty: Quality of mothering and grandmothering. *Child Development, 65*(2), 373–393.

Cherlin, A. J., & Furstenberg, F. F. (1986). *The new American grandparent: A place in the family, a life apart.* New York: Basic Books.

Clarke, L., & Cairns, H. (2001). Grandparents and the care of children: The research evidence. In B. Broad (Ed.), *Kinship care: The placement choice for children and young people* (pp. 11–20). Dorset, UK: Russell House.

Copen, C., & Silverstein, M. (2007). The transmission of religious beliefs across generations: Do grandparents matter? *Journal of Comparative Family Studies, 38*, 497–510.

Fingerman, K. L. (2004). The role of offspring and in-laws in grandparents' ties to their grandchildren. *Journal of Family Issues, 25*(8), 1026–1049.

Friedman, D., Hechter, M., & Kreager, D. (2008) A theory of the value of grandchildren. *Rationality and Society, 20*(1), 31–63.

Hagestad, G. O., & Oppelaar, J. A. (2004). *Grandparenthood and intergenerational context.* Paper presented at the annual meeting of the Gerontological Society of America. Washington, DC.

Hank, K. (2007). Proximity and contacts between older parents and their children: A European comparison. *Journal of Marriage and the Family, 69*(1), 157–173.

Hughes, M. E., Waite, L. J., LaPierre, T. A., & Luo, Y. (2007). All in the family: The impact of caring for grandchildren on grandparents' health. *The Journals of Gerontology: Psychological Sciences and Social Sciences, 62*(2), S108–S119.

Hunter, A. G. (1997). Counting on grandmothers: Black mothers' and fathers' reliance on grandmothers for parenting support. *Journal of Family Issues, 18*(3), 251–269.

Kennedy, G., & Kennedy, C. E. (1993). Grandparents: A special resource for children in stepfamilies. *Journal of Divorce and Remarriage, 19*, 45–68.

King, V., & Elder, G.H., Jr. (1995). American children view their grandparents: Linked lives across three rural generations. *Journal of Marriage and the Family, 57*(1), 165–178.

King, V., Silverstein, M., Elder, G. H., Jr., Bengtson, V. L., & Conger, R. D. (2003). Relations with grandparents: Rural Midwest versus urban southern California. *Journal of Family Issues, 24*(8), 1044–1069.

Mitchell, W. (2007). Research Review: The role of grandparents in intergenerational support for families with disabled children: A review of the literature. *Child & Family Social Work, 12*, 94–101.

Neugarten, B. L., & Weinstein, K. K. (1964). The changing American grandparent. *Journal of Marriage and the Family, 26*(2), 199–204.

Park, J. M., Hogan, D. P., & D'Ottavi, M. (2005). Grandparenting children with special needs. *Annual Review of Gerontology and Geriatrics, 24*, 120–149.

Ralitza, D., & Wolff, F.-C. (2007, December 20). Grandchild care transfers by aging immigrants in France: Intra-household allocation and labor market implications. *European Journal of Population.*

Ruiz, S. A., & Silverstein, M. (2007). Relationships with grandparents and the emotional well-being of late adolescent and young adult grandchildren. *Journal of Social Issues, 63*(4), 793–808.

Silverstein, M., & Chen, X. (1999). The impact of acculturation in Mexican-American families on the quality of adult grandchild-grandparent relationships. *Journal of Marriage and the Family, 61*(1), 188–198.

Silverstein, M., & Marenco, A. (2001). How Americans enact the grandparent role across the family life course. *Journal of Family Issues, 22*(4), 493–522.

Simmons, T., & Lawler Dye, J. (2003). Grandparents living with grandchildren: 2000. Washington, DC: U.S. Census Bureau. Retrieved May 25, 2008, from http://www.census.gov/prod/2003pubs/c2kbr-31.pdf

Solomon, J. C., & Marx, J. (1995). "To grandmother's house we go": Health and school adjustment of children raised solely by grandparents. *The Gerontologist, 35*(3), 386–394.

Szinovacz, M. E. (Ed.). (1998). Grandparent research: Past, present, and future. In *Handbook on grandparenthood*. Westport, CT: Greenwood Press.

Uhlenberg, P. (1996). Mortality decline in the 20th century and supply of kin over the life course. *The Gerontologist, 36,* 681–685.

Uhlenberg, P., & Hammill, B.G. (1998). Frequency of grandparent contact with grandchild sets: Six factors that make a difference. *The Gerontologist, 38,* 276–285.

Merril Silverstein

H

HEAD START

SEE Volume 1: *Child Care and Early Education; Policy, Education.*

HEALTH BEHAVIORS, CHILDHOOD AND ADOLESCENCE

Adolescence is a transitional phase of development that begins at the onset of puberty and continues into early adulthood (Nielsen, 1996), whereas *health behaviors* have been broadly defined as the voluntary activities an individual undertakes to promote or enhance health, to prevent or detect disease, and/or to protect from risk of disease, injury, or disability (Alonzo, 1993).

Health behaviors have often been discussed from a *health-enhancing* or *health-damaging* perspective (Spear & Kulbok, 2001). It is necessary to understand the main themes and theories (e.g., social learning theory) in research on youths' health-damaging behavior (e.g., smoking), the predictors of health-damaging behavior (e.g., substance-using peers), the illnesses (e.g., obesity, cancer), and the costs involved in order to design and implement effective prevention and intervention policies and programs and thus reduce health-damaging behaviors and the illnesses associated with them. Among others, primary health-damaging behaviors include smoking, sexual risk behavior, violence and aggression, poor diet, and lack of exercise (Spear & Kulbok, 2001). This review explores three of these behaviors: exercise, diet, and smoking. It also seeks to include factors influencing health-damaging behaviors, consequences of health-damaging behavior, and theories on youth health behaviors, as well as current trends in major health behaviors.

HEALTH-DAMAGING BEHAVIORS

The majority of adolescent and early adult morbidity and mortality can be attributed to preventable risk factors such as smoking, poor nutritional habits, and lack of exercise (Irwin, Burg, & Cart, 2002). All of these behaviors are initiated during adolescence and young adulthood and often form behavior patterns lasting into adulthood (Irwin et al., 2002; Irwin, Igra, Eyre, & Millstein, 1997). The following offers detailed explanation on how health-damaging behaviors determine adolescent health status.

Cigarette Smoking Cigarette smoking is a major health-damaging behavior in the United States. Studies by Tomeo, Field, Berkey, Colditz, and Frazier (1999) suggested that three-quarters of American youth have tried at least a few puffs of a cigarette before attaining their 18th year. A more recent study by D. B. Wilson et al. (2005) found that 23% of high school students in the United States were currently smoking. Although the average age of smoking initiation was earlier reported to be 14.5 years, children younger than 12 have reported experimenting with cigarettes (Tomeo et al., 1999).

Effects of Smoking Early onset of smoking has been shown to be predictive of sustained adolescent and adulthood smoking, with lung cancer mortality being highest among adults who began smoking before age 15 (Tomeo

211

et al., 1999; D. B. Wilson et al., 2005). Interestingly, studies also have found smoking to be associated with greater consumption of fatty foods, lower consumption of fresh foods, less exercise, and reduced intake of important vitamins, minerals, and fiber (Margetts & Jackson, 1993). For example, milk consumption has been found to be declining nationally among adolescents, often being replaced by soft drinks (Jacobson, 1998). This pattern may be even more prevalent in teens who smoke.

Females who smoke appear more likely to exhibit compromised food intake in both middle and high school when compared to males. This may be due to concerns about body weight, which may be a motivator to begin smoking as well as restricting food intake. It has also been found that smoking coupled with low intake of calcium-rich food combines risk factors that are associated with poor bone health, particularly among girls. If poor dietary behavior and/or decreased physical activity among smokers persist beyond adolescence, it may significantly lead to obesity or overweight and other risks for chronic diseases beyond those solely attributable to smoking (D. B. Wilson et al., 2005).

Diet and Exercise/Sedentary Behavior Increased attention has been focused on exactly what children are eating. Many school districts now grapple with the problems of unhealthy lunch options and nutritionally deficient items in school vending machines (Suarez-Balcazar et al., 2007). Although national objectives of promoting healthy dietary behavior and physical activity suggests an increase in consumption of fruits and vegetables and participation in vigorous physical activity (U.S. Department of Health and Human Services, 2001), many American adolescents engage in only minimal physical activity, spending much of their time in sedentary pursuits such as viewing television and playing computer games (Lowry, Wechsler, Galuska, Fulton, & Kann, 2002; D. B. Wilson et al., 2005).

A study by D. B. Wilson et al. (2005) documented that a higher prevalence of health-damaging behavior is found among African American and Hispanic American adolescents. Numerous studies suggest that there is greater TV viewing and less participation in physical activity mostly among Black and Hispanic youth compared to White youth (U.S. Department of Health and Human Services, 2001). Black and Hispanic youth are also less likely to regularly eat breakfast.

Exercise/Sedentary Behavior by Grade Level, Gender, and Ethnicity Additional study by Jorge Delva, Patrick O'Malley, and Lloyd Johnston (2006) found consistent gender differences in exercise levels, with more males than females in each racial/ethnic group reporting that they get vigorous exercise. Along racial/ethnic lines, a greater percentage of White females than Black or Hispanic American females reported frequent, vigorous exercise at each of the three grade levels (eighth, tenth, and twelfth). No racial/ethnic differences were found, however, among males across all three grades.

Regarding TV viewing, Delva et al. (2006) again reported significant racial/ethnic differences in the number of hours youths watched TV on an average weekday. Black students consistently reported the highest number of TV viewing hours compared to Whites. For example, from 2001 to 2003 the number of hours that eighth-grade males viewed TV on an average weekday was 2.5, 3.7, and 3.2 hours for Whites, African Americas, and Hispanic Americans, respectively. Among females during this same period, the respective numbers were 2.3, 3.7, and 3.1 hours; there were, however, no appreciable gender differences in hours of television watched within each racial/ethnic group. At all grades and ethnicities, White youth spent less time watching TV overall than either African American or Hispanic American adolescents.

Dietary Behavior by Grade Level, Gender, and Ethnicity Delva et al. (2006) also found dramatic disparities in breakfast eating by grade level, gender, and ethnicity, with African American youths eating the least breakfast, followed by Hispanic Americans, then Whites. Among male students in eighth grade, 58% of the Whites ate breakfast frequently, compared to 42.6% of the African Americans and 43.7% of the Hispanics. The corresponding figures for eighth-grade females who ate breakfast were 40.3% for the Whites, 28.3% for the African Americans, and 31.2% for the Hispanics. Among males in twelfth grade, only 33.8% of the Whites ate breakfast frequently versus 20.3% of the African Americans and 29.8% of the Hispanics. Of twelfth-grade females, 30.4% of the Whites, 17.8% of the African Americans, and 22.3% of the Hispanics ate breakfast frequently.

TRENDS IN CONSEQUENCES OF HEALTH-DAMAGING BEHAVIOR

Studies have suggested that health-damaging behaviors such as smoking, poor dietary behavior, and decreased activity/sedentary behaviors have adverse health consequences such as obesity, overweight, and other risks for chronic diseases (D. B. Wilson et al., 2005). In a study of children and adolescents ages 2 to 19 years, Cynthia Ogden and colleagues (2006) found that the prevalence of overweight in U.S. children and adolescents was on the rise. Reflecting the overall growth of obesity in the United States during the same period, this study found that since the mid-1960s, the proportion of children considered obese has increased from about 3.5% to approximately 16%. Whereas health-damaging behaviors

Variable	Male Nonsmokers	Smokers	Female Nonsmokers	Smokers	Total Nonsmokers	Smokers
Middle school (6th–9th grade)						
n (%)	3711 (95.3)	185 (4.7)	3909 (94.7)	217 (5.3)	7620 (95)	402 (5)
Eat fruit (%)						
≥1 serving per day	45.5	48.6	48	44.2	46.8	46.3
<7 times during past week	54.5	51.4	52	55.8	53.2	53.7
Eat vegetables (%)						
≥1 serving per day	48.1	49.7	53.7	45.6	51	47.5
<7 times during past week	51.9	50.3	46.3	54.4**	49	52.5
Consume milk dairy (%)						
≥1 serving per day	62.8	61.6	60.5	56.7	31.6	59
<7 times during past week	37.2	38.4	39.5	43.3	38.4	41
Exercise frequency (%)						
≥3 times per week	79.4	74.6	76.6	69.1	77.9	71.6
<3 times per week	20.6	25.4	23.4	30.9**	22.1	28.4**
High school (9th–12th grade)						
n (%)	967 (78.7)	262 (21.3)	1100 (79.5)	284 (20.5)	2067 (79.1)	546 (20.9)
Eat fruit (%)						
≥1 serving per day	42.8	42	45.5	41.9	44.2	41.9
<7 times during past week	57.2	58	54.5	58.1	44.1	58.1
Eat vegetables (%)						
≥1 serving per day	51.2	46.9	56.5	47.2	54	47.1
<7 times during past week	48.8	53.1	43.5	52.8**	46	52.9**
Consume milk dairy (%)						
≥1 serving per day	66.4	56.5	63.6	54.6	64.9	55.5
<7 times during past week	33.6	43.5**	36.4	45.4**	35.1	44.5***
Exercise frequency (%)						
≥3 times per week	78.3	71	70.2	56	74	63.2
<3 times per week	21.7	29**	29.8	44***	26	36.8***

** *P* < 0.01, all comparing smokers to nonsmokers.

*** *P* < 0.01, all comparing smokers to nonsmokers.

Table 1. *Food intake and exercise level by smoking status and gender.* CENGAGE LEARNING, GALE.

(e.g., poor dietary behavior, inadequate exercise, sedentary behavior) result in adolescent obesity and being overweight, these consequences subsequently lead to both contemporaneous and long-term physical and mental health outcomes and higher health care costs (Thulitha Wickrama, Wickrama, & Bryant, 2006). Studies from the early 21st century have demonstrated that obese adolescents are at high risk for mental and physical health problems such as depression, lower-body disability, diabetes, hypertension, heart disease, increased blood lipid levels, and glucose intolerance (Thulitha Wickrama et al., 2006).

GENDER AND ETHNIC DIFFERENCES IN CONSEQUENCES OF HEALTH-DAMAGING BEHAVIOR

Nationwide it is apparent that the United States is becoming a "wide nation," and the consequences of health-damaging behavior become even clearer when gender and race/ethnicity are taken into account. Research by K.A.S Wickrama, Glen Elder, and Todd Abraham (2007) found the prevalence rate of these health consequences (e.g., chronic diseases) among minorities (e.g., Latino adolescents) to be higher than that among Whites. In addition, minorities (e.g., African Americans or Hispanic Americans) rather than Whites, because of structural constraints both at the individual and the community level, are more likely to live in low socioeconomic status neighborhoods (D. B. Wilson et al., 2005). Thus, the prevalence and persistence of chronic illnesses associated with adolescent health-damaging behavior might be explained partially by the disparities in racial/ethnic minority backgrounds (Godoy-Matos et al., 2005). Such associations between adolescents' health outcomes and factors at the neighborhood level (e.g., neighborhood socioeconomic status) and individual level (e.g., being a racial/ethnic minority) have been explored elsewhere (e.g., Thulitha Wickrama et al., 2006; Wickrama & Bryant, 2003; Wilson,1987).

FACTORS INFLUENCING HEALTH-DAMAGING BEHAVIOR

The paragraphs that follow use social learning processes, prototype models of health risk behavior, life-course perspectives on development, and social structural theory to explain when, why, and how adolescent health outcomes are influenced, specifically by unhealthy dietary practices, smoking, and lack of exercise.

Individual-Level Influences The social learning perspective provides a useful framework for understanding how adolescents develop positive attitudes and expectations regarding specific health-damaging behaviors (Bandura, 1977). According to this perspective, substance-using peers model and reinforce health-damaging behaviors such as smoking among peer group members (Conger & Rueter, 1996).

Both older and newer studies (e.g., Gibbons & Gerrard, 1995; Gerrard, Gibbons, Stock, Lune, & Cleveland, 2005) propose a prototype model of health-damaging behavior. This model assumes that people maintain a prototype, or image, of the type of person who engages in a particular health-damaging behavior (e.g., poor diet, smoking), and that their attitudes and propensity toward the behavior reflect the favorability of this prototypic individual. The more favorable the image, the more willing the individual is to engage in the health-damaging behavior.

Intergenerational studies from a life-course perspective (Wickrama, Conger, Wallace, & Elder, 1999) showed the effects of parental health-damaging behavior on adolescent health-damaging behavior both in terms of overall lifestyle and specific behaviors. These researchers also suggested that this intergenerational transmission of health behaviors may have gender symmetry (i.e., fathers' health risk lifestyles affected only boys' health risk lifestyles, and mothers' health risk lifestyles affected only girls' health risk lifestyles). From a social learning perspective, it makes sense that behavioral similarities may be greatest between parents and children who spend time interacting with one another, as often occurs within same-gender pairs.

Community-Level Influences Social structural theory suggests that structural constraints among socially disadvantaged individuals (e.g., minorities) dominate their choices of health-related behaviors and accordingly result in an unhealthy lifestyle. That is, socially disadvantaged adolescents may have less autonomy to choose healthy behavior because of reduced access to health information and limited control over sleeping hours and food choices (Wickrama et al., 1999).

Community studies from the late 20th and early 21st centuries have attempted to place individual-level theoretical explanations within the community context. A study by Thulitha Wickrama et al. in 2006 integrated these multilevel theoretical perspectives and suggested several mechanisms through which community poverty influences

adolescent health-damaging behavior. The paragraphs that follow offer detailed explanations of such mechanisms.

First, the influence of community poverty on adolescent health-damaging behavior may operate through several structural constraints that limit availability of health resources in the community (Sorensen, Emmons, Hunt, & Johnston, 1998). As a result, poor communities are unable to meet their residents' dietary health needs. The structural constraints in poor communities include such factors as unaffordable prices that limit access to proper food, a greater number of unhealthy fast-food restaurants than in higher income communities, a lack of recreational activities and safe areas for physical activities, and unavailability or inaccessibility of health care services. These factors may then contribute to a higher prevalence of unhealthy dietary practices, inadequate exercise, and sedentary behaviors, which subsequently may lead to health consequences such as obesity for adolescents in disadvantaged communities (U.S. Department of Health and Human Services, 1999).

Second, community poverty may influence adolescent dietary practices through the erosion of community norms and values (Wickrama & Bryant, 2003). If community norms and values do not have adequate power to enforce healthy dietary practices, an emergence of "health-related subcultures" associated with increased community-level tolerance for risky lifestyles may adversely influence or affect the motivation for adolescents to properly manage their weight (Browning & Cagney, 2003; Kowaleski-Jones, 2000).

Third, community influence may also manifest itself through adolescent learning, emulating, and cognitive processes. That is, adolescents who live in poor communities are less likely to find positive role models who support and promote healthy activities among youth. Instead, they often may find negative role models who exert negative influences on their health behaviors (Kowaleski-Jones, 2000). This influence can be explained by both the social learning theory (Bandura, 1977) and the prototype perspective. For example, adolescents who are exposed to community members who engage in smoking or unhealthy dietary practices may model or emulate such community members (Wickrama et al., 1999). Similarly, adolescents who operate with favorable attitudes and perceptions of smoking may engage in behaviors that promote smoking (Thornton, Gibbons, & Gerrard, 2002). Thus, this willingness to engage in such behaviors results from opportunities presented at both the community and the individual level (Wickrama et al., 1999).

Finally, community poverty contributes to the erosion of social trust and social cohesion among residents, which in turn results in lower collective efficacy to acquire health-promoting and preventive services. Collective efficacy emphasizes the willingness of residents to engage in and the capacity to take collective action toward community goals, regardless of preexisting social ties (Browning & Cagney, 2003; Thulitha Wickrama et al., 2006).

HEALTH BEHAVIOR INTERVENTION

Better understanding of the correlates of physical activity and sedentary behavior in children and adolescents can support the development of effective interventions that promote an active lifestyle and prevent a sedentary lifestyle. These interventions can occur on the individual, family, and community levels.

Individual-level interventions can focus on cognitive factors (e.g., images and willingness) that can be altered to reduce adolescents' willingness to engage in health-damaging behaviors. In addition, active lifestyles involving walking, biking, camping, martial arts, and other ranges of activity (excluding TV viewing, video games, and other screen-based activities) can have some favorable effects on adolescents' health-related behaviors. These physical activities may also contribute to reductions in other health-damaging behaviors through a variety of mechanisms such as providing role models, peer networks, opportunities for teamwork, social development, problem solving, and effective outlets for energy (Gerrard et al., 2005).

Family-level interventions, such as positive parental health behavior, can also contribute to a reduction in adolescents' health-damaging behaviors (e.g., providing role models to adolescents, acting as a buffer against the deleterious effects of low socioeconomic status neighborhoods) and unhealthy peer associations, especially among minority youth (Ellickson & Morton, 1999).

Community-level interventions aimed at eliminating significant health disparities in the neighborhoods (Delva et al., 2006) may focus on distal factors such as homes where parents smoke and neighborhoods where illegal substances are available (Browning & Cagney, 2003; Wickrama et al., 1999). For positive changes in physical activity and dietary behaviors to occur, however, other strategies, including broad-based community efforts, are needed (U.S. Department of Health and Human Services, 2001). Communities must create environments with safe playgrounds and parks, walking and bicycle trails, and neighborhood recreation centers with sports facilities and supervised activities for youth (Lowry et al., 2002). Nutrition education campaigns can be sponsored by public health and community-based organizations. Schools can have a major impact through comprehensive high-quality health and physical education programs that prepare students for physical activity and healthy eating, and by sponsoring after-school programs that provide youth with safe and active alternatives to watching television (Lowry et al., 2002).

LIMITATIONS

It should be noted that there are a number of common limitations overlaying these studies. The studies by Wickrama et al. (1999) and Tomeo et al. (1999) sampled predominantly Whites and therefore cannot be extrapolated to other races/ethnicities. The studies by D. B. Wilson et al. (2005) and Tomeo et al. (1999) were cross-sectional, and as such, causation cannot be inferred from them. The studies by D. B. Wilson et al. (2005) and Thulitha Wickrama et al. (2006) had self-reporting measures of obesity, which could create bias in the sample.

CONCLUSION

Despite the aforementioned limitations, these studies collectively call for innovative strategies to promote resilience and protective factors (e.g., proper eating behaviors and regular exercising in addition to encouraging disapproving attitudes toward smoking and obesity) in order to overcome the challenges faced by adolescents and their families. Reduction of health-damaging behavior and disease prevention in adolescents calls for a comprehensive approach that engages the entire community and family as participants and positive role models for adolescents.

SEE ALSO Volume 1: *Drinking, Adolescent; Drug Use, Adolescent; Sexual Activity, Adolescent;* Volume 2: *Health Behaviors, Adulthood;* Volume 3: *Health Behaviors, Later Life.*

BIBLIOGRAPHY

Alonzo, A. A. (1993). Health behavior: Issues, contradictions, and dilemmas. *Social Science and Medicine, 37,* 1019–1034.

Bandura, A. (1977). *Social learning theory.* Englewood Cliffs, NJ: Prentice Hall.

Browning, C. R., & Cagney, K. A. (2003). Moving beyond poverty: Neighborhood structure, social processes, and health. *Journal of Health and Social Behavior, 44,* 552–571.

Conger, R. D., & Rueter, M. A. (1996). Siblings, parents, and peers: A longitudinal study of social influences in adolescent risk for alcohol use and abuse. In G. H. Brody (Ed.), *Sibling relationships: Their causes and consequences* (pp. 1–30). Norwood, NJ: Ablex.

Delva, J., O'Malley, P. M., & Johnston, L. D. (2006). Racial/ethnic and socioeconomic status differences in overweight and health-related behaviors among American students: National trends, 1986–2003. *Journal of Adolescent Health, 39,* 536–545.

Ellickson, P. L., & Morton, S. C. (1999). Identifying adolescents at risk for hard drug use: Racial/ethnic variations. *Journal of Adolescent Health, 25,* 382–395.

Gerrard, M., Gibbons, F. X., Stock, M. L., Lune, L. S., & Cleveland, M. J. (2005). Images of smokers and willingness to smoke among African American preadolescents: An application of the prototype/willingness model of adolescent health risk behavior to smoking initiation. *Journal of Pediatric Psychology, 30*, 305–318.

Gibbons, F. X., & Gerrard, M. (1995). Predicting young adults' health-risk behavior. *Journal of Personality and Social Psychology, 69*, 505–517.

Godoy-Matos, A., Carraro, L., Vieira, A., Oliveira, J., Guedes, E. P., Mattos, L., et al. (2005). Treatment of obese adolescents with sibutramine: A randomized, double-blind, controlled study. *The Journal of Clinical Endocrinology & Metabolism, 90*, 1460–1465.

Irwin, C. E., Burg, S. J., & Cart, C. U. (2002). America's adolescents: Where have we been, where are we going? *Journal of Adolescent Health, 31*, 91–121.

Irwin, C. E., Igra, V., Eyre, S., & Millstein, S. (1997). Risk-taking behavior in adolescents: The paradigm. In M. S. Jacobsen, J. M. Rees, N. H. Golden, & C. E. Irwin (Eds.), *Adolescent nutritional disorders prevention and treatment* (pp. 1–35). New York: New York Academy of Sciences.

Jacobson, M. F. (1998). *Liquid candy: How soft drinks are harming America's health.* Washington, DC: Center for Science in the Public Interest.

Kowaleski-Jones, L. (2000). Staying out of trouble: Community resources and problem behavior among high-risk adolescents. *Journal of Marriage and the Family, 62*, 449–464.

Lowry, R., Wechsler, H., Galuska, D. A., Fulton, J. E., & Kann, L. (2002). Television viewing and its associations with overweight, sedentary lifestyle, and insufficient consumption of fruits and vegetables among U.S. high school students: Differences by race, ethnicity, and gender. *The Journal of School Health, 72*, 413–421.

Margetts, B. M., & Jackson, A. A. (1993). Interactions between people's diet and their smoking habits: The dietary and nutritional survey of British adults. *British Medical Journal, 307*, 1381–1384.

Nielsen, L. (1996). *Adolescence: A contemporary view* (3rd ed.). Fort Worth, TX: Harcourt Brace College Publishers.

Ogden, C. L., Carroll, M. D., Curtin, L. R., McDowell, M. A., Tabak, C. J., & Flegal, K. M. (2006). Prevalence of overweight and obesity in the United States, 1999–2004. *Journal of the American Medical Association, 295*, 1549–1555.

Sorensen, G., Emmons, K., Hunt, M. K., & Johnston, D. (1998). Implications of the results of community intervention trials. *Annual Review of Public Health, 19*, 379–416.

Spear, H. J., & Kulbok, P. A. (2001). Adolescent health behaviors and related factors: A review. *Public Health Nursing, 18*, 82–93.

Suarez-Balcazar, Y., Redmond, L., Kouba, J., Hellwig, M., Davis, R., Martinez, L. I., et al. (2007). Introducing systems change in the schools: The case of school luncheons and vending machines. *American Journal of Community Psychology, 39*, 335–345.

Thornton, B., Gibbons, F. X., & Gerrard, M. (2002). Risk perception and prototype perception: Independent processes predicting risk behavior. *Personality and Social Psychology Bulletin, 28*, 986–999.

Thulitha Wickrama, K. A., Wickrama, K. A. S., & Bryant, C. M. (2006). Community influence on adolescent obesity: Race/

ethnic differences. *Journal of Youth and Adolescence, 35*, 641–651.

Tomeo, C. A., Field, A. E., Berkey, C. S., Colditz, G. A., & Frazier, A. L. (1999). Weight concerns, weight control behaviors, and smoking initiation. *Pediatrics, 104*, 918–924.

U.S. Department of Health and Human Services, Public Health Service, Office of the Surgeon General. (1999). Overview of cultural diversity and mental health services. In *Mental health: A report of the Surgeon General* (pp. 80–92). Retrieved April 16, 2008, from http://www.surgeongeneral.gov/library

U.S. Department of Health and Human Services, Public Health Service, Office of the Surgeon General. (2001). *The Surgeon General's call to action to prevent and decrease overweight and obesity.* Washington, DC: U.S. Government Printing Office.

Wickrama, K. A. S., & Bryant, C. M. (2003). Community context of social resources and adolescent mental health. *Journal of Marriage and the Family, 65*, 850–866.

Wickrama, K. A. S., Conger, R. D., Wallace, L. E., & Elder, G. H., Jr. (1999). The intergenerational transmission of health-risk behaviors: Adolescent lifestyles and gender moderating effects. *Journal of Health and Social Behavior, 40*, 258–272.

Wickrama, K. A. S., Elder, G. H., Jr., & Abraham, W. T. (2007). Rurality and ethnicity in adolescent physical illness: Are children of the growing rural Latino population at excess health risk? *The Journal of Rural Health, 23*, 228–237.

Wilson, D. B., Smith, B. N., Speizer, I. S., Bean, M. K., Mitchell, K. S., Uguy, L. S., et al. (2005). Differences in food intake and exercise by smoking status in adolescents. *Preventive Medicine, 40*, 872–879.

Wilson, W. J. (1987). *The truly disadvantaged: The inner city, the underclass, and public policy.* Chicago: University of Chicago Press.

Dan Nyaronga
K. A. S. Wickrama

HEALTH CARE USE, CHILDHOOD AND ADOLESCENCE

Health care is an important pathway to improving and sustaining well-being over the life course. Children and adolescents have different needs and concerns in regards to health care than older individuals and these needs differ by factors such as socioeconomic status (SES), race and ethnicity, gender, sexual orientation, and disability, as well as by the diverse contexts in which young people live. Health care use among children and adolescents is important because early detection and intervention has the potential to greatly improve health outcomes in later life. By contrast, inappropriate, insufficient, or poor quality care at this critical early stage in the life course may exacerbate inequities in later life outcomes.

Childhood and adolescence represent two important periods of the life course, generally covering children and youths from ages 0 to 19. Childhood and adolescence are also periods of tremendous growth and accelerated development where, because of the rapid rate of change, the impacts of disadvantage and inequity experienced are amplified (Stein, 1997). Health care access and use have been identified as important intervening factors between social stratification and poor health outcomes across the life course.

DEFINING AND STUDYING HEALTH CARE USE

Although *health care* is usually discussed as a unified concept, this term actually represents a varied set of interventions that are delivered in different ways, settings, and for different goals to diverse groups of people. Common categorizations divide health care into specific kinds of care, such as preventive, acute, and long-term; primary versus secondary care; or ambulatory versus inpatient care. Preventive care encompasses immunizations and routine checkups, acute care refers to treatment for severe but short-term injuries and illnesses, and long-term care covers treatments for chronic health conditions lasting at least several months. Primary care is a broad term describing the first point of entry into a health care system, which also includes secondary care (specialists and hospitals), as well as a kind of care that is comprehensive and longitudinal in nature (Donaldson et al., 1996). Ambulatory care simply refers to care delivered on an outpatient basis, in contrast to inpatient care, which requires hospital admission. Beyond more instrumental classifications of health care, however, health care is also understood in terms of its symbolic meanings. For example, good motherhood is evaluated, in part, on the basis of how healthy one's child is—and quality of health has increasingly been linked with appropriate health care use. Yet, what is appropriate varies over time, by context, and is, in part, culturally determined.

An important distinction should also be made between health care *access* and health care *use*. Good access is broadly understood to mean an individual's capacity to obtain timely, appropriate, and high-quality medical care at the hands of competent medical providers in an efficient and culturally sensitive manner (American Academy of Pediatrics, 2002). Thus, uninsured sick children using the emergency room as their primary source of care might have comparably higher rates of use but poorer access than healthy, privately insured children who visit a physician once per year. Access also entails how easily one can reach providers, the quality of providers in one's community, and the content and quality of patient-provider interactions. In order to untangle health care use from health care access, many researchers use the concept of *unmet need*, which refers to the gap between the care individuals need and the care they are able to get.

Social scientists study health care use differentials and trends in two main areas. First, they examine health care disparities as either a manifestation of other kinds of inequality, along the lines of race or class in particular, or as a pathway through which early life structural inequalities are translated into differential rates of morbidity and mortality in later life. Although health care access and use inequalities are still important for older children and adolescents, health care is also important in a second way. Social scientists have identified the provision of medical care as a potential site of social control, and for adolescents in particular, the gatekeeping role played by health care professionals over access to information about sexuality, sexually transmitted diseases (STDs), contraceptives, and pregnancy become a significant factor in timely and appropriate reproductive health care delivery.

INEQUALITY AND HEALTH CARE

Most research in the area of health care inequalities in early life use survey data to document inequalities for specific groups of young people, generally racial minorities and poor youths. Conceptual frameworks linking poor health care access to disparities in child survival and health outcomes have existed at least since the late 1970s, but the data to explicitly link health and health care have been sparse. Despite methodological weaknesses, early life disadvantage has been convincingly linked to poor health outcomes later in life, including shortened life expectancy, heightened risk of obesity, and increased risk for heart disease (Marmot & Wilkinson, 2006). Recent work uses newly available, longitudinal data to link health care inequalities explicitly with health outcomes. For example, in their study using the Fragile Families and Child Well-being Study, a national, longitudinal survey of almost 5,000 children, Erin Hamilton and colleagues (2006) link the reversal of the well-documented Mexican-American infant health advantage to health care disparities. They note that Mexican-American infants experience significantly lower rates of low birth weight and mortality relative to other infants, but lose this health advantage by age 3 as a result of their consistent disadvantage in health insurance coverage, health care access, and health care use.

In the United States, health insurance coverage is one of the most important routes to ensuring good health care access. Consequently, a significant chunk of research on the health care use of American children compares uninsured to insured children and publicly to privately insured children. Health insurance is vital because of the assistance it provides in covering the frequently expensive costs of even routine medical care. However, the presence and kind of insurance coverage is also related to the likelihood that children experience unmet need for reasons other than cost, the kinds of providers children see, and the quality of medical care children receive. For example, Medicaid, a public health insurance program for low-income individuals and families, covered 40% of all U.S. births in 2002 (Matthews, 2007), but its receipt has been associated with poorer health care access compared with privately insured children because of the systematically lower payments to providers (Smedley et al., 2003).

Problems with health care access are not limited only to uninsured and publicly insured children; the presence and type of health insurance coverage affects children differentially. Janet Currie and Duncan Thomas (1995) found that privately insured African-American children did no better in securing good access to health care than uninsured white children, suggesting that other factors, such as residential concentration in neighborhoods with relatively few or low-quality providers, hostility on the part of health care workers toward Blacks and other racial minorities, and institutional racism embedded in the organization and financing of health care also matter in patterning health care disparities by race and class (Smedley et al., 2003).

The United States is a special case: It is the only developed country that does not provide some form of universal health care. Despite the lack of universal health care, the United States has the highest per capita expenditures on health care among developed countries, as well as some of the lowest indicators of national health—most significantly, the highest rates of infant mortality in the developed world (Stein, 1997). Countries with universal health care consistently do better in providing citizens with access to primary care, which has been linked to reductions in differentials across population subgroups (Starfield, Shi, & Macinko, 2005).

Over the past decade, the United States has made strides to reduce the number of uninsured children. In 1997, Congress enacted the State Children's Health Insurance Plan (SCHIP) legislation, allocating $40 billion over the subsequent 10-year period for the expansion of health insurance among children. Although the passage of SCHIP legislation has been credited with significant declines in uninsurance among children in poor households (incomes less than the poverty line) and near-poor households (incomes between 100% and 200% of the poverty line), an estimated 2.7 million children eligible for Medicaid or SCHIP remain uninsured (Dubay, Hill, & Kenney, 2002). In 2007, efforts to reauthorize SCHIP failed, but Congress extended SCHIP

funding to maintain current coverage levels through March 2009.

ADOLESCENT REPRODUCTIVE HEALTH CARE

Although adolescents face the same health concerns as younger children, reproductive health care becomes an increasingly important component of primary care. The delivery of reproductive health care also marks the point at which health care differentiates along gender lines. Reproductive health care is frequently associated with family planning, although it also encapsulates STD prevention. Two major research areas on reproductive health care delivery for adolescents are STD transmission and teen pregnancy. While HIV and AIDS infection also looms large as a significant issue, the advent of an effective vaccine against human papillomavirus (HPV) has propelled the prevention of HPV transmission among adolescents to the forefront of public health agendas. Specific strains of HPV are associated with heightened risk of cervical cancer. Worldwide, HPV prevalence rates among women aged between 15 and 74 range from a low of 1.4% in Spain to a high of 25% in Nigeria (Clifford et al., 2005). An estimated 74% of all new HPV infections in 2000 were among American youths aged 15 to 24, resulting in 4.6 million new cases during 2000 in this age group alone (Weinstock, Berman, & Cates, 2004). Following the 2006 approval of Gardasil (a vaccine against HPV) by the Food and Drug Administration, at least 20 states considered bills to make inoculation mandatory for girls in elementary school. The quick movement to adopt mandatory inoculations has spurred a backlash against the vaccine, however, for reasons ranging from fears of increased teen sexual activity, to fears of potential medical complications stemming from vaccination.

Teen pregnancy is a long-standing concern of health care providers. Relative to other developed countries, the United States has the highest teen pregnancy rates (Darroch, Singh, & Frost, 2001), with 68.5 births per 1,000 women aged 15 to 19 in 2006 (Hamilton, Martin, & Ventura, 2007). Although the high rates of U.S. teen pregnancy have been attributed to low and inconsistent contraceptive use among U.S. teens (Committee on Adolescence, 2007), the lion's share of efforts to reduce teen pregnancy focus on transmitting the values of abstinence and waiting, despite the inefficacy of or even heightened risk associated with such programming (Santelli et al., 2006). Another barrier to adolescent contraceptive access is privacy concerns; one national study found that more than 30% of adolescents did not receive needed care because they did not want a parent to find out, yet 29 states require some form of parental consent for the

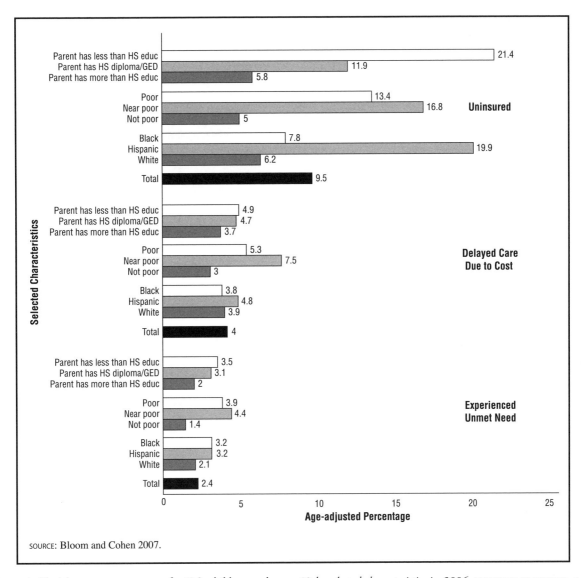

Figure 1. *Health care access measures for U.S. children under age 18 by selected characteristics in 2006.* CENGAGE LEARNING, GALE.

delivery of contraceptive services to minors (Monasterio, Hwang, & Shafer, 2007).

CONTRIBUTIONS OF RESEARCH TO HEALTH CARE POLICY AND PRACTICE

Scholarly research has proven valuable in evaluating policies and programs, and in identifying how health care inequities in early life impact important outcomes. Nationally representative surveys allow researchers to regularly generate estimates of uninsurance and unmet need among children and adolescents in order to keep tabs on the scope of the problem and measure its progress. Investigations of the impact of un- or underinsurance on

short- and long-term well-being allow one to understand how health insurance can ameliorate the increased risks of poor health among minority and low-income children, but also demonstrate that health insurance expansion alone is not a catchall solution. Studies on doctor-patient interactions and work on the efficacy of current interventions targeting teen pregnancy and STD transmission illuminate how the content of health care delivery can impact whether interventions work, and why the same interventions may work differently for diverse groups of young people. Continued research in this area is imperative to maintaining health gains among children and adolescents and new research developments, such as prospective studies of inequality in early life and long-term health, promise to enhance one's understanding of the relationships between health care inequities and important health outcomes over the life course.

SEE ALSO Volume 1: *Health Behaviors, Childhood and Adolescence; Health Differentials/Disparities, Childhood and Adolescence; Illness and Disease, Childhood and Adolescence;* Volume 2: *Health Care Use, Adulthood;* Volume 3: *Health Care Use, Later Life.*

BIBLIOGRAPHY

American Academy of Pediatrics. (2002). The medical home. *Pediatrics, 110*(1), 184–186.

Bloom, B., & Cohen, R. A. (2007). Summary health statistics for U.S. children: National health interview survey, 2006. National Center for Health Statistics. *Vital Health Stat 10*(234). Retrieved April 18, 2008, from http://www.cdc.gov/nchs/data/series/sr_10/sr10_234.pdf

Clifford, G. M., Gallus, S., Herrero, R., et al. (2005). Worldwide distribution of human papillomavirus types in cytologically normal women in the International Agency for Research on Cancer HPV prevalence surveys: A pooled analysis. *Lancet, 366*(9490), 991–998.

Committee on Adolescence. (2007). Contraception and adolescents. *Pediatrics, 120*(5), 1135–1148.

Currie, J., & Thomas, D. (1995). Medical care for children: Public insurance, private insurance, and racial differences in use. *Journal of Human Resources, 30*(1), 135–162.

Darroch, J. E., Singh, S., & Frost, J. J. (2001). Differences in teenage pregnancy rates among five developed countries: The roles of sexual activity and contraceptive use. *Family Planning Perspectives, 33*(6), 224–250, 281.

Donaldson, M. S., Yordy, K. D., Lohr, K. N., & Vanselow, N. A. (Eds.). (1996). *Primary care: America's health in a new era.* Washington, DC: National Academy Press.

Dubay, L., Hill, I., & Kenney, G. (2002). Five things everyone should know about SCHIP. Washington, DC: The Urban Institute. Retrieved April 18, 2008, from http://www.urban.org/url.cfm?ID=310570

Hamilton, B. E., Martin, J. A., & Ventura, S. J. (2007). Births: Preliminary data for 2006. *National Vital Statistics Reports, 56*(7). Retrieved April 18, 2008, from http://www.cdc.gov/nchs/data/nvsr/nvsr56/nvsr56_07.pdf

Hamilton, E. R., Hummer, R. A., You, X. H., & Padilla, Y. C. (2006). Health insurance and health care utilization of U.S.-born Mexican-American children. *Social Science Quarterly, 87*(1), 1280–1294.

Marmot, M., & Wilkinson, R. G. (Eds.). (2006). *Social determinants of health.* Oxford, UK: Oxford University Press.

Matthews, L. (2007). *Maternal and child health (MCH) update 2005: States make modest expansions to health care coverage.* Washington, DC: National Governor's Association Center for Best Practices. Retrieved April 19, 2008, from http://www.nga.org/Files/pdf/0609MCHUPDATE.PDF

Monasterio, E., Hwang, L. Y., & Shafer, M. (2007). Adolescent sexual health. *Current Problems in Pediatric and Adolescent Health Care, 37*(8), 302–325.

Santelli, J., Ott, M.A., Lyon, M., Rogers, J., Summers, D., & Schleifer, R. (2006). Abstinence and abstinence-only education: A review of U.S. policies and programs. *Journal of Adolescent Health, 38*(1), 72–81.

Smedley, B. D., Stith, A. Y., & Nelson, A. R. (Eds.). (2003). *Unequal treatment: Confronting racial and ethnic disparities in health care.* Washington, DC: National Academy Press.

Starfield, B., Shi, L., & Macinko, J. (2005). Contribution of primary care to health systems and health. *The Milbank Quarterly, 83*(3), 457–502.

Stein, R. E. K. (Ed.). (1997). *Health care for children: What's right, what's wrong, what's next.* New York: United Hospital Fund of New York.

Weinstock, H., Berman, S., & Cates, W., Jr. (2004). Sexually transmitted diseases among American youth: Incidence and prevalence estimates, 2000. *Perspectives on Sexual and Reproductive Health, 36*(1), 6–10.

Julia A. Rivera Drew

HEALTH DIFFERENTIALS/ DISPARITIES, CHILDHOOD AND ADOLESCENCE

Health disparities refers to gaps in the quality of health and health care across racial, ethnic, and socioeconomic groups. The term itself has become very important in medical sociology as well as in public health and medicine more broadly over the past several decades in the United States. Outside of the United States, the term used the most is *health inequalities*. Within the United States, this area of research has received more attention because a reduction in health disparities is listed as one of the major goals of the *Healthy People 2010* report, a major goal setting and reporting effort within the U.S. government. As specified by the goals of that project, the United States strives to eliminate health disparities among segments of the population, including differences that occur by gender, race or ethnicity, education or income, disability, geographic location, or sexual orientation. Collecting and reporting about data on health care disparities is complex, and this entry will focus on U.S. efforts and data.

EARLY SOCIOLOGICAL BACKGROUND

Sociological research on inequalities and health has been a long-standing concern in medical sociology, and this concern within sociology is much older than specific U.S. government policy efforts in the area. A large number of studies conclude that people of lower socioeconomic status (SES) have worse health and lower life expectancies than those from a higher socioeconomic position. While much of this research in sociology has been conducted among adults, there has also been research on children growing up in poverty demonstrating that growing up in such families, as compared to growing up in more advantaged families, leads to poorer health status as a child. More recent research has also begun to focus on differentials in health during childhood as a factor in understanding adult disparities in health, arguing that health disparities linked to socioeconomic position in childhood may have very important, long-lasting impacts. More recent and sophisticated studies still find that people with incomes below the poverty level in the United States have a higher (some studies find two to three times higher) chance of dying early, even when other basic factors such as age, race, and gender are controlled.

U.S. GOVERNMENT EFFORTS RELATED TO HEALTH DISPARITIES

One of the first attempts by the U.S. government in addressing health disparities was a 1985 report by the Secretary of Health and Human Services (HHS), Margaret Heckler (b. 1931). This landmark report described large, persistent gaps in health status among Americans of different racial and ethnic groups. Because of this report, the Office of Minority Health (OMH) within the HHS was created with a mission to address these disparities. The Centers for Disease Control (CDC), another U.S. federal government agency that focuses on public health and epidemiology, also established its own Office of the Associate Director for Minority Health (ADMH) in 1988 in response to the same report. This unit became the CDC's OMH in 2002, with the mission to promote health and quality of life by preventing and controlling the disproportionate burden of disease, injury, and disability among racial and ethnic minority populations. This unit was expanded in 2005 to create the new Office of Minority Health and Health Disparities (OMHD) in the CDC. The mission of the renamed agency was broadened, with a focus on reducing health disparities experienced by populations defined by race and ethnicity, SES, geography, gender, age, disability status, and risk status related to sex and gender. There have been special programs developed with CDC to focus on adolescent health. One push has been the development of culturally appropriate school programs that address risk behaviors among youth, especially when coordinated with community efforts. The CDC's Division of Adolescent and School Health (DASH) has a mission of preventing the most serious health risks among youths, and tries to incorporate efforts to address health disparities among at-risk communities.

During the same time period, one of the pushes for more research on health care inequalities came from the passage of the Healthcare Research and Quality Act of 1999. This law directed the Agency for Healthcare Research and Quality (AHRQ) to develop two annual reports, one focused on quality and one on disparities. The AHRQ was directed to track prevailing disparities in health care delivery as they relate to racial and socioeconomic factors among priority populations. Priority populations include low-income groups, racial and ethnic minorities, women, children, the elderly, individuals with special health care needs, the disabled, people in need of long-term care, people requiring end-of-life care, and those living in rural communities.

The first National Healthcare Disparities report in 2003 built on some previous efforts in the federal government, especially *Healthy People 2010* and the Institute of Medicine's (IOM) 2002 report, *Unequal Treatment: Confronting Racial and Economic Disparities in Health Care* (Smedley, Stith, & Nelson, 2003). *Unequal Treatment* extensively documents health care disparities in the United States, with a focus on those related to race and ethnicity. One of the weaknesses of this report is that there is not any focus on disparities related to SES. The *National Healthcare Disparities Report* (2003) does have a focus on the ability of Americans to access health care and variation in the quality of care. This report will be discussed in further detail later in the entry.

DESCRIPTION OF CHILD AND ADOLESCENT HEALTH DISPARITIES AND TRENDS

Most experts agree that for children, as for adults, health disparities can be found in both health and access to health care. The disparities are linked to racial and ethnic impacts and SES. The importance of these differences is one of the reasons why *Healthy People 2010* has listed the elimination of health disparities as one of its two overarching goals and has spent much effort on documenting differences by a variety of factors, with great attention to racial and ethnic differences. Other research also documents the importance of SES for children. Both low SES and minority status have been associated with poorer health in childhood. Children of lower SES are less likely to have contact with physicians at early ages and have poorer health behaviors as they become adolescents, such as higher rates of smoking. In addition, children of lower SES have higher rates of injury and more sedentary behaviors that lead to obesity and poor overall health. They also have more chronic health problems.

Some research now argues that poor health in childhood leads to more health problems in later life and may impact the entire trajectory of a person's life. Similar findings hold true for many aspects of health and access to health care by minority status. Most of the data in this area has dealt with problems of Black children, although some newer datasets have information on other minority groups, such as Hispanics, as well. In the following sections, most of the data are reported from *America's Children: Key National Indicators of Well-Being, 2007*, a major federal data source on the health of children and a source that incorporates data from the *Healthy People 2010* project.

One of the most important issues regarding health disparities occurs at the very beginning of life, with variation in birth weight. Black, non-Hispanic infants have a much higher percentage of low birth weight as compared to other racial and ethnic groups. From 1990 to 2003, the percentage of low birth weight in that group ranged from 13.1 to 13.6%. The percentage rose in 2005 to 14%. An increase in the percentage of babies with low birth weight was true in other groups also from 1990 to 2005, going from 5.6 to 7.3% among White, non-Hispanic infants and from 6.1 to 6.9% among Hispanic infants. Within the Hispanic category, variation also occurs. Women of Mexican origin had the lowest percentage of infants with low birth weight (6.4%) and Puerto Ricans had the highest (9.8%). One explanation given for better birth outcomes among women of Mexican origin is better nutrition practices in the immigrant generation and stronger family support in that same generation.

Related to this variation in birth weight is another indicator that occurs at the beginning of life and is often thought of as one of the most important ways to compare health status among population groups, states, and countries: the infant mortality rate, which is the number of deaths before an infant's first birthday per 1,000 live births. Black, non-Hispanic and American Indian and Alaska Native infants have consistently had a higher infant mortality rate than that of other racial or ethnic groups. In 2004, the Black, non-Hispanic infant mortality rate was 13.6 infant deaths per 1,000 live births and the American Indian and Alaska Native rate was 8.4, both higher than the rates among White, non-Hispanic (5.7), Hispanic (5.5), and Asian and Pacific Islander (4.7) infants. As with infants who have low birth weight, there is important variation within the Hispanic category. In 2004, the infant mortality rate ranged from 4.6 deaths per 1,000 live births for infants of Cuban origin to a high of 7.8 for Puerto Rican infants.

In the national databases on health problems and health status, only a few other major health indicators have available data for children—these include behavioral and emotional difficulties, activity limitations, obesity, and asthma. Most of the data about behavioral and emo-

Race and Hispanic origin	Percentage
White, non-Hispanic	7.9
Black, non-Hispanic	13.1
Asian	6.5
Hispanic	8.6
Mexican	7.4
Puerto Rican	19.9

SOURCE: Centers for Disease Control and Prevention, National Center for Health Statistics, National Health Interview Survey.

Table 1. *Percentage of children ages 0–17 with current asthma by race and Hispanic origin, 2005.* CENGAGE LEARNING, GALE.

tional difficulties come from parents' answers to survey questions. Generally these types of difficulties are first noticed by parents or by teachers in school. Children with emotional or behavioral difficulties may have problems managing their emotions, focusing on tasks, or controlling their behavior. These difficulties often persist throughout a child's development and can lead to lifelong disability. In 2005, children 4 years of age and older from low-income families were slightly more likely to have these problems, with 7% of children living below the poverty level having serious emotional or behavioral difficulties, compared with 5% of children in near-poor families (family incomes between 100 and 199% of the poverty level) and 4% of children in nonpoor families (family incomes of 200% or more of the poverty level). For this health issue, boys have a higher rate of problems as compared to girls. There has been very little variation in this since 2003.

Activity limitation is a broad measure of health and functioning, generally reported for children from 5 through 17 years of age. It is affected by a variety of chronic health conditions, and thus is a good overall measure of health limitations for children. Activity limitation refers to a person's inability, due to a chronic physical, mental, emotional, or behavioral condition, to participate fully in age-appropriate activities, such as attending school. Using the same categories again of poor, near-poor, and nonpoor families as a measure of SES, 11% of children in poor families, 9% of children in near-poor families, and only 7% of children in nonpoor families had activity limitations. Hispanic children were less likely than White, non-Hispanic and Black, non-Hispanic children to have a parental report of activity limitation. Looking at trends, percentages have increased a small amount, going from about 6% in 1999 to 8% in 2005. As with emotional and behavioral difficulties, percentages for boys are slightly higher than for girls.

Obesity and asthma are also both very important areas of health disparities. Asthma is a disease with important variation by race and ethnicity and obesity is a health problem that has been increasing in American society overall. It is difficult to compare asthma trends over time because the method of measurement in national data changed in 1997. Rates doubled from 1980 to 1995, but were stable from 1997 to 2005. In 2005, about 13% of Black, non-Hispanic children were reported to have asthma, compared with 8% of White, non-Hispanic children and 9% of Hispanic children. Within the Hispanic population, Puerto Rican children have the highest percentage with asthma (20%), as compared with 7% of children of Mexican origin. Being overweight changed little among U.S. children from 1960 to 1980, with steady increases since then. While only 6% of children between 6 and 17 were overweight in 1980, the figure increased to 18% during 2003 and 2004. In 2003 and 2004, Black, non-Hispanic females aged 6 through 17 were at a much higher risk of being overweight (25%), compared with White, non-Hispanic and Mexican-American females (16% and 17 %, respectively). For males between 12 and 17, there were virtually no differences existing between racial and ethnic groups.

In addition to disparities in health status, disparities exist in access to health care services. Two important factors that impact access to care are whether children are covered by health insurance and whether children have a regular source to use when they need health care services. Tracking whether children have health insurance coverage is often encountered with difficulty because the federal government changed some aspects of their data collection in 2004, making the 2004, 2005, and later data not comparable to earlier data. In 2005, 89% of children had health insurance coverage at some point during the year, a figure very similar to the 2004 figure (90%). All health insurance coverage is not the same, and much health policy work contrasts coverage by private health insurance and coverage by government programs.

Looking first at the earlier data to understand trends, the proportion of children covered by private health insurance decreased from 74% in 1987 to 66% in 1994. The figures then increased to 70% in 1999, and then dropped to 66% in 2003. Looking at the figures for public health insurance coverage, this type of coverage grew from 19% in 1987 to 27% in 1993. Public health insurance decreased until 1999, the year the State Children's Health Insurance Program (SCHIP) began. By 2003, due to the impact of this program's focus on providing health care insurance to children of the working poor (in contrast to Medicaid, which provides coverage to many but not all poor children), public health insurance coverage climbed to 29% in 2003. Racial and ethnic differences do occur in health insurance coverage rates. Percentages from 2005 show that Hispanic children are less likely to have health insurance (only 79%),

as contrasted with White, non-Hispanic children (93%) and Black children (88%).

In addition to having health insurance, having a usual source of care makes it more likely that a child will be able to use pediatric services in a timely and appropriate manner. A usual source of care is defined as a particular person or place a child goes to for sick and preventive care. Emergency rooms are not considered a usual source of care in most data sources, including the ones used here, because their focus on emergency care generally excludes the other elements of health care. Overall, only 5% of children in the United States had no usual source of care in 2005. Insurance status is important in whether children have a regular source of care, as is income. Uninsured children are almost 16 times as likely as those with private insurance to have no usual source of care. There are variations by type of health insurance coverage. If children have public coverage such as Medicaid, 4% had no usual source of care versus only 2% of children with private health insurance. Nine percent of children in poor families do not have a usual source of care, as contrasted with only 3% of children in nonpoor families. Children of the near poor report figures of not having a usual source of care that are closer to poor children, with a figure of 8%.

Two other important measures of use of health care services as indicators of access to health care are whether children are immunized and whether children have seen a dentist recently. For immunization, there are data available on whether children have received the doses of five vaccinations that have been recommended for them since 1991 or earlier, by the appropriate ages of 19 to 35 months. These include such well-known vaccinations for polio; diphtheria, pertussis, tetanus (DPT); measles; and chicken pox. In 2005, 81% of children had received the recommended five-vaccine series. Disparities in coverage rates are found both by income and race and ethnicity. Children living in families below the poverty level report coverage rates of 77%, whereas children living above the poverty level have a coverage rate of 83%. White, non-Hispanic children have the highest rates of coverage (82%) as compared with Hispanic children (79%) and Black, non-Hispanic children (79%).

For dental care, some comparative data are available both on whether a child between 2 and 17 years old has seen a dentist in the past year and whether the child has any untreated cavities. Looking first at use of dental care, the percentage of children having seen a dentist in the past year has been fairly stable from 1997 to 2005, with about three-quarters of all children between 2 and 17 having visited a dentist in the past year. For children in poor families, only 66% had seen a dentist in the past year, as compared to 69% of children in near-poor

Ages 2–17	Percentage
Total	76.2
Poverty status[a]	
Below 100% poverty	66.2
100-199% poverty	68.6
200% poverty or above	82.0
Type of insurance[b]	
Private insurance[c]	82.1
Public insurance[c,d]	71.4
No insurance	49.5

[a]Family income was imputed for data years 1997 and beyond. Missing family income data were imputed for 22–31 percent of children ages 5–17 in 1997–2005.

[b]Children with health insurance may or may not have dental coverage.

[c]Children with both public and private insurance coverage are placed in the private insurance category.

[d]As defined here, public health insurance for children consists mostly of Medicaid or other public assistance programs, including State plans. Beginning in 1999, the public health insurance category also includes the State Children's Health Insurance Program (SCHIP). It does not include children with only Medicare, Tricare, or CHAMP-VA.

SOURCE: Centers for Disease Control and Prevention, National Center for Health Statistics, National Health Interview Survey.

Table 2. *Percentage of children ages 2–17 with a dental visit in the past year, 2005.* CENGAGE LEARNING, GALE.

families and 82% in nonpoor families. Coverage for dental insurance is different, in many cases, than is coverage for health insurance, but there is more information available about health insurance coverage. Very few families who have no health insurance will have dental insurance. If children do not have health insurance, only 50% have visited a dentist in the past year, whereas 82% of children with private health insurance have visited their dentist. For children with Medicaid, which generally does include dental care coverage, 71% of children have visited the dentist in the past year. Untreated dental caries (cavities) are the most common health care problem in children. In 2003 and 2004, 25% of all children had untreated cavities. For children living in poor and near-poor families, 29% of children had untreated cavities, compared with 18% of children from nonpoor families. The percentage of untreated cavities is higher for Mexican-American children than for White, non-Hispanic and Black, non-Hispanic children.

This finding of poor dental health among children of Mexican origin has been found in some other recent work that uses the National Health Interview Survey data. Research in the scholarly literature raises some questions about looking in more detail at racial and ethnic groups. A number of studies that have tried to look at disparities within minority racial and ethnic groups and to look more closely at the Hispanic population have concluded that Hispanic children of Mexican origin have poorer access to health care than other Hispanic groups, and that this is true for such factors as physician visits, emergency room visits, and delay in accessing care. Often data are limited for looking at smaller minority groups (such as Asians, American Indians, and even Hispanics if the research tries to break the data down into the specific subgroups within the Hispanic population). Data limitations are often particularly true for information about children.

COMPLEXITY OF RACE AND SES DISCUSSIONS

One limitation in many of the federal government data sources is that they look at disparities in health and access to health care for race and ethnicity, SES, or by age of the child, but these reports do not take into account the complexity of the interrelationship between race and SES. Some more scholarly articles have used some of the federal data sources, such as the National Health Interview Survey, to try to determine whether childhood health disparities are best understood as effects of race, of SES, or of the synergistic effects of the two. Because SES and race and ethnicity are closely intertwined, researchers need to understand how these factors interact. There are several possibilities. The effect of low SES could be particularly pronounced among minority groups, especially if there is discrimination, which has sometimes been termed the *double jeopardy* hypothesis. A different argument is that low SES could be particularly negative among the native born because immigrants generally have better health than the people left behind or even some of those in the country to which they move, especially if compared to those of lower SES. (This is known as the *healthy immigrant effect.*) Whites and Blacks are less likely to be immigrants but this could impact Hispanics and Asians.

In a 2006 article in the *American Journal of Public Health*, Edith Chen, Andrew Martine, and Karen Matthews analyzed race and SES interactions, looking at SES as measured by parental education. In this study, the traditional relation of fewer years of parental education with poorer health was true for White and Black children for overall health, activity, school limitations, and (for Blacks only) chronic circulatory conditions. The relationship with education was often not found for Hispanic and Asian children, and in some situations (respiratory illnesses) was reversed. One suggestion of the authors is that the different gradients across racial groups may be linked to social and cultural values of that group. Perhaps both Asian and Hispanic families share health beliefs and practice certain health behaviors across education groups, thus diminishing the effect of low education in those groups. A different possible explanation is that Asian and Hispanic children are more likely to be from immigrant families and that there is a healthier immigrant impact, making children in these families less susceptible to the negative impact of lower parental education.

MAIN THEMES AND THEORIES AND FUTURE TOPICS

More recent government reports raise many concerns about inequalities in all population groups, including children and adolescents. The *National Healthcare Disparities Report* (2003) explores the relationship between race and ethnicity and socioeconomic position. There are seven key findings from the report: First, inequality in quality of care continues to exist and often these are true for particularly serious health care problems; second, disparities come at a personal and societal price; third, differential access to health care often leads to disparities in quality of care actually received; fourth, opportunities to provide preventive care are often missed; and finally, the last three points all relate to the need for more data, more research, and the linkage of those to policy within the United States.

The knowledge about why disparities continue to exist is still limited, and data limitations may limit improvement efforts. A different government report has raised some concerns about data limitations. Measurement issues are of importance for data about health disparities. This is true both in general and in government disparity reports, which this article has referred to a great deal. A recent government report by the National Center for Health Statistics, the part of the federal government charged with collection of health-related data, has argued that there are six important decisions that impact disparities data, including selecting a reference point from which to measure disparity, whether to measure disparity in absolute or in relative terms, and deciding whether to consider any inherent ordering of the groups. These types of choices can impact the size and direction of disparities reported, and can therefore influence conclusions. These are some of the reasons that research on health disparities is complex.

Some recent research suggests that, for children, there may be important differences between the experiences in childhood versus adolescence. SES gradients may not be static across the life span. Though SES gradients were found for global health measures at all ages, there were variations by age on SES gradients for specific acute health conditions. Future research will need to examine pathways emerging in adolescence and that may be linked to development and peer relationships, which

may be important in shaping health gradients related to SES. The importance at all ages of SES on global health measures reinforce the importance of understanding the role that social and environmental factors, not linked to developmental processes in childhood, also may play.

There are some new approaches that consider an inequality paradox linked to population level approaches to this topic and some considerations about how to incorporate disparities frameworks into newer work from a health services perspective. These new approaches and critiques provide a summary of what types of approaches that disparities research on children and adolescents may cover in the future. Discussions about how to incorporate disparities frameworks into an overall health services perspective have focused on three phases. The first is detection. This phase emphasizes the definition of health disparities, the identification of vulnerable populations, and an important methods concern, the development of valid measures. The second phase focuses on understanding why disparities exist, including the incorporation of research to identify what factors explain gaps in health and health care disparities between more and less vulnerable groups. The last phase argues for the importance of developing evaluations to reduce health disparities and then implementing and evaluating the success of those approaches.

The inequality paradox, as described by Katherine Frohlich and Louise Potvin (2008), also discusses the idea of vulnerable groups. They argue that disparities in health may be exacerbated by population approach interventions that focus on improving the health of the overall population. The objective of improving population health may not necessarily reduce health disparities and could leave vulnerable populations behind. Vulnerable populations are different than a population at risk. A population at risk generally has a higher exposure to a risk factor. Usually all the individuals in the population will have that higher risk exposure. In contrast, a vulnerable population is a subpopulation or subgroup who is at higher risk, most often because of shared social characteristics. This can be linked to the important idea of fundamental causes in medical sociology. Vulnerable populations concentrate numerous risk factors throughout their life course because of shared fundamental causes that are linked to position in the social structure. Vulnerable populations may be least able to respond to population level interventions, which then improve the health of many, but leave the health of the vulnerable population as it was, thus increasing health disparities.

The Robert Wood Johnson Foundation (RWJF), a major national health foundation, has just issued a new report on how education, race, and ethnicity impact the health of Americans. This report has concluded that the relationship between how Americans live their lives and the surrounding economic, social, and physical environment may be more important than access to health care in the determination of future health. RWJF has formed a commission to examine these factors and identify innovative ways to improve health for both children and adults, an important future direction in research and policy on health disparities. The commission will pursue strategies for reducing illness, preventing early death, and extending life. In doing this, attention should also be paid to the inequality paradox and the complexity of interrelationships between factors, including the complexity of the interrelationship between child and adolescent health and health in late life.

SEE ALSO Volume 1: *Attention Deficit/Hyperactivity Disorder (ADHD); Autism; Birth Weight; Health Care Use, Childhood and Adolescence; Illness and Disease, Childhood and Adolescence; Infant and Child Mortality; Mental Health, Childhood and Adolescence; Obesity, Childhood and Adolescence;* Volume 2: *Health Differentials and Disparities, Adulthood;* Volume 3: *Health Differentials and Disparities, Later Life.*

BIBLIOGRAPHY

Centers for Disease Control. *Health topics: Six critical health behaviors.* Retrieved April 4, 2008, from http://www.cdc.gov/Healthyyouth/healthtopics/index.htm

Chen, E., Martine, A.D., & Matthews, K.A. (2006). Understanding health disparities: The role of race and socioeconomic status in children's health. *American Journal of Public Health, 96*(4), 702–708.

Frohlich, K.L., & Potvin, L. (2008). Transcending the known in public health practice: The inequality paradox: The population approach and vulnerable populations. *American Journal of Public Health, 98*(2), 216–220.

Interagency Forum on Child and Family Statistics. *America's children: Key national indicators of well-being, 2007.* Retrieved April 4, 2008, from http://childstats.gov/americaschildren/index.asp

Keppel K., Pamuk E., Lynch J., et al. (2005). Methodological issues in measuring health disparities. National Center for Health Statistics. *Vital Health Statistics 2*(141). Retrieved April 4, 2008, from http://www.cdc.gov/nchs/data/series/sr_02/sr02_141.pdf

Robert Wood Johnson Foundation. *Overcoming obstacles to health: Stories, facts, and findings.* Retrieved April 4, 2008, from http://www.rwjf.org/files/research/obstaclestohealth highlight.pdf

Smedley, B.D., Stith, A., & Nelson, A.R. (Eds.). (2003). *Unequal treatment: Confronting racial and ethnic disparities in health care.* Washington, DC: Institute of Medicine, National Academies Press.

U.S. Department of Health and Human Services. (2000). *Healthy people 2010: Understanding and improving health.* (2nd ed.). Washington, DC: Author. Retrieved April 7, 2008, from http://www.healthypeople.gov/Document

U.S. Department of Health and Human Services. (2003, July). *National healthcare disparities report.* Washington, DC: Author. Retrieved April 4, 2008, from http://www.ahrq.gov/qual/nhdr03/nhdr2003.pdf

Jennie Jacobs Kronenfeld

HIGH SCHOOL DROPOUT

The No Child Left Behind (NCLB) legislation of 2001 has triggered a renewed interest in school dropout. The NCLB legislation is particularly salient because federal law now requires that high schools be held accountable for graduation rates. On a personal level, dropouts are more likely to lack the skills necessary for successful employment and further education, be unemployed, use food stamps and welfare programs, lack health insurance, suffer health problems, and be involved in crime (Belfield & Levin, 2007). Dropouts present numerous challenges at a societal level, including a serious socioeconomic problem, given that most are undereducated and ill equipped to meet the rapidly advancing technological needs of society's workforce. As a result of the increase risk of deleterious individual outcomes, the costs to the society also include (a) the vast loss of taxable income because of unemployment (or underemployment, holding jobs that do not require high school graduation), (b) increased participation of dropouts in social welfare programs, and (c) higher prison funding, as a disproportionate number of individuals incarcerated in U.S. prisons are high school dropouts. Costs due to unemployment, welfare status, and the incarceration of dropouts have been estimated at $240 billion per year (Dryfoos, 1990). Considering inflation and the advancing technological demands of contemporary society, current financial costs would be significantly more.

DEFINING SCHOOL DROPOUT

Precisely measuring the number of students who do not graduate from high school has been very controversial. Typically, a school dropout is defined as an individual who quits school before graduation and has not enrolled in or completed an educational equivalency program. Although this seems like a simple definition, it is not currently possible to know exactly how many students drop out of school because most states do not follow individual students over time.

The most common method used to determine the educational outcomes of students in the United States is the *status dropout rate*, which is the percentage of an age group in the civilian, noninstitutionalized population who were not enrolled in a high school program and had not received a high school diploma or obtained an equivalency certificate (i.e., general education degree [GED]). In the United States in 2002, 10% of 16- to 24-year-olds were out of school without a high school credential. Dropout rates of young people ages 16 to 24 have gradually declined between 1972 and 2004, from 15% to a low of 10% in 2004 (Jimerson, Reschly, & Hess, 2008). The status dropout rate remained relatively stable during the 1990s, and there continue to be millions of youths who do not complete high school. For instance, in 2000 it was estimated that about 3.8 million young adults were not enrolled in, nor had completed, a high school program.

Status dropout rates provide a national indicator of school completion rates. However, these methods also count individuals who earn GEDs and other certificates as *completers*, thus important inequities may be obscured. For instance, the rate of GED completion has increased dramatically for Hispanic and African American youth, creating the appearance of a narrowing gap in school dropout rates. However, wages, hours of work, unemployment experiences, and job tenure are similar for dropouts and GED recipients. Moreover, individuals who are incarcerated are not counted; thus, the corresponding decrease in the dropout rate for African Americans may be partially accounted for by the increasing incarceration rate for this group since the 1980s.

Another measure that may be used to estimate school completion rates is the Cumulative Promotion Index (CPI), which approximates the probability that a student entering the ninth grade will complete high school on time with a regular diploma. Consistent with NCLB, students who receive nonstandard diplomas (e.g., certificates of attendance) or a GED are not considered graduates. An advantage of this approach is that it requires only two years of data collection, and because this approach adheres to the definition of the high school graduation rate provided in NCLB, it can be used for accountability purposes. Essentially, this technique estimates the rate of yearly school-leaving for each grade, and the potential pool of graduates is reduced by this percentage for each subsequent year, providing the district with a probability of graduation or estimated graduation rate. Applying this method to U.S. data for 2001, the graduation rate was 68%. As a result of the definition and methodology used, this figure is three times lower than indicated by the status count. Specifically, the criteria of the CPI examine the graduation rate among all ninth-grade students, focusing on receiving the regular diploma, whereas the status count examines the population of 16- to 24-year-olds who were out of school without a high school diploma or an equivalency certificate.

The lack of agreement regarding how to estimate school completion rates has resulted in challenges in comparing rates across states or districts, determining whether schools are meeting accountability standards, and examining the relative effectiveness of prevention and intervention programs.

DEMOGRAPHIC CHARACTERISTICS OF THE YOUTH WHO DROP OUT

Despite the diverse methods used to measure school dropout, it is clear that in the United States some groups of students are at higher risk for dropping out. Students that are particularly overrepresented among dropouts come from backgrounds of low socioeconomic status (SES) and are often Hispanic, African American, Native American, or have disabilities. U.S. data reveals that dropout rates among the Hispanic youth remain much higher than other ethnic groups, although the rate has declined in more recent years from 30% in 1998 to 24% in 2004 (Child Trends Databank, 2006). Higher dropout rates among recent Hispanic immigrants partially account for the elevated rates. For example, 44% of Hispanics between16 and 24 years old who were born outside the United States were not enrolled in school and had not earned a certificate of high school completion, a percentage more than double the rates for first- or second-generation Hispanic youths born in this country and approximately six times the rate for non-Hispanic immigrant populations (7.4%).

Factors that appear to be most directly related to school dropout include demographic (e.g., male, lower SES, and minority status), academic (e.g., high absenteeism, history of retention, and low-achieving), social (e.g., poor peer relationships), familial (e.g., low family involvement or family stress), and individual domains (e.g., behavior problems and substance abuse). Previous efforts to target students in high-risk groups have not been successful. Instead, prevention and intervention efforts have been found to be most successful at meeting the needs of all students when those efforts are focused on creating school and community systems that help all students to graduate. The change in focus from reducing the incidence of school dropout to increasing the rate of school completion reflects an important shift in contemporary thinking, which encourages practitioners to address this important issue through a wide range of services that include targeted interventions to broad systemic reforms.

HISTORICAL AND CONTEMPORARY THEORIES INFORMING SCHOOL DROPOUT SCHOLARSHIP

Early research exploring high school dropout mostly focused on student and family variables that appeared to be antecedents associated with dropping out. Demographic, individual, academic achievement, behavior problems, peer relations, and family factors are those most often considered the major risk factors for dropping out. For instance, male students who are ethnic and racial minorities, from disadvantaged family backgrounds, are low-achieving, display problem behaviors, and have poorer peer relations are more likely to drop out relative to other students. Longitudinal studies designed to understand the phenomenon have mostly examined high school and elementary school predictors of dropping out. Predictors of dropout have been categorized as proximal (e.g., attendance and homework completion) and distal (e.g., SES) variables and, more recently, according to amenability to intervention.

Increasingly, contemporary scholarship has focused on questions of why or how students drop out. Thus, recent studies attempt to articulate multidimensional models to explain the process influencing students' decisions to drop out instead of descriptions of correlates. These models also suggest developmental pathways to dropout that involve family expectations and involvement, early school difficulties, poor peer relations, lack of school engagement, drug use, and cumulative family stress. It is important to recognize that only some of the risk factors for dropping out of high school are characteristics of the students, whereas others are characteristics of the schools these students attend or their family's origin. Dropping out of school is at least partially a product of school practices that are ineffective in promoting the success of all students and of community pressures that fall disproportionately on underprivileged families. Early childhood experiences have been related to school adjustment and dropout or completion years later (Jimerson, Egeland, Sroufe, & Carlson, 2000). The confluence of empirical evidence illustrates that school dropout is best conceptualized as a process that occurs over many years.

Understanding why students drop out of school is important for developing effective prevention and intervention approaches. A comprehensive review of the extant literature reveals that school dropout is typically influenced by social, behavioral, and academic problems in school. Thus, several conceptual models of processes and pathways leading to early school withdrawal warrant consideration. Given the diversity among dropouts, each of these models is helpful in considering the multiple influences and pathways that may result in school withdrawal.

An early model of school dropout developed by Ruth Ekstrom, Margaret Goertz, Judith Pollack, and Donald Rock (1986) described a multifaceted pathways model of a student's decision to drop out or stay in school. This model included demographic factors (e.g.,

SES and ethnicity), the family educational support system, the student's school performance, and the previous behaviors warranting discipline as influences on the student's decision to drop out or stay in school. This model also emphasized that problem behaviors and grades were partially determined by the home educational support system. Although this path model was used specifically to examine high school data, these factors are also important in elementary school.

Jeremy Finn's (1989) participation-identification model emphasized that students' active participation in school and in classroom activities and feeling of identification with the school affected school completion. "Identification with school" referred to an internalized conception of belonging and valuing school success. From this perspective, lack of school engagement was central to the process of dropping out. Engagement is composed of student behavior (involvement with classroom and school activities) and identification with school. The participation-identification model explained dropout in terms of a behavioral antecedent (lack of participation) and a psychological condition (lack of identification). It portrayed dropping out as a process of disengagement over time rather than as a phenomenon that occurs in a single day or even a single school year. Participating in the school environment includes attending school, being prepared to work, and responding to the teacher's directions and questions. Other levels of participation include students' initiative to be involved in the classroom and school, participation in social and extracurricular activities, and involvement in decision making. This model has a developmental emphasis in that it reflects how participation in the school environment changes as students progress through school with greater opportunities to become involved in the nonacademic aspects of the school environment.

Ian Evans and Adria DiBenedetto (1990) provided four possible pathways that focus on the interaction of the individual and school factors that lead to early school withdrawal: (a) unexpected events, (b) long-term underlying problems, (c) early skill deficits, and (d) entry problems. Consistent with Finn (1989), these authors suggested that dropouts can be better identified by examining behaviors rather than searching for predetermined characteristics of students. Moreover, they proposed that dropping out may be characterized by a snowballing effect, wherein events that occurred early impact subsequent events.

The first pathway emphasized unexpected events such as a pregnancy or the death of a loved one occurring that subsequently influences school enrollment. Such unforeseen events may be more likely to appear in certain contexts, and thus these events are not completely unexpected. Moreover, adaptation and coping following these unexpected events will also be influenced by the context and support available to the student.

The second pathway focuses on long-term underlying problems. Students on this pathway may not display any psychological difficulties, but over time the student engages in deviant behaviors, perhaps begins to associate with maladjusted peers, or possibly begins using drugs, which ultimately influences school enrollment. Clearly this pathway takes time as the student follows a deviant pathway that may ultimately lead to school dropout.

In the third pathway, a student may possess cognitive or social deficits that interact over time and influence school enrollment. For example, a child who experiences early reading difficulties may subsequently lack the motivation to continue to struggle with reading. The student may also be shy or may be actively neglected by peers because of his or her academic performance, and over time both dimensions may interact and result in early withdrawal.

The fourth and final pathway recognizes that some children begin school with emotional or behavioral problems. For instance, if a student is immature and overactive, this is likely to lead to problems with classroom behavior and possible struggles with teachers, and over time the student may dislike school and ultimately choose to withdraw.

As illustrated in the above models, numerous pathways may potentially lead to school dropout, and thus it is particularly important to consider the multiple influences that may facilitate school completion. Each of the models emphasizes the combined impact of diverse influences across the individual's development (including social, behavioral, and academic considerations). However, many of these models fail to recognize the role that school environmental factors play in school dropout. For example, large school size is positively correlated with decreased attendance, lower grade point averages and standardized test scores, higher dropout rates, and higher crime than smaller schools serving similar children. School practices, such as tracking and grade retention, have a negative correlation with school completion rates independent of the students' ability level. Other school-related factors such as high concentrations of low-achieving students and less qualified teachers are also associated with higher dropout rates.

FUTURE SCHOLARSHIP PROMOTING SCHOOL COMPLETION

Considering the multiple pathways that may lead to school dropout, an array of potential intervention strategies have been developed to facilitate the academic success of students who may be at risk of dropping out. It is important that scholars and professionals seek further understanding and knowledge of the individual strengths and needs of youths and be prepared to provide appropriate prevention and interventions to promote their success. There are currently numerous studies across the United States that aim to promote school completion and reduce school dropout. A multifaceted California Dropout Research Project (2008) aims to synthesize existing research and undertake new research to inform policy makers and the larger public about the nature of—and potential solutions to—the dropout problem. In addition, the Institute of Education Sciences (2008) has posted a report on the effectiveness of programs to prevent school dropout. Such contemporary science aims to enhance understanding and identify effective strategies to promote school success and school completion among students in the United States.

SEE ALSO Volume 1: *Cognitive Ability; College Enrollment; Employment, Youth; High School Organization; High-Stakes Testing; Policy, Education; Racial Inequality in Education; School Tracking; Socioeconomic Inequality in Education; Vocational Training and Education.*

BIBLIOGRAPHY

Belfield, C. R., & Levin, H. M. (2007, August). *The economic losses from high school dropouts in California* (Policy Brief No. 1). Retrieved May 30, 2008, from http://www.lmri.ucsb.edu/dropouts

Cairns, R., & Cairns, B. (1994). *Lifelines and risks: Pathways of youth in our time.* New York: Cambridge University Press.

California Dropout Research Project. (2008). Retrieved April 4, 2008, from http://www.lmri.ucsb.edu/dropouts

Child Trends Databank. (2006). *High school dropout rates.* Retrieved April 4, 2008, from http://www.childtrendsdatabank.org/indicators

Dryfoos, J. (1990). *Adolescents at risk: Prevalence and prevention.* New York: Oxford University Press.

Ekstrom, R., Goertz, M., Pollack, J., & Rock, D. (1986). Who drops out of high school and why? Findings from a national longitudinal study. *Teachers College Record, 87*(3), 356–373.

Evans, I., & DiBenedetto, A. (1990). Pathways to school dropout: A conceptual model for early prevention. *Special Services in School, 6*, 63–80.

Finn, J. (1989). Withdrawing from school. *Review of Educational Research, 59*(2), 117–142.

Institute of Education Sciences. *What Works Clearinghouse.* Retrieved April 4, 2008, from http://ics.ed.gov/ncee

Jimerson, S. R., Egeland, B., Sroufe, L. A., & Carlson, E. (2000). A prospective longitudinal study of high school dropouts: Examining multiple predictors across development. *Journal of School Psychology, 38*(6), 525–549.

Jimerson, S. R., Reschly, A., & Hess, R. (2008). Best practices in increasing the likelihood of school completion. In A. Thomas & J. Grimes (Eds.), *Best practices in school psychology* (5th ed.). Bethesda, MD: National Association of School Psychologists.

Lehr, C. A., Clapper, A. T., & Thurlow, M. L. (2005). *Graduation for all: A practical guide to decreasing school dropout.* Thousand Oaks, CA: Corwin Press.

Orfield, G. (Ed.). (2004). *Dropouts in America: Confronting the graduation rate crisis.* Cambridge, MA: Harvard Education Press.

Rumberger, R. (1995). Dropping out of middle school: A multilevel analysis of students and schools. *American Educational Research Journal, 32*(3), 583–625.

Shane Jimerson

HIGH SCHOOL ORGANIZATION

One critical element of examining the life course of adolescents centers on examining those experiences held in common—the defining features or events that characterize being a teen in the United States. Whereas many of these common experiences, such as church attendance, volunteer organizations, or scouting, have fragmented over the period from 1960 to 2000, the role played by the modern American high school has become more common and defining (Dorn, 1996). In many ways, going to high school is one of the few experiences that a wide variety of American teens have in common, and, although there are certainly variations in that experience, the typical structure and organization is surprisingly similar across the United States.

BASIC STRUCTURE OF THE AMERICAN HIGH SCHOOL

According to U.S. Census data, by the year 2007 there were 15.1 million students enrolled in public high schools and 1.46 million students in private high schools, the highest point projected during a steady increase since 1970. This division identifies the two primary classifications of high school: They are either (a) publicly funded through tax dollars, a designation that includes comprehensive high schools, schools of choice and magnet schools, as well as Department of Defense and Bureau of Indian Affairs high schools; or (b) privately funded, through religious organizations, private institutions, or private funding supplemented with individual tuition. The two most common configurations of high school are those including grades 9 through 12 and those including only grades 10 through 12. Using the most recent national database (the Educational Longitudinal Study of 2002) as a reference, the average size of American high schools as of 2002 was about 1,400 students, with 22% of the high schools smaller than 800 students and 18% larger than 2,000 (Planty & DeVoe, 2005).

Typically, the local decision to expand or restrict grade levels in a high school has more to do with the building resources relative to student population than any overarching difference in educational approach or student success.

Curriculum offerings in the modern high school are measured by content and amount. Content designations break broadly into "academic" subjects—mathematics, sciences, English language arts, social sciences, and foreign languages—and "nonacademic" subjects—physical education, music and arts education, technology education, career and vocational education, and life skills. Across these different subject areas, one may find any number of different individual courses, typically ranging from 100 to 300 individual classes in the overall course offerings for one high school. The amount of instruction can be measured by credit, by semester, or by hour, but the standard form used to report how much curriculum a student receives (especially in college applications) is based around a Carnegie unit, which currently corresponds to 120 hours or 7,200 minutes of instruction. The determination of Carnegie units originated in 1906 by the Carnegie Foundation, with 14 units deemed to be the minimum number required for college preparation (Boyer, 1983). Since its institution, a far more complex formula has become standard, based on determination of the content over an 18-week semester (regardless of actual length of the term) as lecture, lab with homework, or lab without homework.

Finally, the modern high school curriculum continues to be differentiated by difficulty level, ranging from "basic" to "advanced" in most academic courses. In some areas (particularly in social sciences and English language arts), the division by difficulty uses the same content for each level but modifies the work demands by complexity. For example, three different levels of an American history course might all cover the Revolutionary War, but at the lowest level students are answering fact-based questions from the text, whereas at the highest level students are writing essays based on original sources. In other areas (particularly in mathematics and sciences), the division by difficulty dictates different content, as the material taught builds sequentially by knowledge complexity. In these situations, lower-level students may never be introduced to higher-level concepts.

This division has recently come under criticism, as the focus of high school is shifting more toward preparing all students to attend college. Although the specific criticism relates to differentiating students across topics at one difficulty level, the problems are inherent even when only one content area offers content separated by difficulty level. In particular, private high schools tend to emphasize a more constrained curriculum, with fewer choices and a general orientation toward college-preparatory material. In general, the content of a high school curriculum is quite complex, with an ongoing emphasis toward individual selection even in the face of pressures for a more constrained curriculum.

With noted exceptions, classes in high school tend to average around an hour in length, allowing for five to seven classes in a day, five days per week. Within this paradigm, an hour typically includes time to travel between classes, transition time, and orientation, leaving between 20 and 40 minutes of instructional time (Planty et al., 2007). Some content areas (such as the performing arts) may cover more than one period, but this arrangement is commonly found only in private high schools. The more common public school exception to this structure schedules around blocks of time, trading class length for class frequency. There are no fewer than five common block structures, varying the length of the block (from 80 minutes to 4 hours), the frequency of classes in a week, and the configuration across the school year. This alternative to traditional scheduling began with criticism in 1994 from the National Education Commission on Time and Learning (Education Commission of the States, 1994). However, subsequent research on the success (or even the successful implementation) of blocks of time in improving learning at the high school level is at best mixed (Zepeda & Mayers, 2006). Typically, high school course offerings in most settings operate in five top seven 1-hour blocks of time, providing about 200 minutes per week of instruction in any given subject area (Education Commission of the States, 1994).

HIGH SCHOOL ORGANIZATIONAL STRUCTURE

Although schools vary extensively across the spectrum of organizational structure, high schools typically most closely adhere to an organizational style described as bureaucratic or mechanistic (Rowan, 1990). The defining features characterizing the work environment of a high school consist most typically of the following components: (a) top-down authority and decision-making concerning school-wide decisions, centering on the principal (in public schools) or director (in private schools), combined with (b) departmental structures governing classroom curriculum and (to some extent) instruction, with (c) little if any daily contact between teachers or administrators across classroom boundaries, originally termed a "loosely coupled" organizational structure (Weick, 1976) and now more generally understood as "organic" in nature (Rowan, 1990). These components are perhaps best understood in terms of their associated political and social pressures, navigating internal and external requirements.

Of any grade-span configuration, high schools (regardless of size) typically have the most complex administrative structure of the different grade-level configurations. For example, it is not uncommon for a high school to have, in addition to a principal (public) or director (private), a set of vice or assistant principals/administrators, each with their own secretarial organization and separate responsibilities. The purpose of this complex administrative structure is to negotiate the outside pressures and concerns that regularly affect high schools. The administrators in a high school must coordinate several different external connections between the high school and the community (including all extracurricular activities, especially sports events), the high school and the state requirements, and finally the high school and advanced opportunities for students (including the military, different college recruiting programs, and vocational programs that place students with outside agencies). In general, the primary role of a high school administrative team is not to provide instructional leadership; rather, it is to negotiate the needs of the high school teachers and students within the larger community and state context.

Given these external requirements placed on high school administrators, much of the school's instructional leadership develops within the specific content area(s), coordinated by a department leader or chair. The formality of this departmental structure varies by the size of the institution, but typically it is within the small group organization that curriculum and (to some extent) instructional decisions are made (Daniels, Bizar, & Zemelman, 2001). Individual teachers within a department commonly bring back new ideas, content, instruction, or assessment innovations from conferences or external training experiences to shape innovation in the material taught in that content area. The focus of attention might be guided by accreditation requirements, external testing requirements, or even the goals held by postsecondary institutions. Whatever the focus, however, instructional leadership at the high school level typically takes place within specific content areas.

However, even within these more tightly knit organizational units, the actual decision making concerning daily instruction typically takes place within the classroom itself. Teachers at the high school level most commonly operate largely unobserved and unmonitored within their own walls (Boyer, 1983; Rowan, 1990). Unless curriculum units undertake a common assessment of classes, the material covered by any individual instructor may be unrelated to that of any other instructor, even within the same department. Even when tests are taken together, the process of delivering instruction may differ dramatically from classroom to classroom, hinging largely on the teacher's own views of student learning (Menlo & Poppleton, 1999; Wilen, Hutchison, & Bosse, 2008).

RESEARCH AND STUDY OF AMERICAN HIGH SCHOOLS

Ongoing efforts to study and ultimately reform American high schools typically orient toward one of two basic goals: to increase the opportunity for students to succeed across social strata or, alternately, to promote individual excellence in educational paths that will not end at 12th grade (Hammack, 2004). Primarily because of compulsory education laws, but also because of external limitations placed on the job market, high school is generally the first exit point from one's educational path. High school dropouts are a serious concern not only of the school system, held accountable for each failure to completely educate and graduate a student, but also for the society carrying the burden of a potentially undereducated citizen. Employers are increasingly requiring higher education credentials for positions, even where higher education is not necessary for the function of the job (Bracey, 1996). Since the 1960s, the nation's economy has slowly and steadily shifted away from the manufacturing sector toward a service sector. The loss of manufacturing jobs, accelerating as a result of technological advances and global competition, means that well-paying jobs that require limited formal education are quickly disappearing and are being replaced by careers in the technological and professional field, arguably requiring higher formal education (Hammack, 2004).

In addition to the struggle between individual and social opportunity, high schools in the early 21st century also must deal with the increased accountability mechanisms that have been a keystone reform effort of the start of the 21st century. In 2001, the No Child Left Behind Act (NCLB) was initiated and had an immense impact on educational reform at all levels. Its mainstay was a reliance on standardized testing at the local, state, and national levels in order to continuously track student progress (Planty et al, 2007). The impact of this increase in school accountability has had serious ramifications at the high school level. Current research investigates whether this level of external federal policy pressure has had an impact on raising test scores, reducing the dropout rate, or increasing 4-year on-time graduation rates. In addition, research continues to explore the possible impact such efforts have had on students who are not able to earn credits at the scheduled pace, students who have learning disabilities, or students who experience family, health, psychological, monetary, or legal difficulties. These challenges confronted by high schools, converging in the face of budget cuts and underfunded federal mandates, will require specialized programs and sufficient funding in order to adequately meet the graduation requirements set out by NCLB.

SEE ALSO Volume 1: *Academic Achievement; High School Dropout; High-Stakes Testing; Policy, Education; School Culture; School Tracking; School Transitions; School Violence; Stages of Schooling.*

BIBLIOGRAPHY

Boyer, E. L. (1983). *High school: A report on secondary education in America.* New York: Harper & Row.

Bracey, G. W. (1996). International comparisons and the condition of American education. *Educational Researcher, 25*(1), 5–11.

Daniels, H., Bizar, M., & Zemelman, S. (2001). *Rethinking high school: Best practice in teaching, learning, and leadership.* Portsmouth, NH: Heinemann.

Dorn, S. (1996). *Creating the dropout: An institutional and social history of school failure.* Westport, CT: Praeger.

Education Commission of the States. (1994). *Prisoners of time: Report of the National Education Commission on Time and Learning.* Education Commission of the States Education Reform Reprint Series. Denver, CO.

Hammack, F. M. (Ed.). (2004). *The comprehensive high school today.* New York: Teachers College Press.

Menlo, A., & Poppleton, P. (1999). *The meanings of teaching: An international study of secondary teachers' work lives.* Westport, CT: Bergin & Garvey.

Orfield, G. (2004). *Dropouts in America: Confronting the graduation rate crisis.* Cambridge, MA: Harvard Education Press.

Planty, M., & Devoe, J. (2005). *An examination of the conditions of school facilities attended by 10th-grade students in 2002.* NCES Publication 2006–302. Washington, DC: U.S. Department of Education.

Planty, M., Provasnik, S., Hussar, W., Snyder, T., Kena, G., Hampden-Thompson, G., et al. (2007). *The condition of education 2007.* NCES Publication 2007–064. Washington, DC: U.S. Department of Education.

Rowan, B. (1990). Commitment and control: Alternative strategies for the organizational design of schools. *Review of Research in Education, 16,* 353–389.

Weick, K. E. (1976). Educational organizations as loosely coupled systems. *Administrative Science Quarterly, 21*(1), 1–19.

Wilen, W., Hutchison, J., & Bosse, M. I. (2008). *Dynamics of effective secondary teaching* (6th ed.). Boston: Pearson/Allyn & Bacon.

Zepeda, S. J., & Mayers, R. S. (2006). An analysis of research on block scheduling. *Review of Educational Research, 76*(1), 137–170.

Julia B. Smith

HIGH-STAKES TESTING

Notwithstanding a tawdry history that tracks testing back to the eugenics movement (Lemann, 1999), "the call for relevance, for clarifying the relationship between what is taught in high schools and later life" (Grubb & Oakes, 2007) is at the root of contemporary standardized testing in the United States. In essence, the motivation for standardized testing for the country is anchored in the "excellence movement" driven primarily by an economic labor-market frame. The history of massive standardized student testing in the United States began as a response to international competition in the fields of science and technology with progressive countries such as Japan, South Korea, and Germany. While some states such as New York have been implementing standardized tests since the 1800s, the national agenda in the United States for standardized tests began in earnest after the launching of an unmanned satellite mission called Sputnik by the Soviet Union in 1957. Testing was again reinforced after the publication of the landmark *A Nation at Risk* (1983), which claimed that the poor quality of the nation's public school system explained why the United States had not advanced as far as other countries in the areas of industry, commerce, science, and technology.

In order to enhance the academic rigor of American schools and thereby improve student performance, several unique but loosely interconnected policies such as the Elementary and Secondary Education Act of 1965 and its successive reauthorizations in 1994 as the Improving America's School Act and in 2001 as the No Child Left Behind Act, together with the birth in 1969 of the National Assessment of Educational Progress, collectively gave rise to the high-stakes testing policies of the early 21st century (Lemann, 1999). Although the 1950s scare of the United States being left behind in the fields of science and technology drove these policies, achievement gaps between the United States and other industrialized countries on international exams such as the Programme for International Student Assessment persist. That is, in comparison to students from other Organisation for Economic Co-operation and Development countries, American 15-year-olds consistently score below average (Baldi, Jin, Skemer, Green, & Herget, 2007). In any case, international performance and the move toward a more global market keeps the issue of high-stakes standardized testing in the forefront as a viable, if controversial, educational policy direction (Heubert & Hauser, 1999) as researchers (Darling-Hammond, 2006; Valenzuela, 2002, 2004) argue that multiple criteria measures offer a more holistic and valid measure of student ability and performance.

DESCRIPTION OF THE HIGH STAKES ASSOCIATED WITH STANDARDIZED TESTING

High-stakes standardized tests are exams administered to students with "stakes" or "penalties" that are levied on the student, teacher, school, school district, or state as a result of student performance. These examination systems are designed to simultaneously achieve four goals: induce teachers to set high standards, motivate students to learn what is being taught, recognize and reward them when they do, and assist in the sorting of students across different post-secondary programs and employment options (Bishop, 1998). Proponents of high-stakes exams purport that the rewards and penalties that are attached to student performance on tests strengthens the accountability mechanisms that are in place and help to ensure that teachers are on task and actively engaged in student learning and that students take schooling seriously (Nichols & Berliner, 2007). In contrast, opponents suggest that these stakes exacerbate existing gaps in opportunity and equity across varying demographic groups (Heubert & Hauser, 1999; Valenzuela, 2004). Using a single test for high-stakes decisions is also against the guidelines of assessment use put forth in the ethical and professional standards maintained by the American Educational Research Association, the American Psychological Association, and the National Council on Measurement in Education (AERA, APA, & NCME, 1999). These opposing views help sustain the continuing debate on the purpose and appropriate use of high-stakes tests.

The exams used for high-stakes testing vary from state to state, and in some cases from school district to school district. The most recent iterations of high-stake testing policies point to those mandated by individual state legislatures that require students at a particular promotion gate (or gates) to pass a standardized exam and, possibly, as a condition of graduation as well. Students in states where such policies exist must pass an exam (or a series of exams) in order to advance either to a particular grade (or grades) or pass an exam in order to receive a high school diploma. These policies place the "stake" at the student level and students are often held back a grade or denied a high school diploma as a result of not meeting the testing requirements. High-stakes testing policies and results are often highly contested by parents and educators.

There are high-stakes policies, particularly at the local and federal levels, such as those instituted by the No Child Left Behind Act, that attach "stakes" to test scores at the school, district, and state levels. Schools are penalized through corrective-action measures when students either in the aggregate or as part of one of several disaggregated groups fail to achieve performance benchmarks from year to year. The "stakes" at the school level range from allowing students to transfer to another public school, which can have the effect of decreased enrollment and, as a consequence, decreased per-pupil funding, to more drastic measures such as removing all of the school faculty and administrators and closing down the school.

Some school districts are implementing their own versions of high-stakes policies. For example, in 2006 the Houston Independent School District's Board of Education unanimously approved a merit pay program that gives teachers additional monetary bonuses if their students perform well on standardized tests. Merit pay bonuses are contested by national teachers unions such as the National Education Association and the American Federation of Teachers, which cite evidence that test results should not be sole measures for teacher performance, and by scholars such as Darling-Hammond (2000), who suggests that merit pay discourages teachers from teaching students who are at the highest risk of failing.

WHAT DO THE EXAMS LOOK LIKE?

About half of the states in the United States have passed legislation that requires all high school students to take and pass an exam in order to receive a high school diploma. These take one of three forms: minimum competency exams (MCEs), standards-based exams (SBEs), and end-of-course exams (EOCs). Minimum competency exams generally measure a student's attainment of basic educational knowledge and skills (Heubert & Hauser, 1999). As early as 1971, two states mandated the use of MCEs, and this number gradually increased to 15 by 1976 (Jaeger, 1982). Many states are moving away from MCEs to more rigorous exams—what Bishop (1998) refers to as curriculum-based exit exams.

In 2006 the Center on Education Policy reported that 10 of the 18 states that were implementing exit exams in 2002 were using MCEs, whereas in 2006 the same study reported that only 3 out of 22 states were using MCEs. Unlike most MCEs, SBEs are generally aligned with state academic standards. The number of states using SBEs as a graduation requirement increased from 7 in 2002 to 15 in 2006. Like SBEs, EOC exams are generally aligned with course content, but students take these after completing their course work. States are also showing an upward trend toward the use of EOCs, primarily because these exams tend to have a more direct link to what students are learning in the classroom. In 2002 only two states, New York and Texas, used EOCs as part of the high school graduation requirement. The number doubled by 2006 and is expected to be as many as nine states by 2012.

DEMOGRAPHIC DISPARITIES AND THE LINK TO GRADE RETENTION AND DROPOUT

General demographic performance trends on standardized tests indicate that the racial and ethnic gaps in achievement still exist. The growing disparities that lie between White and Asian students and their Hispanic and Black counterparts, those who are fluent in English and those who are not, and students with disabilities and those without are extreme (Heubert, 2001). Valenzuela (2002) found that because of their language-dependent nature, the tests are neither valid nor reliable for English language learners.

One such example of the negative impact of high-stakes exams on race, ethnicity, and language minorities can be found in California, which has a large percentage of students who are Hispanic and English language learners. Results from the 2006–2007 California High School Exit Examination (CAHSEE) show that the disparities persist. A random coefficient model analysis conducted by the Human Resources Research Organization (HumRRO) indicates that Black students on average scored 22 points lower than White students in math and 18 points lower in English language arts. In this same analysis, Hispanic students scored 11 points lower than White students in math and 10 points lower than White students in English language arts (HumRRO, 2007). This same report showed that the pass rate for English learners dropped from 31% to 27%.

States have taken varying approaches to how they deal with the disparities in student pass rates. In Georgia, the cut scores are adjusted to allow for lower failure rates, and this adjustment thus accounts for some of the declining gaps between demographic groups (Fordham Foundation, 2005; Scafidi & Robinson, 2006). Other states adhere to similar practices. The state of Washington has pushed back the dates on which the graduation penalties for the new tests are implemented in order to give students an opportunity to learn the material before the requirement to withhold their diploma is implemented (Senate Democratic Caucus, 2007). California followed a similar phased-in approach for the CAHSEE. These approaches do not necessarily predict student success, and some argue that the pressure itself of having high-stakes testing policies predicts failure (Clarke, Haney, & Madaus, 2000; Nichols & Berliner, 2007).

There has been an ongoing debate about whether states that put in place more stringent graduation requirements based on performance on high-stakes tests obtain higher rates of retention and dropping out. In their analyses of national data, researchers Nichols and Berliner (2007) demonstrate that these high-stakes assessments have the effect of increasing dropout rates and contributing to higher rates of retention in grade. These effects are notably pronounced and pose long-term negative consequences for poor and minority students. Valencia and Villarreal (2005) show that dropout levels are likely to increase the more times a student is retained. Their research also finds that although all groups of children start school at about the same age, by ages 15

Testing Gap. *Tenth grade students take a chemistry test while in class at Springfield High School in Springfield, Ill. An Associated Press analysis of new state data found an average 28% gap statewide between the percentage of elementary pupils meeting or exceeding standards on tests and high school students doing the same.* AP IMAGES.

to 17, approximately 45% of African-American and Hispanic youth are below the expected grade level for their age.

RESEARCH AND DIRECTION OF HIGH-STAKES EXAMS

In addition to recent studies that show a link between high-stakes exams and retention and dropout, there is a growing amount of interest and literature around opportunity to learn and value-added approaches to measurement. These discourses are anchored in an equity critique of how students learn and are educationally prepared, as opposed to how well they are tested. Scholars argue that these equity concerns are more fundamental to long-term success than the technical debates on how to alter tests and accountability systems to get improved results (Scheurich, Skrla, & Johnson, 2000). Future research is likely to take into account individual student growth over time and not static measurements at policy time points.

SEE ALSO Volume 1: *Academic Achievement; College Enrollment; Gender and Education; High School Organization; Policy, Education; Racial Inequality in Education; Socioeconomic Inequality in Education.*

BIBLIOGRAPHY

American Educational Research Association, American Psychological Association, & National Council on Measurement in Education (AERA, APA, & NCME). (1999). *Standards for educational and psychological testing.* (3rd ed.). Washington, DC: American Educational Research Association.

Baldi, S., Jin, Y., Skemer, M., Green, P. J., & Herget, D. (2007). *Highlights from PISA 2006: Performance of U.S. 15-year-old students in science and mathematics literacy in an international context* (NCES 2008-016). Washington, DC: U.S. Department of Education, National Center for Education Statistics. Retrieved November 2, 2007, from http://nces.ed.gov/pubsearch/pubsinfo.asp?pubid=2008016

Bishop, J. H. (1998). The effect of curriculum-based external exit exam systems on student achievement. *The Journal of Economic Education, 29,* 171–182.

Bishop, J. H., & Mane, F. (2001). The impacts of minimum competency exam graduation requirements on high school graduation, college attendance, and early labor market success. *Labour Economics, 8,* 203–222.

Clarke, M., Haney, W., & Madaus, G. (2000). *High stakes testing and high school completion* (The National Board on Educational Testing and Public Policy statements, vol. 1, no. 3). Chestnut Hill, MA: Boston College, Lynch School of Education.

Darling-Hammond, L. (2000). Teacher quality and student achievement: A review of state policy evidence. *Education Policy Analysis Archives, 8*(1). Retrieved June 22, 2007, from http://epaa.asu.edu/epaa/v8n1

Darling-Hammond, L. (2006). Assessing teacher education: The usefulness of multiple measures for assessing program outcomes. *Journal of Teacher Education, 57,* 120–138.

Fordham Foundation. (2005, October 19). *Gains on state reading tests evaporate on 2005 NAEP* (press release). Retrieved March 24, 2008, from http://www.edexcellence.net/foundation/about/press_release.cfm?id=19

Grubb, W. N., & Oakes, J. (2007, October). *"Restoring value" to the high school diploma: The rhetoric and practice of higher standards.* Retrieved November 7, 2007, from the Arizona State University Web site: http://epsl.asu.edu/epru/documents/EPSL-0710-242-EPRU.pdf

Heubert, J. P. (2001). High-stakes testing: Opportunities and risks for students of color, English-language learners, and students with disabilities. In M. Pines (Ed.), *The 21st Century challenge: Moving the youth agenda forward.* Baltimore, MD: Sar Levitan Center for Social Policy Studies, John Hopkins University Press.

Heubert, J. P., & Hauser, R. M. (Eds.). (1999). *High stakes: Testing for tracking, promotion, and graduation* (National Research Council report). Washington, DC: National Academy Press.

Human Resources Research Organization (HumRRO) (2007). *Independent evaluation of the California High School Exit Examination (CAHSEE): 2007 evaluation report* (D. E. Becker & C. Watters, Eds.). Sacramento: California Department of Education. Retrieved February 14, 2008, from http://www.cde.ca.gov/ta/tg/hs/documents/evalrpt07.pdf

Jaeger, R. M. (1982). The final hurdle: Minimum competency achievement testing. In G. R. Austin & H. Garber (Eds.), *The rise and fall of national test scores* (pp. 223–246). New York: Academic Press.

Lemann, N. (1999). *The big test: The secret history of the American meritocracy.* New York: Farrar, Straus and Giroux.

National Commission on Excellence in Education. (1983). *A nation at risk: The imperative for educational reform.* Retrieved November 7, 2007, from http://www.ed.gov/pubs/NatAtRisk/index.html

Nichols, S. L., & Berliner, D. C. (2007). *Collateral damage: How high-stakes testing corrupts America's schools.* Cambridge, MA: Harvard Education Press.

Scafidi, B., & Robinson, H. (2006, June 2). *A new day for Georgia education.* Retrieved June 22, 2007, from the Georgia Public Policy Foundation Web site: http://www.gppf.org

Scheurich, J. J., Skrla, L., & Johnson, J. F. (2000). Thinking carefully about equity and accountability. *Phi Delta Kappan, 82,* 293–299.

Senate Democratic Caucus of Washington Legislature. (2007, April 22). *Legislators reach compromise on WASL changes* (press release). Retrieved July 1, 2007, from http://senatedemocrats.wa.gov/2007/releases/McAuliffe/waslfinal.htm

Valencia, R. R., & Villarreal, B. J. (2005). Texas' second wave of high-stakes testing: Anti-social promotion legislation, grade retention, and adverse impact on minorities. In A. Valenzuela (Ed.), *Leaving children behind: How "Texas-style" accountability fails Latino youth* (pp. 113–152). Albany: State University of New York Press.

Valenzuela, A. (2002). High-stakes testing and U.S.–Mexican youth in Texas: The case for multiple compensatory criteria in assessment. *Harvard Journal of Hispanic Policy, 14,* 97–116.

Valenzuela, A. (Ed.). (2004). *Leaving children behind: How "Texas-style" accountability fails Latino youth.* Albany: State University of New York Press.

Madlene Hamilton
Angela Valenzuela

HOME SCHOOLING

Home educating children has become increasingly popular in the United States. The home schooling movement originated in the 1960s within the countercultural or libertarian political left. By the mid-1980s, however, the religious right was leading the movement. Throughout the 1990s and into the 21st century, the number of home schooled children in the United States grew tremendously. The movement has come to be considered mainstream as home schooling is advocated for a variety of reasons and has become much more publicly acceptable. This topic is important to life course scholars given that education is a major social institution that transmits societal norms, values, and a knowledge base. As an alternative to public and private schools, home education provides children with a different socialization experience. This specific educational alternative should be examined for its impact on society.

STUDYING HOME SCHOOLERS

The social scientific study of home schooling in the United States began to take off in the late 1980s as at least 250,000 children were being educated at home. Over the next 20 years numerous studies of this population were published. Nonetheless, although the quantity of evidence is notable, the quality of the scholarship is less than ideal. As many scholars have noted, home schoolers are a difficult population to study. They are geographically dispersed, and there are no reliable lists from which a representative, random sample can be drawn. Some states do not require home schooling

Lesson Plan. *Alison Pittman, right, and her daughter, Adrienne, 10, work on reading skills while Daniel Pittman, 15, left, gets ready to get to work on his English lesson at the Pittman's home in Petal, MS.* AP IMAGES.

families to register, and some parents refuse to either way. Moreover, many home schoolers hold alternative world-views and are unwilling to participate in studies by unknown researchers.

Estimating the number of home schooled children in the United States is politically contentious. Critics have interests in portraying the movement as marginal, whereas advocates seek to stress its prevalence. Thus, government estimates are considerably lower than those given by advocacy groups. Data from the 2003 National Household Education Survey (NHES) indicates that an estimated 1.1 million children were home educated that year (up from an estimated 850,000 in 1999). Given a similar rate of growth, the estimate for 2007 can be increased to 1.4 million. By 2007 some advocacy groups were estimating that the number had reached more than 2 million.

The majority of research studies on home schoolers is qualitative in nature, based on interviewing and/or observing relatively small numbers of parents and their

children. These studies tend to find participants through local networks and associations and then ask them for referrals to other local home schooling parents. Several studies are quantitative and employ sophisticated statistical analyses. These are often based on surveys of members of home schooling organizations. Response rates in these types of surveys tend to be problematically low. Other quantitative studies (such as the NHES) are based on telephone surveys of large-scale, random samples of American adults with school-age children. Home-educating parents are part of this population, but it is difficult to assess whether they differ from parents of publicly and privately schooled children in regard to having a landline telephone, answering it, and actually participating in the survey.

MAJOR FINDINGS ABOUT HOME SCHOOLING

There are only a few major areas of inquiry in the home schooling scholarship. Most important, scholars (as well

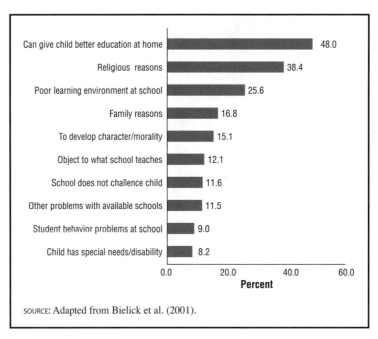

Figure 1. *Ten reasons for home schooling and the percentage of home schooled students whose parents gave each reason, 1999.* CENGAGE LEARNING, GALE.

as policymakers and the public) want to know the reasons why parents choose to home school their children (the inputs), who it is that does it (the demographics), and what is accomplished in doing so (the outputs). The literature indicates that the motivations for home schooling as well as the demographics of who home schools have changed across time. Regarding outputs, the focus has been on three primary topics: the socialization and academic performance of home schooled children as well as the long-term effects into adulthood.

Early studies of motivations to home school identified two major groups: *pedagogues* and *ideologues*. The pedagogues tended to be leftists on the political spectrum and stood against the bureaucratization and professionalization of public schools. These do-it-yourselfers sought personalization and decentralization under family control. The ideologues came largely from the political right, crusading against the secular forces of modern society, and seeking to impart religious values to their children. Although many commentators and much of the public continue to associate home schooling with the religious right, this segment of home schoolers has been declining. The majority of home schooling parents are not motivated by religious reasons.

As the movement grew in the 1990s and early 21st century, it became much more diverse. Home schooling is now advocated for a variety of reasons from average, mainstream Americans. Overall, there is a general consensus among researchers that there are four broad

categories of motivation (and considerable overlap). Academic/pedagogical concerns and religious values certainly continue to be prevalent. In regard to the former, 48.9% of the home schooling parents who responded to the 1999 NHES survey indicated that they were motivated by the ability to give their child a better education. This type of response reflects the fact that, for many parents, home schooling is a positive choice rather than just a reaction against public schools. Also, according to the 1999 NHES, only 38.4% of respondents reported that they were motivated by religious reasons. In the 2003 NHES, 29.8% of respondents chose "to provide religious or moral instruction" as their most important reason for home schooling.

In addition to pedagogical and ideological reasons, home schooling parents also cite their general dissatisfaction with the public schools and family lifestyle reasons. The public school criticism usually takes two forms. First, there are environmental concerns such as safety issues and the potential for negative peer influences. Second, there are curricular concerns surrounding standards-based education and high-stakes testing resulting from state and national government initiatives (such as California's Public Schools Accountability Act of 1999 and the national No Child Left Behind Act of 2001). Finally, although they are the least common of the four, family lifestyle reasons are very important for some parents who decide to home school. This diverse set of motivations includes the inflexibility of the public school schedule, having children with special learning needs, and having children with unique abilities.

As the home schooling movement has grown, the demographic profile of its participants has changed. The pedagogues of the 1960s and 1970s and the ideologues of the 1980s were fairly homogeneous groups. Research from the mid-1990s through the middle of the first decade of the 21st century is quite consistent in regard to the demographics of who home schools. These families differ from the average American family in that they are more likely to be White, to be headed by a married couple, to have greater numbers of children, to be headed by college-educated parents, to have larger annual incomes, and to be state-certified to teach. Mothers usually provide about 90% of the home instruction, and most are not in the paid labor force. The fathers are more likely to work in professional/technical occupations or be self-employed. The importance of religion for some home schoolers also makes this population more likely to be socially and politically conservative and have stronger religious values than average Americans.

Even though home schooling has become less polarizing as an issue, research concerning its effects is still likely to be of the greatest interest. Many studies have documented that home schooled children do not suffer in terms of self-esteem or self-concept (an indicator of positive socialization). A few studies provide evidence that these children are better socialized than their public school counterparts. Indeed, some argue that it is in formal education settings that students experience negative socialization and peer pressure. Much of the research in this area highlights the fact that home educators do not usually act in isolation. They work together through networks and organizations. By sharing teaching materials and ideas, taking their children on group field trips, and engaging in other social activities, home schooling parents and their children build a community. Some note that the children also benefit from the age diversity of their social contacts—they are not segregated in age-based classrooms.

A multitude of studies concerning the student achievement of home-educated children have been conducted. More than 25 different studies indicate that these students score above national averages. Only two studies have demonstrated otherwise. Although home schooled students consistently score higher on various student achievement measures, the specific determinants of their achievement are less well known. Parental demographic factors have had inconsistent effects across these studies. Of all the areas of research, the long-term effects are the least studied. The available evidence does indicate that home-educated children have high college completion rates and go on to lead successful careers and lives.

The social scientific study of this growing movement is certain to continue. The existing literature illustrates that the least is known about the long-term effects, so this area should be a priority for researchers. Although home schooling has been a very polarizing issue, this is changing as it continues to become more popular and more publicly acceptable. The methodological problems of identifying and studying this population are likely to diminish in the future, providing the grounds for additional rigorous analysis. As a whole, the existing research indicates that those who are home schooled match or exceed their public school counterparts in terms of socialization, academic performance, and success in later life. Despite its many critics, there is no consistent or major evidence in the social scientific literature suggesting that home schooling is in any way detrimental to those involved or to society as a whole.

SEE ALSO Volume 1: *Parental Involvement in Education; Religion and Spirituality, Childhood and Adolescence.*

BIBLIOGRAPHY

Bielick, S., Chandler, K., & Broughman, S. P. (2001). *Home schooling in the United States, 1999.* Washington DC: U.S. Department of Education, Office of Educational Research and Improvement.

Collom, E. (2005). The ins and outs of home schooling: The determinants of parental motivations and student achievement. *Education and Urban Society, 37,* 307–335.

Green, C. L., & Hoover-Dempsey, K. V. (2007). Why do parents home school? A systematic examination of parental involvement. *Education and Urban Society, 39,* 264–285.

Mayberry, M., Knowles, J. G., Ray, B., & Marlow, S. (1995). *Home schooling: Parents as educators.* Thousand Oaks, CA: Corwin Press.

Medlin, R. G. (2000). Home schooling and the question of socialization. *Peabody Journal of Education, 75,* 107–123.

Princiotta, D., & Bielick, S. (2006). *Home schooling in the United States, 2003.* Washington DC: U.S. Department of Education, National Center for Education Statistics.

Stevens, M. L. (2001). *Kingdom of children: Culture and controversy in the home schooling movement.* Princeton, NJ: Princeton University Press.

Van Galen, J. A. (1991). Ideologues and pedagogues: Parents who teach their children at home. In J. Van Galen & M. A. Pitman (Eds.), *Home schooling: Political, historical, and pedagogical perspectives* (pp. 67–92). Norwood, NJ: Ablex.

Ed Collom

HOMELESS, YOUTH AND ADOLESCENTS

Estimating the number of runaway and homeless youths in the United States is challenging because homelessness is not a steady state, particularly for adolescents. Even if returning home is not an alternative, minors have the options of temporary shelters, foster care, group homes, or other institutional living arrangements. Most chronic runaways and homeless adolescents live a "revolving door" existence, alternating between various housed living arrangements punctuated by time directly on the streets. The amount of time unsupervised and unsheltered may vary from episodes of a single night to several months. Definitions of homelessness for adolescents have distinguished between *runaways* and *thrownaways*.

According the National Incidence Studies of Missing, Abducted, Runaway, and Thrownaway Children (NISMART), a "runaway episode" is defined by meeting one of the following three criteria: (a) a child leaves home without permission and stays away overnight; (b) a child 14 years old or younger who is away from home chooses not to come home when expected to and stays away overnight; (c) A child 15-years old or older who is away from home chooses not to come home and stays away two nights. A "thrownaway episode" is defined by either of the following two criteria: (a) a child is asked or told to leave home by a parent or other household adult, no adequate alternative care is arranged for the child by a household adult, and the child is out of the household overnight; (b) a child who is away from home is prevented from returning home by a parent or other household adult, no adequate alternative care is arranged for the child by a household adult, and the child is out of the household overnight (Hammer et al. 2002, p. 2).

Usually the term "chronic runaways" refers to those who have run away three or more times. "Homeless youth" refers to those 18 years or older who cannot return home, who have chosen to never return home, and who have no permanent residence (GAO, 1989). Here we focus on chronic runaways and homeless adolescents.

Estimates based on data from the three components of the Second National Incidence Studies of Missing, Abducted, Runaway, and Thrownaway Children (NISMART-2)—the National Household Survey of Adult Caretakers, the National Household Survey of Youth, and Juvenile Facilities Study annual reporting period of 1999—indicate that 1.68 million youth had experienced a runaway/thrownaway episode. Youths aged 15 to 17 years accounted for two-thirds (1.15 million) of the runaway/thrownaway episodes. Only 4% were under 11 years (Hammer et al. 2002, p. 6). Nineteen percent of the episodes were less than 24 hours, 58% percent of the episodes 24 hours to less than one week, and 22% were more than one week. About equal numbers of boys and girls run away (Hammer et al. 2002).

Historically, Americans have had a tendency to romanticize runaway adolescents. They were the Huckleberry Finns, the cowboys, the hobos of the Great Depression, and even the hippies of the 1960s who left home by choice seeking their fortune, adventure, rebelling against

society, or "dropping out" altogether. Much of the research on runaways from the 1950s, 1960s, and early 1970s at least partially reflects this view. This early research argued there were two types of runaways: those running to something and those running from something. Other typologies of the period also included adventure seekers. By the late 1970s and early 1980s, with the advent of more systematic empirical research, the emphasis on adventure and fortune-seeking largely disappears.

Subsequent research has shown convincingly that the vast majority of chronic runaways and homeless youth are running from or drifting out of disorganized and troubled family situations. Numerous studies have documented problems in the caretaker–child relationship ranging from control group studies of bonding, attachment, and parental care to combined caretaker and runaway child reports on negative parenting behaviors and mutual violence. Studies based on adolescent self-reports have indicated high levels of caretaker physical and sexual abuse among chronic runaways and homeless youth.

There is a growing literature on sexual minorities among homeless and runaway adolescents; however, the estimates vary extensively by type of sample. Widely cited estimates reported by Kruks (1991) of 25% to 40% were based on reports from street outreach agencies in Los Angeles. Numbers from larger, more systematic studies are lower and vary by region. For example, a Hollywood study reported 18% of the youths interviewed self-identified as gay or bisexual (Unger, Kipke, Simon, Montgomery, et al, 1997); a Seattle study reported 22% self-identified bisexual, gay, lesbian or transgender youths (Cochran, Stewart, Ginzler, & Cauce, 2002); and a four-state longitudinal study in the Midwest reported 15% self-identified bisexual, gay, or lesbian adolescents (Whitbeck, Chen, Hoyt, Tyler, et al, 2004). Sexual minority runaways tend to be especially at risk. They are more likely to be "thrownaways," more highly victimized, to engage in survival sex (males), and have higher rates of psychopathology.

Youths' troubled backgrounds affect behaviors and experiences on the street. At the point a runaway first leaves home, he or she is already in acute distress. Experiences on the street may either amplify existing behavioral problems and psychological symptoms or result in new symptoms. For runaway and homeless adolescents the street environment is essentially a combat zone. Rates of physical and sexual assaults are extremely high and, at least in the case of sexual assault, apt to be underreported. Kipke and colleagues found that more than 50% of their sample of Hollywood street kids had been beaten up on the streets, 45% had been chased, 26% had been shot at, 9% had been stabbed, and 15% had been sexually assaulted. Their risk was such that they lived in constant fear. More than one-half feared being shot or stabbed and nearly one-half feared sexual and physical assault (Kipke, Simon, Montgomery, Unger, et al, 1997).

In a four-state Midwest sample, 19.7% reported they had been robbed, 26.8% beaten up, 31.5% threatened with a weapon, and 15.6% assaulted with a weapon. Eighteen percent of the young women had been sexually assaulted when on the streets (Whitbeck & Hoyt, 1999). In a three-year longitudinal study of runaway and homeless youth in the Midwest, 91% of the men and 85% of the women who had remained in the study had been physically victimized when on the streets by the final wave of interviews. Forty-two percent of the women who had remained in the study had been sexually assaulted by the final wave, a 40% increase over 3 years (Whitbeck, 2009).

Running away is often preceded by or accompanied with other types of serious violations of rules (e.g., staying out at night despite parental prohibitions, truancy), and other behavioral problems such as substance abuse or delinquent behaviors. These behaviors are carried onto the streets, where the need for self-protection and the lack of legitimate means for self-support increases the likelihood of engaging in deviant survival strategies (Hagen & McCarthy, 1997). Once on the streets, social networks teach survival strategies. Because runaway minors have very few legitimate means of independent economic support, they often rely on the street economy to get by. Conservative estimates of survival strategies based on a sample of runaway and homeless adolescents in large to moderate sized cities in the Midwest indicate that nearly one-half had sold drugs, 14% had panhandled, and 16% had shoplifted or stolen to get money or food (Whitbeck, 2009).

Estimates of sexual survival strategies (i.e. trading sex for money, food, drugs or a place to stay) vary widely by how questions are asked, location, and methods of sampling. Although it is likely underreported most studies indicate that relatively few adolescents engage in prostitution. In a sample of New York City street youth, Rotheram-Borus and associates reported that 13% of males and 7% of females engaged in survival sex (Rotheram-Borus, Meyer-Bahlburg, Koopman, Rosario, et al, 1992). Greene, Ennett, and Ringwalt (1999) in a national sample of sheltered youths reported 28% of "street youths" and 10% of "shelter youths" had engaged in survival sex. Only about 7% of males and 6% of females reported survival sex in a 1999 study of Midwestern homeless youth in small- to moderate-sized cities, although about one-fifth said that they had considered doing so (Whitbeck & Hoyt, 1999).

Nearly all runaways are sexually experienced and along with high rates of sexual activity comes risk for sexually transmitted infections (STI). Studies show self-reported rates of STIs ranging from 20% to 37% (Noell, Rohde, Ochs, Yovanoff, et al, 2001). There is some evidence that rates of STIs vary by sexual orientation, though this is likely a function of survival sex and number of sexual partners.

HIV-risk among runaway and homeless youths has been a serious concern since the late 1980s. Homeless youths are subject to multiple risks for HIV infection including high rates of sexual activity, multiple partners, survival sex, and exposure to IV drug users (Solario, Milburn, Rotheram-Borus, Higgins, et al, 2006). Estimates of HIV infection among this population vary widely by site and sample. The 1990 Office of the Inspector General reported prevalence rates of 3% to 31.5% depending on the shelter where the testing was done (U.S. Office of Inspector General, 1990). Stricof and colleagues (Stricof, Kennedy, Nattell, Weisfuse, & Novick, 1991) reported an infection rate of 5.3% among a sample of New York City homeless youths; Pfeifer and Oliver (1997) reported a rate of 11.5% in Hollywood, California.

Street survival strategies do not always work well and runaway adolescents often go hungry. Obtaining food regularly is precarious and it is often necessary to turn to street survival strategies to obtain it. Based on current estimates of 500,000 homeless and runaway adolescents on the streets or in shelters in the United States on a typical day approximately one-third or 165,000 homeless young people in the United States went hungry (Whitbeck, Chen, & Johnson, 2006).

The distress that builds prior to leaving home and that incurred when on the streets exact an emotional toll. Although prevalence estimates vary due to measurement and sampling, both symptom and diagnostic studies indicate high rates of externalizing and internalizing problems among runaway and homeless youth. For example, prevalence rates for conduct disorder (CD) in three samples of adolescents who could be confidently classified as "chronic runaways" or "homeless" were very similar: 54% in the Booth and Zhang study (1996), 59% in a New York City study (Feital, et al, 1992) and 53% in a Seattle area sample (Cauce, et al, 2000). Studies reporting prevalence rates for major depressive episode (MDE) indicate less agreement than those for CD, with prevalence rates ranging from 49% (Feital, et al, 1992) to 21% (Cauce, et al, 2000). Studies of substance abuse also yield varying estimates for self-reported prevalence and meeting diagnostic criteria for substance use disorders (SUD). Prevalence rates of SUDs among homeless and runaway adolescents range from 71% (Kipke, Montgomery, Simon, Unger, & Johnson, 1997) to 24% for drug abuse (Cauce, et al, 2000). The variation in prevalence rates reflects differences in age groups and sampling strategies that beleaguer research efforts with this population.

Homeless Adolescents. *Research has shown that the vast majority of chronic runaways and homeless youth are running from or drifting out of disorganized and troubled family situations.* © **TONY ARRUZA/CORBIS.**

A 2004 diagnostic study indicates that 21.3% of Midwestern runaway adolescents aged 16 to 19 years (mean age was 17.4 years, with standard deviation of 1.05) met diagnostic criteria for SUD, CD, MDE, or posttraumatic stress disorder (PTSD), and 66.3% met criteria for two or more of these disorders (Whitbeck, Johnson, Hoyt, & Cauce, 2004). Runaways were six times more likely to meet criteria for lifetime comorbid mental disorders than were similarly aged young people in the National Comorbidity Survey (67.3% vs. 10.3%). The most prevalent lifetime disorder was CD (75.7%) followed by alcohol abuse (43.7%), drug abuse (40.4%), PTSD (35.5%), and MDE (30.4%) (Whitbeck, et al, 2004).

It should be apparent from this review that chronic running away and periods of adolescent homelessness have serious life course consequences by interrupting or otherwise modifying the timing, context, and completion of fundamental developmental tasks and important life transitions. Ideally, adolescence is a time for learning and rehearsing adult roles characterized by forays into independence with retreats to home and depend-ency. For homeless youth, the transition to adult behaviors tends to be more abrupt and off time. Running away, with concomitant academic disruptions, loss of conventional adult mentors and conventional peer affiliations, early alcohol and drug use, and participation in the street economy interrupts learning of conventional behaviors essential for successful adulthood. Important pro-social developmental pathways have been missed or at minimum delayed, and there is little on the streets to replace them.

SEE ALSO Volume 1: *Aggression, Childhood and Adolescence; Child Abuse; Drinking, Adolescent; Drug Use, Adolescent; Foster Care; Gays and Lesbians, Youth and Adolescence; Health Behaviors, Childhood and Adolescence; Poverty, Childhood and Adolescence; Sexual Activity, Adolescent;* Volume 2: *Homeless, Adults.*

BIBLIOGRAPHY

Booth, R. E., & Zhang, Y. (1996). Severe aggression and related conduct problems among runaway and homeless adolescents. *Psychiatric Services, 47*, 75–80.

Cauce, A. M., Paradise, M., Ginzler, J. A., Embry, L., Morgan, C. J., Lohr, Y., & Theofelis, J. (2000). The characteristics and mental health of homeless Adolescents: Age and gender differences. *Journal of Emotional and Behavioral Disorders, 9,* 220–239.

Cochran, S., Stewart, A., Ginzler, J., & Cauce, A. M. (2002). Challenges faced by homeless GLB: Comparison of gay, lesbian, bisexual, and transgender adolescents with their heterosexual counterparts. *American Journal of Public Health, 92,* 773–777.

Feitel, B., Margetson, N., Chamas, J. & Lipman, C. (1992). Psychosocial background and emotional disorders of homeless and runaway youth. *Hospital and Community Psychiatry, 43,* 155–159.

General Accounting Office (GAO). (1989). Homeless and runaway youth receiving services at federally funded shelters. Report HRD–90–45. Washington, DC: Author.

Greene, J. M., Ennett, S. T., & Ringwalt, C. L. (1999). Prevalence and correlates of survival sex among runaway and homeless youth. *American Journal of Public Health, 89,* 1406–1409.

Hagen, J., & McCarthy, B. (1997). *Mean streets: youth crime and homelessness.* New York: Cambridge University Press.

Hammer, H, Finkelhor, D., & Sedleck, A. (2002). Runaway/Thrownaway children: National estimates and characteristics. *Bulletin released by the National Incidence Studies of Missing, Abducted, Runaway, and Thrownaway Children [NISMART],* October, 1–12. Retrieved May 13, 2008, from http://criminal justice .state.ny.us/missing/graphics/nismartrunthrowaway2002.pdf

Kipke, M., Montgomery, D., Simon, T., Unger, J. & Johnson, L. (1997). Homeless youth: Drug use patterns and HIV risk profiles according to peer group affiliation. *AIDS and Behavior, 1,* 247-257.

Kipke, M. D., Simon, T. R., Montgomery, S. B., Unger, J. B., & Iverson, E. (1997). Homeless youth and their exposure to violence while living on the streets. *Journal of Adolescent Health, 20,* 360–367.

Kruks, G. (1991). Gay and lesbian homeless/street youth: Special issues and concerns. *Journal of Adolescent Health, 12,* 515–518.

Noell, J., Rohde, P., Ochs, L., Yovanoff, P., Alter, M. J. Schmid, S., et al. (2001). Incidence and prevalence of chlamydia, herpes and viral hepatitis in a homeless adolescent population. *Sexually Transmitted Diseases, 28,* 4–10.

Pfeifer, R. W., & Oliver, J. (1997). A study of HIV seroprevalence in a group of homeless youth in Hollywood, California. *Journal of Adolescent Health, 20,* 339–342.

Rotheram-Borus, M. J., Meyer-Bahlburg, H. F. L., Koopman, C., Rosario, M., Exner, T. M., Henderson, R., et al. (1992). Lifetime sexual behaviors among runaway males and females. *Journal of Sex Research, 29,* 15–29.

Solario, M. R., Milburn, N. G., Rotheram-Borus, M. J., Higgins, C., & Gelberg, L. (2006). Predictors of sexually transmitted infection testing among sexually active homeless youth. *AIDS and Behavior, 10,* 179–184.

Stricof, R. L., Kennedy, J. T., Nattell, T. C., Weisfuse, I. B., & Novick, L. F. (1991). HIV seroprevalence in a facility for runaway and homeless adolescents. *American Journal of Public Health, 81,* 50–53.

Unger, J., Kipke, M., Simon, T., Montgomery, S., & Johnson, C. (1997). Homeless youths and young adults in Los Angeles: prevalence of mental health problems and the relationship between mental health and substance abuse disorders. *American Journal of Community Psychology, 25,* 371–394.

U.S. Office of Inspector General. (1990). *HIV infection among street youth.* Washington, DC: Department of Health and Human Services.

Whitbeck, L. B. 2009. *Emerging adulthood and mental disorders among homeless and runaway adolescents.* New York: Psychology Press.

Whitbeck, L. B., Chen, X., Hoyt, D. R., Tyler, K. A., & Johnson, K. D. (2004). Mental disorder, subsistence strategies, and victimization among gay, lesbian, and bisexual homeless and runaway adolescents. *Journal of Sex Research, 41,* 329–342.

Whitbeck, L. B., Chen, X., & Johnson, K. (2006). Food insecurity among runaway and homeless adolescents. *Public Health Nutrition, 9,* 47–52.

Whitbeck, L. B., & Hoyt, D. R. (1999). *Nowhere to grow: Homeless and runaway adolescents and their families.* New York: Aldine de Gruyter.

Whitbeck, L. B., Johnson, K., Hoyt, D. R. & Cauce, A. (2004). Mental disorder and comorbidity among runaway and homeless adolescents. *Journal of Adolescent Health, 35,* 132–140.

Les B. Whitbeck

HUMAN CAPITAL

Education and training are important determinants of life course outcomes. Across individuals, no variable better predicts lifetime earnings and mental and physical well-being than the number of years of schooling completed. Conceived by economists as *human capital investment* (Becker, 1964; Hanushek & Welch, 2006), the study of the determinants and consequences of variation in years of schooling and training completed has emerged as one of the dominant paradigms in social scientific research. Sociologists have made their own contribution to economists' ideas, adding *social capital* (resources available via social networks and group membership) and *cultural capital* (skills and habits available via family socialization and group culture) to the list of investment goods yielding a later life return (Coleman, 1988; Farkas, 1996, 2003; Portes, 2000). In concert with the individual's genetic and wealth inheritance, achieved human, social, and cultural capital strongly affect life course attainment of occupational standing, earnings, and wealth, as well as correlated outcomes such as health and well-being.

HUMAN, SOCIAL, AND CULTURAL CAPITAL

The *human capital* investment paradigm has proven particularly durable for the study of education and training as assets useful in production. Economists have expanded their discussion to include noncognitive skills and habits as well as cognitive skills, and the effects of intervention programs and schools on student values, habits, and behaviors as well as academic skills (Duncan & Dunifon, 1998; Heckman & Lochner, 2000). Even economists, however, seem to realize that the notion of a carefully calibrated cost–benefit calculation regarding how much to invest in one's children's education does not explain why low-income families are generally unable to help their children achieve school success.

This is better explained by the second research literature, concerned with *cultural capital*. In this perspective, Swidler (1986) sees individuals as using strategies that build on their culturally shaped skills and habits to organize their life. Low-income parents sometimes fail to help their children succeed at school not because they see too low a payoff to such action, but because they lack the skills, habits, and knowledge needed to do so. Lacking these skills and habits themselves, they are unable to help their children obtain them.

These skills and habits include the usual academic skills of language (including vocabulary and grammar), reading and mathematics, and the teacher-demanded work habits of homework, class participation, effort, organization, appearance and dress, and lack of disruptiveness (Farkas, 1996, 2003). They may also include participation in elite cultural activities such as ballet and piano lessons, although parental assistance with more mundane skills (e.g., reading) is more consequential for student success. Of course, the parents' own cultural capital (school-related skills and habits) is central to the provision of such parental assistance.

Finally, a third research tradition follows the cultural capital tradition by emphasizing the resources parents use to assist their children toward school success but focuses on *social capital*—resources stemming from parental and neighborhood social networks (Coleman, 1988; Portes, 2000). Central to these is the concept of community closure—parents' relationships with other adults in the neighborhood. "Intact families double the supervisory and supportive capacity of parents, while closure expands these capacities further by involving other adults in the rearing and supervision of children" (Portes, 2000, p. 6).

These three perspectives contain a common core—the application of resources to child rearing, that is, to building skills and habits in children. These resources are primarily provided by the child's parents but may also be provided by extended family members, neighborhood adults, and adults paid with public funds. Assistance from the latter two groups, however, must typically be gained by parental or family actions. Thus, all three

theoretical perspectives—human, cultural, and social capital—combine to constitute what may be referred to as *family resource theory*. Children raised in families with high levels of human, cultural, and social capital resources tend to develop high levels of these resources themselves. It is variation in these resources that create the mechanisms leading to inequality in life course outcomes.

HUMAN CAPITAL AND LIFE COURSE INEQUALITY

The most obvious mechanism of human capital's effect is that those who complete more schooling use it to attain occupational employment with better pay and working conditions than would otherwise be available to them. Because better paying occupations typically also provide better fringe benefits, including health insurance and retirement benefits, as well as lower unemployment rates, the advantages of increased position within the occupational structure accumulate rapidly, leading to very substantial differences in full earnings (all compensation included) over the life cycle.

This effect of schooling on earnings is central to the human capital paradigm—workers who go further in school are thereby investing in skills that increase their productivity, and their increased life course earnings represent a fair market return on this investment. But the effects of schooling on later-life inequality potentially encompass more than the simple fact that increased years of schooling increase work productivity and, therefore, pay. Rather, greater education is also associated with increased social and cultural capital, and these affect not only pay but also the individual's social-psychological resources, health lifestyle, physical functioning, and perceived health and happiness.

Perhaps the key social-psychological resource is the sense of personal control. This has been operationalized as internal locus of control (Rotter, 1966), mastery (Pearlin, Lieberman, Menaghan, & Mullan, 1981), instrumentalism (Wheaton, 1980), self-efficacy (Gecas, 1989), and personal autonomy (Seeman & Seeman, 1983). Individuals scoring high on this dimension believe that they can achieve their goals through their own efforts. By contrast, lack of control or powerlessness is the belief that one is relatively helpless against powerful external forces. Not surprisingly, the sense of personal control is correlated with other measures of well-being. A sense of personal control is positively associated with lifestyle behaviors that improve health and prevents the suppression of the immune system that is associated with personal demoralization. It is also positively related to the individual having a strong network of social support. Higher schooling attainment pushes all of these variables upward.

The effects of human, social, and cultural capital are intertwined as they increase schooling, employment, earnings, perceived personal control, health-related behaviors, and social support networks. Positive feedback loops are common. Over the life course, the likely effect is that (as has been shown for early reading skill [Stanovich, 1986] and progress though the educational system [Kerckhoff & Glennie, 1999]), "the rich get richer," where wealth is measured not just monetarily but also includes both physical and psychological health (and, therefore, happiness). This notion of cumulative advantage has become a major paradigm in the study of life course outcomes (O'Rand, 1996; Ross & Wu, 1996), one that has been explicitly linked to feedback loops in human, social, and cultural capital (O'Rand, 2001).

The mechanisms underlying cumulative advantage are straightforward. During K–12 schooling, families with greater human, social, and cultural capital resources (such as two parents as opposed to one; higher parental education, occupation, and earnings; greater parenting focus and skill; better psychological and physical health; more extensive social networks; and a more positive neighborhood environment) translate these advantages to their children by instilling skills and habits that help the child have positive engagement and success with school and peers. The efforts and activities these parents and their children engage in lead to the children's development of positive skills and habits, and the success of these efforts leads to positive outcomes and a sense of personal control for the children. These in turn encourage further effort and optimism: High goals are set for the future because present goals have been met, and the experience has been pleasant and rewarding.

Children who have been provided with these advantages are disproportionately likely to enroll in, and complete, college. Many go further, receiving training and a degree in business, law, medicine, engineering, education, or other professional field. At each stage, human, social, and cultural capital increase further. Skills and habits of social interaction and productive work are learned and practiced. Network connections are expanded. The sense of personal control increases. New horizons become visible, and goals are adjusted upward. Completing more schooling affects these outcomes positively because the individual has a greater stock of productive skills. Positive effects may also flow from credentialing and from the selectivity (prestige) of the schools attended (Ross & Mirowsky, 1999).

The process continues through the period of working life and the employment career. Better educated workers pursue careers in a national labor market. They choose spouses from a larger pool of individuals, with career experiences and personal strengths similar to their own. They are geographically mobile and build social networks across multiple geographic locations. They travel more, both for business and pleasure, and adopt a cosmopolitan outlook. Once again, multiple feedback loops are in operation. Education increases earnings and economic security. The sense of personal control is strengthened. A healthy lifestyle is more likely to be adopted. (This includes a greater emphasis on exercise and the avoidance of obesity and a lower likelihood of excessive drinking and smoking.) Social support networks are denser and more extensive, as well as more likely to overlap both professional and personal lives.

Higher levels and more effective use of these economic, psychological, and social assets are characteristic of better educated workers as they progress through the life course (Mirowsky & Ross, 1999). This results in a higher standard of living, less psychological distress, a greater sense of personal control, and greater happiness.

THE ROLE OF INSTITUTIONAL ARRANGEMENTS

Thus far the description of life course behavior and outcomes has focused primarily on individuals, their families, and their social networks. Yet individual-level outcomes are crucially determined by the incentive and reward structures within which these individuals operate. This *social/economic structure* is embodied in the institutional arrangements of society in general and of individual firms in particular.

Central to these arrangements is the system of privately financed retirement and health insurance fringe benefits, provided by employers largely to mid- to upper-level, as well as to unionized lower-level, employees. In many cases, long-term employment with the same firm resembles an implicit contract, in which workers are paid less than their actual productivity when they are young and more than their actual productivity when they are old (England & Farkas, 1986). Lower wages when workers are young help compensate the firm for the cost of training, and they provide an incentive for those workers with long time horizons and the greatest amount of firm-specific human capital to be loyal over the long run. This "back-loaded" compensation scheme includes health and retirement fringe benefits that become most valuable later in life. In concert with monetary wages that rise faster than productivity in later life, this scheme shows strong patterns of cumulative advantage for the better educated and most highly placed employees.

Other forces driving cumulative advantage include the historic increase in economic opportunity for women and the associated rise in rates of divorce. One result has been greater diversity in household types. With continued strong patterns of assortative mating by educational level, combined with decreased marriage and increased nonmarital childbearing among the least well-educated, particularly inner-city African Americans, total household incomes at the top and bottom of the social class hierarchy are diverging ever more widely.

Also important has been the long-term decline in unionization, as employment has shifted out of manufacturing and into the service sector. The resulting decline in "blue-collar elite" jobs in unionized manufacturing such as autos and steel has been exacerbated by globalization and foreign competition. Added to this has been the increased computerization of the economy. A consequence has been a dramatic increase in the wage premium paid to college graduates and stagnant or declining real (adjusted for inflation) wages for jobs requiring lesser skills (Bernhardt, Morris, Handcock, & Scott, 2001). These technological changes, and their effects on institutional arrangements, have also acted to increase life course patterns of cumulative advantage and disadvantage in employment and household life.

HUMAN CAPITAL AND THE LIFE COURSE

Years of schooling attained is *the* key stratifying variable determining life course outcomes. Given the positive correlation between parents' educational attainment and that of their offspring, the process of cumulative advantage/disadvantage begins at birth. Where income and wealth are concerned, this process operates particularly strongly toward the high end of the occupational distribution, with individuals in high-paying occupations accumulating wealth quite rapidly as they age. Where health is concerned, even high occupational standing, income, and wealth cannot prevent an eventual decline with age. Nevertheless, the best educated individuals are able to put this decline off longer, and they experience a more gradual slope of decline. This is at least as much attributable to the healthier lifestyle and greater psychological resources and sense of control as to the greater income of the better educated.

In sum, there is a pattern of human capital based cumulative disadvantage, as the physical and psychological pressures of a harder life, with fewer economic and psychological resources, take their toll on the less educated. As the economy becomes increasingly based on high technology, individual differences in human capital accumulation will continue to determine inequality at all ages, as well as cumulative inequality over the life course.

SEE ALSO Volume 1: *Academic Achievement; Cognitive Ability; Coleman, James; College Enrollment; Cultural Capital; Social Capital.*

BIBLIOGRAPHY

Becker, G. S. (1964). *Human capital.* New York: National Bureau of Economic Research.

Bernhardt, A., Morris, M. S., Handcock, M., & Scott, M. A. (2001). *Divergent paths: Economic mobility in the new American labor market.* New York: Russell Sage Foundation.

Coleman, J. S. (1988). Social capital in the creation of human capital. *American Journal of Sociology, 94,* S95–S120.

Duncan, G. J., & Dunifon, R. (1998). "Soft-skills" and long-run labor market success. *Research in Labor Economics, 17,* 123–149.

England, P., & Farkas, G. (1986). *Households, employment, and gender: A social, economic, and demographic view.* New York: Aldine Publishing.

Farkas, G. (1996). *Human capital or cultural capital? Ethnicity and poverty groups in an urban school district.* New York: Aldine de Gruyter.

Farkas, G. (2003). Cognitive and noncognitive traits and behaviors in stratification processes. *Annual Review of Sociology, 29,* 541–562.

Gecas, V. (1989). The social psychology of self-efficacy. *Annual Review of Sociology, 15,* 291–316.

Hanushek, E. A., & Welch, F. (Eds.). (2006). *Handbook of the economics of education.* Amsterdam: North-Holland.

Heckman, J. J., & Lochner, L. (2000). Rethinking education and training policy: Understanding the sources of skill formation in a modern economy. In S. Danziger & J. Waldfogel (Eds.), *Securing the future: Investing in children from birth to college* (pp. 47–83). New York: Russell Sage Foundation.

Kerckhoff, A. C., & Glennie, E. (1999). The Matthew effect in American education. *Research in Sociology of Education and Socialization, 12,* 35–66.

Mirowsky, J., & Ross, C. E. (1999). Economic hardship across the life course. *American Sociological Review, 64,* 548–569.

O'Rand, A. M. (1996). The cumulative stratification of the life course. In R. H. Binstock & L. K. George (Eds.), *Handbook of aging and the social sciences* (4th ed., pp. 188–207). San Diego, CA: Academic Press.

O'Rand, A. M. (2001). Stratification and the life course. In R. H. Binstock & L. K. George (Eds.), *Handbook of aging and the social sciences* (5th ed., pp. 197–213). San Diego, CA: Academic Press.

Pearlin, L. I., Lieberman, M. A., Menaghan, E. G., & Mullan, J. T. (1981). The stress process. *Journal of Health and Social Behavior, 22,* 337–356.

Portes, A. (2000). The two meanings of social capital. *Sociological Forum, 15,* 1–12.

Ross, C. E., & Mirowsky, J. (1999). Refining the association between education and health: The effects of quantity, credential, and selectivity. *Demography, 36,* 445–460.

Ross, C. E., & Wu, C.-L. (1996). Education, age, and the cumulative advantage in health. *Journal of Health and Social Behavior, 37,* 104–120.

Rotter, J. B. (1966). Generalized expectancies for internal versus external control of reinforcement. *Psychological Monographs, 80,* 1–28.

Seeman, M., & Seeman, T. E. (1983). Health behavior and personal autonomy: A longitudinal study of the sense of control in illness. *Journal of Health and Social Behavior, 24,* 144–160.

Stanovich, K. E. (1986). Matthew effects in reading: Some consequences of individual differences in the acquisition of literacy. *Reading Research Quarterly, 21*, 360–407.

Swidler, A. (1986). Culture in action: Symbols and strategies. *American Sociological Review, 51*, 273–286.

Wheaton, B. (1980). The sociogenesis of psychological disorder: An attributional theory. *Journal of Health and Social Behavior, 21*, 100–124.

George Farkas

HUMAN DEVELOPMENT

SEE Volume 1: *Bronfenbrenner, Urie; Elder, Glen H., Jr.; Erikson, Erik; Piaget, Jean.*

HUMAN ECOLOGY

SEE Volume 1: *Bronfenbrenner, Urie.*

I

IDENTITY DEVELOPMENT

Identity is a term often employed in the social sciences, typically referring to some individual-level attribute that a researcher seeks to explore. As such, it runs the risk of being defined so generally as to be analytically useless. The relevant definitions of identity are summarized in two domains: *ego* and *social* approaches to identity. Ego identity is traditionally associated with classic psychological scholars, namely Jean Piaget (1896–1980) and Erik Erikson (1902–1994), and represents an integration of the self as a unitary construct that typically is largely developed by adolescence. This is often what is referred to in the colloquial idea of "finding yourself." In contrast, the social identity approach is a broad umbrella term encompassing more recent sociological and psychological orientations linking the individual to socially meaningful groups and societal positions. In this perspective, everyone has a set of identities ranging from one's gender, religion, occupation, and so on. Social identities situate an individual in relation to others in situations and can develop differentially across the life course.

EGO IDENTITY

In developmental literature, identity is a concept employed to capture the sense of coherence and unity in the self across life experiences and is most commonly associated with Erikson (1963); he wrote that a "sense of identity provides the ability to experience one's self as something that has continuity and sameness and to act accordingly" (p. 42). Erikson's work focused on the stages of individual development that people encounter as they develop understandings

of themselves in relation to others. Current scholarship builds on Erikson's psychosocial model of a single pattern of stages by focusing on the place of narrative (i.e., the story of oneself) in building a sense of identity, as well as the importance of historical and social contexts on this process. Within this tradition, there is a tension between analytical models of identity imposing academic definitions on individual experience and allowing an individual's notions of identity to be voiced and validated. Some, both professionals and laypeople, view identity as static across the life span, whereas others see it as contingent and shifting based on situational circumstances. Current models attempt, with differential success, to simultaneously capture different multidimensional levels to conceptualize and measure identity.

SOCIAL IDENTITY

In the more sociologically oriented life-course literature, identity refers to the various meanings an individual attaches to oneself and by others within social relationships. Two theories organize the relationship of identities to the social world: the more sociological identity theory and the more psychologically oriented social identity theory. Identity theory has traditionally focused on the self as a set of internalized social roles and how identities are primarily (but not exclusively) linked to hierarchically organized roles (such as employee, sister, and volunteer) that an individual prioritizes. Social identity theorists study how individuals conceptualize group boundaries. Specifically, identities focus on the commonalities among people who share a group membership (ethnicity, gender, and club membership) and the differences between that group and other related groups. Recent research in each tradition moves beyond this simple dichotomy to focus on how membership in roles and groups combine to produce an organized sense of self.

Identities are meanings that guide behavior and organize a personal sense of self. They frame perception within situations and offer intuitive guidelines for how well one fits the expectations they (and others) have for themselves in a particular social encounter. They also circumscribe social allegiances. The self is made up of one's conglomeration of identities. Over time, the expectations associated with each identity become internalized and become part of one's expectations for guiding and interpreting behavior. More recently, identity and social identity theorists have incorporated a social notion of the personal identity that might usefully be compared to the ego identity notion found in developmental work. Typically, however, these models are studied in isolation.

IDENTITY DEVELOPMENT

Both ego identity and other important social identities form early in a person's development. When discussing the development of a person, however, one must be careful to differentiate between the child and childhood. The former focuses on the human organism and the latter on a set of cultural ideas. Identity scholars typically focus on childhood, but any discussion should be bracketed by an understanding of the historically contingent ideals, norms, and ideologies that influence individual development.

There are multiple theoretical perspectives on identity development within the ego identity tradition. Cognitive theory is particularly influential, largely pioneered by Piaget, who argued that children develop schemas, or classification and categorical systems for interpreting the social world. Young children understand themselves as discrete entities, separate from their primary caregiver, and possess simplistic schemas (e.g., viewing oneself simply as a girl or boy). As cognitive abilities develop and the child engages in more social experience, schemas become more complex and integrated. Children are able to build a more substantively complex self-understanding rather than simply thinking of themselves in terms of discrete entities. Children understand that they can view themselves as student, friend, and soccer player simultaneously, and that others inhabit multiple identities as well. Children also learn how and when to evaluate aspects of themselves, forming judgments about what is good and bad. These evaluations become more sophisticated and nuanced as children become adolescents. Lev Vygotsky (1978) extended the cognitive approach into the interactional realm, focusing on the child as an independent problem solver within parental and peer interactions. This focus on the agency of the child as an active shaper of his or her socialization is gaining currency in current social science approaches.

Symbolic interactionist approaches bridge more psychologically oriented approaches to identity development with more sociological notions. George Herbert Mead (1863–1931) discussed children's development of a sense of themselves through the process of taking the role of significant others and developing what he termed *reflected appraisals*, whereby the children imagines how others evaluate their behavior. Roles are expectations for behavior attached to social positions such as mother, sister, and student. Role-taking involves the conscious process of anticipating the responses of actors in the situation and directing one's behavior accordingly. Developmentally, role-taking begins early when a young child begins to imitate the behaviors of his or her primary caregiver. Through imitating significant others, a child learns to act out one role one at a time. For example, a child playing with a baby doll displays mirrored behaviors such as holding, rocking, and patting the doll and slowly learns to adopt the perspective of a caregiver.

The developing child learns to take on multiple roles simultaneously, understanding that the same person might possess a variety of roles with potentially irreconcilable expectations. As part of this process, Mead focused on the way the child abstracts a sense of a *generalized other*, a representation of the perceived attitudes of the larger community. Developing a conventional sense of the generalized other is tantamount to learning the culture the child was born into. Developing a stable sense of self-identity involves properly understanding the wider community the child inhabits. At the root of this development is the idea of reflexivity, a self-evaluative process that children use to imagine how their behaviors are perceived and to form standards through which they gauge if they are enacting roles properly.

Bar Mitzvah Ceremony. By adolescence, individuals situate their identities along a variety of dimensions, including multifaceted understandings of cultural meanings of race, ethnicity, religion, and gender. © ROBERT MULDER/GODONG/ CORBIS.

As children advance cognitively, they understand that different situations may call for different role expectations. These differential sets of expectations become internalized as identities, and the imagined evaluation of others leads to self-evaluation that consciously and nonconsciously guides behavior within a situation. By adolescence, individuals situate their identities along a variety of dimensions, including multifaceted understandings of cultural meanings of race, ethnicity, religion, and gender. Symbolic material for constructing these various identities also comes from a larger variety of sources than early childhood, including media, schools, and peer culture. One orienting principle for researchers involves the extent that they weigh internal processes (e.g., personality dispositions and maternal attachment) versus environmental influences (e.g., school structure and the occupational stratification system) in shaping the development of identity as a child ages and transitions into adulthood.

These more individually oriented perspectives on identity development can be subsumed under the life-course perspective, a larger framework focused on the interplay of individual lives within a broader sociological and historical context. As identity theory highlights, an individual's life involves overlapping trajectories in a variety of domains, from occupation to family life, that set out expected sequences of identities. People develop socially and biologically over time within socially specified, normative trajectories. People who have a child as a teenager or who marry for the first time in their 40s, for example, are experiencing these events at periods in their life that are not in step with normative cultural expectations. Embedded within these trajectories are sets of typical transitions, such as leaving high school, that orient people's lives.

Part of the individual-level maintenance of a trajectory involves the internalization of the appropriate identity that reflects and guides an individual's behavior in the appropriate situations. Transitions, or turning points, signify a reconstruction of an identity, such as *college graduate*, or the adoption of a new identity, such as *parent*. A key point of this perspective is that historical events channel and shape the possible identities an individual might develop, even as there is a concurrent focus on the importance of individual agency to shape his or her own life course within societal bounds. For example, scholarship surmises that the culturally sanctioned stage of adolescence may be extending in the United States. Jeffrey Arnett (2000) suggested a normative life stage termed *emerging adulthood*, whereby people in their late teens and early 20s do not believe they have reached adulthood but also no longer see themselves as adolescents.

IDENTITY RESEARCH AND MEASUREMENT

The varying concepts of identity have led to challenges translating theoretical approaches about identity into empirical practice. One needs to differentiate between the study of different life-course stages (e.g., childhood and adulthood) and the study of identity development across the life course. An entire branch of psychology is focused on the study of childhood development, largely based on systematic observation, as very young children are not particularly facile with surveys. As children age, psychologists shift focus to cognitive development, often employing laboratory experiments. Much of this work is focused on the evaluative aspects of the self and what a person illustrates when focusing on mental health outcomes. For example, self-esteem, evaluative feelings about the self, has received a particularly large amount of attention. Young children at, for example, 8 years old will typically report relatively high self-esteem, but as they enter adolescence their self-esteem often dips to the lowest levels of their lifetimes. Self-esteem often rebounds as a person enters adulthood, and stays rather stable until older age when the individual's self-esteem is more likely to decrease. Such processes are affected by an individual's location in social structure.

Notable empirical treatments of identity in the Eriksonian tradition include the identity status paradigm (Marcia, 1966), which focuses on identity statutes based on differential dimensions of exploration and commitment, and Michael Berzonsky's (1989) derivation focused on how people confront issues of identity. Other perspectives draw on the coherency element of Erikson's work, such as James Côté's (1997) attempt to develop a multidimensional measure based on established scales measuring self-esteem, locus of control, and purpose in life. In particular, the extended transition to adulthood is gaining currency as a topic of interest among identity scholars. Erikson posited a developmental stage in which the young are able to explore potential identities without having to make permanent commitments. Given the current historical period of an extended transition time, a focus on individual agency in the development of identity formation is prevalent.

Social psychological approaches to identity bridge the psychological focus on direct observation—often in the experimental setting—and the more sociologically oriented focus on large-scale analyses, which often relies on survey data. For example, Alison Bianchi and Donna Lancianese (2005) used nationally representative longitudinal survey data of kindergartners to find that children's developing student identity is affected by their positions in societal stratification systems such as family wealth and personal beauty. In contrast, researchers studying social identities based on group membership typically rely on experimental settings. Others, focused on the range of identities developed and enacted in a human life, can rely on instruments ranging from computer surveys to daily diary studies. Scholars focused on narrative constructions of identity can employ in-depth interviews or content analyses, whereas those focused on identity performance can draw on participant observation.

Sociological identity researchers typically focus on adults as their unit of analysis, often studying identity transitions (e.g., from student to worker), including negotiating the transition out of a role identity (e.g., retirement). Identity researchers study how people manage multiple identities at the same time (i.e., being a spouse, parent, and worker). Again, a focus on mental health outcomes is a common area for research in this approach. Sociological study of children is rare and typically involves observation of naturally occurring groups of children. Although the typical focus has been on how society shapes the individual, the more recent trend has been to simultaneously focus on how children are active shapers of their own socialization— a capacity that extends across the life course.

FUTURE DIRECTIONS

The term *identity* has great potential utility for developmental and life-course researchers, although it is often used haphazardly and without theoretical precision. In many cases, people use the term *identity* as a placeholder for other social processes, such as self-understanding, identification, categorization, commonality, or a feeling of group belonging. Scholars appeal to different traditions and presuppositions when speaking of ego identity, social identity, role identity, ethnic identity, gender identity, and so on. All of these approaches share a necessary focus on the social, interrelated nature of individuals, which is important for conceptualizing the developing person across the life course. Future work should both specify the particular use of the term as well as build linkages across literatures to best advance understanding of the mechanics and importance of individual self-conceptions across time and situation.

The notion of identity is useful for developing a proper conception of social actors and points to fundamental ways humans understand themselves, understand others, and align with meaningful social groups and categories. A plausible thesis involves the increasing importance of identity within highly differentiated, varied social systems, whereby individuals are increasingly likely to encounter distinct situations calling forth different aspects of self. To the extent that multiple potential interaction partners recognize different aspects of a person, identity in both of its academic uses—as a unifying ego and as representing internalized meanings represented by a series of social locations and categories— becomes even more important than in the less differentiated past. To interact with others, a standpoint is needed through which to understand oneself, others, and the situation. Identity serves that purpose, both within situations and as a coherent self-understanding develops across the life course. Current research attempts to bridge these two approaches to identify and disentangle this apparent paradox, to the extent that individual consistency is present or overshadowed in different situations or at different times in the life course. Facilitating well-being requires elements from both literatures: helping individuals develop a stable sense of self while also allowing for the multiple situation, influences, and life stage appropriate roles that comprise a human life.

SEE ALSO Volume 1: *Activity Participation, Childhood and Adolescence; Erikson, Erik; Media Effects; Piaget, Jean; Self-Esteem; Social Development; Socialization; Socialization, Gender; Socialization, Race;* Volume 3: *Self*

BIBLIOGRAPHY

Arnett, J. J. (2000). High hopes in a grim world: Emerging adults' views of their futures and "Generation X." *Youth & Society, 31*(3), 267–286.

Berzonsky, M. D. (1989). Identity style: Conceptualization and measurement. *Journal of Adolescent Research, 4*(3), 268–282.

Bianchi, A. J., & Lancianese, D. A. (2005). No child left behind?: Identity development of the "good student." *International Journal of Educational Policy, Research, and Practice, 6*, 3–29.

Brubaker, R., & Cooper, F. (2000). Beyond "identity." *Theory and Society, 29*, 1–47.

Côté, J. (1997). An empirical test of the identity capital model. *Journal of Adolescence, 20*, 421–437.

Erikson, E. (1963). *Childhood and society.* New York: Norton.

Hogg, M. A., Terry, D. J., & White, K. M. (1995). A tale of two theories: A critical comparison of identity theory with social identity theory. *Social Psychology Quarterly, 58*(4), 255–269.

Marcia, J. E. (1966). Development and validation of ego identity status. *Journal of Personality and Social Psychology, 3*(5), 551–558.

Stryker, S., & Burke, P. J. (2000). The past, present, and future of an identity theory. *Social Psychology Quarterly, 63*(4), 284–297.

Tajfel, H., & Turner, J. C. (1979). An integrative theory of intergroup conflict. In W. G. Austin & S. Worchel (Eds.), *The social psychology of intergroup relations* (pp. 33–47). Monterey, CA: Brooks/Cole.

Vygotsky, L. (1978). *Mind in society: The development of higher psychological processes.* Cambridge, MA: Harvard University Press.

Steven Hitlin
Donna A. Lancianese

ILLNESS AND DISEASE, CHILDHOOD AND ADOLESCENCE

Sufferers of sickness face difficulties, and children are no exception. For ill children, the ordeals and challenges can be especially complex. The impact of pediatric conditions is far-reaching, because not only individual children but also their families are affected. Another complexity lies in the disjuncture between how children and families experience *illness*, and how biomedical professionals conceive of a condition as a predominantly physical *disease* (Kleinman, 1988). The disease construct used in biomedicine is not a template for understanding illness as a felt, life-embedded experience. In contrast to biomedical disease, illness entails the social context of the ordeal (including the patient's and family's experiences) and the ways in which young sufferers live with, understand, and respond to sickness. Childhood pediatric conditions can be approached in both ways: as biomedical diseases to be tracked (and treated) clinically, but also as instances of illness that families and children experience within everyday human exchange.

TRACKING CHILDHOOD DISEASE

U.S. trends since the beginning of the 20th century reveal dramatic changes in childhood sickness. The long-troubling infectious diseases such as measles, smallpox, poliomyelitis (polio), scarlet fever, tuberculosis, and cholera have receded. Concurrently, leading causes of child death and hospitalization have shifted from infectious to chronic conditions, which now affect 15% of American children (Wise, 2004). Asthma prevalence in those up to age 17, attributable to many factors including indoor and outdoor pollution, rose by an average of 4.3% per year from 1980 to 1996 (Akinbami and Schoendorf, 2002). Whereas 5 million American children suffer from asthma, only 8,000 children have cancer—mainly leukemia and brain malignancies. Death rates from childhood cancer have plummeted. Overall, the majority (55%) of children under age 17 hospitalized in the United States in 2000 were admitted for some sort of chronic condition—double the proportion seen in 1962 (Wise, 2004). Figure 1 shows one ranking of the diagnoses of hospitalized children.

Children living in developing nations face grave health risks, including specific diseases often prevented or effectively treated in wealthier countries. UNICEF has estimated that 2.3 million children under age 15 are living with HIV, and many of these children die before their second birthday (UNICEF, 2007a). Although overall child mortality declined in developing countries between 1990 and 2006, children continue to die of

Asthma	43%
Dehydration/gastroenteritis	16%
Pneumonia	11%
Seizure disorder	8%
Skin infection	8%
Urinary tract infection/pyelonephritis	4%
Failure to thrive	3%
Severe ear-nose-throat infections	2%
Pelvic inflammatory disease	2%
Diabetes mellitus	2%

SOURCE: Flores, Abreu, Chaisson, and Sun 2003.

Figure 1. *U.S. children admitted to urban hospitals, conditions presented.* CENGAGE LEARNING, GALE.

pneumonia (the leading cause of child death worldwide), diarrhea, and complications following birth (UNICEF 2007b). Worldwide, disease is more deleterious for poor children than for more economically advantaged children. Health disparities also are prevalent in the United States, where socioeconomic gaps persist in the prevalence and severity of pediatric asthma, among other conditions. Increasingly, epidemiologists have begun to examine the connection between children's health and the physical environment. Such studies trace how changes in the built environment (e.g., lack of space for active play) might relate to disease risk factors (e.g., lessened play and obesity-linked diabetes). Asthma and its links to indoor and outdoor air pollution pose another topic of interest, as does cancer in children exposed to carcinogens.

THE FAMILY'S ROLE IN CHILDHOOD ILLNESS

Many have suggested that families, rather than children as isolated individuals, should be the unit of assessment and intervention in childhood conditions. According to *family systems theory*, there is a reciprocal relationship between childhood illness and family functioning. A systems approach focuses on the strategies families use to adjust to stressors or strains. Strains may include financial burdens, loss of privacy, problems with service providers, personal distress, feelings of isolation, worries about the future, medical decision making, and responsibility for medication.

Families use various coping behaviors and resources to adjust to these strains and maintain balance. *Adaptability* refers to the family's capacity to change in response to external stressors. Notably, behaviors that may seem maladaptive can in fact be protective for the family when systemic functioning is taken into account. For example, families may embrace some forms of medical treatment that are consistent with maintaining a "normal" life, while rejecting or neglecting other clinically ordained prescriptions that disrupt routines or shared meanings. Moreover, family conflict or poor parental mental health can interfere with disease management. In dealing with chronic conditions, families endure multiple cycles of crisis and adaptation, enacting change and making adjustments many times over. Family systems theorists posit that failures in family functioning overall are more likely to *result from* a childhood illness than to *cause* such illness.

Family life, as a system, reorganizes around the illness through a series of transactional mechanisms. In some cases, a family may establish increased regimentation, structuring and organizing their interactions at the expense of warmth and communication. In other instances, siblings of the ill child may develop adjustment problems (see Bluebond-Langner, 1996, for a contrasting view). Known protective factors for siblings of ill children include parental marital satisfaction, positive sibling relationships, and a cohesive family environment (Bellin and Kovacs, 2006).

Adjustments in the family system in response to chronic illness need not be negative. Myra Bluebond-Langner (1996) identified possible strategies that families of children with cystic fibrosis use to contain the intrusion of illness into everyday family life. Positive steps that mitigated stress included routinizing tasks for managing illness, reinterpreting what was considered "normal," reassessing priorities, and reconceptualizing the future. Protective measures also can be taken in families of children with asthma. Regular practices that hold symbolic meaning and contribute to shared family identity can serve a protective function for the child's health and overall family wellbeing. Cindy Dell Clark (2003) offers numerous examples of how play and ritual can knit together a family of an ill child and lend positive meaning to needed care. For example, a parent and child might play a game together that makes treatment into a fun routine, such as pretending together that a syringe used for injection is a zebra (with lines), or by enjoying how a child sings upon injection.

Social ecological theory further broadens the family systems approach to include other important systems—including peer networks, school, and healthcare systems—that affect the health and development of children facing illness. Within these contexts, particular risks for children with chronic illness have been pinpointed, such as more difficult peer relationships due to illness-related stigma, and poorer school performance, traced to missed school and/or cognitive impairment associated with illness or treatment, as in childhood cancer (Madan-Swain, Fredrick, & Wallander, 1999). Factors such as peer support and parental satisfaction with health providers

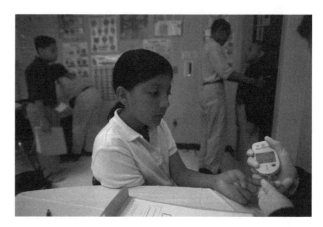

Diabetes. *A fourth grade student receives a glucose test.* © **KAREN KASMAUSKI/CORBIS.**

enhance illness management and outcomes, as has been documented for diabetes (Naar-King, Podolski, Ellis, Frey, et al., 2006).

COGNITIVE DEVELOPMENTAL MODELS OF ILLNESS

One influential approach to children's illness experience has been based on developmental distinctions, presuming that there are stage-based differences in how children at each phase of development approach illness. In other words, cognitive developmental models presume that children's understandings of illness gradually mature and transform over time.

R. Bibace and M. Walsh (1980) and E. C. Perrin and P. S. Gerrity (1981) exemplify the cognitive-developmentalist approach, which draws from theory developed by the Swiss psychologist Jean Piaget (1896–1980). Using questionnaires to elicit children's understanding and beliefs about the causes of disease, treatment, and prevention, these studies found that children between the ages of 4 and 7 hold concepts of illness laden with irrational themes such as punishment or blame, magic, or even witchcraft. Moreover, among children this age, the physical spread of disease (as conceived biomedically) is poorly understood. With advancing maturity and cognitive development, children in middle childhood are better able to understand causative factors in illness. Children's understandings of AIDS and a number of other conditions likewise have been viewed as developmentally determined.

A restrictive focus on developmental stages in children's understandings of illness has become subject to debate. Greater emphasis is now placed on the role of personal experience in children's knowledge (e.g., Eiser, 1989). At the same time, researchers studying childhood and childhood illness using qualitative methods have focused less on a developmental trajectory and more on youngsters' first-hand, present-day experience (how children themselves make sense of and approach illness in everyday contexts) (e.g., Bearison, 1991; Sourkes, 1995).

CHILDREN'S EXPERIENCES OF ILLNESS

Theoretical and methodological advances in sociocultural studies of children have drawn attention to how children actually live with and experience illness. Bluebond-Langner's (1978) ethnography of a pediatric ward in a Midwestern hospital was a foundational investigation, designed to consider children's perspectives, thoughts, and feelings about their leukemia. She documented how these young patients came to know of their impending death, even as they concealed this awareness to protect their parents. The study highlighted the fact that children's experiences may be distinctly their own and not fully disclosed to adults. As attention to children's agency and personal meanings has gathered interest, more is being learned about how children cope with illness. Clark (2003) showed how children use fantasy, imagination, and play to constructively deal with diabetes and asthma. Such research suggests that what developmentalist theorists construe as irrational or immature understandings (fantasy) may have positive emotional and social value for making sense of and managing illness.

Additional research highlighting children's vantage points on illness and disease has reframed scholarly thinking about children's ongoing experiences and understandings. I. Clemente (2007) has shown that Catalan youth with cancer exercise subtle agency in doctor-patient communication to overcome institutionalized evasiveness and nondisclosure. In New Zealand, Helen Mavoa (1999) found that even at age 3 or 4, preschool children with asthma sometimes displayed awareness of symptom onset prior to more overt signs (cough, wheeze, and shortness of breath). Although their adult caregivers were inattentive to such leading cues, the children could describe them. Through studies attentive to children's perspectives, new vistas are opening into their active responses to illness.

Finding investigational avenues that confront children's views and social contexts has involved methodological challenges. Studies with very young children increasingly use ethnographic or other child-centered methods that are comfortable and amenable to children. The use of play, props, photography, and drawings are ways for children to show and tell about their experiences in a manner that is child-attuned. As a case in point, researchers at Children's Hospital in Boston have captured children's perspectives on chronic illness by giving chronically ill children video camcorders and asking them to record aspects of their daily lives. This approach has been valuable for teaching clinicians what it is like to live with illness and to vividly set forth youthful perspectives on illness management and coping (Buchbinder, Detzer, Welsch, Christiano, et al., 2005; Rich, Lamola, and Chalfen, 1998).

CONCLUSIONS AND EXPECTED FUTURE DIRECTIONS

In the American pediatric context, sickness has increasingly taken the form of chronic conditions that children and families experience over time. Asthma, a prime example of the contemporary surge in chronic pediatric conditions, is not only chronic but also life-threatening (for it entails disrupted breathing). The challenges for asthma treatment, with its often intrusive remedies, are pronounced. Families are prone to overlook prescribed regimens when they conflict with established routines and practices, creating a nagging problem for biomedical treatment.

Children with chronic illnesses have more than doubled chances for developing emotional disorders, relative to other children, which raises the stakes for reaching out to children in pediatric contexts. Anxiety and depression can occur in children with chronic illnesses either as a direct result of the illness or its treatment, or as a consequence of the family's response to new challenges posed by illness management. For example, parental stressors such as worry and the burden of care-giving may negatively affect children's emotional well-being. Furthermore, the implications of emotional problems may be long-lived, since adults who had cancer in childhood incorporate that earlier experience, for better or worse, into their adult identity. Chronic illness thus challenges clinicians to better comprehend the experiences facing families and children, and to form lasting partnerships that will, over time, provide for effective management of children's symptoms and the whole family's well-being.

To relieve children's suffering, then, is a complex and socially encumbered challenge. Contemporary research trends offer some optimism for confronting these challenges. In biomedical research, the National Institutes of Health set forth policy in 1998 to increase the extent of child participation in research, a step taken to be sure that knowledge from clinical trials will be relevant to pediatric treatment. Beyond this biomedical policy, information that promises both theoretical and practical utility is accumulating on children's own experience of illness and their ways of dealing with treatment.

In an era when chronic disease weighs on the younger generation in industrialized countries, research on children's illness holds profound implications. Treating chronic illness competently is a task that requires both families and young patients to be part of the solution. In turn, clinicians must understand children's experiences with illness in the here and now, rather than merely as a demarcated, maturational "stage." Even as the U.S. population ages, the study of childhood conditions will be pressing, because diseases (and lifestyle habits) originating in childhood can leave significant traces on illness over the life course.

SEE ALSO Volume 1: *Birth Weight; Bronfenbrenner, Urie; Disability, Childhood and Adolescence; Health Differentials/Disparities, Childhood and Adolescence; Health Care Use, Childhood and Adolescence; Infant and Child Mortality.*

BIBLIOGRAPHY

Akinbami, L. J., & Schoendorf, K. C. (2002). Trends in childhood asthma: Prevalence, health care utilization, and mortality. *Pediatrics, 110*(2), 315–322.

Bearison, D. (1991). *"They never want to tell you": Children talk about cancer.* Cambridge, MA: Harvard University Press.

Bellin, M., & Kovacs, P. (2006). Fostering resilience in siblings of youths with a chronic health condition: A review of the literature. *Health & Social Work, 31*, 209–216.

Bibace, R., & Walsh, M. (1980). Development of children's concepts of illness. *Pediatrics, 66*(6), 912–917.

Bluebond-Langner, M. (1978). *The private worlds of dying children.* Princeton, NJ: Princeton University Press.

Bluebond-Langner, M. (1996). *In the shadow of illness: Parents and siblings of the chronically ill.* Princeton, NJ: Princeton University Press.

Buchbinder, M. H., Detzer, M. J., Welsch, R. L., Christiano, A. S., Patashnick, J. L., & Rich, M. (2005). Assessing adolescents with insulin-dependent diabetes mellitus: A multiple perspective pilot study using visual illness narratives and interviews. *Journal of Adolescent Health, 36*: 71e9–71e13.

Clark, C. D. (2003). *In sickness and in play: Children coping with chronic illness.* New Brunswick, NJ: Rutgers University Press.

Clemente, I. (2007). Clinicians' routine use of non-disclosure: Prioritizing "protection" over the information needs of adolescents with cancer. *Canadian Journal of Nursing Research, 39*, 18–34.

Eiser, C. (1989). Children's concepts of illness: Towards an alternative to the "stage" approach. *Psychology and Health, 3*(2), 93–101.

Kleinman, A. (1988). *The illness narratives: Suffering, healing and the human condition.* New York: Basic Books.

La Greca, A. M., Bearman, K. J., & Moore, H. (2002). Peer relations of youth with pediatric conditions and health risks: Promoting social support and healthy lifestyles. *Journal of Developmental and Behavioral Pediatrics, 23*: 1–10.

Madan-Swain, A., Fredrick, L. D., & Wallander, J. (1999). Returning to school after a serious illness or injury. In R. T. Brown (Ed.), *Cognitive aspects of chronic illness in children* (pp. 312–332). New York: Guilford Press.

Mavoa, H. (1999). Tongan children with asthma in New Zealand. *Pacific Health Dialog 6*(2), 236–239.

Naar-King, S., Podolski, C. L., Ellis, D. A., Frey, M. A., & Templin, T. (2006). Social ecological model of illness management in high-risk youths with Type 1 diabetes. *Journal of Consulting and Clinical Psychology, 74*: 785–789.

Perrin, E. C., & Gerrity, P. S. (1981). There's a demon in your belly: Children's understanding of illness. *Pediatrics, 67*(6), 841–849.

Rich, M., Lamola, S., & Chalfen, R. (1998). Video Intervention/Prevention Assessment (VIA): An innovative methodology for understanding the adolescent illness experience. *Journal of Adolescent Health, 22*, 128.

Sourkes, B. (1995). *Armfuls of time: The psychological experience of the child with a life-threatening illness.* Pittsburgh: University of Pittsburgh Press.

UNICEF. (2007a). *Children and AIDS: A stocktaking report—actions and progress during the first year of Unite for Children Against AIDS.*

UNICEF. (2007b). *Progress for children.*

Wise, P. (2004). The transformation of child health in the United States. *Health Affairs, 23*(5), 9–25.

Mara Buchbinder
Cindy Dell Clark

IMMIGRATION, CHILDHOOD AND ADOLESCENCE

Over the coming decades, immigrant children will have a significant impact on the social, cultural, and economic future of the United States. Immigrant children have a powerful effect on U.S. population growth and its composition. As a diverse group, they bring important cultural assets, but, at the same time, they are considered a high-risk group because of their relatively low socioeconomic status, lack of English skills, and unfamiliarity with U.S. culture. Moreover, a migration event in a child's or his or her parent's life brings about changes that will alter the life course, with social and economic ramifications that will reverberate throughout the rest of the child's life, as well as those of future generations. Thus, the well-being of immigrant children and their adaptation into U.S. society are key issues of analysis for researchers from a broad range of disciplines.

This entry presents information about the demographic characteristics of immigrant children in the United States, as well as a discussion of their resources and obstacles and the main theories used to explain their adaptation to U.S. society and its institutions. The norm is followed in demographic research by using generational categories to discuss immigrant children. Immigrant children discussed in this entry are both first- and second-generation. First-generation children are foreign-

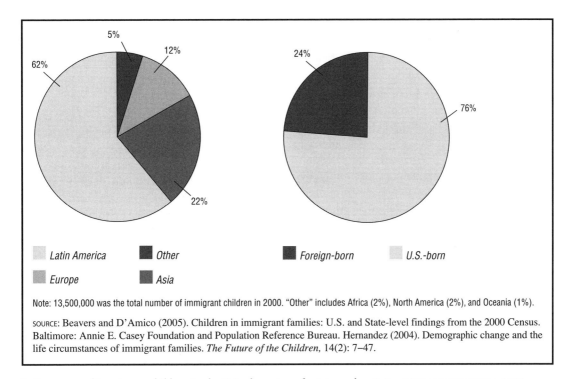

Note: 13,500,000 was the total number of immigrant children in 2000. "Other" includes Africa (2%), North America (2%), and Oceania (1%).

SOURCE: Beavers and D'Amico (2005). Children in immigrant families: U.S. and State-level findings from the 2000 Census. Baltimore: Annie E. Casey Foundation and Population Reference Bureau. Hernandez (2004). Demographic change and the life circumstances of immigrant families. *The Future of the Children,* 14(2): 7–47.

Figure 1. *Percentage of immigrant children in the U.S., by region of origin and nativity.* CENGAGE LEARNING, GALE.

born and have foreign-born parents. Second-generation children are U.S.-born but have at least one foreign-born parent.

THE DEMOGRAPHY OF IMMIGRANT CHILDREN

According to population projections, by 2050, one in five individuals in the United States will be an immigrant, compared to one in eight in 2005. Part of this trend is because of the increase in the number of migrants to the United States, but it is also because of slightly higher fertility rates among recent immigrants, especially Hispanics. According to 2005 data, the average Hispanic woman gives birth to 2.5 children over her lifetime, whereas for Whites the figure is 1.9. Consequently, immigrant children are one of the fastest growing populations in the United States. Between 1990 and 2000, the number of immigrant children increased by 63%, and, in 2000, about 15 million immigrant children accounted for 20% of the U.S. child population, with most of them (68%) living in California, Texas, New York, Florida, Illinois, and New Jersey. However, immigrant settlement patterns are rapidly becoming more dispersed throughout the United States (Beavers & D'Amico, 2005).

Immigrant children are an ethnically and culturally diverse subpopulation. Following current overall immigration patterns and differentiating sharply from the immigrant children arriving in the early 1900s, who primarily came from Europe, most of immigrant children in the early 21st century arrive from Latin America and the Caribbean (62%) or Asia (22%). By far, most immigrant children have Mexican origins (39%), followed by Filipino (4%) and Chinese (3%; Beavers & D'Amico, 2005).

IMMIGRANT CHILDREN'S RESOURCES AND OBSTACLES

Compared with children of native-born parents, immigrant children overall are more likely to experience poverty. The 2000 poverty rate for immigrant children was 21%, whereas it was 14% for children in native-born families (Hernandez, 2004). Also, in 2003, 54% of immigrant children lived in families with mean incomes at only half of the federal poverty threshold, 31% had parents with less than a high school diploma, and median household income was $44,600—which is 11% lower than the median income of native families (Capps et al., 2005). However, there are important racial and ethnic differences in the economic challenges they face. For instance, Latino immigrant children are more likely to be poor (32%) than immigrant children from Asia or Africa (15% and 19%, respectively; Lichter, Qian, & Crowley, 2007). Furthermore, the poverty rate for immigrant children with Mexican origins is 31%, compared to 11% among Filipino immigrant children and 5% among Chinese immigrant children (Beavers & D'Amico, 2005).

In addition to economic hardship, another potential obstacle to positive immigrant adaptation is the lack of English skills. Nationally, 18% of the U.S. child population, and 72% of immigrant children, speak a language other than English at home (Hernandez, 2004). Moreover, in 2000 about 25% of children in immigrant families lived in households where no one age 14 or older spoke English only or spoke English very well. Although prior studies suggest that most children of immigrants report they prefer to use English, retention of a native language varies by country of origin (Portes & Hao, 1998). Children from Latino backgrounds, especially Mexicans, appear most likely to maintain foreign language proficiency, presumably because Latino children all share a common linguistic background and thus have more social opportunities to use their native language.

Concerning family structure, immigrant children are more likely to live in two-parent families (78%) than children of native-born parents (65%; Beavers & D'Amico, 2005). Moreover, immigrant children are more likely to live in a multigenerational household that may include grandparents. About 27% of children in immigrant families live with a grandparent or have another relative present, whereas about 8% of nonimmigrant children live with grandparents and 12% with other relatives (Hernandez, 2004). Living in two-parent families may facilitate access to resources, and, therefore, these children's parents may provide greater amounts of time and attention.

At the same time, immigrant parents are reportedly more optimistic about the future, have higher educational expectations for their children than native parents, and provide positive home environments. Immigrant parents and children maintain healthy communication, and parents act as successful role models for their children, promoting strong family values and respect for authority figures (Pong, Hao, & Gardner, 2005).

Further, immigrant parents tend to cultivate strong ethnic communities and useful social networks, participating in ethnic organizations that encourage positive outcomes among their children. Being part of a strong ethnic community yields possibilities for valuable information about jobs and educational opportunities, helpful social contacts, or financial support (Zhou & Bankston, 1994). Cohesive ethnic communities facilitate social control among adolescents, affirm cultural values, and may provide exposure to positive role models. Strong family and community ties are protective factors that can counterbalance the economic disadvantage immigrant children experience.

IMMIGRANT CHILDREN'S OUTCOMES

Given the concerns about the impact of immigration on the growth and composition of the U.S. population, sociologists have focused many of their studies on the assimilation experiences of immigrant children. Based on the assimilation patterns observed among European immigrants in the early 1900s and because of the notion of U.S. society as a melting pot, it was expected that immigrant adults would quickly experience the benefits of the U.S. society's opportunity structure, and thus their children would mature to become nearly indistinguishable from the mainstream U.S. population.

However, empirical research during the 1990s yielded mixed results and some paradoxical conclusions. In particular, although immigrant families face financial challenges, their children often have better health status and higher educational achievement than their native-born counterparts. Yet some research has found troubling negative outcomes among immigrant youth, showing lower rates of secondary and postsecondary educational completion and declining health status over time.

THE HEALTH STATUS OF IMMIGRANT CHILDREN

Research has shown that immigrant children are less likely to have health insurance or a primary care physician than children of U.S.-born parents. In 2004, 20% of immigrant children between the ages of 6 and 17 were uninsured, compared to 8% of children of native-born parents. At the same time, children of immigrants were less likely than children of native-born parents to be covered under employer-based insurance in 2002 (22% and 31%, respectively; Capps et al., 2004).

Although insurance coverage is less likely among immigrants, immigrant mothers tend to give birth to healthier babies compared to U.S.-born mothers (Finch, Lim, Perez, & Do, 2007). Mexican immigrant mothers tend to have healthier behaviors during pregnancy than comparable native-born mothers, and they are more likely to breast-feed their infants (Padilla, Radey, Hummer, & Kim, 2006). Also, asthma and obesity are less common among first-generation children than among those born in the United States (Fuligni & Hardway, 2004), and research pertaining to adolescent risk behavior demonstrates that first-generation immigrant youths are less involved in risky behaviors—such as smoking or drug and alcohol abuse—than native-born adolescents (Cavanagh, 2007).

Yet many immigrant children experience increasingly negative health outcomes across generations and over time. As measured by body mass index, the health status of immigrant children decreases after arriving to the United States, suggesting that exposure to American culture may be associated with increasing rates of being overweight and obesity (Van Hook & Balistreri, 2007). Further, in 2002, immigrant parents of children under age 6 were more likely to report their children were in fair or poor health than native-born parents, and this difference even increased among low-income immigrant families (Capps et al., 2004).

Importantly, both individual differences in ethnicity and generation status affect the health outcomes of immigrant children and adolescents. Low-income immigrant parents in rural areas are less likely to seek preventive medical and dental care. Macro-level factors also drive immigrant and native-born differences in health, such as residential concentrations of immigrant groups. Neighbor-hood segregation can have protective effects by facilitating the informal flow of information about health resources for immigrant families, such as those from Mexican origins, but deleterious effects for some ethnic minorities, especially Puerto Ricans (Lee & Ferraro, 2007).

EDUCATIONAL OUTCOMES AMONG IMMIGRANT CHILDREN

Most immigrant children are considered to be educationally at risk because of relatively low levels of school readiness, a lack of English skills due to a non-English home background, socioeconomic disadvantage, and their families' lack of familiarity with U.S. school systems. Overall, immigrant children, especially from Latino families, are less likely to enroll in formal preschool and center-based childcare that could help prepare them for the demands of formal schooling (Takanishi, 2004).

English Learners. *Terri Gehman, center, assists English-as-a-Second-Language students as they shape letters at Stone Spring Elementary School in Harrisonburg, VA. A decade of explosive growth in immigrants has given Harrisonburg's schools a high percentage of English-as-a-Second-Language students.* **AP IMAGES.**

Research has provided little consensus as to how immigrant children perform in the U.S. educational system. Some research suggests that immigrants actually have better educational outcomes than similar native-born students; in contrast, some findings have pointed to troubling outcomes among immigrant students. Thus, mixed findings regarding immigrant school performance are reflected in the variety of approaches represented in the literature on immigrant educational outcomes.

When analyzing generational differences in achievement outcomes with standardized test scores and school grades, research demonstrates that first- and second-generation Mexican children arrive at kindergarten with lower cognitive skills than third- and higher generation Mexican children (Crosnoe, 2005), and, accordingly, by fifth grade, immigrants are found to have lower math and reading scores than children from U.S.-born families (Reardon & Galindo, 2008). Yet research examining performance during the middle-school grades has found that native-born Hispanics have higher tests scores but lower grade point averages than foreign-born Hispanics (Portes & Rumbaut, 2001) and that first- and second-generation eighth-graders have higher grade point averages and test scores than children in native-born families. When comparing educational outcomes between White students and the two most predominant immigrant groups, research shows that Latino students' educational outcomes lag far behind White students but that Asian immigrant students outperform White children.

In terms of high school or college completion, first-generation students demonstrate lower educational attainment than the second generation. However, the same patterns of improvement in attainment have not been found between first or second generations or third and higher generations.

In sum, and as with immigrant children's health outcomes, the differences in educational outcomes between immigrant groups, from different social strata, and from varied cultural backgrounds highlight the necessity of acknowledging the vast diversity among the immigrant population and the correspondingly complex picture of their educational experience in the United States.

THEORETICAL FRAMEWORKS

A number of theoretical arguments have been used to explain the adaptation experiences of immigrant children. Most of them point to individual, family, community, and societal factors to explain differences in outcomes. Some theories attempt to explain broad patterns of assimilation among all immigrants, but cultural arguments are also commonly cited to explain the outcomes of certain groups of immigrant children. Three primary approaches to a theoretical understanding of immigrant assimilation are particularly prominent in the literature: (a) classical theories of assimilation, (b) segmented assimilation theory, and (c) the immigrant optimism hypothesis. Each of these predicts different patterns of improvement or deterioration in outcomes across time or generations.

First, classical theories of assimilation define *assimilation* as the acquisition of the receiving country's cultural values and historical memories. Adherents to this perspective argue that immigrants, regardless of place of origin or characteristics, inevitably become assimilated into American society. As a result, they become more similar to participants in the mainstream culture over time through social incorporation and economic upward mobility. This theory predicts that immigrant children experience improvement in outcomes based on generational succession, length of residence in the country, and concomitant integration into local social institutions.

Second, the segmented assimilation theory, developed by Alejandro Portes and Min Zhou (1993), emphasizes that immigrants from different origins assimilate into distinct sectors of U.S. society. The authors identified three alternate scenarios for assimilation. First, immigrants may experience upward assimilation with rapid integration into and acceptance by the mainstream, White middle class, sharing their cultural customs, economic success, and social advantages. Second, immigrant children may experience downward assimilation by becoming part of an existing minority underclass culture, thereby destined for permanent poverty and disadvantage. Finally, immigrant children may achieve upward assimilation with biculturalism (or assimilation without accommodation) by achieving similar economic status as the White middle class while at the same time preserving their own culture. Proponents of segmented assimilation theory argue that the assimilation of immigrant children is a function of the interaction of four main factors: the immigration experience of the first generation, the differences in the pace of acculturation between immigrant children and their parents, available family and community resources, and macro-level barriers to integration, such as racial discrimination or local residential segregation.

Finally, Grace Kao and Marta Tienda (1995) proposed the immigrant optimism thesis, specifically to explain differences in educational outcomes across immigrant groups. This hypothesis predicts that U.S.-born children of foreign-born parents will have better educational outcomes than first- or third- and higher generation students. Second-generation students benefit from strong parental support; it is through high expectations and encouragement that immigrant parents reinforce educational success. However, this theory also predicts that second-generation children will experience improvement in outcomes not only because of their positive home environments but also because of their relatively stronger English skills, which are required to succeed in schools.

Cultural arguments also have been applied in explaining immigrant children's outcomes, particularly their educational outcomes. From a cultural perspective, John Ogbu (1987) proposed the cultural ecology model for explaining educational failure and success and categorized immigrants into two groups: voluntary and involuntary. Involuntary immigrants are those incorporated into the United States through slavery, conquest, or colonization, and voluntary immigrants left their home countries with the belief that greater opportunities awaited them in the United States.

Both voluntary and involuntary immigrants develop different perceptions of opportunities, adaptive strategies, and responses to treatment by the dominant group. Involuntary immigrants compare their situation and experiences to that of the mainstream, perceiving the dominant mainstream group as oppressive, resisting assimilation into this dominant mainstream group and developing an oppositional culture to preserve their own cultural identity. Further, they perceive schools to be dominated by mainstream, White, middle-class culture and thus may not cite education as a legitimate route to economic success—resulting in their limited effort and involvement in school. Voluntary immigrants compare their current situation with experiences "back home." Therefore, they perceive fewer obstacles to upward mobility, optimize the opportunities the receiving country provides, and have a positive frame of reference.

Also, the cultural discontinuity theory departs from explanations of minority students' lower performance using individual or group variables but instead claims differences are a consequence of mismatches in communication, behaviors, or language (Suárez-Orozco & Suárez-Orozco, 1995). Consequently, ethnic minorities' low school achievement is the result of curriculum designs that are not responsive to cultural or language differences between the dominant group and the immigrant group.

Most of the cultural arguments reviewed in this section bring important insights into how macro-social dynamics affect minority students' individual behaviors. However, these models are commonly considered to be highly deterministic because they do not allow for the possibility of school success or only portray schools as institutions engaged in the social reproduction of inequality.

FUTURE DIRECTIONS

During the latter part of the 20th century, a number of perspectives were offered from a variety of disciplines toward understanding immigrant children's experiences and outcomes, and valuable theoretical formulations and empirical understanding have been achieved. However, the study of immigrant children remains immersed in conflicting theoretical discussions, mixed empirical findings, and unanswered fundamental questions, especially the question of whether immigrant children will successfully integrate into U.S. society, experiencing upward social and economic mobility.

The answer to this question is elusive because of at least three key issues: First, immigrants are not homogeneous; second, the characteristics of the immigrant population are still changing over time; and third, researchers grapple with the question of *who* migrates. A great deal of evidence suggests that the families who are able to migrate to the United States tend to be more educated, have greater financial resources, and have better health statuses than their counterparts who remain in the native country. This pattern is referred to as *immigrant selectivity*, which implies that those who migrate to the United States are the healthiest and most able of the sending country's population. Typically, immigrants arriving with higher levels of human or financial capital have greater access to gainful employment, better neighborhoods, and superior schools. The issue of immigrant selectivity raises another question: To which group does one compare the outcomes of immigrants in the United States to gauge their success—to the White middle class, their counterparts in their home country, or others?

In the future, research will also need to take into greater account the timing of immigration and age at arrival. A number of studies have shown that students arriving earlier in life tend to have better outcomes than those arriving in later adolescence, due in part to the fact that social adjustment depends on the age at which they arrive in the United States. Children who arrive in their preschool years, having more time to mature, can better adapt to the U.S. educational system, master English, and be less identifiable as foreign, which is potentially stigmatizing, relative to children who arrive during adolescence.

Another issue receiving increased attention is transnational movement between the United States and immigrants' home countries and how this type of movement affects assimilation. Additionally, the increasing diversity of the U.S. population inherently calls into question traditional racial and ethnic identities. Many immigrant youths assert biracial or binational identities, which also have implications for social incorporation into American society. In sum, as immigration continues to be a politically charged topic, questions regarding how immigrants fare once they arrive in the United States are unlikely to abate. Immigrant children are the fastest growing population in the United States, and most of them are native-born

U.S. citizens, which has implications for social and educational policy in terms of essential and beneficial service programs (e.g., English as a second language course offerings at schools, health information and medical services). Therefore, the future of the United States will benefit from a thorough, comprehensive, and multifaceted approach to understanding immigrant children's experiences and outcomes.

SEE ALSO Volume 1: *Assimilation; Bilingual Education; Racial Inequality in Education.*

BIBLIOGRAPHY

Beavers, L., & D'Amico, J. (2005). *Children in immigrant families: U.S. and state-level findings from the 2000 Census.* Baltimore, MD: Annie E. Casey Foundation; Washington DC: Population Reference Bureau.

Capps, R., Fix, M., Murray, J., Ost, J., Passel, J. S., & Herwantoro, S. (2005). *The new demography of America's schools: Immigration and the No Child Left Behind Act.* Washington, DC: The Urban Institute. Retrieved April 30, 2008, from http://www.urban.org

Capps, R., Fix, M., Ost, J., Reardon-Anderson, J., & Passel, J. S. (2004). *The health and well-being of young children of immigrants.* Washington, DC: The Urban Institute. Retrieved April 30, 2008, from http://www.urban.org

Cavanagh, S. E. (2007). Peers, drinking, and the assimilation of Mexican-American youth. *Sociological Perspectives, 50*(3), 393–416.

Crosnoe, R. (2005). Double disadvantage or signs of resilience? The elementary school context of children from Mexican immigrant families. *American Educational Research Journal, 42*(2), 269–303.

Finch, B. K., Lim, N., Perez, W., & Do, D. P. (2007). Toward a population health model of segmented assimilation: The case of low birth weight in Los Angeles. *Sociological Perspectives, 50*(3), 445–468.

Fuligni, A. J., & Hardway, C. (2004). Preparing diverse adolescents for the transition to adulthood. *The Future of the Children, 14*(2), 99–119.

Hernandez, D. J. (2004). Demographic change and the life circumstances of immigrant families. *The Future of the Children, 14*(2), 17–47.

Kao, G., & Tienda, M. (1995). Optimism and achievement: The educational performance of immigrant youth. *Social Science Quarterly, 76*(1), 1–19.

Lee, M., & Ferraro, K. F. (2007). Neighborhood residential segregation and physical health among Hispanic Americans: Good, bad, or benign? *Journal of Health and Social Behavior, 48*(2), 131–148.

Lichter, D. T., Qian, Z., & Crowley, M. L. (2007). Poverty and economic polarization among children in racial minority and immigrant families. In D. R. Crane & T. Heaton (Eds.), *Handbook of families and poverty: Interdisciplinary perspectives* (pp. 119–143). Los Angeles, CA: Sage.

Ogbu, J. U. (1987). Variability in minority school performance: A problem in search of an explanation. *Anthropology and Education Quarterly, 18*(4), 312–334.

Padilla, Y. C., Radey, M. D., Hummer, R. A., & Kim, E. (2006). The living conditions of U.S.-born children of Mexican immigrants in unmarried families. *Hispanic Journal of Behavioral Sciences, 28*(3), 331–349.

Pong, S., Hao, L., & Gardner, E. (2005). The roles of parenting styles and social capital in the school performance of immigrant Asian and Hispanic adolescents. *Social Science Quarterly, 86*(4), 928–950.

Portes, A., & Hao, L. (1998). *E pluribus unum*: Bilingualism and loss of language in the second generation. *Sociology of Education, 71*(4), 269–294.

Portes, A., & Rumbaut, R. (2001). *Legacies: The story of the immigrant second generation.* Berkeley: University of California Press.

Portes, A., & Zhou, M. (1993). The new second generation: Segmented assimilation and its variants among post-1965 immigrant youth. *The Annals of the American Academy of Political and Social Science, 530*(1), 74–98.

Reardon S. F., & Galindo C. (2008). The Hispanic–White achievement gap in math and reading in the elementary grades. Stanford, CA: Stanford University Institute for Research on Education Policy and Practice. Retrieved April 30, 2008, from http://irepp.stanford.edu/publications

Suárez-Orozco, M. M., & Suárez-Orozco, C. E. (Eds.). (1995). *Transformations: Immigration, family life, and achievement motivation among Latino adolescents.* Stanford, CA: Stanford University Press.

Takanishi, R. (2004). Leveling the playing field: Supporting immigrant children from birth to eight. *The Future of Children, 14*(2), 61–80.

Van Hook, J., & Balistreri, K. S. (2007). Immigrant generation, socioeconomic status, and economic development of countries of origin: A longitudinal study of BMI among children. *Social Science and Medicine, 65*(5), 976–989.

Zhou, M., & Bankston, C. L., III. (1994). Social capital and the adaptation of the second generation: The case of Vietnamese youth in New Orleans. *International Migration Review, 28*(4), 821–845.

Claudia Galindo
Rachel E. Durham

INCARCERATION, CHILDHOOD AND ADOLESCENCE

SEE Volume 1: *Crime, Criminal Activity in Childhood and Adolescence; Juvenile Justice System.*

INFANT AND CHILD MORTALITY

The death of an infant or child is always a tragic and costly event. The costs incurred include the trauma experienced by family and friends, the loss of years of expected life, and the monetary expenditures incurred in the attempt to preserve the life of the infant or child. Moreover, infant/child mortality is often taken as a useful indicator of quality of life across the nations of the world (Caldwell, 1996). Infant mortality, in particular, is viewed as "a synoptic indicator of the health and social condition of a population" (Gortmaker & Wise, 1997, p. 147). Lowering infant and child mortality and eliminating inequalities in rates that exist by socioeconomic status (SES), race/ethnicity, and urban/rural residence are prominent goals for both national and international health organizations.

MEASURING INFANT AND CHILD MORTALITY

The infant mortality rate (IMR) is defined as deaths of children before 1 year of age per 1,000 live births in a given year (or a set of contiguous years). The IMR may be computed on either a calendar year or a birth cohort basis. Using the calendar year approach, if an infant is born in October of year t and dies in January of year $t + 1$, the birth will enter the denominator in year t, but the death will enter the numerator in year $t + 1$. Thus, only the birth cohort approach provides a true probability based on the population at risk, but it requires that birth and infant death records be linked from one year to the next. Because the two measurement approaches usually yield fairly similar results, and because a number of countries do not link records, international comparisons tend to be based on calendar year computations.

Infant mortality is often divided into the neonatal period (deaths before the 28th day of life) and the post-neonatal period (deaths from the 28th day through the 364th day). Although timing of death is important, at times it is useful to partition infant mortality by cause-of-death structure into endogenous and exogenous deaths. The first category consists of deaths that are due largely to natal and antenatal factors whereas exogenous deaths are more apt to be due to environmental factors—both natural and social.

As used here, the child mortality rate refers to deaths of children younger than 5 years of age per 1,000 live births in a given year or time period. Other definitions might be used, such as deaths of children 1 to 4 or 1 to 14 years of age. However, the largest amount of data available from international organizations (e.g., United Nations Children's Fund [UNICEF], 2006) is for deaths to children less than 5 years of age. Although this definition combines deaths of children aged 1 through 4 with infant deaths, it is useful, whenever possible, to maintain a distinction between infant mortality and child mortality. Especially in developed countries, the etiologies of infant mortality and child mortality are distinct and may

require different specific preventative approaches. At a more general level, reducing rates of both infant and child mortality often require approaches that are fundamentally similar.

Precise and comprehensive international comparisons of infant and child mortality rates are often difficult to achieve because of differentials in definitions, coverage, accuracy, and availability of vital statistics. In such cases, researchers may derive estimates using birth histories from nationally representative surveys (e.g., the Demographic and Health Survey) or indirect estimates using census data.

CONCEPTUAL FRAMEWORK

Mosley and Chen (1984) have provided a conceptual framework, used primarily in developing countries, to study both social and biological determinants of child mortality. Working under the premise that "socioeconomic determinants must operate through more basic proximate determinants that in turn influence the risk of disease and the outcome of disease processes" (p. 27), these authors developed a multitiered framework consisting of three levels of determinants. The first—individual-level determinants—includes individual characteristics (e.g., maternal education, maternal health), as well as cultural factors such as norms and attitudes that dictate power relationships in the household, the value placed on children, and beliefs about the causes of disease. Next, household-level factors, mainly consisting of income and wealth, influence child mortality through the availability of food, potable water, clothing, housing, fuel and energy, transportation, hygienic and preventive measures, care for the sick, and health information. Finally, Mosley and Chen identified a number of community-level variables that potentially affect child mortality. These include ecological factors, such as rainfall and climate (which can influence food and water availability, sewage disposal, and the existence of parasites and bacterium), the political economy (organization of production, infrastructure, and political institutions), and heath system variables (such as vaccination programs, public information and education, cost subsidies, and technology).

Complementing Mosley and Chen's (1984) work, the Link and Phelan's (2002) conceptualization of "fundamental" social causes is often applied in more developed countries (e.g., Frisbie, Song, Powers, & Street, 2004). The term *fundamental* implies that the ability of individuals to reduce the risk of disease and death "is shaped by resources of knowledge, money, power, prestige, and beneficial social connections" (Link & Phelan, p. 730). Wise (2003) made the case that the widening relative racial gap in infant mortality in the United States is largely a function of differential access to health care at a time when the means of beneficial intervention have

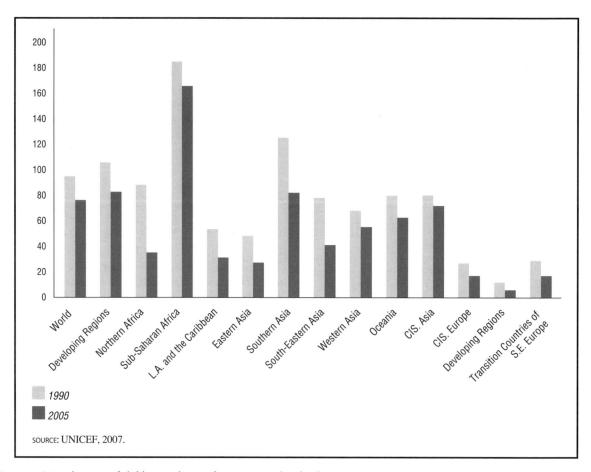

Figure 1. Mortality rate of children under age five per 1,000 live births. CENGAGE LEARNING, GALE.

been greatly expanded, and Gortmaker and Wise (1997) warned that greater racial disparities in mortality may follow advances in health services technology because of inequality of access to health care, which, in turn, is attributable to social inequality.

In both developing and developed nations, empirical research (e.g., Hummer et al. 1999; Wang, 2003) has identified a large number of risk factors that are associated with higher risk of infant and child mortality. Among the maternal risk factors are very young or old maternal age, low education, disadvantaged minority status, lack of adequate prenatal care, no breast-feeding, short birth spacing, risky behaviors (smoking, alcohol consumption, and drug use), low weight gain during pregnancy, and previous pregnancy loss and morbidity. Family risk factors include low SES (e.g., income and wealth), rural residence, and a large number of siblings. Among the community (or regional) risk factors are poor sanitation levels, lack of availability of health care, and absence of health-promoting programs. Characteristics of the infant itself, including low birthweight, preterm birth, gestational age, fetal immaturity, male sex, and

plurality are important and may be regarded as the most proximate risk factors for infant mortality.

INFANT AND CHILD MORTALITY: RATES AND TRENDS

Rates of infant and child mortality differ substantially around the world, and progress in reducing these rates varies dramatically by level of societal development. Figure 1 displays estimated under-5 mortality rates per 1,000 live births by world regions for the years 1990 and 2005. For the world as a whole, the relative decline during this period was approximately 20% (from 95 to 76 deaths per 1,000 live births). Whereas the relative rate of decline (22%) slightly outpaced that for the world, child mortality levels were much higher in developing regions (106 in 1990 and 83 in 2005). In sub-Saharan Africa, the rates were 185 and 166 in 1990 and 2005, respectively—a drop in the rate of only 10%. In sharp contrast, in developed countries, children died at a rate of 12 per 1,000 at the first time point, with a decline to 6 per 1,000 in 2005 (United Nations, 2007).

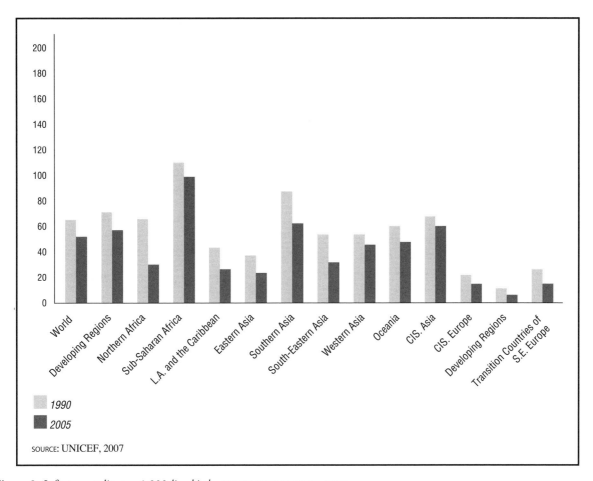

Figure 2. *Infant mortality per 1,000 live births.* CENGAGE LEARNING, GALE.

Comparable values for infant mortality appear in Figure 2. Across all countries, the relative decline between 1990 and 2005 (20%) matched that for child mortality, and the same was true in developing regions, which recorded a reduction in the IMR from 71 per 1,000 to 57 per 1,000 over the time period. Once again, sub-Saharan Africa was the most disadvantaged. The IMR for that region was 110 in 1990 and 99 in 2005—again, an improvement of only 10%. Developed regions were once more able to achieve a 50% decrease (from 10 to 5 infant deaths per 1,000 live births). Although their rates do not approach the high levels of sub-Saharan Africa, other regions, particularly northern Africa, southern Asia, and the Asian portion of the Commonwealth of Independent States (essentially the countries that emerged from the breakup of the Soviet Union) are also greatly disadvantaged with respect to both child and infant mortality. An important distinction here is that northern Africa and southern Asia were able to achieve substantial reductions in child and infant mortality rates between 1990 and 2005, whereas the decline in the Asian portion of the Commonwealth of Independent States was relatively small (United Nations, 2007).

In addition to between-country differences in infant and child mortality, profound disparities exist within nations as well. The most affluent or otherwise more advantaged segments of the population tend to have lower rates of infant and child mortality, whereas less advantaged groups, on average, consistently have higher mortality. Such a polarization has been observed both for entire countries (Frenk, Bobadilla, Sepúlveda, & López Cervantes, 1989) and for local areas (e.g., Forbes & Frisbie, 1991). Infant and child mortality have been found to vary substantially across groups based on characteristics such as education, income, race/ethnicity, and urban/rural residence.

Within-country disparities in infant and child mortality exist for both developed and developing nations. For example, in the United States the mortality rate for non-Hispanic White infants in 2004 stood at 5.7 per 1,000 live births, as compared to a rate of 13.6 for non-Hispanic Black infants. Other U.S. minorities, including Native Americans (IMR = 8.5) and Puerto Ricans (IMR = 7.8), also had elevated rates (Mathews & MacDorman, 2007). One likely reason for the race/ethnic disparities in the United States is the social and economic inequality that exists between the more affluent White population and disadvantaged minorities, such as differences in access to health care, health insurance, and nutritional foods.

SPECIFIC CAUSES OF DEATH

Research on specific causes of death is crucial because such research can uncover factors that underlie infant and child mortality rates and thus informs attempts to lower these rates. The advances most responsible for lowering early childhood and postneonatal mortality are not the same as those for achieving lower neonatal mortality. Half of the deaths of children under 5 in less developed nations are due to five specific conditions (pneumonia, diarrhea, malaria, measles, and AIDS), and more than half of child deaths from all causes are associated with malnutrition (UNICEF, 2006). According to UNIICEF,

> Most of these lives could be saved by expanding low-cost prevention and treatment measures. These include exclusive breast-feeding of infants, antibiotics for acute respiratory infections, oral rehydration for diarrhoea, immunization, and the use of insecticide-treated mosquito netting and appropriate drugs for malaria. (p. 3)

One general conclusion regarding less developed nations is that low- (or no-) cost measures such as oral rehydration and breast-feeding are effective in driving down IMRs, whereas improvements in the economic and health infrastructures are needed to substantially reduce child mortality. This implies that, in developing nations at least, it is easier to achieve reductions in the IMR than in child mortality (Wang, 2003).

Looking only at the neonatal component of under-5 mortality for the world as a whole, three causes—preterm birth (28%), severe infections (26%), and birth asphyxia (23%)—account for more than three-fourths of all neonatal deaths. Adding congenital anomalies (8%) and neonatal tetanus (7%) increases the total to 92%. Although improved living standards have a beneficial effect on neonatal survival, research has suggested that lowering neonatal mortality depends more heavily on improvements in maternal and obstetrical care and on medical advances.

Specific causes of child mortality are different in developed, as compared to developing, countries. To illustrate, in the United States in 2004, the five leading causes of child mortality (using data on children ages 1 to 4 and excluding infant deaths) were unintentional injuries, birth defects, cancer, homicide, and heart disease, in descending order. Unintentional injuries were far and away the most prominent cause (10.3 deaths per

100,000 children ages 1 through 4). The next most frequent cause was birth defects (3.6 per 100,000), whereas the rates for each of the other three causes were less than 3.0 (Federal Interagency Forum on Child and Family Statistics, 2007).

FUTURE DIRECTIONS

One area in which future research is clearly needed involves the stubborn disparities in rates of infant and child mortality across subpopulations within countries. Theoretically, if one subpopulation of a country is able to achieve low rates of infant and child mortality, others should be able to do so as well. Yet, this is often not the case:

> The fact that a particular health intervention is used to prevent or treat a disease that is more prevalent among the poor does not mean that the poor will be the ones who benefit from increased spending on that intervention. In fact, without specific attention, just the opposite is likely to happen. (Freedman et al., 2005, p. 10)

Thus, rate disparities signal continued need for improvement. Disparities in rates of mortality by urban/rural residence, socioeconomic status, and race/ethnicity may be largely driven by differentials in access to health-promoting resources across subpopulations. To illustrate, when pulmonary surfactant replacement therapy was approved by the U.S. Food and Drug Administration (FDA) in 1990 for treatment of respiratory distress syndrome (RDS), individuals with higher incomes, health insurance, and greater access to the health care system were clearly best positioned to take advantage of this beneficial intervention. After FDA approval of surfactant therapy, the IMR for RDS declined for both Black and White infants, but the decline was much more substantial for Whites than for Blacks. Both clinical research (Hamvas et al., 1996) and a nationwide study using linked birth and infant death records (Frisbie et al., 2004) demonstrated that what had been a survival advantage in the risk of infant death from RDS for Blacks as compared to Whites before surfactant therapy was approved became a Black survival disadvantage after the approval of this innovation in perinatal technology. As suggested by Link and Phelan (2002), it is likely that more fundamental societal changes in resource distribution will have to occur before such discrepancies can be eliminated.

There are also paradoxical findings that need further investigation. For example, the U.S. Hispanic population, which is quite disadvantaged according to socioeconomic characteristics and health care access, is nevertheless characterized by an IMR that is much lower than that of non-Hispanic Blacks (with whom they share social disadvantages) and similar to, or modestly lower than, that of non-Hispanic Whites. Further, the mortality rate of infants born to Hispanic immigrant women is even lower than among the native-born segment of this population (Hummer, Powers, Pullum, Gossman, & Frisbie, 2007). Several hypotheses have been offered as potential explanations for the paradoxically low IMR of U.S. Hispanics (especially those of Mexican origin). These include positive health selection of migration from Mexico and other countries of Latin America (that is, the healthiest persons migrate to the United States), culturally based social and family support systems that make for more positive pregnancy outcomes among this population, and healthier behaviors among Hispanics (Frisbie, 2005). Further understanding of positive mortality outcomes in the context of social disadvantage can provide valuable insights for programs and policies designed to assist other populations that currently experience higher rates of mortality.

Similarly, in developing countries, Caldwell (1986) has identified populations that are "health achievers" in the area of infant and child mortality in the face of resource disadvantages. Focusing on areas within countries (e.g., the state of Kerala within India) and entire countries (e.g., Costa Rica) that have lower rates of mortality than one might expect based on their overall wealth, Caldwell identified factors such as maternal education and autonomy, political will on the part of local/national governments, and equitable access to health care that can and do make a difference in health achievements. It is important that such achievements continue to be documented and that initiatives be implemented by policy makers as part of the effort to both lower rates of infant and child mortality and reduce resource-based disparities in these outcomes. Given the lack of information on infant and child mortality in a number of countries, realization of the latter goal means that more adequate procedures are needed for collection of data and dissemination of information on mortality in "data-poor" countries.

SEE ALSO Volume 1: *Birth Weight; Health Differentials/ Disparities, Childhood and Adolescence; Illness and Disease, Childhood and Adolescence;* Volume 3: *Death and Dying; Life Expectancy; Mortality.*

BIBLIOGRAPHY

Caldwell, P. (1996). Child survival: Physical vulnerability and resilience in adversity in the European past and the contemporary Third World. *Social Science and Medicine, 43,* 609–619.

Federal Interagency Forum on Child and Family Statistics. (2007). *America's children: Key national indicators of well-being, 2007.* Washington, DC: U.S. Government Printing Office. Retrieved March 5, 2008, from http://www.childstats.gov

Forbes, D., & Frisbie, W. P. (1991). Spanish surname and Anglo infant mortality: Differentials over a half-century. *Demography, 28,* 639–660.

Freedman, L. P., Waldman, R. J., de Pinho, H., Wirth, M. E., Chowdhury, A. M. R., & Rosenfield, A. (2005). *Who's got the power? Transforming health systems for women and children.* London: Earthscan. Retrieved March 5, 2008, from http://www.unmillenniumproject.org/documents

Frenk, J., Bobadilla, J. L., Sepúlveda, J., & López Cervantes, M. (1989). Health transition in middle-income countries: New challenges for health care. *Health Policy and Planning, 4,* 29–39.

Frisbie, W. P. (2005). Infant mortality. In D. L. Poston & M. Micklin (Eds.), *Handbook of Population* (pp. 251–282). New York: Kluwer Academic/Plenum.

Frisbie, W. P., Song, S.-E., Powers, D. A., & Street, J. A. (2004). Increasing racial disparity in infant mortality: Respiratory distress syndrome and other causes. *Demography, 41,* 773–800.

Gortmaker, S. L., & Wise, P. H. (1997). The first injustice: Socioeconomic disparities, health services technology, and infant mortality. *Annual Review of Sociology, 23,* 147–170.

Hamvas, A., Wise, P. H., Yang, R. K., Wampler, N. S., Noguchi, A., Maurer, M. M., et al. (1996, June 20). The influence of the wider use of surfactant therapy on neonatal mortality among Blacks and Whites. *The New England Journal of Medicine, 334,* 1635–1641.

Hummer, R. A., Biegler, M., De Turk, P. B., Forbes, D., Frisbie, W. P., Hong, Y., et al. (1999). Race/ethnicity, nativity, and infant mortality in the United States. *Social Forces, 77,* 1083–1118.

Hummer, R. A., Powers, D. A., Pullum, S. G., Gossman, G. L., & Frisbie, W. P. (2007). Paradox found (again): Infant mortality among the Mexican-origin population in the United States. *Demography, 44,* 441–457.

Link, B. G., & Phelan, J. C. (2002). McKeown and the idea that social conditions are fundamental causes of disease. *American Journal of Public Health, 92,* 730–732.

Mathews, T. J., & MacDorman, M. F. (2007). Infant mortality statistics from the 2004 period linked birth/infant death data set. *National Vital Statistics Reports, 55*(14). Retrieved March 5, 2008, from http://www.cdc.gov/nchs

Mosley, W. H., & Chen, L. C. (1984). An analytical framework for the study of child survival in developing countries. *Population and Development Review, 10*(Suppl.), 25–45.

United Nations. (2007). *Millennium development goals indicators.* Retrieved November 15, 2007, from http://mdgs.un.org/unsd

United Nations Children's Fund. (2006). *Progress for children: A child survival report card.* Retrieved June 13, 2006, from http://childinfo.org/areas/childmortality

Wang, L. (2003). Determinants of child mortality in LDCs: Empirical findings from demographic and health surveys. *Health Policy, 65,* 277–299.

Wise, P. H. (2003). The anatomy of a disparity in infant mortality. *Annual Review of Public Health, 24,* 341–362.

W. Parker Frisbie
Robert A. Hummer
Sarah McKinnon

The authors gratefully acknowledge the support provided by the National Institute of Child Health and Human Development through Grant No. RO1 HD49754.

INTELLIGENCE/IQ

SEE Volume 1: *Cognitive Ability.*

INTERGENERATIONAL CLOSURE

A social network with "intergenerational closure" is characterized by social ties between (a) parents and their children's friends and (b) parents and their children's friends' parents (Coleman, 1988; Coleman & Hoffer, 1987). To illustrate the concept, consider a hypothetical social network of four parent–child dyads (A through D). The sociogram in Figure 1 represents both nodes and social ties within this social network. Dyads A and B have intergenerational closure because all four people know each other and regularly interact. Intergenerational closure requires social ties within and between families and generations. If any one of the social ties between the nodes in dyads A and B were missing, the social network would lack intergenerational closure. Parent–child dyads C and D illustrate the part of the network that lacks intergenerational closure. Child C is friends with child A, but parents A and C do not know each other. In addition, parents A and C do not know each others' children. For family D, it is seen that parent D knows parent B, but child D is not friends with child B. The absence of these ties across families and generations leaves parents C and D in a network without intergenerational closure.

Coleman (1988) theorized that intergenerational closure was critical in facilitating the creation of social capital for individuals within a social network. Network closure is a necessary condition for the formation and enforcement of norms that produce social capital. Parents' connections with their children's friends and their children's friends' parents facilitate: (a) the flow of information about their children's activities, (b) the communication of expectations to students, and (c) the establishment of norms that shape behavior (Coleman, 1988; Coleman & Hoffer, 1987). For Coleman, intergenera-

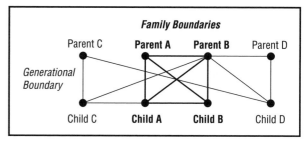

Figure 1. *Intergenerational closure within a social network.*
CENGAGE LEARNING, GALE.

tional closure can lead to the creation of a "functional community" that allows parents to shape and constrain their children's behavior (Coleman & Hoffer, 1987). In short, intergenerational closure is valuable because it produces social capital that helps students. Social networks that lack intergenerational closure make it difficult for parents "to discuss their children's activities, to develop common evaluations of these activities, and to exercise sanctions that guide and constrain these activities" (p. 226). Thus, a lack of intergenerational closure impedes the creation of social capital.

DEVELOPMENT OF THE CONCEPT

Coleman first studied youth and adult cultures in *The Adolescent Society* (1961). In that study, Coleman examined the social relationships and peer culture of students in 10 Illinois high schools in 1957–1958. He found that adolescents of the 1950s were "cut off . . . from the adult society." They were "still oriented toward fulfilling their parents' desires, but they look[ed] very much to their peers for approval as well" (p. 9). The adolescents in Coleman's study were more interested in popularity (via athletics and dating) than academic success. Subsequent research on educational achievement and attainment indicated that both parents' and peers' expectations predicted students' own aspirations and educational attainment (Sewell, Hauser, Springer, & Hauser, 2004). Researchers hypothesized that "significant others" influenced student outcomes by providing role models for students to emulate. These seminal studies inspired many studies that examined how (a) parent–student interactions and (b) peer interactions affect academic outcomes.

Coleman (1988) believed that the heavy emphasis on human capital within the family (i.e., parental education) in status attainment research (Blau & Duncan, 1967; Jencks et al., 1972; Sewell & Hauser, 1975) provided an incomplete picture of how families helped their children succeed in school. Coleman introduced his concept of "social capital" to redirect attention from individual characteristics toward the importance of social relationships in the status attainment process. Social networks with intergenerational closure could empower adults to impose their value system upon young people and direct them toward academic success (in contrast with the adults in *The Adolescent Society*).

Empirical research on Coleman's theory led to several subsequent theoretical advances in the concept of intergenerational closure. Morgan and Sørensen (1999a) used Coleman's theory to propose two ideal typical schools. "Norm-enforcing" schools have social networks with high levels of intergenerational closure among children and parents, as well as overlapping ties with teachers and school administrators. The dense social networks of norm-enforcing schools can lead to the creation of social capital (as Coleman believed), but they could also become "suffocating communities in which excessive monitoring represses creativity and exceptional achievement" (Morgan & Sørensen, 1999a, p. 663). In contrast, parents in "horizon-expanding" schools invest their time and energy in forming relationships with adults outside of the school setting. Morgan and Sørensen disagreed with Coleman's assumption that norm-enforcing schools are more beneficial for students and argued that an equally compelling case can be made for the merits of horizon-expanding schools.

Carbonaro (1999) noted an important (but underappreciated) issue in Morgan and Sørensen's (1999a) operationalization of Coleman's theory: Intergenerational closure may operate at either the individual or the school level. Carbonaro argued that Coleman's writings reflected an individual level explanation: Intergenerational closure provided a mechanism for sharing information, setting expectations, and enforcing norms, and if students were not nested within such a network, they would not enjoy the benefits of this normative system. When viewed in this way, intergenerational closure is a mechanism of social control that benefits only students connected to the network.

In Carbonaro's (1999) view, group level explanations were plausible—students could benefit from attending a school with high levels of intergenerational closure even if they themselves were not connected to an intergenerationally closed network. Carbonaro argued, however, that more theorizing about the specific mechanisms involved in a group level explanation was needed to guide future research in this area.

Finally, it is unclear which characteristics of the social network matter most when judging the importance of intergenerational closure for student outcomes. Carbonaro (1999) argued that the quantity of "closed" ties is likely less important than the quality of the relationships between individuals within the network. Membership in a closed social network that creates a "dysfunctional" community (i.e., a social network that discourages academic achievement) will likely depress academic performance. In short, the number of social ties in a network may be less important than how they are used by actors within the social network.

MAIN RESEARCH FINDINGS

Despite their lengthy discussion of intergenerational closure and functional community, Coleman and Hoffer (1987) lacked direct measures of these concepts in their study and thus provided no evidence to support their claims. The National Education Longitudinal Survey of 1988 (NELS:88) was the first data set that collected information on parents' relations with their children's friends and the friends' parents. There was general agreement that the NELS:88 data set had some important limitations (see Carbonaro, 1999; Hallinan & Kubitschek, 1999; Morgan & Sørensen, 1999b). Most notably, NELS:88 lacked measures of important attributes of social networks (e.g., network density, centrality). In addition, there was almost no information regarding how actors within the social network interacted with one another—a key component of Coleman's argument. Despite these limitations, several important empirical studies of the NELS:88 emerged.

Teachman, Paasch, and Carver (1996) examined which students had high levels of intergenerational closure in eighth grade. As Coleman predicted, children in Catholic schools had higher levels of closure than public school students. Their findings also revealed that parents with more education and income were connected to networks with higher levels of closure. Finally, Black students had lower levels of closure than White students.

Muller (1993) analyzed the cross-sectional eighth-grade NELS:88 data and found that students in social networks with (parent-reported) higher levels of intergenerational closure had higher test scores and grades. Pong (1998) also analyzed the NELS:88 data to examine the effects of intergenerational closure on achievement. Pong, however, used a longitudinal design (eighth to tenth grade) as well as student reports of intergenerational closure. She found that neither individual- nor school-level measures of "parents know friends' parents" were related to math or reading achievement. Carbonaro (1998) also examined how intergenerational closure was related to achievement using parent measures of closure and a longitudinal design that spanned eighth to twelfth grade. He found a positive relationship between closure and math achievement but no significant relationship between closure and cumulative grade point average in twelfth grade.

Morgan and Sørensen (1999a) took a very different approach in their analysis. First, unlike the aforementioned studies, Morgan and Sørensen included information regarding student friendships within the school as well as children's friends' parents in their measure of school closure (labeled "social closure"). Second, Morgan and Sørensen aggregated their intergenerational closure measure to the school level. They found no significant differences in average math achievement levels among schools with differing levels of social closure. When they examined "friends in school" and "parents known by parents" separately in the analysis, they found that (a) in-school friendships had a positive relationship with math achievement, and (b) "parents known by parents" had a negative relationship with achievement. Carbonaro (1999) reanalyzed his data by adding controls for "friends in school" (at the individual level) and school-level closure to his models. He found that these new controls, particularly friends in school, made the relationship between closure and math achievement statistically insignificant. Pribesh and Downey (1999) also found that controlling for many different types of social ties among parent, friends, and friends' parents yielded closure coefficients that were either insignificant or negative.

A more recent study of the relationship between intergenerational closure and achievement analyzed data from the Educational Longitudinal Study of 2002 (ELS:2002). Rosenbaum and Rochford (in press) found that the number of in-school friends' parents known by a parent was positively related to tenth grade math (but not reading) scores. The analyses in this study, however, were cross-sectional and therefore susceptible to bias due to self-selection. A second wave of ELS data is now available for analysis to address this limitation.

Researchers have also examined whether intergenerational closure is related to student outcomes other than achievement. One outcome that received considerable attention was high school dropout. Teachman et al. (1996) found that (net of other controls) intergenerational closure was unrelated to a student's chances of dropping out of high school by the end of 10th grade. However, Teachman, Paasch, and Carver (1997) and Carbonaro (1998) found that by the end of twelfth grade, students with higher levels of intergenerational closure were less likely to drop out of high school. Teachman et al. (1997) also found that the closure–dropping out relationship grew stronger as parental income increased.

Finally, a few studies have examined how intergenerational closure is related to outcomes that mediate the closure relationship with achievement and attainment. Muller and Ellison (2001) found that students with higher levels of closure had higher levels of religious involvement, felt more in control of their individual circumstances (i.e., locus of control), spent more time on homework, and were less likely to cut class. Inter-generational closure was unrelated, however, to students' own educational expectations and their chances of enrolling in an advanced math course. Fletcher, Hunter, and Eanes (2006) collected data on 400 third-grade children and found that intergenerational closure was related to students' self-efficacy but not to internalizing or externalizing problem behaviors. However, these relationships varied based on the types of friends students had (e.g., in school, neighborhood, family), as well as the student's race or ethnicity.

FUTURE RESEARCH

The concept of intergenerational closure remains an important theoretical contribution because it highlights the role that social relationships play in the status attainment process. Although Coleman's ideas about network characteristics provided sociologists with an ambitious research agenda, empirical research has not fulfilled this initial promise. There are several ways to improve future research on intergenerational closure and bring about a greater understanding of how social relationships affect student outcomes.

First, Coleman's theory about intergenerational closure needs further conceptual development. Coleman's writings are too general to provide useful, testable hypotheses. Researchers must work harder to identify how closure might affect students' academic outcomes. Coleman pointed to information flows, shared expectations, and collective norms as key features of closed networks that led to the creation of social capital. Future researchers need to specify what information, expectations, and norms matter for student outcomes. In addition, researchers need to identify specific mediating mechanisms that link intergenerational closure with educational outcomes. Morgan and Sørensen (1999a) identified (theoretically but not empirically) student effort as the main mediating mechanism that linked intergenerational closure to academic achievement. Further theorizing about additional mediating mechanisms will help yield more testable hypotheses for researchers to examine and ultimately create a deeper understanding of how intergenerational closure actually works.

Surprisingly, there is almost no discussion of what constitutes a social tie in this literature. When parents say they "know" another parent or child, what do they mean? Is there some minimal threshold for "knowing," or are there important dimensions of "knowing" that can be measured and quantified? This is a major conceptual gap that must be addressed. Researchers who study intergenerational closure would benefit from a closer reading of the literature on social networks and stratification (see Lin, 1999, for a review). Concepts such as network centrality, homophily, bridging ties, and strong/weak ties can strengthen the underlying theoretical framework that motivates researchers' analyses.

Future research on intergenerational closure also requires better data. Ideally, data that maps the full set of social relationships among parents and students within several schools would provide much greater insight into whether and how social ties between students and parents affect student outcomes. A major limitation of available data is that individuals and schools remain the sole units of analysis. Individuals are nested within social networks, both inside and outside school, and social networks must become units of analysis to advance empirical work in this area.

Analyses of social network data could address some interesting and important questions in this area. Returning to Figure 1, it is clear that the social network is "closed" for some nodes but not others. Do students C and D benefit from the closure enjoyed by A and B? If so, do these students benefit equally? How much would increasing the degree of intergenerational closure within the network affect student outcomes? Is there a point of diminishing returns to additional closed ties because of the circulation of redundant information? Do bridging ties with other social networks with more closure have effects on student outcomes? This is just a sampling of the many questions that researchers could address if they had access to network data for parents and students. With more detailed and better quality information on social networks, it is likely that further research on intergenerational closure will fulfill its initial promise.

SEE ALSO Volume 1: *Coleman, James; Cultural Capital; Human Capital; Parental Involvement in Education; Parent-Child Relationships, Childhood and Adolescence; Social Capital.*

BIBLIOGRAPHY

Blau, P. M., & Duncan, O. D. (1967). *The American occupational structure.* New York: Wiley.

Carbonaro, W. J. (1998). A little help from my friend's parents: Intergenerational closure and educational outcomes. *Sociology of Education, 71,* 295–313.

Carbonaro, W. J. (1999). Opening the debate on closure and schooling outcomes: Comment on Morgan and Sørensen. *American Sociological Review, 64,* 682–686.

Coleman, J. S. (1961). *The adolescent society: The social life of the teenager and its impact on education.* New York: Free Press of Glencoe.

Coleman, J. S. (1988). Social capital in the creation of human capital. *American Journal of Sociology, 94,* S95–S120.

Coleman, J. S., & Hoffer, T. (1987). *Public and private high schools: The impact of communities.* New York: Basic Books.

Fletcher, A. C., Hunter, A. G., & Eanes, A. Y. (2006). Links between social network closure and child well-being: The organizing role of friendship context. *Developmental Psychology, 42,* 1057–1068.

Hallinan, M. T., & Kubitschek, W. N. (1999). Conceptualizing and measuring school social networks: Comment on Morgan and Sørensen. *American Sociological Review, 64,* 687–693.

Jencks, C., Smith, M., Acland, H., Bane, M. J., Cohen, D., Gintis, H., et al. (1972). *Inequality: A reassessment of the effect of family and schooling in America.* New York: Basic Books.

Lin, N. (1999). Social networks and status attainment. *Annual Review of Sociology, 25,* 467–487.

Morgan, S. L., & Sørensen, A. B. (1999a). Parental networks, social closure, and mathematics learning: A test of Coleman's social capital explanation of school effects. *American Sociological Review, 64,* 661–681.

Morgan, S. L., & Sørensen, A. B. (1999b). Theory, measurement, and specification issues in models of network effects on learning: Reply to Carbonaro and to Hallinan and Kubitschek. *American Sociological Review, 64,* 694–700.

Muller, C. (1993). Parent involvement and academic achievement: An analysis of family resources available to the child. In B. Schneider & J. S. Coleman (Eds.), *Parents, their children, and schools* (pp. 77–113). Boulder, CO: Westview Press.

Muller, C., & Ellison, C. G. (2001). Religious involvement, social capital, and adolescents: Evidence from the National Educational Longitudinal Study of 1988. *Sociological Focus, 34,* 155–183.

Pong, S. (1998). The school compositional effect of single parenthood on 10th-grade achievement. *Sociology of Education, 71,* 23–42.

Pribesh, S., & Downey, D. B. (1999). Why are residential and school moves associated with poor school performance? *Demography, 36,* 521–534.

Rosenbaum, E., & Rochford, J. A. (in press). Generational patterns in academic performance: The variable effects of attitudes and social capital. *Social Science Research.*

Sewell, W. H., & Hauser, R. M. (1975). *Education, occupation, and earnings: Achievement in the early career.* New York: Academic Press.

Sewell, W. H., Hauser, R. M., Springer, K. W., & Hauser, T. S. (2004). As we age: A review of the Wisconsin Longitudinal Study, 1957–2001. *Research in Social Stratification and Mobility, 20,* 3–111.

Teachman, J. D., Paasch, K., & Carver, K. (1996). Social capital and dropping out of school early. *Journal of Marriage and the Family, 58,* 773–783.

Teachman, J. D., Paasch, K., & Carver, K. (1997). Social capital and the generation of human capital. *Social Forces, 75,* 1343–1359.

William Carbonaro

INTERPRETIVE THEORY

Since the early 1980s, theoretical advances in the sociology of childhood and youth have broken free from the individualistic doctrine that regards children's social development solely as the private internalization of adult skills and knowledge. Central to interpretive theory is the appreciation of the importance of collective, communal activity—how children and youth negotiate, share, and create culture with adults and each other (Corsaro, 2005; James, Jenks, & Prout, 1998; Thorne, 1993). This emphasis on collective activity within a cultural context in interpretive theory is clearly in line with the life-course perspective.

INTERPRETIVE REPRODUCTION

Interpretive theory builds on Anthony Giddens's (1984) notion of the duality of social structure. In his theory of structuration Giddens argued that "the structural properties of social systems are both medium and outcome of the practices they recursively organize" (p. 25). A central concept in interpretive theory is *interpretive reproduction* (Corsaro, 2005). The term *interpretive* captures innovative and creative aspects of children's participation in society. Children and youth produce and participate in their own unique peer cultures by creatively taking or appropriating information from the adult world to address their own concerns. The term *reproductive* captures the idea that children and youth are not simply internalizing society and culture but are also contributing to cultural production and change. The term also implies that children and youth are, by their very participation in society, constrained by the existing social structure and by social reproduction.

Although many developmental psychologists have stressed the importance of children's agency and now recognize that social and cultural context is important, they often view context in static terms, as a variable affecting individual development. From the perspective of interpretive reproduction, however, cultural context is not a variable that affects development. Rather cultural context is a dynamic that is continually constituted in routine practices collectively produced at various levels of organization. Even when developmental psychologists visualize context as something that is collectively produced, primary concern usually remains with how these collective processes get inside the individual child and not the collective processes themselves.

Children and youth do, of course, develop individually, but throughout this development the collective processes they are always part of are also changing. These processes are most accurately viewed as occurring in the interwoven local cultures making up children's worlds. When discussing these collective processes developmentally or longitudinally, one must consider the nature of *membership* of children and youth in these local cultures and the changes in their degree or intensity of membership and participation over time. One also must consider how different structural and institutional features constrain and enable the collective processes of interest.

From this view human development, or perhaps better phrased *the development of humans*, is always collective, and transitions are always collectively produced and shared with significant others.

ROUTINES, DEVELOPMENT, AND PEER CULTURES

Children and youth participation in cultural routines is an essential aspect of the interpretive approach. Routines, such as greetings, run and chase games, role play, insult exchanges, and gossip, are recurrent and predictable activities (Corsaro, 2005; Eder & Nenga, 2003). Thus, they provide actors with the security and shared understandings of belonging to a cultural group. Furthermore, this very predictability empowers routines, providing frames with which a wide range of sociocultural knowledge can be produced, displayed, and embellished. In this way cultural routines serve as anchors that allow children and all social actors to deal with ambiguities, the unexpected, and the problematic comfortably within the friendly confines of everyday life.

The interpretive approach views development over the life course as reproductive rather than linear. From this perspective, children enter into social groups and, through interaction with others, establish social understandings that become fundamental knowledge upon which they continually build. Thus, interpretive theory extends the notion of stages of cognitive and emotional maturity by viewing development (or the evolving membership of children or youth in their culture) as a productive–reproductive process of increasing density and reorganization of knowledge that changes with children's developing cognitive and language abilities and with changes in their social worlds. A major change in children's worlds is their movement outside the family. By interacting with peers first in child-care centers and preschools and then in age-graded schools, children produce a series of peer cultures.

Interpretive theory stresses that the production of peer culture is not a matter of simple imitation of the adult world. Children and youth creatively take or appropriate information from the adult world to produce their own unique peer cultures. Such appropriation is creative in that it both extends or elaborates peer culture (transforms information from the adult world) and simultaneously contributes to the reproduction of the adult culture. Thus, the peer cultures of children and youth have an autonomy that makes them worthy of documentation and study in their own right.

From the interpretive perspective peer culture is defined as a stable set of activities or routines, artifacts, values, and concerns that children and youth produce and share with peers (Corsaro, 2005; Corsaro & Eder,

1990). Corsaro (1985, 2005) has discovered that children produce and participate in peer cultures as young as 3 years of age in his studies of preschools in the United States and Italy. A good example of an aspect of children's peer cultures can be seen in their attempts to evade adult rules through their collaboratively produced secondary adjustments, which enable children to gain a certain amount of autonomy over their lives in the preschool.

For example, the children employed several concealment strategies to evade the rule that prohibited bringing toys or other personal objects from home to school. This rule was necessary in the preschool because personal objects were attractive to other children simply because they were different from the everyday materials in the preschools, and as a result the teachers were constantly settling disputes about these items. Therefore, such objects were not to be brought to school; if they were and were discovered by a teacher, they were taken away and stored in the child's locker until the end of the day. In both the American and Italian schools Corsaro studied, the children evaded the rule by bringing small, personal objects that they could conceal in their pockets. Particular favorites were toy animals, race cars, and small dolls. Of central importance here is that the children did not get the idea of bringing these small objects from parents, who surely did not tell their children: "The teacher says you can't bring toys, so just take your toy race car and hide it in your pocket!" No, the children come up with these strategies, share them with their peers, and delight in feeling they are fooling the teachers.

The teachers, of course, often knew what was going on but simply ignored the transgressions. The teachers overlooked these violations because the nature of the secondary adjustments often eliminated the organizational need to enforce the rule. Children shared and played with smuggled objects surreptitiously to avoid detection by the teachers. If the children always played with personal objects in this way, there would be no conflict and hence no need for the rule. Thus, the children's secondary adjustments (which are innovative and highly valued aspects of the peer culture) often contribute to the maintenance of the adult rules.

The story does not end there, however. The children's secondary adjustments to school rules often led to the teachers' selective enforcement of the rules and, in some cases, to changes in the rules and in the organizational structure of the preschool. Corsaro found that teachers relaxed the enforcement of school rules because they recognized the creativity of features of the peer culture. For example, in an American school, teachers first overlooked a rule prohibiting children from moving objects from one play area to another; they allowed the children to use string and blocks from a worktable to

create a "fishing" game by dangling the string from an upper-level playhouse to their peers below, who then attached the blocks. The teachers then actually endorsed the secondary adjustments by joining in the play. In these instances the teachers themselves engaged in secondary adjustments to their own rules and exposed the children to a basic feature of all rules—that is, knowledge of the content of a rule is never sufficient for its application; rules must be applied and interpreted in social context.

LIFE TRANSITIONS AND PRIMING EVENTS

Interpretive theory focuses not only on children's production and participation in a series of peer cultures but also on the nature of critical transitions in children's lives. Transitions are seen as collective events that are often embedded in routine activities that signify that one is part of a group. At the same time, cultural practices in these routines prepare or prime members for future transitions. Along these lines, William Corsaro and Luisa Molinari (2005) developed the notion of *priming events*. Priming events involve activities in which children, by their very participation, attend prospectively to ongoing or anticipated changes in their lives. Some priming events Corsaro and Molinari (2005) identified in their research on Italian children's transition from preschool to elementary school were formal and organizational. These included the preschool children's visits to the elementary school to see the school and meet their new teachers and an end-of-the-year party to celebrate their three years in the preschool. These priming events are much like rites of passage (van Gennep, 1960).

Corsaro and Molinari (2005) also discovered other more subtle events that were part of familiar routines in the children's peer culture. For example, two children got into a teasing routine, or what is often referred to in Italian as *discussione*. Angelo accused Marina of telling a lie about him and said that "her nose would grow longer than Pinocchio's." Marina responded that she was not a liar, but the dispute continued with Angelo saying his brother in first grade would beat up Marina. Marina said she was not afraid because her brother in second grade would beat up Angelo's brother in first grade. Angelo then countered that his brother in third grade would beat up Marina's brother in second grade. Mariana then started to laugh and say her cousin in fourth grade would beat up Angelo's brother in third grade. Both children then began laughing and the dispute ended. The use of humor by turning to fanciful threats (the impossibility of having siblings or relatives at every grade level) lightened the seriousness of the discussion and deftly connected a typical peer spat to the children's ongoing concern about ending their time together as a group and moving on to elementary school, where they would join one of four first-grade classes. Thus the children's thinking about age in terms of where they are and where they are going in the educational system is anchored in the everyday teasing and *discussione* of the peer culture.

The concept of priming events shares certain features with an often referred to concept in sociology: Robert Merton's (1968) "anticipatory socialization." However, priming events, because they are part of identifiable collective actions, provide empirical grounding to Merton's more abstract concept.

Merton (1968) saw anticipatory socialization as a function of reference groups and defined it as "the acquisition of values and orientations found in statuses and groups in which one is not yet engaged but which one is likely to enter" (p. 438). Although Merton noted that anticipatory socialization can occur through formal education and training, he argued that much of such preparation "is *implicit, unwitting,* and *informal*" (p. 439).

Anticipatory socialization is frequently cited in work on childhood and adult socialization. Yet there was no inductive empirical grounding of the concept in Merton's presentation, nor has there been a tradition of empirical research on the concept. In Merton's discussion children themselves are not mentioned, but one can infer that the socialization he implied even if not didactic is still something that is completed with the appraisals of those with more power; in the case of children these appraisals come primarily from adult caretakers. Merton's contention that anticipatory socialization is unwitting and informal does imply some control and power to children or adults in the transitions in which they are participating, but determining the nature of this power or agency depends on investigation of empirical events. Such events and their characteristics are captured in the notion of priming events, which defines the processes more precisely as having interactive and communicative features that Merton only speculated about. As mentioned earlier, Corsaro and Molinari (2005) found that priming events often involve the innovative productions of children within their peer cultures as well as input from adults.

A major strength of life-course research is its insistence on situating the developing individual in historical time and structural or cultural place. Glen Elder (1994) defined the life course as a "multilevel phenomenon, ranging from structural pathways through social institutions and organizations to the social trajectories of individuals and their developmental pathways" (p. 5). The concept of priming events, or collective activities that impel social actors to attend prospectively to ongoing and anticipated changes in their lives, clearly can aid in the conceptualization and empirical investigation of the form and meaning of transitions over the life course (Corsaro & Molinari, 2005, p. 22).

SEE ALSO Volume 1: *Friendship, Childhood and Adolescence; Peer Groups and Crowds; Socialization; Youth Culture.*

BIBLIOGRAPHY

Corsaro, W. A. (1985). *Friendship and peer culture in the early years.* Norwood, NJ: Ablex.

Corsaro, W. A. (2005). *The sociology of childhood.* (2nd ed.). Thousand Oaks, CA: Pine Forge Press.

Corsaro, W. A., & Eder, D. (1990). Children's peer cultures. *Annual Review of Sociology, 16,* 197–220.

Corsaro, W. A., & Molinari, L. (2005). *I compagni: Understanding children's transition from preschool to elementary school.* New York: Teachers College Press.

Eder, D., & Nenga, S. K. (2003). Socialization in adolescence. In J. Delamater (Ed.), *Handbook of social psychology* (pp. 157–182). New York: Kluwer Academic/Plenum.

Elder, G. H., Jr. (1994). Time, human agency, and social change: Perspectives on the life course. *Social Psychology Quarterly, 57,* 4–15.

Giddens, A. (1984). *The constitution of society: Outline of the theory of structuration.* Berkeley: University of California Press.

James, A., Jenks, C., & Prout, A. (1998). *Theorizing childhood.* New York: Teachers College Press.

Merton, R. K. (1968). *Social theory and social structure.* (Rev. ed.). New York: Free Press.

Thorne, B. (1993). *Gender play: Girls and boys in school.* New Brunswick, NJ: Rutgers University Press.

van Gennep, A. (1960). *The rites of passage* (M. B. Vizedom & G. L. Caffee, Trans.). Chicago: University of Chicago Press.

William A. Corsaro

J–L

JUNIOR HIGH SCHOOL

SEE Volume 1: *Stages of Schooling.*

JUVENILE JUSTICE SYSTEM

Adolescence is a time of excitement and possibility; during the teenage years, lives and personalities take shape and individuals make the first set of choices that will affect them in adulthood. Adolescence is a turbulent time for many young people, and some struggle as they make the difficult transition between childhood and the adult world. They find out the hard way that mistakes made in adolescence can change the trajectory of a young life dramatically. Some spend years in a correctional facility for a criminal decision made as a teenager.

The juvenile justice system is charged with handling a difficult segment of the population: children and adolescents who have committed crimes and may be considered a danger to the community or to themselves. Although each state has a slightly different vision for dealing with juvenile offenders, the juvenile justice system generally attempts to correct wayward youth at a critical point in the life course. Administrators and staff members who work in the system try to intervene and redirect the life trajectories of troubled adolescents, pointing them toward more conforming futures. As B. C. Feld explained, "One premise of juvenile justice is that youths should survive the mistakes of adolescence with their life chances intact" (Feld 1996, pp. 425–426). Unfortunately, virtually since its inception the juve-

nile justice system has struggled to meet these high ideals and to merge the interests of the community with the best interests of the young people in the system.

JUVENILE JUSTICE: BEGINNINGS

The first juvenile court was created in Chicago in 1899. Although there were houses of refuge and reformatories to house dependent and delinquent youth before that time (Krisberg, 2005), Chicago at the turn of the 20th century was the home of the reform movement that led to the modern juvenile justice system. The "child savers" were feminist reformers who helped pass special laws and create new institutions for juveniles (Platt, 1977, p. 75). The child saving movement was dominated by middle and upper class women who "regarded their cause as a matter of conscience and morality ... the child savers viewed themselves as altruists and humanitarians dedicated to rescuing those who were less fortunately placed in the social ordre" (Platt, 1977, p. 3). They built the juvenile justice system on the philosophy of *parens patriae*: the belief that the state—in the form of the juvenile court—should act as a "superparent" to its troubled youth, treating them sternly but kindly and always attempting to act in the best interests of the individual child.

Initially, until concerns with labeling and the "criminal contamination" of nondelinquent youth arose in the 1960s and 1970s, delinquent children were treated and housed alongside neglected children and those who were dependent on the state for their basic needs. These children were generally sent to reformatories or reform schools in the country which were designed using a cottage system, and were meant to teach working class

275

skills and simulate family life. In the original juvenile court judges had full discretion in deciding how to sentence the children who came before them. Because the court was expected to act in the best interest of each individual child, children had few rights; delinquent and dependent youth were subject to judges' personal biases and sometimes discriminatory practices.

Studies of juvenile "reform schools" and state training schools through the 1970s revealed conditions that in the early 21st century would be viewed as unacceptable: Staff members and cottage "parents" frequently used physical means to punish the young people in their care, striking, shaking, and shoving them (Weber, 1961, Wooden, 1976); boys were housed in the "tombs," an extreme form of isolation, and were not allowed to speak in their cottage living units (Feld, 1999, Miller, 1998); and younger and weaker youths were victimized by their tougher peers (Bartollas, Miller, & Dinitz, 1976, Feld, 1977, Polsky, 1962).

The juvenile corrections system was altered significantly when the Juvenile Justice Delinquency and Prevention (JJDP) Act was passed in 1974. In the 1960s and 1970s concerns about the labeling of minor offenders as delinquents and a focus on rehabilitation led to widespread attempts at deinstitutionalization and community alternatives. The JJDP Act offered financial incentives for states to decriminalize status offenses—acts that would not be crimes if committed by adults—and deinstitutionalize status offenders. Secure juvenile institutions became the agency of last resort, reserved for the most serious juvenile offenders, and alternative programs were developed to provide supervision in the community and care for delinquent children who had committed less serious crimes.

In the contemporary period agencies charged with supervising juvenile offenders after they have been adjudicated delinquent, the equivalent to an adult being convicted of a crime, fall under the umbrella term of *juvenile corrections*. Juvenile offenders may be treated in the community or may be placed in secure facilities; in essence, they may get probation. They may be placed in a group home, halfway house, or foster care; they also may be sentenced to serve time in secure facilities such as training schools and wilderness programs. Together, these agencies share the responsibility for "correcting" or attempting to reform delinquent youth. In addition, new alternatives have been created to try to divert minor offenders from the system. Juvenile drug treatment courts and peer courts, where teens hold each other accountable for minor offenses by holding their own hearings and deciding appropriate sanctions, are two examples of programs that focus on treatment and accountability without subjecting offenders to the stigma of being labeled delinquent and processed through the formal juvenile justice system.

CHANGES IN JUVENILE JUSTICE: MORE PUNITIVE TIMES

In the last decade of the 20th century and the first decade of the 21st the United States made a clear effort to "get tough" on juvenile crime, and the sentencing of juvenile offenders became increasingly punitive (Feld, 1999). The possibilities for punishment shifted to the point where secure training schools seemed to be the kinder, gentler option; the two alternatives facing serious juvenile offenders were usually confinement in juvenile correctional facilities and confinement in adult prisons (Inderbitzin, 2006b).

This shift in attitude perhaps is best symbolized by the public fear of juvenile superpredators that developed in the mid-1990s. Popular magazines and media outlets ran prominent features warning the public of a new epidemic of young villains. Some scholars fueled the fear by making statements that demonized delinquents. For example, W. J. Bennett, J. J. DiIulio, and J. P. Walters stated, "The problem is that today's bad boys are far worse than yesteryear's, and tomorrow's will be even worse than today's" (Bennett, DiIulio, and Walters, 1996, pp. 26–27). Those authors went on to describe the juvenile offenders of people's worst fears: "America is now home to thickening ranks of juvenile 'super-predators'—radically impulsive, brutally remorseless youngsters, including ever more preteenage boys, who murder, assault, rape, rob, burglarize, deal deadly drugs, join gun-toting gangs, and create serious communal disorders. They do not fear the stigma of arrest, the pains of imprisonment, or the pangs of conscience" (Bennett et al., 1996, p. 27).

In response to the fear engendered by such vivid and hard descriptions of America's teenage population, public sentiment shifted, and the pendulum of juvenile justice swung from rehabilitation to punishment. The argument seemed to be that as long as young offenders commit serious crimes, they should be prepared to face the consequences. Indeed, it seemed that "recent reforms in juvenile justice have placed the notion of youth itself on trial" (Grisso & Schwartz, 2000, p. 5). By the mid-1990s public fear translated into political action and most states had passed new laws that made it easier to transfer juvenile offenders to the adult system; there was a public movement to adjudicate more and more juveniles in adult criminal courts and confine them in adult prisons (Howell, 1998). In addition, many states enacted mandatory minimum sentences that were designed for both adult and teenage offenders; that led to long sentences in adult prisons without the incentive of being able to earn time off for good behavior for some adolescent offenders.

The rationale behind having a separate system for delinquents rests on two distinct assumptions: the belief that juveniles are less responsible for their crimes than their adult counterparts are and the idea that because adolescents are thought to be more malleable than adults

Juvenile Detention. *Inmates at the Department of Youth Services juvenile boot camp wait to go outside for physical training.* AP IMAGES.

and still developing as individuals, there is generally more hope for rehabilitation with younger offenders.

Developmental psychologists have argued that "adolescence in modern society is an inherently transitional time during which there are rapid and dramatic changes in physical, intellectual, emotional, and social capabilities ... other than infancy, there is probably no period of human development characterized by more rapid or pervasive transformations in individual competencies" (Steinberg & Schwartz, 2000, p. 23). Because they work primarily with adolescents, juvenile justice agencies have the potential to exert enormous influence over the rapidly changing lives of their captive populations. Because most delinquents will return to the community in months or a few years, it is imperative to try to find effective treatment programs for adolescent offenders.

JUVENILE JUSTICE CHALLENGES

Since the 1980s scholars have been questioning the utility of a separate court system for juvenile offenders, pointing

to "its transformation from an informal, rehabilitative agency into a scaled-down second-class criminal court" (Feld, 1996, p. 418). Given the current focus on the punishment of young offenders and the lack of resources in juvenile courts, Feld argued that the United States should "abolish juvenile court jurisdiction over criminal conduct, try all offenders in criminal courts, and introduce certain procedural and substantive modifications to accommodate the youthfulness of younger offenders" (Feld, 1999, p. 289). All resources would be concentrated into an integrated criminal court system, and states could provide extra procedural safeguards for juvenile offenders. In addition, Feld suggested, an integrated court system should incorporate a youth discount—recognizing youthfulness as a mitigating factor—into sentencing decisions. Adolescents tried and convicted in an integrated criminal court still could be sentenced to separate placement in juvenile correctional facilities rather than adult prisons; only the courts would be combined.

An integrated court system might help reduce racial disproportionally and the double standard of juvenile justice, two problems that have plagued the juvenile justice system since its inception. Even with sentencing guidelines in place in many jurisdictions, racial minorities continue to be overrepresented in secure confinement for their offenses, and girls continue to face a double standard of juvenile justice. Along with females' adolescent criminality, the moral conduct of girls has long been a concern of the larger community (Chesney-Lind & Shelden, 2004). Juvenile courts historically treated girls paternalistically: Girls were sanctioned more severely than boys for early sexual behavior and for disobeying their parents; in the early part of the 19th century they were institutionalized away from males, learning domestic skills until their release. The 1992 reauthorization of the JJDP Act focused attention on gender bias and called for more services to girls in the juvenile justice system, encouraging gender-specific programming.

While race, gender, and social class have always had at least an indirect impact on justice decisions, The U.S. system of juvenile corrections arguably has been becoming a two-tiered system: a state-run system for poor minorities and private facilities for children whose parents can afford alternative placements. In this emerging system, gender, race, and social class interact and lead to clear patterns of incarceration: White girls are more likely to be channeled into private facilities whereas girls of color are concentrated in state institutions (Chesney-Lind & Shelden, 2004).

JUVENILE JUSTICE: IMPORTANCE IN ADOLESCENT DEVELOPMENT AND THE LIFE COURSE

Growing up is difficult in the best circumstances, and spending one's adolescent years in the juvenile justice system increases the problem immeasurably. The experiences adolescent inmates miss while incarcerated cannot be replaced; their teenage lives are part of the price they pay for their crimes. The boys and girls who enter juvenile correctional facilities often return to their communities as young women and men with little preparation but all the responsibilities of adulthood (Inderbitzin, 2006a).

Although incarceration appears to be a turning point for some offenders who "desist" from crime, for nearly all who pass through a correctional facility it adds to the cumulative disadvantage and the obstacles they will face after their release. Incarceration may weaken community bonds, contribute to school failure and unemployment, and ultimately increase the likelihood of committing adult crime (Laub & Sampson, 2003). Young adults returning to the community from juvenile institutions generally face the outside world with little to no money or savings, few marketable skills, and no history in the legitimate labor market to help their employment prospects. Many juvenile offenders have the advantage of being "adjudicated delinquent" rather than convicted of

a felony, so they do not have to "check the box" on employment applications. They do, however, face the difficult task of explaining the gaps that accumulated during their incarceration in their education and work experience to prospective employers and college admissions officers. As L. Steinberg, H. L. Chung, and M. Little (2004) found: "Despite its putatively rehabilitative aims, it is all too often the case that young offenders finish their time with the justice system and move into the adult world with just as many, if not more, problems than when they first entered" (Steinberg, Chung, & Little, 2004, p. 23).

Although the juvenile justice system has become more punitive, many working in the system continue to search for a better way to address the needs of both delinquents and the community. In considering the state of juvenile justice at the turn of the 21st century, J. A. Butts and D. P. Mears (2001) argued that "get-tough policies weakened the integrity of the juvenile justice system, but growing evidence about the effectiveness of new ideas in prevention and rehabilitation may save the system yet" (Butts & Mears, 2001, p. 171). Consistent with the original philosophy of *parens patriae*, guiding the troubled young people in their care through "the immense journey of adolescence, a journey of peril and possibility" (Ayers, 1997, p. 138) remains an important goal for many people who work in the juvenile justice system. The perils are clear for delinquent youth; ideally, their experiences in the juvenile justice system will help them see the possibilities as well.

SEE ALSO Volume 1: *Crime, Criminal Activity in Childhood and Adolescence; Gangs; Theories of Deviance;* Volume 3: *Incarceration.*

BIBLIOGRAPHY

Ayers, W. (1997). *A kind and just parent: The children of juvenile court.* Boston: Beacon Press.

Bartollas, C., Miller, S. J., & Dinitz, S. (1976). *Juvenile victimization: The institutional paradox.* Beverly Hills, CA: Sage.

Bennett, W. J., DiIulio, J. J., Jr. & Walters, J. P. (1996). *Body count: Moral poverty—and how to win America's war against crime and drugs.* New York: Simon & Schuster.

Butts, J. A., & Mears, D. P. (2001). Reviving juvenile justice in a get-tough era. *Youth & Society, 33*(2), 169–198.

Chesney-Lind, M., & Shelden, R. G. (2004). *Girls, delinquency, and juvenile justice* (3rd ed.). Belmont, CA: Wadsworth.

Feld, B. C. (1977). *Neutralizing inmate violence: Juvenile offenders in institutions.* Cambridge, MA: Ballinger Pub. Co.

Feld, B. C. (1996). Juvenile (in)justice and the criminal court alternative. In J. G. Weis, R. D. Crutchfield, & G. S. Bridges, (Eds.), *Juvenile delinquency: Readings,* pp. 418-428. Thousand Oaks, CA: Pine Forge Press.

Feld, B. C. (1999). *Bad kids: Race and the transformation of the juvenile court.* New York: Oxford University Press.

Grisso, T. & Schwartz, R. G. (2000). Introduction. In T. Grisso and R. G. Schwartz (Eds.), *Youth on trial: A developmental perspective on juvenile justice*, pp. 1–7. Chicago: The University of Chicago Press.

Howell, J. C. (1998). NCCD's survey of juvenile detention and correctional facilities. *Crime & Delinquency, 44*(1), 102–109.

Inderbitzin, M. (2006a). Growing up behind bars: An ethnographic study of adolescent inmates in a cottage for violent offenders. *Journal of Offender Rehabilitation, 42*(3), 1–22.

Inderbitzin, M. (2006b). Lessons from a juvenile training school: Survival and growth. *Journal of Adolescent Research, 21*(1), 7–26.

Krisberg, B. (2005). *Juvenile justice: Redeeming our children.* Thousand Oaks, CA: Sage.

Laub, J. H., & Sampson, R. J. (2003). *Shared beginnings, divergent lives: Delinquent boys to age 70.* Cambridge, MA: Harvard University Press.

Miller, J. G. (1998). *Last one over the wall: The Massachusetts experiment in closing reform schools* (2nd ed.). Columbus: Ohio State University Press.

Platt, A. M. (1977). *The child savers: The invention of delinquency* (2nd ed.). Chicago: University of Chicago Press, 1977.

Polsky, H. W. (1962). *Cottage six: The social system of delinquent boys in residential treatment.* New York: Russell Sage Foundation.

Steinberg, L., Chung, H. L, & Little, M. (2004). Reentry of young offenders from the justice system: A developmental perspective. *Youth Violence and Juvenile Justice, 2*(1), 21–38.

Steinberg, L., & Schwartz, R. G. (2000). Developmental psychology goes to court. In T. Grisso & R. G. Schwartz (Eds.), *Youth on trial: A developmental perspective on juvenile justice*, pp. 9-32. Chicago: University of Chicago Press.

Weber, G. H. (1961). Emotional and defensive reactions of cottage parents. In D. R. Cressey (Ed.), *The prison: Studies in institutional organization and change*, pp. 189-228. New York: Holt, Rinehart and Winston.

Wooden, K. (1976). *Weeping in the playtime of others: America's incarcerated children.* New York: McGraw-Hill.

Michelle Inderbitzin

KOHLBERG, LAWRENCE

SEE Volume 1: *Moral Development and Education.*

LEARNING DISABILITY

The Learning Disabilities Association of America (2006) describes learning disabilities as "neurologically based processing problems. These processing problems can interfere with learning basic skills such as reading, writing, or math. They can also interfere with higher level skills such as organization, time planning, and abstract reasoning." The association estimates that 15% of the U.S. population has a learning disability of some kind and that 6% of the public school population receives special education sup-

port because of an identified learning disability. Definitions of learning disabilities in the United States and Canada all emphasize that it is an umbrella term that covers a wide range of difficulties and that these difficulties are not indicative of low intelligence or of lack of adequate instruction. Strictly speaking the term *specific learning disabilities* (U.S. Department of Education, 2004) should be used, which underscores the particular rather than generic nature of the cognitive processing difficulties assumed to lead to basic skills difficulties. The Individuals with Disabilities Education Act (IDEA) of 2004 specifies that an individual can be identified as having a learning disability only if their difficulty is not attributable to any of the following causes:

- a visual, hearing, or motor disability
- mental retardation
- emotional disturbance
- cultural factors
- environmental or economic disadvantage
- limited English proficiency

MODELS AND TYPES OF LEARNING DISABILITY

Learning disabilities can be characterized either by difficulties with a specific skill such as reading (hence the term *reading disability*) or by the processing difficulties thought to underlie a particular skill, such as the phonological impairments that are at the heart of the majority of reading difficulties. Most estimates reveal that approximately 70–80% of children identified as having a learning disability are designated as having a reading disability. Other specific difficulties involve writing, mathematics, speech, and language, and also include dyspraxia (motor planning disorder), but beyond reading, writing, and mathematics what should be included or excluded is more debatable with differing opinions among leading authorities in the area. Critics such as Kavale (2005) have argued that the term *learning disabilities* now covers too broad a range of difficulties and in doing so renders itself too vague to be useful in practice. A complicating factor is that individuals may have more than one specific learning disability or have difficulties that do not fall into a single clear-cut category. Most would agree that learning disabilities represent a spectrum of difficulties, and the debate focuses on whether and how they should be divided into subcategories.

CAUSES OF LEARNING DISABILITIES

The causes of learning disabilities are complex. In some cases several factors may interact to cause a difficulty, whereas in other cases several different causes may lead to the same kind of basic skill difficulty. Studies of identical twins indicate that if one twin has difficulty with the phonological processing aspects of reading (detecting and manipulating the sounds in words), the other twin is also likely to have similar difficulties (Shaywitz & Shaywitz, 2003). Such studies suggest that genetic factors are implicated in a significant proportion of learning disabilities. For other children, by contrast, infections, traumas, or damage to the brain pre- or postnatal, during birth, or in early childhood are the likely cause. In all these instances the way in which the initial processing difficulties interact with the wider social and learning environment will affect the degree and manner in which they are expressed. A child with mild phonological processing difficulties may develop reasonably well with good reading support and high exposure to reading activities, whereas a child with identical difficulties but with poor reading support and little exposure to reading may struggle.

READING DISABILITIES AND DYSLEXIA

Because of the central role of reading in children's learning and the apparent predominance of these difficulties, reading disabilities probably comprise the most intensively researched learning disability area. The problem of using a seemingly deficient skill as a label is that it can emphasize one skill—in this case, reading—and may draw attention away from the fact that most children with reading disabilities also have persistent problems with spelling, writing, and some aspects of arithmetic. Some prefer the term *dyslexia* because this suggests a syndrome with a range of interlinked difficulties with reading at the core. This allows one label to be used to summarize a range of specific difficulties rather than having to give an individual several labels. It also allows a developmental perspective to be taken, one acknowledging that different skills may be highlighted as problematic at different stages in an individual's learning.

Whereas some treat reading disabilities and dyslexia as synonymous, others argue that dyslexia is a major reading disability subcategory. The International Dyslexia Association (2007) provided the following definition of dyslexia:

> It is characterized by difficulties with accurate and/or fluent word recognition and by poor spelling and decoding abilities. These difficulties typically result from a deficit in the phonological component of language that is often unexpected in relation to other cognitive abilities and the provision of effective classroom instruction.

Such individuals often have relatively good reading comprehension skills in comparison to their reading fluency and accuracy skills, revealing that the problem is with the technical rather than the conceptual aspect of reading. Whereas many children with reading disabilities fit this description, some do not; for example, a child with comprehension difficulties but no phonological processing difficulties has a very different cognitive profile—one that would indicate a different course of intervention. Fletcher, Morris, and Lyon (2003) presented evidence that highlights the need to separate learning disabilities into subgroups based on different processing difficulties. The advantage of this approach is that a much clearer understanding of why some children are struggling with particular skills can be gained. In the case of dyslexia this led to a considerable body of research on the difficulties such children have in detecting the sounds (phonemes) that make up words (Snowling, 2000) and in developing effective interventions to address these difficulties. Researchers are also focusing on the role of the magnocellular system in the phonological, visual, and timing aspects of reading.

IDENTIFICATION OF LEARNING DISABILITIES

A combination of approaches can be used to identify all learning disabilities, including specific tests of the basic skill in question, tests designed to identify the underlying processing difficulties, case histories and observations of the child, and noting the child's responses to specific interventions. According to the U.S. National Center for Health Statistics survey, 8% of children between the ages of 3 and 17 (10% of boys and 6% of girls) had a learning disability (Bloom & Cohen, 2006). The survey also reported that twice as many children (12%) from poor economic backgrounds as compared to economically well-off backgrounds (6%) were identified as having a learning disability. Some claim that girls are under-identified as having a learning disability in comparison with boys (Fletcher, Lyon, Fuchs, & Barnes, 2007), possibly because they internalize rather than act out their feelings in response to learning difficulties and are therefore less likely to be noticed and subsequently diagnosed. Although there may be some truth to this, others argue that there are sound biological reasons (linked to genetics) that account for the predominance of boys in the learning disability category (Rutter et al., 2004).

Past definitions of learning disability stressed the discrepancy between children's general intelligence and their unexpected difficulty with specific skills such as basic literacy. Critics point out that definitions of this sort could discriminate against less able and less socially advantaged children, for whom such discrepancies may be less obvious (Vaughn & Fuchs, 2003). For this reason the IDEA stipulates that a severe discrepancy between achievement and intellectual ability is not required to identify a child as having a learning disability. Vaughn and Fuchs suggest that an alternative criterion for specific learning disabilities should be response to intervention. This is based on the argument that children with specific learning disabilities fail to respond at the same rate as other children to specific and well-structured reading programs. Although there is considerable interest in this approach, it does have serious limitations. Who is to decide what counts as lack of response to intervention, and precisely what kind of intervention at what particular age? Others argue that tests of basic skills and underlying processing difficulties should continue to be refined and improved to create clearer cognitive profiles that can inform interventions. The ideal for all approaches is to identify children before they have "failed" in the school system so that appropriate interventions can be put in place to prevent failure from arising, particularly because the specific information-processing difficulties underlying learning disabilities are considered to be lifelong. The hope is that targeted interventions can be used to improve or create alternative processing strengths through which persons with learning disabilities can develop adequate basic skills.

IMPACT OF LEARNING DISABILITIES

An issue for all learning disabilities is distinguishing between the primary processing and basic skill difficulties that an individual has and the possible secondary difficulties that can arise as a consequence. In the case of reading disabilities, it is well documented that children who struggle with reading end up reading considerably less than children who are competent readers; thus, lack of practice exacerbates their slow reading development. Individuals with learning disabilities report a range of personal reactions to their difficulties including shame, guilt, anger, and frustration, especially in situations in which they feel their difficulties are not understood. A number of studies have reported that children and adult students with learning disabilities, reading disabilities, and/or dyslexia have lower self-esteem and self-efficacy than their peers (Riddick, 1996). With improvements in identification, interventions, attitudes, and environmental accommodations, such findings should become less common.

Cognitive approaches to understanding learning disabilities and particularly reading disabilities have been very helpful in suggesting specific forms of intervention, but they have their limitations. One is that many children with severe reading disabilities (or dyslexia) still have considerable difficulties with spelling and writing and poorer reading fluency and accuracy than the general population, despite improvements in their reading. Another difficulty is that these processing models tend to focus on individual deficits and how they should be fixed rather than on challenging or questioning the environment within which individuals with learning disabilities must function. From a social model of disability perspective, it can be argued, for example, that reading disabilities are exacerbated by the fact that written language is the main medium through which school education takes place.

Research has demonstrated that there is a gradient of language difficulty, with English being the most difficult to read and write because of the disjunction between how it is written and spoken. Words such as *yacht* or *colonel* are highly irregular, and many words are not phonologically transparent in that they are not written the way they sound. It is estimated that phonologically transparent languages such as Italian, in which nearly all words are written the way they sound, present only half the difficulty of English for children who have phonological processing problems (Seymour, Aro, & Erskine, 2003). It can be argued that English is a disabling language for some children and that in a genuinely inclusive culture English would be reformed to make it simpler to read and spell.

The importance of early, appropriate, and well-targeted intervention for learning disabilities cannot be underestimated, but for children and adults with persistent difficulties the environment also needs to change. This can be seen as part of the wider philosophy of inclusion in which it is recommended that organizations adapt to meet the needs of individuals rather than expecting the individual to fit into the organization's preexisting approach to education or work. More specifically, there is interest in the development of dyslexia-friendly schools and colleges where in addition to individual interventions a number of changes are made in classroom practices and school policies and organization (British Dyslexia Association, 2005; Riddick, 2006). For adults the Americans with Disabilities Act of 1990 stipulates that employers must make reasonable accommodations for individuals with disabilities in the workplace.

A national survey in Canada has compared the life experiences of people with and without learning disabilities at various ages (Learning Disabilities Association of Canada, 2007). Adults with learning disabilities left school with poorer qualifications, they were less likely to work, those who worked earned less, and they were two to three times as likely to report poor physical or mental health. This reinforces a number of previous findings showing that adults with learning disabilities fare less well as a group than those without learning disabilities. Poor basic skills at school appear to lead to fewer educational qualifications, and these in combination lead to fewer employment opportunities and the risks associated with a low income or unemployment. This pathway is not inevitable, and many examples exist of adults with learning disabilities who have been very successful.

In his seminal research on disability and stigma, Goffman (1963) distinguished between what he called evident and not-evident disabilities. A difficulty with not-evident disabilities such as learning disabilities is that the wrong reasons can be ascribed to why an individual is behaving in a certain way. The classic case has been with poor reading and spelling, which have been attributed to carelessness, laziness, or lack of intelligence. Although some have questioned the use of labels, arguing that they are stigmatizing and focus on individuals' deficits, a counterargument has been that they allow for better understanding of individuals' difficulties and legitimize the reasons why they are struggling with a skill such as reading (Riddick, 1996, 2001). In addition, it is only when learning disabilities are recognized and acknowledged as legitimate that individuals with such difficulties can criticize environmental conditions that exacerbate their difficulties or create barriers for them.

SEE ALSO Volume 1: *Academic Achievement; Attention Deficit/Hyperactivity Disorder (ADHD); Autism; Cognitive Ability; Disability, Childhood and Adolescence; Gender and Education.*

BIBLIOGRAPHY

Bloom, B., & Cohen, R. A. (2006). *National health statistics survey.* Retrieved March 13, 2008, from http://www.cdc.gov/nchs.htm

British Dyslexia Association. (2005). *Achieving dyslexia friendly schools* (5th ed.). Retrieved March 13, 2008, from http://www.bdadyslexia.org.uk

Fletcher, J. M., Lyon, G. R., Fuchs, L. S., & Barnes, M. A. (2007). *Learning disabilities: From identification to intervention.* New York: Guilford Press.

Fletcher, J. M., Morris, R. D., & Lyon, R. (2003). Classification and definition of learning disabilities: An integrative perspective. In H. L. Swanson, K. Harris, & S. Graham (Eds.), *Handbook of learning disabilities.* (pp. 514–531). New York: Guilford Press.

Goffman, E. (1963). *Stigma: Notes on the management of spoiled identity.* Englewood Cliffs, NJ: Prentice-Hall.

International Dyslexia Association. (2007). *What is dyslexia?* Retrieved March 6, 2008, from http://www.interdys.org/FAQWhatIs.htm

Kavale, K. A. (2005). Identifying specific learning disability: Is responsiveness to intervention the answer? *Journal of Learning Disabilities, 38,* 553–562.

Learning Disabilities Association of America. (2006). *Types of learning disabilities.* Retrieved March 6, 2008, from http://www.ldanatl.org/aboutld

Learning Disabilities Association of Canada. (2007). *Putting a Canadian face on learning disabilities (PACFOLD).* Retrieved March 6, 2008, from http://www.pacfold.ca/

National Dissemination Center for Children with Disabilities. (2004). *Learning disabilities* (Disability Fact Sheet No. 7). Retrieved March 6, 2008, from http://www.nichcy.org/pubs

Riddick, B. (1996). *Living with dyslexia: The social and emotional consequences of specific learning difficulties.* London: Routledge.

Riddick, B. (2001). Dyslexia and inclusion-time for a social model of disability perspective? *International Studies in Sociology of Education. 11*(3), 223–236.

Riddick, B. (2006). Dyslexia-friendly schools in the UK. *Topics in Language Disorders, 26,* 144–156.

Rutter, M., Caspi, A., Fergusson, D., Horwood, L. J., Goodman, R., Maughan, B., et al. (2004). Sex differences in developmental reading disability: New findings from four epidemiological studies. *Journal of the American Medical Association, 291,* 2007–2012.

Seymour, P. H. K., Aro, M., & Erskine, J. M. (2003). Foundation literacy acquisition in European orthographies. *British Journal of Psychology, 94,* 143–174.

Shaywitz, S., & Shaywitz, B.(2003) Neurobiological indices of dyslexia. In H. L. Swanson, K. Harris, & S. Graham (Eds.), *Handbook of learning disabilities* (pp. 514–531). New York: Guilford Press.

Snowling, M. J. (2000). *Dyslexia* (2nd ed.). Malden, MA: Blackwell.

U.S. Department of Education. (2004). *Building the legacy: IDEA 2004.* Retrieved March 6, 2008, from http://idea.ed.gov/

Vaughn S., & Fuchs, L. S. (2003). Redefining learning disabilities as inadequate response to instruction: The promise and potential problems. *Learning Disabilities Research and Practice, 18,* 137–146.

Barbara Riddick

LONELINESS, CHILDHOOD AND ADOLESCENCE

SEE Volume 1: *Friendship, Childhood and Adolescence; Mental Health, Childhood and Adolescence.*

M

MATERNAL EMPLOYMENT

In the United States, the labor force participation rate of mothers with children under the age of 18 increased dramatically from 47% in 1975 to 71% in 2006. This rise was particularly sharp for mothers with children under the age of 3, increasing from 34% to 60%. The corresponding figures for mothers with children between ages 3 and 5 are from 45% to 68% and for mothers with children between 6 and 17, 55% to 77% (see Figure 1). Consequently, as many as two-thirds of infants and toddlers received some kind of nonmaternal child care in 2006 (U.S. Bureau of the Census, 2007). These trends are changing family life in significant ways, and increasing attention is being paid to how maternal employment affects child well-being.

THEORETICAL PERSPECTIVES ON MATERNAL EMPLOYMENT AND CHILDREN'S OUTCOMES

To understand how parental employment might impact children's outcomes, it is first necessary to understand child development and its influencing factors (Bornstein, 2002; Bronfenbrenner, 2005). Theorists from a variety of disciplines have underscored the importance of the parent–child relationship in developing trust and a sense of self. Parents are also invaluable in helping children understand and express language, develop a variety of skills, and solve cognitive tasks. Furthermore, parents aid in the development of children's emotional capacities, such as regulating emotions, dealing positively with frustration, and delaying gratification. Building on this base,

psychologists, sociologists, and economists have provided insights into the relationship between maternal employment and child outcomes.

Theory from Psychology Psychological theories of child development generally assume the mother is the primary caregiver. As women increasingly entered the labor force, theorists began to speculate about how this might affect children's development.

The effects of maternal employment are believed to fluctuate over the child's life course. For example, mothers working in the early life course of children's lives were first seen as potentially detrimental to attachment and the mother–child relationship (see review in Belsky, 2001). For instance, one theory argued that working mothers may be more fatigued and less sensitive to their children, thus interfering with children's secure attachment. Early empirical studies, however, found mixed results, most likely because of small sample sizes. More recently, studies have used a longitudinal data set, the NICHD Study of Early Child Care and Youth Development (NICHD SECCYD), to indirectly examine this issue through the use of nonmaternal child care and have generally found negative impacts on attachment only for children who spend a great deal of time in such care and return home to less sensitive mothers. Nevertheless, direct analyses of maternal employment are needed to more accurately examine the effects of maternal employment on attachment.

Scholars have also suggested that maternal employment could negatively affect the quality of the home environment, maternal sensitivity and responsiveness, and the amount of cognitive stimulation offered to

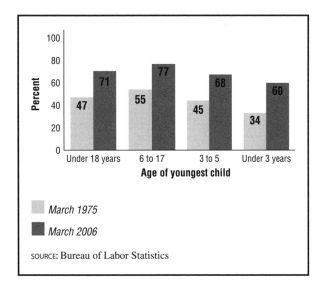

Figure 1. *Labor force participation rates have increased dramatically among mothers over the past 31 years.* CENGAGE LEARNING, GALE.

children (e.g., talking to, playing with the child), or increase the risk of placing their children in poor quality care (Belsky, 2001). It is also possible of course that working mothers might provide higher quality parenting and home environments, or might place their children in care that benefited their development. There is, however, little empirical evidence to support or reject any of these hypotheses. Another hypothesis is that working mothers may experience stress when balancing work and family responsibilities and as a result use harsher parenting tactics. However, a number of small studies of working mothers who had recently exited welfare have generally found no effects in this regard (Chase-Lansdale et al., 2003). Yet another concern is that working mothers may spend less time with their children, and reduced time may mean less intellectual or emotional stimulation. An analysis of NICHD SECCYD data found that working mothers did spend less time with their infants and appeared to provide poorer quality home environments (but not less sensitivity; Huston & Rosenkrantz Aronson, 2005).

These mixed effects may be due to the diversity of families of working mothers, and thus a more nuanced analysis of the impacts of maternal employment on child outcomes is necessary. For instance, theorists suggest that child temperament might moderate such relationships (Damon, Lerner, & Eisenberg, 2006). Additionally, child gender has been found to explain some of maternal employment's effects on externalizing behavior problems such as aggression or hyperactivity. Other potential moderating factors may be the quality and complexity of the mother's job (Bianchi, Casper, & King, 2005) and, given

that childrearing practices are heavily culturally dependent, the mother's racial or ethnic group membership (Damon et al., 2006). For instance, maternal employment may be viewed as more normal and acceptable in communities in which mothers have historically participated in the workforce, such as the African American community. Indeed, researchers have long noted the greater involvement of kin in the rearing of African American children, and thus the impact of mothers going to work may be less for families that are already accustomed to sharing care responsibilities.

Theory from Sociology Sociologists emphasize the important role that environmental factors in the home and work spheres play in child development and how such influences vary across individuals and families (Bianchi et al., 2005). Similar to the role strain and spillover theories developed by psychologists (which emphasize the potential of parents bringing work stress home that may spill over on the family and home environment and in turn affect children's well-being), sociologists have also outlined how home and work roles may complement or conflict with one another (Presser, 2003). For instance, positive work experiences may help a parent be more relaxed and responsive at home, whereas negative or draining work experiences may do the opposite. Together, therefore, both psychologists and sociologists point to the critical role played by moderating factors, whether at the level of child, family, or parental job.

Theory from Economics Economic perspectives complement those from psychology and sociology (Bianchi et al., 2005). First, theorists have noted the benefits of having one parent work and the other parent manage the home to avoid breaks in employment and part-time schedules that are usually penalized in the labor market (Bianchi et al., 2005). This suggests that there will be trade-offs involved in decisions to work or not work in the labor market or to work full-time versus part-time. Second, parental employment may have positive (e.g., increased income/resources), negative (e.g., overburdened parents, heavy reliance on nonparental child care), or neutral effects on children's health and development (Bianchi et al., 2005). This point is crucial as it suggests the importance of considering the role of key mediators such as income, the home environment, and the quality of parenting and nonmaternal child care. Third, early investments in children may lead to more positive development and learning over time. Economic theory, like psychological theory, thus suggests that the effects of maternal employment persist over time and points to the importance of conducting longitudinal analyses.

RESEARCH ON MATERNAL EMPLOYMENT AND CHILDREN'S OUTCOMES

Generally drawing on the National Longitudinal Survey of Youth–Child Supplement (NLSY–CS), studies of the relationship between maternal employment and infant/toddler outcomes have uncovered positive associations, negative associations, and important moderating factors such as child gender and the timing and amount of employment in the child's early years. Some NLSY–CS analyses have found negative associations between full-time maternal employment in the first year of a child's life and child cognitive outcomes up to age 8 (see review in Brooks-Gunn, Han, & Waldfogel, 2002). Additionally, a study using early NICHD SECCYD data found a negative relationship between full-time maternal employment during the child's first year and children's school readiness at age 3 (Brooks-Gunn et al., 2002). A follow-up study that considered several mediating factors found negative relationships with cognitive outcomes from age 3 up to the first grade, that the effect was partially explained by the mother's earnings and the use of center-based care, that lower maternal sensitivity and low-quality home environments helped explain the poorer cognitive outcomes, and that negative associations were stronger for mothers with nonprofessional as opposed to professional jobs. In all of the above-mentioned studies, however, the significant effects applied only to non-Hispanic White children.

Early studies of NLSY–CS data also found negative associations between maternal employment in the child's first two years and problem behaviors during the preschool and school-age years. Some studies, however, have found fewer early behavioral problems and greater compliance for 2-year-olds (see review in Belsky, 2001).

Moreover, several studies have looked at the relationship between parental work schedules and children's outcomes. For instance, one qualitative study found that school-age children had lower math scores if their parents worked evenings and higher school suspension rates if their parents worked nights (Heymann, 2000). In other studies, parents reported that their nonstandard schedules (e.g., early mornings, evenings, nights, or variable hours) interfered with their being able to help their school-age children with homework or participate in after-school activities (Heymann, 2000). Similarly, an analysis of national data revealed lower interest and less willingness to do schoolwork for 6- to 11-year-old children whose parents worked nonstandard hours (see review in Han, 2007).

With respect to socioemotional outcomes, an analysis of NLSY–CS data indicated that nonstandard schedules were related to poorer behavioral outcomes for 4- to 10-year-old children, especially if their mothers were single,

receiving welfare, earning a low income, working as a cashier or in some other service-sector occupation, or working a nonstandard shift full-time (Han, 2007). Children whose mothers and fathers both worked full-time, nonstandard shifts had the most behavioral problems of all (Han, 2007). Studies from Canada have also found a negative relationship between parental nonstandard shifts and school-age children's engagement in property offenses and that parental depression and ineffective parenting behaviors partially explained this finding (see review in Han, 2007).

In sum, early 21st century research indicates that maternal employment during the child's early years, particularly the first year, has some negative effects on cognitive, social, and emotional development during the preschool and possibly the early school years. It is not clear, however, whether these effects last into adolescence or "fade out" over time. In addition, it is important to understand whether there is any link between contemporaneous maternal employment and adolescents' well-being. The next section summarizes this line of research.

MATERNAL EMPLOYMENT AND ADOLESCENTS' OUTCOMES

Adolescence presents children with biological, psychological, and social challenges that can be resolved positively or, if their developmental needs are not met at home and/or school, can lead to negative developmental trajectories (Eccles & Gootman, 2002). Accordingly, adults, especially parents, continue to play guiding roles throughout the teenage years (Eccles & Gootman, 2002). Indeed, adolescents tend to fare better on a range of outcomes when they have open and intimate relationships with their parents, spend more time with them, receive more supervision, and have parents who are more aware of their activities and whereabouts. Unsupervised time with peers is especially problematic when the peers themselves engage in problematic behaviors, when parental monitoring is low, or when the parent–child relationship is poor.

Parents' abilities to provide monitoring and foster healthy relationships with their children are affected by a number of factors, one of them being how their work demands fit with family responsibilities. Psychological theory (e.g., Belsky, 2001; Bronfenbrenner, 2005) and the research on work–family balance (e.g., Presser, 2003) both stress the importance of positive relationships with the home and surrounding environments in fostering positive child development. Economic and sociological theories also point to the significance of income and parent–child relationships in developing adolescents' human, cultural, and social capital. To the extent that employment reduces parent–child contact and parents' physical and emotional energy, they may be less able to nurture their children's development (Heymann, 2000). This discussion suggests that the impact of parental employment may depend on how it affects the time parents have for their children, the quality of the adolescent–parent relationship and the home environment, and the financial resources available to the family. Certainly, parental employment can have benefits, such as greater family resources and reduced financial stress, but longer hours may make it difficult for parents to talk with, monitor, or spend time with their children. These factors make it difficult to predict the effects of parental employment.

PRIOR RESEARCH ON THE RELATIONSHIP BETWEEN PARENTAL EMPLOYMENT AND ADOLESCENT OUTCOMES

A series of experimental studies of welfare-to-work programs by the Manpower Demonstration Research Corporation found some negative effects of maternal employment on achievement and school behavior for adolescents (Gennetian et al., 2002). For instance, adolescents with younger siblings were more likely to be babysitting or working than enrolled in after-school activities, and these same adolescents experienced lower grades, more special education services, and higher expulsion and dropout rates. It has been suggested that taking on adult responsibilities may have some negative ramifications for adolescents' continued development (Bianchi et al., 2005). The results on the behavioral outcomes were mixed (Gennetian et al., 2002). Only two of the six studies found significant results on behavioral outcomes (Gennetian et al., 2002), but maternal employment led to increased drinking and minor (but not major) delinquent activity such as skipping school and increased police involvement.

Most observational studies on the links between maternal employment and adolescent outcomes have focused on middle-income, dual-earner families and have tended to find no significant associations with behavioral problems, social adjustment, drug or alcohol use, or early sexual activity. Effects on school achievement are rare, but some studies have found poorer outcomes for boys (but not girls). One analysis of low-income families from the so-called Three-City Study of welfare reform found that adolescents whose mothers shifted from welfare to work showed improved mental health, especially lessened anxiety (Chase-Lansdale et al., 2003).

A more recent study investigated the long-term effects of early maternal employment and found that children in low-income families had higher verbal, mathematics, and reading test scores in adolescence if their mothers had worked part-time rather than full-time in their infant and toddler years (Ruhm, 2006). Relatively advantaged children (e.g., non-Hispanic White, higher socioeconomic status, college-educated mothers, mothers married at child's birth), however, had poorer cognitive outcomes and an increased risk of obesity in the preadolescent years. In contrast, an earlier study found few long-term effects of early maternal employment on adolescent's risky behaviors such as engaging in sex, crime, and alcohol, cigarette, or substance use.

Another study used the NLSY–CS to examine how adolescents' outcomes were related to parents' current work schedules. For single-mother families, rotating shifts were strongly related to increases in adolescents' delinquent behavior, whereas for two-parent families, mothers' and fathers' nonstandard work schedules were correlated with more parental monitoring but less adolescent–parent closeness. Another study found lower father monitoring and adolescent–parent closeness with both parents when fathers worked nonstandard hours but increased adolescent–mother closeness when only mothers worked nonstandard hours. The mixed findings on adolescent–parent closeness may be explained by how parents' work shifts affect their ability to spend time with their children, such as having dinner with them or helping them with their homework (see review in Han, 2007, for above studies).

AVENUES FOR FUTURE RESEARCH

Both theory and research point to complex links between maternal employment and child and adolescent outcomes, operating through such factors as the home environment, child care, the family's economic resources, and child and maternal characteristics. However, there is still much to learn about these relationships, and fine-grained analyses are needed to adequately explain them. Given the existing research and that a greater number of fathers are participating in child-rearing and household responsibilities, future studies should consider the role that maternal employment plays within the family's division of work and child-care responsibilities. In other words, mothers may work a given shift for a given number of hours in order to increase the family income, to provide more care for their children, or for a number of other reasons. Additionally, considering family members' employment patterns together will provide a better understanding of the father's role in children's development.

Furthermore, more needs to be learned about the changes that arise in the home environment when parents work full- or part-time. In this regard, analyses using time-use data for both parents would be particularly important.

Finally, future research should also consider how the nature, type, and quality of the job (such as flexibility, job satisfaction, and work schedules) affect family dynamics and children's outcomes. In addition, little is known about how mothers make decisions about their employment in the child's first year, and how such decisions relate to the flexibility and financial support received through their jobs. Mothers in the United States have fewer choices than mothers in Canada and many European countries, which offer more governmental financial support to mothers in the early years of a child's life and more subsidized child care to families during the preschool years. Given that long-standing research has shown that longer paid parental leave can improve child health by reducing infant mortality and by leading to more well-baby health care, such research could have very important policy implications given the lack of paid parental leave and subsidized child care offered to parents in the United States.

SEE ALSO Volume 1: *Attachment Theory; Child Care and Early Education; Parent-Child Relationships, Childhood and Adolescence; Parenting Style; Poverty, Childhood and Adolescence; Socialization, Gender;* Volume 2: *Gender in the Workplace.*

BIBLIOGRAPHY

Belsky, J. (2001). Developmental risk (still) associated with early child care. *Journal of Child Psychology and Psychiatry, 42,* 845–859.

Bianchi, S. M., Casper, L. M., & King, R. B. (Eds.). (2005). *Work, family, health, and well-being.* Mahwah, NJ: Erlbaum.

Bornstein, M. H. (Ed.). (2002). *Handbook of parenting.* Mahwah, NJ: Erlbaum.

Bronfenbrenner, U. (Ed.). (2005). *Making human beings human: Bioecological perspectives on human development.* Thousand Oaks, CA: Sage.

Brooks-Gunn, J., Han, W.-J., & Waldfogel, J. (2002). Maternal employment and child cognitive outcomes in the first three years of life: The NICHD Study of Early Child Care. *Child Development, 73,* 1052–1072.

Chase-Lansdale, P. L., Moffitt, R. A., Lohman, B. J., Cherlin, A. J., Coley, R. L., Pittman, L. D., et al. (2003). Mothers' transitions from welfare to work and the well-being of preschoolers and adolescents. *Science, 299,* 1548–1552.

Damon, W., Lerner R. M. (Series Eds.), & Eisenberg, N. (Vol. Ed.). (2006). *Handbook of child psychology: Vol. 3. Social, emotional, and personality development* (6th ed.). Hoboken, NJ: Wiley.

Eccles, J. S., & Gootman, J. A. (2002). *Community programs to promote youth development.* Washington, DC: National Academy Press.

Gennetian, L. A., Duncan, G. J., Knox, V. W., Vargas, W. G., Clark-Kauffman, E., & London, A. S. (2002). *How welfare and work policies for parents affect adolescents: A synthesis of research.* New York: Manpower Demonstration Research Corporation. Retrieved January 18, 2008, from http://www.mdrc.org/publications

Han, W.-J. (2007). *Nonstandard work schedules and work–family issues.* Sloan Work and Family Research Network

Encyclopedia. Retrieved November 20, 2007, from http://wfnetwork.bc.edu

Heymann, J. (2000). *The widening gap: Why America's working families are in jeopardy and what can be done about it.* New York: Basic Books.

Huston, A. C., & Rosenkrantz Aronson, S. (2005). Mothers' time with infant and time in employment as predictors of mother–child relationships and children's early development. *Child Development, 76,* 467–482.

Presser, H. B. (2003). *Working in a 24/7 economy: Challenges for American families.* New York: Russell Sage Foundation.

Ruhm, C. J. (2006). *Maternal employment and adolescent development* (NBER Working Paper No. 10691). Retrieved November 10, 2007, from http://www.uncg.edu/eco

U.S. Bureau of Labor Statistics. (2007). *Charting the U.S. labor market in 2006.* Retrieved November 15, 2007, from http://www.bls.gov/cps

U.S. Bureau of the Census. (2007). *Statistical abstract of the United States.* Retrieved November 15, 2007, from http://www.census.gov/compendia

Wen-Jui Han

MEAD, MARGARET
1901–1978

Anthropologist Margaret Mead was born in Philadelphia, Pennsylvania, on December 16. The eldest of five children (four of whom lived to adulthood) of sociologist Emily Fogg Mead (1871–1950) and economist Edward Sherwood Mead (1874–1956), she first became interested in childhood development by learning to observe her siblings, as her mother and grandmother, a teacher, had observed her. This interest continued throughout her life.

Margaret Mead became well known as a pioneering social scientist while still in her 20s, with her earliest work devoted to the cultural and psychological aspects of childhood and adolescence. She paid particular concern to the ways in which children learn to be members of their cultures.

Mead studied children's thought and enculturation and how an individual's development intersected with and was shaped by cultural emphases on particular temperamental traits. Along with her mentor and friend, anthropologist Ruth Benedict (1887–1948), Mead was a prominent popularizer of anthropology in the United States. Both she and Benedict were associated with the *culture and personality* approach to anthropology.

Mead studied the children of seven cultures in the South Pacific and Indonesia intensively in the 1920s and 1930s. She also studied children in the United States, including her own daughter (Mary Catherine Bateson, born 1939) and granddaughter (Sevanne Margaret Kassarjian, born 1969).

By the post–World War II (1939–1945) years, Mead had switched much of her focus from preliterate cultures to American culture, and she was quoted extensively as an expert on youth culture and the *generation gap* in the 1960s and 1970s, publishing *Culture and Commitment: A Study of the Generation Gap* in 1970.

Using varied—and often groundbreaking—methods, Mead explored the role of culture in personality formation by observing children in various environments. Some of the methods she used by herself, or in collaboration with others, were observation and note-taking; interviewing; photography (still and moving picture); psychological testing (including standardized intelligence tests); and the collection of children's drawings from the study participants. The use of these methods culminated in a collaborative project with her third husband, Gregory Bateson (1904–1980), and other colleagues, in Bali between 1936 and 1939.

Mead earned her master's degree in psychology in 1924, and the research for her master's thesis examined the effect of culture on the intelligence test scores of Italian immigrant children in Hammonton, New Jersey. She used similar psychological testing materials to investigate the thought of girls in American Samoa on her first anthropological field trip in 1925 and 1926. Mead's professor, Franz Boas (1858–1942), known as the father of modern American anthropology, thought she was particularly well equipped to investigate the degree to which adolescence was culturally determined rather than a universal given. Granville Stanley Hall (1844–1924) had argued in his work, *Adolescence* (1904), that adolescence was a time universally marked by "storm and stress."

In 1928 William Morrow (d. 1931) published Mead's first book, *Coming of Age in Samoa*, a study of adolescent girls in Samoa and the implications of their experiences for American education. Mead found that the emphases of Samoan culture led to adolescent girls having a carefree experience of sexuality; it lacked the turbulence of adolescence expected in Western culture.

Although *Coming of Age* brought Mead rapid fame, other researchers over the years disagreed to varying degrees with her findings. The most prominent of these, New Zealand–born Australian anthropologist Derek Freeman (1916–2001), first publicly took on Mead after her death in *Margaret Mead and Samoa: The Making and Unmaking of an Anthropological Myth* (1983). The resulting debate over field methods and the respective roles of nature and nurture is often referred to as the *Mead–Freeman controversy*. The controversy was never definitively resolved, in part because Freeman's work in Samoa,

MARGARET MEAD. AP IMAGES.

on which his conclusions were based, had been conducted in a different area and time period than Mead's.

On Mead's second field trip in 1928 and 1929—to the Manus (or Admiralty) Islands (now part of Papua New Guinea)—with her second husband, Reo Fortune (1903–1979), she investigated the thought of the children through various methods, most prominently through collecting their drawings but also through such projective testing methods as inkblot tests. Mead collected approximately 35,000 children's drawings in Manus and did not find evidence of spontaneous animism (the attribution of spiritual qualities to objects) in their depictions of the world, though Swiss psychologist Jean Piaget (1896–1980) had argued that animism was a universal feature in the development of children's thought. The lack of animism among Manus children was especially striking given a heavy emphasis on the supernatural in the adult Manus culture. Mead continued to collect children's drawings in most of the other cultures she studied, including several return visits to Manus in the post–World War II years.

In the 1950s, Mead and her colleague, Rhoda Metraux (1914–2003), collected drawings from American children depicting images of scientists and from American and Balinese children depicting a Sputnik Soviet satellite. In the postatomic world, Mead felt that children knew more about living in the time of rapid social change than did their parents, in contrast with an older, more slowly changing type of society.

A major figure in American anthropology—especially in the areas of culture and personality, gender, and child development—Mead kept detailed and voluminous notes on her work and that of her colleagues, including those who worked extensively in childhood development, and those papers have been preserved in the Manuscript Division of the Library of Congress for scholarly reference.

SEE ALSO Volume 1: *Puberty; Sexual Activity, Adolescent; Youth Culture.*

BIBLIOGRAPHY

Côté, J. (Ed.). (2000). The Mead–Freeman controversy in review. *Journal of Youth and Adolescence, 29*(5), 525–538.

Francis, P. A. (2005). Margaret Mead and psychology: The education of an anthropologist [Special issue]. *Pacific Studies, 28*(3–4), 74–90.

Library of Congress (2001). *Margaret Mead: Human nature and the power of culture.* Retrieved June 8, 2008, from www.loc.gov/exhibits/mead

Mead, M. (1970). *Culture and commitment: A study of the generation gap.* Garden City, NY: Natural History Press.

Mead, M. (1972). *Blackberry winter: My earlier years.* New York: Morrow.

Mead, M. (2001). *Coming of age in Samoa: A psychological study of primitive youth for Western civilization.* New York: HarperCollins. (Original work published 1928)

Mead, M. (2001). *Growing up in New Guinea: A comparative study of primitive education.* New York: HarperCollins. (Original work published 1930)

Patricia A. Francis

MEDIA AND TECHNOLOGY USE, CHILDHOOD AND ADOLESCENCE

It becomes more and more evident with each generation that young people's media use is more than simple entertainment; it provides them with tools for understanding their world, for presenting their own personality, and for escape and distraction. In the past, when television was the only electronic, audiovisual media in the home, children's and adolescents' media use was a purely receptive experience. Their involvement with the medium was limited to responding to the onscreen content and turning the channel. The world they witnessed was one they could not interact with or influence. In the media environment of the early 21st century, this is far from the case. Modern Internet applications and inexpensive consumer electronics have merged the roles of producer and consumer so that young people create, shape, and own parts of their media environment.

Broadcast and mass-appeal media content, however, is still very much a part of the lives of children and adolescents. On any given day, a young person in the United States is likely to watch television, play video games, use the Internet, listen to music, talk on the phone, and use numerous other forms of electronic media and technology. All of these activities have the potential to influence them in numerous and varied ways. Television and movies can expose young people to violent behavioral options and frightening images or to multiple worldviews and valuable information. Aspects of the Internet can help young people connect with friends, learn about hobbies, and volunteer, or they can be used to harass schoolmates, illegally acquire software, and gamble. All these are tools for young people, and their overall impacts on their lives are determined by how they are used. Understanding these outcomes, therefore, starts with understanding patterns and differences in usage across groups of young people.

PATTERNS OF MEDIA USE

On average, young people aged 8 to 18 spend 6 hours and 21 minutes per day using media (Roberts, Foehr, & Rideout, 2005). Use is primarily composed of electronic media, including 3 hours and 4 minutes of television, 49 minutes of video games, 1 hour and 2 minutes of computers, and 1 hour and 44 minutes of music. Young people also spend 43 minutes per day with leisure time print media. Each type of media use, however, does not occur independently of the others. Portable music players, wireless Internet routers, handheld video game units, and other similar technology allows users to participate in numerous mediated activities at one time. When this type of use, known as multitasking, is considered, young people are exposed to 8 hours and 33 minutes of media per day.

Television Viewing Age is a strong predictor of the amount of television children and adolescents view. As very young children age, they show a fairly consistent increase in television viewing until age 3 after which use remains fairly constant. At the age of school entry, typically age 6, weekday viewing decreases slightly and is met with an increase in weekend viewing (Wright et al., 2001). The amount of viewing is fairly level from ages 8 to 14 (approximately 3

Guitar Hero. *Boy plays Guitar Hero video game.* AP IMAGES.

hours and 15 minutes per day [Roberts et al., 2005]), with some studies finding a peak around age 12 (Comstock, Chaffee, Katzman, McCombs, & Roberts, 1978). Upon entry into high school, television viewing tends to decrease as social and school activities take on higher priorities in the lives of young people.

Gender does not differentiate the amount of viewing, as boys and girls tend to watch similar amounts of television. There are differences, however, in what is watched by gender: Girls are heavier viewers of comedy, whereas boys watch more sports, noneducational cartoons, and fantasy programming (Wright et al., 2001).

Ethnicity and socioeconomic status are both predictors of viewing. African Americans tend to watch the most television, followed by Hispanic Americans and European Americans. Lower levels of income and parental education tend to be associated with more viewing. While these variables are highly related to each other,

economic indicators predict television viewing within each group (Bickham et al., 2003).

Interactive Media Video game play and computer use show opposite relationships with age. As young people progress through childhood and into adolescence, their video game play reduces while their nongame computer use increases (Roberts et al., 2005). Gender differences also exist in use of these media. Boys spend more time playing video games than girls (1 hour 12 minutes versus 25 minutes a day), whereas girls spend slightly more time using the computer.

As young people age, they spend more time using the computer for instant messaging, e-mailing, and visiting Web sites. Some studies show small differences in Internet use between the genders with girls spending slightly more time e-mailing, instant messaging, and visiting Web sites (Roberts et al., 2005). Other research has found no gender differences with these activities

(Gross, 2004). As these media continue to change in form and popularity, it is likely that these time estimates and differences will change as well.

EARLY MEDIA USE RESEARCH: TELEVISION AND OTHER ACTIVITIES

While the newer forms of media are unique in many ways, the existing research on television is relevant to understanding media's influences on young people in the early 21st century for at least two reasons: (a) television continues to be the medium that is used most by children and adolescents and (b) numerous aspects of the theoretical explanations for television's influence on development are applicable to newer media.

In the 1950s as television spread throughout U.S. households, it was met with concerns about how its presence would impact young people. Critics of the medium painted a picture of a fractured family where dinner conversations, homework, and other productive pursuits give way to hours of time in front of the screen. Researchers interested in exploring the effects of the introduction of television on individuals, families, and communities had to move very quickly, and many focused on remote areas of the United States and other countries that received television later in its adoption. These are some of the most effective and unique studies to answer the question of how media impact individuals' everyday lives.

One of the most influential studies to examine the introduction of television into a region was conducted in the Canadian mountains where towns geographically close to each other received various amounts of television (Williams, 1986). Data were collected before one town received television and again 2 years after television was available. With the introduction of television, children attended fewer community activities and adults attended fewer social clubs and activities. Young people did not reduce their participation in youth-oriented social organizations. While television may have replaced some out-of-home activities, those that served important functions (such as social events for adolescents) were less likely to be affected.

In a study in South Africa that examined the effects of the introduction of television, researchers found that television mainly displaced other types of media such as radio and movies (Mutz, Roberts, & van Vuuren, 1993). A study performed in the 1950s found that young adolescents (6th to 10th graders) in communities with television spent less time listening to the radio, going to movies, and reading comic books than their peers with no access to the medium (Schramm, Lyle, & Parker, 1961). Overall, these studies point to the same general conclusion: Television is a force that restructures young people's free time but tends not to greatly influence activities that serve a more important function than occupying unscheduled time.

Displacement These naturalistic studies explored how television may replace other activities. This relationship, known as displacement, plays a central role in many of the concerns that media are detrimental for young people. For example, the use of media could potentially be linked to obesity, poor school performance, and underdeveloped language skills by replacing physical activities, homework, and reading. Many findings from the large naturalistic studies as well as other investigations show limited support for such dramatic displacement by television and other media.

While not fully conclusive, a number of research studies do support the notion that television viewing is related to lower levels of reading achievement (Beentjes & van der Voort, 1988). One potential explanation for this relationship is the displacement of book reading by television viewing. Whereas some studies have found a negative relationship between TV viewing and reading (Koolstra & van der Voort, 1996), others have not (Vandewater, Bickham, & Lee, 2006). Similarly, there are mixed findings concerning whether or not reading increases when an intervention reduces children's television viewing (Gadberry, 1980; Robinson & Borzekowski, 2006). Given these inconsistent and complex findings, an explanation as simple as television displacing reading appears insufficient. Almost certainly, individual family characteristics apart from electronic media availability play a role in determining whether or not a child is a heavy reader.

Similarly, the relationship between television viewing and obesity is more complex than it might initially appear. As indicated by various studies, the more television young people watch, the more likely they are to be overweight (Dietz & Gortmaker, 1985; Rey-López, Vicente-Rodríguez, Biosca, & Moreno, 2008). Additionally, successful media reduction campaigns have been shown to have positive, healthy effects on children's body mass index. It does not appear, however, that screen use and exercise have a simple time-exchange relationship; the reduction of media use does not correspond to an increase in physical activity (Robinson, 2000). Again, displacement is not sophisticated enough to explain fully how television influences young people's weight status. Two other mechanisms have been put forth to explain this relationship: (a) Seeing advertisements for unhealthy foods encourages poor nutritional choices (Dixon, Scully, Wakefield, White, & Crawford, 2007), and (b) children eat more and worse foods while they watch television (Matheson, Killen, Wang, Varady, & Robinson, 2004). More research is still necessary to further reveal the process through which television viewing and other screen media can influence weight status.

Uses and Gratifications The time-exchange of displacement may explain the relationship between media and some other activities, but it does not provide a theoretical structure for understanding why young people use media. The uses and gratifications approach, a theoretical perspective exploring the motivations that drive media use, focuses on the functions that media serve for their users. Individuals are seen as involved participants in media use who make active choices that are motivated by their specific needs.

In terms of television viewing, the main gratifications associated with its use are escape and avoidance, with information seeking also occasionally being mentioned. Similar functions are served by video games, with boys listing as their primary reasons for using the medium as fun, excitement, and the challenge of figuring out the game (Olson et al., 2007). When media meet similar goals, users, according to this approach, will choose one over the other. Take, for example, the desire to learn about the weather. An individual could turn to television, the Internet, the newspaper, or other media sources for this information, but he or she is unlikely to seek more than one.

The Internet serves specific functions that many other media are incapable of fulfilling. While young people go online for many reasons, socializing and connecting with friends is a primary one (Borzekowski, 2006). Uses and gratifications would predict that this behavior might replace telephone use because it serves a similar function, but it is also possible that online interactions could replace actual face-to-face interactions (Kraut et al., 1998). Given the broad spectrum of uses available on the Internet and of motivations for these uses, this approach may be in a unique position to help researchers study and understand how and why young people spend their time online.

Cultivation While the uses and gratifications theory addresses the motivations behind media use, the cultivation theory posits that viewing television contributes to users' perception and development of social reality (Gerbner, Gross, Morgan, & Signorielli, 1994). The social universe portrayed on television is a limited and consistent one that uses recurrent images and messages to convey stereotypical ideas about topics ranging from violence to gender roles. These characteristics are common across all types of programming. Young people who spend more time with the medium will have a world perspective and a belief system that more closely resembles the world as it is presented on television.

Explorations into the cultivation paradigm tend to use survey methodology because they are attempting to illustrate that widespread beliefs about society are linked to television viewing. Results have fairly consistently supported this perspective, with heavy viewers conceiving of the world as a scary, dangerous place and holding racial and gender stereotypes that are consistent with those presented on television. Television viewing has been found to shape an exceptionally broad spectrum of world beliefs. For example, dramas and situational comedies often depict sexual relationships as casual and risk free. Frequent viewers, therefore, are likely to hold more permissive views about sex. Similarly, the popularity of crime shows such as *Law and Order* overrepresent the number of police officers, judges, and lawyers in society. Frequent viewers of these types of programs believe that such occupations are more common than they actually are.

This approach has been met with some criticism. By considering the content of television as monolithic and homogenous, this approach does not address the specific impact of watching certain types of programming. Viewing violent content on television, for example, has been consistently linked to aggressive thoughts and behaviors. Proponents of this perspective argue that the pre- and post-viewing testing methodology of media effects research does not capture the lifelong experience with television or the role it plays in the overall development of individuals' social understanding.

CURRENT ISSUES IN THE AREA OF MEDIA USE

Since the late 1980s, shifts in the media landscape have brought new issues to the forefront of parents' and researchers' minds. Dramatic structural changes in the space media occupy in homes as well as the types of new media available to families contribute to the modern role that media play in the lives of young people. What had once been seen as an occasional entertaining diversion is now a major presence in the home and is often used to occupy, inform, and socialize even very young children.

Media in the Bedroom With the advent of television, sets were expensive and families had one located in a central location of their house. As prices fell, working televisions were replaced by newer models and moved to more private parts of the home. In the early 21st century, it is more common for a family to have three or more televisions than to have one television, and approximately 68% of children aged 8 to 18 have one in their bedroom (Roberts et al., 2005). Furthermore, 30% of children under 3 are sleeping in rooms with televisions in them (Rideout, Vandewater, & Wartella, 2003).

Research has repeatedly demonstrated that having a television in one's bedroom is linked to higher levels of viewing and puts young people at risk for multiple negative outcomes. These include obesity, poor school performance, and poor social skills (Borzekowski & Robinson, 2005; Delmas et al., 2007; Mistry, Minkovitz, Strobino, & Borzekowski, 2007). While less research has explored the relationships between these outcomes and other types of media in the bedroom, there are sound reasons to believe that having a computer or video game system occupy a similar space could lead to similar consequences. The lack of parental oversight in the media these children use is potentially the mechanism that links bedroom televisions and negative outcomes. With additional modes of electronic media available in their private space, young people with computers and game systems in their rooms are likely to experience high levels of use and exposure to age-inappropriate material.

Very Young Children's Media Use Until fairly recently, it was assumed that children under the age of 2 did not watch television. No content was created for this audience, and the limited research in this area indicated that infants and toddlers were unable to understand and learn from television. Between 1998 and 2000, however, Baby Einstein videos, designed to be viewed by very young children, became immensely popular. Their educational value, while implied by the marketing, was not backed by research findings. The great success of these videos spawned a new market with competing video releases as well as a digital television network with constant content for infants and toddlers.

Early research in the area of young children's television viewing indicated that it was difficult for children under the age of 2 to apply information conveyed to them through a television to a real-world task. Coining the phrase "video deficit," investigators found that it was more difficult for young children to imitate actions demonstrated by a researcher on a screen than those demonstrated face-to-face (Anderson & Pempek, 2005; Barr & Hayne, 1999). By 2007 the first academic reports had been released evaluating the educational and developmental usefulness of videos such as those created by Baby Einstein and other companies. Some of this early research has found evidence of a relationship between viewing baby-targeted videos and smaller vocabularies (Zimmerman, Christakis, & Meltzoff, 2007). As this research area grows and children become media users at younger and younger ages, social scientists will begin to have a more complete understanding of the effects of early viewing on brain development and lifelong cognition.

Internet Safety Childhood and adolescence have been forever changed with the advent of the Internet. Along with its potential to expose young people to previously inaccessible information and broad cultural perspectives come concerns about the medium's impact on well-being and overall safety.

As young people began to utilize the social components of the new technology, researchers examined links between virtual connectivity and social outcomes. Early investigations received broad attention and indicated that Internet use contributed to feelings of depression and loneliness as well as reductions in family communication and social support (Kraut et al., 1998). Research from the early 21st century has identified online communication with unknown or lesser known individuals as linked to feelings of loneliness, while no relationship to well-being has been found for general use and communication with friends (Gross, 2004; Gross, Juvonen, & Gable, 2002). While interactions with strangers do occur online, it is much more common for young people to use the Internet to communicate with their local friends (Gross, 2004). Children of the early 21st century will grow up interacting online, and as they enter adulthood, it will be possible for research to track the long-term effects of early Internet use on social adjustment and well-being.

A major concern with young people's online interaction is the potential for them to be targets of sexual predators. Apart from the news attention given to extreme and rare crimes, there is evidence from survey research that a true, day-to-day threat exists. A survey from 2005 found that approximately 17% of 16- to 17-year-olds experienced some type of unwanted or inappropriate sexual solicitation online. Promisingly, this was a drop from 23% in a similar survey performed five years earlier (Mitchell, Wolak, & Finkelhor, 2007). If young people are adequately prepared for the dangers present online, then the Internet has the potential to be an environment in which they supplement their real-world relationships with valuable virtual interactions.

CONCLUSION

From the mid-20th century to the early 21st century, electronic media has been playing a role of ever-increasing importance in the lives of children and adolescents. Researchers and theorists have continually sought to understand the motivations behind the use of media as well as its influences on young people's social, physical, and cognitive well-being. The next 50 years will most certainly see advances in technology that will offer young people both promise and peril. It is the continual challenge of researchers to adapt their theories of media effects to these new technologies and seek to understand their contributions to the lives of children and adolescents.

SEE ALSO Volume 1: *Activity Participation, Childhood and Adolescence; Media Effects; Obesity, Childhood* *and Adolescence; Youth Culture;* Volume 2: *Time Use, Adulthood;* Volume 3: *Time Use, Later Life.*

BIBLIOGRAPHY

Anderson, D. R., & Pempek, T. A. (2005). Television and very young children. *American Behavioral Scientist, 48,* 505–522.

Barr, R., & Hayne, H. (1999). Developmental changes in imitation from television during infancy. *Child Development, 70,* 1067–1081.

Beentjes, J. W. J., & van der Voort, T. H. A. (1988). Television's impact on children's reading skills: A review of research. *Reading Research Quarterly, 23,* 389–413.

Bickham, D. S., Vandewater, E. A., Huston, A. C., Lee, J. H., Caplovitz, A. G., & Wright, J. C. (2003). Predictors of children's electronic media use: An examination of three ethnic groups. *Media Psychology, 5*(2), 107–137.

Borzekowski, D. L. G. (2006). Adolescents' use of the Internet: A controversial, coming-of-age resource. *Adolescent Medicine Clinics, 17,* 205–216.

Borzekowski, D. L. G., & Robinson, T. N. (2005). The remote, the mouse, and the no. 2 pencil: The household media environment and academic achievement among third grade students. *Archives of Pediatrics & Adolescent Medicine, 159,* 607–613.

Comstock, G., Chaffee, S., Katzman, N., McCombs, M., & Roberts, D. (1978). *Television and human behavior.* New York: Columbia University Press.

Delmas, C., Platat, C., Schweitzer, B., Wagner, A., Oujaa, M., & Simon, C. (2007). Association between television in bedroom and adiposity throughout adolescence. *Obesity, 15,* 2495–2503.

Dietz, W. H., Jr., & Gortmaker, S. L. (1985). Do we fatten our children at the television set? Obesity and television viewing in children and adolescents. *Pediatrics, 75,* 807–812.

Dixon, H. G., Scully, M. L., Wakefield, M. A., White, V. M., & Crawford, D. A. (2007). The effects of television advertisements for junk food versus nutritious food on children's food attitudes and preferences. *Social Science & Medicine, 65,* 1311–1323.

Gadberry, S. (1980). Effects of restricting first graders' TV-viewing on leisure time use, IQ change, and cognitive style. *Journal of Applied Developmental Psychology, 1,* 45–57.

Gerbner, G., Gross, L., Morgan, M., & Signorielli, N. (1994). Growing up with television: The cultivation perspective. In J. Bryant & D. Zillmann (Eds.), *Media effects: Advances in theory and research* (pp. 17–41). Hillsdale, NJ: Erlbaum.

Gross, E. F. (2004). Adolescent Internet use: What we expect, what teens report. *Journal of Applied Developmental Psychology, 25,* 633–649.

Gross, E. F., Juvonen, J., & Gable, S. L. (2002). Internet use and well-being in adolescence. *Journal of Social Issues, 58,* 75–90.

Koolstra, C. M., & van der Voort, T. H. A. (1996). Longitudinal effects of television on children's leisure-time reading: A test of three explanatory models. *Human Communication Research, 23,* 4–35.

Kraut, R., Patterson, M., Lundmark, V., Kiesler, S., Mukophadhyay, T., & Scherlis, W. (1998). Internet paradox: A social technology that reduces social involvement and psychological well-being? *American Psychologist, 53,* 1017–1031.

Matheson, D. M., Killen, J. D., Wang, Y., Varady, A., & Robinson, T. N. (2004). Children's food consumption during television viewing. *The American Journal of Clinical Nutrition, 79,* 1088–1094.

Mistry, K. B., Minkovitz, C. S., Strobino, D. M., & Borzekowski, D. L. G. (2007). Children's television exposure and behavioral and social outcomes at 5.5 years: Does timing of exposure matter? *Pediatrics, 120,* 762–769.

Mitchell, K. J., Wolak, J., & Finkelhor, D. (2007). Trends in youth reports of sexual solicitations, harassment, and unwanted exposure to pornography on the Internet. *Journal of Adolescent Health, 40,* 116–126.

Mutz, D. C., Roberts, D. F., & van Vuuren, D. P. (1993). Reconsidering the displacement hypothesis: Television's influence on children's time use. *Communication Research, 20,* 51–75.

Olson, C. K., Kutner, L. A., Warner, D. E., Almerigi, J. B., Baer, L., Nicholi, A. M., II et al. (2007). Factors correlated with violent video game use by adolescent boys and girls. *Journal of Adolescent Health, 41,* 77–83.

Rey-López, J. P., Vicente-Rodríguez, G., Biosca, M., & Moreno, L. A. (2008). Sedentary behaviour and obesity development in children and adolescents. *Nutrition, Metabolism, and Cardiovascular Diseases, 18,* 242–251.

Rideout, V. J., Vandewater, E. A., & Wartella, E. A. (2003). *Zero to six: Electronic media in the lives of infants, toddlers, and preschoolers.* Menlo Park, CA: Henry J. Kaiser Family Foundation. Retrieved May 13, 2008, from http://www.kff.org/entmedia/3378.cfm

Roberts, D. F., Foehr, U. G., & Rideout, V. J. (2005, March). *Generation M: Media in the lives of 8–18 year-olds.* Menlo Park, CA: Henry J. Kaiser Family Foundation. Retrieved May 13, 2008, from http://www.kff.org/entmedia/entmedia030905pkg.cfm

Robinson, T. N. (2000). Can a school-based intervention to reduce television use decrease adiposity in children in grades 3 and 4? *The Western Journal of Medicine, 173,* 40.

Robinson, T. N., & Borzekowski, D. L. G. (2006). Effects of the SMART classroom curriculum to reduce child and family screen time. *Journal of Communication, 56,* 1–26.

Schramm, W., Lyle, J., & Parker, E. B. (1961). *Television in the lives of our children.* Stanford, CA: Stanford University Press.

Vandewater, E. A., Bickham, D. S., & Lee, J. H. (2006). Time well spent? Relating television use to children's free-time activities. *Pediatrics, 117,* e181–e191. Retrieved May 13, 2008, from http://pediatrics.aappublications.org/cgi/content/full/117/2/e181

Williams, T. M. (1986). *The impact of television: A natural experiment in three communities.* Orlando, FL: Academic Press.

Wright, J. C., Huston, A. C., Vandewater, E. A., Bickham, D. S., Scantlin, R. M., Kotler, J. A. et al. (2001). American children's use of electronic media in 1997: A national survey. *Journal of Applied Developmental Psychology, 22,* 31–47.

Zimmerman, F. J., Christakis, D. A., & Meltzoff, A. N. (2007). Associations between media viewing and language development in children under age 2 years. *The Journal of Pediatrics, 151,* 364–368.

David S. Bickham

MEDIA EFFECTS

Public concern about media influence is as old as media forms themselves. The debates and controversies of the early 21st century about video game violence and sexual images on the Internet can be understood not merely as contemporary issues but rather as the latest in a long-standing series of critical conversations about the capacity of the media to affect audiences of all ages. From the comic books and pulp fiction of yesteryear to the electronic and interactive media of the early 21st century, media effects have raised the ire of parent and watchdog groups, inspired research scrutiny by scholars, triggered inquiry by the U.S. Congress, and captured the imagination of the public at large.

Although media influence has been documented in college-age and adult populations, much attention to the topic has centered on media influence on children and adolescents. Young people are often considered particularly vulnerable to media effects because their notions of social roles and cultural norms—and their own place within roles and norms—are in flux. The firsthand knowledge that older audience members use to question or counter media messages is limited in children and adolescents by age and experience. The media, thus, become important sources of cultural and social information for the young. Media effects are an important area of study also because of the central role of media in the lives of young people, with the typical 8- to 18-year-old spending more than 6 hours per day with various forms of media (Roberts & Foehr, 2004).

Some media effects have been shown to occur in audience members of all ages or within the broad stages of childhood or adolescence, but for other media effects, there are developmental periods in which an influence is more or less likely. Huesmann, Moise-Titus, Podoloski, and Eron (2003), for example, found that heightened exposure to violent media between the ages of 6 and 9 poses a particular risk for long-term effects on aggression. In another example, brand loyalty has been shown to develop as early as age 2 in children, and preschoolers are especially likely to request items they have seen advertised. Because diet and exercise habits are formed between the ages of 2 to 4 years, and given the amount of advertisements for high-sugar or salty foods that children are exposed to and the paucity of advertisements for healthy foods, media exposure by children in this young age range can contribute to the development of poor dietary habits.

MAIN THEMES

The potential for violent images in film, on television, and in video and computer games to influence individuals is the single topic that has spurred the largest number of studies and inquiries among media researchers. More than 200 studies to date have used experimental (both lab- and field-based) and survey (including longitudinal) methodologies to explore exposure to violent television and film and aggressive outcomes. Among most social scientists of the early 21st century, the conclusion drawn from this research is unequivocal. Media violence does contribute to the learning of aggression, with convincing statistical links established between various forms of media exposure and attitudes about aggression, aggressive thoughts, and, perhaps most critically, aggressive behavior. The conclusions from these studies have been confirmed through meta-analysis as well. Similar research concerning video game violence is growing, with dozens of experiments and surveys that collectively and decisively point, once again, to the conclusion that video game violence can influence the aggression (in thoughts, feelings, attitudes, or behavior) of individuals young and old. Once again the technique of meta-analysis has synthesized these studies and documented a statistically significant relationship between exposure to violent games and aggression.

Yet the effects of media violence on aggression are just one way to conceptualize the influence of media violence. Other studies, admittedly much smaller in number and many now rather outdated, have supported a desensitization influence, or the potential for repeated exposure to media violence to lower the emotional and/or physiological (e.g., heart rate, blood pressure) response that would usually attend such exposure. Still another manner in which media violence has been shown to affect individuals is through cumulative exposure to television's violent stories (present, in large part, regardless of time slot, genre, or other distinction) and its ability to cultivate perceptions of social reality. Survey research supporting George Gerbner's cultivation theory, as well as more recent experimental inquiries to attempt to explain how individuals process violent television content, has provided ample evidence that heavy television viewers tend to perceive the world around them as dangerous, violent, and threatening compared to lighter television viewers.

Members of the public often express concern about the amount of sexual content in media as well, and the potential for that content to influence audience members. Researchers have not examined this topic as thoroughly as the topic of violence, but nonetheless a few key patterns emerge in the research evidence. A couple of high-profile studies from the early 21st century have used longitudinal survey design to determine that exposure to sexual content on television among early adolescents is a significant contributor to whether they initiate sexual intercourse over the time span of the 2-year study (e.g., Martino, Collins, Kanouse, Elliott, & Berry, 2005). Previous to these studies of direct effects, research in this area had largely been confined to evidence from surveys that showed heavy exposure to television was associated with perceptions of the number of peers who are having sex and other perceptions of sexual behavior in the "real world."

The ability of the media to relay information to audiences regarding gender roles is another key aspect of media effects research. For decades, researchers have conducted experiments and performed surveys to investigate the messages that individuals, especially children and adolescents, receive from television and commercials regarding the "proper" and acceptable ways of being for women and men and for girls and boys. The surveys, once again often but not always conceived within the theoretical tradition of cultivation theory, and undertaken largely in the 1970s and 1980s, have shown the potential of heavy television viewing to be associated with stereotypical or otherwise narrow ways of considering gender (e.g., heavy viewers would be more likely to think women should cook and clean or that men should be scientists or doctors). Experiments have shown the capacity for commercials geared toward children to teach gendered ways of interacting with toys (e.g., boys play in a "rough and tumble" way outdoors with trucks, whereas girls play in a more subdued manner indoors with dolls). Importantly, however, the research evidence has long established that if a depiction on television is counterstereotypical (e.g., a female character fixes things with tools), then the effect on the child audience can be characterized as a widening of conceptions of gender roles rather than a narrowing.

Finally, and relating to this last point, there has been a vast amount of research undertaken with the goal of identifying how audience members young and old are influenced by advertising and commercial content in media. Scholarly inquiry in this area has addressed a broad range of pressing social issues. For instance, experiments have documented dissatisfaction with one's body among adolescents (especially young women) after exposure to magazine advertisements featuring "flawless" and unrealistically slender models. Other experiments have shown that children learn erroneous information about nutrition and can develop preferences for nonnutritious foods through exposure to commercials for fast foods, soda and other sugary beverages, and candy and salty snacks. Still other studies have examined the potential for public service announcements and other aspects of public communication campaigns—the kind that might admonish teens to refrain from smoking, for instance—to change attitudes and behavior in a manner that would advance public health and safety.

KEY THEORIES

A review of all or even most of the theories that have been advanced to explain media effects is well beyond the scope of this entry, but two theories are particularly noteworthy, in addition to the already discussed cultivation theory. Social learning/social cognitive theory is one of the most widely recognized and supported media effects theories, whereas the general aggression model can be understood as a combination of key aspects of a number of previous theories. These theories provide insight regarding why and how the media have the effects reviewed above.

Social learning theory, more recently referred to as social cognitive theory to emphasize the mental processing of social information, proposes that people learn how to act by observing others (Bandura, 1986, 2001). The ability to learn vicariously from real-world others and those presented on television is a unique human capacity, as well as a particularly common and important pursuit for young people because of their developmental tendencies. The theory predicts that the consequences that ensue from an observed model's actions—positive or negative—play a vital role in whether the child will adopt or emulate those actions. For example, if a child is in an environment in which violent behaviors are modeled and are rewarded or simply go unpunished, it follows that the child will then learn to act aggressively. Conversely, if a child is provided with a positive model, the likelihood that that child will emulate that positive behavior increases. Although real-life models (i.e., family and peers) may have the most influence on behavior, exposure to characters in different media forms has been shown to influence aggression levels as well as prosocial behaviors.

One critique of this theory is that there is no empirical evidence to suggest that the short-term effects observed in laboratory situations extend outside the laboratory or later on in life. Moreover, some believe that the theory does not sufficiently take into account the unique factors that each person brings to the viewing situation (e.g., baseline levels of aggression, prior exposure to violent media, socioeconomic status). Beginning in the 1990s, researchers turned their attention more closely to the study of the influence of individual differences with respect to media effects. In addition, with the introduction of the general aggression model, the possible effects of individual differences have come to the forefront.

The general aggression model was conceptualized by Craig Anderson and Brad Bushman and introduced in 2002. With regard to violence and aggression it is perhaps the most comprehensive media effects framework to date. This model takes into account both internal (or personological) and external (environmental or situational) factors when looking at the effects of media. Several characteristics are considered in the person-related variables that an individual brings to the media exposure situation, such as personality traits, gender, beliefs, attitudes, long-term goals, and scripts (or previously learned, processed, and stored information). Research has shown that individually, all of these elements play a role in predicting whether a person will act aggressively. By taking all of them into consideration at once, the model is better able to assess how susceptible or resistant an individual is to acting aggressively in response to violent media. The situational factors that the model explores include aggressive cues present in the physical environment (such as weapons), provocation, frustration, and incentive to aggress. Together these two sets of factors affect cognition, affect, and arousal. Therefore, a person with certain personality traits or who is experiencing a particularly stressful or frustrating situation may be predisposed to interpreting violence differently than another person; that is, depending on individual experiences and the larger context, viewing violent media may produce strikingly different effects or even no effects at all.

The general aggression model has been applied only to violent media, but it may be a useful tool when considering other types of media effects as well. Although there is sufficient evidence to show that various types of media content can affect audience members, the extent to which internal and external states quell or exacerbate these effects remains underresearched. To fully understand the effects of the media, more thorough theoretical propositions (such as that in the general aggression model) are needed.

FUTURE DIRECTIONS

Future scholarly attention is likely to place the individual differences that media audience members bring to a media exposure situation and situational factors into further account through theory and research. Past research has arguably focused much of its attention on whether media effects occur at all, perhaps driven by a desire to be sure media are not dismissed as "just entertainment" but rather are taken seriously for their potential contribution to human behavior and social life. Once clear patterns of whether effects occur have been revealed (as is the case with violence and aggression, for instance), future inquiries are likely to continue attempts to address under what circumstances and for whom effects are most (and least) likely. In doing so, the complexities of media influence will be further revealed.

Media effects research will also look increasingly at the role of new technologies in everyday life, and judgments will be made regarding whether previously existing theories that are based largely on television and print media forms continue to apply convincingly in a digital and interactive media environment. For example, what new social and cultural issues are raised through interacting with the Internet? The scholarly attention in the early 21st century to social networking sites such as MySpace and their role in adolescent life is an example of this new direction for media effects research. Does video game violence influence individuals in similar ways as television and film violence, or are there unique aspects of video game technology that suggest otherwise? What are the implications, more broadly, of a world in which children are spending ever larger amounts of time with media? These are other critical questions likely to guide future media effects research.

Another new direction for media effects research is to study ways of reducing the likelihood of negative effects, particularly for children and adolescents. Parental mediation, which includes the ability of the parent or caregiver to actively counter media messages by speaking up during co-viewing occasions, is likely to gain even more research attention in the years that come. Similarly, a growing focus on media literacy research, which tests the effectiveness of critical and analytical discussions of media in the K–12 classroom, is likely to continue.

Finally, links between media effects research and public policy will continue to be exceptionally important. A unique feature of scholarship in this area is that it is of the utmost interest and relevance to parents, citizen groups, and policy makers. Media effects researchers have been called upon to testify in Congress, for example, regarding whether advertising to young children is ethical. The V-chip (created through the Telecommunications Act of 1996), which can be programmed to block out television content deemed objectionable by parents, is the result, in large part, of the evidence accumulated by those studying the effects of media violence. Media have a huge and growing presence in the lives of children and teens, and parents and policy makers are quite concerned with the health and well-being of children and teens. Thus, the need to draw on this sort of research in governmental decision making as well as decision making within individual households is likely to grow. Valid and reliable media effects research, therefore, will continue to be exceptionally important.

SEE ALSO Volume 1: *Aggression, Childhood and Adolescence; Cultural Images, Childhood and Adolescence; Media and Technology Use, Childhood and Adolescence; Sexual Activity, Adolescent; Socialization, Gender; Youth Culture.*

BIBLIOGRAPHY

Anderson, C. A., & Bushman, B. J. (2002). Human aggression. *Annual Review of Psychology, 53*(1), 27–52.

Bandura, A. (1986). *Social foundations of thought and action: A social cognitive theory.* Englewood Cliffs, NJ: Prentice-Hall.

Bandura, A. (2001). Social cognitive theory: An agentic perspective. *Annual Review of Psychology, 52,* 1–26.

Gerbner, G., Gross, L., Morgan, M., & Signorielli, N. (1994). Growing up with television: The cultivation perspective. In J. Bryant & D. Zillmann (Eds.), *Media effects: Advances in theory and research* (pp. 17–41). Hillsdale, NJ: Lawrence Erlbaum.

Huesmann, L. R., Moise-Titus, J., Podoloski, C.-L., & Eron, L. D. (2003). Longitudinal relations between children's exposure to TV violence and their aggressive and violent behavior in young adulthood: 1977–1992. *Developmental Psychology, 39*(2), 201–221.

Martino, S. C., Collins, R. L., Kanouse, D. E., Elliott, M., & Berry, S. H. (2005). Social cognitive processes mediating the relationship between exposure to television's sexual content and adolescents' sexual behavior. *Journal of Personality and Social Psychology, 89*, 914–924.

Roberts, D. F., & Foehr, U. G. (2004). *Kids and media in America*. Cambridge, U.K.: Cambridge University Press.

Erica Scharrer
Lindsay Demers

MENARCHE

SEE Volume 1: *Puberty*.

MENTAL HEALTH, CHILDHOOD AND ADOLESCENCE

Mental health has been defined as the "successful performance of mental function, resulting in productive activities, fulfilling relationships with other people, and the ability to adapt to change and cope with adversity" (U.S. Department of Health and Human Services, 1999, p. vii). Conversely, mental illness or mental health problems have been described as "alterations in thinking, behavior, and mood with concomitant impairments in social, educational, and psychological functioning" (Roe-Sepowitz & Thyer, 2004, p. 67). Those authors further suggested that "to the extent that so-called mental disorders have etiologies related to environmental, biological, familial, or peer factors, they may not be justifiably construed as *mental* disorders at all, but rather behavioral, affective, and intellectual disorders" (Roe-Sepowitz & Thyer, 2004, pp. 68–69). In this entry mental health disorders of childhood and adolescence are defined as problems that interfere with the way young people think, feel, and act and that when left untreated can lead to school failure, family conflicts, drug abuse, violence, and suicide.

Mental health problems affect one in every five young people at any specific time (U.S. Department of Health and Human Services, 1999). Although the diagnosis of a particular disorder is often difficult, contradictory, or impossible, it is helpful to categorize the various types of abnormal behavior more commonly seen in youth as anxiety disorders (e.g., phobias, chronic anxiety, obsessive–compulsive disorder), mood disorders (e.g., depression), developmental disabilities (mental retardation, autism), organic brain disorders (diseases, brain injury), conduct disorders, eating disorders, and attention deficit hyperactivity disorder (ADHD; American Psychiatric Association, 2000).

The prevalence rate differs for each condition, and trends are often difficult to interpret as they are affected by changes in diagnostic criteria and classification, improved assessment techniques, the increased availability of services and treatment interventions, and the requirement of a diagnosis for receipt of services. For example, the current prevalence rate for autism, estimated to be as high as 1 in 250, represents more than a 500% increase in diagnoses in one decade. The cause of that dramatic increase has not been determined, but no doubt it has been influenced by the factors mentioned above (Calahan & Peeler, 2007).

In line with the thinking of Dominique Roe-Sepowitz and Bruce Thyer (2004), the mental health of children and adolescents is discussed in this entry in terms of the development, both physical and psychosocial, of the individual from preconception through adolescence. Because these disorders typically occur at particular developmental stages of childhood and adolescence, each condition is discussed according to the stage at which it generally is noted.

Human development is a complex process that is influenced by multiple genetic and environmental factors. From the moment when the egg is fertilized by the sperm, the being that will be transformed from embryo to fully developed infant will proceed through the various stages of growth, development, maturation, and aging with a predetermined blueprint that will undergo various changes as the fetus confronts the environment in which it lives. Thus, the infant is born with a physiology determined by prenatal influences, and throughout its life will be influenced by the context in which it will live. If children and adolescents are to develop satisfactory mental health—behaviors conducive to productive activities, fulfilling relationships, and the ability to adapt to change and cope with adversity—they need a healthy physiology that is fostered by a supportive physical and psychosocial environment. When one or the other of these influences goes awry, remediation efforts will be required of parents and/or caregivers and mental health and social service professionals.

Dwyer and Hunt-Jackson (2002) stated that to understand child and adolescent mental health from the perspective of human growth and development, knowledge must be derived and subsequently drawn from a variety of sources. Those sources include psychoanalytic theories and stage theories, biological facts, human behavior theories, economic reports, legal issues, and specific cultural information. Thus, assessment becomes "somewhat of an *art,* and the more available choices, the more opportunities there are for adequate assessment and intervention" (p. 84). In their overview of the life span perspective in human growth and development, Dwyer and Hunt-Jackson noted that although Freud and other psychoanalytic and psychodynamic theorists proposed developmental stages in early life, the life span perspective expands on those ideas. They go on to state that "according to Lefrancois (1993), this unique theoretical view studies the developmental changes that transpire in the individual from conception through old age" and that "key to this study is the consideration given to the biological, psychological, and social processes and influences that account for changes in human behavior" (p. 84).

Dwyer and Hunt-Jackson (2002) noted that although the works of earlier developmental psychologists, including Piaget, Bowlby, and Erikson, suggested that human developmental ceases when adulthood begins, current theorists agree that human development is a continuous process. Because human development is continuous, it also is bidirectional, meaning that children influence their parents as much as parents influence their children and that all development is influenced by the environmental context.

One might conclude that children and parents influence each other along life's continuum and that environment functions pre- and postnatally to determine how, when, and if genetic predispositions will evolve. Peter Gluckman and Mark Hanson (2006) concluded that "environmental factors acting during the phase of developmental plasticity interact with genotypic variation to change the capability of the organism to cope with its environment in later life … and because the postnatal environment can change dramatically, whereas the intrauterine environment is relatively constant over generations, it may well be that much of humankind is now living in an environment beyond that for which we evolved" (p. 4). Thus, the study of child and adolescent mental health must begin with genetic and environmental prenatal influences and move through each developmental period to address both biological and external influences on mental health.

PRENATAL INFLUENCES ON HUMAN GROWTH AND DEVELOPMENT

In spite of advances in the field of human life span development, there are inherent difficulties in deciphering clues about the relative roles of nature and nurture in the origins of health and disease. People are familiar with the saying "you are what you eat" but sometimes do not recognize that individuals are also products of what their grandmothers may or may not have eaten. People also may not be familiar with the role of the father in prenatal as well as postnatal development. It has been discovered, for example, that deficiencies in folic acid during pregnancy can result in neural tube defects in the fetus. Further, children born to mothers whose own mothers might have been exposed to nutritional deficiencies or toxins such as certain medications may be affected by genetic or physiological aberrations in the womb. Similarly, fathers exposed to toxins may have altered sperm that influences prenatal development. However, not all mothers deficient in folic acid will bear children with neural tube defects, nor will all children with neural tube defects be the offspring of nutritionally deprived mothers. Similarly, nutritional and other biological insults in grandmothers or grandfathers may or may not result in defects in subsequent offspring.

Genetic factors link a fetus to the family ancestry and determine the individual characteristics that influence subsequent development. These physiological changes occur within a psychosocial context. As outlined by Keith Godfrey (2006), determinants of fetal growth and functioning include biological, psychological, and social factors. Those factors include the pregnant woman's age at the time of conception, hereditary characteristics, nutrition, alcohol consumption, smoking habits, and ingestion of prescribed or illicit drugs. Physical neglect or abuse plays a role in the development of a baby, as do environmental factors such as access to medical care, financial and emotional support, and exposure to hazardous chemicals. The emotional status of the mother and of her partner and other family members has an effect on the baby. Chemical and hormonal levels in both the brain and the bloodstream affect the biological processes of the mother and the baby when the mother experiences high amounts of stress; this may have an adverse effect on the child's physical and mental health.

INFANTS AND TODDLERS (BIRTH TO 2 YEARS)

During the first 6 months, a normally developing baby will progress from random ineffectual motions to accomplishments such as rolling over and beginning to crawl. The baby will begin to reach for objects, grasping them and learning about them from touch and taste. Newborns are active, constantly interacting with their environment and seeking stimulation and opportunities to improve their competencies (Dwyer & Hunt-Jackson, 2002). During this period an infant expends significant amounts of energy trying to understand and master his or her world.

Within the next few months, infants gain pleasure from their increasing ability to affect what happens to them and the world around them. Erik Erikson (1964) termed the next stage the trust versus mistrust stage, the time during which babies learn to trust their environment and caregivers. When his or her cries are responded to with cuddling, food, and affection, an infant soon realizes that he or she has control over a portion of the immediate environment and is able to develop a sense of security that allows further emotional growth. Without this caring response, however, infants become mistrustful and form an insecure or indifferent attachment to his or her parents. Thus, parents must be educated about the needs of and appropriate responses to infants during the various developmental stages. Parents must be able to facilitate emotional and cognitive development by providing consistency and warmth for the infant. Consistency has been found to be important in that infants who are separated from a consistent caregiver for long periods tend to have retarded motor development, delayed language development, and cold emotional responses. There is also an increased probability of the development of delinquent behavior and an inadequate self-image.

The environment should be structured so that it stimulates the infant's sensory, motor, and social exploration of his or her surroundings and meets the biological needs. In terms of positive mental health, infants must learn to trust or mistrust adults on the basis of the care that is provided. It has been posited that the critical period of attachment and trust occurs between the ages of 6 months and 2 years. The primary role of the health practitioner is to assist parents in providing the infant with an environment conducive to the successful resolution of the issue of trust versus mistrust. Parents and caregivers are responsible for providing a stimulating environment so that the child can grow intellectually and develop in terms of cognitive and social competencies, both of which are essential to good mental health. Parental effectiveness courses should focus on helping parents develop better communication and consistent child management skills, two variables that have been shown to be necessary conditions for successful child rearing (Thyer & Wodarski, 1998, 2007).

Developmental Disabilities Developmental disabilities are defined as significant deficits in intellectual or cognitive functioning, a significant deficit in adaptive behavior, and the presence of these deficits during the developmental period of an individual's life (before age 18; American Psychiatric Association, 2000). Although most babies are born healthy, 2–5% of live-born infants have a major developmental defect (Heindel & Lawler, 2006).

Children and adolescents with developmental disabilities have been found to be at a higher risk of psychiatric comorbid disorders such as ADHD, depression, learning disorders, and childhood psychoses (Andreason & Black, 2006; Filho et al., 2005). Thus, the differential diagnoses for developmental disabilities often are complex.

EARLY CHILDHOOD (2 TO 6 YEARS)

Young children continue to enlarge their repertoires of behaviors between the ages of 2 and 6, and their success at those behavioral tasks determines their future mental health. Physical coordination improves, and the child uses locomotion to explore the environment while learning to master independence-producing tasks such as dressing and toilet training. The use of language increases, as do expressive social interactions with family members and friends. The child becomes more efficient at accumulating and processing new information, providing a foundation for the development of intellectual attributes (Bartsch & Wellman, 1989; Moses & Flavell, 1990).

During this developmental period children begin to see themselves as individuals, separate from others. Erik Erikson (1964) described the two psychosocial crises of this stage as initiative versus guilt and autonomy versus shame and doubt. Initiative results when a child responds actively to his or her environment and is eager to investigate it. The central process by which this occurs is identification. It is necessary for children to explore their environments, and they must have confidence in their ability to control themselves. If adults limit investigation and experimentation, the child develops an overwhelming sense of guilt.

A toddler usually passes through a period when his or her primary word is *no;* this represents a child's attempts to control and order the environment. The establishment of autonomy requires tremendous effort by the child as well as patience and support from the parents. The child begins to develop feelings of self-confidence and independence. The critical point in the development of mental health is the process of moving from dependence to independence. If a toddler develops symptoms such as temper tantrums, aggression, withdrawal, phobias, and school phobia, interventions are available to help parents and caregivers alter the maladaptive behaviors (Thyer & Wodarski, 2007).

There is some controversy about the cross-cultural universality of these developmental crises. Some theorists suggest that independence is valued more highly in European and American cultures (Thomas, 1990). Leonore

Adler (1989) warned that on a worldwide basis individual as well as group differences occur when behavior is observed and studied. The child grows within a social context and during this phase begins to learn the knowledge, values, and skills necessary for effective functioning when interacting in the group setting. These socially sanctioned ways of life (culture) are transmitted to the child through the socialization process, and parents are the first agents of that process. Hence, it is within the family that the child first experiences the requirements of group life. Family circumstances and parenting styles are key variables in a child's adaptation to society (Dwyer & Hunt-Jackson, 2002). Ultimately, the successful adaptation of a child to societal rules, norms, and mores will determine his or her mental health.

A young child's playmates augment the socialization process. By providing cognitive stimulation, play helps prepare children for interaction with their environments and lets them learn and rehearse adult roles. Using play therapy, practitioners allow a child to create imaginary situations in which problems can be expressed and resolved. Play therapy often is used by mental health practitioners as a mechanism for interpreting children's behaviors and helping them learn to cope better (Ganda & Pellegrini, 1985).

The development of good mental health in children requires the provision of appropriate and consistent limits as the child begins to learn self-control. Parents' form of discipline often includes the power-assertive style that involves the use of physical punishment and threats. Psychological discipline may be characterized by love withdrawal and guilt (Patterson, 1982). During this stage the child is beginning to develop moral standards through a gradual internalization of parental values and standards. Numerous researchers have reported that females experience more guilt than males during this developmental stage, and it has been suggested that this results from the unclear messages that society presents to females about their roles and appropriate behaviors.

The development of good mental health requires parental discipline to help the child interpret the undesired behavior. Parents also need to generate alternative actions, explain the reason behind discipline if any is given, and stimulate empathy for the victim of the behavior. Parents should be consistent in their discipline patterns and should set suitable limits that result in appropriate behavior (Quinn, 1998). During early childhood children learn primarily by imitation. If parents model deviant, aggressive, and/or uncontrolled behavior at home, the toddler is likely to imitate that behavior (Wodarski & Wodarski, 1998). Parents can be instructed in ways to exhibit appropriate socialization behaviors.

Because young children are in the early stages of language development, they concretely interpret verbal mes-sages such as "God will punish you" and "big boys don't cry"; that is, those statements are accepted at face value. The development of sex roles, morality, and competence is especially vulnerable to such labels, and parents should be encouraged to use direct communication rather than all-encompassing statements (Dwyer & Hunt-Jackson, 2002).

During this stage of development children have their first experiences with structured early childhood education in the form of preschool, kindergarten, and the first grade of elementary school. With support and patience most children successfully make the transition from the security of the family to the larger school social system. Children who do not experience early school success are at greater risk for dropping out and becoming involved in delinquency (Feldman, Caplinger, & Wodarski, 1983; Rapp-Paglicci, Roberts, & Wodarski, 2002). All children who appear to be struggling with school adjustment require early assessment and intervention; if this is done early, those children can be identified and responded to with appropriate planning.

Separation Anxiety Separation anxiety disorder (SAD) is a severe and disabling form of a maturational experience that all children have (Andreasen & Black, 2006). The fundamental feature of SAD is excessive anxiety about separation from an attachment figure at home. Children with this disorder typically become socially withdrawn and display apathy, sadness, or difficulty in concentration and attention to work or play when separated from the parent or other figure (Sowers-Hoag & DiDona, 1998). The symptoms include distress at being separated from home, worry that harm will come to the parents, and worry that the child will be lost or separated from them. The behaviors include school refusal, sleep refusal, and clinging, and the physiological symptoms include nightmares and physical complaints such as headache and nausea.

Common differential diagnoses include but are not limited to panic disorder, mood disorders, generalized anxiety disorder, and social phobia. Depressed mood is typically concurrent and may intensify over time, usually precipitating an additional diagnosis of dysthymic disorder or major depressive disorder. SAD often is comorbid with school phobia. Approximately 4% of children and young adolescents present with SAD, and the prevalence decreases from childhood through adolescence. It may develop after a life stressor such as the death of a relative or pet, a change in schools, or an illness of the child or a relative. Young people with chronic illnesses may be at higher risk for SAD (Sowers-Hoag & DiDona, 1998). Onset may be as early as preschool age and may occur at any time before age 18 years, although onset in late adolescence is uncommon. The incidence is apparently equal in males and females, although some research has shown that it may be more common in girls than in boys (Sowers-Hoag & DiDona, 1998).

SAD generally is treated through individual therapy with a variety of models of intervention. Karen Sowers-Hoag and Toni DiDona (1998) identified these models as psychodynamic models, play therapy models, cognitive-behavioral treatments, exposure-based procedures, contingency management procedures, and real-life exposure techniques.

LATER CHILDHOOD (7 TO 12 YEARS)

During later childhood the influence of school life becomes paramount. This is a critical time during which someone other than the parents exerts a major influence on the socialization process. Children begin to assess their self-worth through a comparison and evaluation of their academic abilities, athletic skills, physical appearance, and social acceptance. Erikson's (1964) stage of industry versus inferiority describes the way children mentally compare themselves with peers, parents, and other role models to see how much they resemble those others (Dwyer & Hunt-Jackson, 2002). Characteristics of children who have high self-esteem include maternal certainty of child rearing, minimal daily family conflict, closeness with siblings, closeness with peers, parental warmth, consistent and firm discipline, and involvement in family decisions. Inconsistency, harsh discipline, minimal parental attention, and continuous moving are factors that dispose a child to low self-esteem, a contributing factor to conditions such as conduct disorder.

Susan Harter (1987) noted that the most important sources of support for children's self-esteem are parents and classmates, not teachers and friends. Teachers have expectations that when transmitted to the child can affect school performance and consequently self-worth. Hence, schools are major contributors to the maturation of a child as they provide opportunities for meaningful and prolonged interactions with significant adults and peers.

Children's interactions with their peer group during this period serve several functions. As mentioned previously, peers influence an individual's sense of self-esteem. Additionally, peers reinforce key cultural norms, thus affecting the formation of values and attitudes in the child. They are also important sources of information about appropriate behavior. Gender roles and the ensuing behaviors receive strong reinforcement from the child's peer group. During the early school years a child engages in sex-role identification, which includes the understanding of gender labels, sex-role standards, sex-role preference, and identification with the same-sex parent. For the most part children want to be like their peers; they want to wear the same clothes, use the same slang terms, and play the same games. However, illness,

hospitalization, a different physical appearance, and different speaking styles can cause a child to be less acceptable to peers (Quinn, 1998). Most children become accustomed to these differences if they have enough contact with adults who do not teach the prejudice that continues to exist in society.

Peer group structures provide opportunities for growth in a school-age child, and in growth there can be pain. Teachers, counselors, and school social workers frequently help children negotiate small-group power structures and thus learn social skills that foster broader societal adjustment. Teaching communication, conflict resolution, and assertiveness skills can be accomplished by these professionals in a small-group setting.

At this life stage children are in almost daily contact with individuals (e.g., peers, teachers, neighbors) outside the family. At this time signs of problems in adjusting socially, psychologically, or academically are noted. Here the mental health practitioner's role is to design and implement a treatment plan with the child, the family, and relevant collaterals. Caution must be used in working with children this age because frequently the behavior they present has been labeled (e.g., behavior disordered, emotionally disturbed, ADHD) by the classroom teacher or family member, and this label can have a detrimental impact on self-concept and social functioning (Thyer & Wodarski, 1998).

Child abuse is most likely to be detected during the early school years. The child no longer is restricted to family contact or family-sanctioned contacts, and teachers and social workers who have been trained to recognize signs of abuse and neglect are expected to report cases of abuse. According to the U.S. Bureau of the Census (1992), the average age of an abused child is 7, with higher risk ratios for children who are younger (under 3) or older (teens). More than 2 million cases of child maltreatment, including physical injuries, neglect, and sexual and emotional abuse, are reported annually (Lauer, 1995).

In determining actual abuse and neglect professionals must take the cultural aspects relevant to each case into consideration. The United States has a diverse population with many different types of parenting styles. Kathleen Sternberg (1993) gave examples of parenting styles that seem to contradict each other but make sense in a sociological context. She included a study conducted by Diana Baumrind (1991) that found that authoritative parenting (including children in decision making and thus encouraging independence) is most productive with White middle-class children and authoritarian parenting (emphasizing compliance as a virtue and punishment as an appropriate way to enforce compliance) fosters social competence in Mexican American children. Although laws direct certain behaviors and children must be protected, it is best to keep cultural differences in mind when determining the course to take in cases of suspected abuse and neglect.

School Phobia School phobia is a significant anxiety disorder in which children develop a fear of going to school and may develop methods for staying home. School phobia often is associated with school refusal, and there is evidence that SAD may be present in as many as 80% of cases of school phobia (Sowers-Hoag & DiDona, 1998). In younger children school phobia may stem from a fear that something will happen to them or their caregiver while they are at school. In older children school phobia usually has its roots in academic problems, social challenges, or bullying (Mayes & Cohen, 2002).

During the early childhood stage children become part of a larger system, and so they are exposed not only to family influences but also to influences from school officials, teachers, and peers. Separation may be traumatic for the child as well as the parents. School phobia can result in eventual school dropout. If a child has unsuccessful school experiences, the likelihood that he or she will drop out of school is greatly increased. School dropout is highly correlated with later delinquent activities (Feldman, Caplinger, & Wodarski, 1983). A child who does not complete this stage successfully will have numerous problem areas as an adult. Besides the obvious problems of low self-esteem and a sense of incompetence, the individual will be faced with the pressures of needing to earn a living at an early age.

Conduct Disorder Conduct disorder is a repetitive and persistent pattern of behavior in which the basic rights of others or major age-appropriate societal norms or rules are violated. There are four major domains of relevant behavior: aggression toward people and animals, destruction of property, deceitfulness or theft, and serious violations of rules (Andreasen & Black, 2006). Conduct disorder is one of the most frequently diagnosed conditions in outpatient and inpatient mental health facilities for children. According to the American Psychiatric Association (2000), the prevalence of conduct disorder ranges from less than 1% to more than 10%. The prevalence is higher among males than females, but the rate in females may be increasing (Andreasen & Black, 2006).

The disturbance in behavior that defines conduct disorder causes clinically significant impairment in social, academic, and occupational functioning. Conduct disorders have considerable comorbidity with childhood disorders such as learning disorders, ADHD, and mood disorders. At least 10% of children with conduct disorder have specific learning disorders. Twenty to 30% of children who present with ADHD also meet the criteria for conduct disorder (Andreasen & Black, 2006). The most

commonly used and most effective treatments are behavior modification, family therapy, and pharmacotherapy (Rapp & Wodarski, 1998).

ADHD ADHD affects an estimated 3–5%, or 2 million, children in the United States and is considered the most common neurobehavioral disorder among school-age children (Dupper & Musick, 2007). Children with ADHD demonstrate inattention, hyperactivity–impulsivity, or both. These children have difficulty with details, instructions, and organization; are easily distracted and forgetful; and have trouble engaging in mental effort tasks such as schoolwork. The prevalence rate for males is six times higher than that for females and nine times higher among male versus female school-age children. Theories about causation include environmental agents, brain injury, food additives and sugar, and genetics, although genetics appears to be the only consistent link to the condition.

Early diagnosis and treatment, including medication and family behavioral therapy, are imperative as the condition can lead to poor peer relations that can lead to aggression, delinquency and school maladjustment, and family dysfunction that results in higher divorce levels, marital discord, and difficulty coping (Dupper & Musick, 2007). Behavior modification is the only nonmedical psychosocial treatment for ADHD and generally is prescribed in combination with psychopharmacology.

ADOLESCENCE (13 TO 19 YEARS)

In American culture adolescence is the life stage that marks the transition from childhood to adulthood. As in the developmental phases that precede it, distinct biological, psychological, and social changes occur within the individual. Physically, an adolescent experiences a growth spurt associated with the onset of puberty. In the early 21st century individuals tend to reach adult height and sexual maturity faster than they did in the past. Known as the secular trend, this tendency to mature earlier has been attributed to better health care, better nutrition, and a higher standard of living (Chumlea, 1982; Lefrancois, 1993).

One of the important tasks for an adolescent is the development of a sense of identity. According to Erikson (1964), this is achieved only after a period of questioning, reevaluation, and experimentation. During the teen years, individuals experiment with various roles that represent possibilities for future identity development. Experimentation in academic pursuits, athletic endeavors, part-time jobs, hobbies, and dating relationships all contribute to identity formation in a prosocial fashion. Fergus Hughes and Lloyd Noppe (1985) reported that teens look ahead to a time when they will be independent adults. Described traits of disliked adolescents include physically handicapped, shy, timid, withdrawn, quiet, lethargic, listless, passive (a nonjoiner), reclusive, pessimistic, and complaining. For a child with a physical disability, it is not uncommon for traits of and statements about physical limitations to be mistaken for complaints. Similarly, during this period role confusion can lead to poorly thought-out actions and behaviors that appear irresponsible, childish, or rebellious. Some adolescents become entrenched in this acting-out behavior and seek negative reinforcement from it that can delay the transition to adulthood (Dwyer & Hunt-Jackson, 2002).

INFLUENCES ON MENTAL HEALTH DURING ADOLESCENCE

Two influences on adolescent mental health are the family and delinquency.

Family Dysfunction Socially, adolescence provides a period for moving from dependence on one's parents to adult independence. Parent–adolescent relationships frequently are strained as the individual struggles to assert independence. Conflicts often arise about performing chores, studying, using time appropriately, dating, choosing friends, and spending money (Kaluger & Kaluger, 1984). These conflicts test family socialization and communication patterns.

Mental health practitioners encounter adolescent clients for a variety of reasons. Families often seek intervention because of conflicted relationships at home. As Lawrence Shulman (1992) pointed out, teenagers are frequently the scapegoats for family dysfunction; thus, family therapy is the treatment of choice. Baumrind (1991) stated that parents who are responsive yet demanding combine authority with reason and have frequent communication with a child tend to have adolescents who are assertive, responsible, and independent. These qualities should be fostered in families receiving treatment.

Adolescent Delinquency Despite the relatively static delinquency rates of the 1980s and early 1990s, the adolescent homicide rate more than doubled between 1988 and 1994 (Kroshus, 1994). The Centers for Disease Control and Prevention (CDC; 1994) reported that almost half of homicide victims in 1991 were males 15 to 34 years old, with adolescents accounting for the greatest change in rate. Among males ages 15 to 19, homicide surpasses suicide as the second leading cause of death. Thus, there is an escalating trend of juveniles being both the perpetrators and the victims of violence. The CDC suggested that factors influencing this trend include poverty, inadequate educational and economic opportunities, social and familial instability, and expo-

sure to violence as a preferred technique for settling disputes (Dwyer & Hunt-Jackson, 2002). Related to adolescent delinquency is the problem of school dropout. John Alspaugh (1998) reported that approximately one fourth of ninth-grade students in the United States drop out before graduating from high school.

Anxiety Disorders and Depression Anxiety disorders seen during the childhood and adolescent years include phobias, obsessive–compulsive disorders, generalized anxiety, separation anxiety, and panic disorders. School phobias and separation anxiety generally manifest earlier in childhood. The anxiety disorders rank highest among all mental disorders of childhood and adolescence (Costello et al., 1996) and can impair cognitive and social functioning. Several risk factors have been identified, including early childhood temperament, negative life events, parents' behavior, and adolescents' coping styles (Roe-Sepowitz & Thyer, 2004).

Childhood and adolescent depression is closely linked to anxiety disorders. It has been estimated that as many as 1 in 33 children and 1 in 8 adolescents are clinically depressed (Wodarski, Wodarski, & Dulmus, 2003). Depression in children and adolescents differs from depression in adults (Roe-Sepowitz & Thyer, 2004). Children and adolescents who are depressed complain more of anxiety symptoms such as separation anxiety, reluctance to meet people, and somatic symptoms of headaches and stomachaches, whereas adults more often exhibit psychotic behaviors (Roe-Sepowitz & Thyer, 2004). If children and adolescents exhibit psychotic symptoms, they tend to be more auditory than delusional (U.S. Department of Health and Human Services, 1999).

Five risk factors identified by the Institute of Medicine are believed to influence the probability of adolescent depression (Roe-Sepowitz & Thyer, 2004):

1. Having a parent or other close biological relative with a mood disorder;

2. Having a severe stressor such as a loss, divorce, marital separation, or unemployment; job dissatisfaction; a physical disorder such as a chronic medical condition; a traumatic experience; or, in children, a learning disability;

3. Having low self-esteem, low self-efficacy, and a sense of helplessness and hopelessness;

4. Being female;

5. Living in poverty.

Eating disorders, most commonly anorexia and bulimia, have emerged as a major adolescent mental (and physical) health concern and are often comorbid with depression. The American Academy of Family Physicians

---

GENDER DIFFERENCES IN DEPRESSION

Before adolescence, boys tend to have higher rates of depression than girls. By age 13 or 14, however, girls are much more likely than boys to be depressed (Nolen-Hoeksema, 1990). Nolen-Hoeksema and Girgus (1994) proposed an integrative developmental model that incorporates biological, psychological, and social explanations for why the gender differences in depression emerge during adolescence. According to this model, before and after early adolescence, girls tend to demonstrate a more cooperative interaction style, less aggression, and a more ruminative, self-focused style of responding to distress, whereas boys tend to be more competitive, domineering, and aggressive. These preexisting differences then interact with biological challenges (e.g., pubertal development and dysregulation of ovarian hormones) and social challenges (e.g., parental and peer expectations and attitudes) of early adolescence to create gender difference in depression. Although the causes of depression are the same in girls and boys, boys are less likely than girls to become depressed in early adolescence because (a) they tend to be exposed to fewer negative and distressing biological and social challenges, and (b) they are less likely to have the preexisting tendencies that make them unassertive in responding to the challenges that they do face (Nolen-Hoeksema & Girgus, 1994).

Other types of mental health problems, including conduct disorder and substance abuse, are more common among boys and men (Kessler et al., 1994). Some researchers have argued that gendered behavioral expectations lead girls to internalize symptoms of distress and boys to externalize them (e.g., Rosenfield, Lennon, & White, 2005). In this way, studies that focus only on depression may underestimate distress among boys and men (Aneshensel, Rutter, & Lachenbruch, 1991).

BIBLIOGRAPHY

Aneshensel, C. S., Rutter, C. M., & Lachenbruch, P. A. (1991). Social structure, stress, and mental health: Competing conceptual and analytic models. *American Sociological Review, 56,* 166–178.

Kessler, R. C., McGonagle, K. A., Zhao, S., Nelson, C. B., Hughes, M., Eshleman, S., et al. (1994). Lifetime and 12-month prevalence of *DSM–III–R* psychiatric disorders in the United States: Results from the National Comorbidity Survey. *Archives of General Psychiatry, 51*(1), 8–19.

Nolen-Hoeksema, S. (1990). *Sex differences in depression.* Stanford, CA: Stanford University Press.

Nolen-Hoeksema, S., & Girgus, J. S. (1994). The emergence of gender differences in depression during adolescence. *Psychological Bulletin, 115*(3), 424–443.

Rosenfield, S., Lennon, M. C., & White, H. R. (2005). The self and mental health: Self-salience and the emergence of internalizing and externalizing problems. *Journal of Health & Social Behavior, 46,* 323–340.

Belinda L. Needham

(1999) reported that 3% of women develop anorexia and another 8% develop bulimia. Treatment options must be comprehensive and multifaceted, directed at resolving the psychosocial issues, medical concerns, and nutritional needs of the client.

Depression and Suicide Adolescent depression and suicide are two of the most pressing problems in the American culture, and the evidence for the link between them is convincing. It has been suggested that more than 90% of children and adolescents who commit suicide have a mental disorder. Suicide attempts are most common among teens who have had previous psychiatric symptoms such as depression, impulsivity, aggression, antisocial behavior, and substance abuse (Berman & Jobes, 1991).

The CDC (2001) noted that in 1999 teen suicide was the third leading cause of adolescent death. In 1994 teen suicide ranked as the second leading cause of adolescent death, eclipsed only by accidents (Hoffman, Paris, & Hall, 1994). The need for accurate and timely assessments is clear. Because mood disorders such as depression substantially increase the risk of suicide, suicidal behavior is a serious concern for mental health practitioners, who must assess suicide risk when treating almost all teens and be ready to render treatment to adolescents experiencing a variety of problems.

Treatment for depression and attempted suicide should be provided in the least restrictive environment that is safe and effective for the child or adolescent. The selection of treatment depends on the severity of the illness, the motivation of the client and/or the client's family to receive treatment, and the severity of additional psychiatric or medical conditions. Cognitive-behavioral therapy, social skills training, and relaxation techniques have been used successfully (Wodarski & Feit, 1995).

The Link between Depression and Substance Use and Abuse Substance use and abuse is another problem that leads to mental health problems in teens. The U.S. Public Health Service (1998) found that the leading cause of death for adolescents is unintentional injuries, and 40% of those injuries are related to alcohol use and abuse. Mental disorders and substance abuse often coexist. Studies have shown that having depression or anxiety disorder in adolescence doubles the risk for later drug abuse and dependency (Bukstein, 1995). Mental health practitioners who work with chemically dependent adolescents and their families include social work and psychological counselors, group facilitators, and educators. Those roles can be used proactively with adolescents to prevent substance abuse.

RISK FACTORS FOR MENTAL HEALTH DISORDERS

Although the determination of the causes of mental health disorders in children and adolescents is often elusive, certain situations and conditions predispose children and adolescents to abnormal behavior. These conditions generally are biological (genetics, chemical imbalances in the body, or damage to the central nervous system, e.g., through head injury or pre- or postnatal drug abuse) or environmental (exposure to high levels of lead; exposure to violence, e.g., witnessing or being a victim of physical or sexual abuse, shootings or muggings, or other disasters; stress related to chronic poverty, discrimination, or other serious hardships; maternal deprivation; faulty parental role models; physical or emotional abuse; or the loss of significant others through death, divorce, or broken relationships; U.S. Department of Health and Human Services, 1999). Two of the more common risk factors of childhood and adolescence that warrant special mention here are poverty and drug abuse.

The Effect of Poverty Although children from all socioeconomic and background types are vulnerable to mental disorders and problems, poverty increases the risk of those problems (Roe-Sepowitz & Thyer, 2004). Poverty places children at increased risk physically (inadequate health care and nutrition), psychologically (stress and self-esteem issues), and socially (homelessness and increased exposure to violence). In 1991 one in five children in the United States lived below the poverty line. This rate is expected to continue well into the 21st century (U.S. Department of Health and Human Services, 1999). Studies have demonstrated that poverty is correlated with a variety of negative outcomes, such as delinquency, academic underachievement, and poor physical and mental health (Thyer & Wodarski, 1998).

Interventions to ensure a healthy physical and psychosocial environment for a developing child may take several forms. Initially, interventions involve the expectant parents and/or other significant others, all of whom must deal with the many changes that occur with the birth of a baby. Expectant parents may need assistance in accessing and securing the resources needed (e.g., prenatal care, good nutrition, adequate financial support, and genetic counseling). Low-income families may be particularly in need of this support relative to information and advocacy because they may have the least access to high-quality prenatal, postnatal, and other medical care. The stresses associated with poverty put an added burden on a pregnant woman and her family, making the prebirth environment for the baby even more hazardous (Streever & Wodarski, 1984). Economic, social, and medical support must continue throughout childhood and adolescence to ameliorate the devastating effects of poverty on physical and mental health.

Effects of Prescription and Illegal Drug Use on Pre- and Postnatal Development A common problem in prenatal development and birth involves street and prescription drug use. The American Academy of Pediatrics (1998) reported that cocaine and marijuana use is most common among people ages 18 to 34 years, the ages at which pregnancy is most likely to occur. It is estimated that up to 40% of developmental defects result from maternal exposures to environmental or drug agents that affect intrauterine development. Adverse effects include death, structural malformation, and functional alteration of the developing infant (Heindel & Lawler, 2006). Despite the knowledge that these drugs also adversely affect the health of a child, the decision to stop taking drugs to protect a baby is a difficult one; sometimes it is a danger to the mother whether the drugs are legal or illegal. Withdrawal from street drugs can cause convulsions, and ending the use of prescription drugs may put the mother and child in danger of seizures, toxemia, diabetic complications, and other disorders.

IMPLICATIONS FOR THE FUTURE

Although the life span perspective is generalist in its application to all human development, individuals belonging to vulnerable groups in society may have different developmental experiences. Women, people of color, and persons with disabilities progress through the same life stages but may process their experiences differently (Dwyer & Hunt-Jackson, 2002). Developmental researchers and practitioners have considered these differences, but more work in this area is necessary.

The universality of adolescent problems across cultures and countries has been documented in the Youth

Self-Report Form that was administered to adolescents from 24 countries in Europe, Asia, the Middle East, Africa, Australia, the Caribbean, and the United States (Rescorla et al., 2007). A striking degree of similarity was found across adolescents around the world. Within-country comparisons revealed that girls consistently and significantly reported more anxious/depressed problems and internalizing problems than did boys. Older adolescents reported significantly more rule-breaking behaviors than did younger adolescents, and adolescents reported significantly more problems than their parents reported. Mood swings, distractibility, self-criticism, and arguments were the most commonly noted individual items.

Mental health practitioners can benefit from these findings. Because the findings suggest the universality of problems, one might conclude that empirically based treatments for problems could share universal qualities. Also, as adolescents in most countries reported more problems than did their parents, those treating adolescents may find that facilitating adolescents' communication with parents could help family dynamics (Rescorla et al., 2007).

Relative to Roe-Sebowitz and Thyer's (2004) viewpoint that mental disorders whose etiologies include environmental, biological, familial, or peer factors may not be actual mental disorders, it is important to note that many obstacles and crises that affect children's and adolescents' mental health are as much a function of societal factors as they are of an individual's development. Therefore, it is imperative that responsible mental health practitioners combine their micro and macro practice skills to enhance individual functioning for their clients and societal functioning for future clients. Adjustments and accommodations must be made within the individual and within society to maximize each person's potential and the overall potential of American society. The Healthy People Consortium, representing an alliance of more than 350 national organizations and 250 state public health, mental health, substance, and environmental agencies, proposed a systematic approach to health improvement. The Consortium recognizes the determinants of health as individual biology and behaviors interacting within the individual's social and physical environments that in turn are affected by policies and interventions that may or may not result in improved health by targeting factors related to individuals and their environments, including access to quality health care (U.S. Department of Health and Human Services, 2000). Thus, the macro-level perspective appears to be most appropriate for research and practice in child and adolescent mental health.

The life span developmental approach to mental health in children and adolescents views the health, both physical and mental, of the individual as resulting from influences along the continuum that extends on either side of birth (Gluckman & Hanson, 2006). Regardless of their theoretical orientation (behavioral, cognitive, constructivist, or psychodynamic), practitioners should design and implement interventions that build on the strengths present in the individual. They are charged, moreover, with the responsibility to consider the individual within his or her social and political environmental context.

SEE ALSO Volume 1: *Drinking, Adolescent; Drug Use, Adolescent; Eating Disorders; Erikson, Erik; Freud, Sigmund; Health Differentials/Disparities, Childhood and Adolescence; Parent-Child Relationships, Childhood and Adolescence; Peer Groups and Crowds; Self-Esteem; Resilience; Socialization.*

BIBLIOGRAPHY

Adler, L. L. (1989). *Cross-cultural research in human development: Life span perspectives.* New York: Praeger.

Alspaugh, J. (1998). The relationship of school and community characteristics to high school dropout rates. *Clearing House, 71,* 184–188.

American Academy of Family Physicians. (1999). Can the development of eating disorders be predicted? *American Family Physician, 60*(2), 623.

American Academy of Pediatrics, Committee on Substance Abuse. (1998). Tobacco, alcohol, and other drugs: The role of the pediatrician in prevention and management of substance abuse. *Pediatrics, 101*(1), 125–128.

American Psychiatric Association. (2000). *Diagnostic and statistical manual of mental disorders* (4th ed.). Washington, DC: Author.

Andreasen, N. C., & Black, D. W. (2006). *Introductory textbook of psychiatry* (4th ed.). Washington, DC: American Psychiatric Publishing.

Bartsch, K., & Wellman, H. (1989). Young children's attribution of action to beliefs and desires. *Child Development, 60*(4), 946–964.

Baumrind, D. (1991). The influence of parenting style on adolescent competence and substance use. *Journal of Early Adolescence, 11,* 56–95.

Berman, A. L., & Jobes, D. A. (1991). *Adolescent suicide: Assessment and intervention.* Washington. DC: American Psychological Association.

Bukstein, O. G. (1995). *Adolescent substance abuse: Assessment, prevention, and treatment.* New York: Wiley.

Calohan, C. J., & Peeler, C. M. (2007). Autistic disorder. *Morbidity and Mortality Weekly Report, 3,* 725–727.

Centers for Disease Control and Prevention. (2001). *Suicide prevention fact sheet.* Atlanta, GA: Author.

Chumlea, W. (1982). Physical growth in adolescence. In B. W. Wolman (Ed.), *Handbook of developmental psychology.* Englewood Cliffs, NJ: Prentice-Hall.

Costello, E. J., Angold, A., Burns, B. J., Stangl, D. K., Tweed, D. L., Erkanli, A., et al. (1996). The Great Smoky Mountains Study of Youth: Goals, design, methods, and the prevalence of *DSM–III–R* disorders. *Archives of General Psychiatry, 53*(12), 1129–1136.

Dupper, D. R., & Musick, J. B. (2007). Attention-deficit/hyperactivity disorder. In B. Thyer & J. S. Wodarski (Eds.), *Social work in mental health: An evidence-based approach* (pp. 75–96). New York: Wiley.

Dwyer, D., & Hunt-Jackson, J. (2002). The life span perspective. In J. Wodarski & S. Dziegielewski (Eds.), *Human behavior and the social environment: Integrating theory and evidence-based practice* (pp. 84–109). New York: Springer.

Erikson, E. (1964). *Childhood and society.* New York: Norton.

Feldman, R. A., Caplinger, T. E., & Wodarski, J. S. (1983). *The St. Louis conundrum: The effective treatment of antisocial youths.* Englewood Cliffs, NJ: Prentice-Hall.

Filho, A. G. C., Bodanese, R., Silva, T. L., Alvares, J. P., Paglioza, J., Aman, M., et al. (2005). Comparison of risperidone and methylphenidate for reducing ADHD symptoms in children and adolescents with moderate mental retardation. *Journal of the American Academy of Child and Adolescent Psychiatry, 44*(8), 748–755.

Ganda, L., & Pellegrini, A. D. (Eds.). (1985). *Play language and stories: The development of children's literate behavior.* Norwood, NJ: Ablex.

Gluckman, P. D., & Hanson, M. A. (Eds.). (2006). *Developmental origins of health and disease.* Cambridge, U.K.: Cambridge University Press.

Godfrey, K. (2006). The "developmental origins" hypothesis: Epidemiology. In P. D. Gluckman & M. A. Hanson (Eds.), *Developmental origins of health and disease* (pp. 6–32). Cambridge, U.K.: Cambridge University Press.

Harter, S. (1987). The determinants and mediational role of global self-worth in children. In N. Eisenberg (Ed.), *Contemporary topics in developmental psychology.* New York: Wiley.

Heindel, J. J., & Lawler, C. (2006). Role of exposure to environmental chemicals in developmental origins of health and disease. In P. D. Gluckman & M. A. Hanson (Eds.), *Developmental origins of health and disease* (pp. 82–97). Cambridge, U. K.: Cambridge University Press.

Hoffman, L., Paris, S., & Hall, E. (1994). *Developmental psychology today* (6th ed.). New York: McGraw-Hill.

Hughes, F. P., & Noppe, L. D. (1985). *Human development across the life span.* St. Paul, MN: West.

Kaluger, G., & Kaluger, M. F. (1984). *Human development: The span of life.* St. Louis, MO: Times Mirror/Mosby College.

Kroshus, J. (1994). Preventing juvenile violence. *Juvenile Justice Digest, 2*(2), 5–6.

Lauer, R. H. (1995). *Social problems: The quality of life.* Madison, WI: Brown & Benchmark.

Lefrancois, G. R. (1993). *The life span.* Belmont, CA: Wadsworth.

Mayes, L. C., & Cohen, D. J. (2002). *The Yale Study Center guide to understanding your child: Healthy development from birth to adolescence.* New York: Little, Brown.

Moses, L., & Flavell, J. (1990). Inferring false beliefs from actions and reactions. *Child Development, 61*(4), 929–945.

Patterson, G. R. (1982). *Coercive family process.* Eugene, OR: Castalia.

Quinn, P. (1998). *Understanding disability: A life span approach.* Thousand Oaks, CA: Sage.

Rapp, L. A., & Wodarski, J. S. (1998). Conduct disorder. In B. A. Thyer & J. S. Wodarski (Eds.), *Handbook of empirical social work practice.* New York: Wiley.

Rapp-Paglicci, L. A., Roberts, A. R., & Wodarski, J. S. (Eds.) (2002). *Handbook of violence.* New York: Wiley.

Rescorla, L., Achenbach, T. M., Ivanova, M. Y., Dumenci, L., Almqvist, F., Bilenburg, N., et al. (2007). Epidemiological comparisons of problems and positive qualities reported by adolescents in 24 countries. *Journal of Consulting and Clinical Psychology, 75*(2), 351–358.

Roe-Sepowitz, D. E., & Thyer, B. A. (2004). Adolescent mental health. In L. A. Rapp-Paglicci, C. N. Dulmus, & J. S. Wodarski (Eds.), *Handbook of preventive interventions for children and adolescents* (pp. 67–99). Hoboken, NJ: Wiley.

Shulman, L. (1992). *The skills of helping: Individuals, families, and groups.* Itasca, IL: F.E. Peacock.

Sowers-Hoag, K., & DiDona, T. M. (1998). Separation anxiety disorder. In B. A. Thyer & J. S. Wodarski (Eds.), *Handbook of empirical social work practice: Mental disorders* (pp. 157–177). New York: Wiley.

Sternberg, K. J. (1993). Child maltreatment: Implications for policy from cross-cultural research. In D. Cicchetti & S. L. Toth (Eds.), *Child abuse, child development, and social policy* (pp. 191–211). Norwood, NJ: Ablex.

Streever, K. L., & Wodarski, J. S. (1984). Life span development approach: Implications for practice. *Social Casework, 65*(5), 267–278.

Thomas, R. M. (Ed.). (1990). *Encyclopedia of human development and education: Theory, research and studies.* New York: Pergamon.

Thyer, B. A., & Wodarski, J. S. (1998). *Handbook of empirical social work practice. Vol. 2: Social problems and practice issues.* New York: Wiley.

Thyer, B. A., & Wodarski, J. S. (2007). *Social work in mental health: An evidence-based approach.* New York: Wiley.

U.S. Bureau of the Census. (1992). *Statistical abstract of the United States* (112th ed.). Washington, DC: U.S. Government Printing Office.

U.S. Department of Health and Human Services. (2000). *Healthy people 2010: The cornerstone to prevention.* Retrieved June 16, 2008, from http://www.healthypeople.gov/document

U.S. Department of Health and Human Services, Office of the Surgeon General. (1999). *Mental health: A report of the surgeon general.* Washington, DC: U.S. Government Printing Office.

U.S. Public Health Service. (1998). Alcohol and other drug abuse in adolescents. *American Family Physician, 56,* 1737–1741.

Wodarski, J. S., & Feit, M. D. (1995). *Adolescent substance abuse: An empirically based group preventive health paradigm.* Binghamton, NY: Haworth Press.

Wodarski, J. S., & Wodarski, L. A. (1998). *Adolescent violence: An empirically based school/family paradigm.* New York: Springer.

Wodarski, J. S., Wodarski, L. A., & Dulmus, C. (2003). *Adolescent depression and suicide: A comprehensive empirical intervention for prevention and treatment.* Springfield, IL: Charles C. Thomas.

John S. Wodarski

MENTORING

The word *mentor* originates from Homer's *Odyssey*. As Odysseus departed for the Trojan War, he charged his infant son, Telemachus, to the care of his good friend Mentor. Mentor's role in Telemachus's development during Odysseus's many years away is typically portrayed as a wise protector, nurturer, and role model (Colley, 2003). It is this portrayal that serves as a foundation for current conceptions of mentoring relationships—as concerned, experienced individuals who care for and guide the development of someone younger and/or less experienced.

Mentoring has long been championed as a cure-all to the increasing complexity of life course trajectories because of its potential to buoy up and provide skills and training for youth, particularly the disadvantaged. This perspective is the basis of 21st-century mentoring movements such as Big Brothers Big Sisters, GEAR UP, the National Mentoring Partnership, and the emphasis on mentoring in U.S. President George W. Bush's Faith-Based Initiative. In short, many consider the promise of mentoring to be the achievement of the American Dream. Despite this ancient beginning, it has been relatively recently that mentors gained attention by those concerned with adolescent development.

WHAT IS A MENTOR?

What exactly is a mentor? How can a mentor be identified? With the proliferation of mentoring programs in diverse settings and addressing a variety of issues, it is increasingly difficult to answer these questions. Current research does not provide a concise, coherent, and consensual definition of mentoring (Allen & Eby, 2007; DuBois & Karcher, 2005). Definitions range from very broad descriptions of functional benefits (e.g., emotional vs. instrumental) to more exact specifications about the details of the relationships (e.g., relative age of mentor and protégé or mentee, frequency of contact, or one-on-one vs. group contexts).

One helpful way to bring some clarity to the definition of mentoring is to categorize mentors along two dimensions: formal versus informal, and social versus functional role (see Table 1). *Formal* mentors are most often adults that the staff of mentoring programs match with adolescents and then provide a venue and/or schedule for their interaction (Rhodes, 2002). The key element that makes a mentoring relationship formal is that it exists as a result of the mentor and protégé coming together in the context of a formal organization. Thus, peer mentors who are older students paired with younger students through school-based programs, e-mentors who interact primarily with their mentees electronically, and group mentors who interact in settings involving more than one youth all fit this description.

	Social Role	Functional Role
Formal	Yes	?
Informal	No	Yes

Table 1. *Defining Mentoring Relationships.* CENGAGE LEARNING, GALE.

Informal mentors encompass youths' influential relationships with adults who are part of their naturally occurring social networks. They have been variously operationalized as important unrelated adults, individuals outside of the immediate family, very important persons, or role models. Importantly, a young person must identify an adult as influential to be classified as a mentor. The distinction of social and functional roles elucidates why this is so.

Hamilton, Darling, and Shaver (2003) introduced this distinction to studies of mentors. A social role is defined by the structure that connects two individuals. Thus, mentoring as a social role consists of a mentor and protégé participating in a formal organization. Any two individuals in such a relationship identify themselves accordingly, regardless of whether or not knowledge and skills are transferred or an emotional connection is established. On the other hand, a functional role is simply defined by the content of the relationship. Mentoring as a functional role involves doing what a mentor is supposed to do.

Consequently, mentoring as a social role necessarily goes along with formal mentoring, and as a functional role it corresponds with informal mentoring (see diagonal in Table 1). However, informal mentoring cannot coincide with mentoring as a social role (bottom left quadrant in Table 1). The social roles of informal mentoring consist of aunt–niece, teacher–student, and so on, and because these relationships develop a mentoring component through natural processes, the original social roles remain salient. Also, though an organization pairs two individuals, there is no guarantee that the relationship will accomplish what is intended. Informal relationships are successful by definition, but formal relationships require concerted effort and can exist without the relationship providing the intended functions (upper right quadrant of Table 1).

HISTORY OF MENTORING RESEARCH

Late in the 19th and into the 20th century, the industrialization and urbanization of society disrupted existing pathways into adulthood for youth. New forms of antisocial behaviors in youth accompanied these disruptions, resulting in the development of the juvenile justice system. It was in this context that the mentoring movement was originally born with the establishment of Big Brothers and Big Sisters organizations (Allen & Eby, 2007; DuBois & Karcher, 2005).

There has been a resurgence of interest in mentoring as globalization has once again changed the landscape against which young people come of age. Although organizations such as mentor/National Mentoring Partnership and Public/Private Ventures have fueled much of this recent interest in mentoring, two particular studies represent benchmarks of this renewed interest: Daniel J. Levinson and colleagues' *The Seasons of a Man's Life* (Levinson, Darrow, Klein, Levinson, & McKee, 1978) and Emmy E. Werner and Ruth S. Smith's (2001) Kauai Longitudinal Study.

The Season's of a Man's Life was an attempt to create a developmental perspective on adulthood in men and focused particularly on early and middle adulthood. Although a small number of men were studied, developing mentoring relationships emerged as a major task of early adulthood with the danger of future difficulties if they were not formed. Mentors served as sponsors and advisers. They were older than their protégés, but not so old to be out of touch with contemporary issues or to be in danger of being perceived as a parent figure. However, if they were too similar in age, their relationships could become too peer-like. Combining the best of both parental and peer relationships, mentors served to support the achievements necessary to successfully navigate the changing developmental demands of early adulthood.

Werner and Smith (2001) studied children from the 1955 cohort born on the island of Kauai who faced substantial prenatal, childhood, and adolescent risks to development. Despite experiencing these risks for extended periods throughout the life course, the majority of these children became well-adapted adults. Among the adults who managed to achieve "normal" levels of functioning in adulthood, each one identified support from at least one influential non-parental adult figure or mentor who was part of their network of informal relationships as an important factor in their resilience.

DEMOGRAPHICS OF YOUTH MENTORING

Following these two studies, most research on mentoring relationships tends to focus on small samples that often have a particular risk profile or are disadvantaged in some particular way. This should not be surprising considering mentoring programs target these youth. Thus, the demographic profile of formal mentoring is a definitional issue—advantaged youth tend not to have formal mentors. The implication is that the advantaged youth do not need mentors because they already have access to rich resources in their social environment (e.g., parents)

Studies of informal mentoring have followed suit, typically using small, community-based samples of at-risk youth. Thus a broad picture of informal mentoring has been unavailable in the literature. Erickson and Elder (2007) reported a description of youth who identified informal mentoring relationships using the National Longitudinal Study of Adolescent Health (Add Health). Overall, three-quarters of young people reported having a mentor during adolescence. Those who were female, White, and lived with both biological parents during their adolescent years, and whose parents had more education, were more likely to report a mentor.

THEMES AND THEORIES

A recurring question addressed in the mentoring literature is whether mentoring relationships provide emotional or instrumental benefits. This is an important issue for formal mentoring relationships: Organizations need to know whether to train their volunteers to develop emotionally supportive, therapeutic relationships or to engage youth in activities that will expand their knowledge and challenge them to develop new skills (Allen & Eby, 2007).

Researchers emphasize one or the other of these perspectives depending on which theoretical model they prefer. Those who stress emotional benefits typically draw on attachment theory or the risk and resilience perspective. Attachment theory (Bowlby, 1982) suggests that children construct relatively stable expectations of their environment or working models based on their early experiences with caregivers. Poorly attached children are anxious in social situations and tend to withdraw from social relationships. However, when a young person feels emotionally connected with a mentor, this can alter the working model, making the social world more of a welcoming place (DuBois & Karcher, 2005).

Within the risk and resilience tradition, resilience is defined as "a dynamic process encompassing positive adaptation within the context of significant adversity" (Luthar, Cicchetti, & Becker, 2000, p. 543). The search for mechanisms of resilience has led, among other things, to mentors. The social support provided by mentors is seen as a protective factor for at-risk youth that serves to buffer the potential effects of stress associated with risk.

Those who consider the instrumental aspects of mentoring relationships tend to emphasize concepts such as social capital. Although its definition and empirical applications vary, social capital can be defined as the ability of individuals to obtain some type of benefit through their social connections (Portes, 1998). Being connected to a mentor provides access to information and opportunities that young people might be unaware of or unable to take advantage of. For example, a student who wants to go to college but whose parents never did may depend on a mentor for information on taking the right courses or filling out a college application. Mentoring relationships that are less emotionally laden may provide some benefits in instances such as these. Young people may see these mentors as more credible sources of information (e.g., providing more of an unbiased opinion), and they may feel more at ease exploring different aspects of the self without the risk of alienating an important source of emotional support.

Identifying whether emotional or instrumental support is what a particular youth might need seems a daunting task for large mentoring programs and could explain why program interventions show only modest levels of success (Hall, 2006; Rhodes, 2002). In informal mentoring relationships, this should be less of an issue: When youth identify influential people, the criteria they use depend on their own particular needs. Most likely, both emotional and instrumental aspects of mentoring relationships are important regardless of whether they are formal or informal, even if one may be more important than the other for different youth or at different points in the life course.

Pierre Bourdieu's ideas of habitus and field (Bourdieu & Wacquant, 1992) provide a foundation for understanding if, when, and how mentors might provide youth with their particular needs and can be applied to both formal and informal mentoring relationships (Colley, 2003). Habitus is essentially a system of dispositions that individuals develop throughout their lives in response to their experiences or position in various fields. A *field* represents any particular context in which an individual is given a position relative to other individuals and in which individuals have a stake in their position because it provides them with some advantage.

Formal mentors who attempt to improve youths' skills engage with them in a field that is in many ways foreign to those from disadvantaged backgrounds. A protégé may misunderstand a mentor's attempts to teach them skills that will be useful to them in situations to which they are not accustomed. For instance, problem-solving techniques (habitus) learned in gang-related settings (field) will likely not transfer to the educational realm, and it may be difficult for a mentor to successfully convey this to youth. Further, such efforts may be interpreted as disempowering by young people.

Informal mentoring relationships between youth and the kinds of adults that might be most helpful to them may never form because of dissimilarities in their habitus. Also, despite continuous social contact, the most helpful adults are likely located in different locations in a field, complicating arriving at shared meanings or understandings. However, it is precisely the fact that a potential mentor sees the world from a different (advantaged) position that adds value to the life of the adolescent. Therefore, if disadvantaged youth develop informal mentoring relationships at all, it may be with adults who could further socialize them into disadvantaged positions, whereas advantaged youth are socialized to achieve further advantage. A reproduction or even exacerbation of social inequalities could result.

TO THE FUTURE

Because mentoring has been a particular focus of attention at the beginning of the 21st century, the future holds many possibilities for research, of which only a few are mentioned here. Very little is known about long-term consequences of mentoring relationships, although Werner and Smith's Kauai Longitudinal Study provides a notable exception. Although it did not focus on mentoring per se, it did identify mentors as an important component of resilient pathways well into adulthood. However, their study began with a particularly disadvantaged population. Later studies hold the promise of identifying long-term consequences of mentoring in normative populations.

Despite Levinson et al.'s (1978) endorsement of mentors, they also recognized the potential harm that mentors could have on development. However, until recently, most researchers ignored this possibility. Yet when it comes to formal mentoring relationships, neither mentoring as an abstract concept nor actual practice guarantee positive outcomes for youth. In particular, mentoring relationships that are of short duration tend to do damage to a youth's self-concept (Rhodes, 2002).

It has been more difficult to identify whether and when informal mentoring might have no influence or even negative consequences for youth. This is in part because informal mentors by definition have had a positive influence on youth, at least from the youth's perspective. However, preliminary evidence suggests that informal mentors within the school setting contribute more to the educational attainment of youth from disadvantaged backgrounds compared to more advantaged youth, whereas mentors who are relatives are associated with gains in attainment for more advantaged youth (Erickson, McDonald, & Elder, 2007).

With regard to both informal and formal relationships, there is a need for more research that examines the intersection of the characteristics of mentors and youth and how their relationships might help or hinder development across a variety of domains. For example, although informal mentors who are relatives may help advantaged youth more in terms of educational attainment, it is possible that they may be more help to disadvantaged compared to advantaged youth in terms of emotional well-being.

SEE ALSO Volume 1: *Cultural Capital; Resilence; Social Capital.*

BIBLIOGRAPHY

Allen, T. D., & Eby, L. T. (2007). *The Blackwell handbook of mentoring: A multiple perspectives approach.* Malden, MA: Blackwell Publishers.

Bourdieu, P., & Wacquant, L. J. D. (1992). *An invitation to reflexive sociology.* Chicago: University of Chicago Press.

Bowlby, J. (1982). *Attachment.* (2nd ed.). New York: Basic Books.

Colley, H. (2003). *Mentoring for social inclusion: A critical approach to nurturing mentor relationships.* New York: RoutledgeFalmer.

Darling, N., Hamilton, S. F., & Shaver, K. H. (2003). Relationships outside the family: Unrelated adults. In G. R. Adams & M. D. Berzonsky (Eds.), *Blackwell handbook of adolescence* (pp. 349–370). Malden, MA: Blackwell Publishing.

DuBois, D. L., & Karcher, M. J. (2005). *Handbook of youth mentoring.* Thousand Oaks, CA: Sage.

Erickson, L. D., & Elder, G. H., Jr. (2007). Informal mentors during adolescence and young adulthood: A national portrait. Brigham Young University. Accessed from http://sociology. byu.edu/Faculty

Erickson, L. D., McDonald, S., & Elder, G. H., Jr. (2007). Informal mentors and educational achievement: Complementary or compensatory resources? Brigham Young University. Accessed from http://sociology.byu.edu/Faculty

Hall, H. R. (2006). *Mentoring young men of color: Meeting the needs of African American and Latino students.* Lanham, MD: Rowman & Littlefield Education.

Levinson, D. J., Darrow, C. N., Klein, E. B., Levinson, M. H., & McKee, B. (1978). *The seasons of a man's life.* New York: Knopf.

Luthar, S. S., Cicchetti, D., & Becker, B. (2000). The construct of resilience: A critical evaluation and guidelines for future work. *Child Development, 71*(3), 543–562.

Portes, A. (1998). Social capital: Its origins and applications in modern sociology. *Annual Review of Sociology, 24,* 1–24.

Rhodes, J. E. (2002). *Stand by me: The risks and rewards of mentoring today's youth.* Cambridge, MA: Harvard University Press.

Werner, E. E., & Smith, R. S. (2001). *Journeys from childhood to midlife: Risk, resilience, and recovery.* Ithaca, NY: Cornell University Press.

Lance D. Erickson

MIDDLE SCHOOL

SEE Volume 1: *Stages of Schooling.*

MORAL DEVELOPMENT AND EDUCATION

Moral development is an area of study in the field of human development that deals with the ways in which people treat one another. Like any scientific endeavor, the study of human development has been characterized by a series of theoretical and empirical debates. One of the most important is the classic debate about nature versus nurture. Are people shaped ultimately by their internal biological components (nature) or by their social experiences in the world (nurture)? Most scientists would agree that neither is the sole propellant of human development but that an intricate and complex interaction of biological, psychological, and social factors shapes people's trajectory.

The same thing can be said about the way people develop morally. According to most traditions, moral development concerns the individual's socialization into the mores and practices of a family and a culture (nurture). From a cognitive developmental perspective, however, moral development has to do with an individual's construction of understanding through phases of development that are based on greater experience and maturation (nature and nurture). Further, from a developmental systems perspective the individual develops in interaction with the various networks of experience—family, school, neighborhood—adapting to local contexts (nature and nurture in a complex, multilayered interaction).

HISTORICAL PERSPECTIVES

Moral development has been studied systematically since Lawrence Kohlberg (1927–1987) advanced developmental psychology by extending the work of the then-unknown Swiss psychologist Jean Piaget (1896–1980). Like Piaget, Kohlberg focused on moral judgment as part of a long-standing focus on reasoning in European and American approaches to moral philosophy and behavior. Adopting a deontological (duty-based) framework, Kohlberg expanded Piaget's two orientations to six stages that progress in an invariant sequence. According to Kohlberg, moral reasoning develops though a leveled, stagelike progression in thinking over time. Individuals progress from preconventional thinking through conventional and on to postconventional thinking (preconventional level: (a) avoid punishment, (b) prudence and simple exchange; conventional level: (c) interpersonal harmony and concordance, (d) law and order; postconventional level: (e) social contract, and (f) universal moral principles).

Kohlberg used an interview method that included the presentation of moral dilemmas as prompts to ascertain the level or stage of moral judgment. In the classic Heinz dilemma, a poor man named Heinz cannot afford

Eleven Principles of the Character Education Partnership

(Lickona, Schaps & Lewis, 2003)
Effective character education...

Principle 1. Promotes core ethical values as the basis of good character.

Principle 2. Defines "character" comprehensively to include thinking, feeling and behavior.

Principle 3. Uses a comprehensive, intentional, proactive and effective approach to character development.

Principle 4. Provides students with opportunities for moral action.

Principle 5. Creates a caring school community.

Principle 6. Includes a meaningful and challenging academic curriculum that respects all learners, develops their character and helps them succeed.

Principle 7. Strives to foster self-motivation.

Principle 8. Engages school staff as a learning and moral community that shares responsibility for character education and attempts to adhere to the same core values that guide the education of students.

Principle 9. Fosters shared moral leadership and long range support of the character education initiative.

Principle 10. Engages families and community members as partners in the character building effort.

Principle 11. Evaluates the character of the school, the school staff's functioning as character educators, and the extent to which students manifest good character.

Table 1. *Eleven Principles of the Character Education Partnership.* CENGAGE LEARNING, GALE.

a costly medicine for his ailing wife. Because Heinz cannot persuade the druggist to discount the price, he considers breaking into the pharmacy and stealing the medicine to save his wife. After thinking about the dilemma, students articulate why Heinz should or should not take the medicine from the pharmacy.

Over many decades, the work of Kohlberg and his colleagues as well as neo-Kohlbergian research programs have uncovered a cross-cultural trajectory of the development of moral reasoning over the course of early life through early adulthood. Progress through Kohlberg's moral judgment stages is influenced primarily by age and cognitive development throughout primary and secondary schooling. Adolescence offers a unique developmental context for studying the development of moral judgment. As an adolescent develops a sense of society and obligations to it, several key transitions can be noted. In high school the individual transitions from a personal orientation (doing what is convenient for oneself or for maintaining friendships) to a societywide view (follow laws to maintain order).

In college individuals move toward postconventional thinking (determine together what laws should govern

people; Rest, Narvaez, Bebeau, & Thoma, 1999). Those with more higher education, especially postgraduate education in philosophy, tend to receive higher scores on interviews and tests such as the Defining Issues Test (DIT). The DIT presents a dilemma and then a series of statements that represent different Kohlberg stages for participants to rate and rank in importance (e.g., for the Heinz dilemma: Isn't it only natural for a loving husband to care so much for his wife that he'd steal?). Higher scores on the DIT are related to more effective professional behavior, such as a democratic classroom orientation in teachers (Rest & Narvaez, 1994).

Kohlberg and Piaget were criticized for underestimating the degree to which children understand morality. Elliott Turiel (1983) and colleagues discovered an understanding of social convention (e.g., use a fork to eat) versus natural morality (e.g., do not hurt people) in children as young as 3 years. Kohlberg's findings also were criticized for weak correlations with actual behavior; individuals often do not act in a manner that is congruent with their moral reasoning or judgment. In contrast, Nancy Eisenberg and colleagues (1999) mapped empathy and prosocial development through childhood, finding significant links to altruistic and prosocial behavior.

The narrowness of Kohlberg's approach to moral development (as moral reasoning about justice) has given way to a greater diversity of views and topics. One of the best-known critiques was that of Carol Gilligan (1982), who proposed that females take a different path in responding to moral dilemmas. She asserted that females are more oriented toward relationships and the care of others, an orientation that she claimed was not represented adequately in Kohlberg's justice-oriented framework. Although research findings have not supported her claims for gender differences in reasoning, the view that justice reasoning is sufficient to explain moral development in all people was problematic for multiple philosophical, psychological, and cultural reasons (Rest et al., 1999).

PROMOTING MORAL DEVELOPMENT

The field of moral development is diverse; empirical and theoretical evidence highlights the complex interaction of the social, psychological, and biological components of moral formation and development. For example, evolutionary and primate psychology has provided a new view of moral development that includes evolved propensities for reciprocity, empathy, and violence that also are found in other primates (de Waal, 1996). Jonathan Haidt (2001) underscored the importance of emotion and intuition in making moral judgments about others. Darcia Narváez (2008) integrated neurobiology into a moral psychological theory called triune ethics that highlights the importance of early care for optimal and evolutionarily expected "moral brain" development.

One of the most promising areas of research on moral development is moral motivation. Although individuals may know the right and good choice to make in any specific situation, they still may fail to act accordingly. Augusto Blasi (1994) attempted to explain this by pointing to the importance of moral identity and moral personality in driving moral action. Anne Colby and William Damon (1992) noted that those who were considered moral exemplars for their contributions to their communities merged their personal with their moral goals. Their moral goals were central to their sense of self. For example, in one case, after becoming aware of her racist background as unjust, one exemplar made working for social justice her life's work. Other work shows how adolescent moral exemplars also include moral goals and moral traits in their self-descriptions (Hart, Yates, Fegley, & Wilson, 1995). In studying the emergence of morality in childhood, Kochanska (2002) mapped the development of conscience, showing the importance of warm, responsive parenting. In his view the source of commitment to moral concerns is mutual, positive, secure relationships with caregivers. Moral identity is deeply relational and is strongest among students who bond to school, work in caring school communities, and form strong attachments to teachers.

MORAL EDUCATION

Social scientists who study moral development often are interested in the application of their findings to moral education. For example, Kohlberg did not think only about outlining moral development; he also was interested in formulating "just community" schools in which students would practice democratic rule, facing everyday dilemmas together. In his application of moral development to education Kohlberg joined a long tradition. Educators, philosophers, and researchers have been interested in the appropriate structure and content of moral education for thousands of years. Early Greek philosophers such as Cicero, Plato, and Aristotle debated the definition and best practice of moral formation. Educators in 18th- and 19th-century Europe and the United States assumed that it was best to teach moral values based in Christian theology and Greek philosophy; in fact, the Christian Bible formed the primary text for that moral curriculum.

The religious-based method of moral education in schools was challenged by social realities (i.e., increased religious heterogeneity) and scientific studies (i.e., the Character Education Inquiry of 1928–1939, which did not find cross-situation consistency in moral traits such as honesty). That resulted in a shift in the debate on moral education away from the primary school classroom and into academia: What works for developing moral character?

As science is brought to bear on the question, key factors are being uncovered and integrated into educational interventions. An example of positive integration between the empirical study of the academy and the real-world concerns of the classroom is the Character Education Partnership, a coalition of individuals and organizations that provides resources, support, and advocacy for individuals and institutions committed to integrating character education in the school curriculum. The partnership has constructed the Eleven Principles for Character Education, which represent the core principles for designing and enacting school communities committed to character education and are embraced by many schools and districts across the country.

SPECIFIC MORAL EDUCATION PROGRAMS

Social scientists have developed innovative and empirically tested programs to promote healthy moral development in educational contexts. For example, Rest's (1985) four-component model delineates three processes beyond moral judgment that are required for moral behavior: moral sensitivity (noticing and interpreting what is happening and imagining alternative choices), moral motivation or focus (prioritizing the moral action), and moral action skills (knowing what steps to take and persevering). This model has been useful for envisioning what effective moral character education might entail. The first component—sensitivity, empathy, and perspective taking—became particularly important in U.S. education at the end of the 20th century and the beginning of the 21st. Many programs emphasize a caring and developmentally supportive classroom and social and emotional learning or address empathy development directly. For example, the Collaborative for Academic, Social, and Emotional Learning (CASEL) advocates five core social and emotional competencies: self-awareness, social awareness, self-management, responsible decision making, and relationship skills. These types of programs emphasize the development of emotional intelligence.

One of the premier moral character education programs of the late 20th century is the Child Development Project (CDP). Developed in the 1970s and 1980s by social scientists and educators, CDP is a comprehensive, multicontext program designed to foster moral and prosocial development for elementary school age children. The CDP focuses on the structure of the home and school communities to promote peer collaboration, inclusive membership, and value sharing. Adult exemplars model caring, supportive relationships, foster a caring community among children, and guide children through the CDP curriculum (perspective taking, values sharing, moral discussions, and so on). Participants in CDP programs have demonstrated increased positive prosocial skills and dispositions such as sense of community, care for others, and academic engagement (Battistich, Solomon, Watson, & Schaps, 1997). The work of the CDP is carried on by the Developmental Studies Center.

The most useful viewpoint in approaching any aspect of children's development is a developmental systems orientation that "draws attention to embedded and overlapping systems of influence that exist at multiple levels; to the fact that dispositional coherence is a joint product of personal and contextual factors that are in dynamic interaction across the life course" (Lapsley & Narvaez, 2006, p. 271). The most effective character education programs will be situated within the framework of developmental science but also will work within a dynamic model of ecological systems in which the child is embedded (e.g., family, school, neighborhood, and global communities). Such programs not only will help children develop skills and resiliency but also will strengthen the positive networks in which a child resides. In fact, positive youth development programs often take that perspective.

With an emphasis on thriving as a basis for taking an ecological systems approach, Richard Lerner, Elizabeth Dowling, and Pamela Anderson (2003) wrote that "an integrated moral and civic identity and a commitment to society beyond the limits of one's own existence enables thriving youth to be agents both in their own healthy development and in the positive enhancement of other people and of society" (p. 172). The Search Institute, an independent nonprofit organization, has developed a list of 20 external assets (e.g., caring school climate, service to others, and adult role models) and 20 internal assets (e.g., bonding to school, honesty, and sense of purpose).

Communities have adopted asset-building practices. Young people with a greater number of assets are less likely to engage in risky and delinquent behavior (e.g., alcohol use, violent behavior) and more likely to thrive (get good grades in school) and serve others. In fact, service learning can be a potent force for moral development. More than isolated acts of volunteerism, service learning aims to coordinate multiple contexts (school, home, neighborhood) and perspectives (personal, other, social justice) within a series of experiences in which individuals do something good for someone else. When embedded in a caring classroom, aligned with educational objectives, and enacted within the larger community, service learning has been shown to increase student agency, responsibility, and awareness across all grade and age levels and continue into adulthood (Youniss, McLellan, & Yates, 1997). Programs inspired by positive youth development such as thriving, positive asset development and service learning are congruent with the aims of moral character education.

The developmental, cognitive, and educational sciences provide an important foundation for moral character education. In fact, the cognitive sciences allow people to see that a false dichotomy is created when one separates reasoning from virtue, as rational moral education and traditional character education are perceived to have done. To develop expertise in moral functioning (sensitivity, judgment, focus, action) both deliberative reasoning and virtues are required. That view is integral to the Integrative Ethical Education model (Narvaez, 2006), which draws together the direct methods of virtue cultivation associated with traditional character education and the deliberative reasoning development associated with rational moral education.

Within a context of high expectations and support, teachers implement five steps (ideally, simultaneously):

1. Educators establish a caring bond and secure attachment with each student, which is one of the most important protective factors that mitigate poor outcomes for a child.

2. Educators establish a caring classroom climate supportive of achievement and ethical character.

3. Ethical skills for each of the four components (ethical sensitivity, judgment, focus, action) are taught across the curriculum and through extracurricular activities; educators design instruction according to the following four levels of novice-to-expert pedagogy: (a) immersion in examples and opportunities, (b) attention to facts and skills, (c) practice procedures, (d) integration of knowledge and procedures.

4. Educators foster student self-regulation, promoting self-authorship for purposeful development.

5. Educators facilitate democratic and asset-building communities by linking to parents and the local community, promoting an integrated civic identity.

This research-based framework offers a comprehensive and empirically derived integration of contemporary developmental and educational sciences.

FUTURE DIRECTIONS

Future research on moral development will incorporate the neurosciences. For example, attachment is deeply neurobiological and is more fragile than was understood previously, yet it is vital for lifetime brain development and emotion regulation. As an example, care-deprived infants develop aberrant brain structures and brain-behavioral disorders that lead to greater hostility and aggression toward others (Kruesi et al., 1992). When caregivers respond and become attuned to the child's needs and moods and when parents coregulate their moods, the child is likely to be more cooperative and develop a good conscience. Neuroscience research opens a window onto the physical results of child neglect that are related to problems with moral behavior.

Moral formation and development are affected by a myriad of social, psychological, and biological factors. Innovative research on those interrelated factors is essential for the continued study of moral development and the advancement of moral education.

SEE ALSO Volume 1: *Developmental Systems Theory; Genetic Influences, Early Life; Piaget, Jean; Religion and Spirituality, Childhood and Adolescence.*

BIBLIOGRAPHY

Battistich, V., Solomon, D., Watson, M., & Schaps, E. (1997). Caring school communities. *Educational Psychologist, 32,* 137–151.

Blasi, A. (1994). Moral identity: Its role in moral functioning. In E. Puka (Ed.), *Fundamental research in moral development* (pp. 168–179). New York: Garland Press.

Colby, A., & Damon, W. (1992). *Some do care: Contemporary lives of moral commitment.* New York: Free Press.

De Waal, F. (1996). *Good natured: The origins of right and wrong in humans and other animals.* Cambridge, MA: Harvard University Press.

Eisenberg, N., Guthrie, D. K., Murphy, B. C., Shepard, S. A., Cumberland, A., & Carlo, G. (1999). Consistency and development of prosocial dispositions: A longitudinal study. *Child Development, 75,* 1360–1372.

Gilligan, C. J. (1982). *In a different voice.* Cambridge, MA: Harvard University Press.

Haidt, J. (2001). The emotional dog and its rational tail: A social intuitionist approach to moral judgment. *Psychological Review, 108,* 814–834.

Hart, D., Yates, M., Fegley, S., & Wilson, G. (1995). Moral commitment in inner-city adolescents. In M. Killen & D. Hart (Eds.), *Morality in everyday life* (pp. 317–341). Cambridge, MA: Cambridge University Press.

Kochanska, G. (2002). Mutually responsive orientation between mothers and their young children: A context for the early development of conscience. *Current Directions in Psychological Science, 11*(6), 191–195.

Kruesi, M., Hibbs, E., Zahn, T., Keysor, C., Hamburger, S., Bartko, J., et al. (1992). A 2-year prospective follow-up study of children and adolescents with disruptive behavior disorders: Prediction by cerebrospinal fluid 5-hydroxyindoleacetic acid, homovanillic acid, and automatic measures. *Archives of General Psychiatry, 49,* 429–435.

Lapsley, D., & Narváez, D. (2006). Character education. In W. Damon & R. Lerner (Eds.), *Handbook of child psychology* (6th ed., pp. 248–296). Hoboken, NJ: John Wiley.

Lerner, R. M., Dowling, E. M., & Anderson, P. M. (2003). Positive youth development: Thriving as a basis of personhood and civil society. *Applied Developmental Science, 7*(3), 172–180.

Narváez, D. (2006) Integrative ethical education. In M. Killen & J. G. Smetana (Eds.), *Handbook of moral development* (pp. 703–733). Mahwah, NJ: Lawrence Erlbaum.

Narváez, D. (2008). Triune ethics: The neurobiological roots of our multiple moralities. *New Ideas in Psychology, 26,* 95–119.

Rest, J. (1985). Morality. In J. Flavell (Ed.), *Cognitive development.* Volume 3 of P. Mussen (Ed.), *Carmichael's manual of child psychology* (pp. 556–629). New York: Wiley.

Rest, J. R., & Narváez, D. (Eds.). (1994). *Moral development in the professions: Psychology and applied ethics.* Hillsdale, NJ: Lawrence Erlbaum.

Rest, J., Narváez, D., Bebeau, M., & Thoma, S. (1999). *Postconventional moral thinking: A neo-Kohlbergian approach.* Mahwah, NJ: Lawrence Erlbaum.

Turiel, E. (1983). *The development of social knowledge: Morality and convention.* Cambridge, U.K.: Cambridge University Press.

Youniss, J., McLellan, J. A., & Yates, M. (1997). What we know about engendering civic identity. *American Behavioral Scientist, 40,* 620–631.

Anthony C. Holter
Darcia Narváez

N

NEIGHBORHOOD CONTEXT, CHILDHOOD AND ADOLESCENCE

Research on the importance of neighborhood contexts examines how the characteristics of the places people live matter for their well-being. While social research has long shown that individual characteristics and family resources and social processes are associated with child and adolescent outcomes, since the late 20th century researchers have increasingly focused on the neighborhood of residence as an additional, meaningful determinant of child and adolescent development (Elliott et al., 2006). Neighborhoods are important contexts to consider for youth, because children and adolescents are transitioning through stages of development that may make them vulnerable to contextual influences. Consequently, neighborhoods may be key contexts for youthful residents, whose negotiation of these transitions may make them more receptive to both the beneficial and the detrimental aspects of their residential environments.

NEIGHBORHOOD EFFECTS RESEARCH

Neighborhood effects research examines how emergent properties of residential contexts can have effects on people above and beyond their own characteristics. Initial contextual research was based on Clifford Shaw's and Henry McKay's (1942) theory of social disorganization, which focused on the demographic and political-economic characteristics of neighborhoods, such as the roles of residential segregation, ethnic diversity, and relative neighborhood advantage. This research found considerable evidence that aspects of neighborhood disadvantage had negative effects on child and adolescent outcomes, such as child maltreatment, delinquency, teenage childbearing, and low IQ (Brooks-Gunn, Duncan, & Aber, 1997).

Subsequent researchers have sought to identify the social properties of neighborhoods that link disadvantage or demographic characteristics to child and adolescent well-being (Sampson, Morenoff, & Gannon-Rowley, 2002). This research draws on the work of William Julius Wilson (1987; 1996), who noted that social change in inner-city neighborhoods, such as the withdrawal of local industry (and its consequent effects on local employment opportunities), brought about decline and social disadvantage in urban neighborhoods. Disadvantaged neighborhoods—those with high rates of poverty, unemployment, residential instability, family disruption, and racial or ethnic heterogeneity, for example—are hindered in their ability to achieve levels of social organization that are thought to increase the life chances of child and adolescent residents. Similarly, in such neighborhoods, orientations toward crime, risk behavior, and problem behavior may be culturally transmitted, affecting child and adolescent behavioral choices. Neighborhood effects research considers how these aspects of social organization and cultural transmission serve as mechanisms that link neighborhood levels of economic disadvantage with child and adolescent outcomes.

NEIGHBORHOOD SOCIAL ORGANIZATION

Social organization approaches examine the factors that limit neighborhood residents' ability to work together

Changing Neighborhood. *Alkali Flat, once known for drug dealing and transiency, is one of many core city neighborhoods across California that have experienced a revitalization as long commutes and soaring rents have brought new home buyers to areas that would have been unthinkable a few years ago.* AP IMAGES.

toward common goals, including healthy child and adolescent development. Much of the research on social organization has its basis in the literature on social capital, which is the capacity for action on behalf of an individual or group that inheres through networks of social relationships (Coleman, 1990). In the context of the neighborhood, social capital can be employed for the benefit of local youths in a number of ways: it can facilitate the informal social control of children and adolescents; it can be used to draw beneficial institutional resources into the neighborhood; and it can be employed to prevent or lower residents' exposure to harmful agents, such as environmental hazards (Leventhal & Brooks-Gunn, 2000). Residents of economically and socially disadvantaged neighborhoods are less able to develop these networks of social relationships and as a result are hindered in their ability to amass social capital.

Collective Efficacy Robert Sampson and colleagues (1997) identified the concept of collective efficacy to explain how social capital may be recognized and then employed to the benefit of neighborhood residents. Collective efficacy captures the extent of mutual trust among neighborhood residents, along with shared expectations for action in support of neighborhood goals. In the case of children and adolescents, collective efficacy is most often employed through the supervision of youth, and intervention in support of informal neighborhood social control. Researchers have examined the role of collective efficacy with regard to a variety of child and adolescent outcomes, such as sexual risk behavior, problem behavior, and health outcomes (Browning, Leventhal, & Brooks-Gunn, 2005). This research has found evidence for a negative relationship between neighborhood levels of collective efficacy and problem behavior and poor health outcomes.

Institutional Resources and Environmental Hazards Researchers have also identified the importance of social capital as a means to attract and retain institutional resources that are beneficial for neighborhood residents (Elliott et al., 2006). Neighborhoods that are highly socially organized and are able to exploit extended social ties to decision makers outside the neighborhood may be able to influence the placement of desirable institutions (such as health care facilities, community centers, and police and fire substations) within the neighborhood. As a result of the proximity of these desirable institutions, child and adolescent well-being may increase. In addition, residents of socially organized neighborhoods may be better able to work together to reduce the number and type of hazards and risks to which local youths are subject. These types of risks include physical disorder (such as the presence of broken glass, litter, or biohazards), social disorder (such as public intoxication or prostitution), or the placement and regulation of harmful institutions proximate to the neighborhood (such as industries that pollute).

CULTURAL TRANSMISSION

An alternative to the social capital approach is one that focuses on the concentration or proliferation of behaviors or influences in neighborhoods as detrimental to child and adolescent well-being—in some neighborhoods, delinquency and problem behavior may be so prevalent as to seem normal, or an expression of a neighborhood subculture, and may be taken up by children and adolescents through a process of social learning. These approaches are typically thought of as subcultural or epidemic approaches (Anderson, 1990), and are rooted in early theories of crime and gang delinquency (Cloward & Ohlin, 1960). Under the cultural transmission perspective, the proliferation of problem behaviors and crime in a neighborhood setting provides access to delinquent and problem behavior opportunities, and thus serves as an illegitimate opportunity structure for the local youth that stands in opposition to legitimate opportunity structures (such as schooling and legal employment) in mainstream society. In disadvantaged neighborhoods, these illegitimate opportunity structures present an alternative path for children and adolescents who are blocked from mainstream opportunities for success.

NEIGHBORHOOD MEASUREMENT AND ANALYSIS

Methodological Considerations Several methodological considerations are key to the examination of neighborhood effects. The first of these is the definition of *neighborhood*. Researchers have employed different definitions of this term, such as administrative boundaries (census tracts or blocks) (Sastry et al., 2006); areas delimited by streets, railroad tracks, and other ecological boundaries (Sampson et al., 1997); people's social networks (Wellman & Leighton, 1979); and residents' subjective definitions of their socio-spatial neighborhood (Lee & Campbell, 1997).

A second key consideration is endogeneity, or selection bias. Parents of children and adolescents select the neighborhoods in which their families will live. The same characteristics that lead parents to select certain neighborhoods in which to live may also influence their children's development. As a result, observed neighborhood effects may be attributable instead to these unmeasured parental

characteristics (Brooks-Gunn, Duncan, & Aber, 1997). Researchers attempt to address this issue through careful control of individual and family characteristics associated with neighborhood selection, more advanced methods for dealing with selection bias in statistical models (such as the use of propensity scores), and quasi-experimental research designs that attempt to randomly assign families to neighborhoods with certain characteristics.

One such quasi-experimental design is that used in the Moving to Opportunity (MTO) experiments, sponsored by the U.S. Department of Housing and Urban Development, in which some residents of disadvantaged neighborhoods were given housing vouchers to live in more advantaged contexts. The goal of this experiment was to assess whether moving residents to more advantaged residential contexts improved outcomes for them and their children. The results of this research are, to date, mixed with regard to child and adolescent outcomes. The most striking finding is that girls whose families "moved to opportunity" were more likely than girls in a control group to graduate from high school and to refrain from delinquency and problem behavior. These same benefits were not experienced by male children, however, who fared no better or worse on most measures of well-being than boys in a control group, indicating that there may be gender differences in the ways that children experience and make use of the opportunities presented to them in their residential contexts (Kling & Liebman, 2004).

Neighborhood Analysis Neighborhood effects research considers both structural aspects of neighborhoods (disadvantage, residential stability, and racial and ethnic diversity or heterogeneity) and social aspects of neighborhoods (the extent of collective efficacy and the presence of delinquent or problem behavior subcultures). One key concern that researchers must address, then, is the measurement of these concepts. With regard to structural characteristics, researchers often use demographic information derived from the census. For example, disadvantage may be measured by the proportion of households in the neighborhood that are below the poverty line, the proportion of single- or female-headed households, the percent unemployed, and the percent on public assistance. Residential stability is often measured by the percentage of residents living in the same house for a set period (often 5 or 10 years) prior to the study, as well as the percent of housing occupied by owners. Finally, researchers use information on the racial and ethnic composition of the neighborhood to measure racial and ethnic heterogeneity (diversity).

In contrast to these structural measures, measures of social processes are often derived from residents' responses to questions on community-based surveys. These responses

Impoverished Neighborhood. *Research has found considerable evidence that aspects of neighborhood disadvantage has negative effects on child and adolescent outcomes, such as child maltreatment, delinquency, teenage childbearing, and low IQ.* AP IMAGES.

are aggregated by neighborhood to form a composite neighborhood measure of these social processes. Collective efficacy, for example, has often been measured using information provided by survey respondents on both informal social control processes in their neighborhoods and the extent of social cohesion among residents (Sampson, Raudenbush, & Earls, 1997). Informal social control of children and adolescents has been measured by the extent that survey respondents agree that neighborhood residents would intervene if they saw children skipping school and hanging out on the street corner, spray painting graffiti on a local building, showing disrespect to an adult, or if there was a fight in the neighborhood and someone was beaten or threatened. Social cohesion is often measured by information on the extent to which people in the neighborhood are willing to help neighbors, can be trusted, get along, and share the same values, as well as whether the neighborhood is close-knit.

In addition to these measures, researchers have noted the importance of considering the composition and form of neighborhood social networks, or whether residents have parties and get-togethers, visit with one another, ask one another advice about important matters such as job openings and child-rearing practices, and do favors for one another. Another important network of social relationships is described by the term *intergenerational closure*, or the extent to which parents in a neighborhood know their children's friends, and the parents of their children's friends, and can be counted on to watch out so that children are safe and do not get into trouble. Intergenerational closure can ensure that information about children's behavior may be shared among parents, increasing opportunities for effective supervision and monitoring, and addressing behavioral problems early. At the same time, however, neighborhood networks and relationships may also have the opposite effect—complex relationships and competing social obligations may complicate residents' willingness to intervene when children and adolescents are misbehaving (Browning, Feinberg, & Dietz, 2004).

Measures of local subcultures and cultural transmission processes are also derived from community surveys. These measures often include information on the extent to which residents are tolerant of various forms of deviance and problem behavior, such as residents' opinions regarding how wrong it is for adolescents to smoke cigarettes, use marijuana, drink alcohol, and get into fistfights (Browning, Feinberg, & Dietz, 2004). These measures allow researchers to take into account whether problem behaviors are more or less acceptable in different neighborhoods, which may affect children's and adolescents' decisions to engage in delinquency and problem behavior and the opportunities for such behavior.

THE FUTURE OF NEIGHBORHOOD EFFECTS RESEARCH

The upsurge in research on neighborhood effects has been driven, in part, by improvements in both statistical software and survey design. First, multilevel statistical models can be executed readily through the use of a variety of software programs that allow researchers to account for the clustering of residents within neighborhoods (Raudenbush & Bryk, 2002). Second, surveys have been specifically designed to examine the effects of sociospatial contexts on individual residents, such as the Project on Human Development in Chicago Neighborhoods (PHDCN) and the Los Angeles Family and Neighborhood Survey (L.A.FANS). Third, researchers have tried to capture both objective and subjective aspects of neighborhood contexts. For example, PHDCN researchers videotaped the neighborhoods in their Chicago study, and used a systematic rating system to record the extent of physical and social disorder in the neighborhoods. This method of systematic social observation (SSO) resulted in rich data that researchers are using to assess a broader range of neighborhood effects on resident well-being (Sampson & Raudenbush, 1999).

Research on neighborhood effects is proceeding in ways that consider places other than residential location that are important to children and families, such as school, work, and child-care contexts. Certainly, there is considerable overlap between school and neighborhood contexts; schools often draw students from the neighborhood or proximate neighborhoods. As a result, neighborhood disadvantage is often mirrored by school disadvantage—further concentrating disadvantage for children and adolescents in these contexts. However, school contexts can and do differ from neighborhood contexts, and researchers are now beginning to consider cross-classified models that allow for the simultaneous examination of the school and the neighborhood, even when the school is not nested within the child's or adolescent's neighborhood of residence.

Finally, researchers are considering a more fluid definition of *neighborhood*, and are taking into account the social influence of adjacent neighborhoods and shared ties across neighborhoods (Pattillo-McCoy, 1999). This research has shown the importance of ties across neighborhoods and spatial dependencies, as crime and problem behavior in adjacent or related neighborhoods can affect crime and problem behavior in the focal neighborhood. Through these and other approaches, researchers hope to capture a better understanding of how residential contexts and collective social processes serve to promote child and adolescent well-being.

The insights gleaned from neighborhood effects research are thus important for purposes of shaping future social policy. Researchers are increasingly able (as in the MTO study) to provide insight on the types of community interventions that may help to shore up deficits in disadvantaged neighborhood contexts and among poor families. Efforts to increase social capital and collective efficacy for residents, such as increasing opportunities for community residents to come together in support of neighborhood goals, may be of particular significance if they allow residents to recognize shared values and develop a common view that they can work together to effect community change and foster resident well-being. Community-based programs can be implemented that bring about these types of relationships among adults, and can also produce opportunities for intergenerational relationships in which neighborhood adults supervise children's sports or cultural activities. As policies and programs informed by neighborhood effects research are put into place, future evaluative research may be used to identify and refine the most successful community-based approaches for healthy child and adolescent development.

SEE ALSO Voume 1: *Poverty, Childhood and Adolescence; Social Capital; Residential Mobility; Youth;* Volume 2: *Neighborhood Context, Adulthood;* Volume 3: *Neighborhood Context, Later Life.*

BIBLIOGRAPHY

Anderson, E. (1990). *Streetwise: Race, class, and change in an urban community.* Chicago: University of Chicago Press.

Bourdieu, P. (1986). The forms of capital. In J. G. Richardson (Ed.), *Handbook of theory and research for the sociology of education* (pp. 241–258). New York: Greenwood Press.

Brooks-Gunn, J., Duncan, G. J., & Aber, J. L. (Eds.). (1997). *Neighborhood poverty: Context and consequences for children.* New York: Russell Sage Foundation.

Brooks-Gunn, J., Duncan, G. J., Klebanov, P. K., & Sealand, N. (1993). Do neighborhoods influence child and adolescent development? *American Journal of Sociology, 99*(2), 353–395.

Browning, C. R., Feinberg, S. L., & Dietz, R. D. (2004). The paradox of social organization: Networks, collective efficacy, and violent crime in urban neighborhoods. *Social Forces, 83*(2), 503–534.

Browning, C. R., Leventhal, T., & Brooks-Gunn, J. (2005). Sexual initiation during early adolescence: The nexus of parental and community control. *American Sociological Review, 70*(5), 758–778.

Cloward, R. A., & Ohlin, L. E. (1960). *Delinquency and opportunity: A theory of delinquent gangs.* Glencoe, IL: Free Press.

Coleman, J. S. (1990). *Foundations of social theory*. Cambridge, MA: Belknap Press of Harvard University Press.

Elliott, D. S., Menard, S., Rankin, B., Elliott, A., Wilson, W. J., & Huizinga, D. (2006). *Good kids from bad neighborhoods: Successful development in social context*. Cambridge, U.K.: Cambridge University Press.

Kling, J. R., & Liebman, J. B. (2004). Experimental analysis of neighborhood effects on youth. Princeton University working paper 483. Retrieved May 5, 2008, from http://www.irs.princeton.edu/pubs/pdfs/483.pdf

Lee, B. A., & Campbell, K. E. (1997). Common ground? Urban neighborhoods as survey respondents see them. *Social Science Quarterly, 78*(4), 922–936.

Leventhal, T., & Brooks-Gunn, J. (2000). The neighborhoods they live in: The effects of neighborhood residence on child and adolescent outcomes. *Psychological Bulletin, 126*(2), 309–337.

Pattillo-McCoy, M. (1999). *Black picket fences: Privilege and peril among the black middle class*. Chicago: University of Chicago Press.

Raudenbush, S. W., & Bryk, A. S. (2002). *Hierarchical linear models: Applications and data analysis methods*. Thousand Oaks, CA: Sage.

Sampson, R. J., Morenoff, J. D., & Gannon-Rowley, T. (2002). Assessing neighborhood effects: Social processes and new directions in research. *Annual Review of Sociology, 28*, 443–478.

Sampson, R. J., & Raudenbush, S. W. (1999). Systematic social observation of public spaces: A new look at disorder in urban neighborhoods. *American Journal of Sociology, 105*(3), 603–651.

Sampson, R. J., Raudenbush, S. W., & Earls, F. (1997). Neighborhoods and violent crime: A multilevel study of collective efficacy. *Science, 227*, 918–923.

Sastry, N., Ghosh-Dastidar, B., Adams, J. L., Pebley, A. (2006). The design of a multilevel survey of children, families, and communities: The Los Angeles family and neighborhood survey. *Social Science Research, 35*(4), 1000–1024.

Shaw, C. R., & McKay, H. (1942). *Juvenile delinquency and urban areas*. Chicago: University of Chicago Press.

Wellman, B., & Leighton, B. (1979). Networks, neighborhoods, and communities: Approaches to the study of the community question. *Urban Affairs Quarterly, 14*(3), 363–390.

Wilson, W. J. (1987). *The truly disadvantaged: The inner city, the underclass, and public policy*. Chicago: University of Chicago Press.

Wilson, W. J. (1996). *When work disappears: The world of the new urban poor*. New York: Knopf.

Lori A. Burrington
Christopher R. Browning

NO CHILD LEFT BEHIND
SEE Volume 1: *Policy, Education.*

NURSERY SCHOOL
SEE Volume 1: *Child Care and Early Education.*

O

OBESITY, CHILDHOOD AND ADOLESCENCE

Childhood and adolescence are generally very healthy life-course stages in the United States, and relatively few children suffer from major, life-threatening diseases. However, one serious health condition has grown increasingly common since the mid-1970s: childhood and adolescent obesity. Data analyzed for a study published in the *Journal of the American Medical Association* in 2006 indicates that slightly more than half of 2- to 19-year-olds in the United States weigh more than what is medically considered to be a normal weight: 33.6% are overweight and another 17.1% are obese (Ogden et al., 2006). Obesity places young people at risk for a wide range of physical, psychological, and interpersonal problems during childhood, adolescence, and adulthood. It also places an economic burden on society.

DEFINING OVERWEIGHT AND OBESITY

Calculating body mass index (BMI) from children's and adolescents' weight and height is the most common way that clinicians and researchers determine whether young people are overweight or obese. BMI is computed using one of the following formulas: BMI = weight (lb) / [height (in)]^2 × 703 or BMI = weight (kg) / [height (m)]^2. BMI is generally considered a reliable measure of body fatness among young people.

For children and adolescents, BMI assessments are age- and sex-specific to take into account the fact that the amount of body fat children have varies by their sex, age, development, and growth. Therefore, BMI scores are converted into age- and sex-specific percentile rankings of BMI to represent boys' and girls' BMI relative to other children of the same sex and age using the Centers for Disease Control and Prevention's (CDC) BMI-for-age growth charts. The CDC publishes two separate growth charts to calculate BMI percentiles for girls ages 2 to 20, and boys ages 2 to 20.

The CDC classifies children's and adolescents' age- and sex-specific BMI percentile according to four weight categories that represent underweight (BMI scores that fall below the 5th percentile of age- and sex-specific scores), normal weight (BMI scores in the 5th through and 84th percentile), at risk of overweight (BMI scores that fall above the 85th percentile and below the 95th percentile), and overweight (BMI scores that fall above the 95th percentile). The percentile cutoffs for at risk of overweight and overweight correspond to physical and psychological health risks known to be associated with the body fatness levels represented by these high BMI percentiles.

The CDC labels for children's and adolescents' weight classifications do not directly correspond to CDC labels for adults' weight classifications. The adult equivalent of the label "at risk of overweight" is "overweight," and the adult equivalent of the label "overweight" is "obese." The conscious decision to use different labels for young people and adults who weigh more than what is clinically considered normal is the result of concerns that young people who are labeled as obese will face more stigmatization than those labeled as overweight. This decision is controversial, and many experts are now calling for continuity in the labels used to describe children, adolescents, and adults whose weight is above a medically normal range to better reflect the gravity of the health problems associated with body fatness levels that are this high. In 2005 the Institute of Medicine (IOM) recommended that children and adolescents with BMI scores above the 95th percentile be classified as obese. In December 2007 an expert panel of pediatricians endorsed the IOM's recommendation in one of the top U.S. academic pediatric health journals and also added that children and adolescents with BMI scores that fall at or between the 85th and 94th percentiles should be classified as "overweight," not "at risk of overweight" (Barlow & Expert Committee Panel, 2007).

ASSESSING BODY FATNESS

BMI scores are generally more reliable when they are determined by actual measurements of height and weight rather than children's and adolescents' self-reported height and weight. Self-reported weight is more subjective than measured weight, is confounded with weight perceptions (Brener, Eaton, Lowry, & McManus, 2004), and is prone to misreports by respondents dissatisfied with their weight (Elgar, Roberts, Tudor-Smith, & Moore, 2005). Girls are also likely to underreport their weight (Ge, Elder, Regnerus, & Cox, 2001).

Estimating BMI is the most common way that clinicians and researchers assess body fatness among children and teens because it is cost-effective and simple. Other methods of measuring body fatness include:

1. measuring waist to hip ratios to determine body fat concentrated in the abdomen;

2. skin-fold tests, which involve pinching skin and fat tissue at designated body sites to gauge body fat;

3. hydrostatic underwater weighing, a procedure that compares individuals' weight measured under water in a specialized water tank to their weight on land to gauge their percentage of body mass that is body fat;

4. dual-energy x-ray absorptiometry (DEXA), which uses x-ray technology to perform a total body scan on individuals to determine the proportion of their body mass that is bone, lean muscle, and body fat;

5. other medical screening procedures that differentiate between body fat and lean muscle mass such as magnetic resonance imaging (MRI) and computed tomography (CT) scans.

The more complex strategies for assessing body fatness such as hydrostatic underwater weighing, DEXA, MRI and CT scans are capable of more accurately differentiating between children and adolescents who have high BMI scores because they have a high concentration of lean muscle mass versus body fat, but they are not practical technologies for large-scale use because of their complexity and expense. Measuring body fatness using skin-fold tests and measurement of waist to hip ratios also have drawbacks. Skin-fold testing is prone to error, and measurements of waist to hip ratios only provide information about the fatness in one area of the body.

TRENDS IN CHILDHOOD AND ADOLESCENT OBESITY

Paralleling trends among U.S. adults, the prevalence of childhood and adolescent obesity has been rising steadily since the 1970s. Analysis of the National Health and Nutrition Examination Surveys (NHANES), one of the most comprehensive sources of data on U.S. adult and childhood health and nutrition, indicates that the prevalence of obesity has doubled among boys and girls ages 2 to 5 and tripled among boys and girls ages 6 to 11 and 12 to 19 from the early 1970s to 2000 (Ogden, Carroll, & Flegal, 2003). More specifically, the study found that only 5% of boys and girls ages 2 to 5 were obese in 1970–1974 but by 2000, roughly 10% of boys and girls ages 2 to 5 were obese. Among 6- to 11-year-olds and 12- to 19-year-olds, roughly 5% of boys and girls were obese in the early 1970s, whereas roughly 15% were obese in 2000.

The most recent estimates of childhood and adolescent obesity using NHANES data from 2004 indicate that 18.2% of boys and 16.0% of girls ages 2 to 19 are now obese (Ogden et al., 2006). Estimates published in the same source also indicate that the prevalence of obesity among boys and girls varies by age and race/ethnicity, with Black and Hispanic children more likely than their White peers to be heavy. Among boys and girls, the prevalence of obesity is lowest among the youngest children. Among 2-to 5-year-olds, 15.1% of boys and 12.6% of girls are obese; among 6- to 11-year-olds, 19.9% of boys and 17.6% of girls are obese; and among 12- to 19-year-olds, 18.3% of boys and 16.4% of girls are obese. Exact estimates of racial/ethnic differences in prevalence depend on how one defines racial/ethnic categories. Comparisons of young people from the three largest racial/ethnic groups in the U.S indicate that a higher proportion of non-Hispanic Black (20.0%) and Mexican American (19.2%) youth ages 2 to 19 are obese than White (16.3%) youth in the same age range (Ogden et al., 2006).

CONSEQUENCES OF OBESITY

Research on childhood and adolescent obesity tends to focus on either its consequences or its causes. Studies document a range of physical, psychological, and social consequences associated with the condition. These consequences are the primary reason why policy makers, researchers, and clinicians view childhood and adolescent obesity as a major social and public health problem. The

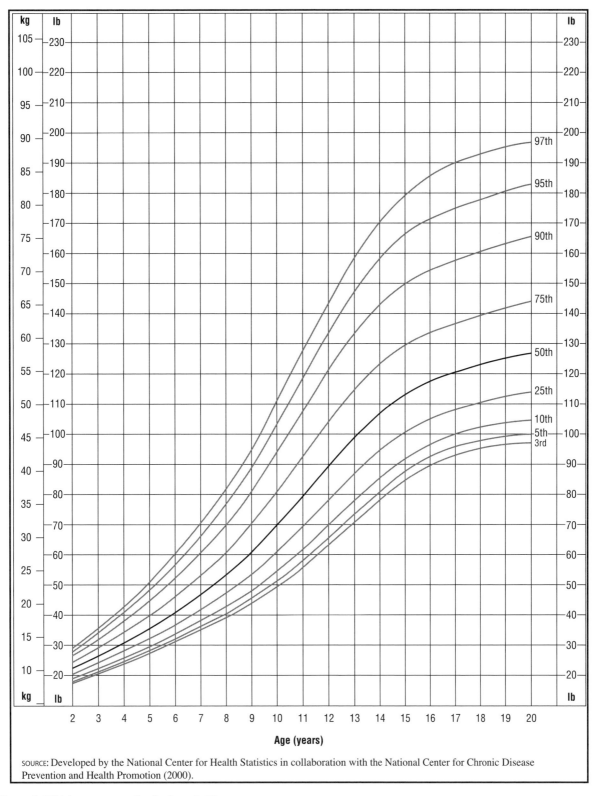

Figure 1. *Weight–age percentiles for boys, 2–20 years.* CENGAGE LEARNING, GALE.

serious consequences of childhood and adolescent obesity also explain why research on the correlates of obesity and obesity prevention among youth has flourished since 2003.

Obese children and adolescents are at elevated risk for serious physical health problems during childhood, adolescence, and adulthood. They are more likely than normal-weight children to become obese adults. They also are likely to develop serious medical conditions that decrease quality of life and life expectancy, including Type II diabetes, high blood pressure, coronary heart disease, asthma, obstructive sleep apnea, and liver disease (Daniels, 2006). Obese female adolescents also are at risk of developing polycystic ovary syndrome, which is a condition that causes problems with women's reproductive systems, including irregular (or no) menstrual cycles, ovarian cysts, and infertility.

Because weight is a physical characteristic that is easily observed by others, obese children and adolescents are also likely to be stigmatized or teased. Experiments show that young people attach many negative characteristics to overweight individuals and are likely to rate them as being unhealthy, ugly, lazy, dirty, and stupid (Crandall & Schiffhauer, 1998). Children and adults also report a bias against befriending an obese person relative to people with other stigmatizing physical characteristics (Latner, Stunkard, & Wilson, 2005). Thus, it is unsurprising that obese adolescents are likely to report a range of psychological problems, including poor self-esteem, social isolation, loneliness, nervousness, anxiety, and suicidal thoughts and attempts.

Socially, obese adolescents report problems establishing and maintaining relationships with friends and romantic partners. In a study of adolescents attending a single high school in New England, obese adolescents were more likely than normal-weight adolescents to be mistreated by peers, obese boys and girls reported dissatisfaction with dating relationships, and obese girls were more likely than normal-weight girls to report that they were not dating (Pearce, Boergers, & Prinstein, 2002). Other research using data from two nationally representative samples of adolescents suggests that obese adolescent boys and girls are less likely to date than normal-weight boys and girls and that they are also less likely to have sex (Cawley, Joyner, & Sobal, 2006).

The consequences of childhood and adolescent obesity are not limited to individuals. It also takes a toll on U.S. society. In 2001, the U.S. Surgeon General indicated that obesity may be responsible for as many preventable illnesses and deaths as cigarette smoking. A recent assessment indicates that the number of hospital discharges of children between the ages of 6 and 17 that involved an obesity-related condition increased dramatically from 1979 and to 1999 in the United States and that hospital costs associated with treating these conditions rose from $35 million to $170 million (Wang & Dietz, 2002). As currently obese children and adolescents grow up and the prevalence of obesity increases, the burden that obesity places on the U.S. health care system and health care costs is expected to rise.

CAUSES OF OBESITY

The range of negative consequences of being obese as a child or adolescent has led to research focused on understanding which children and adolescents are most likely to become obese. On the surface, the answer is simple. A pound of body weight is equivalent to approximately 3,500 calories. Young people who consume 3,500 more calories than they can burn through regular physical activity and body functioning gain weight. The more excess calories that young people consume, the greater the likelihood that they will eventually become obese. Therefore, the formula for decreasing the prevalence of childhood and adolescent obesity is ensuring that young people eat less and are more physically active. When the factors implicated in the rising prevalence of childhood and adolescent obesity are examined, though, it becomes clear that this is not an easy task.

Although the role of genetics in one's likelihood of becoming obese has received increasing attention and certain genes have been associated with body weight regulation, the nature of the relationship between genetics and obesity is not well understood. Furthermore, there is increasing agreement that only a small proportion of children and youth are genetically predisposed to becoming obese and that the rise in childhood and adolescent obesity has occurred too quickly to be explained by genes alone given the relatively slow nature of human evolution. Therefore, it is acknowledged that children and adolescents with obese parents are more likely to be obese, but there is a growing recognition that this family resemblance in weight likely reflects a family similarity in physical activity, diet, and lifestyle.

Large-scale societal changes also have contributed to the rising prevalence of childhood and adolescent obesity. For example, most Western nations, including the United States, have experienced a nutrition transition. This refers to major changes in a population's diet and levels of physical activity (Popkin, 2002). The result is a population that has ready access to a food surplus and a food supply that contains a greater abundance of fatty foods, highly processed foods, preprepared food, restaurant foods, and soft drinks. There is also a trend toward more sedentary versus active leisure time activities for children and adults; video games are replacing baseball as an afterschool activity, for example. These changes in

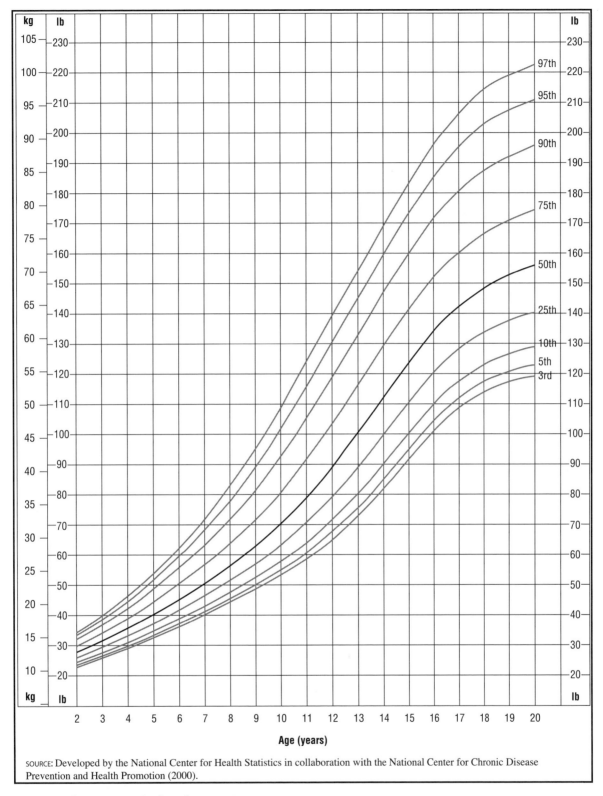

Figure 2. *Weight–age percentiles for girls, ages 2–20.* CENGAGE LEARNING, GALE.

diet and physical activity largely result from economic development, modernization, and technological advances that make food production, transportation, work, and sedentary leisure activity easier. The result of these societal changes in diet and activity is an increasing prevalence of obesity and obesity-related diseases.

The United States has not only experienced the nutrition transition; it also has little government regulation over the types of foods that food companies place on supermarket shelves and how foods are marketed to consumers compared to many other Western countries. Many advocacy groups argue that marketing food to children in the media and in schools should be limited or banned altogether because it influences food choices in ways that lead to a high consumption of calorie-dense food with low nutritional quality. Review articles and books provide excellent overviews of food advertising in the media and schools to youth and its link to childhood and adolescent obesity (e.g., Nestle, 2003). These pieces indicate that corporations spend billions of dollars on advertising food to children and that more than half of all food advertisements youth view on television are marketing sweets, fats, soda, or fast food. These studies also outline the marketing techniques that food corporations use to target children in schools. They include direct advertising via television programming; advertising in school publications; ads placed on school equipment, supplies, buses, and facilities; and product sales in school vending machines, in cafeterias, at school events, and as fundraisers.

Such societal-level trends contribute to high population-level prevalence of childhood and adolescent obesity, but research has also shown that within the U.S. population, some children and adolescents are at a greater risk of becoming obese than others. Factors such as where children live and their socioeconomic and racial/ethnic background are also important.

With respect to U.S. geography, the prevalence of obesity among children and adolescents is higher in some geographic areas than others. Analysis of data from a nationally representative sample of children ages 10 to 17 in 2003 indicated that a greater proportion of children in southeastern states are overweight or obese, whereas the lowest prevalence of childhood overweight and obesity is evident in the central Rocky Mountain states (Tudor-Locke, Kronenfeld, Kim, Benin, & Kuby, 2007). When smaller geographic areas are compared, different analysis of the same data source indicates that youth living in rural areas also are more likely to be obese than youth living in metropolitan areas even when confounders indicating race, gender, income, leisure activities, and health care are controlled (Lutfiyya, Lipsky, Wisdom-Behounek, & Inpanbutr-Martinkus, 2007).

Researchers also have begun to investigate how physical characteristics of neighborhoods, such as the presence of parks, sidewalks, fast-food restaurants, grocery stores, convenience stores, crime, and violence, influence the prevalence of childhood and adolescent obesity. A review of this literature indicates that these neighborhood characteristics influence children's physical activity levels and diet, but neighborhood characteristics have not yet conclusively been tied to youth obesity.

Children's demographic characteristics such as their race/ethnicity, family income, and parents' education level also are related to their likelihood of being an obese child or adolescent. Generally, this research shows that ethnic minority children and adolescents and those who are economically disadvantaged are at greater risk of obesity than their more advantaged peers. For example, data presented above indicates that Mexican Americans and non-Hispanic Blacks are more likely to be obese than Whites. Compared to Whites, the prevalence of childhood and adolescent obesity is also higher among Native Americans and U.S. Pacific Islanders and lower among other Asian American groups. Among ethnic minority youth, those who are immigrants are less likely to be obese than those born in the United States. With respect to income, youth living in poverty are more likely to be obese than youth not living in poverty, but the relationship between income and childhood obesity is not the same for White, Mexican American, and Black youth. Among Whites and Mexican Americans, the likelihood of obesity among 2- to 19-year-olds declines as family income rises, but among Blacks, the likelihood of obesity rises as family income rises (Freedman et al., 2007). In general, parents' education is also inversely related to weight.

FUTURE RESEARCH DIRECTIONS

Future research will continue to focus on identifying the causes and consequences of childhood and adolescent obesity. The primary purposes of these studies will be documenting the multifaceted ways in which obesity influences children's lives and the reasons why some children are at greater risk of becoming obese than others. Because obesity is a social problem, a public health problem, and a medical condition, more research is needed that focuses on both the causes and consequences of childhood and adolescent obesity and that draws from the expertise of multiple disciplines. Future research also must explore the reasons why some young people—especially those from disadvantaged backgrounds—are more likely to become overweight, so that effective interventions can target these children and adolescents. Research attention must also be given to effective ways to prevent childhood and adolescent obesity in a culture in which overconsumption of food and sedentary behavior are normal.

SEE ALSO Volume 1: *Body Image, Childhood and Adolescence; Eating Disorders; Health Behaviors, Childhood and Adolescence; Health Differentials/Disparities, Childhood and Adolescence; Illness and Disease, Childhood and Adolescence;* Volume 2: *Obesity, Adulthood.*

BIBLIOGRAPHY

Barlow, S. E., & Expert Committee Panel. (2007). Expert committee recommendations regarding the prevention, assessment, and treatment of child and adolescent overweight and obesity: Summary report. *Pediatrics, 120,* S129–S192.

Brener, N. D., Eaton, D. K., Lowry, R., & McManus, T. (2004). The association between weight perception and BMI among high school students. *Obesity Research, 12,* 1866–1874.

Cawley, J., Joyner, K., & Sobal, J. (2006). The influence of adolescents' weight and height on dating and sex. *Rationality and Society, 18*(1), 67–94.

Crandall, C., & Schiffhauer, K. (1998). Anti-fat prejudice: Beliefs, values, and American culture. *Obesity Research, 6,* 458–460.

Daniels, S. R. (2006). The consequences of childhood overweight and obesity. *The Future of Children, 16*(1), 47–67.

Elgar, F., Roberts, C., Tudor-Smith, C., & Moore, L. (2005). Validity of self-reported height and weight and predictors of bias in adolescents. *Journal of Adolescent Health, 37,* 371–375.

Freedman, D. S., Ogden, C. L., Flegal, K. M., Khan, L. K., Serdula, M. K., & Dietz, W. H. (2007). Childhood overweight and family income. *Medscape General Medicine, 9,* 26.

Ge, X., Elder, J., Glen H., Regnerus, M., & Cox, C. (2001). Pubertal transitions, perceptions of being overweight, and adolescents' psychological maladjustment: Gender and ethnic differences. *Social Psychology Quarterly, 64,* 363–375.

Latner, J. D., Stunkard, A. J., & Wilson, G. T. (2005). Stigmatized students: Age, sex, and ethnicity effects in the stigmatization of obesity. *Obesity Research, 13,* 1226–1231.

Lutfiyya, M. N., Lipsky, M. S., Wisdom-Behounek, J., & Inpanbutr-Martinkus, M. (2007). Is rural residency a risk factor for overweight and obesity for U.S. children? *Obesity 15,* 2348–2356.

Nestle, M. (2003). *Food politics: How the food industry influences nutrition and health.* Berkeley: University of California Press.

Ogden, C., Carroll, M. D., & Flegal, K. M. (2003). Epidemiologic trends in overweight and obesity. *Endocrinology & Metabolism Clinics of North America, 32,* 741–760.

Ogden, C. L., Carroll, M. D., Curtin, L. R., McDowell, M. A., Tabak, C. J., & Flegal, K. M. (2006). Prevalence of overweight and obesity in the United States, 1999–2004 *Journal of the American Medical Association, 295,* 1549–1555.

Pearce, M. J., Boergers, J., & Prinstein, M. J. (2002). Adolescent obesity, overt and relational peer victimization, and romantic relationships. *Obesity Research, 10,* 386–393.

Popkin, B. M. (2002). An overview of the nutrition transition and its health implications: The Bellagio meetings. *Public Health Nutrition, 5,* 93–103.

Tudor-Locke, C., Kronenfeld, J. J., Kim, S. S., Benin, M., & Kuby, M. (2007). A geographical comparison of prevalence of overweight school-aged children: The National Survey of Children's Health 2003. *Pediatrics, 120,* e1043–e1050.

Wang, G., & Dietz, W. H. (2002). Economic burden of obesity in youths aged 6 to 17 years: 1979–1999. *Pediatrics, 109,* e81–e89.

Michelle L. Frisco

OPPOSITIONAL CULTURE

The oppositional culture theory is a framework used to explain group differences in academic performance among youth. It is more formerly known as the *cultural-ecological theory* or *resistance model*. The theory has received a great deal of attention among academic scholars, teachers, other school personnel, and the mainstream press. It was developed by the Nigerian American cultural anthropologist John U. Ogbu (1939–2003), who was interested in exploring differences in academic outcomes between minority and nonminority groups in the United States and other societies. Ogbu (1978) went on to refine the oppositional culture theory to include an explanation for achievement differences among minority groups.

Ogbu's framework has two major components and several tenets. Prior to a summary of the research on the oppositional culture theory, the major components of the theory are discussed, followed by a description of how groups are classified within the framework, which is a prerequisite for understanding the theory. This entry concludes with a brief discussion on directions for future research. Because most research on the oppositional culture theory has focused on samples within the United States, the summary of research is limited to the American context.

COMPONENTS OF THE THEORY

The oppositional culture theory relies on two components to explain racial or group differences in achievement: (a) societal and school forces and (b) community and individual-level forces. These factors coincide, respectively, with the *ecological* (environment) and *cultural* portions of Ogbu's cultural-ecological framework. The first factor captures the unfair treatment members of minority groups typically face within a given society. Ogbu noted that minority groups are systematically denied access to educational opportunities equal to those received by the dominant group. They also experience barriers to success in future employment and earnings because of racial or ethnic discrimination and structural inequalities, which Ogbu referred to as the *job ceiling*. Thus, minorities are precluded from competing for the most desirable roles in society on the basis of individual training and abilities. Even after a society abandons formal mechanisms responsible for the unequal treatment of minority groups, vestiges of past discriminatory policies in the labor market and education remain in effect.

Differential access to opportunities triggers community forces among minority groups, which corresponds with the second component of the oppositional culture theory. Ogbu (1978) noted that motivation for maximizing school achievement results from the belief that more education leads to better jobs, higher wages, higher social status, and more self-esteem. When members of minority groups encounter barriers within the opportunity structure—the system by which people acquire resources that determine their socioeconomic mobility (up or down)—they develop the perception that they receive lower rewards for education than the dominant group. These perceptions give rise to community and individual-level forces characterized by an oppositional culture that includes resisting educational goals. Ogbu described a *cultural inversion* that takes place whereby some minority groups define certain behaviors, in this case achievement, as inappropriate for them because they are the domain of their oppressors. This repudiation is marked by truancy, lack of serious efforts and attitudes toward school, delinquency, and even early school withdrawal altogether.

The effect of societal and community forces are not the same for all minority groups. Their effects are especially acute for members of subordinate minority groups—groups who have historically been specific targets of exclusionary policies. These groups occupy a subordinate status in stratification systems—the societal systems in which people are arranged hierarchically according to their social class and access to resources—more rigid than systems of stratification based entirely on acquired social class. Ogbu referred to these groups as *castelike minorities*. Ogbu (1978) published a comparative assessment of differences in school achievement among minority groups in six countries (Britain, India, Israel, Japan, New Zealand, and the United States) and found that subordinate minorities exist in each context.

The societal-level (ecological) and community-level (cultural) forces provide a general understanding of how societal forces influence individual-level behaviors to produce achievement differences between dominant and minority groups across numerous contexts. These components, however, do not allow for distinctions to be made between different minority groups. As discussed above, although all minority groups in general experience and perceive some form of barriers to upward mobility, not all minority groups respond similarly. As such, Ogbu refined his framework to include a minority classification scheme useful for understanding the variability among different minority groups.

GROUP CLASSIFICATION

Within the oppositional culture framework, minority groups are placed into three distinct classifications: autonomous minorities, voluntary minorities, and involuntary minorities. Autonomous minorities are groups that may be small in number and/or different in race, ethnicity, religion, or language from the dominant group. Voluntary or immigrant minorities are groups who willingly move to a host country. Typically, these groups immigrate in search of better opportunities in employment or greater political or religious freedom. In contrast, involuntary minorities are groups who have been historically enslaved, colonized, or conquered, and thereby interpret the incorporation of their group into their host country as forced by the dominant group.

Because this framework is used to understand racial differences in achievement within the United States, race is often used as a proxy for group classification. Examples of autonomous minorities include the Amish, Jews, and Mormons; examples of voluntary minorities include Asian Americans and West-Indian Americans; and examples of involuntary minorities include African Americans, Hawaiians, and Native Americans. That African Americans are involuntary minorities and immigrant Blacks (e.g., Caribbean Americans) are voluntary minorities shows that not all Blacks have the same minority status. Group classification does not necessarily correspond with race. Nevertheless, researchers typically use Asian Americans, Blacks, and Whites to represent voluntary, involuntary, and dominant groups, respectively.

A group's place within the classification scheme is not always obvious. Although most Hispanic groups within the United States are either immigrants or descendants of immigrants, Ogbu claimed that Mexican Americans—who comprise nearly two-thirds of the Hispanics living in the United States (Therrien & Ramirez, 2000)—are involuntary minorities. He argued that they feel alienated from American society because they have bitter memories of their incorporation into the United States via American imperialistic expansion in the 1840s. Nevertheless, roughly 50,000 Mexican nationals remained within the newly acquired U.S. territory, a small fraction of the more than 35 million people of Mexican ancestry currently in the United States; most Mexican Americans are therefore immigrants or descendants of immigrants who arrived after the Mexican revolution of 1910 (see Jaffe, Cullen, & Boswell, 1980). Because nearly all Hispanic children in American schools derive from voluntary immigration, Ogbu's classification of this group as involuntary minorities is highly implausible.

GROUP CLASSIFICATION AND SCHOOLING RESPONSES

The minority classification scheme is useful for understanding why ethnic minority groups differ in their schooling behaviors and, subsequently, school achievement. With regard to autonomous minorities, despite experiencing some discrimination they are not dominated or oppressed and have levels of school achievement similar to the dominant group. Therefore, this group is not discussed within the oppositional culture literature.

Members of voluntary minority groups view education as the primary mechanism through which the opportunities that led them to immigrate can be realized. Therefore, they fail to adopt counterproductive schooling behaviors or attitudes and often overcome experiences of discrimination and difficulties in school to do well academically. Their distinction from the dominant group in culture, language, and social or collective identity are characterized by *primary cultural differences*—differences that existed prior to their immigration and acquisition of minority status. These groups understand that their cultural ideals are distinct and view learning aspects of the dominant group's mainstream culture as necessary for success. Thus, they attribute the barriers they initially experience to cultural differences and interpret them as temporary obstacles to overcome. In addition, with no history of being targeted for oppression by the dominant group, they are more trusting of them and their institutions for upward mobility.

As noted earlier, involuntary minorities hold caste-like minority status, which means that they experience greater and more persistent forms of discrimination. They perceive themselves as direct targets of the barriers to success with regard to future employment and earnings. As such, they become disillusioned about the future and begin to doubt the value of schooling. They hold *secondary cultural differences*—differences that emerge after two groups have been in continuous contact, particularly when the contact involves one group's domination over another. Whereas cultural differences from the dominant group for immigrant minorities are based on *content*, for involuntary minorities they are based on *style*. Therefore, they see these differences as markers of group identity to be maintained.

In sum, the oppositional culture theory claims some minorities adopt counterproductive schooling behaviors because of the knowledge or belief that the system of social mobility has been rooted in educational and occupational discrimination based on group membership. Ogbu's group classifications are useful for identifying which minority groups adopt counterproductive schooling behaviors and for understanding why minority groups differ in their responses to societal barriers. Thus, the major premise of the oppositional culture theory is that the prevailing system of social mobility greatly determines achievement motivation and behavior largely through students' beliefs about the opportunity structure.

Ogbu viewed the lower academic performance of involuntary minorities as an adaptation to the barriers they encounter. As discussed below, the scientific evidence on the oppositional culture theory is mixed.

PREVIOUS RESEARCH

There are three primary strands of research on the oppositional culture theory. One focuses on whether societal conditions affect individuals' achievement via perceptions about the opportunity structure (Ford & Harris, 1996; Mickelson, 1990). These studies find that students who believe in the achievement ideology (i.e., education leads to status attainment) experience academic successes, whereas those who challenge this belief do not. Studies also show, however, that Blacks are capable of maintaining high academic achievement or orientation despite beliefs in structural barriers within the opportunity structure for members of their group (O'Connor, 1999; Tyson, Darity, & Castellino, 2005). Also, whereas some studies find Blacks perceive fewer returns to education than Whites (Fordham & Ogbu, 1986; Mickelson, 1990; Ogbu, 2003), others find that Blacks believe in the achievement ideology (e.g., Ainsworth-Darnell & Downey, 1998; Harris, 2006; O'Connor, 1999).

A second strand of research focuses on the community or cultural component of the theory. Specifically, these studies examine whether minorities differ from Whites on oppositional schooling attitudes and behaviors. With regard to attitudes, a common finding of previous studies is that Blacks express greater pro-school attitudes than Whites (e.g., Ainsworth-Darnell & Downey, 1998; Cook & Ludwig, 1997; Harris, 2006). Some scholars argue, however, that these studies do not distinguish between academic (i.e., abstract) and practical (i.e., concrete) attitudes (Mickelson, 1990). Whereas abstract attitudes (e.g., education is important) reflect dominant ideology regarding the ideal role of education, concrete attitudes are rooted in life experiences and inform achievement behavior. Research by Mickelson (1990) shows that (a) abstract attitudes have no effect on grades, (b) concrete attitudes have a positive effect, and (c) Blacks hold less positive concrete attitudes toward education than Whites, which she attributes to the material realities they experience that "challenge the rhetoric of the American Dream" (p. 59). Thus, Mickelson's research raises the importance of distinguishing between different types of attitudes.

With regard to schooling behaviors, studies generally find that racial differences exist on oppositional schooling behaviors. Specifically, these studies find that relative to the dominant group (Whites), voluntary minorities (Asian Americans) have more productive schooling behaviors (i.e., exert more effort in school, spend more time on homework, get in trouble less, and are less disruptive), and involuntary minorities (African Americans) have less productive school behaviors (Ainsworth-Darnell & Downey, 1998; Harris & Robinson, 2007). They also find that these behaviors are related to achievement in the expected manner and partially account for racial differences in achievement. Harris and Robinson (2007) found, however, that the hypothesized and estimated effects of oppositional schooling behaviors on academic achievement are overestimated because of the omission from the framework of students' skill level prior to the theory's applicability—around middle school when youths begin to learn about and understand the opportunity structure (see Ogbu, 2003, pp. 41, 154).

The third strand of research is on the tenet that has received the most attention among researchers: the *acting White* hypothesis. This hypothesis posits that the primary mechanism by which Blacks resist educational goals is by equating academic success with acting White. There is some evidence for this hypothesis (Fordham & Ogbu, 1986; Ogbu, 2003). Fryer and Torelli (2005) found that the phenomenon has a significant effect on Black student achievement in schools with high interracial contact and among high-achieving students but little or no effect in predominantly Black or private schools. Other studies show Blacks do not experience greater social cost for high achievement than Whites; these studies suggest Blacks' peer groups are not monolithic and allow space to affirm academic identity (e.g., Carter, 2005; Horvat & Lewis, 2003; O'Connor, 1999; Tyson, Darity, & Castellino, 2005). In fact, Carter (2005) found that negative sanctioning for acting White is driven by a rejection of generic American, "White" middle-class behavioral styles with regard to such things as interaction, speech, dress, and musical tastes rather than academic achievement. In sum, people interested in employing the oppositional culture theory should carefully weigh the evidence on both sides of the debate. The lack of consensus within the scientific community suggests further research is necessary before the theory can be adopted or rejected.

DIRECTIONS FOR FUTURE RESEARCH

There are several directions for future research on the oppositional culture theory. First, although adequate schooling remains a serious obstacle for many Hispanics, researchers have not incorporated them into the oppositional culture literature. Their lower school achievement as immigrants or descendents of immigrants makes their group classification within the framework unclear. Thus, future research is necessary to determine why, despite their reality as immigrant minorities, their schooling experiences resemble that of involuntary minorities within the United States.

Second, few studies assess the oppositional culture theory among younger students. The theory was developed with reference to high school students. Studies among Black children in elementary school show they begin school very much engaged and achievement-oriented but that the schooling experience plays a central role in the development of negative schooling attitudes (Tyson, 2002, 2003). In addition to poor achievement, a strong emphasis on transforming many aspects of Black children's culture by school officials inadvertently communicates inadequacy associated with *Blackness* (Tyson, 2003). Black children's negative attitudes toward schooling may reflect desires to avoid further failure in school rather than the maintenance of an oppositional culture. Similarly, other studies have found that students' academic skill-sets determine both their schooling behaviors and later achievement (Harris & Robinson, 2007). Future research should determine when perceptions about the opportunity structure become consequential for school achievement and schooling behaviors.

Finally, more research is needed on the oppositional culture theory in non-U.S. contexts. Although the framework was partially developed based on Ogbu's (1978) comparative assessment across several countries, few researchers have attempted to replicate his original conclusions. Studies based on non-U.S. populations may help to further refine the theory and generate new propositions for the framework.

SEE ALSO Volume 1: *Cultural Capital; Immigration, Childhood and Adolescence; Peer Groups and Crowds; Racial Inequality in Education; Socialization, Race.*

BIBLIOGRAPHY

Ainsworth-Darnell, J. W., & Downey, D. B. (1998). Assessing the oppositional culture explanation for racial/ethnic differences in school performance. *American Sociological Review, 63*, 536–553.

Carter, P. L. (2005). *Keepin' it real: School success beyond Black and White.* Oxford, U.K.: Oxford University Press.

Cook, P. J., & Ludwig, J. (1997). Weighing the "burden of 'acting White'": Are there race differences in attitudes toward education? *Journal of Policy Analysis and Management, 16*, 256–278.

Ford, D. Y., & Harris, J. J., III. (1996). Perceptions and attitudes of Black students toward school, achievement, and other educational variables. *Child Development, 67*, 1141–1152.

Fordham, S., & Ogbu, J. U. (1986). Black students' school success: Coping with the burden of "acting White." *The Urban Review, 18*, 176–206.

Fryer, R. G., Jr., & Torelli, P. (2005). *An empirical analysis of "acting White."* Cambridge, MA: National Bureau of Economic Research.

Harris, A. L. (2006). I (don't) hate school: Revisiting "oppositional culture" theory of Blacks' resistance to schooling. *Social Forces, 85*, 797–834.

Harris, A. L., & Robinson, K. (2007). Schooling behaviors or prior skills? A cautionary tale of omitted variable bias within oppositional culture theory. *Sociology of Education, 80*, 139–157.

Horvat, E. M., & Lewis, K. S. (2003). Reassessing the "burden of 'acting White'": The importance of peer groups in managing academic success. *Sociology of Education, 76*, 265–280.

Jaffe, A. J., Cullen, R. M., & Boswell, T. D. (1980). *The changing demography of Spanish Americans*. New York: Academic Press.

Mickelson, R. A. (1990). The attitude-achievement paradox among Black adolescents. *Sociology of Education, 63*, 44–61.

O'Connor, C. (1999). Race, class, and gender in America: Narratives of opportunity among low-income African American youth. *Sociology of Education, 72*, 137–157.

Ogbu, J. U. (1974). *The next generation: An ethnography of education in an urban neighborhood*. New York: Academic Press.

Ogbu, J. U. (1978). *Minority education and caste: The American system in cross-cultural perspective*. New York: Academic Press.

Ogbu, J. U. (2003). *Black American students in an affluent suburb: A study of academic disengagement*. Mahwah, NJ: Lawrence Erlbaum Associates.

Therrien, M., & Ramirez, R. R. (2000). *The Hispanic population in the United States: Population characteristics, March 2000* (Current Population Report P20-535). Washington, DC: U.S. Bureau of the Census.

Tyson, K. (2002). Weighing in: Elementary-age students and the debate on attitudes toward school among Black students. *Social Forces, 80*, 1157–1189.

Tyson, K. (2003). Notes from the back of the room: Problems and paradoxes in the schooling of young Black students. *Sociology of Education, 76*, 326–343.

Tyson, K., Darity, W., Jr., & Castellino, D. (2005). "It's not a Black thing": Understanding the burden of acting White and other dilemmas of high achievement. *American Sociological Review, 70*, 582–605.

Angel L. Harris

P

PARENT-CHILD RELATIONSHIPS, CHILDHOOD AND ADOLESCENCE

The parent-child relationship is unique among human relationships. Parent-child interactions are among the most common interactions during a child's formative years. This relationship is nonvoluntary and endures over the life course. Because of the obligatory interdependence involved in child rearing, even if the relationship is stressed, parents and children rarely sever their ties completely, as sometimes happens with friendships (Maccoby, 2003).

Scholars have examined parent-child relations at various developmental stages, including infancy, early childhood, middle childhood, adolescence, and even adulthood. Research has focused on the ways children are affected by diverse aspects of parent-child relations, such as parenting styles, parenting practices, parental characteristics and behaviors, child characteristics and behaviors, mutual influences between parents and children, parent-child interactions, mother's versus father's influence, and cultural and ethnic influences. Researchers widely agree that the quality of parent-child relationships plays a significant role in children's intellectual, social, emotional, and behavioral development (Cox & Harter, 2003). Parent-child relationships not only have an impact on children's development during childhood and adolescence but also affect their socioeconomic success and psychological well-being throughout the life course.

HISTORICAL TRENDS IN RESEARCH ON PARENT-CHILD RELATIONS

Research on parent-child relationships can be divided into two eras: before bidirectionality and after bidirectionality (Kuzynski, 2003). The before bidirectionality period encompasses studies conducted before the late 1960s, when research focused on identifying parenting strategies, practices, behaviors, styles, and traits that might influence children's outcomes, such as competence, healthy development, school achievement, and problem behaviors. Although these topics continue to be of interest to scholars, early studies largely ignored mutual interactive processes in which parents and children influence each other. In the period since the late 1960s, the after bidirectionality period, researchers started to pay attention to the reciprocal or mutual influences between parents and children. In that period scholars began to recognize that both parents and children act as "agents" in socialization processes; that is, both generations are capable of being goal-oriented, initiating intentional behaviors, reflecting about interactions, interpreting situations, resisting requests, and blocking each other's goals (Bell, 1968, Kuczynski & Parkin, 2007). In contrast to the earlier, more simplistic top-down approach that emphasized parental effects on children's outcomes, a bidirectional perspective recognizes that parents and children together contribute to the processes that shape children's outcomes.

Research and theory adopting the unidirectional approach experienced a second surge in the 1980s and early 1990s. That resurgence, led by behavioral geneticists, turned earlier unidirectional research on its head by exploring the ways in which children's traits elicited specific responses and practices from parents (Harris,

1998, Rowe, 1994). Behavioral geneticists asserted that children's different genetics caused parents to respond to them differently, which affected parent-child interactions and subsequent child outcomes. Those studies compared and contrasted siblings who differed in the degree of genetic relatedness, such as identical versus fraternal twins. Researchers found that genetics plays an important role in shaping children's characteristics and explaining similarities in how brothers and sisters turn out. At the same time they found a powerful influence of the siblings' unshared environment, that is, the unique environments and experiences of each sibling. Unshared environment may encompass factors such as parents' distinctive responses to each child's unique traits and peer or teacher influences on each child. Interestingly, behavioral geneticists reported that the influence of shared environmental factors on child outcomes was minimal; those factors encompass aspects of family life that brothers and sisters share, such as the parents' economic status and attitudes toward child rearing.

On the basis of those studies the behavioral geneticists concluded that parents have little or no long-term influence on children's behavior or personality. This bottom-up orientation elicited strong reactions from many researchers, who challenged the assumptions and methodologies used in behavior-genetic studies; those researchers then set out to document evidence that parents do in fact matter for child outcomes (Collins, Maccoby, Steinberg, Hetherington, & Bornstein, 2000).

Although the findings are often controversial, the key ideas of behavioral genetics have had a strong influence on research on parent-child relationships. Contemporary scholars have accepted several of their assertions, especially the observation that children's genetic makeup plays an important role in shaping parent-child interactions and consequently the types of parent-child relationships that ultimately develop. To evaluate parental influence fully, however, researchers must take into consideration the ways in which parenting may be affected by children's original differences in temperament and abilities (Maccoby, 2007).

Contemporary scholars generally agree that a bidirectional view provides a more compelling explanation of parent-child relations than does either of the unidirectional approaches. In particular, researchers are calling for efforts to study both children's influence on parents and parents' influence on children. This is based on the assumption that in any long-term relationships both parties develop expectations for each other. Through repeated actions and reactions a culture emerges that includes mutually shared views, values, memories, goals, expectations, and rules that worked and rules that did not work. Ongoing daily interaction is based on this shared

culture. Thus, the contextually based mutually influential relationship between parents and children is the major force influencing a child's outcomes (Maccoby, 2003).

PARENT-CHILD RELATIONSHIPS AND CHILD OUTCOMES

Early parent-child relationships are an important topic of inquiry for developmental psychologists. The overall quality of early parent-child relationships reflects the levels of warmth, security, trust, positive affect, and responsiveness in the relationship. Warmth is recognized almost universally as a fundamental component of parent-child relationships. Children who have a warm relationship with their parents report feeling loved and thus develop a strong sense of self-confidence. They typically trust and enjoy participating in shared activities with their parents. Their confidence facilitates their exploration of the environment and development of competence. Children who are in warm relationships with their parents also tend to identify with the parents and thus usually are willing to cooperate or comply with parental demands. Warmth in the relationship creates a context of positive affect, which tends to enhance the mood of those who are involved and makes them more attentive and responsive to each other (Laible & Thompson, 2007).

Security is another central dimension of parent-child relationships, especially among young children. Attachment theory (Bowlby, 1969), one of the most important theories of child development, was developed to characterize parent-infant relationships. Its basic premise is that the primary function of an infant's attachment behavioral system is to obtain and keep the caregiver at times of need. The caregiver's availability, sensitivity, and responsiveness to an infant's signals or calls for help make the infant feel secure. Infants' repeated interactions with caregivers lead to the development of an internal working model; this refers to an internal representation of attachment, including images of the self and others as well as expectations for interaction in close relationships. A child's confidence grows from his or her feelings of security, feeling that the environment is safe, viewing others as loving and kind, and valuing one's self. Subsequent research has confirmed the importance of the attachment relationship, which continues beyond infancy and early childhood. Attachment ties with parents during the later stages of childhood and adolescence continue to provide a supportive foundation for the development of competence and psychological well-being (Armsden & Greenberg, 1987).

An important area of research that developed during the bidirectional period focuses on the connection between parent-child interactions and the relationship they form. Robert Hinde's work has had an important influence on research on parent-child relations. According to Hinde, the basic components of a relationship are interaction and time. Two individuals who interact over a period of time form a relationship, with each interaction having an impact on the ones that follow. Repeated interactions lead to the establishment and development of a relationship, and that relationship influences the frequency and nature of subsequent interactions (Hinde & Stevenson-Hinde, 1988).

Several main principles form the base of interaction models of parent-child relations. The three core principles are as follows:

1. Interaction. Parents and children interact over time and thus create their relationship. At any point in time the relationship consists of only one interaction that occurs in the present moment, together with many memories of past interactions and anticipations of future interactions.

2. Mutual contribution. Parents and children both play a part in their interactions and thus the relationship.

3. Distinctiveness. Each parent-child relationship is unique for the involved dyad and cannot be replicated by a relationship with another parent or child. Two additional principles characterize the "time" dimension: (a) Past expectations. Prior interactions between the parent and the child are building blocks for their current expectations. On the basis of his or her past experiences and observations, a parent has a solid understanding of how a child is likely to act in a specific situation. The child also has expectations about how the parent will act. (b) Future anticipations. Because parent-child relationships are enduring, both parents and children expect a future, which is built into their interactions (Hinde & Stevenson-Hinde, 1988, Lollis, 2003).

Establishing a healthy parent-child interaction pattern in early childhood has important implications for subsequent stages, especially for the potentially problematic parent-child interactions that often emerge during adolescence.

PARENT-CHILD RELATIONSHIPS AND OUTCOMES DURING ADOLESCENCE

In many families, parent-child relationships undergo substantial changes when a child reaches adolescence. A common perception in American society is that adolescence is a period of storm and stress, that is, a period of biologically determined upheaval and disturbance. Classical psychoanalytic perspectives view adolescent storm

and stress as normative (Freud, 1969, Blos, 1962). Anna Freud (1969) claimed that adolescent upheaval is both necessary and desirable for personal growth. Peter Blos (1962) viewed adolescence as a process of individuation; at this time teenagers must experience an emotional disengagement from their parents before establishing new social and emotional bonds outside the family. However, empirical research reveals that adolescence is not necessarily a stage of upheaval. To the contrary, the actual proportions of conflict-ridden families found in nonclinical populations are much lower than what advocates of the psychoanalytic perspective predicted (Hill & Holmbeck, 1986).

The view of adolescent storm and stress as entirely biologically based, universal, and inevitable has been rejected by most scholars. Empirical data provide more persuasive support for what might be called a modified version of the storm and stress hypothesis. This perspective acknowledges that conflicts with parents, mood disruptions, and risky behaviors are more likely to occur during adolescence than in any other stage of the life course (Arnett, 1999). Adolescent arguments with parents usually focus on mundane matters such as curfews, bedtimes, dating, hair, clothing, friends, and chores rather than on fundamental values. However, parents are likely to be upset by daily bickering and often remain distressed, whereas adolescents tend to brush it off and recover quickly from the negative interactions (Steinberg, 2001). One reason for this pattern is that parents and adolescents define conflict differently. Whereas parents tend to think about an issue such as keeping the room clean in moral or social conventional terms of right and wrong, adolescents tend to see it merely as an indicator of personal choice (Smetana, 1989).

Researchers note that conflicts may serve as a warning signal for parents to modify and adjust their interactions with their teenage children. For example, parental monitoring is still necessary when children are adolescents, yet parents need to modify the degree and extent of supervision to adapt to a child's developmental changes. Monitoring is also more effective if it is conducted within a close parent-child relationship and if the adolescent is cooperative and willing to let the parents know his or her whereabouts and with whom he or she is associating (Kerr & Stattin, 2003).

Adolescent behaviors such as being secretive may be a result of earlier parental reactions. For example, an adolescent may have shared with his or her parents worries about problems encountered at school or with peers, and the parents may have responded with unsympathetic criticism, mockery, or disciplinary actions. Even worse, parents may have used the disclosed information against the adolescent at a later time. As a result, the adolescent becomes increasingly guarded and refrains from disclosing further information. This makes parents even more suspicious about what might be concealed and less likely to trust the adolescent, who in turn may become more deceptive. That negative cycle is often a result of earlier failed interactions (Maccoby, 2007).

PARENTING STYLES

Research on parenting styles also attempts to identify patterns of child rearing that are consistently related to children's outcomes. The well-established distinction among parenting styles is a fourfold typology that involves different combinations of the two core dimensions of responsiveness and demandingness. Responsiveness entails parental warmth, acceptance, support, and sensitivity. Demandingness includes high standards, demands for maturity, and control. Authoritative parenting is characterized by high levels of responsiveness and demandingness, authoritarian parenting is high on demandingness but low on responsiveness, indulgent parenting is high on responsiveness but low on demandingness, and rejecting-neglecting parenting is neither responsive nor demanding (Baumrind, 1991, Maccoby & Martin, 1983). Authoritative parenting, in which the parents are warm and firm, is reported to be more successful than any of the nonauthoritative parenting styles, as evidenced by its consistent association with higher academic achievement, better mental health, and fewer problem behaviors for both children and adolescents.

PARENT-CHILD RELATIONSHIPS IN CONJUNCTION WITH PEER GROUPS

The parent-child relationship needs to be considered against a broader social context that encompasses influences such as schools, teachers, media, and peer groups. Peers and parents may appear to represent two vastly different social worlds for a child. However, research has provided mounting evidence on the significant and complex interplay between the two worlds. Studies have reported that parents directly influence their children's friendship formation by selecting residential areas and schools, enrolling children in extracurricular programs, and serving as gatekeepers in the management of children's peer relationship by initiating, arranging, monitoring, and facilitating their children's contacts with potential friends (Parke, MacDonald, Burks, Carson, Bhavnagri, Barth, et al., 1989). Parents also have an influence on children's selection of friends through socialization, that is, by passing on their values in a way that facilitates their children's choice of friends who have similar values and orientations (Chen, Dornbusch, & Liu, 2007).

PREDICTORS OF PARENTAL BEHAVIOR

Since the 1980s scholars increasingly have explored the antecedents of parental behavior. Specifically, researchers have investigated the extent to which a broad range of factors shape parental behavior; those factors include early upbringing, personality, mental health, child characteristics, marital satisfaction, social network support, economic hardship, and parenting beliefs (Belsky & Jaffee, 2006)

Many longitudinal and multigenerational studies have found that parents of young children tend to use the same parenting practices that their parents used in raising young children. The intergenerational transmission of parenting has been documented persuasively in studies of dysfunctional parenting, specifically, child abuse. These findings also shed light on the intergenerational transmission of parenting in the general population. Early exposure to harsh or abusive parenting is related to subsequent adoption of coercive parenting practices with one's own children, whereas early satisfying experiences in the parental household predict supportive parenting behavior. Scholars also have investigated specific mechanisms that might have mediated the transmission of parenting practices across generations. Through modeling or social learning processes, children may internalize parenting beliefs. An early harsh upbringing also may contribute to their development of psychopathology or antisocial traits, which can lead to the adoption of similar harsh parenting behavior toward their own children (Belsky & Jaffee, 2006).

The marital relationship between parents also is an important predictor of parental behavior. Family researchers generally view a family as a social system, with marital, parent-child, and sibling interdependent subsystems (Minuchin, 1974). The quality of the marital relationship is believed to be the leading factor affecting the quality of family life. The spillover hypothesis, which proposes that moods, emotions, and behaviors are transferred directly from the marital situation to the parent-child context, has received a great deal of empirical support. A satisfactory marriage is likely to enhance parents' supportive and constructive behavior, whereas a poor marital relationship tends to hinder parenting performance and hurt the parent-child relationship (Erel & Burman, 1995).

Socioeconomic status is one of the major structural factors that predict parenting practices. Economic difficulties have negative effects on parental sensitiveness and responsiveness because financial strains may add pressure and stress to a family's daily life. For example, higher rates of child abuse are reported among families with lower levels of education, in poverty, or with blue-collar occupations (Belsky & Jaffee, 2006).

CULTURAL INFLUENCES

European and North American scholars have focused on the effects of reciprocity and mutuality on parent-child relationship, yet those processes may not be generalizable to all cultures. One of the guiding concepts in cross-cultural research for several decades has been individualism-collectivism. In individualistic societies such as the United States, Canada, and Western European countries, individual goals and freedom are honored. In collectivist societies such as Asian, African, and Latin American countries, group goals are predominant and seniority and authority are respected. In collectivist societies parents tend to have more absolute authority, demanding obedience from children and granting them little autonomy. Because this is the norm in such societies and because there are societywide assumptions that legitimize parental authority and power as fair, children may take it for granted, do not expect much autonomy, and grow up into responsible adults as expected by their society.

In individualistic societies, in contrast, individual rights are valued in all arenas, with a weaker social emphasis on parental authority and power. From a young age children are encouraged to develop independent and critical thinking and display their individuality. Parents in democratic societies will be more successful when they establish a cooperative relationship with their children so that the children will be more willing to accept parental guidance (Maccoby, 2003, Trommsdorff & Kornadt, 2003). Immigrant families that move from collectivist societies to the United States may face a special challenge when the parents continue to exercise their more absolute parental authority and control as their children are being immersed in an individualistic culture that leads them to expect a democratic relationship.

WHERE IS RESEARCH GOING?

Although scholars have made progress in conceptualizing bidirectionality in studies of parent-child relationships, those theoretical advances have not been accompanied by commensurate advances in empirical research. Most empirical research has continued to be dominated by a persistent top-down emphasis on parental influence over children's outcomes. Research guided by the bidirectional framework is still expanding. The challenge scholars face is not a lack of evidence documenting the effect of the parent or that of the child; instead, studies need to investigate the complexity of those effects. Numerous contextual factors need to be explored further so that researchers can specify under what conditions an effect or combined effects are more likely to play out.

Whereas researchers who study parent-child relations may have overemphasized the unidirectional influence from parents to children, there is no reason to underestimate parental influence and deny the existence of a parent-child hierarchy. The important message is how parents should adapt to the situation in a way that is consistent with their parental responsibilities. In a completely egalitarian household without any parental authority or in a reversed hierarchy in which the child runs the house the outcome is usually detrimental. However, at the same time parents need to realize that the assertion of parental power will not be successful unless children's power is taken into consideration. This approach requires warm, sensitive, and responsive parenting practices to build a healthy parent-child relationship so that children will be willing and cooperative as the parents undertake their child-rearing tasks (Maccoby, 2003).

SEE ALSO Volume 1: *Attachment Theory; Child Abuse; Cultural Capital; Family and Household Structure, Childhood and Adolescence; Parental Involvement in Education; Parenting Style; Policy, Child Well-Being; Poverty, Childhood and Adolescence; Racial Inequality in Education; Social Capital; Socioeconomic Inequality in Education.*

BIBLIOGRAPHY

Armsden, G., & Greenberg, M. T. (1987). The inventory of parent peer attachment: Individual differences and their relation to psychological well-being in adolescence. *Journal of Youth and Adolescence, 16*(5), 427–454.

Arnett, J. J. (1999). Adolescent storm and stress, reconsidered. *American Psychologist, 54*(5), 317–326.

Baumrind, D. (1991). The influence of parenting style on adolescent competence and substance use. *Journal of Early Adolescence, 11*(1), 56–95.

Bell, R. Q. (1968). A reinterpretation of the direction of effects in studies of socialization. *Psychological Review, 75*(2), 81–95.

Belsky, J., & Jaffee, S. R. (2006). In D. Cicchetti & D. J. Cohen (Eds.), *Developmental psychopathology* (2nd ed.) Vol. 3: *Risk, disorder, and adaptation* (pp. 38–85). Hoboken, NJ: John Wiley & Sons Inc.

Blos, P. (1962). *On adolescence.* New York: Free Press of Glencoe.

Bowlby, J. (1969). *Attachment and loss.* Vol. 1: *Attachment.* New York: Basic Books.

Chen, Z, Dornbusch, S. M., & Liu, R. X. (2007). Direct and indirect pathways in the link between parental constructive behavior and adolescent affiliation with achievement-oriented peers. *Journal of Child and Family Studies, 16*(6), 837–858.

Collins, W. A., Maccoby, E. E., Steinberg, L., Hetherington, E. M., & Bornstein, M. H. (2000). Contemporary research on parenting: The case for nature and nurture. *American Psychologist, 55*(2), 218–232.

Cox, M. J., & Harter, K. S. M. (2003). Parent-child relationships. In M. H. Bornstein, L. Davidson, C. L. M. Keyes, & K. A. Moore (Eds.), *Well-being: Positive development across the life course* (pp. 191–204). Mahwah, NJ: Lawrence Erlbaum Associates.

Erel, O., & Burman, B. (1995). Interrelatedness of marital relations and parent-child relations: A meta-analytic review. *Psychological Bulletin, 118,* 108–132.

Freud, A. (1962). Adolescence. In *The writing of Anna Freud: Research at the Hampstead child-therapy clinic and other papers* (vol. 5, pp. 136–166). New York: International Universities Press (Orig. publ. 1958).

Harris, J. R. (1998). *The nurture assumption: Why children turn out the way they do.* New York: Free Press.

Hill, J. P. & Holmbeck, G. N. (1986). Attachment and autonomy during adolescence. *Annals of Child Development, 3:* 145–189.

Hinde, R. A., & Stevenson-Hinde, J. (1988). *Relationships within families: Mutual influences.* Oxford: Clarendon Press.

Kerr, M., & Stattin, H. (2003). Parenting of adolescents: Action or reaction? In A. C. Crouter & A. Booth (Eds.), *Children's influence on family dynamics: The neglected side of family relationships* (pp. 121–151). Mahwah, NJ: Lawrence Erlbaum Associates.

Kuczynski, L. (2003). Introduction and overview. In L. Kuczynski (Ed.), *Handbook of dynamics in parent-child relations* (pp. ix–xv). Thousand Oaks, CA: Sage Publications.

Kuczynski, L., & Parkin, C. M. (2007). Agency and bidirectionality in socialization: Interactions, transactions, and relational dialectics. In J. E. Grusec & P. D. Hastings (Eds.), *Handbook of socialization: Theory and research* (pp. 259–283). New York: Guilford Press.

Laible, D., & Thompson, R. A. (2007). Early socialization: A relationship perspective In J. E. Grusec & P. D. Hastings (Eds.), *Handbook of socialization: Theory and research* (pp. 181–207). New York: Guilford Press.

Lollis, S. (2003). Conceptualizing the influence of the past and the future in present parent-child relationships. In L. Kuczynski (Ed.), *Handbook of dynamics in parent-child relations* (pp. 67–87). Thousand Oaks, CA: Sage Publications.

Maccoby, E. E. (2003). Epilogue: Dynamics viewpoints on parent-child relations—their implications for socialization processes. In L. Kuczynski (Ed.), *Handbook of dynamics in parent-child relations* (pp. 439–452). Thousand Oaks, CA: Sage Publications.

Maccoby, E. E. (2007). Historical overview of socialization research and theory. In J. E. Grusec & P. D. Hastings (Eds.), *Handbook of socialization: Theory and research* (pp. 13–41). New York: Guilford Press.

Maccoby, E. E., & Martin, J. A. (1983). Socialization in the context of the family: Parent-child interaction. In E. M. Hetherington (Ed.), P. H. Mussen (Series Ed.), *Handbook of child psychology* (vol. 4, pp. 1–101). New York: Wiley.

Minuchin, S. (1974). *Families and family therapy.* Cambridge, MA: Harvard University Press.

Parke, R. D., MacDonald, K. B., Burks, V. M., Carson, J., Bhavnagri, N. P., Barth, J. M., et al. (1989). Family and peer systems: In search of the linkages. In K. Kreppner & R. M. Lerner (Eds.), *Family systems and life-span development* (pp. 65–92). Hillsdale, NJ: Lawrence Erlbaum Associates.

Rowe, D. C. (1994). *The limits of family influence: Genes, experience, and behavior.* New York: Guilford Press.

Smetana, J. G. (1989). Adolescents' and parents' reasoning about actual family conflict. *Child Development, 60*(5), 1052–1067.

Steinberg, L. (2001). We know some things: Parent-adolescent relationships in retrospect and prospect. *Journal of Research on Adolescence, 11*(1), 1–19.

Trommsdorff, G., & Kornadt, G. (2003). Parent-child relations in cross-cultural perspective. In L. Kuczynski (Ed.), *Handbook of dynamics in parent-child relations* (pp. 271–306). Thousand Oaks, CA: Sage Publications.

Zeng-yin Chen

PARENTAL INVOLVEMENT IN EDUCATION

Research in parental involvement in education is founded on the premise that families and communities—as well as schools—play a crucial role in children's social and academic development. Hypothesizing that children are more likely to succeed when their schools, families, and communities work together, parental involvement research has contributed to a broad-based effort in U.S. educational policy and practice to bridge gaps between home and school. Although important questions persist regarding the independent effects of parental involvement, research clearly indicates that affluent and highly educated parents are more involved than less advantaged parents and that this school involvement is an important mechanism underlying the reproduction of educational inequality. By encouraging all parents to take a more active role in their children's schooling, parental involvement efforts seek to level this source of educational inequality, improving the life chances of all youth.

THEORIES OF PARENTAL INVOLVEMENT

Although sociologists of education and developmental psychologists have long been interested in the ways in which schools and families interact to influence children's educational experiences, this work has traditionally viewed the school and the home as separate, isolated spheres of activity. Parental involvement efforts, by contrast, blur the boundaries between family and school. Much of the work around parental involvement began in the 1980s when the sociologist Joyce L. Epstein (1987) developed a theory of parental involvement that conceptualized the school and the family as "overlapping spheres of influence" in children's lives.

Epstein's (1992) widely cited typology of parental involvement draws attention to five areas for collaboration between schools and families: (a) basic parental obligations such as providing a safe home environment that is conducive to learning; (b) communication between school and home; (c) school meetings, volunteering, and other activities that bring parents into schools; (d) home learning and homework help; and (e) parental participation in school decision making.

Drawing on work in developmental psychology, Epstein argues that collaboration between families and schools increases the overlap between these two spheres of influence, hypothesizing that parental involvement helps families and schools more effectively advance shared goals related to children's development and educational attainment. This developmental theory leads Epstein to assume that increasing parental involvement can have positive consequences for all children, regardless of the social context in which it occurs. As a result, most of Epstein's work focuses on developing strategies to improve parental involvement.

Subsequent sociological work, however, draws this assumption into question by situating parental involvement as a mechanism through which class advantages are reproduced. The research of Lareau (2000, 2003) indicates that affluent and highly educated parents tend to take a more active role in their children's schooling than poor and working-class parents. Whereas highly educated parents are at ease in schools, can interact comfortably and effectively with teachers, and have the confidence to help their children with homework, less highly educated parents lack this ease and confidence and tend to view schools as having the exclusive role in their children's education. Furthermore, Lareau argues that the involvement of parents of high socioeconomic status (SES) may be more effective than the involvement of less advantaged parents, because highly educated parents have a better understanding of teachers' expectations and the unspoken rules at work in parent–school interactions. Lareau's work, which is grounded in sociological reproduction theory, suggests that raising the involvement of low-SES parents may do little to narrow class-based gaps in academic achievement.

METHODS IN PARENTAL INVOLVEMENT RESEARCH

Parental involvement research draws widely from among the methods commonly used in social science and educational research, including structured teacher interviews, ethnography, and quantitative analysis of parent survey and student test data.

Much of Epstein's research in parental involvement revolves around teacher interviews and surveys, in which teachers are asked about their willingness to collaborate with parents, their efforts to stimulate parental involvement, the extent to which parents participate in their children's schooling, and the consequences of this partic-

ipation. This research demonstrates that many teachers are very interested in raising parental involvement levels and provides rich data on the approaches that teachers and schools use to do so (c.f. Epstein & Becker, 1982; Epstein & Dauber, 1991). These teacher interviews, however, offer little direct insight into the experience of parents and students. Therefore, they are arguably less reliable sources for information about how parents approach their children's schooling and the efforts that parents make to help their children, particularly when these efforts take place outside of the school.

Most research in parental involvement conducted in the past 20 years has focused on data gathered directly from parents and students. In her ethnographic research, Lareau (2000) carefully tracked a small number of families, observing the practices of a diverse sample of parents and discussing these practices with parents, children, and, in select cases, school officials. These observations provide a nuanced description of the relationships between families and schools and considerable insight into *why* educationally and economically advantaged parents tend to be more highly involved in their children's schooling than less advantaged parents.

Although this nuance is lost in survey data analyses, quantitative research methods have become increasingly central to parental involvement research. Often based on random samples of the population of U.S. schoolchildren, these quantitative studies improve the generalizability of findings derived from the qualitative studies that Epstein, Lareau, and others have conducted. Furthermore, by linking parental descriptions of involvement activities to data describing family's racial, ethnic, and socioeconomic background, as well as standardized measures of students' academic achievement and socio-development, these quantitative studies make it possible to assess the causal effects of parental involvement.

CLASS-BASED INEQUALITIES IN PARENTAL INVOLVEMENT

Survey data gathered by the U.S. Department of Education clearly indicates that the extent to which parents are involved in their children's education is closely related to their socioeconomic background (Vaden-Kiernan & McManus, 2005). Class-based inequalities are particularly pronounced in intensive forms of parental involvement, such as school volunteering. The 2003 National Household Education Survey indicates that 55% of parents with a B.A. volunteered in their children's school or served on a school committee. By comparison, just 30% of parents whose formal education ended with a high school diploma and just 16% of parents who dropped out of high school were similarly involved in their children's schools. In a similar manner, parents in poverty households are approximately half as likely to report volunteering in their children's schools relative to parents in non-poverty households (27–45%). Language barriers also discourage parental involvement. Children from households in which no parent speaks English are less than half as likely to have a parent who volunteers at school.

Although class-based inequalities are less pronounced in other areas of parental involvement, they remain substantial. Most parents attend general school meetings and parent–teacher conferences, regardless of their educational attainment or poverty status. For example, 93% of parents with college degrees report attending a school meeting in the 2002–2003 school year, compared to 84% of parents whose formal educated ended with a high school diploma. The gap separating the school meeting and parent–teacher conference attendance rates for parents above and below the poverty level is similar in magnitude.

Despite these class-based inequalities, racial and ethnic inequalities are relatively muted. In 2003, for example, African American parents were slightly more likely to attend parent–teacher conferences and school events than White parents. Furthermore, Lareau's (2003) ethnographic research as well as multivariate analyses based on survey data suggest that among parents with similar levels of educational attainment and family income, racial and ethnic gaps in parental involvement are small (Fan, 2001; Muller & Kerbow, 1993).

PARENTAL INVOLVEMENT IN EDUCATIONAL POLICY

The effort to eliminate these class-based inequalities in parental involvement has been central to federal efforts to improve the education of poor and minority youth since 1965, when the Elementary and Secondary Education Act set aside funds for parental involvement efforts in high-poverty schools. The Clinton administration's 1994 Goals 2000: Educate America Act made universal parental involvement a central federal goal. The 1994 Clinton administration reauthorization of the Elementary and Secondary Education Act added a new provision requiring the nation's poorest schools to spend at least 1% of their Title I supplementary federal funds to develop educational "compacts" between families and schools.

These federal efforts, as well as state- and district-level parental involvement efforts, have particularly focused on providing opportunities for parents to visit their children's schools and meet with teachers and administrators (Domina, 2005). A 1995–1996 survey by the National Center for Education Statistics showed that nearly all public elementary and middle schools in the United States sponsored activities that were designed

to foster parental involvement. According to the survey, 97% of schools invited parents to attend an open house or back-to-school night, 96% hosted arts events, 92% scheduled parent–teacher conferences, 85% sponsored athletic events, and 84% had science fairs (Carey, Lewis, Farris, & Burns, 1998).

The No Child Left Behind Act of 2001 took the parental involvement effort further. The law defined parental involvement as "the participation of parents in regular, two-way, and meaningful communication involving student academic learning and other school activities." It requires schools to regularly distribute information about teacher qualifications and student test scores to parents. In addition, the law sets aside federal funds to encourage parents in high-poverty schools to help their children with homework, attend scheduled parent–teacher conferences and other events at school, and volunteer in the classroom. In addition, the law instructs schools to give parents advisory roles in school governance decisions. Many schools have used these funds to hire full-time parental involvement coordinators, who are responsible for maintaining communications with parents and organizing school events.

TRENDS IN PARENTAL INVOLVEMENT

These policy efforts seem to have helped boost parental involvement levels at U.S. schools, and a growing proportion of parents report participation in school-sponsored involvement activities. In 1999, 78% of the nationally representative sample of parents with school-age children surveyed in the National Household Education Survey reported that they attended a general meeting at their child's school in the past year (Nord & West, 2001). By 2003, when the survey was replicated with a new sample of U.S. parents, that figure had risen to 88% (Vaden-Kiernan & McManus, 2005). Furthermore, the 1999 and 2003 National Household Education Surveys point to smaller, but still statistically significant, improvements in other areas of parental involvement. Between 1999 and 2003, the portion of parents who had attended a parent–teacher conference increased from 73% to 77%; the portion of parents who had attended a school athletic event or performance increased from 65% to 70%; and the portion of parents who had volunteered at school or served on a school committee increased from 37% to 42%.

The 2003 National Household Education Survey also indicated that 85% of parents had helped their children with their homework in the previous year. The 1999 survey, however, did not collect data on homework help, so it is impossible to identify trends in this common form of parental involvement.

THE CONSEQUENCES OF PARENTAL INVOLVEMENT

Given the popularity of parental involvement initiatives as a tool for school reform, it is surprising to note that research on the link between involvement and school success has been inconclusive. In the past decades, dozens of studies have attempted to isolate the consequences of parental involvement. Many of these studies use multivariate analysis techniques to isolate the consequences of parental involvement from the potentially confounding effects of family background, in effect asking whether the children of highly involved parents do better in school than children from families with similar SES whose parents are less involved.

Whereas some studies indicate that the children of highly involved parents tend to do better in school than their peers with less involved parents, others report that parental involvement in education is negatively related to children's educational outcomes. Even within studies the results are often uneven, with the observed effects of parental involvement depending on which aspects of involvement and which educational outcomes are being considered (Fan & Chen, 2001).

Evaluations of programs that attempt to improve student educational outcomes by stimulating parental involvement yield similarly mixed results. In 2002 Mattingly and colleagues reviewed the evaluations of 41 school- and district-level parental involvement programs and found little empirical evidence to support the claim that schools' efforts to improve parental involvement ultimately improve students' outcomes. Only half of the parental involvement programs that provided adequate data to reach conclusions about program effectiveness actually improved student outcomes.

These mixed results are difficult to interpret. Epstein's developmental theory of parental involvement predicts that involvement should substantially improve children's odds of school success. Furthermore, although Lareau's reproduction theory suggests that the positive consequences of parental involvement are closely linked to other advantages that highly involved parents offer their children, even this more skeptical theory suggests that involvement's effects should be generally positive. Why, then, do parental involvement researchers so frequently fail to find positive results? One possible explanation is that the relationship between parental involvement and children's educational outcomes is complex. Parents often intervene in their children's education only when their children are experiencing difficulties in school; conversely, they often relax their involvement when children are succeeding in school. If this is the case, many estimates of the consequences of parental involvement are likely biased, because they confuse the effects that parental involvement has on children's educational success with the cross-cutting effects that children's educational success has on parental involvement.

A handful of studies have attempted to address these causal complexities by studying the consequences that parental involvement activities have on children's test scores over time. These studies draw on cohort-based longitudinal studies, which follow the same students as they progress through school, to produce unbiased estimates of the consequences that parental involvement has on children's schooling. Muller (1998) and Fan (2001) examined parental involvement activities that occurred when students were in eighth grade, measuring the consequences on student math and reading achievement score gains between the eighth and twelfth grade. Although neither study found unequivocal evidence for the effectiveness of parental involvement, Muller found that children who discuss school with their parents experience relatively rapid gains in math achievement. Similarly, Fan found that children who work on homework with their parents show large gains on standardized tests of reading, math, science, and social studies skills.

Hypothesizing that parental involvement may have greater consequences for younger students and for student's social and behavioral outcomes, Domina (2005) built on these findings, studying a cohort of elementary school children and their mothers. Although this study indicates that involvement has no measurable impact on student improvement on academic achievement exams, it did find that parental involvement could have strong causal effects on a student's behavior. Domina found that parents protect their children from behavioral problems when they volunteer at schools and check on their homework.

FUTURE OF PARENTAL INVOLVEMENT RESEARCH AND POLICY

After more than 40 years of government activity dedicated to using parental involvement as a tool for narrowing persistent educational inequalities, and more than 20 years worth of social science research about the involvement of parents in children's educations, a striking gap between the state of parental involvement research and practice has emerged. It is clear that parental involvement practices are becoming increasingly common in American schools. It is less clear, however, that these involvement activities are improving children's chances of success in school. In the near future, researchers, policymakers, and practitioners should work together to bridge this gap, by investigating and implementing efforts to improve the efficacy of parental involvement activities, particularly for disadvantaged parents.

SEE ALSO Volume 1: *Cultural Capital; Data Sources, Childhood and Adolescence; Intergenerational Closure; Parent-Child Relationships, Childhood and Adolescence; Policy, Education; Racial Inequality in Education; Social Capital; Socioeconomic Inequality in Education.*

BIBLIOGRAPHY

Carey, N., Lewis, L., Farris, E., & Burns, S. (1998). *Parent involvement in children's education: Efforts by public elementary schools* (NCES 98-032). Washington, DC: U.S. Department of Education, National Center for Education Statistics.

Domina, T. (2005). Leveling the home advantage: Assessing the effectiveness of parental involvement in elementary school. *Sociology of Education, 78,* 233–249.

Epstein, J. L. (1987). Toward a theory of family–school connections: Teacher practices and parent involvement across the school years. In K. Hurrelmann, F.-X. Kaufmann, & F. Lösel (Eds.), *Social intervention: Potential and constraints* (pp. 121–136). Berlin: Walter de Gruyter.

Epstein, J. L. (1992). School and family partnerships. In M. C. Alkin (Ed.), *Encyclopedia of educational research* (pp. 1139–1151). New York: Macmillan.

Epstein, J. L., & Becker, H. J. (1982). Teachers' reported practices of parent involvement: Problems and possibilities. *The Elementary School Journal, 83,* 103–113.

Epstein, J. L., & Dauber, S. L. (1991). School programs and teacher practices of parent involvement in inner-city elementary and middle schools. *The Elementary School Journal, 91,* 289–303.

Fan, X. (2001). Parental involvement and students' academic achievement: A growth modeling analysis. *Journal of Experimental Education, 70,* 27–61.

Fan, X., & Chen, M. (2001). Parental involvement and students' academic achievement: A meta-analysis. *Educational Psychology Review, 13,* 1–22.

Lareau, A. (2000). *Home advantage: Social class and parental intervention in elementary education.* (2nd ed.). London: Falmer Press.

Lareau, A. (2003). *Unequal childhoods: Class, race, and family life.* Berkeley: University of California Press.

Mattingly, D. J., Prislin, R., McKenzie, T. L., Rodriguez, J. L., & Kayzar, B. (2002). Evaluating evaluations: The case of parent involvement programs. *Review of Educational Research, 72,* 549–576.

Muller, C. (1998). Gender differences in parental involvement and adolescents' mathematics achievement. *Sociology of Education, 71,* 336–356.

Muller, C., & Kerbow, D. (1993). Parent involvement in the home, school, and community. In B. Schneider & J. S. Coleman (Eds.), *Parents, their children, and schools* (pp. 13–42). Boulder, CO: Westview Press.

Nord, C. W., & West, J. (2001). *Fathers' and mothers' involvement in the children's schools by family type and residence status* (NCES 2001-032). Washington, DC: U.S. Department of Education, National Center for Education Statistics.

Vaden-Kiernan, N., & McManus, J. (2005). *Parent and family involvement in education, 2002–2003* (NCES 2005-043).

Washington, DC: U.S. Department of Education, National Center for Education Statistics.

Thurston Domina

PARENTING STYLE

Parenting styles and parenting practices are often considered synonymous, but each term describes a distinct aspect of the parenting role. *Parenting practices* refer to specific parenting behaviors such as parental involvement, monitoring, or spanking. *Parenting styles,* by contrast, refer to a set of childrearing attitudes or orientations, which are partially expressed through parenting practices (Darling and Steinberg, 1993). For several decades scholars have sought to identify parenting styles that each individual parent displays. These styles have been found to consistently relate to various outcomes of children and adolescents. Scholars also have studied connections between parenting styles and children's outcomes cross-culturally. Most research concurs that one particular parenting style, *authoritative* parenting, is consistently associated with positive child outcomes such as higher levels of academic performance, social competency, self-confidence, mental health, and responsible behaviors. Authoritative parenting is characterized by parental warmth, acceptance, having high standards for one's children, age-appropriate maturity demands, and two-way communication. These positive outcomes are believed to benefit children and adolescents throughout their life course.

A BRIEF HISTORY OF RESEARCH ON PARENTING STYLES

A number of research programs pertinent to parenting style were established in the 1930s and early 1940s by several prominent European intellectuals who came to the United States as refugees during World War II (1939–1945) (Maccoby, 2007). Kurt Lewin (1890–1947), a European field theorist who believed in the contextual influences of social behavior, worked with American colleagues on group atmospheres. They studied club groups made up of boys ages 10 to 11 who were participating in recreational activities. By placing adult leaders who displayed three different leadership styles into the groups, Lewin and his colleagues created three kinds of group atmospheres: authoritarian, democratic, and laissez-faire. An authoritarian leader would decide on the group activities, assign roles and partners to the boys, give praise and criticism without explaining the policy, and remain emotionally distant. A democratic leader would let the group members discuss and decide on the activities, roles, and partners themselves, yet would also provide guidance, suggest alternative options, give feedback, and provide explanations. A laissez-faire group leader would be friendly but withdraw to the sideline and let the boys run everything, providing information if asked but not giving any guidance, advice, suggestion, or feedback. Compared to the authoritarian and laissez-faire groups, the boys in the democratic group were observed as more engaged with the activities, more successful in getting the job done, and continuing the work without supervision (Lewin, Lippitt, & White, 1939).

Lewin's work on group atmospheres greatly influenced Alfred Baldwin (b. 1914), a graduate student at Harvard in the 1930s, who later applied Lewin's paradigm to family settings—since a family can be considered a small group, and thus the style of a parent may affect how children function. Baldwin identified several different parenting styles. He found that "warm democracy," which resembled Lewin's democratic leadership and was characterized by parental affection and empathy, was associated with the most successful child outcomes, including the development of intellectual competence and low levels of anxiety in children (Baldwin, 1955). Another European scholar, Else Frenkel-Brunswik (1908–1958), with her psychoanalytical approach, worked with American colleagues and identified a harsh and threatening parental upbringing as a critical influence on the development of a personality that was prejudiced and intolerant (Adorno, Frenkel-Brunswik, Levinson, & Sanford, 1950).

While these early studies documented the benefits of parental warmth and leniency as well as the detrimental effect of controlling and harshness, developmental psychologist Diana Baumrind (b. 1927) strongly believed that a successful parenting style required both warmth and firm control. Based on these two critical components of warmth and control, Baumrind proposed a three-category typology of parenting style: authoritarian, authoritative, and permissive. The authoritative parenting style was believed to be the most optimal one, which encompassed a high level of warmth, and a high level of firmness; the latter criterion was not an aspect of Baldwin's "warm democratic" parenting (Maccoby, 2007). By contrast, the authoritarian style was distinguished by high levels of firmness and low levels of warmth, and the permissive parent typically showed low levels of firmness.

By the early 1980s Baumrind's typology was solidly established in the field of child development. However, some critics suggested Baumrind's typology was incomplete, in that the three parenting styles were limited to relatively well-functioning families. Because earlier research on parenting had revealed two core clusters or dimensions (Becker, 1964), responsiveness and demandingness, Eleanor Maccoby and John Martin (1983) broadened the scope of Baumrind's threefold typology into four by differentiating the permissive parenting into two distinctive styles: indulgent parenting and indifferent or uninvolved parenting. This fourfold scheme better captures all parenting styles including those in maladapted families, so it is more generalizable to a larger population than was the original three-category scheme (Darling & Steinberg, 1993).

A CLOSER LOOK AT THE FOURFOLD PARENTING STYLE TYPOLOGY

The well-established fourfold typology of parenting styles is characterized by high or low levels of two core dimensions: (a) responsiveness, which refers to warmth, acceptance, support, and sensitivity; and (b) demandingness, which entails consistent discipline, age-appropriate maturity demands, and control. As noted earlier, *authoritative* parenting is characterized by high levels of responsiveness and demandingness. Authoritative parents guide children's activities in a rational manner. They set high standards and will apply firm control in situations where parental guidance is needed. They also recognize children's individual rights, explain the reasoning behind their policies and disciplinary practices, and encourage verbal give-and-take communication. Their disciplinary actions are supportive rather than punitive. Authoritarian parents recognize that they are not invincible and flawless, yet they also do not base their decision-making on their child's desires only. They aim to raise children who are assertive, self-regulated, responsible, cooperative, and considerate.

Authoritarian parenting, by contrast, is high on demandingness but low on responsiveness. Authoritarian parents demand high levels of compliance, discourage verbal give-and-take, and use disciplinary measures that are forceful and punitive. They emphasize obedience, work, respect for authority, and the importance of maintaining order and traditional structures. *Indulgent* parenting is high on responsiveness but low on demandingness. Indulgent parents do not set high standards nor demand conformity from their children. They tolerate their children's impulse to a greater extent and do not require mature and responsible behavior. They offer themselves as a resource to be used, with disciplinary actions that are relaxed and lenient. They grant their children extensive autonomy and tend to avoid asserting control to stay away from confrontations. *Rejective or neglectful* parenting is neither demanding nor responsive. Rejective or neglectful parents do not set standards and structures for their children to follow and do not monitor their children. They are not supportive and often reject their children. Frequently disengaged and withdrawn, they neglect their own responsibilities as parents as well as the development of their children.

Nancy Darling and Laurence Steinberg (1993) propose that parenting style should be regarded as a relationship's emotional climate, which moderates (that is, either amplifies or reduces) the effect of specific parenting practices such as monitoring and involvement. They argue that parenting styles create a context that can be partially expressed through specific parenting practices that convey parental attitudes toward the child. Thus parenting styles will first affect children's willingness to be socialized, and thus they affect how effective a particular parenting practice is.

For example, the ways that a specific parental practice, such as involvement in children's school performance, affects children's behavior may vary based on whether the parent uses an authoritarian, authoritative, or permissive style. An authoritative parent might mold their child's school performance by using explanations, respecting children's perspectives, encouraging discussions, and two-way communication. Consequently parental involvement under such a climate may make children more receptive to parental guidance and, ultimately, make parental involvement more effective. Conversely, an authoritarian parent may make parental involvement an imposed practice, which is likely to lead to children's resistance and thus make their involvement noneffective. Similarly, monitoring conducted with a warm and responsive style is likely to convey parental caring and concern and thus make adolescents more likely to cooperate. Supervision with a cold and unsympathetic manner may arouse adolescents' resentment, which may lead them to look for ways to get away from supervision.

IMPLICATIONS OF PARENTING STYLE FOR CHILD AND ADOLESCENT DEVELOPMENT

Many studies have examined Baumrind's typology among children aged between 3 and 12 and reported that children from authoritative families are more confident, assertive, competent, considerate, cooperative, and mature (Maccoby & Martin, 1983). Parenting style research has been extended to the period of adolescence, with the consistent finding that authoritative parenting is the most successful parenting style, whereas rejective or neglectful parenting is the least effective. Authoritative parenting is consistently related to adolescents' superior academic achievement, higher confidence and self-reliance, lower psychological distress, and less misconduct. Even a study based on a sample of economically disadvantaged, ethnic minority juvenile offenders revealed the beneficial

effects of authoritative parenting (Steinberg, Blatt-Eisengart, & Cauffman, 2006). Adolescents with rejective or neglectful parents tend to struggle academically, have lower confidence, are more vulnerable to distress, and are more likely to get into trouble (Lamborn, Mounts, Steinberg, & Dornbusch, 1991).

Authoritarian parenting and indulgent parenting have received less consistent results regarding their associations with various outcomes. Adolescents with authoritarian parents have lower levels of drug use or problem behaviors. However, they are likely to report lower self-confidence. Adolescents from indulgent families tend to score high in social competence and self-reliance, but do less well in school, and are more likely to get involved in drugs, alcohol, and problem behaviors (Lamborn et al., 1991). These patterns reflect the importance of both parental warmth and acceptance on the one hand, and parental authority and control on the other. Without parental warmth and acceptance, it is difficult for children to develop self-confidence. Without firm parental control, children are less likely to obey authority under necessary circumstances and are more likely to drift into activities valued by peers rather than adults.

One question many people ask is, "What will happen if two parents display different parenting styles?" Traditionally, childrearing tasks are performed by mothers more than fathers. Most parenting research throughout history is based on the assumption that mothers matter more than fathers for childrearing. Many researchers focused exclusively on mothers' parenting styles only, assuming that fathers concur with mothers in this arena. Other researchers did not share this assumption and collected data on both parents' styles, but they took the average of both mothers' and fathers' scores in the analysis or left out from the analysis those families where parents displayed dissimilar parenting styles. To remedy this limitation in prior studies, several researchers have analyzed parenting style scores for mothers and fathers separately and examined interparental inconsistencies in parenting styles. As expected, having two authoritative parents is associated with the best child outcomes. Having one authoritative parent and one nonauthoritative parent is better for children's development than having two nonauthoritative parents. These findings suggest that having one authoritative parent is a resource that may help to buffer against the potentially negative effects of having a nonauthoritative parent. Consistency between two parents is beneficial for children only if both parents are authoritative (Fletcher, Steinberg, & Sellers, 1999).

While authoritativeness, characterized by the two dimensions of responsiveness and demandingness, is predictive of many positive outcomes for children and adolescents, some scholars consider the dimension of demandingness, which includes levels of control, deserves further analysis, especially for adolescents. Brian Barber and his colleagues (1994) propose that the concept of parental control should be disaggregated into two types of control: behavioral control and psychological control. Analyses show that the former is related to fewer externalizing problems such as using drugs or alcohol, whereas the latter is associated with more internalizing problems such as loneliness or depressive symptoms in children.

A study by Marjory Gray and Steinberg (1999) has "unpacked" authoritativeness into three components that represent responsiveness, behavioral control, and psychological control. Behavioral control involves strictness such as child monitoring and limit setting, whereas psychological control entails the extent of psychological autonomy granting and encouragement of adolescent individuality in the family. As expected, the component of responsiveness is consistently related to positive outcomes such as higher academic achievement, healthier identity development, and lower rates of psychological distress and misconduct. The two types of control, behavioral and psychological, have predicted different outcomes. A high level of behavioral control is associated with fewer behavioral problems, whereas a high level of psychological control (and thus a lower level of autonomy granting) is associated with poor emotional health. A combination of high responsiveness with high behavioral control and low psychological control from parents is associated with the best psychosocial development.

CROSS-CULTURAL EXPLORATIONS OF PARENTING STYLES

Although the benefits of authoritative parenting are clear-cut among non-Hispanic White students, researchers have reported inconsistent patterns in samples of Asian, African-American, and Hispanic-American students (Dornbusch, Ritter, Liederman, Roberts & Fraleigh, 1987). Ruth Chao (1994) argues that the Baumrind typology of parenting styles is entirely based on Euro-American cultural ideals that highly value individual freedom. Baumrind's conceptualization of authoritativeness entails encouragement for children's independence and self-expression. These values do not have the same relevance in Asian societies, which have a traditional heritage of Confucianism that emphasizes hierarchies in social structure and children's obligations to respect and obey parents. Strict control and a lack of two-way communication, which are considered potentially harmful to children in Western cultures, may be perceived very differently in Asian cultures, where successful parenting corresponds to high levels of parental control and government involvement in children's lives and high expectations for children's success.

However, some cross-national studies on parenting styles have reported authoritative parenting as the most optimal and authoritarian parenting as being harmful, similar to what has been observed in the U.S. individualistic society (Sorkhabi, 2005). The results from cross-cultural studies are still inconclusive. More studies are needed before generalizing the parenting style findings to all cultures.

CONCEPTUAL AND METHODOLOGICAL CHALLENGES

Standard measures of parenting style are based on the assumption that styles are stable, rather than dynamic, over the life course of both the parent and child(ren). Typologies of parenting style suggest that each individual parent may recurrently display a certain type of parenting style, and that parenting style is relatively stable over time. Some researchers question this assumption of stability. Joan Grusec and Jacqueline Goodnow (1994) propose that parents may not necessarily use one consistent style across time and context, or for all of their children. Rather, parenting styles may differ depending on the child's age, emotional state at the moment, the nature of the misbehavior, and the social context. For example, responsiveness rather than demandingness is more appropriate if a child is upset. Reasoning is used if a child is not considerate of others or does not follow social conventions. Parental power assertion is necessary if a child is defiant to the legitimate parental authority.

FUTURE DIRECTIONS

While parenting research has achieved a great deal in identifying parenting styles and their associated outcomes, many scholars argue that the top-down approach of focusing on parental influence on children is incomplete, and that scholars should also explore horizontal processes that involve mutual influences between parents and children (Kuczynski, 2003). Parents who have raised more than one child must have noticed how children in the same family differ in their temperament, personality, and ability, and how the same parenting strategies may not have the same effect on different siblings within the family. A bidirectional approach to studying parent-child relationship suggests that children are capable of initiating intentional behaviors, reflecting over interactions, interpreting situations, resisting requests and influence, and blocking parental goals.

Certainly, both parental responsiveness and demandingness are powerful influences on child outcomes. However, a parent's behavior is shaped by (and in turn shapes) the child's behavior. While the effect of parents as a critical socialization force is still considered highly important, scholars call for increased attention to children as an important instrumental force that influences parenting. More studies are needed to sort out how parents adjust their parenting styles toward different children, at different developmental stages, and in different social and emotional situations. Moderating effects on parenting styles also need to be further explored; scholars are interested in exploring whether the consequences of parenting style vary based on factors such as the family's socioeconomic status, family structure (e.g., two-parent households versus single-parent households and stepfamilies), and ethnic or cultural influences.

The utility and benefits of authoritative parenting has been consistently demonstrated over the decades. Parents need to be warm and firm, and adjust their parenting to the developmental changes of their children. As to the question of what is the best parenting style, evidence persuasively demonstrates the advantages of authoritative parenting style over any nonauthoritative style. More important future endeavor should be devoted to the practical application of parenting styles. Parenting classes should be offered to parents, foster parents, and primary caregivers of children of all ages to pass along the basic information on parenting styles, so as to educate adults on how to be authoritative in order to facilitate the healthy growth of children.

SEE ALSO Volume 1: *Attachment Theory; Child Abuse; Parental Involvement in Education; Parent-Child Relationships, Childhood and Adolescence.*

BIBLIOGRAPHY

Adorno, T.W., Frenkel-Brunswik, E., Levinson, D.J., & Sanford, R.N. (1950). *The authoritarian personality.* New York: Harper.

Baldwin, A.L. (1955). *Behavior and development in childhood.* New York: Dryden Press.

Barber, B.K., Olsen, J.E., & Shagle, S.C. (1994). Associations between parental psychological and behavioral control and youth internalized and externalized behaviors. *Child Development, 65*(4), 1120–1136.

Baumrind, D. (1968). Authoritarian versus authoritative parental control. *Adolescence, 3*(11), 255–272.

Baumrind, D. (1991). The influence of parenting style on adolescent competence and substance use. *Journal of Early Adolescence, 11*(1), 56–95.

Becker, W.C. (1964). Consequences of different kinds of parental discipline. In M.L. Hoffman & L.W. Hoffman (Eds.), *Review of child development research* (Vol. 1, pp. 169–208). New York: Russell Sage Foundation.

Chao, R.K. (1994). Beyond parental control and authoritarian parenting style: Understanding Chinese parenting through the cultural notion of training. *Child Development, 65*(4), 1111–1119.

Darling, N., & Steinberg, L. (1993). Parenting style as context: An integrative model. *Psychological Bulletin, 113*(3), 487–496.

Dornbusch, S.M., Ritter, P.L., Leiderman, P.H., Roberts, D.F., & Fraleigh, M.J. (1987). The relation of parenting style to adolescent school performance. *Child Development, 58*(5), 1244–1257.

Fletcher, A.C., Steinberg, L., & Sellers, E.B. (1999). Adolescents' well-being as a function of perceived interparental inconsistency. *Journal of Marriage and the Family, 61*(3), 599–610.

Gray, M.R., & Steinberg, L. (1999). Unpacking authoritative parenting: Reassessing a multidimensional construct. *Journal of Marriage and the Family, 61*(3), 574–587.

Grusec, J.E., & Goodnow, J.J. (1994). Impact of parental discipline methods on the child's internalization of values: A reconceptualization of current points of view. *Developmental Psychology, 30*, 4–19.

Kuczynski, L. (Ed.). (2003). Beyond bidirectionality: Bilateral conceptual frameworks for understanding dynamics in parent-child relations. In *Handbook of dynamics in parent-child relations* (pp. 3–24). Thousand Oaks: Sage.

Lamborn, S.D., Mounts, N.S., Steinberg, L., & Dornbusch, D.M. (1991). Patterns of competence and adjustment among adolescents from authoritative, authoritarian, indulgent, and neglectful families. *Child Development, 62*(5), 1049–1065.

Lewin, K., Lippitt, R., & White, R.K. (1939). Patterns of aggressive behavior in experimentally created "social climates." *Journal of Social Psychology, 10*, 271–299.

Maccoby, E.E. (2003). Epilogue: Dynamics viewpoints on parent-child relations: Their implications for socialization processes. In L. Kuczynski (Ed.), *Handbook of dynamics in parent-child relations* (pp. 439–452). Thousand Oaks: Sage.

Maccoby, E.E. (2007). Historical overview of socialization research and theory. In J.E. Grusec & P.D. Hastings (Eds.), *Handbook of socialization: Theory and research* (pp. 13–41). New York: Guilford Press.

Maccoby, E.E., & Martin, J.A. (1983). Socialization in the context of the family: Parent-child interaction. In P.H. Mussen (Ed.), *Handbook of child psychology* (Vol. 4, pp. 1–101). New York: Wiley.

Sorkhabi, N. (2005). Applicability of Baumrind's parent typology to collective cultures: Analysis of cultural explanations of parent socialization effects. *International Journal of Behavioral Development, 29*(6), 552–563.

Steinberg, L. (2001). We know some things: Parent-adolescent relationships in retrospect and prospect. *Journal of Research on Adolescence, 11*, 1–19.

Steinberg, L., Blatt-Eisengart, I., & Cauffman, E. (2006). Patterns of competence and adjustment among adolescents from authoritative, authoritarian, indulgent, and neglectful homes: A replication in a sample of serious juvenile offenders. *Journal of Research on Adolescence, 16*(1), 47–58.

Zeng-yin Chen

PEER GROUPS AND CROWDS

One of the most striking features of children's social development is the evolving character of peer groups.

Elementary school-age children develop small "interaction-based" peer groups or "cliques." In adolescence, however, larger peer groups or "crowds" emerge. Leading peer group researcher B. Bradford Brown has defined crowds as relatively large "reputation-based" collectives of individuals who share a common image but who do not necessarily spend much time together. An adolescent's crowd affiliation provides a social identity and indicates the activities and attitudes with which he or she is associated in the perceptions of peers. Examples of crowds common among American secondary school students are nerds, brains, druggies, and populars.

Though differences in the character of prototypic identities exist across schools, peer crowd membership is typically associated with a number of important behaviors and beliefs, such as substance use and achievement orientations (Kindermann, 2007; La Greca, Prinstein, & Fetter, 2001; Sussman, Pokhrel, Ashmore, & Brown, 2007) as well as peer relations (Eder, 1985). Longitudinal research (Barber, Eccles, & Stone, 2001) indicates that one's high school social identity category even predicts psychological adjustment and educational outcomes through young adulthood.

PEER GROUPS/CROWDS AND DEVELOPMENT

Peer groups become more prominent as a cultural and interpersonal force during adolescence because of changes within individuals and in the social environment. Cognitive developmental shifts associated with adolescence, which have been implicated in the changing character of "person perceptions" and other components of social information processing, probably have a bearing on the emergence of crowds and changes in crowd conceptions across adolescence. Developmental gains in abstract reasoning capacities and hypothetical thinking are especially relevant for understanding adolescent social cognition, but it is important to realize that some negative results can accrue from the same structural changes that allow for more advanced forms of thought. Youth may be more enthusiastic than accurate in hypothesizing about crowds. They may engage in overgeneralization and flagrant exaggeration, especially in early adolescence.

The social environment of the secondary school is more complicated and challenging than that of the primary school. Students are typically faced with a much larger school environment populated by a bewildering number of strangers. Brown, Mory, and Kinney (1994) have suggested that social cognitive theory and research on adult social identity could offer insight into how crowds might serve individuals in coping with these challenges. Social psychologists assert that social categories serve as perceptual organizers and also help individuals anticipate interactions with others based on previous

category knowledge. Social risk can be avoided if individuals can anticipate friendly or unfriendly reception by others. Unfortunately, however, social categorization typically involves stereotyping and in-group favoritism, characterized by a perceptual exaggeration of similarities within categories and differences between categories.

CROWDS AS CARICATURES, CHANNELS, AND CONTEXTS

Brown et al. (1994) suggested that crowds serve as *caricatures*, *channels*, and *contexts* for adolescents. They provide identity caricatures, or prototypic identities recognized by peers, through which adolescents may themselves be known (Stone & Brown, 1999). This knowing may be superficial, but it seems that adolescents prefer to be understood partially to being invisible or mistaken for someone with values the individual rejects. Moreover, theorists (Newman & Newman, 1976) have speculated that adolescents may prefer a "safe" group identity to the commitment of a deeper personal identity in early adolescence as they cope with uncertainties engendered by multiple social and physical changes, such as the demands of dating and romance.

As channels, crowds regulate relationships by serving as perceptual anchorages in one's mental map of the social system, helping adolescents recognize potential friend or foe by signaling who is likely to share values and who is likely to rebuff a friendly text message. It is clear why adolescents may sense the need for an efficient cognitive map (Brown et al., 1994) to guide their movement through their social world.

Finally, crowds serve as contexts for relationships and identity development. Many crowds are identified with leisure activity contexts such as student council, sports, or even risk-taking pursuits. Research centering on extracurricular activities (Barber, Stone, Hunt, & Eccles, 2005; Fine, 1987) suggests that social identities are part of a larger contextual system—including personality and personal identity, activity choices, the peer context existing in each activity, and the cultures of particular activities. Although this dynamic system applies most explicitly to those identities associated with organized community and school activities, one can easily extrapolate to those identities that are not centered on institutionally organized activities. Thus, crowd niches (perhaps even categories based on exclusion from such prominent and accepted niches, such as "nobodies" or "outsiders") can form adolescent cultures in which leisure activities, values, and styles are consensually validated (Youniss & Smollar, 1985).

CROWD MAPPING

Researchers as well as youth have often depicted crowds as mapped in terms of their relationships to each other or to some dimension—for instance, a status hierarchy. Brown and colleagues (Brown et al., 1994; Stone & Brown, 1998) suggested a two-dimensional arrangement of crowds in conceptual social space, based on theoretical notions of formal and informal reward structures in high schools developed by sociologists Rigsby and McDill (1975). The formal dimension was seen to represent school engagement or adherence to the adult reward structure, while the informal dimension was seen to represent peer status. Some of their empirical work has employed multidimensional scaling, a statistical technique in which data regarding respondents' perceptions of the difference or similarity between crowds is analyzed and yields a display of the crowds as points on a map (Brown et al., 1994; Stone & Brown, 1998). In the maps yielded by these analyses, crowds were distributed across the map's quadrants in several "crowd clusters." Jocks and populars were perceived as similar, as were rebels and Blacks. Normals and brains were also perceived as similar, but the wannabe Blacks were remote from all other crowds.

In a 2008 study by Stone, Barber, and Eccles, discriminant function analysis of the reported characteristics of youth who would in coming years identify with particular social identities yielded mappings similar to those derived through multidimensional scaling. In discriminant function analysis, data about group differences on the set of variables hypothesized as predictors are analyzed to indicate which groups differ most and which predictors best explain differences. In these analyses, the most prominent dimensions of difference between social identity groups seem to have represented a blurring of the formal and informal domains of the reward structure. Figure 1 provides an example of such a crowd map derived through discriminant function analysis.

DETERMINANTS OF PEER GROUPS/ CROWDS

The effect of the wider social structure on the adolescent peer group structure has been a frequent theme of research on peer groups (Eckert, 1989). In such accounts, crowd identities are seen to reflect the socioeconomic or ethnic composition of the community, with students being assigned to their crowd on the basis of parental position. Future adult socioeconomic class identities are then seen to be determined through dynamic processes of alignment and polarization operating in the high school social system.

Quantitative research has documented only a modest correlation between parental social class and crowd affiliation. Most theorists and researchers acknowledge, however, that socioeconomic class affects peer relationships at least indirectly through residential patterns, neighborhood schools, and youth activity participation, and that the significance of crowd-based social identities does in part lie in their delineation of varying adolescent orien-

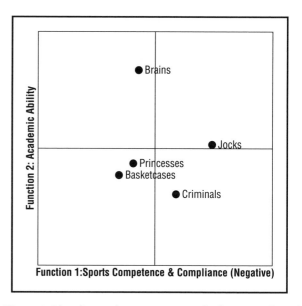

Figure 1. *Plot of mean discriminant scores for future members of each peer group for the first (X–Axis) and second (Y–Axis) functions for 616 participants in the Michigan Study of Life Transitions.* CENGAGE LEARNING, GALE.

tations toward the values of normative adult society (Brown et al., 1994; McLellan & Youniss, 1999). For the most part, however, current research locates the meaning of peer group processes primarily in the social world of adolescents themselves.

THE CHOICE OF A PARTICULAR CROWD NICHE

Social scientists have pondered the question of what future "jocks" or "burnouts" are doing in the years preceding high school. The separation of research traditions regarding children and adolescents delayed quantitative efforts to determine what characteristics of preadolescent children are predictive for the social identities they will develop during adolescence. However, ethnographic studies of preadolescent social development provided relevant evidence (Adler & Adler, 1998; Fine, 1987), especially regarding popular identities, describing such factors as personality, interests, activities, expressive characteristics specifically located in the peer culture, and differences between factors associated with male and female social identities.

Stone and Brown (1998) invited adolescents themselves to describe the differences between crowds. They reported that though there is some contention about the characteristics of crowds, based on the crowd affiliation of the perceiver, six domains were commonly used to describe crowds: dress and grooming styles, sociability or antagonism toward outsiders, academic attitude, the crowd's hangout at school, typical weekend activities, and participation in extracurricular activities.

Findings from one longitudinal study mentioned previously (Stone et al., 2008) suggest that differences between individuals associated with particular high school social identity categories predate adolescence. Discriminant function analysis assessed the predictive relevance of nine characteristics measured in sixth grade for differentiating among social identities claimed 4 years later by more than 600 participants in the Michigan Study of Life Transitions. For females, academic motivation and sports competence accounted for more than 80% of the variability between groups. For males, academic ability, self-concept of appearance, and sports competence together accounted for more than 80% of between-group variability. Over half of the students could be correctly classified into their future crowd based on earlier characteristics.

COMPETITION IN THE SEARCH FOR A CONGENIAL PEER GROUP IDENTITY

According to Newman and Newman's (1976) account, the adolescent who does not find a peer group niche feels alienated from peers. Nevertheless, mixed findings regarding the self-esteem benefits of crowd membership have suggested that it may not be crowd membership in itself that gives a sense of having a congenial place in the world. Brown and Lohr (1987) found that high-status crowds were valuable in that regard, but others were not.

Crowds are typically arranged in a status hierarchy, and young people may prefer to identify with crowds toward the top of the hierarchy (Eder, 1985). Thus, in addition to its role in identity formation, crowd identification can be an exercise in impression management. Though adolescents may attempt to convince their peers that they belong to a particular crowd, peers may either accept or reject this effort at impression management (Brown & Lohr, 1987; Eder, 1985; Merten, 1997). Indeed, teenage social systems often include crowds called the "wannabes," populated by individuals who strive unsuccessfully to be accepted into one crowd or another; their efforts at impression management are acknowledged by peers but not accepted. One area of related peer research explores aspects of aggression in the search for a positive peer group niche (see Cillessen & Rose, 2005, for a review of this literature).

MEASURING PEER GROUP MEMBERSHIP

This conflictual dynamic renders identity negotiation problematic for many youth. It has also engendered controversy about how best to measure peer group membership (Sussman et al., 2007; Urberg, Degirmencioglu, Tolson, & Halliday-Scher, 2000). In keeping with the reputational basis of crowds and to avoid social desirability bias, Brown (Brown et al., 1994) prefers to ask selected knowledgeable youth to identify the crowds that exist in their schools and to assign their peers to one of the crowds thus identified. This "social type rating" method as well as "sociocognitive mapping" (Kindermann, 2007) have provided interesting methodological alternatives but are somewhat difficult to enact in school settings. Other researchers have asked youth to indicate the crowd with which they most identify (Barber et al., 2001; Stone et al., 2008). Despite findings of considerable correlation between self-reported and peer-reported crowd membership (Urberg et al., 2000), it is generally acknowledged that both methods are subject to bias and that multiple measures are beneficial.

Research has documented fluidity in crowd orientation. Strouse (1999) found, in her analysis of a large longitudinal data set, that more than half of participants changed their self-perceived crowd affiliation over a 2-year period. Kinney's ethnographic research (1999) revealed that new crowd cultures may be forged when individuals decide that they no longer are comfortable with a former social identity category and make behavioral changes to signal new interests and values to their peers. The salience of crowd affiliation and of conformity pressures within peer groups has also been shown to decline in later adolescence.

FUTURE DIRECTIONS

Certainly the meaning and function of adolescent peer groups have not been fully explained. Differences in peer group functioning for youth beyond North America and for males and females are two areas much in need of further development. Another important issue for future research to explore is multiple peer group membership. Though researchers acknowledge that students may consider themselves to have multiple memberships, it has been difficult to accommodate this notion in research.

Adults (and adolescents themselves) might be reluctant to emphasize the category labels with which peers are organized in perception, but they are as powerful as adult labels such as "old people" and "academic types" in constraining identities and relations. A realistic understanding of adolescent peer groups could offer adults opportunities to help students examine and question stereotypes and anxieties that impede the formation of friendships across the borders of diverse groups. It will be important for those who promote positive youth development to understand more about how institutional practices such as ability grouping and extracurricular activity promotion or neglect might impede the enhancement of development in peer group contexts. Given the importance of peers in the lives of adolescence, understanding peer groups is clearly essential.

SEE ALSO Volume 1: *Dating and Romantic Relationships, Childhood and Adolescence; Friendship, Childhood and Adolescence; Interpretive Theory; School Culture; Social Capital; Social Development; Youth Culture.*

BIBLIOGRAPHY

Adler, P. A., & Adler, P. (1998). *Peer power: Preadolescent culture and identity.* New Brunswick, NJ: Rutgers University Press.

Barber, B. L., Eccles, J. S., & Stone, M. R. (2001). Whatever happened to the Jock, the Brain, and the Princess? Young adult pathways linked to adolescent activity involvement and social identity. *Journal of Adolescent Research, 16,* 429–455.

Barber, B. L., Stone, M. R., Hunt, J. E., & Eccles, J. S. (2005). Benefits of activity participation: The roles of identity affirmation and peer group norm sharing. In J. L. Mahoney, R. W. Larson, & J. S. Eccles (Eds.), *Organized activities as contexts of development: Extracurricular activities, after-school, and community programs* (pp. 185–210). Mahwah, NJ: Erlbaum.

Brown, B. B., & Lohr, M. J. (1987). Peer group affiliation and adolescent self-esteem: An integration of ego identity and symbolic interaction symbolic-interaction theories. *Journal of Personality and Social Psychology, 52,* 47–55.

Brown, B. B., Mory [Stone], M. S., & Kinney, D. (1994). Casting adolescent crowds in a relational perspective: Caricature, channel, and context. In R. Montemayor, G. R. Adams, & T. P. Gullotta (Eds.), *Advances in adolescent development* (Vol. 6). *Personal relationships during adolescence* (pp. 123–167). Newbury Park, CA: Sage.

Cillessen, A. H. N., & Rose, A. J. (2005). Understanding popularity in the peer system. *Current Directions in Psychological Science, 14,* 102–105.

Eckert, P. (1989). *Jocks and burnouts: Social categories and identity in the high school.* New York: Teachers College Press.

Eder, D. (1985). The cycle of popularity: Interpersonal relations among female adolescents. *Sociology of Education, 58,* 154–165.

Fine, G. A. (1987). *With the boys: Little League baseball and preadolescent culture.* Chicago: University of Chicago Press.

Kindermann, T. A. (2007). Effects of naturally existing peer groups on changes in academic engagement in a cohort of sixth graders. *Child Development, 78,* 1186–1203.

Kinney, D. A. (1999). From "headbangers" to "hippies": Delineating adolescents' active attempts to form an alternative peer culture. *New Directions for Child and Adolescent Development, 84,* 21–35.

La Greca, A. M., Prinstein, M. J., & Fetter, M. D. (2001). Adolescent peer crowd affiliation: Linkages with health-risk behaviors and close friendships. *Journal of Pediatric Psychology, 26,* 131–143.

McLellan, J. A., & Youniss, J. (1999). A representational system for peer crowds. In I. E. Sigel (Ed.), *Development of mental representation: Theories and applications* (pp. 437–449). Mahwah, NJ: Erlbaum.

Merten, D. E. (1997). The meaning of meanness: Popularity, competition, and conflict among junior high school girls. *Sociology of Education, 70,* 175–191.

Newman, P. R., & Newman, B. M. (1976). Early adolescence and its conflict: Group identity versus alienation. *Adolescence, 11,* 261–274.

Rigsby, L. C., & McDill, E. L. (1975). Value orientations of high school students. In H. R. Stub (Ed.), *The sociology of education: A sourcebook* (3rd ed., pp. 53–75). Homewood, IL: Dorsey Press.

Stone, M. R., Barber, B. L., & Eccles, J. S. (2008). We knew them when: Sixth grade characteristics that predict adolescent high school social identities. *The Journal of Early Adolescence, 28,* 304–328.

Stone, M. R., & Brown, B. B. (1998). In the eye of the beholder: Adolescents' perceptions of peer crowd stereotypes. In R. E. Muuss & H. D. Porton (Eds.), *Adolescent behavior and society: A book of readings* (5th ed., pp. 158–169). Boston: McGraw-Hill.

Strouse, D. L. (1999). Adolescent crowd orientations: A social and temporal analysis. *New Directions for Child and Adolescent Development, 84,* 37–54.

Sussman, S., Pokhrel, P., Ashmore, R. D., & Brown, B. B. (2007). Adolescent peer group identification and characteristics: A review of the literature. *Addictive Behaviors, 32,* 1602–1627.

Urberg, K. A., Degirmencioglu, S. M., Tolson, J. M., & Halliday-Scher, K. (2000). Adolescent social crowds: Measurement and relationship to friendships. *Journal of Adolescent Research, 15,* 427–445.

Youniss, J., & Smollar, J. (1985). *Adolescent relations with mothers, fathers, and friends.* Chicago: University of Chicago Press.

Margaret R. Stone

PEER VICTIMIZATION

SEE Volume 1: *Bullying and Peer Victimization.*

PERSON-ORIENTED APPROACHES

At the core of the person-oriented approach is a focus on the person as an organized totality. The person "as a whole" is regarded as the main conceptual unit and is often also the main analytical unit in the statistical analyses. At a theoretical level this implies that key components involved in the theoretical formulation of the system under study are regarded as operating together and influencing each other so that the main expression of the system is in how these components all together emerge and develop. They are then regarded as indivisible, and the configuration of them within an individual is fundamental and cannot be understood by treating the components as isolated entities.

In methodological realizations of the person-oriented approach, the individual's patterns of values in the variables measuring the components are treated as the main analytical units, and typical patterns at a given time point and across the development of the individuals are observed. To give a simple example: In a Swedish study of boys' adjustment, six components were identified as together depicting their adjustment status. The whole six-variable pattern of values for a child was regarded as indivisible and was searched for distinct typical patterns using cluster analysis. Seven clusters (= typical patterns) were found, each with its distinct profile, and each boy belonged to one of these typical patterns. The person-oriented approach can be compared to the standard variable-oriented approach, whereby the variable is regarded as the main conceptual and analytical unit. This approach is far more common but is not addressed in this entry.

The person-oriented approach need not be anchored in the mainstream quantitative research tradition and can instead be carried out by case-oriented research and by using a qualitative approach. However, most commonly the modern person-oriented approach is a quantitative approach, studying a sample of persons and directed at explaining interindividual differences in emerging individual patterns of components. This formulation of the person-oriented approach is the focus of this entry. A related tradition, called p-technique, was introduced by Raymond Cattell and focuses on the study of the single individual. This technique has been further developed by John Nesselroade and others but is not discussed here.

ROOTS OF THE PERSON-ORIENTED APPROACH

Early formulation of the person-oriented approach preceded the variable-oriented approach by thousands of years, in the form of the typological approach by which individuals were categorized into different types. It is seen in the classical categorization of individuals into the four basic temperaments: sanguine, phlegmatic, melancholic, and choleric, whereby several individual properties together defined a type. There the organization of these properties within the individual was regarded as the essential feature that describes the person, and it was claimed that this organization can take only one of a limited number of forms. This approach is a natural first step for the "primitive" scientist to take, considering the basic tendency of people cognition to categorize the world around them. In the life sciences, the typological approach is still strong, especially in the social sciences, as exemplified by taxonomy in biology and diagnosis in medicine (see Misiak & Sexton, 1966, for an overview).

The typological approach has to a certain extent fallen into disrepute in the social sciences. The criticism has mainly been directed toward three aspects of earlier formulations of the approach: the often implicit assumption that the typology was innate, the subjectivity usually involved in constructing a typology, and the inability of the typology to represent all the information about the studied subjects. However, none of these limitations, except perhaps the last one, apply to the modern typological research within the person-oriented approach. For a discussion of the promise and limitations of the typological approach, see Bergman and Magnusson (1997) and Waller and Meehl (1998).

The person-oriented approach has emerged in symbiosis with the new developmental science (Cairns, Elder, & Costello, 1996), where a new type of person-oriented research has emerged (Bergman & Magnusson, 1997). It is focused on the integrated study of individual development and is rooted in the holistic-interactionistic research paradigm, which emphasizes a systemic perspective implying continuous interactions and transactions (Magnusson, 1999).

THE MODERN PERSON-ORIENTED APPROACH

The modern person-oriented approach was first proposed by Magnusson and Allen (1983) and was summarized in the following way: "The person oriented approach to research (in contrast to the variable-centered approach) takes a holistic and dynamic view; the person is conceptualized as an integrated totality rather than as a summation of variables" (p. 372). Thus, individual functioning and development is seen as a dynamic, complex process involving both individual factors (behavioral, mental, biological) and environmental factors (social, cultural). The adaptive properties of the process are stressed, as well as the individual as an active agent in changing his or her environment. In these aspects, the person-oriented view shows similarities to Bronfenbrenner's (1979) multilevel individual-environment system view.

Four basic tenets of the person-oriented approach were presented by Bergman and Magnusson (1997):

1. The process of individual development is partly individual-specific, complex, and contains interacting factors at different levels.

2. A meaningful coherence and structure exists in individual growth and in differences in individuals' process characteristics. It is seen in the development and functioning of all subsystems and in the functioning of the organized totality.

3. Processes are lawful within structures organized and functioning as patterns of operating factors whereby

each factor derives its meaning from its relations to all the others.

4. Theoretically, there exists at a detailed level an infinite variety of differences in process characteristics and observed states, but at a global level there is usually a small number of dominating typical patterns that emerge ("common types").

Three comments are especially pertinent with regard to these tenets: (a) Biological systems tend to be characterized by self-organization, which is a process by which new structures "spontaneously" emerge, even from chaos. In individual development, the operating factors tend to organize in ways to maximize the functioning of each subsystem, and subsystems organize themselves to fulfill their function in the totality. They are organized to ensure survival, and that leads to perpetuation. This principle has been demonstrated in many human biological systems, such as the brain and the respiration system, supporting tenets (3) and (4) above. (b) The tenet concerning common types is also suggested by empirical findings of demonstrated distinct types, for instance often found in biological and ecologic systems (species, ecotypes). Within psychology, Block (1971) discussed "system designs" (p. 110) with more enduring properties and demonstrated longitudinal personality types; Gangestad and Snyder (1985) made a case for the existence of distinct types, each sharing a common source of influence. (c) For the researcher who takes the tenets of the person-oriented approach seriously, using standard variable-oriented methods would normally be seen as questionable, because these methods are not designed to handle patterns of information as undivided units of analysis and or to handle complex interactions and nonlinear relationships (e.g., many variable-oriented methods use the correlation matrix as the data to be analyzed, a matrix that reflects linear relations, not non-linear relations and interactions).

It should be noted that the person-oriented approach is foremost a theoretical perspective that serves as a guide and framework for carrying out empirical research. Of course, the methods most frequently used within this perspective are various forms of pattern analysis, such as cluster analysis, because in a person-oriented context such methods tend to match the theoretical propositions better than conventional variable-oriented methods.

TWO CLASSES OF METHODS COMMONLY USED IN PERSON-ORIENTED RESEARCH

As examples, two classes of methods are presented that are commonly used in person-oriented research (for an overview of different methods, see Bergman, Magnusson, & El-Khouri, 2003).

Classification Using Cluster Analysis or Latent Class Analysis The purpose of these methods is to divide a sample of individuals (or cities, school classes, and so on) into different groups (called "clusters" or "classes") so that each individual belongs only to one group, all individuals in a group are similar, and the different groups are dissimilar. For this purpose, different types of information can be used, but most commonly the information for each individual consists of a pattern or profile of values in a number of different variables. In *cluster analysis*, the classification is exploratory and empirically driven. Most often a dissimilarity matrix among all individuals is analyzed and some type of algorithm is used to cluster individuals so that the three criteria mentioned above are fulfilled as well as possible. This can be done differently; perhaps most commonly, a hierarchical method is used whereby, at the start, each individual is its own cluster, then the two most similar ones are merged, then of the remaining clusters the two most similar ones are merged, and so on until every individual is in the same cluster. Procedures exist for deciding on the most useful number of clusters to accept and for evaluating the quality of the cluster solution (for overviews, see Bergman, 1998; Gordon, 1981).

This method can be extended to a developmental setting, for instance in the following way. For a longitudinal sample with two measurement points, a cluster analysis is performed at each time point, and then the two cluster memberships of each individual are cross-tabulated (see the description of the LICUR method in Bergman et al., 2003). A more model-based approach can also be used. An example of this is latent structure analysis or *latent class analysis,* which was proposed by Paul Lazarsfeld. In this method it is assumed that a number of latent classes (essential, true groups) exist and that, within each latent class, no relationships exist between the variables because, except for errors of measurement, all individuals in a latent class are identical (the assumption of local independence). From this starting point, a classification is searched that comes as close as possible to fulfilling these conditions and the parameters of the model, and the fit to data can then be estimated (Goodman, 1974). This type of model can be extended to a developmental setting, as is done in latent transition analysis (Collins & Wugalter, 1992).

Analyzing All Value Patterns as a Goal in Itself Analyzing all value patterns as a goal in itself is done in *configural frequency analysis.* The variables described by the pattern must be discrete and are often dichotomized to make the number of value patterns to be examined manageable. Configural frequency analysis was originally suggested by Gustav Lienert and has been further developed, foremost, by von Eye (2002). The basic idea is to list all possible value patterns and analyze which occurs more frequently than expected by chance (types) or less frequently than expected by chance (antitypes). This simple idea has been developed to address the needs of a number of different research questions and designs, as well as developmental ones. A method used rather frequently in sociology—for instance in career research—is *sequence analysis* or optimal matching (see Abbott & Tsay, 2000). In this methodology, time sequences of categorized events are observed, and a procedure has been developed for assessing the dissimilarity between two individuals' sequences (e.g., between two persons' sequences of jobs each year between the ages 20 and 40). Optimal matching builds on a calculation of the costs involved in transforming one of the two sequences into the other. Often the results from optimal matching are subsequently subjected to a cluster analysis to find different sequence types.

COMMENTS ON THE USE OF THE PERSON-ORIENTED APPROACH

The following two research examples give some indication of the scope of the modern use of the typological approach: Bergman and colleagues (2003) studied stability and change in typical patterns of boys' adjustment problems. For instance, they found that positive and negative adjustment patterns were quite distinct in that no typical pattern occurred that was characterized by good adjustment in some components and bad adjustment in other components, and in a related study it was found that the boys characterized by a multiproblem syndrome often showed adjustment problems at adult age, but that those with just a single or a few adjustment problems did not have a bad prognosis. This could not be seen by standard variable-oriented analyses. Block (1971) studied personality development using longitudinal data and arrived at a typology of longitudinal personality types (he introduced the term "person approach"). These types referred to the "whole" organization of personality, and this knowledge could not have been obtained by an ordinary mainstream approach. Salmela-Aro and Nurmi (2004) studied employees' motivational orientation and well-being at work using a person-oriented approach. The findings indicated the coupling between these two systems, each reflected by its typical patterning of the involved factors, and provided information unattainable within, for instance, a linear model framework. For an overview of person-oriented research as contrasted to variable-oriented research, see Laursen and Hoff (2006).

Broadly speaking, the person-oriented approach is mainly a meta-theoretical research paradigm that has achieved rather wide acceptance. It undoubtedly paints a more complex picture of reality than that offered by the standard "box and arrow" paradigm, whereby variables tend to be seen as dependent or independent in a linear framework, often assuming that the same model holds for all studied subjects. Clearly, the usefulness of applying a person-oriented approach for a research problem depends on what kind of assumptions hold in the specific case. Sometimes the box and arrow types of assumptions are reasonable, or in some cases the needed information from the statistical analyses is straightforward and simple, such as a comparison of means or a correlation. Then there is no need to apply a person-oriented approach, neither theoretically nor by using, for instance, pattern-based methods of analysis. However, if the tenets of the person-oriented approach are fulfilled in the specific case, implying that dynamic interactions and whole-system properties must be taken seriously, then a person-oriented approach is natural, and this standpoint is now more frequently taken in empirical developmental research.

BIBLIOGRAPHY

Abbott, A., & Tsay, A. (2000). Sequence analysis and optimal matching methods in sociology: Review and prospects. *Sociological Methods & Research, 29*(3), 3–33.

Bergman, L. R. (1998). A pattern-oriented approach to studying individual development: Snapshots and processes. In R. B. Cairns, L. R. Bergman, & J. Kagan (Eds.), *Methods and models for studying the individual* (pp. 83–121). Thousand Oaks, CA: Sage.

Bergman, L. R., & Magnusson, D. (1997). A person-oriented approach in research on developmental psychopathology. *Development and Psychopathology, 9,* 291–319.

Bergman, L. R., Magnusson, D., & El-Khouri, B. M. (2003). *Studying individual development in an interindividual context.* Mahwah, NJ: Lawrence Erlbaum.

Block, J. (1971). *Lives through time.* Berkeley, CA: Bancroft Books.

Bronfenbrenner, U. (1979). *The ecology of human development: Experiments by nature and eesign.* Cambridge, MA: Harvard University Press.

Cairns, R. B., Elder, G. H. Jr., & Costello, E. J. (1996). *Developmental science.* Cambridge, U.K.: Cambridge University Press.

Collins, L. M., & Wugalter, S. E. (1992). Latent class models for stage-sequential dynamic latent variables. *Multivariate Behavioral Research, 27,* 131–157.

Gangestad, S., & Snyder, M. (1985). To carve nature at its joints: On the existence of discrete classes in personality. *Psychological Review, 92,* 317–349.

Goodman, L. A. (1974). Exploratory latent structure analysis using both identifiable and unidentifiable models, *Biometrika, 61,* 215–231.

Gordon, A. D. (1981). *Classification: Methods for the exploratory analysis of multivariate data.* London: Chapman & Hall.

Laursen, B., & Hoff, E. (Eds.). (2006). Person-centered and variable-centered approaches to longitudinal data. *Merrill–Palmer Quarterly, 52*(3), 377–644.

Magnusson, D. (1999). Holistic interactionism: A perspective for research on personality development. In L. A. Pervin & O. P. John (Eds.), *Handbook of personality* (2nd ed., pp. 219–247). New York: Guilford.

Magnusson, D., & Allen, V. L. (1983). Implications and applications of an interactional perspective for human development. In D. Magnusson & V. L. Allen (Eds.), *Human development: An interactional perspective* (pp. 369–387). New York: Academic Press.

Misiak, H., & Sexton, V. S. (1966). *History of psychology.* New York: Grune & Stratton.

Salmela-Aro, K., & Nurmi, J.-E. (2004). Employees' motivational orientation and well-being at work: A person-oriented approach. *Journal of Organizational Change Management, 17,* 471–489.

von Eye, A. (2002). *Configural frequency analysis: Methods, models, and applications.* Mahwah, NJ: Lawrence Erlbaum.

Waller, N. G., & Meehl, P. E. (1998). *Multivariate taxometric procedures: Distinguishing types from continua.* Thousand Oaks, CA: Sage.

Lars R. Bergman

PIAGET, JEAN
1896–1980

Jean Piaget was born in Neuchâtel, Switzerland, on August 9 and died in Geneva on September 17. He was a psychologist, epistemologist, logician, and social theorist and is considered the most famous child psychologist of the 20th century. Although many of his views have been questioned, his empirical work and theoretical conceptions changed the way psychologists and life course scholars think about children's psychological development.

CAREER AND EARLY PUBLICATIONS

Piaget's research program focused on the construction of a genetic (developmental) epistemology, a theory of the way knowledge develops (Kitchener, 1986, in press-a). That theory included child development and the history of science (Piaget, 1950).

Piaget's early interests centered on biology, which provided the backdrop for his subsequent theorizing. After earning a doctorate in science in 1918, he studied psychoanalysis in Zurich and then spent 2 years at the Sorbonne studying psychology and the philosophy of science. He then worked in Alfred Binet's laboratory, standardizing Cyril Burt's reasoning tests for children. That work led him to study children's reasoning and served as the basis for several early articles. During his appointment at the J. J. Rousseau Institute in 1921, Piaget began a systematic study of children's reasoning and then the cognitive development of infants.

In 1925 Piaget occupied the chair in the philosophy of science at Neuchâtel, and in 1929 he was a professor of the history of scientific thought at Geneva. In 1939 he was appointed to a professorship of sociology and then to a chair in experimental psychology at Geneva. In 1955 he established the International Center for Genetic Epistemology at Geneva, where he taught until his death.

THEORIES AND INTERPRETATIONS

Piaget was not just a child psychologist; in addition to scores of books on children's cognitive development, he wrote several volumes on philosophy (genetic epistemology), sociology, and education. Nevertheless, he continues to be thought of as a child psychologist.

Piaget's theory is widely misunderstood. First, his psychological theory is a theory of cognitive development; other aspects of psychological development (e.g., personality, emotion, gender role identity) were of interest to him only as they related to the child's cognition of the world. Second, his theory is about the development of the child and early adolescent. Piaget believed that by age 14 a person's cognitive architecture and processes are securely established as a necessary foundation for further cognition. After that age, few fundamental changes occur. It is at this point that the organism's reasoning power—its logic—is finalized in terms of the ultimate stage of formal logic.

Piaget saw the child as an organism in relation to its environment; those two aspects could not be ignored or separated. Piaget thus was an interactionist. To satisfy his or her many needs, an individual must adapt to his or her milieu. This process of adaptation involves the dual processes of assimilation (i.e., environmental input is interpreted in terms of the individual's cognitive structure) and accommodation (i.e., the individual's cognitive structure must change to reflect the nature of the environment). The balance between the two processes represents a particular state of equilibrium, and the process leading to that outcome is equilibration.

Because environmental input must be interpreted cognitively, the individual's cognitive structure is crucial. That structure consists of a variety of elements, including concepts, categories, schemas, schemes, images, symbols, and operations. Those elements are constructed by the individual from his or her interaction with the environment and integrated to various degrees over time. The subsequent higher level of representation is based initially on a lower plane of motor behavior (praxis) and occurs through a process of abstraction and integration.

In addition to the structural aspects of cognition, there is a dynamic aspect of change—genesis and development—that results in one structure being modified or replaced by another. Neither of these major

Jean Piaget. © FARRELL GREHAN/CORBIS.

radical notion that individuals in some sense can choose (construct) their development freely. Later debates about the precise relationship of genetics and the environment are closely related to this issue. Piaget was clearly a constructivist and interactionist, rejecting both empiricism linked to environmentalism and rationalism linked to nativism. His position that both are involved in the process of constructivism leaves unanswered the question of how, with this approach, one can believe in universal stages and explain the apparent uniformity of behavioral development, at least up to adolescence.

A second issue that has emerged from this position concerns a common criticism of Piaget's theory: that it ignores the social or underestimates its importance or that it is an inadequate theory of the social-cultural. These criticisms often reflect a misunderstanding of Piaget's theory and failure to consider Piaget's (1995) important work, *Sociological Studies.* Running throughout most of Piaget's work is a consistent emphasis on the role of the social in explaining the course of development (Kitchener, in press-b), although the way Piaget conceptualized the nature and importance of the social aspect changed over time. Later scholarship on this question produced a variety of approaches (Carpendale & Mueller, 2004; Smith & Vonèche, 2006).

SEE ALSO Volume 1: *Identity Development; Moral Development and Education.*

BIBLIOGRAPHY

Carpendale, J. I. M., & Müller, U. (Eds.). (2004). *Social interaction and the development of knowledge.* Mahwah, NJ: Lawrence Erlbaum.

Ginsburg, H. P., & Opper, S. (1988). *Piaget's theory of intellectual development* (3rd ed.). Englewood Cliffs, NJ: Prentice-Hall.

Gruber, H. E., & Vonèche, J. J. (Eds.). (1995). *The essential Piaget* (2nd ed.) Northvale, NJ: J. Aronson.

Kitchener, R. F. (1986). *Piaget's theory of knowledge: Genetic epistemology & scientific reason.* New Haven, CT: Yale University Press.

Kitchener, R. F. (in press-a). *Developmental epistemology: Cognitive development and naturalistic epistemology.*

Kitchener, R. F. (in press-b). Piaget's sociological studies. In L. Smith, J. Carpendale, & U. Müller (Eds.), *The Cambridge companion to Piaget.* New York: Cambridge University Press.

Piaget, J. (1950). *Introduction à l'epistémologie génétique.* Paris: Presses Universitaires de France.

Piaget, J. (1995). *Sociological studies,* Trans. L. Smith. London: Routledge.

Piaget, J., & Inhelder, B. (1969). *The psychology of the child,* Trans. H. Weaver. New York: Basic Books.

Smith, L, & Vonèche, J. (Eds.). (2006). *Norms in human development.* Cambridge, England: Cambridge University Press.

Richard F. Kitchener

aspects—structure or genesis—can be ignored because a structure always is generated from an earlier structure. The course of development follows a particular kind of trajectory depicted as a series of universal and global stages of cognitive development: sensory-motor (age 0 to 2), preoperational (2 to 7), concrete-operational (7 to 11), and formal operational (11 to 15). It is unclear how strongly Piaget believed in the existence of a universal stage theory.

An important issue that emerges from this debate concerns the question of developmental determinism: Is the course of psychological development fixed and set by an underlying process or law of psychological development, or does the individual have freedom with respect to his or her developmental trajectory? If there are universal stages of development, that would appear to support developmental determinism because everyone ideally proceeds through these stages. Many individuals would link such determinism to a genetically fixed biological-psychological program of development; those opposed to such determinism tend to favor an open developmental program or the more

POLICY, CHILD WELL-BEING

The life course paradigm recognizes that human development varies depending on its historical time and place. This variation is determined, in part, by the public policies applying to a particular geographic locale in a particular historical period. In the 20th and early 21st centuries in the United States, major social policies affecting children's well-being included cash assistance, child care, early childhood education, and after-school programs. Changes in each policy across historical time were affected by demographic shifts in family composition and maternal labor force participation and changing understandings of children's developmental needs. Tensions associated with these policies reflect the interdependence of children's and parents' lives. For example, traditional welfare policies aimed to directly improve children's financial circumstances through cash transfers but sometimes had unintended consequences such as undermining parental incentives to work for pay or encouraging unmarried parenthood. Likewise, conclusions about how to best design child-care policies often differ when they are viewed from the lens of parental-employment supports versus childhood interventions.

CASH ASSISTANCE

Cash assistance, commonly referred to as *welfare*, provides money to families when an adult provider is unable to do so. Federally funded welfare began as part of the Social Security Act of 1935 with the Aid to Dependent Children (ADC) program, later renamed Aid to Families with Dependent Children (AFDC). Although AFDC was an outgrowth of state mothers' pensions programs that allowed widowed mothers, most of them White, to care for their children, the fraction of AFDC families headed by never-married, divorced, or separated minority mothers quickly rose. In 1938 nearly half of children receiving ADC were living with widowed mothers. By 1961 the percentage had fallen to less than 10. The portion of recipients who were African American increased from 14% to 44% over this same period (Soule & Zylan, 1997, p. 736).

The shifts in demographics of the AFDC recipient population, from White widowed to African-American unmarried women, coincided with broad societal increases in the labor force participation of mothers, including married mothers with young children. These general shifts in maternal employment undermined the premise that the government should support mothers to stay home to raise children when a male breadwinner was unavailable. Divorced and never-married mothers were also seen by conservatives as less deserving of public cash assistance than were families headed by widowed mothers

because a living but absent father might have supported the family or the mother might have delayed childbearing until marriage. In addition, analyses of welfare caseloads by social scientists Bane and Ellwood (1994) found that many recipients received benefits for long spells—close to a decade at a time—and arguments by conservative scholars such as Charles Murray (1984) that an open-ended entitlement to welfare might cause dependency gained momentum at the end of the 20th century.

The Personal Responsibility and Work Opportunity Reconciliation Act of 1996 (commonly referred to as PRWORA or simply as welfare reform) fulfilled President Bill Clinton's 1992 campaign promise to "end welfare as we know it," although PRWORA's specifics were shaped by the Republican-controlled Congress that came into power in 1995. The law sent the clear message that cash assistance would be time limited—the program was now called Temporary Assistance to Needy Families (TANF)—and conditional on employment. Recipients had to start a job within two years (with accompanying sanctions for noncompliance) and could receive cash assistance for a total of 5 years across their lifetime. The law also emphasized marriage, for example by permitting states to keep benefits at their prior level when new children were born to an unmarried mother (a *family cap*).

The new law was controversial. Although most agreed that the welfare system needed reform, some believed the law that ultimately passed was too severe, for example by emphasizing "two years and you're off" rather than "two years and you work." Several high-ranking members of Clinton's administration resigned in protest, including social scientists Mary Jo Bane, Peter Edelman, and Wendell Primus. Many social commentators and scholars forecast that child poverty would increase dramatically as families moved from welfare to work without seeing their incomes rise above the poverty line and when families reached their lifetime limits or saw their grants decreased because of caps or sanctioning. Critics also worried that children's well-being would suffer when mothers went to work, because some children's care arrangements would be inferior to care by their mothers, because of the strains on mothers of combining paid work with childrearing, and because of reduced time, attention, and monitoring devoted to children by employed mothers. Others believed, however, that child well-being would improve after welfare reform because families would ultimately have higher incomes, because parents and children would no longer feel the stigma of receiving welfare, and because families would benefit from the structure associated with employment.

The period leading up to and following the major welfare reforms of 1996 also produced increased state-to-state variation in cash assistance to families. Prior to the

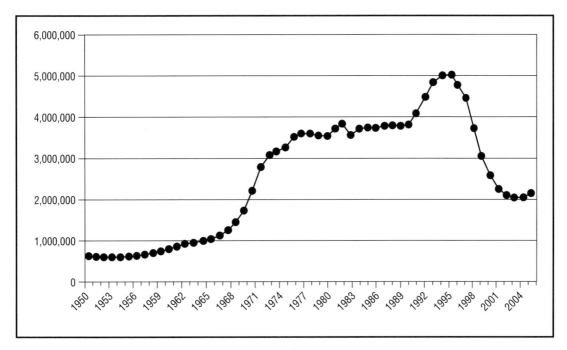

Figure 1. *Number of families receiving AFDC/TANF, 1950–2004.* CENGAGE LEARNING, GALE.

1996 reforms, the federal government granted states waivers that allowed them to experiment with some of the ideas that would become part of the 1996 federal reform, including work requirements, time limits, and family caps. These experiments stimulated research because states were required to evaluate the success or failure of their modified AFDC programs. With PRWORA, TANF funding now came in the form of a *block grant*, giving all states considerable discretion within broad parameters. As a consequence, increasingly over the late 20th and early 21st centuries, a child's well-being vis-à-vis federal cash assistance depended on the state within which he or she lived.

Numerous research efforts—large and small, local and national—have documented how changes in welfare policy affected children and families. These studies all have limitations. Policies typically came in "bundles"—whether as part of randomized or natural experiments—making it difficult to pinpoint which aspect of the policy was associated with child well-being. In addition, state waiver experiments generally preceded the federal welfare reform, and surveys designed specifically to study the federal changes followed it, making it impossible to make statements about the impact of PRWORA itself. The booming economy of the late 1990s and other changes that coincided with the reforms (such as the expansion of the Earned Income Tax Credit [EITC] discussed below) also complicate trend analyses of PRWORA's effects.

Nevertheless, consistent findings across studies strengthen the evidence for some conclusions. Highlighted below are general findings in two areas: (a) moving from welfare to employment and associated changes in family income and (b) losing benefits because of family caps, sanctions, and time limits.

Employment and Poverty As anticipated, the TANF rolls dropped sharply, beginning in 1997 (see Figure 1). Yet critics' fears of a dramatic rise in poverty did not occur. In fact, overall poverty rates dropped and employment rates of single mothers rose steadily throughout the 1990s. Critics' concerns that child well-being would plummet were also not revealed within studies of work requirements, although variation exists across subgroups.

Two major studies that differ on many dimensions—including their historical timing (before or after federal reform), experimental versus nonexperimental designs, and measures—find variation in outcomes by child's age. The reasons for these findings are not well understood, although both studies suggest changes in income and time use may be contributing factors. An important synthesis of welfare waiver experiments and other random-assignment studies conducted by Manpower Demonstration Research Corporation (MDRC) found that adolescents were more vulnerable to negative effects of work requirements than were preschool children, in terms of school achievement. This was especially evident for adolescents with younger siblings, suggesting that they may have taken on more responsibilities within the family.

One of the largest surveys specifically designed to study welfare reform (Welfare, Children, and Families: A Three City Study) also found variation in outcomes by child age, although this study's results showed that mothers' transitions into employment were not correlated with changes in preschoolers' outcomes but were associated with increases in adolescents' mental health. The Three City Study found that mothers of older children, but not preschoolers, compensated for time away from home by cutting back on personal time, leaving no net change in total time spent with adolescents. It also found that transitions to employment were correlated with substantial increases in family income, and MDRC's studies suggest that when families' earnings were supplemented by cash assistance some of the income increment was spent purchasing center child care for preschoolers (the group their study found had more positive school outcomes).

When TANF families move into employment, the extent to which they would see such a beneficial rise in earnings depends on the state in which they live. One reason is because of *earned income disregards*, which allow welfare recipients to supplement their wages with cash assistance during their transition into the labor force. The Urban Institute's Welfare Rules Database, which centralized details about state welfare programs from 1996 to 2006, shows that most states included such disregards, although the amount varied substantially across states and typically declined as parents accumulated time in the labor force. The second reason is the Earned Income Tax Credit (EITC). This federal credit provides low-wage earners with a refund of up to several thousand dollars when they file a tax return. The EITC was expanded in 1993, prior to the 1996 welfare reform, and many see it as an important companion to PRWORA's work requirements. Nearly half of the states offer a similar credit to state income taxes, and the state amounts range from an additional 3.5% to 43% of the federal credit.

Losing Benefits: Sanctions, Time Limits, and Family Caps There are three major reasons a family might involuntarily lose TANF benefits: (a) their case might be closed for procedural reasons (such as failing to produce documents to verify continued eligibility), (b) they might be sanctioned (such as for not following the rules regarding work requirements), or (c) they might reach their lifetime limit for benefits. States have considerable discretion in defining each of these reasons for losing benefits. Their grant might also be capped at its current level on the birth of a new child. For example, the Welfare Rules Database found sanction amounts ranged

from a fraction up to the entire grant, and the duration ranged from the sanction being lifted immediately upon compliance to permanent termination. Regarding lifetime limits, some states permit receipt for just 2 or 3 rather than 5 years, whereas others effectively lengthen the federal limit by *stopping the clock*, for example when families receive cash assistance to supplement earned income.

There is much less evidence on the effect of these specific policies associated with losing benefits than more general research on work requirements. The most general conclusion that emerges is that families who are sanctioned or have their cases closed are more disadvantaged than other families. For example, in the Three City Study, families who lost benefits had lower education, poorer health, and were more likely to report not having enough food than were other families, and sanctioning was correlated with children's behavior problems. Why sanctions and time limits are associated with disadvantage and hardship is unclear. On the one hand, loss of cash assistance might lead to harsher and less consistent parenting or might disrupt children's regular care arrangements. On the other hand, unobserved characteristics that make families more likely to miss an appointment to determine eligibility (and thus lead to procedural case closure) or make it harder for a parent to find and keep a job (and thus lead to violation of work requirements or faster approach of the time limit) may also be associated with poorer child outcomes.

A Look to the Future The most consistent findings from studies of welfare reform are that work requirements increase parental employment and earnings with little evidence of negative impacts on child well-being. These findings support public sentiment for a strong work ethic, and a return to unlimited cash assistance is unlikely. Rather, broad-based support of programs such as the Earned Income Tax Credit—which supplement the earnings of those whose behavior conforms to this work ethic—is likely to continue.

Concerns about how to support working families, and how to help parents with the greatest barriers to employment, will also likely persist, especially in times of economic recession. Given the flexibility of TANF block grants, and the dramatic declines in caseloads in the late 1990s, states were able to spend close to two-thirds of their grants on work supports (such as child care, transportation, education, and job training) and on extensions and exemptions to the lifetime time limit. The recession of the early 21st century made it harder for states to continue these supports and exemptions.

TANF was reauthorized by the Deficit Reduction Act of 2005, which kept most features of the program intact but increased the fraction of TANF recipients that had to be engaged in paid work. The reauthorization also highlighted PRWORA's marriage promotion objectives. These objectives had provided much of the basis for PRWORA and were included in three of the legislation's four stated goals, but were overshadowed by the law's work requirements and time limits in terms of research and policy attention. The reauthorization's annual allocation of $150 million for healthy marriage and responsible fatherhood initiatives reflects a culmination of efforts by the Bush administration to bring marriage promotion to the fore. Prior to the reauthorization, the Bush administration had used several existing funding streams to support programs and research, including waivers in the office of child support and two large-scale random-assignment demonstrations, Supporting Healthy Marriage and Building Strong Families (Ooms, Bouchet, & Parke, 2004). Although the political priority of marriage promotion may change under a new administration, the momentum generated by these demonstrations for research and policy will likely continue.

CHILDREN'S CARE ARRANGEMENTS

One of the most important supports to employed families is child-care assistance. The *cost*, *quality*, and *availability* of child care are salient issues to parents, advocates, and policymakers. In terms of public policy, child-care *subsidies* and *tax credits* offset some of the cost of care. *Regulations* set a minimum bar on quality, although increasingly states are experimenting with strategies to improve quality beyond a minimum, including through *tiered reimbursement programs*. Availability is less often directly addressed through policy, although policies aimed at cost and quality affect supply indirectly. As with cash assistance, the policy supports and constraints for child care vary considerably from state to state. In addition, how children spend their time when neither with their parents nor in school can be seen through the lens of a child-directed intervention rather than a parent-directed work support. Moves toward *universal prekindergarten* (pre-K) and *extended learning opportunities*, which gained momentum during the mid-1990s, exemplify such child-focused policies.

Subsidies, Regulations, and Universal Programs For low-income families, subsidies under the Child Care and Development Fund (CCDF), part of the PRWORA legislation, offset the cost of care. Across states, income eligibility cutoffs range from 34% to 85% of state median income. Parents pay a co-payment, and the state reimburses the provider at a set rate. Typically, the co-pay is higher for higher income families, and the reimbursement rate is higher for center-based than home-based providers, for younger than older children, and in different regions of the state. The federal government recommends that co-payments not exceed 10% of a family's income and that reimbursement rates are at or above the 75th percentile of market rates, but some states set higher co-pays and the majority set lower reimbursement rates. Although this reduces the cost of the state's program and in theory allows more families to be covered, co-payments that are higher than parents can afford reduce participation, while under-market reimbursement rates make the most expensive providers inaccessible, because providers are not required to accept subsidies. The Child and Dependent Care Credit—a federal tax credit—also makes child care more affordable, including for middle-income families. The maximum credit is about half the size of the maximum EITC and is not refundable.

State regulations require group care settings, including all centers and large home-based providers (typically those caring for six or more children), to be licensed. Small home-based providers are license-exempt. Concrete *structural* features of care settings are regulated, such as ratios of staff to children and group sizes. State regulations vary and are often less stringent than standards recommended by child-care accreditation programs. For example, accreditation programs recommend one adult for every four 18-month-olds and one adult for every ten 4-year-olds. Just eight states require ratios these low for toddlers, and only 14 states meet the recommendations for preschoolers. Some states allow one person to care for as many as nine 18-months-olds and as many as twenty 4-year-olds.

States are required to use at least 4% of their TANF funds on initiatives to improve child-care quality. States can meet this objective in a wide variety of ways, from providing resource and referral services to parents to enhanced inspections of licensed providers to investment in caregiver education and training. Tiered reimbursement programs—which provide higher reimbursement rates to providers who rank higher on quality rating systems—have emerged as an innovative mechanism for stimulating quality. Typically the programs include higher program standards for higher rated programs (often through meeting national accreditation standards), accountability measures to assign ratings, outreach to help programs meet higher standards, and parent education to encourage use of the ratings in their child-care decisions.

Most states cover both licensed and license-exempt providers with CCDF subsidies. Beneficially, this means that parents who prefer small home-based providers can receive assistance in paying for them. From the perspective of supporting parental employment, this also means

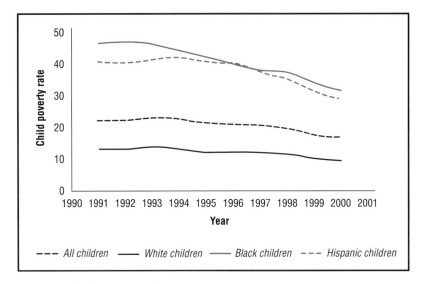

Figure 2. *Child poverty rate by race, 1991–2001.* CENGAGE LEARNING, GALE.

that supply will be maximized, because any adult is a potential caregiver. However, from the perspective of supporting children's well-being, license-exempt and home-based settings score lower on average than centers in terms of providing structured learning activities. Children's safety may also be compromised in some license-exempt settings (or larger home-based settings operating illegally without a license). A unique study by sociologists Wrigley and Dreby (2005) found that although fatality rates in child care were low overall, they were higher in homes than centers. The MDRC synthesis finds that subsidies encourage use of licensed and center care, although direct evidence of associations of subsidies to child well-being is not available.

High-income families have historically used part-day, part-week preschool to enrich children's social and educational experiences. Especially starting in the 1960s, government-funded programs provided similar experiences to lower income children, most prominently through Head Start. Low-income children who participated in model early intervention programs (including the Abecedarian and Perry Preschool programs) with well-designed curricula, skilled staff, and small groups, showed better school achievement and behavioral adjustment than other children. Cost–benefit studies and long-term follow-ups confirm their effectiveness. Historically, Head Start funding levels have not been sufficient to cover all eligible families. Increasingly, states are moving toward funding their own universal preschool programs (see Figure 2). Most state programs target *at-risk* children who are most likely to experience poor school achievement, based on economic disadvantage, disabilities, or other special needs. Just three states aim to cover all children, regardless of risk or income.

Evidence from large-scale efforts, including the citywide Chicago Parent–Child Centers for at-risk children and the statewide Oklahoma universal preschool program, suggests beneficial outcomes can be achieved when programs are scaled up to citywide or statewide levels. Nevertheless, universal programs may shift the costs of preschool from higher income families to the states (in those instances in which these parents would have paid for preschool in the absence of state programs), and debate (and limited knowledge) exists about whether preschool boosts school readiness for middle- and higher income families. There is also concern about whether quality can be maintained in statewide programs. For example, although most state pre-K programs require a staff-to-child ratio of one to ten, only about half require a teacher to have a bachelor's degree. A 2005 study of state pre-K in six states by the National Center for Early Development and Learning found that quality averaged only a minimal rating (Clifford et al., 2005).

Future Directions What does the future hold for early care and education? Many call for increased funding of the subsidy program, especially as state TANF surpluses drop. The tension between quantity and quality of care is likely to continue, as are expansions of tiered reimbursement and universal preschool. Efforts such as the Foundation for Child Development's PK–3 initiative further encourage continuity and sustainability so that gains from early education are maintained through the school years (Bogard & Takanishi, 2005). Debate about strategies to care for infants and toddlers and to support quality in home-based care will likely persist as well.

Increasing the supply of infant and toddler care, especially in centers, is expensive (because structural ratios and group sizes are lower than for older children). Quality is typically lower, on average, for this age group, and parents are more likely to express a preference for home-based care for these youngest children. Most states exempt parents from TANF work requirements when their children are young (generally under 1 year of age), and investing in parental leave may be an important complement to expanding infant and toddler care.

Once children enter full-day kindergarten or first grade, the amount of time parents must arrange for care declines (at least for parents who work weekday hours). Historically, the patchwork of opportunities before and after school and in the summertime has received less research and policy attention than early child care. Yet sociologists Entwisle and Alexander (1992) powerfully documented a *summer setback*, whereby poorer children show growth in mathematics test scores comparable to that of higher income children during the school year, but slower growth during the summer. CCDF subsidies and Child and Dependent Care Credits may be used for the care of school-age children up to age 12. In addition, the federal 21st Century Community Learning Centers support objectives of the No Child Left Behind Act of 2001 through an array of academic enrichment services in out-of-school hours, including tutoring and homework assistance. Some states also fund their own after-school programs. In a 2006 survey, 27 states had funding dedicated explicitly to after-school programs, 8 states included after-school programs as one of a menu of options in certain streams, and 15 states had no dedicated after-school funding (Stedron & Thatcher, 2007).

Extended learning is a new conceptualization of children's time that emphasizes integration and continuity across contexts. In part, this reflects new pressure for after-school programs to help support learning under strict accountability standards, although not all agree that after-school programs can and should meet these expanded purposes. In addition, as children move into adolescence, they are increasingly active decision-makers about where and how they spend their time, and thus their perceptions of a setting's quality become increasingly important. Maintaining steady participation is challenging, especially among older youth. Similar to the situation in early care, innovation and evaluation of strategies to monitor and stimulate the dimensions of quality in after-school programs that best reflect youth interests and support their well-being are likely to continue.

SEE ALSO Volume 1: *Child Care and Early Education; Family and Household Structure, Childhood and Adolescence; Policy, Education; Poverty, Childhood and Adolescence.*

BIBLIOGRAPHY

Bane, M. J., & Ellwood, D. T. (1994). *Welfare realities: From rhetoric to reform.* Cambridge, MA: Harvard University Press.

Bogard, K. & Takanashi, R. (2005). PK-3: An aligned and coordinated approach to education for children 3 to 8 years old. *SRCD Social Policy Report, 19*(3). Retrieved June 18, 2008, from http://www.icpsr.umich.edu/PK3/resources/463961.html

Clifford, R.M., Barbarin, O., Chang, F., Early, D., Bryant, D., Howes, C. et al. (2005). What is pre-kindergarten? Characteristics of public pre-kindergarten programs. *Applied Developmental Science, 9*(3), 126–143.

Entwisle, D. R., & Alexander, K. L. (1992). Summer setback: Race, poverty, school composition, and mathematics achievement in the first two years of school. *American Sociological Review, 57,* 72–84.

Glass, J. L., & Estes, S. B. (1997). The family responsive workplace. *Annual Review of Sociology, 23,* 289–313.

Gormley, W. T., Jr. (2007). Early childhood care and education: Lessons and puzzles. *Journal of Policy Analysis and Management, 26,* 633–671.

Grogger, J., & Karoly, L. A. (2005). *Welfare reform: Effects of a decade of change.* Cambridge, MA: Harvard University Press.

Halpern, R. (1999). After-school programs for low-income children: Promise and challenges. *The Future of Children, 9*(2), 81–95.

Haskins, R. (2006). *Work over welfare: The inside story of the 1996 welfare reform law.* Washington, DC: Brookings Institution Press.

Jencks, C. (1992). *Rethinking social policy: Race, poverty, and the underclass.* Cambridge, MA: Harvard University Press.

Lichter, D. T., & Jayakody, R. (2002). Welfare reform: How do we measure success? *Annual Review of Sociology, 28,* 117–141.

MDRC. *Next Generation project.* Retrieved May 12, 2008, from http://www.mdrc.org/project_8_10.html

Murray, C. (1984). *Losing ground: American social policy, 1950-1980.* New York: Basic Books.

National Center for Children in Poverty. *Improving the Odds for Young Children project.* Retrieved May 12, 2008, from http://www.nccp.org/projects/improvingtheodds.html

Ooms, T., Bouchet, S., & Parke, M. (2004, April). *Beyond marriage licenses: Efforts in states to strengthen marriage and two-parent families.* Washington, DC: Center for Law and Social Policy. Retrieved May 12, 2008, from http://www.clasp.org/publications/beyond_marr.pdf

Scarr, S. (1998). American child care today. *American Psychologist, 53*(2), 95–108.

Soule, S. A., & Zylan, Y. (1997). Runaway train? The diffusion of state-level reform in ADC/AFDC eligibility requirements, 1950–1967. *American Journal of Sociology, 103,* 733–762.

Stedron, J., & Thatcher, D. (2007, August/September). State funding for expanded learning opportunities. *National Conference of State Legislatures Leigisbrief, 15*(34). NCSL: Denver, CO.

U.S. Department of Health and Human Services. *Child Care Bureau.* Retrieved May 12, 2008, from http://www.acf.dhhs.gov/programs/ccb/index.html

U.S. Department of Health and Human Services. *Office of Family Assistance.* Retrieved May 12, 2008, from http://www.acf.hhs.gov/programs/ofa

Urban Institute. *Welfare Rules Database: Assessing the New Federalism Project at the Urban Institute.* Retrieved May 25, 2008, from http://anfdata.urban.org/wrd/wrdwelcome.cfm

Welfare, Children, and Families: A Three City Study. Retrieved May 12, 2008, from Johns Hopkins University Web site: http://web.jhu.edu/threecitystudy

Wrigley, J., & Dreby, J. (2005). Fatalities and the organization of child care in the United States, 1985–2003. *American Sociological Review, 70,* 729–757.

Zuckerman, D. M. (2000). Welfare reform in America: A clash of politics and research. *Journal of Social Issues, 56,* 587–599.

Rachel A. Gordon

POLICY, EDUCATION

The United States public education system is characterized by several features that influence the nature of educational policies, the relative success or failure of their implementation, and policy outcomes.

INVESTMENT AND DECENTRALIZATION

First, in the United States the federal government, states, and localities spend more than most other industrialized nations on public schools but invest relatively less in social welfare programs (social security, housing, health care, etc.) (Hochschild & Scovronik, 2003). This large investment of resources in public education reflects the national faith in the power of the public educational system—and in educational policies specifically—to address a broad range of social, economic, and political problems (Hochschild & Scovronick, 2003). Public schools are expected to provide a means for children to attain a desired occupational level and to equalize the opportunities of children with disadvantages in terms of race, social class, or special needs status and to prepare students to be informed participants in the democratic process (Hochschild & Scovronick, 2003). These multiple and often conflicting goals influence not only the types of policies that are adopted and implemented, but these goals also influence whether policies are considered successful, on what measures, and by whom.

Second, the U.S. public education system is highly decentralized compared with those of other industrialized nations. Whereas most countries centralize school governance, curriculum decisions, and finance at the federal level, the U.S. system vests most authority over those issues in states and localities. Funding for education is largely the responsibility of states and local governments as well; federal spending on education constitutes a small fraction of overall school spending, accounting for 9% of the estimated $1 trillion spent in fiscal year 2007 (U.S. Department of Education, 2008). This reliance on local funding for education and local control has allowed a relatively greater degree of community influence and input than is given to localities in other nations, yet this decentralized structure also has meant that public school students in the United States often find themselves in what Hochschild and Scovronick (2003) termed "nested structures of inequality": in states with dramatically different levels of funding and capacity, in districts with vastly unequal funding levels, within schools that have differential allocations of qualified teachers. Each of these nested structures shapes educational policy outcomes in profound ways.

This entry on educational policymaking, policy implementation, and educational policy research covers three major educational policies that have affected public schools in the late 20th and early 21st centuries: (a) Title 1 and the ensuing accountability requirements through No Child Left Behind (NCLB); (b) school desegregation policy designed to equalize access and opportunity across highly unequal school contexts; and (c) school choice policy, which has been designed to hold schools accountable through competitive market forces. Each of these policies differs dramatically in its origins (the courts or the federal, state, and local governments), goals, and outcomes. Thus, this entry explores the diverse ways in which success has been defined in each of these policies and on what measures.

TITLE 1 AND NO CHILD LEFT BEHIND

One of the most significant policies that has affected public schools since the 1960s has been Title 1, first enacted as part of the Elementary and Secondary Education Act (ESEA) of 1965. It emerged in an era in which great faith had been placed in the power of government, especially the federal government, to address social inequality. Title 1, which granted significant federal funding for public education for the first time, was intended to provide greater equity in resources for higher-poverty urban school districts and ameliorate the disadvantages of poverty.

At the time it was enacted, Title 1 contained few provisions dealing with school accountability for fear that such provisions would be perceived as federal intrusion into local educational decisions and lessen the chances for passage (Halperin, 1975). However, a subsequent amendment written by Senator Robert F. Kennedy required states to adopt "effective procedures, including provision for appropriate objective measurements of educational achievement...for evaluating at least annually the effectiveness of the programs in meeting the special educational needs of culturally deprived children" (cited in Halperin, 1975, p. 8). Those initial efforts to hold districts and schools accountable were entirely procedural and revolved around the appropriate and effective use of federal monies, or "inputs."

It was not until the 1990s that the inputs-driven model of accountability shifted to reflect the growing popularity of systemic reform, a school reform model that is based on the assumption that schools should be given greater regulatory flexibility with day-to-day operations, or "inputs," in exchange for increased accountability for test scores and other "outputs." In the late 1980s the National Governors Association embraced this model of reform, with a strong leading role taken by then Governor Bill Clinton of Arkansas. When Clinton became president, his reauthorized 1994 version of ESEA, known as the Improving America's Schools Act (IASA), was retooled to reflect the principles of systemic reform. States receiving Title 1 monies were required to develop content and performance standards and adopt assessments that were aligned to those standards to hold their schools accountable. This outcome-driven accountability under IASA in 1994 made modest strides: By 2000 most states had some sort of accountability system in place, but most of those systems were not well developed (Sunderman, Kim, & Orfield, 2005). Before the enactment of NCLB only 19 states had fully approved systems under IASA, largely as a result of inadequate capacity at the state and local levels (Sunderman et al., 2005).

When Governor George W. Bush of Texas was elected president in 2000, he arrived from the state that arguably had made the most progress under the outcomes-driven accountability model promoted under IASA, one that involved strict timelines, strong state intervention, and a focus on testing outcomes. His first legislative priority was to reauthorize ESEA under that model, and in 2001 a bipartisan coalition helped him push through his first major legislative achievement, No Child Left Behind, which he signed into law in January 2002.

NCLB marked a dramatic increase in federal intervention into public education (Elmore, 2002). The law requires all states to develop content standards and measure the progress of students through annual testing in English/language arts, mathematics, and science in the third through eighth grades and to test students once in high school. The law also mandates that 95% of students must be tested and that all teachers in core subject areas must be "highly qualified." States also must develop Annual Measurable Objectives (AMOs) to determine whether their schools are making Annual Yearly Progress (AYP). According to the legislation, 100% of schools

must demonstrate proficiency in reading and math by the 2013–2014 school year.

Strict sanctions are imposed for nonperformance: If a school fails to make AYP for two consecutive years on the same indicator, it is identified for the first year of program improvement and districts must offer those students the choice to attend a nonfailing school. Failure to make the required improvement for three consecutive years on the same indicator entitles students to "supplemental services" such as tutoring (paid for out of school funds), failure the fourth year dictates "corrective action" (replacing staff, new curriculum), and failure the fifth year allows more drastic measures, including conversion to charter school status, reconstitution, or takeover by a state or private management agency.

Although some researchers have documented gains in achievement since the enactment of NCLB (National Center for Education Statistics, 2005a), others have found that achievement gaps have not changed on cross-state measures such as the National Assessment of Educational Progress, a cross-state student achievement test (Lee, 2006). Other researchers have raised questions about the effectiveness of this punitive model of school reform; for example, researchers have found that schools that are more diverse and thus have a greater number of student subgroups to be accountable for are disproportionately likely to be identified for improvement status and become the target of NCLB sanctions. A 2008 study by the U.S. Department of Education found that schools identified for improvement were disproportionately urban, high-poverty schools and that "school poverty and district size better predicted existing improvement status than the improvement strategies undertaken by the schools" (U.S. Department of Education, 2008, p. xii). Researchers also have found that schools have narrowed their curricula to focus much more heavily on tested subject areas while cutting time in science, social studies, music, art, and physical education (Center on Education Policy, 2007).

As policymakers take on the task of reauthorizing NCLB, it is likely that the legislation will stay intact but undergo significant modifications. Proposed changes include granting greater flexibility to states and school districts in meeting achievement targets through growth models and granting states more flexibility in the kinds of sanctions they impose on schools in need of improvement.

SCHOOL DESEGREGATION POLICY

School desegregation policy has had an impact on virtually all urban school districts in the United States. Although desegregation is not an educational policy in the formal sense of the term, state and federal court orders mandating desegregation have given way to a wide range of federal, state, and local policies to comply with those decisions.

The implementation of school desegregation varied greatly across local contexts. Ironically, desegregation was most successful in the Southern states, in which district boundaries encompassed entire counties; in those districts suburbs could not escape the reach of court mandates. In contrast, implementation of desegregation in Northern and Midwestern metropolitan areas, in which central-city school districts were usually smaller and surrounded by dozens of smaller self-contained suburban districts, was virtually impossible after the U.S. Supreme Court's *Milliken I* decision in 1974, which made cross-district desegregation legally unfeasible.

Most current data show that schools across the United States largely have resegregated. According to research by E. Frankenberg, C. Lee, and G. Orfield (2003), although significant gains in integration were made in the South through the late 1980s, integration levels everywhere declined since that time and in the first decade of the 21st century were lower than they had been in 1970 for both African Americans and Latinos. At the same time high-minority schools are more likely to be high-poverty than in the past (Frankenberg, Lee, & Orfield, 2003). At the same time that segregation levels have been worsening, the nature of school segregation in the United States has also changed dramatically. However, although segregation levels have been worsening, the nature of segregation has changed dramatically. As Charles Clotfelter's research has found, in 1970, the majority of school segregation was within school distirct boundaries, and just 43% of school segregation was between school districts; by 2000 the proportion of segregation that was between-district jumped to 70.8% (Clotfelter, 2004). Thus, court orders, which usually are intradistrict only, are less likely to ameliorate these between-district inequalities.

Research on the outcomes of school desegregation has focused on short-term effects on achievement and peer relations and on "long-term" effects in terms of social mobility, rates of college attendance, and life after graduation. The literature on short-term effects has shown that while students are in school, desegregation has resulted in slight achievement gains for African-American students and has had no impact on achievement for Whites. This literature has found that desegreation also has resulted in more positive intergroup relations (Schofield, 1995). Research on longer-term impacts has documented more uniformly positive effects: Desegregation has had a positive impact on the aspirations of African-American students, and African-American graduates of desegregated schools are more likely to attend integrated colleges and have higher levels of educational attainment. African-American graduates of desegregated

schools are also more likely to have desegregated social networks and work in white-collar occupations (Wells, 1995).

Although a series of U.S. Supreme Court decisions have led to the dismantling of many mandatory court-ordered desegregation plans (Orfield & Eaton, 1996), a number of localities have maintained integrated schools voluntarily. Those voluntary efforts, however, might have been endangered after the 2007 *Parents Involved in Community Schools v. Seattle School District No. 1 et al.* ruling by the U.S. Supreme Court, which circumscribed the conditions under which school districts may pursue voluntary measures. The ultimate influence of this decision on voluntary desegregation is unclear; in light of the growing problem of between-district segregation, many researchers observe that any integration efforts will need to cross district lines.

SCHOOL CHOICE POLICY

School choice policies are diverse, reflecting the different historical contexts from which they evolved. In the 1960s support was strong for both alternative schools, which were based on progressive education models, and magnet schools, which were designed to promote voluntary desegregation through choice. In the 1980s and 1990s many states adopted intra- and interdistrict open enrollment policies as a way to promote school improvement through competition. These efforts expanded in the 1990s as a growing number of states enacted charter school legislation, and several states enacted voucher legislation; in 2002 school choice became a cornerstone of NCLB.

Proponents of choice believe that schools will perform better if they are subjected to competitive market forces. However, researchers studying achievement results from school choice programs have been faced with a number of methodological challenges, the most significant of which is selection bias: Most students who enroll in school choice programs come from families with measured differences in terms of family background (income and education levels) compared with students who do not enroll as well as unmeasured differences such as greater parental savvy in navigating the school system. Although attempts are made to factor out these differences, the validity of research continues to be challenged as a result of these issues.

Critics of school choice fear that choice will exacerbate racial, ethnic, and income stratification between schools. Research has shown that this depends on the type of choice policy. Many researchers have found that more deregulated choice policies such as vouchers and charter schools, which have fewer controls on choice and fewer requirements for transportation or parent information, typically lead to greater stratification across schools. The more regulated policies, which have strong parental information requirements, racial and economic balance requirements, and provide transportation, tend to lead to reduced stratification. Each of these types of choice policies is examined below.

Vouchers Four publicly funded voucher programs were in operation in the first decade of the 21st century: the Milwaukee Parental Choice Program (MPCP) for low-income students, which began in 1990; the Cleveland Scholarship and Tutoring Program (CSTP), which was authorized in 1995 for low-income families in that city, and a statewide program in Ohio, EdChoice, which began in the 2006–2007 school year; and, in Washington, D.C., the D.C. School Choice Incentive Act, which was authorized in January 2004 for low-income students in that district. Enrollments in the programs are relatively low: The enrollment in Milwaukee reached 19,233 students in 123 schools in the 2007–2008 school year, short of the 22,500 student cap (Wisconsin Department of Public Instruction, 2008). The Cleveland program enrolled 6,300 students in the 2006–2007 school year (National School Boards Association, 2007). Enrollment in the statewide program in Ohio was approximately 7,000 in the 2007–2008 school year, well short of the 14,000-student cap (School Choice Ohio, 2008), and in 2008 the program was slated for elimination by the governor. The Washington, D.C., program reached its enrollment cap of approximately 2,000 students in the 2006–2007 school year, which included 2.3% of the district's students (U.S. Government Accountability Office, 2007). There are also privately funded voucher programs in a number of major U.S. cities.

Evaluating the effect of voucher programs on student achievement has been difficult because private schools generally are not required to administer standardized tests. Limited evidence from voucher programs has found either no overall gains (U.S. Government Accountability Office, 2001, Witte, Stern, & Thorn, 1995) or limited gains for some subgroups of students in some grades but not others; those gains generally have faded over time (Howell & Peterson, 2002, Ladd, 2002).

Few evaluators have examined stratification in voucher programs because most of the programs are restricted to low-income students. However, research on the Cleveland program has found that students of color utilized the CSTP vouchers at a disproportionately low rate and that the majority of students entering the CSTP had been enrolled in private schools (without using vouchers) the prior year. Students entering the CSTP from private schools were more likely to be White (Plucker, Muller, Hansen, Ravert, & Makel, 2006).

The 2002 U.S. Supreme Court ruling in *Zelman v. Simmons-Harris* opened the door for publicly funded voucher programs that allow the use of vouchers at parochial schools in states that do not have constitutions that prohibit such programs. However, these programs have not expanded dramatically since that time; an effort to enact a statewide program in Utah was defeated, and the EdChoice program in Ohio faced elimination after its first year of operation.

Charter Schools Since the early 1990s, 40 states and the District of Columbia have enacted legislation authorizing charter schools, which are schools of choice that are freed from many of the regulations that apply to regular public schools in exchange for greater accountability to their authorizing entity (usually a school board, a university, or a state board of education), which can "revoke" the charter and close a school for nonperformance. In 2004–2005, 1.8% of all students (887,243 students) in the United States were enrolled in 3,294 charter schools (National Center for Education Statistics, 2007). Charter schools are situated largely in urban areas (52%) and largely in the Western states (National Center for Education Statistics, 2007).

Research on the impact of charter schools on student achievement has been mixed. Some studies show positive gains (Hoxby & Rockoff, 2005), but others show no measurable differences (Miron & Nelson, 2000) and still others show poorer performance in comparing charter school students with their regular public school counterparts (National Center for Education Statistics, 2005b).

With respect to the impact of charter schools on school stratification, most data have indicated that at the aggregate level charter schools serve a relatively greater diversity of students than do public schools (National Center for Education Statistics, 2007). However, when data are examined at the local level, charter schools tend to be more segregated by race and social class than their public school counterparts (Frankenberg & Lee, 2003). A number of researchers also have pointed out that charters have relatively greater control over student enrollment through the requirement of parent involvement and behavior contracts and thus are able to select students who may look demographically similar but differ in subtle but important ways (Becker, Nakagawa, & Corwin, 1997).

Despite this mixed evidence, charter schools enjoy widespread support and have been promoted as an option for students in failing schools under NCLB, with significant federal support for their expansion. They are likely to remain a strong component of the choice landscape.

Magnet Schools Magnet schools represent one of the longest-standing school choice policies in the United States, originating in the efforts of central-city school systems to stem middle-class flight and foster voluntary desegregation within city borders (Wells, 1993). Magnet schools, which usually are created around a particular curricular or academic focus, have been included in nearly all school desegregation plans since 1980 (Wells, 1993), and ongoing though fluctuating federal support for magnet schools has existed since the 1970s through the reauthorization of NCLB.

Although recent national statistics are not available, estimates from the 1999–2000 school year indicate that "there were 3,026 magnet schools with explicit desegregation objectives enrolling 2.5 million students. However, if one also counts magnet (or specialty) schools without explicit desegregation objectives, the estimate increases to 5,576 schools and 4.5 million children" (National Center for Education Statistics, 2007, p. 5).

Magnet schools are distinct from the more recent, more deregulated charter and voucher programs in that magnets are designed specifically to foster racial diversity and typically include both requirements and funding for parental information and transportation. Extra resources usually are funneled into those programs, and demands for slots are usually quite high.

Research on academic achievement in magnet schools has been more positive than evidence from voucher and charter school programs (Gamoran, 1996, Goldhaber, 1999). However, some researchers have raised concerns about the within-school segregation that has occurred in magnet schools, as many magnets are "schools within schools" that separate students into specialized programs apart from the regular student population (West, 1994). Other research has raised concerns that lower-income parents utilize magnet schools at relatively lower rates (Saporito, 2003).

Magnet schools have had long-standing political support, and funds were included for their support under NCLB. However, the ability of magnets to achieve their original goal of racial diversity may be curtailed in light of the 2007 *Parents Involved* ruling and the growing restrictions on the use of race in admissions.

SEE ALSO Volume 1: *Academic Achievement; College Enrollment; High-Stakes Testing; Private Schools; Racial Inequality in Education; School Tracking; Segregation, School; Socioeconomic Inequality in Education.*

BIBLIOGRAPHY

Becker, H. J., Nakagawa, K., & Corwin, R. G. (1997). Parent involvement contracts in California's charter schools: Strategy for educational improvement or method of exclusion. *Teachers College Record, 98*(3), 511–536. Retrieved May 2, 2008, from http://www.tcrecord.org

Center on Education Policy (2007). *Choices, changes, and challenges: Curriculum and instruction in the NCLB era.* Washington, DC: Center on Education Policy. Retrieved May 2, 2008, from http://www.cep-dc.org

Clotfelter, C. T. (2004). *After Brown: The rise and retreat of school desegregation.* Princeton, NJ: Princeton University Press.

Elmore, R. F. (2002). Unwarranted intrusion. *Education Next, 2*(1). Retrieved May 2, 2008, from http://www.hoover.org/publications/ednext/3367491.html

Frankenberg, E., & Lee, C. (2003). Charter schools and race: A lost opportunity for integrated education. *Education Policy Analysis Archives, 11*(32). Retrieved May 2, 2008, from http://www.epaa.asu.edu/epaa/v11n32

Frankenberg, E., Lee, C., & Orfield, G. (2003). *A multiracial society with segregated schools: Are we losing the dream?* Civil Rights Project. Retrieved May 2, 2008, from http://www.civilrightsproject.ucla.edu/research/reseg03.php

Gamoran, A. (1996). Student achievement in magnet, public comprehensive, and private city high schools. *Educational Evaluation and Policy Analysis, 18*(1), 1–18.

Goldhaber, D. D. (1999). School choice: An examination of the empirical evidence on achievement, parental decision making, and equity. *Education Researcher, 28*(9), 16–25.

Halperin, S. (1975). ESEA ten years later. *Educational Researcher, 4*(8), 5–9.

Hochschild, J., & Scovronick, N. (2003). *The American dream and the public schools.* New York: Oxford University Press.

Howell, W. G., & Peterson, P. E. (2002). *The education gap: Vouchers and urban schools.* Washington, DC: Brookings Institution Press.

Hoxby, M., & Rockoff, J. (2005). Findings from the city of big shoulders. *Education Next.* Retrieved May 2, 2008, from http://www.vanderbilt.edu/schoolchoice/downloads/articles/ednext-hoxbyrockoff2005.pdf

Ladd, H. F. (2002). School vouchers: A critical view. *Journal of Economic Perspectives, 16*(4), 3–24.

Lee, J. (2006). *Tracking achievement gaps and assessing the impact of NCLB on the gaps.* Civil Rights Project. Retrieved May 2, 2008, from http://www.ucla.edu/research/esea/nc16_naep_lee.pdf

Miron, G., & Nelson, C. (2000). *Autonomy in exchange for accountability: An initial study of Pennsylvania charter schools.* Kalamazoo: Western Michigan University Evaluation Center. Retrieved May 2, 2008, from http://www.wmich.edu/evalctr/charter/pa_reports

National Center for Education Statistics. (2005a). *National assessment of educational progress: The nation's report card.* Washington, DC: Author. Retrieved May 2, 2008, from nces.ed.gov/naep3

National Center for Education Statistics. (2005b). *America's charter schools: Results from the NAEP 2003 pilot study.* Washington, DC: U.S. Department of Education. Retrieved May 2, 2008, from http://www.nces.ed.gov/pubsearch/pubsinfor.asp?pubid=2005456

National Center for Education Statistics. (2006). *The condition of education 2006.* Washington, DC: U.S. Department of Education. Retrieved May 2, 2008, from http://www.nces.ed.gov/pubsearch/pubsinfor.asp?pubid=2006071

National Center for Education Statistics. (2007). *Trends in the use of school choice: 1999 to 2003.* Washington, DC: U.S. Department of Education. Retrieved May 2, 2008, from http://www.nces.ed.gov/pubs2007/2007045.pdf

National School Boards Association. (2007). *Cleveland voucher program.* Retrieved May 2, 2008, from http://www.nsba.org

Orfield G., & Eaton, S.E. (1996). *Dismantling desegregation: The quiet reversal of* Brown v. Board of Education. New York: New Press.

Plucker, J., Muller, P., Hansen, J., Ravert, R., & Makel, M. (2006). *Evaluation of the Cleveland scholarship and tutoring program technical report 1998–2004.* Bloomington, IN: Center on Educational Evaluation and Policy.

Saporito, S. (2003). Private choices, public consequences: Magnet school choice and segregation by race and poverty. *Social Problems, 50*(2), 181–203.

Schofield, J. W. (1995). Review of research on school desegregation's impact on elementary and secondary students. In J. A. Banks (Ed.), *Handbook of research on multicultural education* (pp. 597–616). New York: Macmillan.

School Choice Ohio. (2008). *Tax-supported scholarships in Ohio.* Retrieved May 2, 2008, from http://www.scohio.org/scholarships.php

Sunderman, G. L., Kim, J. S., & Orfield, G. (2005). *NCLB meets school realities: Lessons from the field.* Thousand Oaks, CA: Corwin Press.

U.S. Department of Education. (2008). *Overview: The federal role in education.* Retrieved May 2, 2008, from http://www.ed.gov/about/overview/fed/role.html

U.S. Government Accountability Office. (2001). *School vouchers: Publicly funded programs in Cleveland and Milwaukee.* Washington, DC: Author. Retrieved May 2, 2008, from http//:www.gao.gov/new.items/d0194.pdf

U.S. Government Accountability Office. (2007). *District of Columbia opportunity scholarship program: Additional policies and procedures would improve internal controls and program operations.* Washington, DC: Author. Retrieved May 2, 2008, from http//:www.gao.gov/new.items/d089.pdf

Wells, A. S. (1993). *Time to choose: America at the crossroads of school choice policy.* New York: Hill and Wang.

Wells, A. S. (1995). Reexamining social science research on school desegregation: Long- versus short-term effects. *Teachers College Record, 96,* 691–706.

West, K. C. (1994). A desegregation tool that backfired: Magnet schools and classroom segregation. *Yale Law Journal, 103*(8), 2567–2592.

Wisconsin Department of Public Instruction. (2008). *MCPC facts and figures for 2007–08.* Retrieved May 2, 2008, from http://www.dpi.wi.gov/sms/choice.html

Witte, J. F., Stern, T. D., & Thorn, C. A. (1995). *Fifth-year report: Milwaukee parental choice program.* Madison: University of Wisconsin, Department of Political Science. Retrieved May 2, 2008, from http://www.disc.wisc.edu/choice/choice_rep95.html

Jennifer Jellison Holme

POLITICAL SOCIALIZATION

Political socialization can be defined as the processes by which a person acquires the necessary skills to function in the political world. While political socialization is an ongoing and changing process that occurs over a lifetime, researchers have primarily focused on how children, teenagers, and young adults learn about politics. If individuals can learn about and participate in civic life as children and adolescents, then hopefully they will continue this civic participation throughout the life course.

Since the mid-1990s there has been a resurgence of political socialization research both domestically and internationally. Although each discipline emphasizes different processes and factors, researchers in mass communication, political science, developmental psychology, education, and other fields are focusing on political socialization. Some even suggest that this renewed interest heralds the "rebirth" of political socialization research (Niemi and Hepburn, 1995).

Researchers have identified three primary agents of political socialization: parents, school, and the media. Parents and the family were originally viewed as the primary agents of political socialization, and early research emphasized a transmission model that suggested children would model the political attitudes and behavior of their parents. More recent research has instead found that children are not a blank slate, and the processes of political socialization are affected by variables such as the child's age, ethnicity and socio-economic status, parents' educational level, and family communication patterns (e.g., McDevitt and Chaffee, 2002).

Similar to the role of parents, early research found that school plays a very important role in the political socialization of children and young adults. However, recent research on schools finds their effects to be more complex. The amount of political and civic knowledge learned in high school is generally low, but is more effectively learned when active teaching methods are used. The *entire* school experience is now understood to be contributing to citizen development. This includes both extracurricular activities at school and voluntary civic participation within the community.

Researchers in the early 21st century also understand that mass media variables such as TV news watching, newspaper reading, and Internet news are all important antecedents of political socialization (e.g., Atkin, 1981; McLeod, 2000). Television in particular is a "bridge to politics" for young people that can bring to life political concepts learned in school.

Theoretically, political socialization fits into a larger body of work that attempts to measure youth and adolescence in a changing world (e.g., Flanagan and Sherrod, 1999). Researchers understand that there are external, conditional, and historical factors that often affect youth development. Periods and events such as the 1960s in the United States, the breakup of the Soviet Union, the end of communism in the 1990s, the events of 9/11, and the 2008 presidential campaign of Barack Obama have had profound effects on youth and their political socialization.

SEE ALSO Volume 1: *Activity Participation, Childhood and Adolescence; Civic Engagement, Childhood and Adolescence; Data Sources, Childhood and Adolescence; Socialization;* Volume 2: *Political Behavior and Orientations, Adulthood; Social Movements.*

BIBLIOGRAPHY

Atkin, C. K. (1981). Communication and political socialization. In D. D. Nimmo & K. R. Sanders (Eds.) *Handbook of political communication* (pp. 299–328). Beverly Hills: Sage.

Flanagan, C. A., & Sherrod, L. R. (1999). Youth political development: An introduction. *Journal of Social Issues, 54,* 447–456.

McDevitt, M., & Chaffee, S. (2002). From top-down to trickle-up influence: revisiting assumptions about the family in political socialization. *Political Communication, 19,* 281–301.

McLeod, J. M. (2000). Media and civic socialization of youth. *Journal of Adolescent Health, 27S,* 45–51.

Niemi, R. G., & Hepburn, M.A. (1995). The rebirth of political socialization. *Perspectives on Political Science, 24,* 7–16.

Edward M. Horowitz

POVERTY, CHILDHOOD AND ADOLESCENCE

For several industrialized nations, especially the United States, the proportion of children under age 18 in poverty has been a long-standing concern. In 2000 the United States had the highest child poverty rate among all developed countries with 16.2% of children younger than 18 years old deemed poor. The nations with the next highest child poverty rates in 2000 were Canada and Australia at 14%. Among European Union members in 2000, the United Kingdom had about 10% of children considered poor, whereas about 7% of children in Italy and Germany were poor. Norway and Belgium have some of the lowest rates of child poverty at about 4.5%.

Although the United States was conspicuous by having the highest child poverty rate among developed countries in 2000, that rate was below its highest child poverty rate of 20.8% in 1995. In 1995 children 0 to 5 years of age had the highest rates of poverty at about 22%. Overall, approximately 10% of American children who are poor are extremely poor—that is, they live in households with incomes less than 50% of the amount that distinguishes the poverty line—and approximately 6 million children who are extremely poor are younger than 6. In the United States and other developed nations,

the proportion of children that live in poverty varies by region, educational levels of parents, race and ethnicity, and family structure.

Noticeable fluctuations have occurred in rates of child poverty over time in the United States. Child poverty rates have varied greatly across the 1970s (14.4% in 1973), 1990s (21% in 1990), and mid-2000s (17% in 2006). Overall, for the United States, poverty rates for both non-Hispanic White and minority children have decreased since 1980, but the total child poverty rate has remained constant. The persistence of the overall child poverty rate across time can be attributed to the growing diversity of the child population in the United States alongside the economic disadvantage of minority groups relative to non-Hispanic Whites. For instance, between 1980 and 2006, the proportion of minority children increased from roughly 25% to more than 40%. Also, in 2006 the poverty rate among minority children of 27% was nearly three times the rate for non-Hispanic White children (10%). No doubt, unrelentingly high child poverty rates in the United States stem partly from the fast-changing composition of the child population.

The lack of progress in reducing child poverty in the United States over time contrasts with the progress made in reducing poverty among the American elderly. Since the late 1960s, the proportion of poor persons older than 65 years of age has steadily decreased from 25% in the mid-1960s to 9% in 2006. Evidence suggests that although Social Security and Medicare drove down the poverty rate among the elderly by transferring enough aid, which was adjusted to keep pace with inflation and cost of living increases, the several public transfer programs for poor children have been ineffective at driving down their poverty rate. Unsurprisingly then, the 2006 child poverty rate of 17% in the United States was nearly double the rate for elderly Americans.

Apart from the estimates of child poverty at any given time in a country, the longer a child stays poor in that country, the harder it is for the family to sustain expenditures on goods and services that are important to that child's development. In addition, prolonged low income fuels social exclusion, which can have long-lasting consequences. Historically among developed nations, the United States has recorded some of the highest figures for children remaining in poverty from one year to the next. Around 6–7% of American children in the poorest one-fifth of families in one year are still there the next year. Moreover, in the United States some 5–6% of children were in the poorest one-fifth in each of 10 consecutive years. For the latter group of children, mobility and opportunity appear extremely limited and signals the substantial persistence of low family income.

MEASUREMENT OF CHILD POVERTY

The base of knowledge about child poverty rates and factors associated with variation in those rates are produced mostly from countries' annual national-level census data. Less common but steadily growing is the use of longitudinal data, which follows children as they grow up and then transition into adulthood, to generate knowledge about the consequences of growing up poor. For the latter sources of data, two main types of surveys have been used: general household panels that follow a random sample of households for successive years and studies of birth cohorts that trace a group of children born around the same time. Using the longitudinal data, researchers have sought to explore questions at the core of life course research, such as the links between disadvantage during childhood and outcomes for children during childhood and adolescence as well as in adulthood.

Most of the research using longitudinal data has occurred in the United States using the Panel Study of Income Dynamics, which started in 1968 and is a very rich source of information regarding the different stages of childhood, spells of poverty, and links to later adult outcomes. The United Kingdom has also conducted considerable longitudinal research using birth cohort data from 1958 and 1970 and, more recently, the British Household Panel Study (begun in 1991) and the Millennium Cohort. Other countries, such as Canada, Australia, New Zealand, as well as the European Union, have started household panels, cohort studies, or both so they can not only produce annual estimates of child poverty but estimates of the effects of income, family structure, and deprivation on children's health and later adult outcomes.

Greater use of scientifically generated longitudinal data will undoubtedly occur in the future as scholars and policy makers aim to precisely monitor trends in child poverty, better understand why so many children are poor, and improve estimates of the impact of childhood poverty on children's later adult lives. Moreover, research will continue efforts to broaden the conception of childhood poverty so that the multidimensionality of the problem is recognized and measured rather than remaining reliant on income alone to measure child poverty. Using only income to conceptualize poverty among children is a narrow approach and does not capture the actual experiences of children. Further, this practice reinforces a unidimensional perspective on child poverty that can overlook issues of social exclusion and stratification.

CORRELATES OF CHILD POVERTY

High rates of childhood poverty in the United States are attributable to several demographic and economic factors, including growing wage inequality, increasingly ineffective antipoverty programs, and rising numbers of single-parent families. Certainly, the erosion of earnings among lower educated workers (e.g., both high school dropouts and high school graduates) is associated with higher child poverty rates. In the first decade of the 2000s the least educated workers earn far less than their counterparts of three or four decades ago. Because the economy has shifted from an industrial to a technological and informational one, wages of the least educated have steadily eroded. Free trade agreements, increased capital flows to developing countries, low minimum wages, and more offshore manufacturing have lowered the demand for manual and low-skill workers in the United States and thereby reduced their earning power. Hence, the number of the working poor, or workers whose income is below the poverty line, has risen noticeably. In 2006, two-thirds of all children growing up in poverty in the United States had one or more working parents and one-third had a parent working full-time year round.

When trying to understand why so many children are poor in America, it is important to remember that whereas the items used to measure poverty have stayed the same over time, the items that can impoverish families have changed. In 1965 the poverty line was set at three times the cost of the basic food basket for a family of a given size. Back then, food items drove poverty-line calculations. However, the costs of many other items overlooked at the time, such as housing, childcare, and transportation, have increased much faster than food items. In other words, items once thought unimportant to families' (necessary) expenditures are necessities now; these new items, such as childcare, are significant financial outlays for families in the early 21st century, especially those families that are minorities or headed by single parents.

Currently, about three-quarters of American households spend over half of their income on rent, and about one-quarter of those households are overcrowded. For a child that lives at the 2008 poverty level, his or her family has only about 60% of the purchasing power of a family that lived at the poverty level four decades ago. Moreover, whereas the welfare system has become less generous and benefits have declined in real dollars, more children compared to the 1960s are fully welfare dependent—that is, they have no other sources of income except public income transfers.

Besides dramatic economic transformations, the striking increase in single-parent families over the past 20 to 30 years is also associated with child poverty. In 1970 the number of single-parent families with children under the age of 18 was 3.8 million. By 1990 the number had more than doubled to 9.7 million. Of these single-parent

Family Attempts to Re-settle. *Shanika Reaux's children Da-Vone Lewis (C), 2, O-Neil Lewis (R), 1, and Tatiana Lewis (L), 6 months, wait for a bus in New Orleans, LA.* **GETTY IMAGES.**

families, most are headed by females. These single mothers raising children have the highest rate of poverty across all demographic groups. Although the annual statistics shift from year to year, at least 55% of children who live in single-parent, fatherless families are poor compared with about 10% of children in two-parent families. Fatherless families are more likely to be poor because of the lower earning capacity of women, inadequate public assistance and childcare subsidies, and lack of enforced child support from nonresidential fathers. The median annual income for female-headed households with children under 6 years old is roughly one-fourth that of two-parent families; however, the number of children per family unit is generally comparable at approximately two per household.

EFFECTS OF CHILD POVERTY

Numerous studies report the effects of poverty on an individual's life during his or her childhood or later during his or her adult years. Studies of the effects of poverty on an individual's outcomes during his or her childhood have focused on physical, cognitive, behavio-

ral, and emotional outcomes. Although a multitude of outcomes within these childhood domains have been examined, most studies have sought to discover whether poor children have had higher rates of mortality, morbidity, teenage pregnancy, environmental distress, high school noncompletion, alcohol and drug abuse, school exclusions, and lower self-esteem compared with non-poor children. Although the findings are oftentimes mixed, many of the studies strongly suggest that the magnitude of the effect of child poverty during children and later adulthood is not just a function of the incidence of child poverty but of its duration and severity as well.

Studies of the impact of child poverty on physical outcomes for children have documented that poverty is associated with higher rates of poor health and chronic health conditions among children. National surveys find that compared with parents who are not poor, parents who are poor more often rate their children's health as *fair* or *poor* and are less likely to rate their children's health as *excellent*. Children who are poor have higher rates of hospital admissions, disability days, and death rates. They also have inadequate access to preventive, curative, and emergency care and are affected more frequently by poor nutrition. Finally, many studies have documented the association between child poverty and teenage motherhood. Adolescents who are poor are three times as likely to have a child born out of wedlock than adolescents who are not poor. These births are associated with increased rates of low birth weight and perinatal and postnatal complications. Other studies suggest that childhood poverty may negatively affect children's intelligence and educational achievement and hence their later adult productivity, employability, and welfare use. Many studies using a variety of data sources that have compared poor children with nonpoor children suggest that children's IQ test scores are associated with poverty. Also, as noted earlier, a child's educational attainment is shaped by the incidence and duration of poverty experienced earlier in the life course.

Another body of research that possesses a more ecological perspective has suggested that childhood poverty is associated with many environmental and contextual inequalities. Compared with nonpoor children, poor children are exposed to more family violence, marital instability, parental separation, and household compositional changes. Psychological studies have documented that poor children's parents are more authoritarian (i.e., they use higher levels of control but less parental warmth in their childrearing tactics), read less to their children, and are less involved in their children's school activities. The children also have fewer educational resources at home, such as books and computers, but watch more television. Poor children live in noisier, more crowded homes that need repair and receive fewer services from local organizations. Exposure to lead hazards is yet another example of how the ecology of poverty directly impacts child health. Four to 5 million children, the vast majority of whom are poor, reside in older homes with lead levels exceeding the accepted threshold for safety. More than 1.5 million of these children (younger than 6 years) have elevated blood lead levels. Thus, the environment associated with child poverty includes multiple risk factors that are far from conducive to healthy child development.

Even in the absence of consequences for children either in the short term or in later adulthood, society might be concerned about the effects of child poverty for social justice and equity reasons. Children have little control over their economic circumstances, which might result from their parents' ill-informed decisions. Apart from children representing an investment in the future, society may wish to protect children because it is just and fair. Overall, poverty among children diverts resources that could be used elsewhere, reduces the future stock of human capital, and creates a variety of social problems from which all people might suffer. Hence, as discussed earlier, the effects of prolonged income deprivation among children may only be fully realized when they are adults.

ONGOING RESEARCH

As of 2008, studies can point only to the correlations between childhood poverty and outcomes in later adult life. Ideally, researchers would like to show a causal link between childhood poverty and later adult outcomes based on solid theory.

One such theoretical argument is that income levels represent the level of investments in children. Higher income investments translate into higher child consumption of goods and services and, hence, better outcomes now and later (e.g., improved physical health). An alternative theory suggests that the value of higher income is that higher levels of income lower parents' stress levels, thereby improving their parenting and, hence, their children's development. Yet another theory asserts that income from employment is important not because of the level of income per se but because working parents are valuable role models for their children, and they transfer positive norms about work to their children.

Even if income safeguards a high standard of living for children, some argue that income is still not the key determinant of children's future outcomes. Susan Mayer (1997) contended that parental characteristics in the workplace (e.g., reliability and interpersonal skills) enhance children's later outcomes net of parental income. In other words, parental characteristics independent of their income have a larger effect on IQ scores, educational attainment, and employability. This line of research reveals that low income does indeed have a genuine impact on a range of future events for a child net of other factors associated with income but that the effect of low-income status during childhood on later adult outcomes is not as large as was first thought. For example, the effect of child poverty on high school dropout rates might not be as great as first thought. Mayer (1997) noted that if the poorest one-fifth of families in the United State had their income doubled, then that increase would lower the high school dropout rate only from 17.3% to 16.1%. Other researchers, such as Dan Levy and Greg Duncan (2000), disagreed and found that the degree to which household income level during childhood affects later adult outcomes is far from overstated. Overall, at this stage the debate continues over the causal mechanisms between childhood poverty and later adult outcomes and the magnitudes of the estimated correlations.

Having accurate estimates of the effects of low income during childhood on later adult outcomes and child well-being is fundamental to effective policy interventions. In the United States, the welfare system now places a great emphasis on labor force participation among parents but not necessarily on earnings that raise living standards. The distinction is important because employment without income growth among poor parents may mean that poor children gain little. Indeed, the employment demonstration projects that have been conducted seem to indicate that employment programs for welfare mothers that failed to raise earnings had little impact on child well-being. At the same time, increased work effort among welfare mothers had no ill effects on their children either. In contrast, work programs that increased incomes among single mothers seemed to have positive effects on child well-being. Another factor that policy makers must consider is that even if raising income is a blunt policy instrument for improving poor children's outcomes, this device might still have greater promise in the short and long term than attempts to change parenting skills or values.

Without better research on the immediate and long-term impact of child poverty, social policy is bound to remain in the political arena. The policy prescriptions advocated will reflect that politicization. Some will argue that the eradication of child poverty must be based on parents' choices with respect to work. Others will claim that the solution to child poverty requires government promoting social change. Like other social problems, the solution to child poverty lay somewhere between both opposing viewpoints because no one factor will ever, by itself, totally explain child poverty. Certainly, in nations such as the United States that possess such a diverse population, a multifaceted strategy to reducing child poverty is imperative. That approach for the United States and numerous other developed countries must include macroeconomic policies that raise incomes, social policies that strengthen families, and work-family policies that help low-income families balance work and family responsibilities and seize labor market opportunities.

SEE ALSO Volume 1: *Family and Household Structure, Childhood and Adolescence; Health Differentials/Disparities, Childhood and Adolescence; Policy, Child Well-Being; Neighborhood Context, Childhood and Adolescence; Socioeconomic Inequality in Education.*

BIBLIOGRAPHY

Bhattacharya, J., Currie, J., & Haider, S. (2004). Poverty, food insecurity, and nutritional outcomes for children and adults. *Journal of Health Economics, 23*(4), 839–862.

Bradbury, B., Jenkins, S. P., & Micklewright, J. (Eds.). (2001). *The dynamics of child poverty in industrialized countries.* Cambridge, England: Cambridge University Press.

Brooks-Gunn, J., Duncan, G., & Aber, L. (1997). *Neighborhood poverty: Vol. 1. Context and consequences for children.* New York: Russell Sage Foundation.

Danziger, S., Heflin, C., Corcoran, M., & Wang, H. C. (2002). Does it pay to move from welfare to work? *Journal of Policy Analysis and Management, 21*(4), 671–692.

Duncan, G., & Brooks-Gunn, J. (Eds.). (1997). *The consequences of growing up poor.* New York: Russell Sage Foundation.

Duncan, G., & Brooks-Gunn, J. (2000). Family poverty, welfare reform, and child development. *Child Development, 71*(1), 188–196.

Duncan, G., Yeung, W. J., Brooks-Gunn, J., & Smith, J. R. (1998). How much does childhood poverty affect the life chances of children? *American Sociological Review, 63*(3), 406–423.

Fiscella, K., & Williams, D. R. (2004). Health disparities based on socioeconomic inequities: Implications for urban health care. *Academic Medicine, 79*(12), 1139–1147.

Haveman, R., & Wolfe, B. (1994). *Succeeding generations: On the effects of investments in children.* New York: Russell Sage Foundation.

Hoynes, H. W., Page, M., & Stevens, A. (2006). Poverty in America: Trends and explanations. *Journal of Economic Perspectives, 20*(1), 47–68.

Iceland, J. (2005). Measuring poverty: Theoretical and empirical considerations. *Measurement, 3*(4), 199–235.

Klebanov K., Brooks-Gunn, J., McCarton, C., & McCormick, M. C. (1998). The contribution of neighborhood and family income to developmental test scores over the first three years of life. *Child Development, 69*(5), 1420–1436.

Koen, V., & Smeeding, T. (Eds.). (2000). *Child well-being, child poverty, and child policy in modern nations: What do we know?* Bristol, England: The Policy Press.

Lerman, R. I. (1996). The impact of the changing U.S. family structure on child poverty and income inequality [Suppl.]. *Economica, 63*(250), S119–S139.

Levy, D., & Duncan, G. (2000). *Using sibling samples to assess the effect of childhood family income on completed schooling* Chicago: Joint Center for Poverty Research. Retrieved July, 1, 2008, from http://econpapers.repec.org/paper

Lichter, D. T. (1997). Poverty and inequality among children. *Annual Review of Sociology, 23*, 121–145.

Lichter, D. T., & Jayakody, R. (2002). Welfare reform: How do we measure success? *Annual Review of Sociology, 28*, 117–141.

Mayer, S. E. (1997). *What money can't buy: Family income and children's life chances.* Cambridge, MA: Harvard University Press.

Wood, D. (2003). Effect of child and family poverty on child health in the United States. *Pediatrics, 112*(3), 707–711.

Peter D. Brandon

PRE-SCHOOL

SEE Volume 1: *Child Care and Early Education; School Readiness.*

PRIMARY SCHOOL

SEE Volume 1: *Stages of Schooling.*

PRIVATE SCHOOLS

Contemporary understandings of private or independent schools are the product of the mass institutionalization of public education. Virtually all schooling was once private. It was in the late 1700s when some European governments began to erect mass public education systems. North American jurisdictions initiated their public systems in the early to mid-1800s. Across the Western world, public school systems were institutionalized over the 20th century, and these systems were rapidly built in developing nations after World War II (1939–1945) (Boli, Ramirez, & Meyer, 1985; Meyer, Ramirez, & Soysal, 1992). In the early 21st century, public schools that, in principle, make themselves universally accessible across gender, ethnic, racial, and regional lines are global in nature. As a result, private schools are typically framed as those educational bodies that are not mandated for all children by a national, regional, or local government and that are not fully funded through state-raised taxes. Indeed, though many kinds of religious and elite private schools predate their public counterparts, today they distinguish themselves as independent alternatives to mass public schooling.

The criteria that demarcate private and public schools vary greatly around the world. Different levels and forms of governance, accreditation, and funding coexist across and within nation-states (Forsey, Davies, & Walford, 2008). In North America, private schools are typically defined as those that are not fully governed, or fully funded, by governments. In this essay, the focus is specifically on North American private schools. In the United States and Canada, the common school tradition idealizes neighborhood public schools as cradles of democracy, equalizers of economic opportunity, and the foundation of a shared civic culture. This historical norm shapes policy debates that arise about private schools, such as the degree to which governments should fund such schools. American advocates of school choice, such as authors John Chubb and Terry Moe (1990), for instance, champion voucher programs that would extend public monies to private schools, an initiative they associate with parental rights, equity, and educational quality. Canadian advocates of charter schools or vouchers similarly cite American policy experiments, or the writings of Chubb and Moe, Milton Friedman (1912–2006), or economist Caroline Hoxby to buttress their claims, but then adapt these ideas for their own peculiar landscape, where virtually all provinces already fund religious schools, there are no political principles to separate church and state, and there is less of the antistate rhetoric that is so popular in the United States.

TRENDS IN PRIVATE SCHOOL ENROLLMENT

As of 2006 about 11% of all American students are enrolled in private schools, a decline from previous decades (National Center for Education Statistics, 2006). The comparable figure in Canada is about 8%, a sizable increase from previous decades (Guppy, 2005). Why are nominally private enrollments falling in the United States but rising in Canada? Some important trends are masked by these figures. Notably, the American decline in private enrollments is largely because of dropping attendance in Catholic schools. Further, that decline obscures the fact that semiprivate alternatives in the United States, such as magnet schools, charter schools, homeschooling, and voucher experiments, are rapidly growing. Moreover, there are different forms of choice in Canada. Many provinces fully fund Catholic schools and consider them to be public, and some provinces extend funds to independent schools. Yet Canada has far fewer charter schools and no voucher experiments. In general, many students in both nations are choosing to attend schools other than their local, regular public school, and these numbers appear to be growing.

TYPES OF PRIVATE SCHOOLS

Private schools in North America are not all alike, of course. The demand for these schools is sometimes associated directly with certain social statuses, but sometimes it reflects more purely pedagogical preferences. These varying associations form the basis for three distinct subsectors of private schools. Those schools that incorporate the teaching of religion into their curriculum represent the largest subsector, and in many jurisdictions, they are public schools' main competitors for students— distinguished in the public eye by their disciplinary climate, moral instruction, uniforms, and other symbols of community identity. So-called *elite* schools cater to high socioeconomic groups, often have long histories, and assume boarding school forms. Within the elite sector, school reputations are strongly conditioned by their students' socioeconomic status. The prestige of much-sought-after elite boarding schools stems as much from their symbols of tradition (premodern architecture, plush lawns, classical curricula) as from their particular instruction. Third-sector schools (Davies & Quirke, 2007), in contrast, are characterized less by the social statuses of their students (e.g., socioeconomic status or religion) and more by their brands of pedagogy and curricula, particularly specialities for languages, science, sports, and people with disabilities. These schools forge their identities more on cognitive grounds than on moral or socioeconomic reputations.

IMPACT OF PUBLIC SCHOOL ATTENDANCE

What is the hypothesized impact of private schools on students? Much research has been done on three major facets of private schooling: social organization, social inequality, and socialization.

Advocates such as Chubb and Moe (1990) voice two bold claims about the social organization of private schools. First, they declare that private schools have an

SINGLE-SEX SCHOOLS

In the United States at least 366 schools have single-sex classrooms and 88 public schools are completely single-sex, according to the National Association for Single Sex Public Education. Private schools are far more likely than public schools to offer single-sex education. Many teachers and parents argue that students learn better when not distracted by the opposite sex, but others counter that the real world is not segregated by sex and therefore single-sex schools do not prepare students for the actual work environment. Those in favor of single-sex schools believe that gender differences exist in the brain, which leads to different learning and behavioral styles for boys and girls, and that single-sex schools reduce the amount of unhealthy competition and distraction students of the opposite sex provide. Opponents of single-sex schools believe that coeducational institutions can spark different kinds of learning for every student; the diversity of having both boys and girls in the classroom is beneficial to the understanding of ideas and viewpoints; and schools attended by both genders will better prepare students to interact with a more diverse population. Some also question whether single-sex schools violate the 14th Amendment to the Constitution, which states that schools must be integrated. In general, the majority of the American public does not endorse single-sex schooling (68%), but with mounting concerns over how the education system may advantage or disadvantage one sex, discussions of utilizing single-sex schools continues.

Anne E. McDaniel

advantage over their public counterparts because the former are more directly subject to market pressures and are less constrained by bureaucratic regulation. They argue that state governance encourages public schools to conform to legal conventions instead of providing incentives for effective instruction. Bureaucratic shackles, such as having labor unions demand the hiring of certified teachers, boards impose curricular guidelines, and governments leverage instructional practices by funding formulae, discourage public schools from directly responding to the needs of their clients. Reducing state bureaucracy in education is said to eliminate the need for mindless rule-following, allowing schools to shed ineffective struc-

tures and to redirect their energies to solving pedagogical problems. A competitive marketplace is touted as more sharply rewarding pedagogical success, punishing pedagogical failure, and encouraging schools to clarify their missions, demonstrate their effectiveness, and satisfy customers.

However, scholars who study educational markets have encountered a situation that is considerably more complex than this theory allows. Competitive forces do not always spawn school variety; parental incentives to choose schools can vary greatly, and parents do not always equate school quality with performance on standardized test scores (Belfield & Levin, 2002; Davies & Quirke, 2007).

A second rationale for private schools centers not on market competition but on their alleged ability to foster effective school communities. For a quarter century, American scholars have investigated whether Catholic schools provide learning environments that are superior to those in public schools. Beginning in the early 1980s, author James Coleman (1990) and his colleagues contended that Catholic schools produce higher levels of learning. Their central concern was whether or not cultural and network characteristics of school communities can boost student achievement. Coleman focused not on market competition per se but on functional communities in schools. A functional community, according to Coleman, enjoys durable parental networks, widespread feelings of trust, caring and social responsibility, and effective norms of reciprocal obligation—all of which can enhance student learning. Schools are said to draw these resources from broader functional community ties. Coleman contended that teachers, parents, and students in Catholic schools utilize bonds that are forged through their common affiliation with the Catholic church and that these bonds raise educational expectations among these stakeholders. Coleman also argued that because public schools lack a comparable institution to reinforce such norms, those schools cannot generate similar levels of social capital. In the 1980s critics disputed these claims for a variety of reasons, but a second generation of studies, utilizing superior analytic techniques and forms of data, has tended to confirm the existence of Catholic school effects. The research debate has been redirected from whether or not those effects actually exist toward specifying their particular form and identifying their causal mechanisms (Hallinan, 2006; Morgan, 2000).

The degree to which private schools reinforce inequality may differ by subsector. Traditionally, the demand for elite private schools has come from the wealthiest rungs of society (Cookson & Persell, 1985), and most observers concede that elite schools largely serve

Catholic School. *A seventh-grade class makes the sign of the cross after a class prayer at St. Rose Catholic School, in Newtown, CT.* AP IMAGES.

to prepare the privileged to assume positions of power. However, a livelier American debate focuses on whether religious private schools, particularly Catholic schools, can provide equitable outcomes for disadvantaged populations. Coleman (1990) contended that Catholic schools lessen the effects of poverty and minority status on student achievement, compared to their public counterparts. Likewise, advocates of school choice, such as Chubb and Moe, contend that poorer parents are most motivated to take advantage of voucher programs because they are far more likely to be trapped in substandard schools and less likely to be able to afford private school tuition or to live in neighborhoods with superior public schools. In contrast, opponents claim that wealthier and more educated parents are likelier to seek Catholic schools and choice programs because they are more knowledgeable about schooling, are more comfortable discussing school issues with professionals and other parents, and have better information networks. The empirical literature on choice offers a mixed assessment. Many American studies show that parents who seek private schools, charter schools, and homeschooling generally have above-average levels of education, income, and socioeconomic status within their communities and

that these traits tend to explain most of the achievement advantages they enjoy over public schools (Lubienski & Lubienski, 2006).

Do private schools socialize students in unique ways? Private schools are typically proclaimed to facilitate higher levels of social control, less deviance, and healthier school climates. However, scholars face a major challenge when addressing this question because of the strong possibility that student outcomes may reflect selection effects as well as the independent effects of schools themselves. That is, because private schools are usually smaller, have wealthier and more homogeneous student populations, and are free to offer distinct missions, like-minded families may self-select into those schools. Student outcomes may reflect more readily the characteristics of those self-selecting families than the efforts of private educators. Research suggests that religious schools influence a student's sense of well-being and enhance community participation, over and above their prior levels of religiosity (Schneider, Hoogstra, Chang, & Sexton, 2006). Research on the elite subsector suggests that boarding schools can resemble total institutions, in which students are surrounded by their peers at all hours that, covertly or overtly, teach each other how to dress, speak, act, and think. At the extreme, the intense experience of boarding schools can encourage some students to adapt by resorting to drugs and alcohol, or by committing white-collar crimes, such as purchasing upcoming tests and exams, study notes, and homework assignments, as well as forging notes and passes. This deviance is frequently hidden from adults through a code of silence that is policed in dense peer networks (Cookson & Persell, 1985). But, in regular day private schools, research has not clearly established whether or not student behavior varies drastically between private and public schools, once family and community characteristics are controlled. Overall, although there are clear differences between private schools and public schools, there is no consensus among researchers as to whether private schools have unique impacts on students that are independent of the traits of students themselves.

SEE ALSO Volume 1: *Academic Achievement; Coleman, James; Gender and Education; Policy, Education; Racial Inequality in Education; School Tracking; Social Development; Socioeconomic Inequality in Education; Stages of Schooling.*

BIBLIOGRAPHY

Belfield, C. R., & Levin, H. M. (2002). The effects of competition between schools on educational outcomes: A review for the United States. *Review of Educational Research, 72*(2), 279–341.

Boli, J., Ramirez, F. O., & Meyer, J. W. (1985). Explaining the origins and expansion of mass education. *Comparative Education Review, 29*(2), 145–170.

Chubb, J., & Moe, T. (1990). *Politics, markets, and America's schools*. Washington, DC: Brookings Institution.

Coleman, J. S. (1990). *Equality and achievement in education*. Boulder, CO: Westview Press.

Cookson, P. W., Jr., & Persell, C. H. (1985). *Preparing for power: America's elite boarding schools*. New York: Basic Books.

Davies, S., & Quirke, L. (2007). The impact of sector on school organizations: The logics of markets and institutions. *Sociology of Education, 80*(1), 66–89.

Forsey, M., Davies, S., & Walford, G. (Eds.). (2008). *The globalization of school choice?* Oxford, U.K.: Oxford Series on Comparative Education; Symposium Books.

Guppy, N. (2005). *Parent and teacher views on education: A policy-maker's guide*. Kelowna, British Columbia, Canada: Society for the Advancement of Excellence in Education.

Hallinan, M. T. (Ed.). (2006). *School sector and student outcomes*. Notre Dame, IN: University of Notre Dame Press.

Lubienski, S. C., & Lubienski, C. (2006). School sector and academic achievement: A multilevel analysis of NAEP mathematics data. *American Educational Research Journal, 43*(4), 651–698.

Meyer, J. W., Ramirez, F. O., & Soysal, Y. N. (1992). World expansion of mass education, 1870–1980. *Sociology of Education, 65*, 128–149.

Morgan, S. (2000). Counterfactuals, causal effect heterogeneity, and the Catholic school effect on learning. *Sociology of Education, 74*(4), 341–374.

National Center for Education Statistics. (2006). *Digest of education statistics, 2006*. Accessed February 22, 2008, from http://nces.ed.gov/programs/

Schneider, B., Hoogstra, L., Chang, F., & Sexton, H. R. (2006). Public and private school differences: The relationship of adolescent religious involvement to psychological well-being and altruistic behavior. In M. T. Hallinan (Ed.), *School sector and student outcomes* (pp. 73–100). Notre Dame, IN: University of Notre Dame Press.

Scott Davies
Stephanie Howells

PUBERTY

Puberty, one of the few universals in early development, involves a set of biological events that produce profound change throughout the human body. For both boys and girls, these physiological changes come about through complex interactions among the brain, the pituitary gland, the gonads, and the environment. For girls, the typical sequence of physical change includes a significant growth spurt, weight gain, breast development and pubic hair growth, and the onset of menstruation. Boys also undergo a significant growth spurt and weight gain along with voice changes and penile growth. Although these sequences are fairly standard, the onset and tempo of changes can vary considerably. For girls, puberty typically begins between the ages 8 to 14, and for boys, between ages 9 to 15. The progression or speed with which young people undergo these changes is also variable. As a result, young people who are the same chronological age can vary considerably in terms of physical development. In all, the amount of physical growth experienced during this period is second only to growth experienced during the first year of life.

Because puberty also ushers in adolescence, the significance of this transition extends beyond the physiological or biological. Simply put, the appearance of adult physical characteristics is often linked with adult expectations of behavior (Graber, Petersen, & Brooks-Gunn, 1996). Thus, young peoples' transformed bodies signal their reproductive capacity, which, in turn, often ushers in a more sexualized image of the self and elicits new expectations from others (Caspi, Lynam, Moffit, and Silva, 1993; Martin, 1996). This biopsychosocial process is gendered, according to empirical studies. Although boys' bodies change in dramatic ways and these changes affect the way they understand themselves and are perceived by others, the more pressing social and psychological consequences of puberty are more pronounced for girls than boys. But for both, puberty is a *life course transition*, a physical change that also redefines social roles and brings about new social expectations and obligations.

HISTORICAL AND DEVELOPMENTAL PERSPECTIVES

The timing of the pubertal events as well as the social and psychological implications of puberty have fluctuated throughout different historical periods. At the population level, increased affluence, changing diets, and rising levels of obesity have contributed to a secular decline in age of pubertal onset for American boys and girls (Herman-Giddens, Slora, Wasserman, Bourdony et al., 1999; Herman-Giddens, Wang, & Koch, 2001). For instance, the average age at menarche at the start of the 20th century was around 15 or 16; at the turn of the 21st century the average age is around 12.5 (Brumberg, 1997). Similarly, American boys today are taller, heavier, and show earlier genital maturation and pubic hair growth than did boys in the late 1960s and 1970s (Hermans-Giddens et al, 2001). These changes have spurred a lot of press coverage and public concern about pubescent elementary school children but, by and large, menstruating second graders remain an anomaly.

These changes in the pubertal process, however, remain meaningful. Although the age at onset of puberty has declined, no parallel acceleration in young people's emotional and cognitive skills has occurred, meaning girls and boys in the early 21st century may look adult-like at 12 or 13 but their thinking is still that of a young girl or boy. Thus, a gap or mismatch between the physical and socio-emotional development of young people is evident (Petersen, Crockett, Richards, and Boxer, 1988). At the same time, the salience of the body, especially for girls, and the cultural context in which young people transition into adolescence has shifted in important ways. In *The Body Project* (1997), Joan Jacobs Brumberg examined girls' diaries from the 1830s through the 1990s to understand changes in the social construction of American girlhood. She argues that over time, the female body has become the primary expression of girls' identity; the body is *the* thing to be managed and maintained but also treated with increasing ambivalence. Moreover, contemporary American culture, more so than ever, emphasizes the importance of sex, sexualizes the female adolescent figure, and provides fewer social protections for young people. Together, these biological, psychological, and social changes make puberty an important piece of the story of adolescence.

PUBERTAL STATUS AND TIMING

Two key dimensions of puberty are *pubertal status* and *pubertal timing*. Both have been used to examine the relationship between puberty and adolescent adjustment. Pubertal status, on the one hand, refers to the level of physical development that young people have reached (e.g., prepubertal, midpubertal, postpubertal), without reference to chronological age. On the other hand, pubertal timing refers to whether young peoples' development is early, on-time, or late compared with their same-age, same-sex peers (Graber et al, 1996). Clearly, these dimensions are related but the expected role each plays in adolescent development is different.

Many different methods have been used to measure pubertal status, ranging from intrusive bone x-rays and physical examinations with unclothed adolescents to single item reports of overall development based on observer reports. Probably the most commonly used method for measuring pubertal development is the Pubertal Development Scale, a self-report of pubertal status obtained through interview or questionnaire (Petersen et al, 1988; Ge, Conger, and Elder, 1996). With this instrument, adolescents rate their own level of development, on a scale of 1 (not begun) to 4 (development completed), on a set of pubertal indicators. All are asked about body hair development, growth spurt, and skin changes. Girls also rate breast development and whether they have begun menstruation and boys rate the development of facial hair and voice change.

Pubertal timing is a relative concept, and thus is defined by a measure of pubertal status that is compared across same-age peers. For girls, age at menarche is probably the most extensively used measure of pubertal timing (Stattin and Magnusson, 1990). Although menarche typically occurs later in the sequence of pubertal development, approximately 6 to 12 months following the height spurt and after the development of secondary sex characteristics, it is the event around which girls organize and assimilate the myriad changes that define this transition. It is also the event around which social roles and expectations about behavior, sexual behavior in particular, are cemented by others, parents especially. This indicator also has good reliability, women and girls easily recall this date, and, because it is anchored in time, can be reported on retrospectively. For boys there is no single pubertal event that has either the salience of menarche or its measurement properties. Therefore, scholars often use an age-standardized indicator of development and then define pubertal timing as one standard deviation above or below the average as early or late.

An emerging focus among puberty scholars is an examination of the social and environmental forces that predict pubertal timing. The timing of pubertal changes is influenced by a host of factors including genes, socio-economic status, diet, environmental toxins, prepubertal fat and body weight, and chronic stress. This last factor, chronic stress, has been the focus of recent work on pubertal timing. Operationalized with indicators of maternal and paternal parenting behaviors, chronic stress in the family environment was associated with an accelerated transition to puberty among girls only (Belsky, Steinberg, Houts, Friedman et al, 2007). Others have addressed the question of timing by using twin and sibling design studies to explore genetic and environmental factors. Work by Ge and colleagues (2007) suggests that, for both girls and boys, about half of all individual differences in pubertal timing was explained by genetic factors and the other half by environmental factors that siblings may not necessarily share, such as nutrition, health status, and early family environment.

Research focused on pubertal development is often guided by the *stressful change hypothesis*, which posits that the changes that define puberty, regardless of when they occur, are stressful for young people. This hypothesis rests on the notion that change is inherently stressful and this stress is often expressed with feelings of psychological distress and increased emotional distance in relationships with parents immediately following puberty. For instance, adolescents often become less involved in family activities shortly after the onset of puberty (Collins, 1990).

Much of the current research on puberty, however, centers on its timing. Variability in age at onset can be pronounced; according to the life course perspective, the significance of a life course transition is dependent, in part, on when it occurs. Ample evidence has documented that timing has significant implications for the lives of girls (Caspi et al. 1993; Cavanagh 2004; Cavanagh, Riegle-Crumb, and Crosnoe, 2007; Haynie 2003; Ge et al, 1996; Graber et al, 1996; Stattin and Magnusson, 1990) and, to a lesser extent, boys (Felson and Haynie, 2002).

Two hypotheses have been proposed to understand the relationship between the pubertal timing and young people's social, psychological, and behavioral outcomes: the off-time hypothesis and the early-maturing hypothesis. The *off-time hypothesis* states that, compared to those who transition on-time, young people who are either early or later maturing will experience more psychological and social problems. Based on the deviance proposition, departure from the normative developmental schedule is less socially desirable and even stressful. Because early and later maturing adolescents depart from this timetable and are visibly different from their peers, at a moment when being like everyone else matters a great deal, off-time adolescents are hypothesized to be at heightened risk for psychological and behavioral problems (Ge et al, 1996). The *early-maturing hypothesis*, on the other hand, focuses exclusively on early maturers. Because they are physically different and perceived as older by peers and adults, early maturers enter the social world of adolescence sooner, doing so with neither the support of their larger peer group nor the development time needed to acquire and integrate the skills needed to confront the new tasks in adolescence (Ge et al, 1996).

Empirical tests of these hypotheses indicate that early maturation, more so than off-time maturation, and particularly for girls, is associated with a host of negative outcomes (Stattin and Magnusson, 1990; Ge et al, 1996; Haynie, 2003; Cavanagh, Riegle-Crumb, and Crosnoe, 2007). During adolescence, early maturing girls are more likely to drink, smoke, and engage in minor delinquent behaviors (Haynie, 2003; Stattin and Magnusson, 1990), become sexually active earlier (Udry, 1988), report higher levels of psychological distress (Ge et al., 1996), and do less well in school (Cavanagh et al, 2007) than other girls. There is also evidence that early pubertal timing is associated with how these young women transition to adulthood, with early maturing girls less likely to graduate from high school and also to cohabit and marry sooner.

Often framed as a biosocial model of adolescent development, these associations come about through the *social* interpretation, by the adolescent as well as those around her, of the physiological and biological changes of early puberty (Stattin and Magnusson, 1990). Specifically, three main social psychological processes are often invoked to explain these links. First, early pubertal timing affects girls' perceptions of self. By virtue of their earlier transition to adolescence, early maturing girls are more likely to be physically out-of-step (i.e., greater breast development and curviness) with age mates at a developmental moment when both the body and social comparison increase in significance. Thus, early maturing girls perceive themselves as different and older. This feeling of difference is often associated with a negative self-appraisal, which, in turn, can heighten their risk for psychological distress and depression (Ge et al, 1996).

Second, early pubertal timing is linked with girls' peer relationships. Because early maturing girls and their peers attribute greater maturity to them than is warranted by their age, early maturing girls are more likely to select and/or be drawn into less normative friendship groups, ones that include older boys and girls and are characterized by riskier behavior and lower academic achievement (Haynie, 2003). Finally, early maturing girls are often embedded in social contexts that offer them opportunities to engage in riskier behaviors. And, because these girls had less time to integrate the coping skills needed to manage the new tasks in adolescence, they negotiate these "opportunities" often without the socio-emotional resources they need to make healthier choices (Haynie, 2003).

Two factors are worth noting. To the extent that researchers have had both hormone and social data on girls, the risks associated with early puberty timing for girls is not simply a function of hormones driving both girls' pubertal timing and behavior. Rather, the social psychological interpretation of this transition plays a key role (Udry, 1988). Second, the risks associated with early puberty do not occur as soon as menarche begins, at age 10 or 11. Instead, risks emerge as adolescent girls are granted more autonomy and independence from parents, establish more intense and intimate relationships with peers, and explore both platonic and romantic relationships with boys. In other words, the disruptive social psychological consequences of early puberty become most pronounced between 8th through 10th grades (Stattin and Magnusson, 1990).

Do these associations hold for boys? The research is less definitive. Some of the earliest work exploring the effect of pubertal timing on boys' adjustment found that early puberty was *positively* associated with their social status, such that early maturing boys, who appeared taller and older relative to their same-age peers, were more popular than others. Moreover, researchers who had both hormone and social data for boys found that the observed puberty and sexual behaviors link was largely spurious, or caused by some third factor. Unlike for girls, increases in hormones drove both pubertal development and boys' sexual behavior (Udry, 1988). These findings, combined with difficulty in measuring boys' pubertal timing and the nearly consistent negative findings for girls, limited the scientific inquiry regarding puberty for boys.

More recently, there has been renewed interest in understanding pubertal timing in the lives of boys, but still, the story remains ambiguous. On one hand, consistent with findings for girls, some studies suggest that early puberty is associated with increased psychological distress (Ge et al, 1996), delinquency, and sexual behavior (Felson and Haynie, 2002). On the other hand, there is also evidence that pubertal timing is also associated with better psychological functioning, more friends, and better academic achievement for boys (Felson and Haynie, 2002).

RACE/ETHNIC DIFFERENCES

Much of the empirical work upon which this biosocial model of adolescent behavior was developed is based on non-representative samples of White youth. Yet important race/ethnic group differences in the factors that comprise this model are evident. For instance, African-American girls reach each pubertal marker significantly earlier than do White girls (Herman-Giddens et al, 1997; Obeidallah, Brennan, Brooks-Gunn, Kindlon et al, 2000). Similarly, the median age of onset of pubic hair growth and genital development is earlier for African-American boys compared with White or Hispanic boys (Herman-Giddens et al, 2001). At the same time, the significance of an "ideal" body may vary by race and ethnicity. Latinas and, especially, White girls react more negatively to body changes at puberty than do African-American girls (Brumberg, 1997), while African -American girls seem better able to thwart the negative feelings that unattainable body images can instill. Thus, what defines "early" as well as the some of the social psychological affect of being early varies by race and ethnicity.

With these differences in mind, Cavanagh (2004) examined whether the biosocial model of development predicted girls' transition to first sex across race and ethnic groups. Using a sample of White, African-American, and Latina girls, within-group analyses revealed important race and ethnic differences in the linkages among pubertal timing, friendship groups, and sexual debut. Support for the biosocial model was found for Whites and, to a lesser extent, Latinas. Little support, however, was found for African-American girls. At the same time, Ge and colleagues (2006), studying a sample of African-American children living in Iowa and Georgia, found that pubertal timing was associated with child reports of internalizing symptoms (feeling lonely or depressed) and externalizing symptoms (acting out in class, or engaging in verbal or physical fights with peers) for both boys and girls. A continued focus on race/ethnic differences and similarities in the biosocial model of development remains an important task.

In the future, researchers will likely continue to focus on the social and genetic factors that predict pubertal timing. Research also may more explicitly incorporate genetic data in studies of biosocial development, where puberty is both an outcome and predictor, to define the study of pubertal development in the future.

SEE ALSO Volume 1: *Body Image, Childhood and Adolescence; Dating and Romantic Relationships, Childhood and Adolescence; Sexual Activity, Adolescent; Socialization, Gender.*

BIBLIOGRAPHY

Belsky, J., Steinberg, L. D., Houts, R. M., Friedman, S. L., et al. (2007). Family rearing antecedents of pubertal timing. *Child Development, 78,* 1302–21.

Brumberg, J. J. (1997). *The body project: An intimate history of American girls.* New York: Random House.

Caspi, A., Lynam, D., Moffit, T., & Silva, P. (1993). Unraveling girls' delinquency: biological, dispositional, and contextual contributions to adolescent misbehavior. *Developmental Psychology, 29,* 19–30.

Cavanagh, S. E., Riegle-Crumb, C., & Crosnoe, R. (2007). Puberty and the education of girls. *Social Psychology Quarterly, 70,* 186–198.

Collins, W.A. (1990). Parent-child relationships in the transition to adolescence: continuity and change in interaction, affect, and cognition. In R. Montemayor, G. Adams, & T. Gullato (Eds.), *Advances in adolescent development: From childhood to adolescence: A transitional period?* (pp. 85–106). Beverly Hills, CA: Sage.

Felson, R. B., & Haynie, D. L. (2002). Pubertal development, social factors, and delinquency among adolescent boys. *Criminology, 40,* 967–988.

Ge, X. J., Conger, R. D., & Elder, G. H. (1996). Coming of age too early: Pubertal influences on girls' vulnerability to psychological distress. *Child Development, 67,* 3386–3400.

Graber, J. A., Peterson, A. C., & Brooks-Gunn, J. (1996). Pubertal processes: Methods, measures, and models. In J. A. Graber, J. Brooks-Gunn, & A. C. Peterson (Eds.), *Transitions through adolescence: Interpersonal domains and context* (pp. 23–54). Mahwah, NJ: Lawrence Erlbaum.

Haynie, D. L. (2003). Contexts of risk? Explaining the link between girls' pubertal development and their delinquency involvement. *Social Forces, 82*: 355–397.

Herman-Giddens, M. E., Slora, E. J., Wasserman, R. C., Bourdony, C. J., Bhapkar, M. V., Koch, G. G., et al. (1999). Secondary sexual characteristics and menses in young girls seen in office practice: A study from the Pediatric Research in Office Settings network. *Pediatrics, 99,* 505–12.

Herman-Giddens, M. E., Wang, L., & Koch, G. G. (2001). Secondary sexual characteristics in boys: Estimates from the National Health and Nutrition Examination Survey III, 1988–1994. *Archives of Pediatrics* & *Adolescent Medicine, 155*: 1022–1028.

Martin, K. A. (1996). *Puberty, sexuality, and the self.* New York: Routledge.

Obeidallah, D. A., Brennan, R. T., Brooks-Gunn, J., Kindlon, D., Earls, F. E. (2000). Socioeconomic status, race, and girls' pubertal maturation: Results from the project on human development in Chicago neighborhoods. *Journal of Research on Adolescence, 10*, 443–464.

Petersen, A. C., Crockett, L. J., Richards, M., & Boxer, A. (1988). A self-report measure of pubertal status: Reliability, validity, and initial norms. *Journal of Youth and Adolescence, 17*, 117–133.

Stattin, H., & Magnusson, D. (1990). *Pubertal maturation in female development.* Hillsdale, NJ: Lawrence Erlbaum.

Udry, J. R. (1988). Biological predispositions and social control in adolescent sexual behavior. *American Sociological Review, 53*, 709–722.

Shannon E. Cavanagh

R

RACIAL INEQUALITY IN EDUCATION

Historically, education has been a battleground for racial and ethnic minorities fighting to gain access to valued resources and credentials. The struggle for equal access to quality schooling has been painstakingly slow and the resistance to integration extreme. Nonetheless, people look to schools to promote equality in society at large. Despite the common perception of schools as the "great equalizer," education researchers continually find most racial and ethnic students still lag behind their White peers in terms of achievement, graduation rates, and college completion. The U.S. public educational system remains one of the most unequal of the industrialized nations. This entry provides a summary of research on U.S. racial differences in education, the long-term implications of these differences, and the theories that attempt to explain inequality in public schools, along with an overview of the main cause of these differences, and then concluding with two contemporary discussions within the study of racial inequality in education. As the U.S. population becomes increasingly diverse, understanding racial and ethnic variation in educational attainment and achievement is imperative.

DIFFERENCES IN ATTAINMENT, DROPOUT RATES, PERFORMANCE, AND COURSE TAKING

Since the late 1970s, gaps in educational achievement have narrowed and educational aspirations are consistently high across all racial and ethnic groups. Yet, significant differences in high school and college attainment remain. On average, Asians and Whites have the highest probability of graduation at each level of education, followed by Blacks, Native Americans, and Hispanics/Latinos; but all racial and ethnic groups have increased their average rates of educational attainment over time. Researchers caution, however, that much variation within ethnic groups exists. For example, among Asian Americans, Chinese, Japanese, and Asian Indians have significantly higher graduation rates than Vietnamese, Laotians, and Hmong. Additionally, one's generation affects high school graduation rates. For Latinos, high school completion rates increase with each generation, but for Asian and White ethnic groups they increase only from the immigrant to the second generation (i.e., the children of immigrants) (Kao & Thompson, 2003). Although educational attainment differences across racial and ethnic groups are narrowing for high school completion, figures from the early 21st century show that inequalities still persist and are increasing at the postsecondary level. According to the 2000 U.S. census, approximately 44% of Asians and Pacific Islanders, 28% of Whites, 17% of Blacks, and 11% of Hispanics/Latinos above 25 years of age had earned a bachelor's degree or higher.

Research also finds that Blacks, Latinos, and Native Americans are more likely than Whites or Asians to drop out of high school, and that students who attend schools with high Black or Latino populations experience higher dropout. There are significant differences, however, across and between ethnic groups. For example, Rumberger (1995) found that socioeconomic status (SES), defined as the measure of an individual's or family's relative economic and social ranking, predicts dropout rates for Latinos and Whites, but not for African

Americans. Low grades, behavioral issues, and changing schools increase dropout rates for Blacks and Whites, but not for Latinos. Absenteeism predicts dropout rates for all racial and ethnic groups. Furthermore, immigrants—especially recent immigrants—are more likely to drop out than native-born students (White & Kaufman, 1997). White and Kaufman (1997) find that once factors such as generation, language, and social capital are controlled, ethnicity effects have only a minor impact on dropping out for Latinos. They also advise that substantial ethnic differences in school performance and expectations lead to variations in dropout rates across ethnic groups. But *social capital*, the ability of adult family members to invest attention, support, values, and advice in children, can be an important factor in reducing the odds of dropping out.

Standardized test scores of African Americans continue to lag behind Whites in math and reading, but this gap is narrowing. According to data from the National Assessment of Educational Progress, the Black–White reading gap for 17-year-olds shrank by 50%, and the math gap by nearly 30%, between 1971 and 1996 (Jencks & Phillips, 1998). Parental SES accounted for much but not all of the Black–White test gap (Kao, Tienda, & Schneider, 1996). Similarly, the White–Latino performance gap has narrowed over time. Asians, however, perform above or comparable to that of Whites, particularly on mathematics assessments. Ethnic and racial patterns in grades mirror that of test scores and are heavily influenced by parental SES. Using data on eighth graders from the National Education Longitudinal Study of 1988, Kao et al. (1996) found that Asians had the highest grade point average (3.24), followed by Whites (2.96), Hispanics/Latinos (2.74), and African Americans (2.73).

In many schools, students are sorted by tracks or ability groups. In general, when ability and other background characteristics are controlled, most research on the determinants of ability group assignment provides little evidence of a direct effect of race on initial placement or subsequent reassignment. Research has shown, however, that low-income and racial and ethnic minorities are disproportionately placed in low-ability groups in elementary school and in vocational tracks in middle and high school. Additionally, Hallinan (1994) found that Black students are more likely to drop to a lower track than White students and are less likely to be reassigned to higher ability groups.

LONG-TERM IMPLICATIONS OF EDUCATIONAL INEQUALITY

In many ways, the racial and ethnic inequality in American education parallels inequality in the country's occu-

pational hierarchy. High school diplomas and bachelor's degrees are important credentials that influence potential labor market outcomes, and research shows persuasively that occupational disadvantages experienced by racial minorities often result from unequal access to educational resources and differential educational attainment. Beyond individual social mobility, the United States looks to education to fulfill multiple societal goals: to transmit knowledge and literacy, socialize values and attitudes, monitor children's behavior, and prepare students for higher educational or occupational opportunities. While inequality in educational opportunity has implications for individual employment opportunities and economic well-being over the life course, these differences are also relevant for current social debates on employment, welfare reform, poverty, homelessness, and crime.

Given historical and contemporary patterns of racial and ethnic educational inequality, researchers and policymakers have voiced concerns over the equality of access to educational resources. Additionally, the civil rights and women's movements of the 1960s influenced much of the sociological research concerning equality of educational outcomes. Moreover, many researchers find that educational institutions themselves play a key role in the reproduction of racial inequality. These orientations toward research are reflected in the theoretical traditions of sociology.

THEORETICAL FRAMEWORKS ON EDUCATIONAL INEQUALITY

The Black–White achievement gap has received significant empirical and theoretical attention, and sociologists approach the subject from very different theoretical frameworks. The major theoretical developments in sociology of education reflect the larger discipline's traditions ranging from Marxist to Weberian. Most contemporary research on race and ethnic racial inequality in education falls into three categories. The first has a *functional* or economic foundation, arguing that the education system is neutral and that the lower academic performances of racial and ethnic minorities can be attributed to lower levels of human capital or credentials. The second theoretical framework, rooted in *conflict* perspectives, argues that larger structural factors such as economic or political conditions affect various ethnic groups' achievement. Finally, the third tradition has a *cultural* origin and suggests certain groups adhere to beliefs and practices that encourage academic achievement more than others.

The functionalist tradition emphasizes that members of a society share core beliefs and values, and that education is one of many institutions necessary to create an efficient society. Functionalists incorporate an economic model that describes education as a mechanism for individual mobility. The primary function of schooling is to teach and strengthen the skills and knowledge required to increase one's human capital and future economic capital. Under this model, racial disparities in educational attainment and achievement are a function of family backgrounds with varying levels of human capital. Factors such as language proficiency and ability are commonly used to explain the negative educational outcomes of racial minorities and immigrants. Thus, the low performance of some minority students is attributable to the disadvantaged position of their parents. While human capital measures do account for some of the inequality in achievement between White and racial/ethnic minority groups, the empirical evidence suggests there could be larger structural factors at work.

In contrast, conflict theories of education critique functional models by conceptualizing schools themselves as a hindrance to social mobility and as producers of inequality. Explanations under the broad umbrella of the conflict perspective examine structural and institutional-level factors that affect educational attainment and reproduce society's system of inequality. Conflict theories suggest that elite and nonelite individuals are often in conflict over the resources and curricula that should be made available to students. Social reproduction theories introduce the notion of power when understanding how education reinforces the social structure. Contemporary Marxists represent one school of thought within the social reproduction camp. Schools are seen as locales that reinforce the class structure through differential socialization patterns. Educational institutions are developed to serve the interest of the capitalist elite, with mass education used to socialize and control working-class children. Similarly, status conflict theorists examine individual outcomes within the framework of macro-level factors, suggesting that competition between groups over resources influences the educational outcomes of group members. In sum, under the conflict tradition, elites maintain their status by limiting the access of racial and ethnic minorities to valued educational resources.

Tracing its theoretical lineage to Max Weber's work on religious ethnic groups, cultural explanations suggest that different racial, ethnic, or immigrant groups, with varying cultural value systems, promote or discourage academic and economic success. Fordham and Ogbu (1986), for example, argued that the oppositional attitudes (or attitudes that consciously reject mainstream pro-education beliefs) many African-American students express toward school account for their low achievement. The majority of contemporary explanations about ethnic group differences in educational attainment and achievement fall somewhere in between cultural and conflict orientations.

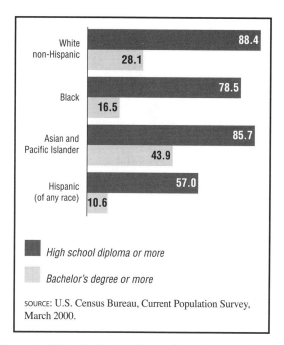

Figure 1. High school and college graduates by race and Hispanic origin, 2000. Percent of the population aged 25 and older. CENGAGE LEARNING, GALE.

CAUSES OF RACIAL INEQUALITY IN EDUCATION

A great deal of educational research focuses on the inequality of educational outcomes. But what are the main causes of these differences? Common thought suggests that the quality of school makes a significant difference in the academic achievement of students. At the extremes, where the average annual per-pupil expenditure can range from $20,000 at an elite independent private school to $3,000 in an inner city public school, the differences between school experiences is fairly obvious. Jonathan Kozol has termed these kinds of comparisons "savage inequalities." Beyond the extremes, however, researchers have attempted to discern the more nuanced inequalities between schools. Since the publication of the Coleman Report in 1966, which showed small improvements in academic performance for minorities in integrated schools, racial contextual effects have received ample attention in the discipline. Studies indicate that, when school is in session, Black and White children in segregated schools learn more than students in integrated schools.

Previous research shows that reading performance is more sensitive to class and ethnic differences in the features of language. Beginning readers draw heavily on their knowledge of spoken language. When spoken language skills do not match the language used in the classroom, learning to read is more difficult. A possible explanation for the relatively slow progress that students in integrated schools made in reading comprehension in

---

STEREOTYPE THREAT

Despite major historical gains in academic achievement, racial and ethnic minorities still face blatant and subtle forms of discrimination in education. In the early 21st century, research in the area of social psychology examined how fear of confirming a stereotype affected the outcomes of non-White students. *Stereotype threat* occurs when individuals perceive others as having low judgments of or expectations about the abilities of members of their own race/ethnic group, regardless of whether they themselves agree with or reject these ideas. Ample evidence indicates that such stereotype threat has a negative impact on the performance of African Americans on standardized tests (Blascovich, Spencer, Quinn, & Steele, 2001). A similar pattern occurs with women's scores on math and science exams (Ben-Zeev, Fein, & Inzlicht, 2005). These stereotype threat effects occur when knowledge about widely held stereotypes result in individuals having anxiety, self-consciousness, and trouble paying attention while test taking.

In experimental settings, social psychologists find students do less well on a task if given the impression that a bad performance would confirm a negative stereotype about their gender, race, ethnic group, or social class (Spencer & Castano, 2007; Steele, 1997). Researchers are less sure, however, about the impact of stereotype threat in everyday settings. Minorities experience prejudice throughout the life course, but researchers warn about the long-term impact of these negative stereotypes, contending that the additional anxiety of stereotype threat remains a psychological challenge for stigmatized groups and may undermine the identity itself (Steele, 1997).

BIBLIOGRAPHY

Ben-Zeev, T., Fein, S., & Inzlicht, M. (2005). Arousal and stereotype threat. *Journal of Experimental Social Psychology*, *41*, 174–181.

Blascovich, J., Spencer, S. J., Quinn, D., & Steele, C. M. (2001). African Americans and high blood pressure: The role of stereotype threat. *Psychological Science*, *12*, 225–229.

Spencer, B., & Castano, E. (2007). Social class is dead. Long live social class! Stereotype threat among low socioeconomic status individuals. *Social Justice Research*, *20*, 418–432.

Steele, C. M. (1997). A threat in the air: How stereotypes shape intellectual identity and performance. *American Psychologist*, *52*, 613–629.

---

winter is that students who come from segregated neighborhoods speak different dialects than their teachers. In summer, Black students who attended integrated schools gained considerably more than their peers who attended segregated schools. The seasonal patterning of reading and math test scores also emphasize that home disadvantages are compensated for in winter because when school is in session children, regardless of race or SES, perform at almost the same level (Downey, von Hippel, & Broh, 2004). Taken together, these between-school effects illuminate interesting patterns and puzzles for future research. It is important to stress, however, that these between-school effects are much smaller than the effects of home background or within-school differences.

A large literature is devoted to examining the impact of teacher quality on student outcomes. But empirical research has repeatedly found limited to no effect of teacher quality on such outcomes. Fuller (1986) found that teacher training, experience, and salary made little difference in student achievement. But more minor differences reveal how teacher behavior could still affect educational trajectories for students. Grant (1994) exam-ined how complex but subtle processes in the classroom encourage Black girls, more than other students, to fill distinctive roles such as helper, enforcer, and go-between.

For example, the "helper role" promotes stereotypical Black female tasks that stress service and nurturance. These positions foster the growth of social skills, but they also limit the academic abilities of Black girls. Skills developed within these roles reflect the occupational roles in which Black women are overrepresented.

Even more controversial than the topic of teacher quality is the literature on the processes and structure of tracking. Curriculum tracking has a substantial and significant influence on students' future educational and occupational outcomes. Tracks, broadly defined, are the divisions that separate students for all academic subjects according to ability for particular subject areas. The vast majority of schools group and/or track students in some manner. In the United States, children in high-ability groups learn more than those in low-ability groups (Oakes, 1985). But students are often sorted by perceived performance tied to race and class stereotypes, not by actual ability. Most research finds that ability grouping (as practiced in the United States) is neither productive nor does it reduce inequality. Research has also found that controlling for ability, minority students and those from low-income families are more likely to be assigned to a low-track class in high school than their high SES or White peers (Hallinan, 1994; Oakes, 1985).

Olsen's research (1997) indicates that tracking does benefit certain students. She found that college-bound track placement, although difficult to gain entrance to, did erase much of the racial disparities in academic achievement for Mexican and immigrant students. Thus, research has repeatedly shown that track allocation plays a key role in the educational process, especially for racial minorities. In all, school resource differences, particularly curriculum tracking, are unequally distributed by class, race, and gender; and these differences help explain educational achievement disparities.

Differences in home background, or what students bring with them into the classroom, have a powerful influence on students' academic achievement and life opportunities. Students vary on a wide array of home background measures, but socioeconomic status has consistently been the most powerful home background measure in predicting educational achievement: The higher the social class is at home, the higher the achievement level of the student. Grades, curriculum placement, college ambitions, dropout rates, and achievement levels are all related to parental SES. In the United States, race and class are strongly linked. Thus, the effects of race on educational attainment tend to operate indirectly. Most studies find that race per se does not have a direct effect, but race does influence resources, test scores, and track placement, which in turn affect attainment (Alexander & Cook, 1982).

Research also examines the role of ethnic support. Sociologists have examined the educational and occupational prosperity of Asian and Jewish immigrants, arguing that their educational and occupational successes are attributed to their skill sets and background rather than any cultural tendency. Literature on immigrants emerging in the 1990s, however, reveals how academic achievement can be fostered beyond the nuclear family and into the larger ethnic social context of the community. Zhou and Bankston (1998) found that ethnic "culture" does matter, but as a proxy for *social capital* for Vietnamese whose ties to the ethnic community and the expectations and social control that go along with membership in the greater Vietnamese community promote educational achievement, despite an economic and socially marginal environment.

Finally, sociologists caution that schools are organizations that do not operate in isolation. Researchers recognize that although local neighborhood settings for schools are often the location where students live, institutional aspects of organizational environments also shape schools. For example, Arum (2000) suggests that within the U.S. federal system, the state level (as opposed to the local level) has become increasingly important because institutional variations in laws, regulations, and court opinions are often structured at the larger level.

CONTEMPORARY RACIAL INEQUALITY ISSUES IN EDUCATION

Trends in resegregation, battles over school funding, and access to higher education continue to be powerful influences on racial and ethnic inequality in U.S. education. But national debates often shape the local ramifications of educational policy. Two areas that have particular relevance to racial inequality in education are implications of schooling alternatives and immigrant educational needs.

Local and state education policy reflects a growing concern for educational choice, voiced in policies that followed enactment of the No Child Left Behind Act of 2001. Under this controversial school-reform act, districts must publicly identify those schools that are evaluated as needing improvement. Voucher and charter programs and homeschooling options have become increasingly popular with specific relevance to racial and ethnic minorities. Historically, public school children have been assigned to the nearest and most available school. Most states have passed or are preparing to pass legislation that increases parental school choice. Proponents argue that voucher programs encourage more choice among schools and subject districts to market forces leading to higher student achievement and equalizing educational opportunity, particularly for minorities and the poor, who can now opt out of their present failing school. Most states also allow organizations or people to form their own school, known as a charter school. Taken together, these trends have the potential to offer contemporary routes for White, wealthy, and elite flight. Critics warn that poor minority students who cannot make the commute to alternative locations seldom use vouchers or alternative schooling options. Additionally, vouchers and charter programs divert tax dollars to support high-status schools and continue to promote a two-class education system (Berliner & Biddle, 1995).

The growth in the immigrant population across the United States has prompted significant cultural, linguistic, and ethnic demographic changes. Apart from political issues surrounding growing diversity, there is increased concern over how schools should address the language obstacles immigrant children face in public schools. As of 2006, one in ten students in the United States was considered an English language learner or limited English proficient (LEP). Furthermore, the language minority population is increasing at a much faster rate than their native-born peers. LEP students continue to lag behind their classmates when it comes to academic achievement, high school graduation, and degree attainment. While educational and linguistic communities defend the effectiveness of bilingual education and English as a second language (ESL) programs, opposition to language assistance has less to do with pedagogical interests than with social and ethnic concerns. Scholars, politicians, and educators remain divided about how to best address language proficiency for LEP students.

Overall, most social scientists find that racial and ethnic gaps in educational attainment have narrowed since the late 1970s across all levels of education, but there is less of a consensus on what factors account for the continued racial differences in educational achievement. Furthermore, much research on minority and immigrant students still considers them as liabilities to overcome and until very recently, racial comparisons have masked the cultural heterogeneity among panethnic groups. Given the increasing importance of global networking, future educators, policy makers, and those concerned with inequality should consider how the growing diversity of the United States can be an educational contribution.

SEE ALSO Volume 1: *Academic Achievement; College Enrollment; Cultural Capital; Human Capital; Immigration, Childhood and Adolescence; Oppositional Culture; Policy, Education; School Culture; School Readiness; School Tracking; Segregation, School; Social Capital; Socialization, Race; Socioeconomic Inequality in Education.*

BIBLIOGRAPHY

Alexander K. L., & Cook, M. A. (1982). Curricula and coursework: A surprise ending to a familiar story. *American Sociological Review, 47,* 626–640.

Arum, R. (2000). Schools and communities: Ecological and institutional dimensions. *Annual Review of Sociology, 26,* 395–418.

Berliner, D. C., & Biddle, B. J. (1995). *The manufactured crisis: Myths, fraud, and the attack on America's public schools.* Reading, MA: Addison-Wesley.

Downey, D. B., von Hippel, P. T., & Broh, B. A. (2004). Are schools the great equalizer? Cognitive inequality during the summer months and the school year. *American Sociological Review, 69,* 613–635.

Entwisle, D. R., & Alexander, K. L. (1994). Winter setback: The racial composition of schools and learning to read. *American Sociological Review, 59,* 446–460.

Fordham, S., & Ogbu, J. U. (1986). Black students' school success: Coping with the burden of "acting White." *The Urban Review, 18,* 176–206.

Fuller, B. (1986). Is primary school quality eroding in the Third World? *Comparative Education Review, 30,* 491–507.

Grant, L. (1994). Helpers, enforcers, and go-betweens: Black females in elementary school classrooms. In M. Baca Zinn & B. T. Dill (Eds.), *Women of color in U.S. society* (pp. 43–64). Philadelphia: Temple University Press.

Hallinan, M. T. (1994). Tracking: From theory to practice. *Sociology of Education, 67,* 79–84.

Jencks, C., & Phillips, M. (Eds.). (1998). *The Black–White test score gap.* Washington, DC: Brookings Institution Press.

Kao, G., & Thompson, J. S. (2003). Racial and ethnic stratification in educational achievement and attainment. *Annual Review of Sociology, 29,* 417–442.

Kao, G., Tienda, M., & Schneider, B. (1996). Racial and ethnic variation in academic performance. *Research in Sociology of Education and Socialization, 11,* 263–297.

Oakes, J. (1985). *Keeping Track: How Schools Structure Inequality.* New Haven, CT: Yale University Press.

Olsen, L. (1997). *Made in America: Immigrant students in our public schools.* New York: New Press.

Rumberger, R. W. (1995). Dropping out of middle school: A multilevel analysis of students and schools. *American Educational Research Journal, 32,* 583–625.

Stewart, D. W. (1993). *Immigration and education: The crisis and the opportunities.* New York: Lexington Books.

White, M. J., & Kaufman, G. (1997). Language usage, social capital, and school completion among immigrants and native-born ethnic groups. *Social Science Quarterly, 78,* 385–398.

Zhou, M., & Bankston, C. L., III. (1998). *Growing up American: How Vietnamese children adapt to life in the United States.* New York: Russell Sage Foundation.

Beth Tarasawa

RELIGION AND SPIRITUALITY, CHILDHOOD AND ADOLESCENCE

Parents, religious communities, and political figures have an intense interest in the transmission of religious knowledge and spiritual practices to children. Thus, they devote extensive resources to this cultural activity, from religious instruction to community festivals to youth-targeted media. In addition, virtually every religious and spiritual tradition has important rituals that signal key transitions in the life course. Some occur within days of birth, such as baptism and circumcision; others in middle childhood, such as catechism and confirmation classes; and still others in adolescence, such as bar and bat mitzvahs, Quinceañera, Amish rumspringa, and Native American vision quests. These rituals require extensive preparation by the youth and his or her family, require the youth to gain a basic understanding of the core tenets and practices of the religious tradition, and culminate in a celebration that frequently includes extended kin, congregations, and entire communities.

Despite the attention that diverse cultures pay to religious transmission, only recently have scholars returned to the study of religion and spirituality among children and adolescents. Researchers have documented the religious involvement of children and teens both in the United States and abroad, explored diverse components of spirituality, and developed several conceptual models to interpret their results. Childhood and adolescent religion and spirituality has become an active field in contemporary social scientific research.

RELIGION IN CHILDHOOD AND ADOLESCENCE

For chiefly practical reasons, scholars have more systematic knowledge about religious involvement than about spirituality and more certainty about adolescents than about children. First, religious organizations are readily identifiable, and participation in them is easy to measure. It is far simpler, for example, to record the number of times a youth enters a house of worship each week than to gauge the piety of his or her prayers. Second, adolescents are relatively easy to survey and observe: They often are clustered together in school or youth organizations, and access is not complicated; they have an adultlike ability to think and communicate; and they are often eager to cooperate with researchers. By contrast, access to children is more difficult, and children often lack the communicative and cognitive skills needed to answer questions about beliefs, meanings, and other abstract matters. Third, there is a wide readership for studies of adolescent religious and moral life, whereas audiences for works on spirituality are smaller and more segmented.

Far more than public fascination drives research into adolescent religious involvements, however. What drives this research is the enormous and significant engagement in religion by youth. The 2000 World Values Survey reported that 84% of Canadian, 93% of U.S., and 97% of Mexican youth ages 18 to 24 believe in God and that 27%, 50%, and 73%, respectively, report that their belief in God is "very important" in their lives (Roehlkepartian, King, Wagener, & Benson, 2006). The 2002–2003 National Longitudinal Survey of Youth and Religion (NLSYR) reported that 40% of U.S. teens ages 13 to 17 attend religious services at least weekly, another 19% do so at least monthly, and only 18%

never attend, with 28% of those in the final category stating that they would attend if they could (Smith, 2005). The NLSYR, a nationally representative survey of more than 3,000 English- and Spanish-speaking adolescents and their parents, showed that belief and attendance are high in the United States because sizable proportions of American teens report a variety of personally meaningful religious experiences (see Table 1). Although involvement in religion declines as teens age, a solid majority (57%) report "frequent" or "occasional" attendance at worship services at the end of the first year of college, the same number report "no change" in their religious beliefs since beginning college, and 35% report "stronger" religious beliefs (Keup & Stolzenberg, 2004).

Surveys provide a useful window into adolescent religious involvement but have some limitations. Gaining a full understanding of adolescent religiosity requires direct, open-ended, and lengthy conversations; entry into and participation in youth worlds; and a close look at the media culture teens consume. Many studies of this type have added nuance and complexity to scholarly understanding of adolescent religious life. From interviews with adolescents one learns that although most U.S. teens have a benign view of religion, matters of belief and practice are distant from the everyday priorities of most adolescents. Teens seem to view religion as "a very nice thing" (Smith, 2005, p. 124) regarding it as something that is "good for you" but rarely something that generates excitement or passion (Clydesdale, 2007, p. 11). There are exceptions: An estimated 10% of U.S. teens rely on their faith to shape their lives and direct their decisions. Also, about 1% of U.S. teens are intentionally nonreligious; they consciously choose to reject religion and explore alternative worldviews. Most U.S. teens, however, regardless of their stated religious self-identification, articulate a belief system that is distinctly American: God exists, fixes personal problems, and wants people to be happy and nice.

Consequently, few U.S. adolescents demonstrate a focused interest in exploring religious life but most grant a wide berth to those who wish to do so. There also is little in U.S. adolescents''' generalized beliefs about the deity that prevents them from considering ghosts,

Teens who ...	Percent
report "a personal commitment to live for God"	55%
report "an experience of spiritual worship that was very moving and powerful"	51%
have experienced "a definite answer to prayer or specific guidance from God"	50%
have "witnessed or experienced what [they] believed was a miracle from God."	46%
SOURCE: The National Longitudinal Study of Youth & Religion, 2002–2003.	

Table 1. *U.S. adolescents report religion is personally meaningful.* CENGAGE LEARNING, GALE.

astrology, extraterrestrials, and reincarnation as possibly real, as many of them do. That openness is facilitated by media such as music, movies, television, and the Internet. In fact, intentional teen exploration of paranormal phenomena, although rare, can be attributed to its labeling as dangerous by mainstream religious groups (Clark, 2003). Some of these teens find the earthy, feminine, and nascent character of Wicca practice an appealing alternate to traditional religious forms (Berger & Ezzy, 2007). However, despite increasing U.S. religious diversity, it is likely that adolescent religious exploration will be of Christianity rather than of any other religious form. Proximity and population prevalence explain that reality.

One of the liveliest areas in the study of adolescent religion involves the sizable, significant, and positive association that teen religiosity has with positive life outcomes such as reduced risk taking, better educational performance, and greater self-confidence. Researchers in psychology, sociology, public health, and economics have investigated this association and concurred that it is both robust and consistent. Scholars have learned that religious adolescents are more likely to stay in school, get better grades, and complete their degrees compared with less religious and especially nonreligious adolescents. They also have learned that religiously devout adolescents have lower rates of risky behaviors, from smoking to alcohol use to sexual promiscuity, than their less religious and nonreligious peers. In addition, they have learned that religiously committed teens are more satisfied with their physical appearance, their relationships with family and friends, and their ability to talk to adults and find support among them (Regnerus, 2003, 2007; Smith, 2005). The explanation for this positive association between religiosity and life outcomes is the subject of considerable debate, but its existence is not disputed.

Scholars know less about religion among children under age 13, and what they do know largely comes through intermediaries. For example, the *Religion & Ethics NewsWeekly* sponsored a "Faith and Family in America" poll in 2005 that surveyed a national sample of U.S. households with children. A solid majority of household heads (62%) reported that religion was "very important" to their families, and an even larger majority (74%) indicated that it was "very" or "somewhat" likely that their children would choose as adults to remain in the faith tradition in which they were raised.

This fits with the little that is known from direct surveys of children. In 1989, the Girls Scouts, the Lilly Endowment, and the Harvard psychiatrist Robert Coles joined forces to sponsor a national survey of U.S. schoolchildren as young as 9 years old. One-half of that survey's elementary school age respondents reported that their religion was "very important" to them, another fifth reported

that religion was "fairly important," and more than three-quarters affirmed belief in God. Moreover, one-third of elementary school age respondents reported a personal religious experience that "changed the direction" of their lives and one-fifth claimed that "a close relationship to God" was their "most important" goal for the future. Children thus largely follow in their parents' footsteps in religious belief and practice. Although that may be their only option, a sizable minority also chooses to embrace that faith and claim it as their own. Both surveys, as well as the NLSYR, are available from the Association of Religion Data Archives (2008; www.thearda.com).

SPIRITUALITY IN CHILDHOOD AND ADOLESCENCE

Perhaps it was the prominence of Harvard's Robert Coles and his 1990 book, *The Spiritual Life of Children*, that made spirituality a safe topic for lesser known scholars. Perhaps it was popular campus lecturer Jonathan Kozol's (1995) bestseller about the spiritual lives of Harlem children, *Amazing Grace: The Lives of Children and the Conscience of a Nation*, published five years later, that piqued academics' attention. Perhaps it was the example of Princeton sociologist Robert Wuthnow, *After Heaven: Spirituality in America since the 1950s* (1998), which not only demonstrated how to study spirituality at the macro level but also outlined a half-century of its history in U.S. culture. Or perhaps it was the influence of an assortment of private and federal funding agencies, such as the Templeton Foundation, the Lilly Endowment, and the National Institutes of Health, whose priorities turned to underwriting spirituality research. Whatever the cause, there is no denying the broad and multidisciplinary nature of scholarly interest in spirituality since 2000.

A major area of attention has been defining spirituality and clarifying its relationship to religion. For historians, sociologists, and anthropologists, spirituality generally is defined as "the beliefs and activities by which individuals attempt to relate their lives to God or to [the sacred]" (Wuthnow, 1998, p. viii); this may or may not overlap with individuals' participation in religious organizations or engagement with religious culture. From this broader perspective spirituality and religion are like circles on a Venn diagram that variously overlap across subjects and over time. Other scholars, however, subsume one within the other, with some scholars defining spirituality within the bounds of religion whereas still others define religion as a type of spirituality and spirituality as a dimension of personality. Childhood and adolescence thus present an important place to test these theories. By examining how spirituality and religiosity develop in children and teens, scholars can refine their definitions and draw implications for religious education, human development, and larger understandings of cultural patterns and change.

CRITICISM AND METHODOLOGICAL DIFFICULTIES

Social scientific research on spirituality is not always well received. Many scholars report that their work is marginalized in their disciplines or that there is outright hostility to it. Some critics question their findings, whereas others are hostile because they hold that religion and spirituality are dying superstitions that have no place in social scientific research. Moreover, the limited sample sizes of most studies of child spirituality and the overwhelming variety of potential spiritual forms make it difficult to draw simple generalizations from this research. There is evidence that this is beginning to change with the publication of three edited volumes (Ratcliff, 2004; Roehlkepartain et al., 2006; Yust, Johnson, Sasso, & Roehlkepartain, 2006) and the emerging scholarly community that these studies represent.

Research on adolescent spirituality is a bit more accepted, as national studies of it in the United States, Australia, and other developed countries have given scholars much to compare (Mason, Singleton, & Webber, 2007). This research, however, has numerous methodological hurdles. In the United States, for example, scholars have documented how many American baby boomers have set out to explore spiritual life (Roof, 1993; Wuthnow, 1998), but few contemporary U.S. adolescents find the term *spirituality* meaningful or describe themselves as spiritual but not religious (Clydesdale, 2007; Smith, 2005). This does not mean that U.S. teens reject or lack a spiritual dimension, only that the term is vague and better ways of researching spirituality need to be devised. Thus, issues of measurement and generalizability will remain at the forefront of this field.

THEORIES AND DEBATES

Beyond the definitional and philosophical debates that surround this field of inquiry, including challenges to the validity of religion and spirituality, there have been a number of scholarly debates. One involves stage theories of spiritual and/or religious development, including arguments among competing stage theorists and between stage and nonstage theorists. Several debates are methodological. One exists between those who give primacy to quantitative methods and those who privilege qualitative research. Another exists between those who view religion as epiphenomenal (i.e., as something that can be explained by other factors, such as social class) and those who view it as a social fact, with the former attempting to demonstrate that religious findings are spurious (i.e., due to chance) and the latter seeking to show that they are nonspurious (i.e., "real"). This debate is most intense over interpretations of the positive association of religious involvement and life outcomes, with one side arguing that this association is a function of self-selection and social location and the other side arguing that religious involvement provides a variety of resources that directly and indirectly contribute to more positive life outcomes.

Because the study of religion and spirituality is multidisciplinary, larger disciplinary differences also shape debates. Psychologists and many economists privilege the individual in their models of human behavior, whereas historians, sociologists, and anthropologists hold that the whole is greater than the sum of its parts and thus prioritize cultural and social factors. Such differences will persist as long as these disciplines exist. Also, no understanding of religion or spirituality research is complete without acknowledging the wide gap between secularization and religious market paradigms: those who hold that modernization necessarily leads to secularization of religion versus those who hold that modernization creates free religious markets in which religion expands and thrives. The impact of the secularization debate cannot be underestimated—those who concur with secularization models point to lower levels of religiosity among adolescents as signaling secularization, whereas those who concur with religious market models point to persistent youth involvement in religion combined with the return to religion or spiritual exploration later in the life cycle as evidence of vital religious markets.

MOVING FORWARD

Many of the above debates are intractable, and no additional evidence will convince partisans one way or the other. Important scholarly gains need to be made, however, in four areas: measurement, method, comparison, and interdisciplinarity. First, a greater variety of measures need to be tested, contrasted, and refined. Also, measures need to go beyond counting (e.g., "How often do you pray?") to matters of content (e.g., "Describe your last prayer"), rationale (e.g., "What outcomes do you seek when you pray?"), and respondent capabilities (e.g., age-gauged questions about prayer). Second, researchers need to diversify their methods. Augmenting a primary method is not expensive—a week spent gathering qualitative data or designing a concise Web survey are well within the resources of most researchers—and such will move scholarship past dead-end methodological debates to more substantive matters.

Third, comparative research is a pressing and widely acknowledged need. Although increasing U.S. religious diversity is attracting scholarly attention to non-Christian religious and spiritual forms, too few non-U.S. scholars or sites are engaged in this issue. Finally, researchers new to the study of religion and spirituality, as well as those long involved, need to do their interdisciplinary "homework." There is much extant scholarship of high quality, and the depth and breadth of religious and spiritual phenomena leaves no room for disciplinary parochialism.

Psychologists need to read beyond psychology of religion, sociologists need to develop a taste for economic scholarship on religion (e.g., Iannaccone, 1998), and so on.

To move forward in scholarly understandings of the religious involvements and spirituality of children and adolescents will require better data, more diverse data, and wider engagement. Although a great deal of good work exists already, much remains to be done.

SEE ALSO Volume 1: *Activity Participation, Childhood and Adolescence; Civic Engagement, Childhood and Adolescence; Home Schooling.*

BIBLIOGRAPHY

Association of Religion Data Archives. (2008). Retrieved April 29, 2008, from http://www.thearda.com

Benson, P. L. (2004). Emerging themes in research on adolescent spiritual and religious development. *Applied Developmental Science, 8*(1), 47–50.

Berger, H. A., & Ezzy, D. (2007). *Teenage witches: Magical youth and the search for self.* New Brunswick, NJ: Rutgers University Press.

Clark, L. S. (2003). *From angels to aliens: Teenagers, the media, and the supernatural.* New York: Oxford University Press.

Clydesdale, T. (2007). *The first year out: Understanding American teens after high school.* Chicago: University of Chicago Press.

Coles, R. (1990). *The spiritual life of children.* Boston: Houghton Mifflin.

Iannaccone, L. R. (1998). Introduction to the economics of religion. *Journal of Economic Literature, 36,* 1465–1495.

Keup, J. R., & Stolzenberg, E. B. (2004). *The 2003 your first college year (YFCY) survey: Exploring the academic and personal experiences of first-year students.* Monograph 40. Columbia: University of South Carolina, National Resource Center for the First-Year Experience and Students in Transition.

Kozol, J. (1995). *Amazing grace: The lives of children and the conscience of a nation.* New York: Crown.

Mason, M., Singleton, A., & Webber, R. (2007). *The spirit of generation Y: Young people"'s spirituality in a changing Australia.* Melbourne, Australia: John Garratt.

Ratcliff, D. (Ed.). 2004. *Children's spirituality: Christian perspectives, research, and applications.* Eugene, OR: Cascade Books.

Regnerus, M. D. (2003). Linked lives, faith, and behavior: Intergenerational religious influence on adolescent delinquency. *Journal for the Scientific Study of Religion, 42*(2), 189–203.

Regnerus, M. D. (2007). *Forbidden fruit: Sex & religion in the lives of American teenagers.* New York: Oxford University Press.

Roehlkepartain, E. C., King, P. E., Wagener, L. M., & Benson, P. L. (Eds.). (2006). *The handbook of spiritual development in childhood and adolescence.* Thousand Oaks, CA: Sage.

Roof, W. C. (1993). *A generation of seekers: The spiritual journeys of the baby boom generation.* San Francisco, CA: HarperSanFrancisco.

Smith, C., with Denton, M. L. (2005). *Soul searching: The religious and spiritual lives of American teenagers.* New York: Oxford University Press.

Smith, C., Denton, M. L., Faris, R., & Regnerus, M. (2002). Mapping American adolescent religious participation. *Journal for the Scientific Study of Religion, 41*(4), 597–612.

Wuthnow, R. (1998). *After heaven: Spirituality in America since the 1950s.* Berkeley: University of California Press.

Yust, K. A., Johnson, A. N., Sasso, S. E., & Roehlkepartain, E. C. (Eds.). (2006). *Nurturing child and adolescent spirituality: Perspectives from the world's religious traditions.* Lanham, MD: Rowman & Littlefield.

Tim Clydesdale

RESIDENTIAL MOBILITY, YOUTH

Residential mobility is a form of migration that is typically thought of and analyzed as occurring within a particular country. Along with changes in the local environment that occur around children without their movement (e.g., gentrification, urban redevelopment), residential mobility is one of two ways that children's local environments can vary over the course of childhood. This variation can occur either within or across counties and states. It typically involves other family members but can occur independently of family as well. Although migration can and often does occur to or from other countries, in-depth studies of residential mobility usually have a national, rather than international, focus. The reasons for this definition are driven primarily by the availability of data: Because the goal of many studies is to understand not just how common movement is, but what changes accompany that movement, researchers often want to link children to data on specific characteristics of each residence. This level of detail is often only possible within one country.

Why is residential mobility an important part of the early life course? As the following discussion will demonstrate, geographic movement among the young population is very common and often coincides with important changes in family composition and organization, the quality of the local environment, the nature of peer groups, and educational quality. Understanding the causes and consequences of residential mobility in children's lives, as well as trends over time, is therefore an important task.

PATTERNS OF RESIDENTIAL MOBILITY AMONG CHILDREN AND ADOLESCENTS

Residential mobility has declined slightly over time but remains a common feature of many societies. In 1948, when the United States Census Bureau began collecting information on geographic mobility, about 20% of the U.S. population moved in a 1-year period. In 2003 that number stood at about 14% (Schachter, 2004). Although overall rates of movement have decreased over time, the distance of moves has increased, with more moves now crossing state boundaries (Schachter, 2004). Families with children are less likely to move than families without children (Long, 1972). Moving rates are higher among children than adults, however: In 2003, for example, 21.4% of children ages 1 to 4 experienced a move, versus 8.6% of adults ages 45 to 54 (Schachter, 2004). Movement among children is also common in other countries, although less so than in the United States (Long, 1992).

THE SOCIAL DETERMINANTS OF YOUTHS' RESIDENTIAL MOBILITY

Despite the frequency of mobility, there are important differences in which children are most likely to move, both within the United States and in other societies. First, there are noticeable age differences in who moves. Rates of movement are higher among very young children and young adults than among older children and adolescents. Between 2002 and 2003, for example, more than 20% of children ages 1 to 4 and almost one-third of young adults ages 20 to 29 experienced a move, versus 16% and 14% of those ages 5 to 9 and 10 to 19, respectively (Schachter, 2004). These age differences are particularly pronounced in the United States. Rates of movement during the early life course, in both early childhood and young adulthood, are much higher in the United States than in other industrialized nations (Long, 1992).

Many social factors determine the likelihood of movement among children, and also partially explain age differences in children's rates of movement. With respect to the greater tendency of very young children to move, it is important to think about their parents' stage in the life course. Childbearing may change the characteristics that make a particular residence desirable, causing parents to place a higher priority in finding a neighborhood with low crime rates, a good educational system, and spacious housing (Long, 1992; Rossi, 1955). High rates of movement among those transitioning out of childhood—young adults ages 20 to 29—also make sense, because these are the ages at which important changes in the life course occur, including the completion of higher education, employment transitions, and the formation of new families.

In addition to life course transitions among young adults and young parents that explain the higher mobility of the oldest and youngest children, a number of other factors determine the likelihood of geographic movement during childhood. Children who live below the national poverty line are more likely to move than their wealthier peers, as are children whose parents are not married. For both groups of children, reduced economic status and access to the resources necessary to live in certain loca-

tions are a source of instability. Children of divorced parents are also more likely to spend time at the residences of multiple family members over the course of childhood, increasing rates of mobility. Poverty and family structure may also influence children's likelihood of movement in combination with one another: Children who live with only one parent are more likely to experience poverty than children in two-parent homes, making their living situation even more economically unstable (South & Crowder, 1998). Particularly high rates of poverty and divorce in the United States, and the economic and occupational instability that accompany them, may also explain higher rates of movement among children in the United States than in other countries.

The type and scale of a child's move are equally important to rates of residential mobility. Although there are not striking differences in rates of movement among Black, Latino, and non-Latino White children, there are large differences in the types of moves that children experience by race/ethnicity. Black children in particular, and Latino children to a lesser degree, are more likely to move from one poor neighborhood to another, whereas non-Latino White children are more likely to transition to wealthier environments (Quillian, 2003; Timberlake, 2007). With respect to the scale of a move, parents' education plays an important role. Children with highly educated parents are not much more likely to move than other children. Among children who do move, however, the distance of the move is greater (Schachter, 2004).

Many factors—including the stage in children's and parents' life course, families' economic well-being, family organization, and parental education—work independently and in combination with one another to determine whether and how far a child will move. These causes also contribute to different patterns of residential mobility across societies. Although the factors discussed are known to be particularly salient in shaping mobility trends, there are other important contributors as well, including urban versus rural residence and housing tenure (Rossi, 1955).

THE CONSEQUENCES OF RESIDENTIAL MOBILITY FOR CHILDREN

Children can be influenced by residential mobility in both positive and negative ways, which are not yet fully understood by researchers. A change in a child's local environment can be thought of as involving two parts: a move (regardless of its direction) and a change in the quality of neighborhood conditions. With respect to a move, residential mobility may adversely influence children, whether or not it involves an improvement in surroundings. Some researchers, for example, define the number of times a child's household moved while growing up as an indicator of family stress, and find that this measure is associated with a lower likelihood of high school completion (Haveman, Wolfe, & Spaulding, 1991).

This way of thinking about children's residential mobility assumes that all moves, whether they involve an upward or downward shift in the quality of surroundings, influence children equally. This may not be the case, however, and relates to the second component of a move: the extent to which it involves a change in the quality of a child's environment. Because children consistently exposed to disadvantage may be subject to more adverse influences than those who only briefly experience poor and unsafe surroundings, residential mobility out of a disadvantaged neighborhood may benefit children if it brings access to better housing, more positive peer role models, and a better educational system (Goering & Feins, 2003). This positive change may outweigh the disruptive influence of moving. On the other hand, if children move frequently between similarly poor or wealthy neighborhoods, then the combination of frequent disruption due to mobility and consistent exposure to disadvantage may be harmful to children. Researchers are trying to understand the extent to which children's exposure to disadvantage varies over time due to residential mobility, with the hope of disentangling the potentially different influences of residential mobility and exposure to disadvantaged environments (Jackson & Mare, 2007; Timberlake, 2007).

IMPLICATIONS AND FUTURE RESEARCH

Much is known about patterns of children's residential mobility in the United States and in other societies. Very young children and young adults are the most likely to experience mobility, and mobility among children of all ages is much higher in the United States than in other industrialized nations. The determinants of residential mobility are also well understood, with clear differences in which children move by families' economic status, marital status, education, and stage in the life course.

Our understanding of the consequences of residential mobility for children's well-being is less developed. Although residential mobility appears to be a meaningful predictor of children's success and well-being, it is not clear if this applies equally to all types of moves. Research in this area is necessary and important in order to gain a thorough understanding of how children's welfare changes over time as they move.

SEE ALSO Volume 1: *Neighborhood Context, Childhood and Adolescence; Family and Household Structure, Childhood and Adolescence;* Volume 2: *Residential Mobility, Adulthood;* Volume 3: *Residential Mobility, Later Life.*

BIBLIOGRAPHY

Goering, J., & Feins, J. D. (Eds.) (2003). *Choosing a better life: Evaluating the moving to opportunity social experiment.* Washington, DC: The Urban Institute Press.

Haveman, R., Wolfe, B., & Spaulding, J. (1991). Childhood events and circumstances influencing high school completion. *Demography, 28,* 133–157.

Jackson, M. I., & Mare, R. D. (2007). Cross sectional and longitudinal measurements of neighborhood experience and their effects on children. *Social Science Research, 36*(2),590–610.

Long, L. (1972). The influence of number and ages of children on residential mobility. *Demography, 9,* 371–382.

Long, L. (1992). International perspectives on the residential mobility of America's children. *Journal of Marriage and the Family, 54,* 861–869.

Quillian, L. (2003). How long are exposures to poor neighborhoods? The long-term dynamics of entry and exit from poor neighborhoods. *Population Research and Policy Review, 22,* 221–249.

Rossi, P. H. (1955). *Why families move.* (2nd ed.). Glencoe, IL: The Free Press. Beverly Hills, CA: Sage, 1980.

Schachter, J. P. (2004). Geographical Mobility: 2002 to 2003. U.S. Census Bureau Current Population Reports. Washington, DC: U.S. Department of Commerce.

South, S. J., & Crowder, K. D. (1998). Avenues and barriers to residential mobility among single mothers. *Journal of Marriage and the Family, 60,* 866–877.

Timberlake, J. M. (2007). Racial and ethnic inequality in the duration of children's exposure to neighborhood poverty and affluence. *Social Problems, 54,* 319–42.

Margot I. Jackson

RESILIENCE, CHILDHOOD AND ADOLESCENCE

Resilience research is guided by the following question: Why do some children experience significant adversity but grow up to be well-adjusted adolescents and adults? Some children are exposed to conditions such as war, poverty, or abuse but are able to overcome these adversities and become healthy, productive members of society, whereas other children exposed to the same circumstances develop mental health problems. Those who achieve positive developmental outcomes in the context of significant risk or adversity are considered to be resilient.

RESILIENCE CONCEPTS

The study of resilience involves three critical components: risk factors, protective factors, and positive adaptation. A risk factor is a condition, context, experience, adversity, or individual characteristic exposure to which has been shown to increase the probability that an individual will experience maladjustment. Researchers have identified risk factors for child maladjustment at the individual, family, and community levels. Individual-level risk factors include factors such as depressogenic cognitive style and negative temperament. Family risk factors include factors such as parental divorce, parental mental illness, and substance use. Community-level risk factors include factors such as neighborhood violence, high neighborhood poverty, and the presence of multiple outlets for selling alcohol to minors.

Risk factors tend to be positively intercorrelated so that children exposed to one risk factor are likely to be exposed to others as well. Risk factors negatively impact child development by disrupting the satisfaction of basic needs and goals or impeding the child's accomplishment of developmental tasks, such as adjusting to school and developing friendships. In general, the more risk factors a child experiences, the greater the chances he or she will develop adjustment problems. For example, using a large sample of members of a health maintenance organization (HMO), Vincent Felitti and colleagues (1998) found that exposure to four or more risk factors in childhood predicted a 4- to 12-fold increase in the odds of developing problems in adulthood, including alcoholism, drug abuse, depression, and suicide attempts. Thus, it is the cumulative effect of exposure to multiple risk factors that makes children most vulnerable.

A protective factor is an individual-, family-, or community-level factor that decreases the likelihood that an individual exposed to significant risk will develop problems or increases the chances he or she will experience positive adaptation. Protective factors facilitate positive adaptation by either promoting children's competencies or by reducing the child's exposure to adversity. Like risk factors, protective factors have been found across levels (e.g., individual-level factors such as high child intelligence; family-level factors such as positive parenting; and community-level factors such as availability of positive mentors). It is important to note that a protective factor is not something that simply produces good outcomes in general but, instead, reduces the effects of a risk factor.

Positive adaptation can be conceptualized as the occurrence of developmental outcomes that are substantially better than would be expected given a child's exposure to specific risk factor(s). For example, children who attend underperforming schools in poverty-stricken neighborhoods are at risk for low academic achievement and behavior problems. Students experiencing positive adaptation in this circumstance would be those demonstrating high academic achievement and prosocial behavior. These students would be considered resilient.

Although resilience research is similar to several other approaches to studying child and adolescent development, such as risk research and the study of child competence, there are important differences as well. Whereas risk research has focused on predicting negative outcomes and the study of competence has focused on positive outcomes, resilience research considers both negative and positive outcomes, as well as the protective factors that account for positive adaptation in the context of risk. Resilience researchers also seek to understand the dynamic processes that explain the processes by which an individual achieves positive outcomes despite being exposed to adversity; including emotional and cognitive mechanisms, gene-environment interactions, and neuroendocrine responses.

HISTORY AND EVOLUTION OF THE RESILIENCE FIELD

The systematic study of resilience began to evolve in the early 1970s as a handful of pioneering researchers identified subgroups of children at risk for psychopathology who were characterized by surprisingly high levels of competence. Whereas most scientists of the time disregarded positive outcomes among high-risk children as unimportant anomalies, researchers such as Norman Garmezy and Michael Rutter described and sought to understand the reasons for these phenomena. For example, Garmezy (1974) observed that most children identified during childhood as at risk for schizophrenia did not ultimately develop mental disorder in adulthood. He laid out a four-stage strategy for studying the development of severe psychopathology that included (a) examining early forms of competence and incompetence; (b) identifying variables that differentiate low- and high-risk children; (c) conducting longitudinal studies to predict patterns of outcomes among vulnerable children; and (d) testing interventions to evaluate whether changing early predictive variables influences later competence.

Around the same time, Rutter (1979) described the cumulative risk phenomenon, in which a child's susceptibility for later disorder increases exponentially as a function of the number of stressors experienced. Using an epidemiological sample of children living in inner-city London, Rutter compared groups of children based on the number of family stressors (e.g., marital conflict, low socioeconomic status [SES], paternal criminality, or maternal psychiatric disorder) they faced. He discovered that children exposed to one stressor were no more likely to have a psychiatric disorder than those who had zero stressors, whereas children with two stressors were 4 times as likely, and those with four or more stressors were nearly 20 times as likely, to have a disorder than children who had only one or no stressors. However, Rutter pointed out that a sizable minority of children who suffered a combination of stressors did not have a mental disorder. He also identified a variety of potential protective factors that were suggested by prior research (e.g., female gender, high-quality education, high levels of parental supervision, and good parent-child relationships).

In the 1980s groundbreaking studies by Rutter and Garmezy, among others, led to new conceptualizations of major constructs, methods, and data-analytic strategies for examining resilience processes and outcomes. In addition, Emmy Werner and Ruth Smith published *Vulnerable but Invincible* (1982), a book describing a longitudinal study that would become a landmark in the resilience field. Over multiple decades, these researchers followed a population of infants born in 1955 on the island of Kauai, Hawaii. They discovered that children with a combination of risk factors including chronic family poverty, poorly educated parents, and perinatal stress (e.g., premature birth) tended to develop learning and behavioral problems in childhood and adolescence. Werner and Smith identified a variety of protective factors, such as early sociability, family cohesiveness, and social support from extended family and friends, that differentiated high-risk, resilient children from those who later developed problems.

The 1990s witnessed an explosion of studies using the concepts and methodologies initiated by these early resilience pioneers. As the field has evolved, researchers have increasingly clarified concepts and terminology, expanded the types of risk and protective factors and processes examined, and recognized that resilience may not be stable across development or different domains of adjustment. In the next section, this entry reviews some of the main protective factors that have been identified by resilience researchers studying children at risk for mental health problems.

PROTECTION IN THE CONTEXT OF RISK

Resilience researchers have identified a variety of factors across multiple domains that buffer children against the negative effects of adverse circumstances.

Family Protective Factors Protective factors within the family have received the most attention in resilience research, because the family is the most proximal and stable of children's contexts. Early resilience research suggested that having a close relationship with at least one parent was protective across various adverse situations (Garmezy, 1974; Rutter, 1979; Werner & Smith, 1982). Subsequent research has shown that the protective effects of the parent-child relationship occur at different developmental stages. For high-risk infants, having a secure attachment with one caregiver is a protective factor for many outcomes throughout the life span (Sroufe, 2002). In childhood and adolescence, warm, close rela-

tionships with mothers or fathers protect children against the negative effects of poverty, parental mental illness, and parental alcoholism (Luthar, 2006). In adolescence, higher levels of parental monitoring also have been shown to protect against the negative effects of risky peer and community contexts (Dishion & McMahon, 1998).

Community-based Organizational Protective Factors Children living in impoverished neighborhoods often face multiple adversities, with each adversity increasing the likelihood of exposure to other adversities. For example, living in poverty increases the likelihood of parental mental health problems and exposure to antisocial peers and community violence. The cumulative exposure to multiple adversities puts children at risk for mental health problems. However, positive experiences in child care, schools, and community contexts can be protective in these environments. For example, in low-income neighborhoods, infants and toddlers in high-quality child care (i.e., those with highly educated and trained child care providers, low child-to-adult ratios, and low staff turnover) tend to be more securely attached to their mothers and less likely to be delinquent later in life (NICHD Early Child Care Network, 2002). For older children, having a warm and trusting relationship with an adult in the community, such as a teacher, coach, or mentor, can lead to positive outcomes in high-risk environments.

Individual Characteristics Individual characteristics of the child have also been the focus of resilience research; but this research typically involves community and family factors as well, because these contexts are instrumental in shaping children's characteristics. One of the most widely studied child characteristics is high intelligence, which has been shown to protect children from negative outcomes and lead to positive outcomes across a variety of adverse circumstances, even after controlling for the association between intelligence and family SES (Masten, 2001). In infancy and toddlerhood, having an easy child temperament or being able to appropriately regulate emotions has been shown to exert protective effects, whereas having a difficult temperament is a risk factor for negative outcomes under stressful conditions (Calkins & Fox, 2002). In childhood and adolescence, active coping strategies, coping efficacy, high self-esteem, internal locus of control, and positive thinking have been found to protect children against the negative effects of adversity (Sandler, Wolchik, Davis, Haine, & Ayers, 2003).

Friendships The quality of peer relationships and social networks are also an important context in which resilience research has been productive. Being accepted by peers and having a supportive relationship with at least one friend protects children from the negative effects of adversity. At the same time, peer rejection and association with deviant peers increase adolescents' vulnerability to behavior problems in high-risk environments. Research has also shown that association with deviant peers is protective against depression among antisocial adolescents living in inner cities (Seidman & Pedersen, 2003). This finding suggests that while having deviant friends may contribute to increased behavior problems, having friends (even if they are deviant) may lead to less depression. This is an example of the complexity involved in identifying risk and protective factors in resilience research because one factor (e.g., association with deviant peers) may be both risky and protective depending on the outcome of interest.

IMPLICATIONS OF RESILIENCE RESEARCH

Resilience researchers have amassed a wealth of knowledge about factors that promote healthy adjustment among children at risk. Prior to the 1970s, researchers had little knowledge of positive influences on children's development in the context of risk; consequently, they lacked information needed to develop interventions to prevent adjustment problems among high-risk children (Rutter, 1979). Since the late 1970s, findings from the resilience literature have provided the impetus for the development and evaluation of a wide range of interventions that build protective factors. Preventive interventions have been developed to counteract the negative effects of risk factors such as neighborhood disadvantage, family poverty, low SES, premature birth, parental divorce, death of a parent, abuse, and trauma. The following sections provide examples of interventions that have been designed to build protective factors to counteract one of these risk factors: parental divorce.

Parental Divorce Each year, more than 1 million children experience parental divorce in the United States. Although the majority of children whose parents divorce do not develop serious problems, these children are at increased risk for a host of difficulties, such as conduct problems, academic underachievement, and increased rates of mental health and substance abuse disorders in adulthood. Several interventions have been shown to reduce children's risk of experiencing significant difficulties following parental divorce by bolstering known protective factors.

Family Programs Two parenting-based preventive interventions have been shown to help children adjust to parental divorce: Parenting Through Change (PTC) and the New Beginnings Program (NBP). Both programs build protective factors by teaching parents effective dis-

ciplinary practices and promoting warm parent-child relationships. Randomized experimental evaluations of these interventions have shown that children whose parents participated in PTC or NBP had fewer conduct problems several years later compared to children in control groups (Martinez & Forgatch, 2001; Wolchik, Sandler, Weiss & Winslow, 2007). In fact, positive effects of the NBP have been found on a wide range of adolescent outcomes 6 years after the intervention, including lower rates of diagnosed mental disorder, fewer mental health problems, lower substance use, fewer high-risk sexual behaviors, higher self-esteem and higher grade point averages. The protective effects of NBP were most pronounced for children who were at the highest risk for problems when the intervention began, which is consistent with research showing that children exposed to multiple risk factors are the most vulnerable and, consequently, most in need of interventions to build protective factors.

Child Programs Two preventive interventions have been shown to promote resilience among children experiencing parental divorce by promoting individual protective characteristics, specifically child coping skills. The Children's Support Group (CSG) and Children of Divorce Intervention Project (CODIP) are group-based programs designed to give children emotional support, correct misconceptions about divorce, and teach coping skills (e.g., emotional expression, problem-solving skills, and anger management) to help them adjust well to parental divorce. Evaluations of both programs have shown increases in child competence and reductions in internalizing (e.g., depressive symptoms) and externalizing (e.g., noncompliance) problems 1 to 2 years following intervention compared to no-treatment control groups (Stolberg & Mahler, 1994; Pedro-Carroll, Sutton, & Wyman, 1999).

FUTURE RESEARCH

The field of resilience has made important contributions to understanding child development within the context of adversity and has laid the groundwork for interventions that can improve child well-being by building protective processes identified through resilience research. In the years ahead, it is expected that scientists will continue to expand the search for protective factors by increasingly attending to genetic and biological factors in addition to psychological mechanisms, as well as examining protective factors that may be specific to ethnic and cultural subgroups (e.g., ethnic pride or collectivist family values). Finally, findings from resilience research will continue to be used to develop theory-based intervention strategies for improving the lives of children at risk.

SEE ALSO Volume 1: *Child Abuse; Family and Household Structure, Childhood and Adolescence; Friendship, Childhood and Adolescence; Neighborhood Context, Childhood and Adolescence; Parent-Child Relationships, Childhood and Adolescence; Peer Groups and Crowds; Poverty, Childhood and Adolescence.*

BIBLIOGRAPHY

National Institute of Child Health and Development (NICHD) Early Child Care Research Network. (2002). Parenting and family influences when children are in child care: Results from the NICHD study of early child care. In J. Borkowski, S. Ramey, & M. Bristol-Power (Eds.), *Parenting and the child's world: Influences on academic, intellectual, and social-emotional development* (pp. 99–123). Mahwah, NJ: Lawrence Erlbaum.

Calkins, S. D., & Fox, N. A. (2002). Self-regulatory processes in early personality development: A multilevel approach to the study of childhood social withdrawal and aggression. *Development and Psychopathology, 14*(3), 477–498.

Dishion, T. J., & McMahon, R. J. (1998). Parental monitoring and the prevention of child and adolescent problem behavior: A conceptual and empirical formulation. *Clinical Child and Family Psychology Review, 1*(1), 61–75.

Felitti, V., Anda, R., Nordenberg, D., Williamson, D., Spitz, A., Edwards, V., et al. (1998). Relationship of childhood abuse and household dysfunction to many of the leading causes of death in adults: The Adverse Childhood Experiences (ACE) study. *American Journal of Preventive Medicine, 14*(4), 245–258.

Garmezy, N. (1974). The study of competence in children at risk for severe psychopathology. In E. J. Anthony & C. Koupernik (Eds.), *Children at psychiatric risk* (pp. 77–97). New York: Wiley.

Luthar, S. (2006). Resilience in development: A synthesis of research across five decades. In D. Cicchetti & D. J. Cohen (Eds.), *Developmental psychopathology, Vol. 3: Risk, disorder, and adaptation* (2nd ed., pp. 739–795). Hoboken, NJ: Wiley.

Martinez, C., & Forgatch, M. (2001). Preventing problems with boys' noncompliance: Effects of a parent training intervention for divorcing mothers. *Journal of Consulting and Clinical Psychology, 69*(3), 416–428.

Masten, A. S. (2001). Ordinary magic: Resilience processes in development. *American Psychologist, 56*(3), 227–238.

Pedro-Carroll, J., Sutton, S., & Wyman, R. (1999). A two-year follow-up evaluation of a preventive intervention for young children of divorce. *School Psychology Review, 28*(3), 467–476.

Rutter, M. (1979). Protective factors in children's responses to stress and disadvantage. In M. W. Kent & J. E. Rolf (Eds.), *Social competence in children* (pp. 49–74). Hanover, NH: University Press of New England.

Sandler, I., Wolchik, S., Davis, C., Haine, R., & Ayers, T. (2003). Correlational and experimental study of resilience in children of divorce and parentally bereaved children. In S. Luthar (Ed.), *Resilience and vulnerability: Adaptation in the context of childhood adversities* (pp. 213–240). New York: Cambridge University Press.

Seidman, E., & Pedersen, S. (2003). Holistic contextual perspectives in risk, protection, and competence among low-income urban adolescents. In S. Luthar (Ed.), *Resilience and vulnerability: Adaptation in the context of childhood adversities* (pp. 318–342). New York: Cambridge University Press.

Sroufe, L. A. (2002). From infant attachment to promotion of adolescent autonomy: Prospective, longitudinal data on the role of parents in development. In J. Borkowski, S. Ramey, & M. Bristol-Power (Eds.), *Parenting and the child's world: Influences on academic, intellectual, and social-emotional development* (pp. 187–202). Mahwah, NJ: Lawrence Erlbaum.

Stolberg, A., & Mahler, J. (1994). Enhancing treatment gains in a school-based intervention for children of divorce through skill training, parental involvement, and transfer procedures. *Journal of Consulting and Clinical Psychology, 62*(1), 147–156.

Werner, E., & Smith, R. (1982). *Vulnerable but invincible: A longitudinal study of resilient children and youth.* New York: McGraw-Hill.

Wolchik, S., Sandler, I., Weiss, L., & Winslow, E. (2007). New beginnings: An empirically-based program to help divorced mothers promote resilience in their children. In J. M. Briesmeister & C. E. Schaefer (Eds.), *Handbook of parent training: Helping parents prevent and solve problem behaviors* (3rd ed.) (pp. 25–62). Hoboken, NJ: Wiley.

Emily B. Winslow
Darya D. Bonds
Irwin N. Sandler

S

SCHOOL CULTURE

Efforts to improve the academic performance of schools typically focus on their formal, structural characteristics, such as curricular offerings and test scores. Yet, social and behavioral research has amassed a good deal of evidence that the more informal, social characteristics of a school—for example, the general quality of relationships among students or between students and teachers—also contribute to its students' academic performance and healthy development. The *culture* of a school matters. School culture is an important topic to consider when studying the life course because it captures how various life course trajectories (e.g., social development and educational attainment) intersect within specific contexts in one stage of life in ways that influence transitions into and through subsequent stages of life.

DEFINING AND STUDYING SCHOOL CULTURE

Culture is an esoteric concept that is not easily defined. A general definition is the established patterns of activity, symbolic structures, and products within a given society. Culture can also exist on the level of the group or organization, such as the school (Coleman, 1961). A school's established patterns of activity can include rituals that are officially organized (e.g., participation in sponsored clubs or the clustering of students by grade or achievement level) and those that arise organically from repeated interactions among students and teachers (e.g., dating rules, sharing a common vocabulary). Its symbolic structures refer to taken-for-granted norms about how students and teachers are supposed to act, what they are supposed to value, and what kinds of emotions are acceptably felt and displayed. For example, drinking alcohol may gain students high social status in one school but low social status in another. Cultural products of the school can be tangible commodities, such as a winning sports team, or they can be more intangible, such as the perceived quality of the young people it is producing. Importantly, a school culture is tapped into broader cultures (e.g., American society) and subsumes different subcultures (e.g., jocks vs. nerds, White vs. Black, young vs. old). Yet, over time, an overarching culture typically arises in the school that is a system of rules, values, and beliefs that creates opportunities for behavior, interpersonal interaction, self- and other-assessments, and social advancement, whether individuals are conscious that this is happening or not (Carter, 2007; McFarland & Pals, 2005; Pianta & Walsh, 1996).

Not surprisingly, then, social scientists have long been interested in school culture. Some school culture researchers conduct in-depth ethnographies of a single school or a small number of schools. Sara Lawrence-Lightfoot (1983) illustrated this approach in *The Good High School*, which focused on six very different high schools widely considered to be successful within their own contexts—for example, an inner-city public school with a higher than expected graduation rate and an elite private school that is a pipeline to Ivy League colleges. Lawrence-Lightfoot constructed a portrait of a good high school based on the cultural commonalities across these schools, such as mutual respect between principals and staff and a general sense of community among students, parents, and school personnel.

Other school culture researchers rely on survey data. For example, Laurence Steinberg, B. Bradford Brown, and Sanford Dornbusch (1996) worked together to survey thousands of students in California and Wisconsin, which allowed them to map out common peer crowds such as the *preps*, *rebels*, and *burnouts*. As another example, the National Education Longitudinal Study is collected and made publicly available by the Department of Education, whereas the National Longitudinal Study of Adolescent Health is made publicly available by the National Institutes of Health. The former study sampled parents, students, counselors, teachers, and principals in thousands of schools across the United States. These data allow researchers to compare how values and expectations of different actors in the school intersect (e.g., commonly shared goals, alienation between young and old) in ways that contribute to student performance (Lee & Smith, 2001). The latter provides a census of 132 schools in the United States. Aggregating student responses to identify prevailing norms in a given school has revealed, for instance, that adolescents avoid risky behavior when they attend schools in which students, on average, are close to their teachers, regardless of their own feelings about teachers (Crosnoe & Needham, 2004).

The ethnographic approach can elucidate important cultural patterns in any one school, and the survey approach facilitates comparisons across diverse schools. Thus, these approaches to studying school culture provide maximum value when partnered.

THREE MAIN AREAS OF SCHOOL CULTURE RESEARCH

Contemporary scientific inquiry into school culture is largely traceable to the enormously influential, occasionally polarizing work of the late sociologist James Coleman (1926–1995). First, Coleman's 1961 book, *The Adolescent Society*, was based on an intensive study of Midwestern high schools. It argued that American high schools house peer cultures that are in direct opposition to conventional society and often undermine the basic goals of school administrators and parents. This idea of a monolithic, defiant teen culture has been largely refuted over the years, but the basic premise that schools are *the* site of youth culture is still an underlying theme of school research.

Schools can be broken down into clearly defined peer groups that share common social and academic positions in the school. For example, those referred to as *geeks* tend to develop a social identity through their repeated interactions in high-level coursework year after year, whereas those generally considered *popular* come together in clubs and teams. In both cases, proximity and shared status create meaningful peer groups (Field et al., 2005). Within a school, disparate groups then aggregate into a larger peer culture, with more powerful

actors better able to define the rules of school social life. *School Talk*, in which Donna Eder, Catherine Evans, and Stephen Parker (1995) detail the hypersexualized nature of middle school, offers a startling glimpse into the power struggles that set the tone of school peer culture. They describe how coaches' denigration of signs of femininity in their athletes eventually infected the entire school culture, primarily because those athletes had such power and status among their peers. Boys who could not live up to this masculine ideal and girls who resisted their subjugated place paid a price in mental health, popularity, and attachment to school and school personnel. In this school, the peer culture of the school was toxic, even if the more formal, structural components of the school were of high quality.

Second, Coleman's 1966 study *Equality of Educational Opportunity*, also known simply as the Coleman Report, was a landmark federal report on inequalities among American schools. Controversially, it concluded that family, community, and school culture—not school funding—were often deciding factors in racial and economic differences in academic achievement. The Coleman Report had many methodological flaws and some regrettable policy applications (e.g., busing and the consequent White flight phenomenon), but its essential message that reforming schools requires transforming values and attitudes, not just pumping in money, has lived on.

This message that efforts to improve schools must start with positive changes to school culture has fueled interest in *caring* schools. In such a school, teachers realize that effectively instructing students in a given curriculum first requires that they and their students see each other as worthy of support, attention, and acceptance (Noddings, 2002). Angela Valenzuela's 1999 book, *Subtractive Schooling*, reveals the pitfalls of school cultures that lack widespread caring. In a Texas high school, the staff had a general tendency to view the cultural heritage of the largely Mexican immigrant student body as a disadvantage rather than a resource, which created an emotional disconnect in the classroom that hampered the ability of even well-trained teachers to motivate their students. Racially divided schools such as this one clearly demonstrate the politics of caring. For example, research has revealed how minority students emphasize modes of talk, dress, and interaction that symbolize their racial heritage, which school personnel then take as evidence that these students do not care about academics—setting in motion a self-fulfilling cycle that can eventually disengage students of color from school (Carter, 2007; Tyson, Darity, & Castellino, 2005).

Fortunately, social scientists have begun to identify ways to build school cultures that are more caring. One strategy, complex instruction, purposely creates student work partnerships across racial lines (Cohen, 1994). These partnerships allow students and their teachers to observe the unique contributions that people from different backgrounds can make to a group, which alters cross-group expectations and feelings of connectedness that can then build into more positive school cultures.

Third, the later years of Coleman's career were spent developing the theoretical concept of social capital, which refers to the different aspects of social connections that serve as resources for school success. It is akin to financial capital, only the currency is not money but something more personal. What do students need to get ahead academically and how can they get it at school? These questions are major foci of school culture research.

High expectations are one resource. Because students often rise to meet the expectations of parents, teachers, and peers who challenge them (Weinstein, 2002), those attending schools in which performance standards for all students are high are immersed in social capital. Another resource is information. Even smart, motivated students get off-track if they do not have practical knowledge about the schooling process. For example, Ricardo Stanton-Salazar (2001) has documented how some bright, goal-oriented Mexican-origin students do not make their way to college because they do not know about college coursework requirements or when to take their Scholastic Aptitude Test (SAT). Thus, a school in which a student has a high probability of tapping into information channels—through contacts with teachers, savvy peers, or the parents of peers—is high in social capital. A third resource is instrumental assistance. A school has social capital, therefore, when it can make available numerous people in the larger school community who are willing and able to help its students with schoolwork, projects, and applications (Morgan & Sorenson, 1999).

For all three resources, the school is an opportunity structure for social contacts and interactions that provide some tangible benefit for academic endeavors. Not surprisingly, socioeconomically advantaged schools are more likely than other schools to provide social capital, which is precisely why socioeconomic integration plans are gaining political momentum (Kahlenberg, 2001). Yet, even socioeconomically disadvantaged schools can develop cultures rich with social capital. Catholic schools, for instance, often serve working-class student populations, but their common curriculum, high performance standards, and more personalized instruction and mentoring have generally resulted in relatively high performance rates. In other words, a surplus of social capital can make up for a deficit of financial capital (Bryk, Lee, & Holland, 1993).

THE IMPORTANCE OF SCHOOL CULTURE AND SCHOOL CULTURE RESEARCH

The commonality among these three themes is the idea that schools can be compared according to the quality and nature of social relations among the people in the school and the common modes of behavior, belief, and values that emerge from these relations over time. These social elements of schooling have traditionally been de-prioritized by educational policies (such as No Child Left Behind of 2001) and evaluations that focus on the more concrete aspects of schooling, such as course offerings, teacher training, per pupil expenditures, and class sizes. Yet, the rich history of school culture research clearly suggests that this is a mistake. The culture of a school helps to determine whether the school is equipped to fulfill its educational mission and whether its students are able to achieve academic success and socioemotional well-being.

SEE ALSO Volume 1: *Intergenerational Closure; Peer Groups and Crowds; Policy, Education; Private Schools; Social Capital; Socioeconomic Inequality in Education; Stages of Schooling; Youth Culture.*

BIBLIOGRAPHY

Bryk, A. S., Lee, V. E., & Holland, P. B. (1993). *Catholic schools and the common good.* Cambridge, MA: Harvard University Press.

Carter, P. L. (2007). *Keepin' it real: School success beyond Black and White.* New York: Oxford University Press.

Cohen, E. G. (1994). *Designing groupwork: Strategies for the heterogeneous classroom.* (2nd ed.). New York: Teachers College Press.

Coleman, J. S. (1961). *The adolescent society: The social life of the teenager and its impact on education.* New York: Free Press of Glencoe.

Coleman, J. S. (1966). *Equality of educational opportunity.* Washington, DC: U.S. Department of Health, Education, and Welfare; Office of Education.

Crosnoe, R., & Needham, B. (2004). Holism, contextual variability, and the study of friendships in adolescent development. *Child Development, 75*(1), 264–279.

Eder, D., Evans, C., & Parker, S. (1995). *School talk: Gender and adolescent culture.* New Brunswick, NJ: Rutgers University Press.

Field, S. H., Frank, K. A., Schiller, K., Riegle-Crumb, C., & Muller, C. (2005). Identifying social contexts in affiliation networks: Preserving the duality of people and events. *Social Networks, 17*(1), 27–56.

Kahlenberg, R. (2001). *All together now: Creating middle-class schools through public school choice.* Washington, DC: Brookings Institution.

Lawrence-Lightfoot, S. (1983). *The good high school: Portraits of character and culture*. New York: Basic Books.

Lee, V. E., & Smith, J. B. (2001). *Restructuring high schools for equity and excellence: What works*. New York: Teachers College Press.

McFarland, D., & Pals, H. (2005). Motives and contexts of identity change: A case for network effects. *Social Psychology Quarterly, 68*(4), 289–315.

Morgan, S. L., & Sorensen, A. B. (1999). Parental networks, social closure, and mathematics learning: A test of Coleman's social capital explanation of school effects. *American Sociological Review, 64*(5), 661–681.

Noddings, N. (2002). *Educating moral people: A caring alternative to character education*. New York: Teachers College Press.

Pianta, R. C., & Walsh, D. J. (1996). *High-risk children in schools: Constructing sustaining relationships*. New York: Routledge.

Schneider, B. L. (2007). *Forming a college-going community in U.S. public high schools*. Seattle, WA: Bill and Melinda Gates Foundation.

Stanton-Salazar, R. D. (2001). *Manufacturing hope and despair: The school and kin support networks of U.S.–Mexican youth*. New York: Teachers College Press.

Steinberg, L. D., Brown, B. B., & Dornbusch, S. M. (1996). *Beyond the classroom: Why school reform has failed and what parents need to do*. New York: Simon & Schuster.

Tyson, K., Darity, W., & Castellino, D. R. (2005). It's not a Black thing: Understanding the burden of acting White and other dilemmas of high achievement. *American Sociological Review, 70*(4), 582–605.

Valenzuela, A. (1999). *Subtractive schooling: U.S.-Mexican youth and the politics of caring*. Albany: State University of New York Press.

Weinstein, R. S. (2002). *Reaching higher: The power of expectations in schooling*. Cambridge, MA: Harvard University Press.

Robert Crosnoe

SCHOOL READINESS

Schooling is a central aspect of childhood in most cultures around the world. In the United States, kindergarten has traditionally been viewed as the beginning of formal schooling, and most U.S. children attend kindergarten at or around age 5. However, educators and researchers have begun to focus more closely than in the past on the skills, knowledge, and habits children have upon entering formal schooling, often referred to as their *school readiness*, which is developed in the early childhood years. Because there is mounting evidence that these early competencies have a significant influence on children's later school success and life trajectories, experts have become increasingly aware of their central role in overall child and human development.

Although different early childhood experts define school readiness in different ways, there is general consensus that skills that promote later reading, writing, and mathematics development are important aspects of school readiness. Some dimensions used by the U.S. Department of Education to measure school readiness include a child's ability to recognize letters of the alphabet, count to 20 or higher, write her or his name, and interact with storybooks (U.S. Census Bureau, 2008). Beyond academic skills, many experts also cite motor skills and physical development, language development, and social and emotional growth as key school readiness components (Rouse, Brooks-Gunn, & McLanahan, 2005). As the National Governor's Association's Task Force on School Readiness (2005) explained:

> Ready children are those who, for example, play well with others, pay attention and respond positively to teachers' instructions, communicate well verbally, and are eager participants in classroom activities. They can recognize some letters of the alphabet and are familiar with print concepts (e.g., that English print is read from left to right and top to bottom on a page and from front to back in a book). Ready children can also identify simple shapes (e.g., squares, circles, and triangles), recognize single-digit numerals, and count to 10. (p. 1)

Pamela High and the American Academy of Pediatrics's Committee on Early Childhood, Adoption, and Dependent Care (2008) classified four conceptualizations of school readiness in the research literature on education and child development:

1. the idealist/nativist view, which emphasizes a child's developmental maturity, particularly in such areas as self-control and the ability to work with peers and follow directions;

2. the empiricist/environmentalist model, which focuses on the knowledge and behavior a child has been taught;

3. the social constructivist perspective, in which the major focus is on the community values and expectations to which a child has been exposed;

4. the interactional relational model, the one most widely recognized by child developmentalists, which emphasizes the individual child, the environment in which he or she learns and develops, and the ongoing interaction between the two.

Researchers have developed and used numerous standardized tests to measure school readiness, and, given the multidimensional nature of school readiness, these tests vary widely (Rock & Stenner, 2005). One of the most commonly used testing instruments, the Peabody Picture Vocabulary Test–Revised, measures a child's vocabulary size by presenting the child with pictures on cards, which he or she is required to match with a stimulus word that is read aloud. The Wechsler Preschool and Primary Scale of Intelligence–Revised is a wider-ranging battery of tests used to assess individual learning patterns in such areas as vocabulary, word reasoning, object assembly, and comprehension. The Stanford–Binet Intelligence Test measures cognitive ability in areas such as verbal reasoning, abstract/visual reasoning, quantitative reasoning, and short-term memory. More recently developed testing instruments, such as those used in the Early Childhood Longitudinal Study–Kindergarten, focus not only on cognitive development and skills but also on social/behavioral and physical/motor skills (Rock & Stenner, 2005). Despite the wider range of school readiness components measured by recent tests, however, many early childhood advocates question the reliability of these assessments to predict school readiness and express concern about the overuse of standardized measures to make educational decisions that may have long-term effects on a child's future schooling (High & Committee on Early Childhood, Adoption, and Dependent Care, 2008).

VARIATIONS IN SCHOOL READINESS

One of the main reasons why school readiness has received increased attention is that research has shown that children enter kindergarten with widely varying levels of many of the skills and competencies that make up school readiness. Early childhood researchers have studied a host of factors to determine whether they contribute to a child's cognitive development and skills in the preschool years, including physical health, family structure, parenting styles, and weight at birth (Rouse et al., 2005). Research has consistently shown that socioeconomic status—a cluster of factors that includes family income as well as a child's access to learning resources and parents' level of educational attainment and expectations—is one of the variables most strongly associated with school readiness. Valerie E. Lee and David T. Burkam (2002), authors of *Inequality at the Starting Gate*, reported that the average cognitive scores of children in the highest socioeconomic category are 60% higher than those of children in the lowest category. Moreover, a frequently cited book by Betty Hart and Todd R. Risley (1995) notes that 3-year-olds whose parents have "professional" jobs have vocabularies that are 50% larger than those of children from working-class families and twice as large as those of children whose families receive public assistance.

Researchers also have found school readiness gaps among students of different racial, ethnic, and language groups. Overall, Black, Hispanic, and American Indian students have been found to have significantly lower prereading, premathematics, and vocabulary skills at school entry than White and Asian American students. A report by Richard Rothstein and Tamara Wilder (2005) noted that the average Black student begins school at the 40th percentile in school readiness, compared to the 57th percentile for the average White student. In addition, data on California kindergartners drawn from the U.S. Department of Education's Early Childhood Longitudinal Study showed that fewer than one-fifth of the children for whom English was a second language scored above average on reading and mathematics tests (Gándara, Rumberger, Maxwell-Jolly, & Callahan, 2003).

As many researchers in education and other fields have noted, race and ethnicity are often closely aligned with socioeconomic status. In an analysis of school readiness gaps by race, ethnicity, and socioeconomics, Greg Duncan and Katherine Magnuson (2005) pointed out that whereas 10% of White children live in poverty, 37% of Hispanic and 42% of Black children do. They also find that, although it is difficult to quantify the effects of poverty and other components of socioeconomic status on school readiness, there is evidence that socioeconomic status differences play a significant role in these early learning gaps by race and ethnicity.

School readiness gaps by race, ethnicity, and socioeconomic status are of particular concern to educators and researchers because of strong evidence that they are associated with what have been called "achievement gaps" in later schooling, all the way up through high school, which in turn affect children's future job choices and overall life trajectories. Throughout K–12 schooling, Black and Hispanic students, on average, score lower on most measures of academic achievement than their White and Asian American peers. A widely cited analysis by Meredith Phillips, James Crouse, and John Ralph (1998) in the Brookings Institution book *The Black–White Test Score Gap* suggests that about half the Black–White gap on standardized test scores at 12th grade can be attributed to gaps that exist at 1st grade. Similarly, Rothstein and Wilder (2005) estimated that Black, Hispanic, and low-income students are, on average, 2 years behind their peers in reading and math by the end of fourth grade, 3 years behind by eighth grade, and 4 years behind by twelfth grade.

EFFORTS TO BRIDGE SCHOOL READINESS GAPS

One of the most widely prescribed solutions to closing socioeconomic and other gaps in school readiness has been improving access to preschool across racial, ethnic, and socioeconomic groups. Data reported by the National Institute for Early Education Research at Rutgers University indicate that more than 70% of 3-year-olds from families with incomes of more than $100,000 attend preschool, compared to less than 40% of 3-year-olds from families with incomes of less than $60,000 (Barnett & Yarosz, 2007). The wide availability of federally funded Head Start preschool programs, which target low-income communities, and the growing number of state-funded preschool programs have closed the gap in preschool attendance between Black and White children: Black children now attend some form of preschool in slightly greater percentages than their White peers; about half of 3-year-olds and more than 75% of 4-year-olds. Hispanic children, however, still attend preschool in the lowest numbers: Only about 30% of Hispanic 3-year-olds and 60% of Hispanic 4-year-olds attend preschool (Barnett & Yarosz, 2007).

Researchers question why the gaps in school readiness persist even as some gaps in preschool attendance have closed, and some point to what they say are differences in the quality and focus of the programs attended by high-income and low-income children. Economist Jane Waldfogel (2006), author of *What Children Need*, emphasizes the difference between *preschool*—which can include a wide range of services such as home care, private day care centers, and Head Start—and *prekindergarten* programs that are usually connected with school districts and are specifically designed to teach children the kinds of skills they will need in kindergarten and the first few years of elementary school. Whereas Head Start is an important resource for many children who might otherwise have no access to preschool, Waldfogel and other researchers have cited evidence that some Head Start programs do not promote school readiness as effectively as the best-quality private programs or school-based prekindergarten.

There also is evidence that access to school-based prekindergarten programs may have the potential to narrow school readiness gaps along racial, ethnic, and socioeconomic lines. One study conducted by Katherine Magnuson, Christopher Ruhm, and Jane Waldfogel (2004) found that students who had attended prekindergarten had higher reading and math skills upon entering school than those who had not; moreover, children from low-income homes, many of whom were Black, Hispanic, or from immigrant families, experienced long-lasting benefits from the prekindergarten programs. Another study of 339 New Jersey children randomly assigned to either full-day or half-day prekindergarten found that those who had attended the full-day program had better literacy and mathematics skills than those in

the half-day program—and that full-day prekindergarten seemed to narrow school readiness gaps between children from upper- and lower-income homes (Robin, Frede, & Barnett, 2006).

GREATER AWARENESS OF SCHOOL READINESS

Recent research such as that highlighted previously, as well as the efforts of early childhood education advocates, have raised both policy makers' awareness of school readiness as an important issue and interest in prekindergarten programs as a possible solution to persistent school readiness gaps. Most recent efforts to improve access to high-quality prekindergarten programs have taken place at the state level. According to National Institute for Early Education Research data, more than 1 million U.S. children now attend state-funded prekindergarten programs, which are offered in 38 states, and interest in the idea of universal prekindergarten has been growing steadily.

Another focus of more recent discussions about school readiness has been not only children's readiness for school but schools' readiness for children. The children who enter school in the United States are an increasingly diverse group. Roughly one in five children currently attending school in the United States is either an immigrant or the child of an immigrant, and this number is rapidly growing. Moreover, U.S. Census Bureau data show that nearly half of all children under 5 in the United States are from groups classified as racial or ethnic minorities. The FPG Child Development Institute, a group based at the University of North Carolina at Chapel Hill, focuses some of its work on examining how schools can be made more ready for children through such practices as addressing the specific learning needs of English-language learners, incorporating "culturally responsive practices" that take into account the diversity of children's ethnic and racial backgrounds, and providing early intervention for students who may be eligible for special services.

Other recent research on issues potentially related to school readiness has focused on such areas as neuroscience (studying the ways early experiences affect brain development and thus cognitive ability); health and wellness, including birth weight and maternal health; and differences in parenting practices (Rouse et al., 2005).

Given the growing diversity of the school-age population in the United States and the evidence that school readiness is strongly associated with success in later schooling (which is, in turn, associated with better future outcomes), school readiness is a concept with far-reaching implications both for the individual life course and for the composition of American society in years to come.

SEE ALSO Volume 1: *Child Care and Early Education; Cognitive Ability; Policy, Education; Racial Inequality in Education; Socioeconomic Inequality in Education; Stages of Schooling.*

BIBLIOGRAPHY

Barnett. W. S., & Yarosz, D. J. (2007). Who goes to preschool and why does it matter? *NIEER Preschool Policy Brief 15.* New Brunswick, NJ: National Institute for Early Education Research. Retrieved June 18, 2008, from http://nieer.org/resources

Duncan, G. J., & Magnuson, K. A. (2005). Can family socioeconomic resources account for racial and ethnic test score gaps? *The Future of Children, 15*(1), 35–54. Retrieved June 18, 2008, from http://www.futureofchildren.org/usr_doc

Gándara, P., Rumberger, R., Maxwell-Jolly, J., & Callahan, R. (2003). English learners in California schools: Unequal resources, unequal outcomes. *Education Policy Analysis Archives, 11*(36). Available from http://epaa.asu.edu/epaa

Hart, B., & Risley, T. R. (1995). *Meaningful differences in the everyday experience of young American children.* Baltimore, MD: P. H. Brookes.

High, P. C., & Committee on Early Childhood, Adoption, and Dependent Care and Council on School Health. (2008). School readiness. *Pediatrics, 121*(4), e1008–e1015. Retrieved June 18, 2008, from http://pediatrics.aappublications.org/cgi

Lee, V. E., & Burkam, D. T. (2002). *Inequality at the starting gate: Social background differences in achievement as children begin school.* Washington, DC: Economic Policy Institute.

Magnuson, K. A., Ruhm, C., & Waldfogel, J. (2004). *Does prekindergarten improve school preparation and performance?* Working Paper No. 10452. Cambridge, MA: National Bureau of Economic Research.

Phillips, M., Crouse, J., & Ralph, J. (1998). Does the Black–White test score gap widen after children enter school? In C. Jencks & M. Phillips (Eds.), *The Black–White test score gap* (pp. 229–272). Washington, DC: Brookings Institution Press.

Robin, K. B., Frede, E. C., & Barnett, S. W. (2006). *Is more better? The effects of full-day vs. half-day preschool on early school achievement.* Working Paper. New Brunswick, NJ: National Institute for Early Education Research. Retrieved June 18, 2008, from http://nieer.org/resources

Rock, D. A., & Stenner, A. J. (2005). Assessment issues in the testing of children at school entry. *The Future of Children, 15*(1), 15–34. Retrieved June 18, 2008, from http://www.futureofchildren.org/usr_doc

Rothstein, R., & Wilder, T. (2005). *The many dimensions of racial inequality.* Paper presented at the Social Costs of Inadequate Education Symposium, Teachers College, Columbia University. Retrieved June 18, 2008, from http://devweb.tc.columbia.edu/manager

Rouse, C., Brooks-Gunn, J., & McLanahan, S. (2005) Introducing the issue. *The Future of Children, 15*(1), 5–14. Retrieved June 18, 2008, from http://www.futureofchildren.org/usr_doc

Task Force on School Readiness. (2005). *Building the foundation for bright futures: A governor's guide to school readiness.* Washington, DC: National Governor's Association. Retrieved June 18, 2008, from http://www.nga.org/cda/

U.S. Census Bureau. (2008). *Statistical Abstract of the United States.* Washington, DC: Author. Retrieved June 18, 2008, from http://www.census.gov/prod

Waldfogel, J. (2006). *What children need.* Cambridge, MA: Harvard University Press.

Michael Sadowski

SCHOOL TRACKING

Curriculum differentiation—the division of a domain of study into segments and/or the division of students into groups—is an important feature of schools in the United States. Tracking is one possible outcome of curriculum differentiation.

Aage Sørenson (1970) identified key dimensions of curriculum differentiation. Two dimensions—*horizontal* differentiation and *vertical* differentiation—concern the basis of the differentiation. Horizontal differentiation divides the curriculum into domains of study. For example, the division of foreign language study into German, French, and Spanish is horizontal differentiation. Vertical differentiation, however, reflects a division of students along lines relevant for learning, such as, for example, prior knowledge. The division of French into introductory (French I), intermediate (French II), and advanced (French III) provides an example of vertical differentiation. In vertical differentiation, students' performance in one course depends on their mastery in the previous course, a dependence reflecting the ordered nature of vertical differentiation.

Another key dimension of curriculum differentiation is *scope*, which reflects the extent to which students share the same peers over the school day. If scope is high, groups are segregated.

Electivity, which refers to the degree to which students' desires matter for their curricular placement, is another key dimension. *Mobility* and *selectivity* are two additional dimensions. Mobility concerns the amount of movement across curricular positions, and selectivity measures the degree of homogeneity within the curricular positions. These dimensions are important tools for understanding curriculum differentiation and, by extension, tracking.

TRACKING: CLASSICAL DEFINITION AND HISTORICAL BASIS

Tracking has been defined in various ways. One restrictive view regards tracking as existing only when a formal process assigns students to explicit, overarching programs encompassing multiple subjects and years. Others define tracking as an association across disparate subjects, regardless of whether or not a formal program assignment process exists. Others claim that an association of course-taking across time is tracking. And the least restrictive definition sees curriculum differentiation *itself* as tracking.

Matters were apparently more straightforward in an earlier epoch, one in which analysts suggest that *classical* tracking existed. Under classical tracking, students are formally assigned to overarching programs that determine the level of all of their academic courses.

Classical tracking was the aim of original tracking advocates who argued that the vast majority of students should be taught "followership," which would lead those students to uncritically parrot the views of their ostensible superiors (Finney, 1928). Further, these advocates contended that some students should be taught leadership, so that they could eventually, unapologetically, take power. In other words, these advocates sought *disparate socialization,* an aim that, they argued, required students to be consistently segregated.

These reformers defended classical tracking using early-20th-century theories of intelligence that asserted the existence of a generalized intelligence. In this view, differences in a person's ability across domains were trivial. Thus, one could identify the top students, provide them with exposure to challenging material in all domains, and avoid spending resources on those deemed constitutionally unable to grasp complex material.

Over time, classical tracking took root in the United States. During a period of high immigration and rising nativist fears, many reformers believed schools could "Americanize" immigrant youth (e.g., Kelley, 1903). In order for schools to do so, however, immigrant youth had to be forced into school. Compulsory schooling laws accomplished this aim, but produced another dilemma. Should schools teach a classical college preparatory curriculum of Latin and Greek to all students? Or should schools elaborate the curriculum to provide different types of training? Classical tracking was the eventual answer (Kliebard, 1995; Spring, 1972; Wrigley, 1982).

Consistent with this historical basis, evidence indicates that students in different tracks receive different amounts and kinds of resources. For example, lower-track classes are less likely to have experienced or capable teachers (Finley, 1984; Kelly, 2004). Further, consistent with the aim of differential socialization, courses of different tracks differ in multiple ways. Researchers have found that low-track classes require students to complete repetitive tasks based on simplified texts that are replete with lists to memorize, while, in contrast, high-track classes at the same school ask students to engage creatively using complex texts that demand students sift and synthesize the material (e.g., Oakes, 1985; Page, 1990).

Historically tracking appears to have been a major cleavage in students' experience. The best evidence suggests that, prior to 1965, most high schools assigned students to explicit, overarching programs that determined their academic course taking and thus governed the socialization to which they would be exposed. Classically, there were three broad tracks: college preparatory, general, and vocational. Students were assigned to one such program upon entering high school (e.g., Hollingshead, 1949; Alexander, Cook, & McDill, 1978), and students in a school spent most of their time with students in the same track.

THE DECLINE OF CLASSICAL TRACKING

Under episodic pressure in the courts (e.g., Hayes, 1990), and in a 1960s context in which the claims of general intelligence were under attack, school districts began to alter their formal practices. Evidence suggests that the classical system of tracking began to erode in the late 1960s. Donald Moore and Suzanne Davenport (1988) studied four urban districts in 1965 and 1975, finding that all four had ended the assignment of students to overarching programs by 1975. Yet the districts maintained within-subject curriculum differentiation; students in the same grade could still take different levels of math, English, and so on. Districts formally decoupled course-taking *across* subjects, but maintained multiple levels of coursework within subjects. This change and stability allowed students to take different course levels in disparate subjects.

Evidence indicates that by the early 1980s most schools had moved similarly, dismantling the formal apparatus of cross-subject tracking (Oakes, 1981). By the early 1990s, 85% of schools lacked formal mechanisms of tracking (Carey, Farris, & Carpenter, 1994).

THE AFTERMATH OF CLASSICAL TRACKING: MULTIPLE DEFINITIONS, MULTIPLE CONTROVERSIES

It is apparent that most schools have no formal process of overarching track assignment. Samuel Roundfield Lucas (1999) contended that an unremarked revolution in tracking had occurred, unremarked in that analysts acknowledged the change in school practice, but research methods did not keep pace with the change. Prior to the end of classical tracking, analysts routinely used students' self-reports or school personnel reports of students' track location (Gamoran & Berends, 1987). Indeed, scholars debated what discrepancies in the two measures signified (e.g., Rosenbaum, 1980 & 1981; Fennessey, et. al, 1981). However, the end of formal program assignment made students' self-reports purely social-psychological measures (Lucas & Gamoran, 2002), and rendered the referent of school personnel reports even more opaque.

In that context, some maintain that curriculum differentiation *is* tracking, because as long as there is curriculum differentiation, students in the same school may be exposed to vastly different levels of rigor and types of socialization. Although some research has studied students' level of placement in individual subjects, this research is based on the inherent importance of those subjects, not on an explicit claim that curriculum differentiation is tracking. However, the view that curriculum differentiation is tracking has been the basis of proposals to end curriculum differentiation and instruct all students in heterogeneous classrooms. This policy proposal has often been called *detracking*, implying thereby that curriculum differentiation is tracking (e.g., Wheelock, 1992).

However, others resist the idea that curriculum differentiation is tracking, pointing to the original basis of tracking in a desire to make socialization consistent for any given student. In this view, the key issue concerns the degree to which courses are associated. Thus, some analysts contend that if students' level of course in one subject predicts their level of course in a disparate subject, this signals de facto tracking. When placement level in one subject predicts placement level in a disparate subject, the courses are associated (e.g., Lucas, 1999).

A great deal of research has been conducted under the assumption that students' academic programs are a collection of associated courses. Some of that research occurred prior to the end of classical tracking, and the questions given attention then have continued to be studied after the unremarked revolution. The research has concerned four principal questions:

1. What are the sources of tracking at the school-level?

2. What are the sources of the assignment of students to various levels of study?

3. What are the effects of students' track location? and

4. What is the source of any effects of students' track location?

PREDICTORS OF SCHOOL-LEVEL TRACK STRUCTURE

The school-level factors that predict the existence or characteristics of tracking are important because any determinant of track structure is thus a determinant of the in-school environment students navigate. Learning the basis of that environment, therefore, is important for understanding the underlying basis of important aspects of adolescent experience and the context within which important life course transitions occur.

Maureen T. Hallinan (1994a, 1994b) maintains that tracking is a technical pedagogical device that allows students to receive education tailored to their prior preparation. In this view, tracking allows the construction of groups homogenous on prior achievement. In contrast, Jeannie Oakes (1994a, 1994b) contends that tracking inescapably involves segregation along lines of race, ethnicity, and class. Indeed, in this view the construction and maintenance of tracking is best understood as a means of buttressing such segregation after the end of de jure school segregation.

Research on detracking and on the factors predicting the strength of tracking systems addresses this debate. Ethnographic research on detracking shows that socio-economically advantaged parents contest detracking until some advantages are preserved for their children (e.g., Wells & Serna, 1996). Yet if advantaged children receive advantageous positions, then both detracking reform and the technical pedagogical basis of tracking are both undermined.

Analysts find that the more race (Braddock, 1990) or class (Lucas, 1999) diversity, the higher scope—that is, the more pronounced de facto tracking is. Socio-demographic diversity is important even after the profile of student achievement is controlled in public (but not private) schools (Lucas & Berends, 2002). Further, the profile of students' achievement matters for track structure; the more highly correlated students' prior achievement is across domains, the higher the association in students' placements across subjects (Lucas & Berends, 2002). This result suggests a technical basis for some amount of de facto tracking. Still, the degree of de facto tracking appears driven, in part, by race and class diversity, consistent with the interest in segregation that provided the historical basis of the implementation of classical tracking.

PREDICTORS OF STUDENT ASSIGNMENTS

Another key question asks what factors place students in different levels. If assignment is connected to factors other than prior achievement, the meritocratic basis of tracking is questioned, and one of the key technocratic rationales for tracking—its ability to produce groups homogenous on prior achievement so as to facilitate instruction—is undermined.

Research indicates that although measured achievement matters for placement, socioeconomic background matters as well (e.g., Gamoran & Mare, 1989; Mickelson, 2001; Jones, Vanfossen, & Ensminger, 1995). Ethnographic research reveals some of the ways social class enters the equation. Elizabeth Useem (1992) found that

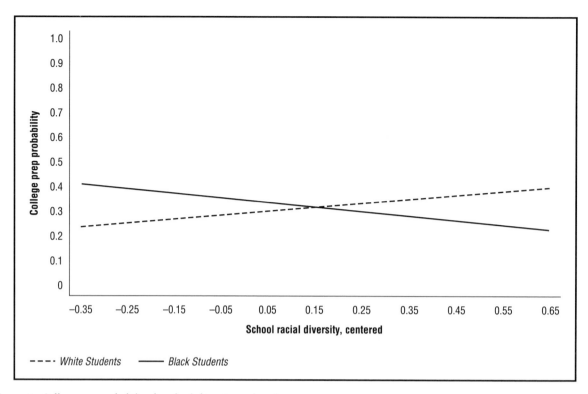

Figure 1. College prep probability by school diversity and student race. CENGAGE LEARNING, GALE.

socioeconomically advantaged parents networked to obtain and distribute information amongst themselves, routinely intervened in placement decisions to assure high track placements for their children, and routinely acted to affect their children's course preferences.

The consensus concerning socioeconomic class and track location is not mirrored for race. Using nationally representative data, Oakes (1985) found that Black and Latino/a students were more likely to be found in vocational or remedial classes. Samuel Lucas and Adam Gamoran (2002), however, found a Latino/a disadvantage but no Black-White difference in college prep track assignment for 1980 sophomores, and parity for Blacks, Whites, and Latino/as amidst an advantage for Asians in their analysis of 1990 sophomores. Michael S. Garet and Brian DeLany (1988) found Blacks and Asians were more likely to be assigned to advanced mathematics classes than were Whites, whereas Roslyn Arlin Mickelson (2001) found Blacks disadvantaged in English class enrollment.

Evidence indicates that one explanation for the disparate findings is that schools differ appreciably in the relationship between race/ethnicity and curricular location (Lucas & Berends, 2007). That variation is not random: The more racially diverse the school, the less

likely Blacks are, and the more likely Whites are, to enter college prep courses, even after test scores and parents' characteristics are controlled. Indeed, the differences resemble one-for-one substitution of Whites for Blacks in college prep courses as one compares more and less diverse schools (Lucas & Berends, 2007).

EFFECT OF STUDENT ASSIGNMENTS

One oft-studied outcome is cognitive achievement. Research indicates an association between measured achievement and track location. Indeed, Gamoran (1987) found the achievement gap between college preparatory and non-college preparatory students exceeded the gap between high school graduates and drop-outs. One challenge in study effects of track location, however, flows from the research on track placement. That research, which finds social class, race, and other factors matter for track placement, implies that the students in any given track are not a random set of students. Thus, it could be that high track students would have higher test scores even if there were no tracking. Yet the difference between high-track and low-track student achievement appears to be caused by track placement, because when researchers control for the non-random assignment of students to tracks, they still find that students in high tracks end up having higher achievement (Gamoran & Mare, 1989; Lucas & Gamoran, 2002).

Tracking could produce the effect on cognitive achievement by raising high-track students' achievement, lowering low-track students' achievement, or both. Alan C. Kerckhoff (1986) compared the outcome of British students of equal levels of achievement prior to their track assignment, and found that high-track students had high achievement growth while equivalent low-track students had less achievement growth. This pattern matches Thomas B. Hoffer's (1992) finding for middle school students in the United States. Hoffer (1992) compared not only high- and low-track students, but also schools with and without tracking. He found the usual advantage of high-track students compared to low-track students. However, comparing schools with and without tracking revealed that schools' overall level of achievement was virtually the same, because tracked schools improved the learning of high-track students but equally hindered the learning of low-track students.

Other outcomes have been studied as well. High-track students are less likely to drop out of school (Gamoran & Mare, 1989), more likely to receive encouragement for college (e.g., Hauser, Sewell, & Alwin, 1976), more likely to enter college (Rosenbaum, 1980), and have higher political efficacy (Paulsen, 1991) and academic engagement (Berends, 1995). General self-efficacy, however, appears unaffected once appropriate pre-track assignment controls are introduced (e.g., Wiatrowski, Hansell, Massey, & Wilson, 1982).

THE SOURCE OF STUDENT ASSIGNMENT EFFECTS

Cognitive effects of track location exist and have been theorized in three ways. Effects may reflect an *instructional* basis, because lower tracks cover less content than do higher tracks (Gamoran, 1989). Effects may be *social,* because tracks provide different social contexts for students to explore and develop their selves and capabilities. Effects may be *institutional,* because tracking is a persistent organizational form and placements are arguably stable and public (Pallas, Entwistle, Alexander, & Stluka, 1994).

Evidence favors a role for instructional effects (Gamoran, 1989; Pallas, et al, 1994). Indeed, although tracking was originally motivated by the aim to teach students differently, some research suggests that low tracks would produce more achievement if they used the same approaches used in high-track classes (e.g., Gamoran, 1993). This is consistent with instruction as a key factor in producing track effects.

FUTURE RESEARCH

Analysts have neglected some dimensions of curriculum differentiation. Electivity has been neglected, in part because students routinely report that they selected their courses when, in fact, researchers know that other factors, such as prerequisites and graduation requirements, play a large role in whether students will be allowed to take what they ostensibly desire (National Center for Education Statistics, 1981). No one has studied, using longitudinal ethnographic means, the coalescence of students' course-taking intentions and eventual behavior from before tracking to the end of high school. Until such research is conducted, carefully attending to students' unfolding preferences and the factors influencing those preferences, it will be difficult to reach any conclusions concerning electivity and tracking.

Almost every research effort would allow one to ascertain the selectivity of tracks. However, few researchers report this information. The little research that has reported on selectivity—the degree to which groups are homogenous—suggests tracks are heterogenous. Mickelson (2001) finds that while some students in the 30th to 39th percentile range are placed in advanced English, other students in the 90th to 99th percentile are placed in regular English. Many such placement inconsistencies existed, suggesting that tracks are often heterogenous on prior achievement.

Some research has studied mobility. One criticism of classical tracking is that it locked students into trajectories from which they could not escape. Early research on track mobility sustained this imagery, describing tracking as a tournament in which upward mobility was impossible (e.g., Rosenbaum, 1976 & 1978). But research after the unremarked revolution suggests that while downward mobility is far more common, a non-negligible amount of mobility is actually upward (Lucas, 1999), and patterns of mobility through the complex curriculum structure appear affected by race and class (Lucas & Good, 2001). Given the importance of mobility for the fairness and implications of tracking, and given the limited understanding we have based on the research completed to date, more research on track mobility would be desirable.

Tracking continues to affect the lives of students in schools throughout the United States. Tracking allows students to be exposed to different levels of cognitive demand and types of socialization. Track structures are most pronounced in racially and socioeconomically diverse schools. Assignment to courses is not meritocratic, as social class has been shown to matter, and race appears to matter as well. That race and class matter is likely consequential, because track location matters for a wide variety of cognitive, affective, and socioeconomic outcomes. The pathways

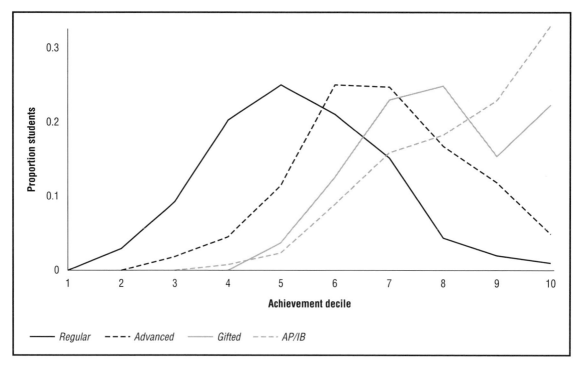

Figure 2. *Distribution of prior achievement in four English courses.* **CENGAGE LEARNING, GALE.**

through which track effects are generated are contested, but instruction appears to be a key conduit for track effects.

SEE ALSO Volume 1: *Academic Achievement; College Enrollment; Gender and Education; High-Stakes Testing; Learning Disability; Racial Inequality in Education; Socioeconomic Inequality in Education; Stages of Schooling; Vocational Training and Education.*

BIBLIOGRAPHY

Alexander, K. L., Cook, M., & McDill, E. L. (1978). Curriculum tracking and educational stratification: Some further evidence. *American Sociological Review, 43,*: 47–66.

Berends, M. (1995). Educational stratification and students' social bonding to school. *British Journal of Sociology of Education, 16,* 327–351.

Braddock, J. H., II, & McPartland, J. M. (1990). Tracking the middle grades: National patterns of grouping for instruction. *Phi Delta Kappan, 71,* 445–449.

Carey, N., Farris, E., and Carpenter, J. (1994). *Curricular differentiation in public high schools: Fast response survey system E.D. tabs.* Washington, DC: U.S. Dept. of Education.

Fennessey, J., Alexander, K. L., Riordan, C., & Salganik, L. H. (1981). Tracking and frustration reconsidered: appearance or reality. *Sociology of Education, 54,* 302–309.

Finley, M. K. (1984). Teachers and tracking in a comprehensive high school. *Sociology of Education, 54,* 233–243.

Finney, R. L. (1928). *A sociological philosophy of education.* New York: Macmillan.

Gamoran, A. (1987). The stratification of high school learning opportunities. *Sociology of Education, 60,* 135–155.

Gamoran, A. (1989). Rank, performance, and mobility in elementary school grouping. *Sociological Quarterly, 30,* 109–123.

Gamoran, A. (1993). Alternative uses of ability grouping in secondary schools: Can we bring high-quality instruction to low-ability classes? *American Journal of Education, 102,* 1–22.

Gamoran, A., & Berends, M. (1987). The effects of stratification in secondary schools: Synthesis of survey and ethnographic research. *Review of Educational Research, 57,* 415–435.

Gamoran, A., & Mare, R. D. (1989). Secondary school tracking and educational equality: Compensation, reinforcement, or neutrality? *American Journal of Sociology, 94,* 1146–1183.

Garet, M. S., and DeLany, B. 1988. Students, courses, and stratification. *Sociology of Education, 61*(2), 61–77.

Hallinan, M. T. (1994a). Tracking: From theory to practice. *Sociology of Education, 67,* 79–84.

Hallinan, M. T. (1994b). Further thoughts on tracking. *Sociology of Education, 67,* 89–91.

Hauser, R. M., Sewell, W. H., & Alwin, D. F. (1976). High school effects on achievement. In W. H. Sewell, R. M. Hauser, & D. L. Featherman (Eds.), *Schooling and achievement in American society* (pp. 309–341). New York: Academic Press.

Hayes, III, F. W. (1990). Race, urban politics, and educational policy-making in Washington, DC: A community's struggle for quality education. *Urban Education, 25,* 237–257.

Hoffer, T. B. (1992). Middle school ability grouping and student achievement in science and mathematics. *Educational Evaluation and Policy Analysis, 14,* 205–227.

Hollingshead, A. B. (1949). *Elmtown's youth: The impact of social classes on adolescents.* New York: Wiley.

Jones, J. D., Vanfossen, B. E., & Ensminger, M. E. (1995). Individual and organizational predictors of high school track placement. *Sociology of Education, 68,* 287–300.

Kelley, F. (1903). An effective child-labor law: A program for the current decade. *Annals of the American Academy of Political and Social Science, 21,* 438–445.

Kelly, S. (2004). Are teachers tracked? On what basis and with what consequences. *Social Psychology of Education, 7,* 55–72.

Kerckhoff, A. C. 1986. Effects of ability grouping in British secondary schools. *American Sociological Review, 51,* 842–858.

Kliebard, H. M. (1995). *The struggle for the American curriculum: 1893–1958.* (2nd ed.). New York: Routledge.

Lucas, S. R. (1999). *Tracking inequality: Stratification and mobility in American high schools.* New York: Teachers' College Press.

Lucas, S. R., & Berends. M. (2002). Sociodemographic diversity, correlated achievement, and de facto tracking. *Sociology of Education, 75,* 328–348.

Lucas, S. R., & Berends, M. (2007). Race and track location in U.S. public schools. *Research in Social Stratification and Mobility, 25,* 169–187.

Lucas, S. R., & Gamoran, A. (2002). Tracking and the achievement gap. In J. E. Chubb & T. Loveless (Eds.), *Bridging the Achievement Gap* (pp. 171–198). Washington, DC: Brookings Institution Press.

Lucas, S. R., & Good, A. D. (2001). Race, class, and tournament track mobility. *Sociology of Education, 74,* 139–156.

Mickelson, R. A. (2001). Subverting Swann: First- and second-generation segregation in Charlotte-Mecklenberg schools. *American Educational Research Journal, 38,* 215–252.

Moore, D. R., & Davenport, S. (1988). *The new improved sorting machine.* Madison: National Center on Effective Secondary Schools, School of Education, University of Wisconsin.

National Center for Education Statistics. (1981). *High school and beyond, 1980: A longitudinal survey of students in the United States.* Ann Arbor, MI: Inter-University Consortium for Political and Social Research.

Oakes, J. (1981). *Tracking policies and practices: School by school summaries: A study of schooling in the United States,* Technical Report Series No. 25. Los Angeles: Graduate School of Education, University of California at Los Angeles.

Oakes, J. (1985). *Keeping track: How schools structure inequality.* New Haven, CT: Yale University Press.

Oakes, J. (1994a). More than misapplied technology: A normative and political response to Hallinan on tracking. *Sociology of Education, 67,* 84–89.

Oakes, J. (1994b). One more thought. *Sociology of Education, 67,* 91.

Page, R. N. (1990). Games of chance: The lower-track curriculum in a college-preparatory high school. *Curriculum Inquiry, 20,* 249–281.

Pallas, A. M., Entwistle, D., Alexander, K. L., & Stluka, M. F. (1994). Ability-group effects: Instructional, social, or institutional? *Sociology of Education, 67,* 27–46.

Paulsen, R. (1991). Education, social class, and participation in collective action. *Sociology of Education, 64,* 96–110.

Rosenbaum, J. E. (1976). *Making inequality.* New York: Wiley.

Rosenbaum, J. E. (1978). The structure of opportunity in school. *Social Forces, 57,* 236–256.

Rosenbaum, J. E. (1980). Track misperceptions and frustrated college plans: An analysis of the effects of tracks and track perceptions in the National Longitudinal Survey. *Sociology of Education, 53,* 74–88.

Rosenbaum, J. E. (1981). Comparing track and track perceptions: Correcting some misperceptions. *Sociology of Education, 54,* 309–311.

Sørenson, A. (1970). Organizational differentiation of students and educational opportunity. *Sociology of Education, 43,* 355–376.

Spring, J. (1972). *Education and the rise of the corporate state.* Boston: Beacon Press.

Useem, E. (1992). Middle schools and math groups: Parents' involvement in childrens' placement. *Sociology of Education, 65,* 263–279.

Wells, A. S., & Serna, I. (1996). The politics of culture: Understanding local political resistance to detracking in racially mixed schools. *Harvard Educational Review, 66,* 93–118.

Wheelock, A. (1992). *Crossing the tracks: How "untracking" can save America's schools.* New York: W. W. Norton.

Wiatrowski, M. D., Hansell, S., Massey, C. R., & Wilson, D. L. (1982). Curriculum tracking and delinquency. *American Sociological Review, 47,* 151–160.

Wrigley, J. (1982). *Class politics and public schools: Chicago, 1900–1950.* New Brunswick, NJ: Rutgers University Press.

Samuel R. Lucas

SCHOOL TRANSITIONS

Common features of virtually all individuals' educational careers are disruptions in organizational contexts, pedagogical approaches, and social relationships during school transitions. Most frequently, these transitions are planned or orchestrated as rites of passage from one level of schooling to the next made by student cohorts each year. Other changes in school enrollment are essentially random events occurring in individuals' lives, which often coincide with other transitions in residence and family status. Regardless of whether they are rites of passage or random events, school transitions involve changes in status and context that have myriad implications for individuals' developmental, social, and academic trajectories. This entry begins by describing the different types of school transitions and then outlines issues to be considered when trying to understand the opportunities and challenges that individuals encounter when changing schools. The entry closes with a discussion of the implications of school transitions for life course research.

TYPES OF SCHOOL TRANSITIONS

Moves between schools can be generally grouped into two types: (a) school transitions marking completion of a program of study and entry into the next stage of schooling and (b) school transfers due to changes in an individual's enrollment or assignment to a particular institution. The former are often called *normative school transitions* in that, annually, student cohorts follow similar patterns of movement between schools as guided by institutional policies and practices. In contrast, the latter are often called *nonnormative school transfers* in that students' movements between schools are unpredictable because they are predicated on individual circumstances, such as a family's move or disciplinary problems (Alexander, Entwisle, & Dauber, 1996). Although sometimes considered together in studies of changing schools, these two types of transitions are fundamentally different in the processes that individuals experience.

Normative school transitions are the most frequent changes in enrollment because they are required by the organizational structures of almost all educational systems. In most developed nations, school systems are organized into levels that can be roughly classified as lower elementary, upper elementary, middle, lower secondary (or junior high), and upper secondary (or senior high; Shavit & Blossfeld, 1993). In the United States, these levels are generally equivalent to kindergarten through third grade, fourth through sixth grade, seventh and eighth grades, ninth and tenth grades, and eleventh and twelfth grades, respectively. Although transitioning between levels reflects a change in program status, students do not necessarily have to enroll in a new school, depending on its grade configuration (e.g., kindergarten through tenth grades compared to a traditional 4-year high school). Although policies on school-leaving ages vary by state, school attendance is generally compulsory from ages 5 to 16 (roughly equivalent to kindergarten through tenth grade) in the United States. Some individuals also enroll in preschool prior to entering the early elementary grades, and others attend postsecondary schooling at colleges and universities. Regardless of the level, all normative transitions occur between academic years as dictated by the grade configurations and academic programs of the schools involved.

In contrast, nonnormative school transfers are generally unpredictable in that these decisions to change schools are the result of an individual's life circumstances. Most frequently, school transfers are prompted by residential moves that cross school attendance zones, whether from one side of the street to the other or across the country. However, many school transfers occur for other reasons, such as a consequence of schools' disciplinary and other administrative actions or a result of parents' exercise of school choice, such as transferring a child from a public to a parochial school (Swanson & Schneider, 1999). Generally, one or a small group of students transfer between two schools at a given time.

Although many school transfers are made in the summer months, they can take place at any time during the school year and may occur with little warning or planning.

Although these two types of changes in school enrollment are distinct, the transition between levels of schooling can coincide with a decision to attend a different school than most of a student's cohort from the prior school. In these cases, the difficulties and opportunities encountered by students during school transitions may share many similarities with those of students transferring between schools during summer months. However, transfer students are usually changing schools at times when few of their classmates, at either their prior or new schools, would be contemplating similar shifts in schools. Thus, as discussed below, the contexts and challenges encountered by transfer students are different than those transitioning between levels of schooling because they tend to make their transition alone and at an unexpected time.

CONTEXTS OF SCHOOL TRANSITIONS AND TRANSFERS

Regardless of whether a transition or a transfer, changing schools usually involves a student's entry into a new building, and it frequently features discontinuity in instructional techniques or approaches. The transition may also entail disruptions in one's teacher and peer networks. The frequency and predictability of students' changes in school enrollments create organizational and educational problems that vary in their intensity and duration. Schools' attempts to address these potential problems shape the structural, pedagogical, and social contexts of students' experiences during transitions and transfers.

Schools experience annual changes in enrollment, with varying proportions of students transitioning between grade levels and others transferring between schools. At one end of the spectrum, new students at schools spanning kindergarten through twelfth grades may be basically limited to a cohort of entering kindergarteners to replace the graduating senior class exiting the year before. At the other end, urban and rural schools serving transient populations may have rates of student mobility so high that the composition of classrooms can change dramatically during an academic year. To the extent that changes in enrollment are due to predictable arrivals of successive cohorts, school districts and systems can institutionalize the transition process by establishing policies and procedures to regulate the flow of students and information about them (MacIver & Epstein, 1991). In contrast, large numbers of transfer students can create chaotic environments for all students as schools must adjust their allocation of classroom and other resources

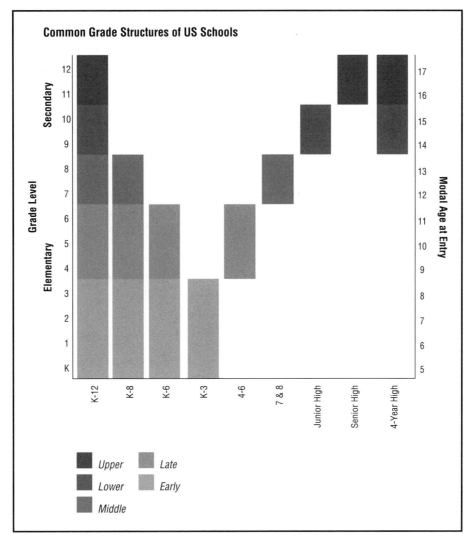

Figure 1. *This figure outlines the major grade configurations in U.S. schools to highlight when transitions between levels of schooling are likely to occur.* **CENGAGE LEARNING, GALE.**

to accommodate their new classmates' needs (Ingersoll, Scamman, & Eckerling, 1989). Thus, from an organizational standpoint, managing the relatively predictable transition of successive cohorts is easier than assisting transfer students arriving under frequently chaotic circumstances.

One major source of organizational uncertainty is created by questions concerning new students' prior academic experiences and intellectual development. Most schools depend largely on official records and teacher recommendations accompanying new students in allocating them to classrooms and courses (Gamoran, 1992). Interpreting this information is easier when district feeder policies establish a consistent flow of students between schools, which facilitates the formation of both formal and informal relationships among school personnel

(Schiller, 2005). For example, many school districts with schools linked by feeder patterns have personnel who meet to both exchange information about specific students and coordinate instructional practices and programs. Such meetings, however, become more problematic as the number of feeder schools increases and are almost impossible when a student transfers in from another district or state. In these situations, schools are more likely to establish internal assessments of prior academic progress to use in placing new students and provide more time in class for review of foundational curricular material (Riehl, Pallas, & Natriello, 1999). Despite schools' efforts to manage transition processes, however, students often encounter significant differences in instructional approaches and programs after changing schools that can disrupt their academic progress.

Changing schools usually requires students to also develop new relationships with peers, teachers, and other school personnel at the same time as they are encountering new academic environments. In addition to finding their way around a new building, new students usually must learn to negotiate the institutional procedures and social environments unique to each school (Eccles et al., 1993). Around the beginning of each academic year, most schools hold at least one orientation session to provide new students and their parents with information about policies they must follow and activities in which they may want to participate. Smaller numbers of schools establish mentoring programs or small learning communities to help new students connect with their teachers and form friendships with their peers. In contrast, transfer students, especially those entering during the middle of the school year, often receive little help in adjusting to the culture of their new school and integrating themselves into established friendship networks. Both members of incoming cohorts and transfer students are at risk of becoming socially isolated, yet they also have opportunities to identify with a new group of peers.

RISK AND RESILIENCY DURING SCHOOL TRANSITIONS AND TRANSFERS

School transitions and transfers may precipitate changes in a student's academic success and social integration. The challenges and opportunities students encounter when entering a new school can dramatically affect their developmental trajectories in both negative and positive ways (Catterall, 1998). As noted earlier, most schools have programs to facilitate the integration of new students into their academic structures and improve the students' adjustment to a new social environment. However, an individual student's family circumstances, personality, and academic status also may influence how he or she is affected by the process of transitioning between schools.

Changing schools may disrupt a student's academic progress if the pedagogical approach—reflected in teaching style and curriculum—differs significantly between the two schools. Students transferring during the academic year are particularly likely to encounter severe discrepancies between what they were learning in the same class at each school (Lash & Kirkpatrick, 1994). Even within the same state or district, each class moves at a different pace and with varying emphases on particular topics based on the teachers' assessment of the students' strengths and weaknesses. These differences are particularly problematic for new students whose prior class was moving at a slower pace, as they will lack a similar

foundation in basic concepts and skills compared to their new classmates.

Programs to coordinate curricula across feeder schools are intended to reduce differences in prior curriculum exposure among entering cohorts. However, these students must still deal with increasing conceptual complexity, more challenging tasks, and less personal attention from teachers with each transition between levels of schooling (Felner, Ginter, & Primavera, 1982). For example, students frequently first experience changing classrooms for some subjects when they enter middle school and do so for each class in high school. How successfully students adjust to new academic challenges and environments after either transfer or transition can have long-term consequences for their educational trajectories. A decrease in grades following a transition or transfer often signals that a student may be on a path to dropping out of school.

Similarly, newly arrived students are at risk of becoming socially isolated and alienated if their peer networks are significantly disrupted in the process of transitioning or transferring between schools. When entering a new school, transfer students are almost always required to build ties to peers who already have established relationships with longer-term classmates. Breaking into established social cliques can be particularly difficult when school transfers coincide with residential moves prompted by other life events, such as a change in parental marital status or a job loss (South & Haynie, 2004). Sometimes, however, school transfers are initiated by parents to distance their child from undesirable peer (or other) influences; such changes may not coincide with residential moves. In contrast, peer networks are likely to remain at least partially intact when cohorts transition between schools together. However, many students encounter difficulties adjusting to a new social status (i.e., being the youngest in a school instead of the oldest) and more diverse social groups when transitioning to larger schools. The larger pools of potential friends allow some previously marginalized individuals to redefine themselves as *normal* by identifying with new subcultures (Kinney, 1993). Individuals differ in their abilities to maintain or develop new connections to peers after entering a new school, which can also affect both their academic and social development.

RESEARCH CHALLENGES AND FURTHER DIRECTIONS

Changing schools is a common feature of most individuals' educational careers, yet it is a complex process involving the interaction of institutional procedures and environments with students' readiness for the academic and social challenges they are likely to encounter.

Researchers are only beginning to understand why many students encounter difficulties that put them on the path to dropping out of school altogether (Roderick, 1993) whereas others find opportunities to blossom both academically and socially (Schiller, 1999). Investigators need to carefully specify the type or types of school transitions being studied to identify the particular processes and experiences that influence students' adjustment to new schools. For example, even if they remain in the same school, high school freshmen often experience similar changes in academic performance and social integration as those who transition into a 4-year high school (Weiss & Bearman, 2007). More studies following significant numbers of students through school transitions are needed to understand whether some changes in context—for example, from an urban to a suburban school—are more challenging for some students than others.

In addition, investigators need to consider what contributes to variation in how individuals are affected by school transitions. Changing schools may provide socially and academically struggling students with windows of opportunities for fresh starts, whereas more successful students may encounter more competition for positions in the elite groups. Studying school transfers can be particularly problematic because the difficulties students encounter may be due as much to the reason (e.g., a divorce or behavioral problems) for changing schools as to the transition process itself. Similarly, prior academic or social difficulties may become problematic again following school transitions and transfers. Disentangling these institutional and individual influences requires longitudinal study designs that allow a detailed examination of the transition process over time to determine what programs and interventions may work for which type of individuals.

SEE ALSO Volume 1: *Academic Achievement; High School Dropout; Peer Groups and Crowds; Policy, Education; Racial Inequality in Education; School Culture; Socioeconomic Inequality in Education; Stages of Schooling;* Volume 2: *School to Work Transition.*

BIBLIOGRAPHY

Alexander, K. L., Entwisle, D. R., & Dauber, S. L. (1996). Children in motion: School transfers and elementary school performance. *Journal of Educational Research, 90*(1), 3–12.

Catterall, J. S. (1998). Risk and resilience in student transitions to high school. *American Journal of Education, 106*(2), 302–333.

Cicourel, A. V., & Kitsuse, J. I. (1963). *The educational decision-makers.* Indianapolis, IN: Bobbs-Merril.

Eccles, J. S., Midgley, C., Wigfield, A., Buchanan, C. M., Reuman, D., & Flanagan, C. (1993). Development during adolescence: The impact of stage-environment fit on young adolescents' experiences in schools and families. *American Psychologist, 48*(2), 90–101.

Felner, R. D., Ginter, M., & Primavera, J. (1982). Primary prevention during school transitions: Social support and environmental structure. *American Journal of Community Psychology, 10*(3), 277–291.

Gamoran, A. (1992). Access to excellence: Assignment to honors English classes in the transition from middle to high school. *Educational Evaluation and Policy Analysis, 14*(3), 185–204.

Ingersoll, G. M., Scamman, J. P., & Eckerling, W. D. (1989). Geographic mobility and student achievement in an urban setting. *Educational Evaluation and Policy Analysis, 11*(1), 143–149.

Kerckhoff, A. C. (1993). *Diverging pathways: Social structure and career deflections.* Cambridge, U.K.: Cambridge University Press.

Kinney, D. A. (1993). From nerds to normals: The recovery of identity among adolescents from middle school to high school. *Sociology of Education, 66*(1), 21–40.

Lash, A. A., & Kirkpatrick, S. L. (1994). Interrupted lessons: Teacher views of transfer student education. *American Educational Research Journal, 31*(4), 813–843.

MacwIver, D., & Epstein, J. L. (1991). Responsive practices in the middle grades: Teacher teams, advisory groups, remedial instruction, and school transition programs. *American Journal of Education, 99*(4), 587–622.

Riehl, C., Pallas, A. M., & Natriello, G. (1999). Rites and wrongs: Institutional explanations for the student course-scheduling process in urban high schools. *American Journal of Education, 107*(2), 116–154.

Roderick, M. (1993). *The path to dropping out: Evidence for intervention.* Westport, CT: Auburn House.

Rumberger, R. W., & Larson, K. A. (1998). Student mobility and the increased risk of high school dropout. *American Journal of Education, 107*(1), 1–35.

Schiller, K. S. (1999). Effects of feeder patterns on students' transition to high school. *Sociology of Education, 72*(4), 216–233.

Schiller, K. S. (2005). School transition programs in organizational context: Problems of recruitment, coordination, and integration. In B. Schneider & L. Hedges (Eds.), *Reflections on the social organization of schooling: A tribute to Charles E. Bidwell.* New York: Sage.

Shavit, Y., & Blossfeld, H.-P. (Eds.). (1993). *Persistent inequality: Changing educational attainment in 13 countries.* Boulder, CO: Westview Press.

South, S. J., & Haynie, D. L. (2004). Friendship networks of mobile adolescents. *Social Forces, 83*(1), 315–350.

Swanson, C. B., & Schneider, B. (1999). Students on the move: Residential and educational mobility in America's schools. *Sociology of Education, 72*(1), 54–67.

Useem, E. L. (1991). Student selection into course sequences in mathematics: The impact of parental involvement and school policies. *Journal of Research on Adolescence, 1*(3), 231–250.

Weiss, C. C., & Bearman, P. S. (2007). Fresh starts: Reinvestigating the effects of the transition to high school on student outcomes. *American Journal of Education, 113*(3), 395–421.

Kathryn S. Schiller

SCHOOL VIOLENCE

Schools are critical in the life course development of children and youth. Success in school is an important developmental outcome in its own right. School performance also has important implications for a range of adulthood outcomes including mental health, suicide, mortality, substance abuse, criminality, employment, and poverty. Research has shown that feeling safe at school is related to success, and even low levels of violence can have detrimental effects on student well-being. On a daily basis, however, many students fear being or are victims of school violence, which may involve either social victimization, such as harassment, social exclusion, and humiliation, or physical violence such as being threatened or physically assaulted.

The study of school violence is complex because such violence stems not only from a broad range of intertwined individual and environmental-level factors, including the social conditions in schools, but also from family and neighborhood factors. Starting in the 1990s, increasing efforts have been focused on the development of effective strategies to reduce and prevent violence in schools. These efforts have been motivated, in part, by a series of particularly violent events in schools such as the now infamous 1999 Columbine school shooting in Colorado, when 12 students and one teacher were killed and 23 students were wounded. Theorizing and research on school violence has moved in recent decades from an individual criminal justice orientation toward a more complex ecological and life course approach, which can be seen in the evolving definition of school violence and the theories informing its study.

DEFINITIONS AND THEORIES OF SCHOOL VIOLENCE

School violence, like most social phenomena, is complex. For example, even defining what is and is not considered school violence is not straightforward. Conventional definitions of school violence have been framed by a criminal justice orientation—the use of force by one person toward another that results in harm to that person or their property—situated within the school, including the student's travel to and from school. Benbenishty and Astor (2005) offered an excellent example of such a definition: "any behavior intended to harm, physically or emotionally, persons in school and their property" (p. 8). *Harm* is in the perception of the harmed, and may encompass verbal, social, and physical forms of violence. Benbenishty and Astor also assert, however, that school violence is shaped by elements of wider social and cultural contexts, including neighborhood and family characteristics, and by "policies, practices, procedures, and social influences within the school setting as well as the impact of the variables external to the school" (p. 7).

Offering a broader definition of school violence, Henry (2000) argued that school violence is perpetrated not just by individuals, but by organizations and policies, and that it is not always by force, but may also involve the misuse of power. Further, the harm may be physical or emotional, but may also be in the areas of economics, opportunity, or equal access. Finally, Henry asserted, as Benbenishty and Astor suggested, that school violence is not limited to the school grounds, or travel to or from school, because that denies the interconnectedness of the school and community. Which definition is more accurate and more compelling? The current trend in research and theory is to use the more complex definition.

A broad range of theories has been applied to study school violence. Social disorganization theory is one traditional approach developed by sociologists Clifford R. Shaw and Henry D. McKay in response to the rapid changes in immigrant and ethnic minority communities in Chicago during the early 1900s. According to social disorganization theory, relationships are seen as organized when individuals actively engage in community institutions that foster social support and connectedness. As a criminology theory, social disorganization locates disorganizing forces within communities by attributing violence to a breakdown in community relationships and institutions. Social disorganization theory says little, however, about "organized" neighborhoods, for example rural and suburban communities, which statistically show school violence rates that are similar to urban schools in terms of teasing and bullying.

Ecological theory, developed by the psychologist Urie Bronfenbrenner, has seen increasing application to the study of school violence. Ecological theory emphasizes the multiple environments of youth, which include school, family, peers, neighborhood, the larger society, and the dynamic interplay of these settings over time. An increasing number of researchers who study school violence apply ecological theory to understand this web of environments and the many factors that contribute to violence in schools. An ecological approach fits well with the increasingly complex definitions of school violence that have emerged since the late 1990s and is especially useful when trying to develop comprehensive efforts to reduce and prevent school violence.

STUDYING SCHOOL VIOLENCE

Although school violence has been the focus of much media attention in recent decades, actual school violence rates have not changed significantly since the 1980s. Overall trend data, however, tell only the surface of the story. As described above, school violence is a function of the wider social context; therefore patterns vary with the characteristics of the students, school, and community. Four approaches typically are used in studying school violence: qualitative interview studies, analyses of school records, surveys of students and school staff, and after-the-fact analyses of violent events in schools such as multiple-victim school shootings.

Qualitative approaches have two advantages: Researchers may "discover" things that were not known, and they may construct "thick" descriptions of school violence issues. For example, in an interview study examining the connection between school violence and the community surrounding a New York City middle school, Mateu-Gelabert and Lune (2003) found that conflicts flow in both directions. The authors found that 18% of the violent events in the neighborhood started at school, while 21% of the violent events in the school started in the neighborhood.

Schools keep records, and such records offer data for the study of school violence. Cantor and Wright (2002) asked a nationally representative sample of school principals about violent events reported to the police and found that most violent events in schools are concentrated in certain schools, with 60% of the serious violence reported occurring in just 4% of the schools from which data were collected. The highest rates of violence occurred in urban schools with a high percentage of minority students and high rates of economic disadvantage. Such findings are consistent with the widespread belief that urban schools experience more violence. Cantor and Wright also found, however, that out of the 4% of schools whose records revealed the highest rates of serious violence a large percentage (36%) were in rural areas. School violence surveys also gather data from school staff such as principals. For example, Crosse, Burr, Cantor, Hagen, & Hantman (2002) surveyed 882 principals from across the United States and found that 72% of middle school principals reported physical fights between students in their schools. By contrast, only 11% of elementary school principals reported such fights, while the figure for high school principals was 56%. Such variation can be looked at from a life course perspective: What is known about children and youth developmentally that would explain such a pattern?

The most widely used approach to studying school violence is with surveys. Such research reveals that a significant percentage of students feel unsafe and report experiencing violence in schools. For example, in a survey of 7th, 9th, and 11th graders in a suburban school district, Cornell and Loper (1998) asked students about experiences of violence in the past 30 days, both "in school" and "out of school," with "in school" defined as the building and grounds and on the way to or from school. They found that 20% of students reported having been in a physical fight at school and 13% reported carrying a weapon to school for protection in the past 30 days. An important survey source of school violence patterns is the *Indicators of School Crime and Safety*, published annually starting in 1998 by the U.S. Departments of Education and Justice (Dinkes, Cataldi, Lin-Kelly, & Snyder, 2007). In this report, data collected for the period from 1992 to 2005 indicated a general decline from 10% to 4% of students aged 12 to 18 who reported being physically victimized during the previous 6 months. Rates varied by gender, however, with 10% of boys being threatened or injured compared to 6% for girls.

Although some survey studies have found that Whites experience lower rates of school violence than African Americans and Latinos, Chandler, Chapman, Rand, & Taylor (1998) found in a national sample of students that after accounting for the impact of socioeconomic conditions, African American students reported only slightly higher rates and Latino students slightly lower rates than White students. Such findings do suggest that some of the variation in school violence may be related to higher rates of poverty and historical patterns and experiences of inequality. Some authors have suggested that school violence increases when social problems—such as poverty, unemployment, drug abuse, crime, and child maltreatment—are left unaddressed. The following question emerges from these suggestions: How might such social conditions relate to violence in schools? One thing is clear: Despite variations in rates in local schools and communities, school violence is a pervasive problem affecting schools across the United States and around the world. Violence within specific schools, however, is linked to the social conditions of the surrounding neighborhoods and larger community.

Researchers also study school violence by retrospectively examining particularly violent events. Although school shootings garner high levels of media attention, violent deaths in schools are quite rare. Much more common are violent deaths of students outside school, although as discussed above community and school violence is connected. The year 1999, that of the infamous Columbine massacre, represented an average year for violent deaths, with a total of 33 murders committed at schools (Dinkes et al., 2007). This statistic suggests that the actual number of murders at schools occurring elsewhere, most typically urban schools, was low that year. Multiple-victim school shootings are unpredictable, but have been increasing in frequency, typically occur in suburban and rural communities, and are most often perpetrated by White males. Examinations of multiple-victim events in schools, such as Columbine or Red Lake (Minnesota) High School have revealed two clear patterns: The shooters were suffering from mental health

Columbine Massacre. *Columbine High School security video shows Dylan Klebold and Eric Harris in the cafeteria.* © REUTERS/CORBIS.

problems and had previously been victims of teasing, bullying, or social rejection (Leary, Kowalski, Smith, & Phillips, 2003). The question then becomes: How are more common types of school violence linked to more rare school shootings? An important consequence of high-visibility school shootings is that policymakers and educators have increased their attention to the widespread problem of more common types of school violence, leading to increased efforts and government funding for prevention and reduction efforts.

PREVENTING AND REDUCING SCHOOL VIOLENCE

Just as the definitions, study methods, and patterns of school violence are varied and complex, so are efforts at reduction and prevention. Such efforts include procedures to make schools safer by keeping out weapons and aggressive students, initiatives to reduce teasing and bullying, the mapping of individual schools to identify where in those schools violence is most likely to occur, and, finally, comprehensive programs to change the social climate in a school.

In regard to the goal of keeping weapons and high-risk students out of schools, one such effort has been the proliferation of "zero-tolerance" policies. Such policies dictate that students who bring weapons to school will be expelled. This policy originated with U.S. Customs to combat drug smuggling in 1988. It was picked up by schools in a few states and then was made a federal law in 1994 with the passage of the Gun-Free Schools Act, which mandated that states punish students bringing guns (most states included all "weapons") to school with yearlong expulsions. By 1998, 91% of public schools had

zero-tolerance policies, and as a result of such policies a total of 2,554 students were expelled for bringing a weapon to school during the 2001–2002 school year. There is little evidence, however, that these policies are effective, and they have been so rigidly enforced that students have been expelled for minor offenses, including bringing into a school such objects as a squirt gun, a 5-inch plastic axe as part of a Halloween costume, a rocket made out of a potato chip can, and nail clippers (Skiba, 2000). Skiba found these overreactions are not rare, whereas the serious infractions these policies were originally intended to address are rare.

Research has found that suspension and expulsion as a disciplinary strategy disproportionally affects poor and African American students. Such policies deny access to educational opportunities in the short- and mid-term, which can negatively impact academic progress. In the long run, such denial of access can negatively impact the educational trajectories of students, which can have life-long consequences in terms of graduating from high school, employment opportunities, and a cascade of other negative adult outcomes as discussed above in the introduction. It is easy to see Henry's point made above—that school policies can do violence to students.

The Bully Prevention Program, developed by the Norwegian psychologist Dan Olweus, is an effective program designed to reduce teasing and bullying in schools. The program encompasses interventions focused on the school, classroom, and individual levels, including interventions for bullies and victims, and to elicit bystanders to intervene to help victims. The intervention process starts with a survey of students to assess the extent of the problem, followed by a schoolwide meeting that includes parents to discuss the results of that assessment survey and get everyone onboard with the need for change. Finally, the interventions are implemented, including clear rules in the school about teasing and bullying and classroom meetings to discuss these rules and ongoing problems.

Mapping the unsafe places in schools, including where and when students are more likely to become victims, was an important development in school violence prevention that emerged from the research of Rami Benbenishty and Ron A. Astor. These authors have also proposed an ecological approach to preventing school violence that takes into consideration the complex interplay between the school and the surrounding environment (Benbenishty, Astor, & Estrada, 2008). For example, they suggest that all schools complete comprehensive school-wide assessments of violence, and they point out that the federal No Child Left Behind Act of 2001 calls for assessments of the level of safety in schools, the results of which are supposed to be made publicly available, although few schools currently complete such assessments. The authors' comprehensive assessment strategy stresses the importance of gathering data from multiple informants—students, teachers, parents, and administrators—and they have developed strategies for how to best use the results to most effectively promote a successful change effort.

Another promising approach to reducing violence is to change the social climate in schools. The term *social climate* refers to the quality and nature of the interpersonal relationships that influence students' behavior and social functioning. This approach, adapted from a public health model, has three levels, termed *universal*, prevention efforts for all students; *selected*, for interventions delivered to students at risk; and *targeted*, for interventions for students who have been aggressive. Bowen, Powers, Woolley, and Bowen (2004), applying an ecological framework while identifying risk factors at the individual, family, peer group, school, and community levels, have described strategies to apply this three-level approach to reduce and prevent school violence. For example, an implementation could include: (a) a universal schoolwide program such as Responding in Peaceful and Positive Ways, which was developed by Albert D. Farrell to advance the social and conflict-resolution skills of all students; (b) a group-based intervention such as the Anger Coping Program, which was developed by John E. Lochman for students with some problematic behaviors; and (c) targeted interventions for students with ongoing problems with aggression such as Multisystemic Family Therapy, which was developed by Scott W. Henggeler.

FUTURE DIRECTIONS

For schools to accomplish their central goal, nurturing the learning and growth of students, they must be and feel like safe places. Schools are most effective when students see the school as welcoming and supportive, and when students look forward to being there (Woolley, 2006). If school is a place where students are teased or bullied, or—even worse—stabbed, shot, or live in fear of a classmate going on a shooting rampage, then the central goal of learning is severely undermined.

With the downturn in reports of victimization in schools that occurred in the period encompassing the last years of the 20th century and the first years of the 21st, it appears that the increased efforts to study this problem and develop strategies to address it may be working. Nevertheless, researchers, policymakers, and school administrators have much work to do to achieve the goal of all students feeling safe at school. The development of more complex definitions and more life course and ecological orientations to this problem have been positive developments. Future research may seek to understand the commonalities found in school violence

between schools in diverse settings in order to inform the development of more robust school violence intervention and prevention efforts.

SEE ALSO Volume 1: *Aggression, Childhood and Adolescence; Bullying and Peer Victimization; Policy, Education.*

BIBLIOGRAPHY

Benbenishty, R., & Astor, R. A. (2005). *School violence in context: Culture, neighborhood, family, school, and gender.* New York: Oxford University Press.

Benbenishty, R., Astor, R. A., & Estrada, J. N. (2008). School violence assessment: A conceptual framework, instruments, and methods. *Children and Schools, 30,* 71–81.

Bowen, G. L., Powers, J. D., Woolley, M. E., & Bowen, N. K. (2004). School violence. In L. A. Rapp-Paglicci, C. N. Dulmus, & J. S. Wodarski (Eds.), *Handbook of preventive interventions for children and adolescents* (pp. 338–358). Hoboken, NJ: Wiley.

Cantor, D., & Wright, M. M. (2002, August). *School crime patterns: A national profile of U.S. public high schools using rates of crime reported by police* (Publication No. 2001-37). Washington, DC: U.S. Department of Education, Planning and Evaluation Service. Retrieved June 5, 2008, from http://www.ed.gov/offices/OUS/PES/studies-school-violence/school-crime-pattern.pdf

Chandler, K. A., Chapman, C. D., Rand, M. R., & Taylor, B. M. (1998, March). *Students' reports of school crime: 1989 and 1995* (NCES Publication No. 98-241/NCJ Publication No. 169607). Washington, DC: National Center for Education Statistics and Bureau of Justice Statistics. Retrieved June 5, 2008, from http://nces.ed.gov/pubs98/98241.pdf

Cornell, D. G., & Loper, A. B. (1998). Assessment of violence and other high-risk behaviors with a school survey. *School Psychology Review, 27,* 317–330.

Crosse, S., Burr, M., Cantor, D., Hagen, C. A., & Hantman, I. (2002, August). *Wide scope, questionable quality: Drug and violence prevention efforts in American schools* (Publication No. 2001-35). Washington, DC: U.S. Department of Education, Planning and Evaluation Service. Retrieved June 5, 2008, from http://www.ed.gov/offices/OUS/PES/studies-school-violence/wide-scope.pdf

Dinkes, R., Cataldi, E. F., Lin-Kelly, W., & Snyder, T. D. (2007, December). *Indicators of school crime and safety: 2007* (NCES Publication No. 2008-021/NCJ Publication No. 219553). Washington, DC: National Center for Education Statistics and Bureau of Justice Statistics. Retrieved June 5, 2008, from http://nces.ed.gov/pubs2008/2008021.pdf

Henry, S. (2000). What is school violence? An integrated definition. *The Annals of the American Academy of Political and Social Science, 567,* 16–29.

Leary, M. R., Kowalski, R. M., Smith, L., & Phillips, S. (2003). Teasing, rejection, and violence: Case studies of the school shootings. *Aggressive Behavior, 29,* 202–214.

Mateu-Gelabert, P., & Lune, H. (2003). School violence: The bidirectional conflict flow between neighborhood and school. *City and Community, 2,* 353–368.

Skiba, R. J. (2000, August). *Zero tolerance, zero evidence: An analysis of school disciplinary practice* (Policy Research Report

No. SRS2). Bloomington: Indiana Education Policy Center. Retrieved June 5, 2008, from http://www.indiana.edu/-safeschl/ztze.pdf

Woolley, M. E. (2006). Advancing a positive school climate for students, families, and staff. In C. Franklin, M. B. Harris, & P. Allen-Meares (Eds.), *The school services sourcebook: A guide for school-based professionals* (pp. 777–783). Oxford, U.K.: Oxford University Press.

Michael E. Woolley
Desmond U. Patton

SCHOOL VOUCHERS

SEE Volume 1: *Policy, Education.*

SEGREGATION, SCHOOL

Social science research in school segregation has both informed and been informed by legal consideration of segregated schools. In the 1954 *Brown v. Board of Education*, the U.S. Supreme Court declared racially segregated schools to be "inherently unequal" and therefore a violation of the Fourteenth Amendment of the U.S. Constitution. Although this ruling initially made very little change in the extent of school segregation in the 17 states that had laws requiring it (Orfield & Lee, 2007), it had an important effect on social scientists because of footnote 11, in which Chief Justice Earl Warren cited social science studies as part of his argument about why segregation was harmful. In particular, the decision and subsequent legal and policy actions to implement the decision created a fertile opportunity for social scientists to examine the extent of segregation (or desegregation), how school segregation related to other types of societal segregation, and the effect of judicial and policy actions, and to investigate the ways in which school segregation and desegregation affected students attending these schools.

WHAT IT MEANS TO STUDY SEGREGATION

The meaning of segregation from a social science perspective is distinct from the legal definition, so it is important to clarify terminology. Social scientists define school segregation in terms of the distribution of students of different racial/ethnic groups, often measured at the school-building level. Social scientists measure segregation in a variety of ways: different dimensions of segregation (e.g., concentration, exposure, dissimilarity),

different levels of analysis (within schools, between schools, between districts), and different groups (usually two groups: White–Black, White–Latino, and so on; Massey & Denton, 1988).

Further, for several decades, there has been a legal distinction between *de jure* segregation, which refers to segregation that occurs as a result of government law or policy, and *de facto* segregation, which occurs in the absence of such law or policy. In subsequent Supreme Court decisions about school desegregation, the Court ruled that unless there is de jure segregation, school districts are not required to—and even may not be able to voluntarily—take action to address the patterns of school segregation that exist (*Freeman v. Pitts*, 1992; *Parents Involved in Community Schools v. Seattle School District*, 2007).

One of the major causes of segregated schools in the absence of overt policies that segregate students is the segregated housing patterns in most communities across the country. There is an interdependent relation between housing and school segregation: Because schools often draw their students from the surrounding neighborhoods, schools will be segregated if neighborhoods are. At the same time, evidence suggests that in areas where there is not complete desegregation across the metropolitan area, communities with segregated schools have higher levels of residential segregation (Frankenberg, 2005).

It is also important to differentiate how K–12 school segregation is different from the higher education context. For the past several decades, discussions of racial diversity in higher education have focused on the design, legality, and effect of affirmative action policies. These policies differ considerably from school desegregation, most notably in that they affect only a fraction of students whereas school districts are required to educate all students who reside in their boundaries.

BASIC TRENDS IN SCHOOL SEGREGATION RESEARCH

The most fundamental area of study is understanding the extent of school segregation. One of the foremost authorities in documenting school segregation trends has been Gary Orfield, a political scientist and founder of the Civil Rights Project. Using national-level data, Orfield and his colleagues have consistently documented segregation for, in particular, Black and Latino students at the national, regional, and state levels (e.g., Orfield & Lee, 2007). Some of his findings include:

1. Black student segregation fell dramatically in the South from the mid-1960s to the late 1980s;

2. Latino segregation has been increasing continuously since the late 1960s;

3. Black segregation has risen since the late 1980s, and Black students nationally are more segregated than they have been since before 1970, when many of the most extensive desegregation plans were implemented;

4. White students are the most isolated students of any racial/ethnic group.

In addition, there are very different percentages of students in intensely segregated minority schools (defined as schools in which 0% to 10% of students are White) by region of the country. Until very recently, Black students in the South were the most desegregated; Latino segregation in the West is high. Quite high percentages of Black and Latino students attend segregated minority schools in the Midwest and Northeast, which is likely a result of the many small school districts that divide students, often by racial background, in these regions' metropolitan areas.

Whereas Orfield's analysis aggregates school-level segregation measures to the state, regional, or national level, most case studies of segregation examine segregation within a particular school district. In general, much of this research dates from the time when school desegregation plans were being implemented. These studies might use a conceptualization such as Orfield's intensely segregated schools to assess whether the number of students in such schools diminishes with the implementation of certain policies. In addition, social scientists may have used measures influenced by court decisions that relate to that school district. For example, in Charlotte, North Carolina, compliance with desegregation as required by the courts—who found Charlotte had illegally segregated students—was that every school had to have a percentage of Black students that was within 15 percentage points of the system-wide percentage (Mickelson, 2001). As a result, case studies of Charlotte often used this definition of segregation.

In addition to examining changes in segregation as desegregation policies were implemented, often under court or administrative supervision and primarily in the South, newer district-level studies examine desegregation after such policies are lifted (e.g., Mickelson, 2001; Orfield & Eaton, 1996). Policies can be ended once the court believes that the district has eliminated all traces of prior segregation policies or are "unitary" systems. However, these studies generally conclude that when race-conscious policies are abandoned there has been a return to segregation—sometimes quite rapidly—and associated educational inequality.

More recent studies suggest that district-level analyses may understate the extent of segregation, because these studies do not incorporate the substantial differ-

ences in racial composition among school districts. One estimate was that 70% of segregation was due to segregation among school districts (Clotfelter, 1998). Additionally, a Connecticut Supreme Court ruling found that the way in which school district boundaries in the case were drawn was unconstitutional because they were the major cause of segregation within the state (Eaton, 2007). Both legally and politically, designing policies to address cross-district segregation is challenging. Yet in the South, countywide districts encompassed much of the metropolitan areas and contributed to lower levels of school segregation (Frankenberg & Lee, 2002; Orfield, 2001).

WHY SEGREGATION MATTERS

The initial argument made in *Brown* about the "harms" of segregation focused largely on potential psychological costs to young people: Those who attend segregated minority schools are harmed in terms of their self-image, whereas those who are the "segregators" or who attended segregated White schools were also found to have a false sense of superiority. In addition, two Supreme Court decisions that paved the way for *Brown* by desegregating higher education relied on sociological evidence about the importance of access to professional networks as part of a graduate education. Since the mid-1950s, research on segregation has confirmed and extended these reasons as to how segregated schools deny their students equal opportunity. This can be divided into three categories: (a) the relation of segregated schools to important educational inputs, (b) cognitive outcomes for students, and (c) democratic outcomes for students.

In 1954 Harry Ashmore released a major study of schooling in the South that documented the persisting inequalities between the funding of "Negro schools" and White schools. Although the gap had closed, in some states schools with White students had three times the funding that schools educating Black students received. The relation between segregation and resource inequality is evident in several major ways. First, minority schools tend to have teachers with lower qualifications and fewer years of experience, both of which relate to lower student achievement. In addition, these schools have higher teacher turnover, particularly among White teachers who comprise the vast majority of the teaching force. Second, because of the relation between poverty and race, segregated minority schools are overwhelmingly likely to have a majority of low-income students (Orfield & Lee, 2007). This is important because research since the 1966 Coleman Report finds that being in classrooms with middle-class students has a powerful effect on student learning. Finally, segregated minority, high-poverty schools generally offer fewer advanced courses and other college preparation opportunities (Solorzano & Ornelas, 2002). These schools may also have facilities that are inferior to schools serving more White and middle-class students.

Cognitive Student Outcomes More research has been focused on understanding the short-term effects on students who attend segregated schools, although recently there has been a shift to also examine long-term student outcomes as well. The bulk of research on short-term effects has examined how student achievement in segregated schools compares to that in desegregated schools. Some of the earlier studies found mixed or inconclusive results on student achievement—usually measured by test scores sometimes after only a year of partially implemented desegregation—although the general consensus is that there are at least modest benefits for Black students' achievement in desegregated schools. There was less focus earlier on the achievement of Latino students, due, at least in part, to the fact that most of the students "desegregated" were in the South, where there were few Latino students at the time. Most of the rationales for desegregation focused on the unequal opportunity for minority students, with less attention on White students. The few studies that have examined White student achievement concluded that as long as desegregated schools remained majority White, the achievement of Whites was not harmed.

With the development of new statistical techniques, more methodologically rigorous research has confirmed that minority students' achievement in segregated minority schools is lower (Hanushek, Kain, & Rivkin, 2006). Further, attending segregated schools is likely to result in lower track placement (Mickelson, 2001). Segregated minority schools have lower graduation rates. In fact, after controlling for other factors that might relate to lower graduation rates, the percentage of minority students in a school was a significant predictor of graduation rates (Swanson, 2004). In addition, attending racially isolated K–12 schools makes students less likely to attend and persist in higher education, even after controlling for factors such as students' standardized test scores (Camburn, 1990). High school and college diplomas are increasingly important for employment opportunities, which affect both the individuals who fail to attain these degrees as well as communities who will have a less-educated employment base.

Noncognitive Student Outcomes Much of the attention, particularly in legal opinions, assessing whether there are benefits to desegregated schools has focused on cognitive outcomes. However, evidence about noncognitive benefits of desegregated schools—psychological, sociological, and democratic—is more robust than that of cognitive outcomes.

In terms of psychological outcomes, several important findings have emerged in prior studies. First, desegregated schools help students to develop cross-racial understanding, which relates to reduced stereotypes and bias (Killen & McKown, 2005). Further, students are more likely to understand racial exclusion is harmful (Killen, Crystal, & Ruck, 2007). Importantly, these gains can only be realized as a result of intergroup contact—which is limited in segregated schools—and the benefits are most significant at early ages before the formation of stereotypes (Hawley, 2007). These outcomes are particularly important because the development of racial stereotypes can affect decisions students make as adults, such as their interracial tolerance and their willingness to live and work in mixed-race settings. In addition, racial stereotypes can affect the achievement of students who perceive themselves as the object of such stereotypes as compared to students who are unaware of stereotypes (Steele & Aronson, 1995).

One of the early arguments for desegregation was the access to networks that minority students lacked in segregated schools. This was particularly evident in the *Sweatt v. Painter* (1950) and *McLaurin v. Oklahoma* (1950) Supreme Court decisions, in which the justices noted that without access to professional colleagues and the reputation of graduate schools, future professional opportunities were likely to be limited for minority students. Likewise, in elementary and secondary schools in particular, networks can be important in helping students think about attending college and know what preparation is important. In addition, the reputation and alumni connections that high schools have with selective colleges and employers are valuable benefits for students who attend such schools. When minority students lack access to such networks because they attend poorly resourced, segregated minority schools, their postgraduate opportunities are more limited solely as a result of the school they attended.

More recent literature has described a democratic or perpetuating effect of attending desegregated schools. As opposed to the other benefits described previously, these accrue to students of all racial backgrounds that attend racially diverse schools. Studies examining the "perpetuation effects" of desegregated schools suggest that Black students who attend desegregated schools are more likely than their peers who attend segregated minority schools to live and work in desegregated environments as adults (Wells & Crain, 1994). This pattern reflects two mechanisms: First, attending diverse, majority White schools allows students to overcome their fears of such institutions, and second, they have access to information about diverse opportunities. Research also suggests that students of all racial/ethnic backgrounds attending diverse high schools are more likely to report that they would feel comfortable living and working in desegregated settings later in life and may have more civic engagement than peers attending segregated schools (e.g., Yun & Kurlaender, 2004). In addition, communities with more thoroughly desegregated schools are likely to experience more integrated neighborhoods. All of these findings point to the conclusion that desegregated schools are more likely to result in desegregated experiences for both students and communities, which are important preparation for citizenship in a pluralistic, multiracial society.

Structuring Desegregation Nearly all the research discussed thus far has focused on segregation at the school level. Although desegregation is important, one expert panel noted that it is only when two (or more) racial groups of students are in a school that the actual process of desegregation, sometimes called *integration*, begins (Hawley et al., 1983). Within-school structures and practices matter, however, as to whether these benefits of desegregated schools accrue to students. In 1954, Harvard psychologist Gordon Allport published *The Nature of Prejudice*. In it, he described his theory of intergroup contact and specified four conditions (equal status for all groups, common goals, no competition among groups, and authority sanction for intergroup contact) that were necessary to fully realize the benefits of contact with people of other groups. A meta-analysis of five decades of research supports the importance of intergroup contact in reducing stereotypes and finds that gains are stronger when the conditions are met (Pettigrew & Tropp, 2006).

Despite the importance of intergroup contact, structures inside schools often segregate students *within* schools even if they are desegregated at the school level (Mickelson, 2001). In addition to preventing interracial exposure, such within-school segregation often disproportionately exposes minority students to novice teachers (Clotfelter, Ladd, & Vigdor, 2005) and contributes to a persistent racial achievement gap (Burris & Welner, 2007). In the early 1970s, the federal government passed a desegregation assistance law to help retrain teachers for desegregated schools and to conduct research to understand what strategies worked best in diverse schools. There is considerably less research focus on how to best structure desegregated schools despite the growing racial complexity of the student enrollment (see Frankenberg & Orfield, 2007).

FUTURE OF SEGREGATION AND SEGREGATION RESEARCH

School segregation research is likely to be shaped by two significant trends: demographic and legal changes. The country is rapidly diversifying, and, by the middle of the 21st century, public schools will have a non-White majority student body, a reality that has already occurred in 10 states (Orfield & Lee, 2007). In many states there are racially changing, complex school contexts (Frankenberg, in press). These new school contexts require research to help teachers and administrators understand how to structure schools in ways that will enhance the learning and social outcomes for all students.

In addition, in 2007 the Supreme Court (in *Parents Involved in Community Schools v. Seattle School District No.1* and *Meredith v. Jefferson County Board of Education*) ruled that commonly used student assignment plans designed to achieve racial integration were illegal. This decision portends a number of important avenues for investigation. First, educators and administrators must understand the possible consequences of abandoning race-conscious voluntary plans. Second, it is important to understand where there are successful race-neutral plans, under what conditions these plans are successful, and how they could be replicated. Third, given the skepticism by Supreme Court justices about the importance of racially diverse schools and the harms of segregated schools, there continues to be the need to research—and compellingly communicate research findings—on these topics.

Thus, whereas research will continue to be informed by social reality (e.g., the persisting high levels of segregation in major urban centers), it is quite likely that the study of school segregation will continue to be influenced by legal and policy decisions to understand their consequences and help inform future policies.

SEE ALSO Volume 1: *Academic Achievement; Neighborhood Context, Childhood and Adolescence; Policy, Education; Racial Inequality in Education; School Tracking; Socioeconomic Inequality in Education; Stages of Schooling;* Volume 2: *Segregation, Residential.*

BIBLIOGRAPHY

Burris, C. C., & Welner, K. G. (2007). Classroom integration and accelerated learning through detracking. In E. Frankenberg & G. Orfield (Eds.), *Lessons in integration: Realizing the promise of racial diversity in American schools.* (pp. 207–227). Charlottesville: University of Virginia Press.

Camburn, E. M. (1990). College completion among students from high schools located in large metropolitan areas. *American Journal of Education, 98*(4), 551–569.

Clotfelter, C. (1998). *Public school segregation in metropolitan areas.* Cambridge, MA: National Bureau of Economic Research.

Clotfelter, C. T., Ladd, H. F., & Vigdor, J. (2005). Who teaches whom? Race and the distribution of novice teachers. *Economics of Education Review 24*(4), 377–392.

Eaton, S. E. (2007). *The children in room E4: American education on trial.* Chapel Hill, NC: Algonquin Books.

Frankenberg, E. (2005). The impact of school segregation on residential housing patterns: Mobile, AL and Charlotte, NC. In J. Boger & G. Orfield, (Eds.), *School resegregation: Must the South turn back?* (pp 164–184). Chapel Hill: University of North Carolina Press.

Frankenberg, E. (in press). School segregation, desegregation, and integration: What do these terms mean in a post-*Parents Involved in Community Schools,* racially transitioning society? *Seattle Journal for Social Justice.*

Frankenberg, E., & Lee, C. (2002). *Race in American public schools: Rapidly resegregating school districts.* Cambridge, MA: Civil Rights Project at Harvard University.

Frankenberg, E., & Orfield, G. (Eds.). (2007). *Lessons in integration: Realizing the promise of racial diversity in our nation's public schools.* Charlottesville: University of Virginia Press.

Hanushek, E. A., Kain, J. F., & Rivkin, S. G. (2006). *New evidence about* Brown v. Board of Education: *The complex effects of school racial composition on achievement.* Working paper. Cambridge, MA: National Bureau of Economic Research.

Hawley, W. D. (2007). Designing schools that use student diversity to enhance learning of all students. In E. Frankenberg & G. Orfield (Eds.), *Lessons in integration: Realizing the promise of racial diversity in American schools* (pp. 31–56). Charlottesville: University of Virginia Press.

Hawley, W. et al. (1983). *Strategies for effective school desegregation.* Lexington, MA: Lexington Books.

Killen, M., Crystal, D., & Ruck, M. (2007). The social developmental benefits of intergroup contact among children and adolescents. In E. Frankenberg & G. Orfield (Eds.), *Lessons in integration: Realizing the promise of racial diversity in American schools* (pp. 57–73). Charlottesville: University of Virginia Press.

Killen, M., & McKown, C. (2005). How integrative approaches to intergroup attitudes advance the field. *Journal of Applied Developmental Psychology, 26,* 612–622.

Massey, D. S., & Denton, N. A. (1988). The dimensions of racial segregation. *Social Forces, 67,* 281–315.

Mickelson, R. A. (2001). Subverting Swann: First- and second-generation segregation in Charlotte, North Carolina. *American Educational Research Journal, 38*(2), 215–252.

Orfield, G. (2001). Metropolitan school desegregation. In J. A. Powell, G. Kearney, & V. Kay (Eds.), *In pursuit of a dream deferred* (pp. 121–157). New York: Peter Lang.

Orfield, G., & Eaton, S. E. (Eds.). (1996). *Dismantling desegregation: The quiet reversal of* Brown v. Board of Education. New York: New Press.

Orfield, G., & Lee, C. (2007). *Historic reversals, accelerating resegregation, and the need for new integration strategies.* Los Angeles, CA: Civil Rights Project.

Pettigrew, T. F., & Tropp, L. R. (2006). A meta-analytic test of intergroup contact theory. *Journal of Personality and Social Psychology, 90,* 751–783.

Solorzano, D. G., & Ornelas, A. (2002). A critical race analysis of advanced placement classes: A case of educational inequality. *Journal of Latinos & Education, 1*(4), 215–226.

Steele, C. M., & Aronson, J. (1995). Stereotype threat and the intellectual performance of African Americans. *Journal of Personality and Social Psychology, 69*(5), 797–811.

Swanson, C. B. (2004). *Who graduates? Who doesn't? A statistical portrait of public high school graduation, Class of 2001.* Washington, DC: Urban Institute.

Wells, A. S., & Crain, R. L. (1994). Perpetuation theory and the long-term effects of school desegregation. *Review of Educational Research, 64,* 531–555.

Yun, J. T., & Kurlaender, M. (2004). School racial composition and student educational aspirations: A question of equity in a multiracial society. *Journal of Education for Students Placed at Risk, 9*(2), 143–168.

Erica Frankenberg

SELF-ESTEEM

Few concepts in the social and behavioral sciences have excited as much interest and spawned as many major research programs dedicated to defining its antecedents and consequences as has self-esteem. This concept is regarded by many social theorists and researchers as a central construct in explanations of human social behavior, over the life course and as a concept that must be attended to in the design and implementation of planned social interventions that are dedicated to the amelioration of individual and societal outcomes (Kaplan, 1986; Kernis, 2006; Mecca, Smelser, & Vasconcellos, 1989).

DEFINITION AND MEASUREMENT OF SELF-ESTEEM

In spite of the central position accorded to self-esteem in theories and research about human social behavior, there are widely divergent positions regarding how to define this concept. Definitions of self-esteem usually involve one or more of four classes of self-referent responses: self-cognition, self-evaluation, self-feeling, and self-enhancing or self-protective mechanisms. *Self-cognition* refers to processes whereby the person conceives of, perceives, or otherwise thinks about him or herself. *Self-evaluation* refers to the process of judging one's self to be close to or distant from more or less salient personal values. *Self-feeling* refers to the more or less positive/negative affective (emotional) responses that are evoked by the ways in which the person thinks about and evaluates him or herself. *Self-enhancing* and *self-protective mechanisms* refer to the person's responses that allow the individual to think about him or herself in more favorable terms, to evaluate and feel more positively about him or herself, and to forestall negative thoughts, evaluations, and feelings about him or herself. Recognizing the conceptual and empirical interrelations among these classes of self-oriented responses, self-esteem is defined here in terms of all four classes: Self-esteem refers to the more or less intense emotional responses that are evoked by self-conceptions and self-evaluations of being close to or distant from relatively important self-values and that motivate responses that are intended and may function to improve positive self-esteem or self-concepts, -evaluations, and -feelings.

Self-esteem may refer to one's momentary situation or to characteristic responses that transcend time and situations for the individual. In the latter case, regardless of the momentary situation, the person generally or chronically tends to conceive of and evaluate him or herself in terms of positively (or negatively) valued qualities, behaviors, and circumstances and to respond with positive (or negative) self-feelings. The stability of level of self-esteem may be accounted for by the stability of life's circumstances (the person continues to evoke rejecting responses from significant others in response to stable socially disvalued qualities), internalized attitudes, or, perhaps, genetic influences. Such stable, characteristic, or chronic levels are often referred to in terms of trait self-esteem.

In the case of momentary (state) or current self-esteem, level of self-esteem is variable and depends on the current circumstances in which the individual finds him or herself. In one social situation the person's attributes and behaviors might evoke accepting responses from significant others that stimulate positive self-conceptions, self-evaluations, and self-feelings and motivate attempts to reproduce these circumstances, and the same or other attributes in other social situations might evoke rejecting responses and concomitant negative self-conceptions, self-evaluations, and self-feelings that motivate responses that might forestall such experiences (e.g., avoidance of those who offered rejecting responses). Trait (characteristic, general) self-esteem and state (momentary, current) self-esteem are mutually influential. Individuals with low trait self-esteem may respond differently to current self-devaluing circumstances than individuals with high trait self-esteem. If a person feels like he or she has had a successful life, a momentary failure would be less traumatic.

Trait or state self-esteem may be based on more or less global (e.g., a worthy person) or specific (e.g., a good student) self-evaluations. A positive global self-evaluation generally is believed to be the most important basis for high self-esteem (Kaplan, 1986).

Defining self-esteem as a trait rather than a state, or in more general rather than specific terms, has implications for measuring self-esteem. The methodologies must be able to permit conclusions regarding cross-time and trans-situational versus current and situational-specific self-conceptions, self-evaluation, and concomitant self-feelings (how does one usually feel about one's self versus how does one feel about one's self now in this situation)

and between more general versus more specific self-conceptions, self-evaluations, and concomitant self-feelings (I am a person of worth versus I am a good student). Further, because traditional self-report measures of self-esteem are vulnerable to distortions motivated by the need to perceive one's self in positive terms or to appear to others as if one perceives one's self in such terms, attention has been turned to the measurement of self-esteem in terms of nonconscious/automatic ways that presumably circumvent such enhancement-motivated distortions.

Such implicit, or "true" self-referent responses might be gauged by asking people to respond to stimuli that are associated with, but do not directly reflect, significant aspects of the self (Koole & Pelham, 2003). Unfortunately, measures of implicit self-esteem are more easily measured in experimental laboratory studies than in large-scale in-community studies. Such studies require alternative methods to determine that responses to self-evaluative scales are valid indicators of the individual's true feelings about him or herself. Among the self-report scales that are frequently used in research on self-esteem is Rosenberg's (1965) Self-Esteem Scale.

SOCIAL ANTECEDENTS OF SELF-ESTEEM

The need for self-esteem is a fundamental human motive that is grounded in even more basic needs related to human survival and the interdependent social relationships that facilitate such survival (Kaplan, 1986). The need for self-esteem is experienced as the distressful emotions that are evoked when people perceive threats to their self-values. Such threats arise developmentally or contemporaneously when individuals anticipate, recall, or currently experience the failure to achieve important goals, evoke positive evaluations from significant others, or in general fail to approximate the self-evaluative criteria according to which they judge themselves to be worthy.

The self-values that comprise the hierarchically and situationally organized criteria for positive/negative self-evaluation may include the ability to perform tasks (contributing to a sense of competence), the ability to influence the environment (perhaps manifested as a sense of autonomy), likeability (perhaps manifested in the evocation of positive responses from significant others), leadership (reflected in the ability to influence others to adopt one's goals), or any number of other criteria. These values are hierarchically organized in the sense that if one could only approximate one value at the cost of failing to approximate another value, precedence would be given to the value that contributed more to one's overall sense of self-esteem. The values are situationally organized in the sense that different values are relevant to different situational contexts. On the battlefield, values relating to loyalty to one's comrades or acting courageously might be salient. In the family, values relating to the ability to give and receive love and to care for those who are dependent might be salient whereas the battlefield-related criteria might be irrelevant. The experiences that contribute to one's negative (positive) self-feelings and, hence, to the need for self-esteem include past developmental circumstances and the current situation.

Other people provide important information to the individual that keys into the person's self-evaluation. Although it is not otherwise apparent as to how successful we are in meeting our self-evaluatively relevant goals, other people provide clues to how well we are doing in this regard. People may offer expression of approval or disapproval by virtue of their evaluation of our performance. Further, our own perception of what we accomplish may be measured against our perceptions of what others like us have accomplished. That is, others provide a normative standard against which we evaluate our own achievements. Other people's responses to us are important to our self-evaluation also insofar as evoking positive responses is intrinsically valued as a criterion for self-evaluation. That is, we think well of ourselves if others like and respect us. One of the criteria for high self-esteem is the self-perception of being liked and respected by others who are highly regarded by us.

It is in the context of emotionally significant interpersonal relationships that individuals develop characteristic levels of self-esteem and contemporaneously respond to what are perceived as threats to one's self-esteem. In the context of relations with parents, siblings, friends, children, spouses, and romantic partners, these interpersonal transactions assume great significance for how one perceives, evaluates, and feels about one's self.

Parenting style, for example, has a profound impact on children's level of self-esteem (DeHart, Pelham, & Tennen, 2006). Young adult children who had more nurturing parents tend to have higher implicit self-esteem. Experiences in secondary social institutions as well contribute to level of self-esteem. Individuals who identify with a positive religious identity tend to display higher self-esteem (Keyes & Reitzes, 2007), particularly during certain life transitions such as retirement, when a religious identity may serve to replace an important loss of some other identity (i.e., worker). Within medicine, better functional health (the ability to perform various activities) anticipates greater levels of self-esteem over a 2-year period (Reitzes & Mutran, 2006); and the success of the therapeutic experience is expected when the patient is able to perceive that the therapist truly places a high value on the patient's worth (Beutler et al., 2004).

In the workplace, a study of abusive supervisory patterns on subordinate's self-esteem (Burton & Hoo-

bler, 2006) confirmed that individuals in the abusive supervision condition manifested lower levels of self-esteem than individuals in a neutral supervision condition. Finally, the relevance of political values is apparent from the observation that for those whose beliefs suggest that socioeconomic status reflects talent and effort, the achievement of higher levels of socioeconomic status is more likely to contribute to one's self-esteem than for those who believe that socioeconomic status is merely a reflection of one's initial placement in a social structure and does not reflect one way or the other on a person's merit (Malka & Miller, 2007).

Experiences of failure/rejection or success/acceptance will affect level of self-esteem depending on the effectiveness of the person's self-enhancing or self-protective responses to the experiences. These mechanisms include increased efforts directed toward achieving valued attributes or performing valued behaviors, distorting one's self-perceptions in a positive direction, changing one's values to give higher priority to those that are achievable and lower priority to those values that are more difficult to achieve, and directly counteracting negative self-feelings by repressing or suppressing them or through the use of mood-enhancing substances. Under some conditions, deviant behaviors that symbolize rejection of the society that engendered low self-esteem serve to decrease self-rejecting attitudes (Kaplan, 1975, 1980). Feelings of being rejected may be reduced by hostility toward the rejecting partners (Murray, Holmes, MacDonald, & Ellsworth, 1998), in effect discrediting the source of the rejection. Overly positive perceptions of acceptance in the context of peer relationships is effective in increasing feelings of self-worth, whether examining a normative sample or a peer-rejected group of grade school students (Kistner, David, & Repper, 2007).

A common mechanism that is employed in the face of probable failure to approximate one's self-evaluatively relevant goals is self-handicapping (Rhodewalt & Tragakis, 2002). This mechanism involves self-imposing barriers to success that would provide self-justification for excuses for failure. By not exerting effort, one can blame one's failure on the lack of effort rather than on any limitations on one's own ability. However, individuals who base their self-esteem on particular conventional values will ordinarily devote greater effort to achieving those values.

SOCIAL CONSEQUENCES OF LOW SELF-ESTEEM

Threats to a person's self-esteem have important positive and negative consequences for the functioning of interpersonal relationships and group memberships. For those who identify with a group, devaluation of the group will lead to increased in-group bias and a subsequent increase in social self-esteem. For individuals whose position in the group is insecure (neophytes, those who are ostracized or peripheral) a common reaction is increased conformity, or at least increased impression management of conforming to group norms (Jetten, Hornsey, & Adarves-Yorno, 2006). Lower status group members are more likely to present themselves as conforming than higher prestige group members, particularly when they are addressing an in-group audience and when they perceive their responses as being made public.

In some groups, paradoxically, conformity may demand the appearance of independence or being a rebel. In such groups, also, one would expect those most in need of group acceptance (that is, new, rejected, or peripheral members) to be most likely to make public displays of being nonconformists. Of course it is to be expected that conformity as a mechanism would be utilized only when the individual feels committed to the group and expects conformity to result in desirable rewards (including recognition) and when no alternatives are offered that lead one to anticipate more self-enhancing outcomes. Absent these contingencies, the loss of motivation to conform to the group and the disposition to seek alternative (perhaps deviant) response patterns becomes a real possibility. These principles play out in a variety of ways in interpersonal and group contexts.

In romantic relationships, low self-esteem on the part of one of the partners may have destructive implications with regard to the maintenance of the relationship. These individuals disbelieve that their partners hold positive attitudes toward them and frequently impute negative attitudes, a self-protective device by low self-esteem individuals that permits them to reject others before they themselves are rejected. Although low self-esteem individuals' partners may initially have positive attitudes toward them, because of their oversensitivity to rejection and their self-defensive responses to expectations of rejection, these individuals create the rejecting attitudes that previously did not in fact exist.

The mechanism of conformity to group norms as an adaptation to self-threatening circumstances has important implications for intergroup behavior because it often takes the form of public displays of derogation or deleterious action against groups that are regarded as competitors or otherwise in conflict with one's own group. It is the individual who has a tenuous relationship with the group rather than the longstanding member that feels the need to derogate outgroups as a mechanism for achieving acceptance within the group (Noel, Wann, & Branscombe, 1995).

As low self-esteem persons associate their self-derogation with participation in the conventional socionormative system, it becomes increasingly likely that individuals will withdraw from participation in conventional social institutions and rather will adopt deviant patterns that offer alternative sources of self-esteem (Kaplan, 1975, 1980, 1986; Kaplan & Johnson, 2001). For example, lower levels of self-esteem during the university years predicted lower levels of work involvement and satisfaction and higher levels of unemployment, weariness, and negative attitudes toward (along with reduced accomplishment at) work (Salmela-Aro & Nurmi, 2007). In the educational realm, studies of adolescents followed into adulthood demonstrate that low self-esteem during adolescence is associated with a greater probability of dropping out of school, and, of course, being less likely to attend a university, findings that are significant because they are frequently observed after controlling for the effects of socioeconomic status, depression, and IQ, among other variables. At the same time, long-term studies confirm that adolescents characterized by low self-esteem predicted higher levels of criminal behavior during adulthood.

To some extent the loss of motivation, and perhaps ability, to conform to conventional role obligations might be accounted for in part by the association of low self-esteem with poor physical and mental health outcomes. For example, over a 2-year period, low self-esteem was associated with decreases in functional health (i.e., the ability to perform various tasks; Reitzes & Mutran, 2006).

LIFE COURSE AND SELF-ESTEEM

The empirical literature clearly suggests that the level, stability, and even capacity for self-esteem is contingent on stage in the life course and its concomitants. Level of self-esteem has been observed to change over the life course. The level of self-approval is relatively high during early childhood but is somewhat lower during middle and late childhood. The level apparently declines during the transitional period between childhood and adolescence. During adulthood, level of self-esteem seems to increase during the latter part of the third decade of life (the 20s) and in the 50s and 60s. However, self-esteem appears to decrease among adults ages 70 or older (Tevendale & DuBois, 2006). More specifically, global self-esteem tends to decrease during adolescence and gradually increase during the third decade of life (between 20 and 30; Kling, Hyde, Showers, & Buswell, 1999). A follow-up of individuals between the ages of 16 and 26 over a 20-year period reported an average increase of self-esteem (Roberts & Bengtson, 1996). Individuals with lower self-esteem ages 18 to 25 manifested a greater increase in self-esteem over this period (Galambos, Barker, & Krahn, 2006). Improvements in self-esteem correlated with receiving greater social support and experiencing less unemployment.

Amount of change/stability in self-esteem, as opposed to change in level of self-esteem, also varies throughout the life course. Amount of change (low stability) tends to be relatively great during childhood. However, stability tends to increase between adolescence and adulthood and decreases between the adult years and old age (Trzesniewski, Donnellan, & Robins, 2003).

The developmental basis for a global sense of high or low self-esteem appears to emerge at an earlier age (3 years) when children begin to use self-referent words "I" and "me" in sentences suggesting that they have particular competencies such as knowing their letters or being a "big boy/girl." Implicit in these statements is the sense that these competencies have evaluative significance. However, very young children do not have a self-conscious concept of global self-esteem that they can verbalize. Nevertheless, a sense of high or low self-esteem becomes manifest in behaviors. Thus, among children ages 4 through 7, children with high and low self-esteem were said to be distinguished by whether or not they showed exploration, confidence, curiosity, and so on (Harter, 1999). Around the age of 8, however, children appear to be able to reflect on themselves as being globally more or less self-accepting. It is around this age that it is possible to conceive of one's self and to communicate a self-conception that is less than positive. The ability to form such global self-evaluations is facilitated by the development at this age of the capacity to compare one's self to a self-evaluative standard and to take the role of another person and perceive one's self from the other person's point of view.

Influences on low or high self-esteem are associated with stage in the life course. Some social changes that are predictably likely to occur at particular stages in the life course may influence level of self-esteem. Thus, the transition from elementary school to middle school appears to adversely affect self-esteem (Seidman et al., 1994). Whether this decrease in self-esteem associated with transition is due to the disruption of normal adaptive/coping/defense mechanisms or increased expectations on the individual that the individual has problems fulfilling or is accounted for by other correlates of the transition remains to be determined. In any case, stage of the life course does appear to moderate level of self-esteem and must, of necessity, be considered in any attempt to fully understand the nature, antecedents, and consequences of processes related to self-esteem.

SEE ALSO Volume 1: *Body Image, Childhood and Adolescence; Identity Development;* Volume 3: *Self.*

BIBLIOGRAPHY

Beutler, L. E., Malik, M. L., Alimohamed, S., Harwood, T. M., Talebi, H., Noble, S., et al. (2004). Therapist variables. In M. J. Lambert (Ed.), *Bergin and Garfield's handbook of psychotherapy and behavior change* (pp. 226–306). New York: Wiley.

Burton, J. P., & Hoobler, J. M. (2006). Subordinate self-esteem and abusive supervision. *Journal of Managerial Issues, 18*(3), 340–355.

DeHart, T., Pelham, B. W., & Tennen, H. (2006). What lies beneath: Parenting style and implicit self-esteem. *Journal of Experimental Social Psychology, 42*(1), 1–17.

Galambos, N. L., Barker, E. T., & Krahn, H. J. (2006). Depression, self-esteem and anger in emerging adulthood: Seven-year trajectories. *Developmental Psychology, 42*(2), 360–365.

Harter, S. (1999). *The construction of the self: A developmental perspective.* New York: Guilford.

Jetten, J., Hornsey, M. J., & Adarves-Yorno, I. (2006). When group members admit to being conformist: The role of relative intragroup status in conformity self-reports. *Personality Social Psychology Bulletin, 32*(2), 162–173.

Kaplan, H. B. (1975). *Self-attitudes and deviant behavior.* Pacific Palisades, CA: Goodyear.

Kaplan, H. B. (1980). *Deviant behavior in defense of self.* New York: Academic Press.

Kaplan, H. B. (1986). *Social psychology of self-referent behavior.* New York: Plenum Press.

Kaplan, H. B., & Johnson, R. J. (2001). *Social deviance: Testing a general theory.* New York: Kluwer Academic/Plenum.

Kernis, M. H. (Ed.). (2006). *Self-esteem issues and answers: A sourcebook of current perspectives.* New York: Psychology Press.

Keyes, C. L. M., & Reitzes, D. C. (2007). The role of religious identity in the mental health of older working and retired adults. *Aging and Mental Health, 11*, 434–443.

Kistner, J., David, C., & Repper, K. (2007). Self-enhancement of peer acceptance: Implications for children's self-worth and interpersonal functioning. *Social Development, 16*(1), 24–44.

Kling, K. C., Hyde, J. S., Showers, C. J., & Buswell, B. N. (1999). Gender differences in self-esteem: A meta-analysis. *Psychological Bulletin, 125*, 470–500.

Koole, S. L., & Pelham, B. W. (2003). On the nature of implicit self-esteem: The case of the name letter effect. In S. J. Spencer, S. Fein, & M. P. Zanna (Eds.), *Motivated social perception: The Ontario symposium* (Vol. 9, pp. 93–166). Hillsdale, NJ: Lawrence Erlbaum.

Malka, A., & Miller, D. T. (2007). Political-economic values and the relationship between socioeconomic status and self-esteem. *Journal of Personality and Social Psychology, 75*(1), 25–42.

Mecca, A. M., Smelser, N. J., & Vasconcellos, J. (1989). *The social importance of self-esteem.* Berkeley: University of California Press.

Murray, S. L., Holmes, J. G., MacDonald, G., & Ellsworth, P. C. (1998). Through the looking glass darkly? When self-doubts turn into relationship insecurities. *Journal of Personality and Social Psychology, 75*(6), 1459–1480.

Noel, J. G., Wann, D. L., & Branscombe, N. R. (1995). Peripheral ingroup membership status and public negativity toward outgroups. *Journal of Personality and Social Psychology, 68*(1), 127–137.

Reitzes, D. C., & Mutran, E. J. (2006). Self and health: Factors that encourage self-esteem and functional health. *Journal of Gerontology, 61B*(1), 844–851.

Rhodewalt, F., & Tragakis, M. (2002). *Self-handicapping and the social self: The costs and rewards of interpersonal self-construction.* Philadelphia, PA: Psychology Press.

Roberts, R. E. L., & Bengtson, V. L. (1996). Affective ties to parents in early adulthood and self-esteem across 20 years. *Social Psychology Quarterly, 59*(1), 96–106.

Rosenberg, M. (1965). Society and the adolescent self-image. Princeton, NJ: Princeton University Press.

Salmela-Aro, K., & Nurmi, J.-E. (2007). Self-esteem during university studies predicts career characteristics 10 years later. *Journal of Vocational Behavior, 70*(3), 463–477.

Seidman, E., Allen, L., Aber, J. L., Mitchell, C., & Feinman, J. (1994). The impact of school transitions in early adolescence on the self-system and perceived social context of poor urban youth. *Child Development, 65*(2), 507–522.

Tevendale, H. D., & DuBois, D. L. (2006). Self-esteem change: Addressing the possibility of enduring improvements in feelings of self-worth. In M. H. Kenris (Ed.), *Self-esteem issues and answers: A sourcebook on current perspectives* (pp. 170–177). New York: Psychology Press.

Trzesniewski, K. H., Donnellan, M. B., & Robins, R. W. (2003). Stability of self-esteem across the life span. *Journal of Personality and Social Psychology, 84*(1), 205–220.

Howard B. Kaplan

SEX EDUCATION/ ABSTINENCE EDUCATION

People learn about their sexuality throughout their lives and from a multitude of sources, including peers, family, religious institutions, online sources, commercial advertising, self-help books, television, and popular music and films. Despite the varied sources and moments at which people glean lessons about sex and sexuality, researchers, educators, policy makers, and community members concerned with sex education routinely focus on school-based instruction for adolescents. This focus often forecloses attention to lifelong learning about sexuality. Sexuality exists in all stages of the life course, and sex education thus has the potential to be a lifelong pursuit.

Most school-based sex education in the United States promotes a narrow conception of not only *when* people learn about sex and sexuality but also *what* they need to learn. Sex education—both policy and practice—routinely approaches sexual health as an issue of reproductive anatomy, disease and pregnancy prevention, and sexual abstinence. The narrow scope of sex education has a long legacy entrenched in moral panics, social purity movements, and disease prevention. This focus obscures other issues that many consider central to sexual well-being, including mental and emotional health, spiritual growth, and a sense of sexual agency.

At the beginning of the 21st century, the debate over sex education in the United States is particularly contentious. Proponents of abstinence-only sex education argue for instruction that insists on the value of abstaining from sexual activity outside of heterosexual marriage. Such definitions of sexual well-being deny lesbian, gay, and bisexual people and other sexual nonconformists access to information that promotes their health. Those advocating comprehensive sex education argue that schools must provide young people with lessons about pregnancy and disease prevention; lesbian, gay, and bisexuality; abortion; and masturbation to address the reality that many people are sexually active outside marriage. Conflicts over school-based sex education are about much more than whether and what young people should learn about sex and sexuality in school; rather, these debates address fundamental issues about gender norms, family formations, sexual possibilities, and life trajectories.

SEX EDUCATION IN THE 20TH CENTURY

Sex education emerged as a public concern in the United States in the late 1800s, when the YMCA, YWCA, and American Purity Alliance, organizations based on Judeo-Christian principles, hosted sex-related panels and lectures. In the early 1900s, U.S. educators and advocates formed organizations to address sexual morality and hygiene, including the American Society of Sanitary and Moral Prophylaxis (founded in 1905), the American Federation for Sex Hygiene (1910), and the American Social Hygiene Association (1914). These groups focused on the dangers of "social evil" especially within the context of venereal diseases. During World War I, policy makers grew concerned that soldiers would return from overseas with venereal diseases such as syphilis and gonorrhea and thus endanger U.S. morality and well-being. In response, organizations such as the American Social Hygiene Association ran a social marketing campaign that brought moral and physical fitness together with the aim of preventing sexually transmitted infections (STIs). During this period, medical information about sexually transmitted diseases (STD) improved and popular media about sexuality increased. As sexuality information escalated in public spaces, parents and teachers worried about adolescents' access to this information. Regulating sex education within schools promised to control how youth talked about sex. Therefore, by the end of the 1920s, consensus was emerging that sex education was primarily schools'—and not families'—responsibility (Rosow and Persell, 1980).

Abstinence. *A billboard in downtown Baltimore promotes teen abstinence.* **AP IMAGES.**

In 1964 the Sexuality Information and Education Council of the United States (SIECUS) formed to promote access to sexual health information. SIECUS founder Mary Calderone explained, "We put sexuality into the field of health rather than the field of morals" (Irvine, 2002, p. 27). SIECUS departed from earlier hygiene groups that combined sexuality and morality rather than sexuality and health. SIECUS continues to be a national organization invested in advocacy for sexual health, sexual rights, the defense of comprehensive sex education, the promotion of sexual pleasure, and the eradication of sexual guilt.

The emergence of the HIV/AIDS epidemic in the 1980s strengthened the link between sexuality and health and had a broad and significant impact on sex education. HIV heightened the fear that lives were at stake in the debate over sex education. For comprehensive-education supporters, HIV provided further evidence that young people needed to be equipped with the tools to fight sexually transmitted diseases. For abstinence-only supporters, HIV affirmed that sexual expression is appropriate and safe only within monogamous, faithful, heterosexual marriage (Irvine, 2002).

CONTEMPORARY DEBATES OVER COMPREHENSIVE AND ABSTINENCE-ONLY SEX EDUCATION

The fight over abstinence-only sex education was well underway before the emergence of the HIV epidemic. In 1978 Senator Edward Kennedy (Democrat-Massachusetts) sponsored the Adolescent Health Services Act, which aimed to address a perceived epidemic of teen pregnancies by funding comprehensive sex education, contraceptive services, and other family planning and reproductive services. Objecting to the Adolescent Health Services Act and to broad shifts in gender and sexual norms, conservatives argued that federal funding should promote sexual abstinence among young people, not suggest that they might engage in sexual activity safely.

Congress passed the Adolescent Family Life Act (AFLA) in 1981. AFLA, also known as the "chastity law," codified and funded a standard of sexual abstinence. The legislation promoted regulation and conventional modesty among youth by funding exclusively abstinence-only education and denying funding to state and local programs that discussed, counseled, or provided abortion.

The 1996 Personal Responsibility and Work Opportunity Reconciliation Act (what many know as "welfare reform") included an amendment that required that programs receiving federal funding adhere to an eight-point definition of "abstinence education" as instruction that asserts mutually faithful, monogamous marriage as "the expected standard of sexual activity" for all people.

All of this legislation contributed to a sustained federal effort to fund education that reasserts conventional gender and sexual norms and morality. In the 1990s and 2000s, conservatives have been unabashed in their aim to reassert heterosexual marriage as fundamental to a healthy society and to establish a taken-for-granted idea that sexual abstinence is the most, if not only, appropriate choice for young people outside of heterosexual marriage. This "common sense" emerged despite consistent support for comprehensive sex education that presents abstinence as one of many ways to prevent sexually transmitted infections and pregnancies and which discusses the benefits, failure rates, and side effects of contraceptive methods. Such instruction usually affirms the value of sexual abstinence in pregnancy and disease prevention. However, comprehensive sex education also argues that, because some youth are sexually active before leaving high school and before marriage and that non-heterosexual youth are invisible in education that relies on marriage, schools must provide sex education that equips them to engage in consensual and safe sexual activity.

SEX EDUCATION IN CONTEXT: CULTURE, CURRICULA, AND CLASSROOMS

Debates and controversies over sex education are about much more than whether, when, and what young people should learn in school about sex and sexuality, and the effects of sex education extend well beyond whether students learn how to prevent pregnancies and disease. Community struggles over sex education policy in the United States reflect, resist, and affirm broad social conflict. Sex educators offer lessons that aim to protect young people from harm but that may have little effect. Further, day-to-day practices in sex education classrooms reflect and reproduce persistent social inequalities such as gender norms, family structures, and sexual expectations.

SEXUAL SENSIBILITIES: SEX EDUCATION AND SHIFTING SOCIAL NORMS

Varying contexts of family values, cultural proscriptions of adolescent sexuality, and religion all have a significant impact on sex education, adolescents' ability to navigate their sexuality, and the role of sexuality in the life course. Amy Schalet's research (2004) on the construction of adolescent sexuality in the Netherlands and the United States reveals the significant role of culture in sexual health. Schalet found that Dutch parents normalize and accept adolescent sexuality whereas U.S. parents dramatize, deny, and fear adolescent sexuality. While Dutch parents see adolescents as sexual agents capable of healthy sexual decision-making and loving relationships, Americans distrust adolescents' ability to navigate sexuality and deny their ability to foster love-based relationships. The U.S practice of denying adolescent sexuality makes it harder for adolescents to gain access to contraceptives, abortion, and relevant sex education material. Young people in the United States also find it difficult to act on healthy sexual choices.

This culture of denial and proscription is also evident in adult talk about young people's sexuality and sex education. Janice Irvine (2002) researched the role of emotions in the politics of sex education debates to unravel the power of emotion and language in the culture wars of sexuality. Irvine realizes that *how* people talk about sex has significant implications on sexual culture. These adult-driven debates ignore adolescents' voices; messages of fear illicit strong emotional responses that further compel abstinence-only support while masking the benefits of comprehensive sex education.

INSTRUCTIONAL IMPACT: EVALUATION RESEARCH AND SEX EDUCATION

Despite proponents' claims that abstinence-only sex education contributes to individual, familial, and social well-being, reports consistently confirm that no studies meeting scientific standards have demonstrated that abstinence-only education reduces teen pregnancies or delays first intercourse (Kirby, 2001). Indeed, few systematic evaluations exist of sex education, broadly defined, and, in particular, of abstinence-only sex education. Those evaluations that are available indicate that participating in abstinence-only programs does not increase the likelihood that youth will abstain from engaging in sex or delay sexual initiation. Some research suggests that many young people who participate in abstinence-only education and who take "virginity pledges" eventually do become sexually active before marriage, are less likely to use condoms at first intercourse, and thus are ultimately at greater risk of contracting an STI.

This finding contradicts the widespread belief that talking with youth about sex education will increase their interest in having sex. Indeed, research shows that participants in comprehensive sex education wait longer to have sex, have fewer sex partners, and use condoms and other contraceptive methods at significantly higher numbers as compared to students not in comprehensive sex

VIRGINITY PLEDGE

Virginity pledges are promises to remain sexually abstinent until marriage. Pledgers tend to be school-aged, evangelical Christians, and from conventional families. Many—though not all—pledgers promise abstinence through formal organizations such as The Silver Ring Thing and True Love Waits. At the turn of the 21st century, virginity pledging, like abstinence-only sex education, contributes to a broad movement to reassert conventional gender and sexual norms.

Definitions of virginity are elusive, often making it unclear precisely what a pledger is committing to. Some believe that engaging in penile-vaginal intercourse constitutes the loss of one's virginity; for others, remaining abstinent may mean avoiding kissing, oral sex, or anal sex. Some assert the possibility of "secondary virginity," in which a person pledges to no longer engage in sexual activity and to remain abstinent until marriage.

Pledgers may postpone sex, but, having had little instruction in disease and pregnancy prevention, they are less likely to use a condom during their first sexual experiences. Identity appears important to pledging's effectiveness: Pledgers successfully delay sexual activity as long as committing to abstinence remains nonnormative among their peers—too many or too few pledging peers results in a less socially significant and therefore less effective pledge.

education. The curricula most effective in reducing teen pregnancies focus on contraceptive use and sexual behavior while also highlighting abstinence as the safest choice to avoid pregnancies and STIs (Kirby, 2001).

These and other findings in the evaluation research have become increasingly politicized. Critics of abstinence-only policies have called on the federal government (a) to desist funding for curricula that contain medical and scientific inaccuracies and (b) to conduct rigorous scientific evaluations of the curricula's effectiveness. These calls grew only louder in 2006, when the nonpartisan Government Accountability Office (GAO) released a report on abstinence education, conducted at the request of Democratic members of the House of Representatives and Senate. According to the GAO, the U.S. Department of Health and Human Services had not required grantees to review their curricula for scientific accuracy or to conduct

scientifically sound evaluations of their programs' effectiveness in meeting their own goal of promoting sexual abstinence among young people. A 2004 report sponsored by Representative Henry Waxman (Democrat-California) demonstrated that abstinence-only curricula contain scientific errors (including false information about contraceptives and the risks of abortion), distort distinctions between religion and science, and provide stereotypical depictions of girls' and boys' behavior.

As of late 2007, approximately one quarter of all states have refused federal funding for sex education that requires abstinence-only curricula. This is a dramatic change from 2006, when only four states rejected this funding.

INSIDE THE CLASSROOM: SEX EDUCATION AND THE REPRODUCTION OF SOCIAL INEQUALITY

While states' refusal of abstinence-only funding is a significant development in the debate over sex education, legislative and administrative decisions are only brief moments in a process of curricular negotiation and contestation. Abstinence-only and comprehensive curricula are never regimes of absolute control. Teachers adapt administration priorities to meet their own instructional aims, and once inside the classroom, negotiations continue as students receive, resist, and revise their teachers' lessons.

Much of the research on sex education focuses on hidden lessons on gender and female sexuality. In one of the earliest and still most influential publications on sex education, Michelle Fine (1988) argued that discourses of female victimization and (im)morality dominated school-based sex education. Fine pointed to a "missing discourse" that recognizes girls' and young women's sexual desires and pleasures. This discourse emerges in whispers from students, but teachers and the formal curriculum routinely mute talk of female sexuality as a site of agency and pleasure, contributing to the broad educational, intellectual, and political disempowerment of girls and women.

This muting continues in the wake of U.S. abstinence-only legislation. Jessica Fields found that whereas abstinence-only lessons may be the formal mandate in sex education courses, multiple curricula abound in a single class meeting. Formal, hidden, and evaded lessons compete for students' attention, and all help to determine the sex education students receive. These lessons are consistently inflected with social differences and inequalities. For example, teachers affirm sexist understandings of sexuality when they present lessons that officially promote open sexual communication while also casting boys and men as sexual aggressors and assigning girls and women responsibility for maintaining sexual boundaries. Often, teachers are complicit in perpetuating racist and able-ist depictions of bodies and sexuality when they present exclusively white, able-bodied, and idealized images of sex organs and anatomies.

SEX EDUCATION'S FUTURE AND POTENTIAL

Building on the work of Michelle Fine, feminist researchers have explored the "whisper" of desire in sex education, finding in these whispers not only the hint of female sexual agency, but also the potential for transformative sex education, and a call for broad social change. Sometimes those whispers occur outside the classroom—in school-based health clinics, in online chat rooms, or over dinner with a trusted adult, for example. Other times those whispers occur inside the classroom during, for example, a discussion of orgasms, clitorises, or heterosexual intercourse. In these moments, the discourse of desire and the challenge to social inequality so often missing from sex education interrupts the prevailing talk about female sexuality as victimization, violence, and individual morality. In these moments, sex education becomes more than an opportunity to teach the skills and norms expected of young people as they move toward adulthood. Instead, sex education becomes a site in which young women and others can struggle against disadvantage and come to a sense of entitlement. This sensibility is crucial to a range of conventional health outcomes—for example, avoiding unwanted pregnancies, STIs, and sexual coercion. It is also entangled with the political, intellectual, and social entitlement that is central to people thriving throughout the life course.

Researchers, advocates, and educators with a critical perspective on sexuality, education, and youth pursue inquiry that "suspends the 'givens' of adolescent sexuality" (McClelland and Fine 2008, p. 67). These givens include the commonsense assumptions that abstinence is the most desirable outcome in young people's sexual lives; that young people are naturally and best heterosexual; that girls are passive in their heterosexual relationships; that boys are inevitably sexually aggressive; that sex education's promise and effects lie primarily in preventing pregnancies and disease and in delaying sexual initiation; and that learning about sex and sexuality ends once we leave adolescence.

Intersections of religion, public policy, school-based education, and media all inform youth and adults of the potential benefits and harms of sexuality. Locating sex education in classrooms that uncritically promote gender and sexuality conventions limits sex education's potential. Sex education represents an opportunity to promote under-

standings of sexuality as a lifelong process; affirm lesbian, gay, bisexual, and heterosexualities; and acknowledge that individuals learn about sex and sexuality across institutions and across their lifetimes.

SEE ALSO Volume 1: *Dating and Romantic Relationships, Childhood and Adolescence; Sexual Activity, Adolescence; Transition to Parenthood.*

BIBLIOGRAPHY

Brückner, H., & Bearman, P. (2005). After the promise: The STD consequences of adolescent virginity pledges. *Journal of Adolescent Health, 36,* 271-278.

Fields, Jessica. 2008. *Risky lessons: Sex education and social inequality.* New Brunswick, NJ: Rutgers University Press.

Fine, M. (1988). Sexuality, schooling, and adolescent females: The missing discourse of desire. *Harvard Educational Review, 58.*

Irvine, J. M. (2002). *Talk about sex: The battles over sex education in the United States.* Berkeley: University of California Press.

Kirby, D. (2001). *Emerging answers: Research findings on programs to reduce teen pregnancy.* Washington, DC: National Campaign to Prevent Teen Pregnancy.

Luker, K. (2006). *When sex goes to school: Warring views on sex—and sex education—since the sixties.* New York: Norton.

McClelland, S. I., & Fine, M. (2008). Embedded science: Critical analysis of abstinence-only evaluation research. *Cultural Studies, Critical Methodologies, 8*(2), 50–81.

Regnerus, M. D. (2007). *Forbidden fruit: Sex and religion in the lives of American teenagers.* Oxford, UK: Oxford University Press.

Rosow, K., & Persell, C. H. (1980). Sex education from 1900 to 1920: A study of ideological social control. *Qualitative Sociology, 3*(3), 186–203.

Schalet, A. (2004). Must we fear adolescent sexuality? *Medscape General Medicine 6*(4), 1–16.

Sexuality Information and Education Council of the United States (SIECUS). (2004). *Guidelines for comprehensive sexuality education: Kindergarten through 12th grade.* (3rd ed.). Retrieved May 22, 2008, from http://www.siecus.org/pubs/guidelines/guidelines.pdf

SIECUS. (2004). *Life behaviors of a sexually healthy adult.* Retrieved May 22, 2008, from http://www.siecus.org/school/sex_ed/guidelines/guide0004.html

Kathleen Hentz
Jessica Fields

SEXUAL ACTIVITY, ADOLESCENT

"It started with a chair," says Juno in the opening line of the 2007 movie of the same name. As quickly becomes apparent, "it" refers to sex between two teenagers that in this case leads to pregnancy, birth, and finally an awakening of what love is all about. In contrast, a very different sequence of life course events is laid out in the traditional children's jump rope rhyme: "first comes love, then comes marriage, then comes a baby in a baby carriage." The chain of events portrayed in Juno, however, in which marriage is not even mentioned, is the one that comes closest to typifying adolescent sexual behavior at the turn of the 21st century. It is also the sequence that continues to worry parents and policy makers alike.

TRENDS IN ADOLESCENT SEXUAL BEHAVIORS

A range of excellent Web sites provide facts and figures pertaining to adolescent sexual behavior, including statistics on pregnancy, birth rates, contraception and sexually transmitted infections (STIs) (for example, www.cdc.gov/nchs; www.childtrends.org; www.guttmacher.org). Since the early 1970s many aspects of adolescent sexual behavior have undergone marked changes; an understanding of the historical trends helps to put current teen activities in context.

Beginning in the late 1960s and continuing throughout the 1980s, increasing proportions of teenagers had nonmarital sexual intercourse. This trend reversed during the late 1990s and into the early 21st century, however. Comparisons of data from the 1988 and 2002 National Surveys of Family Growth (NSFG) show that, whereas 60% of 15- to 19-year-old males in 1988 reported having had intercourse, by 2002 this had dropped to 46%. Girls showed a decrease over the same time period, from 51% to 46%. American teens also appear to be waiting longer to have sex. Whereas in 1995 19% of girls and 21% of boys said they had had sex by age 15, in 2002 the corresponding percentages were 13% and 15% (Abma, Martinez, Mosher, and Dawson, 2004).

Adolescents have always had sex, so why should current figures be of concern, especially given their downward trend? Despite the declines, close to half of American teens report having had sexual intercourse at least once, and roughly 1 in 7 initiated sex prior to age fifteen. Alongside moral disquiet that adolescent sexual behavior overwhelmingly occurs outside of marriage, risk of pregnancy and STIs also lie at the forefront of societal worries.

Looking at the glass half full, a steady drop in teen pregnancy rates from a high point in 1991 has paralleled declines in sexual intercourse among teenagers in the United States. So has the teen birth rate, which in 1991 stood at 62 births per 1000 15- to 19-year-old American girls, but by 2004 had dropped by one third to 41 births per 1000. Although social conservatives have credited much of the drop in teen pregnancy rates to abstinence education programs, the numbers show that improved

contraception has been the principal player (Santelli, Lindberg, Finer, and Singh, 2007). Data from the 1995 and 2002 rounds of the NSFG placed the percentages of 15- to 19-year-old males who used contraceptives the most recent time they had sex at 82% in 1995 and 91% in 2002. Comparable figures for girls were 71% and 83%. Teens are increasingly using condoms, which provide the best protection from STIs, and the birth control pill and injectable hormonal methods such as Depo-Provera to prevent pregnancy. The number of adolescents who used two or more methods simultaneously over this time period also more than doubled, from 11% to 26%, reflecting a desire to protect against both pregnancy and STIs. Contraceptive use at first sex has also become more common. This is important because youth who use contraceptives at first sex are more likely to use them when they have sex later than are youth who do not, and consistent contraception is key to preventing pregnancy and STIs.

The same glass may also be seen as half empty, however, as close to one-third of American girls get pregnant by age 20—a figure that stands at just over 50% for Latina teens. More than four-fifths of teen pregnancies are unplanned and just under half of unintended teen pregnancies end in abortion. Of those teen pregnancies that are carried to term, the proportion of teen births that occur within marriage decreased from just under half in 1982 to less than 1 in 5 in 2002 as fewer unmarried pregnant teens chose to marry prior to giving birth, if at all. Further, despite significant declines in teen sex, pregnancy, abortions and births since 1991, U.S. teen pregnancy rates, birth rates, abortion rates, and rates of STIs remain among the highest in the industrialized world (Singh and Darroch, 2000; Darroch, Singh, and Frost, 2003). Although levels of adolescent sexual activity do not vary much across comparable developed countries such as Canada, Great Britain, France, Sweden, and the United States, American teens are more likely than their foreign counterparts to have sexual intercourse before their 15th birthday and to have two or more sexual partners in a year. They are also less likely to use contraceptives, especially the pill and other long-acting reversible hormonal methods that are the most effective at preventing pregnancy.

Youths who delay sexual initiation are less likely to regret the timing of their first sexual experience, and are less likely to be involved in coercive sexual relationships. Approximately 10% of first sexual encounters that occur prior to age 20 are reported by young women as involuntary or forced; for those who first have sex before age 15, the percentage is higher still. Early initiation of sex is also especially worrisome because young teens are less likely to use contraception at first sex and to use contraceptives

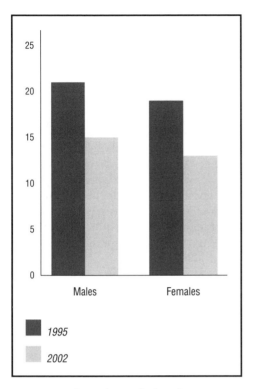

Figure 1. *Percent of teens having had sex by age 15 years.*
CENGAGE LEARNING, GALE.

consistently throughout their first relationship (Ryan et al., 2003). Although most teens who have had sex report none or only one sexual partner in the past year, close to 10% report four or more. Having more partners also is associated with a lower likelihood of contraception and a higher likelihood of acquiring an STI. According to figures published by Child Trends in 2006, young people between 15 and 24 years old account for almost half of all newly acquired STIs in the United States. Figures released by the Centers for Disease Control in 2008 show that at least 1 in 4 teenage girls had an STI.

Why might American teens have sex at an earlier age, have more partners, and use contraception less consistently than teens in other developed countries? A number of reasons have been proposed, including more negative societal attitudes toward teenage sexual relationships, more restricted access to contraceptives, and higher costs associated with reproductive health services. The Alan Guttmacher Institute notes that countries with low levels of adolescent pregnancy, childbearing, and STIs are not only characterized by a greater acceptance of teenage sexual relationships but also provide more comprehensive and balanced information about sexuality, and the prevention of pregnancy and STIs within these relationships.

WHAT DO PEOPLE MEAN BY "SEX"?

At a White House press conference on January 26, 1998, President Bill Clinton denied allegations of prior sexual relations with White House intern Monica Lewinsky, stating: "I did not have sexual relations with that woman." In the sense that sexual relations means male-female intercourse, President Clinton did not lie. However, sexual relations can include behaviors other than intercourse; adolescents in the early 21st century are engaging in both oral and anal sex, sometimes while clinging to the belief that by avoiding intercourse in favor of these behaviors, they officially remain virgins (Remez, 2000). Data from the NSFG show that in 2002, 16% of teens between the ages of 15 and 17 years who had not yet had sexual intercourse had engaged in oral sex.

What teens do or do not consider to be sex is also a subject that has received publicity as a result of research on virginity pledges by Peter Bearman and Hannah Brückner (2001, 2005). Approximately 1 in 10 teens in the United States takes a virginity pledge that is designed to curb adolescent sexual activity (Maynard, 2005). While it appears that taking the pledge delays the initiation of sexual intercourse for some teens, pledgers who remain virgins are more likely to have anal and oral sex. Adolescents view oral sex as less threatening to their values and beliefs, and perceive it to have fewer health, social, and emotional consequences than vaginal sex (Halpern-Felsher, Cornell, Kropp, and Tschann, 2005). However, although both oral and anal sex remove the potential for pregnancy, these activities can put teens at high risk for STIs.

WHY DO TEENS HAVE SEX?

Much research over the past few decades has tried to work out why some teens will have sex while others will refrain, and why among sexually active teens, some will use contraception and others will not. To identify both risk and protective factors, many of these studies have used data from large surveys such as the National Longitudinal Survey of Adolescent Health (Add Health), the 1979 and 1997 National Longitudinal Surveys of Youth (NLSY79 and NLSY97), the NSFG, and the Youth Risk Behavior Survey (YRBS), supplemented with information gleaned from questionnaires targeted at more specific subpopulations of adolescents.

There is no simple answer to the question: What predicts adolescent sexual activity? Myriad structural and contextual influences play a role. For example, African-American youth initiate sexual activity at earlier ages than do European-American or Latino/a youths, and features of the neighborhood environment have been highlighted to account for this racial difference: neighborhood poverty as a factor that elevates the risk of early sexual initiation, and neighborhood social cohesion and social

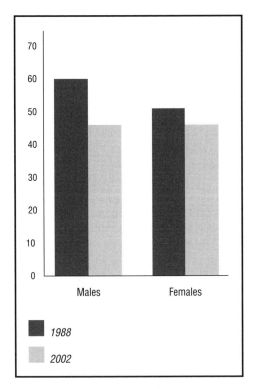

Figure 2. *Percent 15–19 year olds who have had sexual intercourse.* **CENGAGE LEARNING, GALE.**

control that helps delay sexual onset (Browning, Leventhal, and Brooks-Gunn, 2004). Television, movies, music, magazines and the Internet all expose teens to sexual imagery, but rarely depict realistic consequences of risky sexual activities. Although relatively few studies have looked at the effect of media exposure on adolescent sexual knowledge, attitudes, and behaviors, findings generally suggest that greater exposure to sexual content within the media is associated with higher estimates of the extent to which friends are sexually active, and more and earlier sexual behavior among adolescents themselves.

Emphasis has been placed on the role of peers, and studies have found that friends' sexual behaviors, perceptions of friends' behaviors and attitudes, and the degree to which adolescents are involved with their friends all influence adolescent sexual behaviors. For example, in 2006 Renee Sieving and colleagues from the University of Minnesota found that sexually inexperienced kids in 9th to 11th grades had greater odds of having intercourse in the following 18 months when a higher proportion of their friends were sexually active. Youths who believed that they would gain their friends' respect if they were to have sex also had higher odds of initiating intercourse.

Numerous studies also demonstrate the importance of both family structure and parental processes for ado-

lescent sexual behavior. For example, youth who grow up in intact two-parent homes are more likely to delay sexual initiation and to have fewer sexual partners than adolescents in single parent and stepfamily homes. Why? Parental supervision may be higher in two-parent families, reducing the amount of time that a teen can spend at home without a parent, and hence their opportunity to have sex. Parental monitoring also conveys to children that they matter. A 2007 study by Chadwick Menning and colleagues that finds that the more involved adolescent males are with their stepfathers the less likely they are to have sex. This finding also underscores the importance of close family relationships (Menning, Holtzman, and Kapinus, 2007). Further, even if teens cite their peers and the media as primary sources of sexual information and influences on their sexual behavior, parents can play a critical intervening role. Although having sexually active peers tends to accelerate sexual initiation, parental responsiveness to adolescents during discussions about sex can decrease risk-taking behaviors, and can buffer the negative influences of peers (Fasula and Miller, 2006). Conversely, peer effects may be stronger when parental bonds are strained (Jaccard, Blanton, and Dodge, 2005).

Reflecting different theoretical perspectives and disciplinary approaches, other research has been more qualitative in nature, using information from semi-structured, one-on-one interviews that allow teens to more easily tell their own stories surrounding their sexual experiences, and to reflect on how and why they made the decisions they did—choices that do not always appear to be in their best interests. Qualitative studies are a reminder that curiosity plays an important role, and that adolescents, like adults, have sexual desires. Such studies also underscore the importance of understanding the social contexts within which adolescents act. For example, how does the quality of a relationship or perception of romance define the context for sexual decision making? Although a teen may well understand the importance of using contraception to reduce health and social risks, their decision to actually use a condom might instead come down to whether they believe this choice might improve or harm their relationship.

In-depth interviews with inner-city Black adolescents uncovered some important gender differences in the ways in which youth approach sexual relationships (Andrinopoulos, Kerrigan, and Ellen, 2006). Whereas young men reported that sexual relationships helped them feel wanted and that they could gain social status among their peers when they had multiple partners, young women desired monogamous romantic partners for emotional intimacy, although most also reported having had nonmonogamous partners. Understanding contextual factors such as a lack of socioeconomic opportunities for young men, or an imbalanced sex ratio as a result of the incarceration of many young men, is crucial for comprehending the types of sexual relationships that youth engage in, as the nature of the relationship can then affect the sexual activities engaged in within the relationship, and the contraceptive decisions they make (Kaestle and Halpern, 2005).

A 2005 study of adolescent sexual decision making by Tricia Michels and colleagues from the University of California provides evidence that at least some young adolescents set a priori boundaries for how far they are willing to go with a particular partner at a given time (Michels, Kropp, Eyre, and Halpern-Felsher, 2005). The limits that teens mentally set for themselves are influenced by their own personal characteristics and the nature of the relationship they are in. Study participants also acknowledged that the lines they drew now might be crossed in the future as they matured, their relationship changed, or they became more sexually experienced. Their findings also suggested a dynamic, self-regulating process of decision making as teens (especially girls) reflected on their behaviors and how they felt about them, and then either reaffirmed or reappraised where their boundaries lay.

Results from both quantitative and qualitative research methods therefore highlight a range of both personal and contextual factors that play an important role in why teens behave in the ways they do, and help to construct our overall understanding of adolescent sexual behaviors block by block. Adolescents are not only influenced by the social contexts within which they operate, but, as the life course perspective underscores, the decisions they make and pathways they take are influenced by the trajectories of their past. As Lisa Lieberman states in her essay on early predictors of sexual behavior (2006), "the seeds of sexual risk-taking are sown early in adolescence."

Adolescence is a stage in the life course marked by significant developmental growth, and as youth mature physically, emotionally, and socially, they are tempted to explore various sexual behaviors. Further, within their unique social contexts that provide both boundaries and opportunities for sexual activity, adolescents appear to consider both costs such as guilt, embarrassment, and risk of perceived negative outcomes along with benefits such as physical pleasure, social status, and intimacy when making their sexual and contraceptive decisions (Deptula, Henry, Shoeny, and Slavic, 2006). The vast array of factors that play a part in teen decision making are complex and interrelated and each can influence the weight that teens ascribe to the risks and advantages that they perceive. For example, teens who attend religious services frequently might attribute high levels of guilt to premarital sexual activity; youth with lofty academic plans might perceive the risk of pregnancy as especially damaging; and adolescents who are themselves children of adolescent parents might place a much lower significance on avoiding pregnancy.

LOOKING TO THE FUTURE

Although it is no longer accepted that a young girl's life script is mostly written should she, like Juno, get pregnant while still a teenager, the decisions that adolescents make concerning their sexual and contraceptive behaviors still have consequences for the ways in which their life course trajectories will unfold, the futures of the next generation, and costs for society in terms of public assistance, or health care for teens and their children. It is unrealistic to expect all teens to remain virgins until marriage, but interventions that promote the value of responsible behaviors, delay sexual initiation, encourage the consistent use of contraceptives that protect against both pregnancy and STIs when teens do decide to become sexually active, and recognize that even "virgins" engage in potentially risky sexual behaviors are important ones to endorse. One can learn from the successes and failures of other countries, and should involve parents, teachers, and other adult role models as active participants in our programs. Finally, and perhaps most importantly adults must keep the lines of communication open and listen to the voices of adolescents themselves as they attempt to navigate the complex transition from child to young adult within a constantly changing culture.

SEE ALSO Volume 1: *Dating and Romantic Relationships, Childhood and Adolescence; Health Behaviors, Childhood and Adolescence; Sex Education/Abstinence Education; Transition to Marriage; Transition to Parenthood;* Volume 2: *Abortion; Birth Control; Childbearing.*

BIBLIOGRAPHY

Abma, J. C., Martinez, G. M., Mosher, W. E., & Dawson, B. S. (2004). Teenagers in the United States: Sexual activity, contraceptive use and childbearing, 2002. *Vital Health Statistics, 23,* 2–4.

Andrinopoulos, K., Kerrigan, D., & Ellen, J. M. (2006). Understanding sex partner selection from the perspective of inner-city black adolescents. *Perspectives on Sexual and Reproductive Health, 38*(3), 132–138.

Bearman, P. S., & Brückner, H. (2001). Promising the future: Virginity pledges and the transition to first intercourse. *American Journal of Sociology, 106*(N4), 859–912.

Browning, C.R., Leventhal, T., & Brooks-Gunn, J. (2004). Neighborhood context and racial differences in early adolescent sexual activity. *Demography 41*(4), 697–720.

Brückner, H., & Bearman, P. S. (2005). After the promise: The STD consequences of adolescent virginity pledges. *Journal of Adolescent Health, 36,* 271–278.

Darroch, J. E., Singh, S., & Frost, J. J. (2003). Differences in teenage pregnancy rates among five developed countries: The role of sexual activity and contraceptive use. *Family Planning Perspectives, 33,* 244–250, 281.

Deptula, D. P., Henry, D. B., Shoeny, M. E., & Slavick, J. T. (2006). Adolescent sexual behavior and attitudes: A costs and benefits approach. *Journal of Adolescent Health, 38,* 35–43.

Fasula, A. M., & Miller, K. S. (2006). African-American and Hispanic adolescents' intentions to delay first intercourse: Parental communication as a buffer for sexually active peers. *Journal of Adolescent Health, 38,* 193–200.

Halpern-Felsher, B. L., Cornell, J. L., Kropp, R. Y., & Tschann, J. M. (2005). Oral versus vaginal sex among adolescents: Perceptions, attitudes, and behavior. *Pediatrics, 115*(4), 845–851.

Jaccard, J., Blanton, H., & Dodge, T. (2005). Peer influences on risk behavior: An analysis of the effects of a close friend. *Developmental Psychology, 41*(1), 135–147.

Kaestle, C. E., & Halpern, C. T. (2005). Sexual activity among adolescents in romantic relationships with friends, acquaintances, or strangers. *Archives of Pediatrics and Adolescent Medicine, 159,* 849–853.

Lieberman, L. D. (2006). Early predictors of sexual behavior: Implications for young adolescents and their parents. *Perspectives on Sexual and Reproductive Health, 38*(2), 112–114.

Maynard, R. (2005). Interview: Sex and the American teen. Retrieved April 15, 2008, from http://www.pbs.prg/pov

Menning, C., Holtzman, M., & Kapinus, C. (2007). Stepfather involvement and adolescents' disposition toward having sex. *Perspectives on Sexual and Reproductive Health, 39*(2), 82–89.

Michels, T. M., Kropp, R. Y., Eyre, S. L., & Halpern-Felsher, B. L. (2005). Initiating sexual experiences: How do young adolescents make decisions regarding early sexual activity? *Journal of Research on Adolescence, 15*(4), 583–607.

Remez, L. (2000). Oral sex among adolescents: Is it sex or is it abstinence? *Family Planning Perspectives, 32*(6), 298–304.

Ryan, S., Manlove, J., and Franzetta, K. (2003). The first time: Characteristics of teens' first sexual relationships. *Child Trends Research Brief no. 2003-16.* Washington, DC.

Santelli, J..S., Lindberg, L. D., Finer, L. B., & Singh, S. (2007). Explaining recent declines in adolescent pregnancy in the United States: The contribution of abstinence and improved contraceptive use. *American Journal of Public Health, 97*(1), 150–155.

Sieving, R. E., Eisenberg, M. E., Pettingell, S., & Skay, C. (2006). Friends' influence on adolescents' first sexual intercourse. *Perspectives on Sexual and Reproductive Health, 38*(1), 13–19.

Singh, S., and Darroch. J. E. (2000). Adolescent pregnancy and childbearing: Levels and trends in developed countries. *Family Planning Perspectives, 32,* 14–23.

Elizabeth Cooksey

SIBLING RELATIONSHIPS, CHILDHOOD AND ADOLESCENCE

Within the family realm, siblings potentially affect each other in myriad ways. Psychologists generally study the quality of relationships and interpersonal dynamics of sibling relationships (see Cicirelli, 1995; Dunn, 1985). Sociologists, on the other hand, tend to focus on the structure of sibling relationships. Growing up with brothers and sisters in a family is associated with a variety of behavioral outcomes from school accomplishments to misconduct (Blake, 1989; Downey, 1995; Steelman, Powell, Werum, & Carter, 2002). Siblings, as members of the family primary group, can and do affect one another over the life course. Envision a family with more than one child at home. As a child matures any sibling may be a confidante or a rival, an ally or foe, a coresident or occasional housemate. Scholarly study of the role that siblings play in child development ironically commands considerable scholarly attention at the precise moment in history when birth rates in most industrialized societies are dropping. Indeed, the famous "only one child" policy in China epitomizes a worldwide trend. Knowledge about whether and how siblings matter thus is ever more compelling.

Virtually every scholar recognizes that the family is the most enduring influential small group to which individuals will ever belong. Families ordinarily affect their members across the life cycle. Because contemporary families are in the midst of significant and sometimes unforeseen changes, experts express concern that children may be adversely affected by such transformations. Growing up in a home with siblings helps shape a child's growth in both positive and negative ways. We focus on the consequences of sibling structure features insofar as this is where most scholarly attention is directed. The most investigated outcome in which siblings may play a role is educational achievement.

CHARACTERISTICS OF SIBLING RELATIONSHIP RESEARCH

Much of what we suspect about the impact of siblings comes from study of the four characteristics that comprise the sibling matrix: sibship size, birth order, space intervals, and sex composition (Steelman, 1985). Sibship size is simply the number of brothers and sisters within the family. Birth order is an individual's rank in the age hierarchy of the sibling group, for instance, firstborn, second-born, and so forth. Child spacing is the time interval that separates consecutive or adjacent siblings. Sex composition captures whether a child is reared in a

BIRTH ORDER

Popular interest in birth order is as high as ever, and, in 2007, the highly prestigious journal *Science* presented new data that suggested the superiority of the eldest child in academics. Renewed interest in birth order followed the publication of Frank J. Sulloway's 1996 book *Born to Rebel: Birth Order, Family Dynamics, and Creative Lives*. For his data, Sulloway asked academic experts to identify famous scientists and other academic luminaries over the past 500 years. He considered those most referenced as those who essentially leapfrogged over their contemporaries in terms of discovery, innovation, and philosophy and identified their birth orders by bibliographic and other archival sources. He observed that although first-borns are more likely than later-borns to become scientists, the latter group more likely spearheaded transformative ways of thinking. Nicolaus Copernicus, Charles Darwin, and Galileo Galilei, all later-born children, exemplify Sulloway's reasoning. Although his study is often criticized for methodological flaws, it pushes the link between birth order and intelligence back to the front burner. Whether future research will corroborate Sulloway's thesis is uncertain, but his bestselling book almost single-handedly reignited the flame of interest in it.

sibling group composed of same sex or opposite sex siblings. These sibling group characteristics mold the conditions under which children and their siblings' lives develop.

PERSPECTIVES ON CHILD DEVELOPMENT

There are two well-known rival views of how the sibling group as a set affects child socialization. The first is Robert Zajonc and Gregory Markus's (1975) confluence model, which maintains that firstborn and children from small families excel in academic outcomes compared to their counterparts. It purports that a child's intelligence is affected by the dynamic intellectual climate to which he or she is exposed during maturation. The confluence model explains the joint consequences of the number of siblings and birth order on intelligence. The model, moreover, applies to the effects of child spacing, sex composition, and parental absence. At the time the

model was fashionable, conventional wisdom was that children in large families and at later birth orders were at an intellectual disadvantage relative to their counterparts. Follow this logic: Suppose that one considers just raw intelligence, not adjusted for age (the standard IQ test). The intellectual atmosphere is the average of the intellectual levels of all family members divided by the number of people in the household. Thus, as the sibling group expands in the number of children (ordinarily less intellectually stimulating), the weaker the intellectual climate to which any given child is exposed. Note also that child spacing should logically matter. The further children are spaced, the greater the odds that both an older child and a much younger newborn sibling experience a similar intellectually stimulating environment.

Even sex composition of the family can be tied to intellectual development, given sex differences. Boys traditionally outperform girls in math and girls outperform boys in English. Families populated with more boys should accordingly benefit from higher intellectual stimulation in the subjects of math and science whereas the presence of girls should enhance reading and language arts (Steelman, 1985). Despite its elegance, the confluence model has not received consistent empirical support and indeed is quite often disconfirmed. Nonetheless, to date no one has provided a definitive test of the model.

Another perspective, the resource dilution model, is embraced by many social scientists for its elegance (Downey, 1995, 2001). Put simply, the amount of resources available to any child is diminished by the number of siblings with whom he or she must compete. Resources cover a gamut of things that children may require to advance academically and in other realms. They encompass social capital or the extent to which parents spend time with children and have ties with important figures in the school system. Financial resources are very important, especially for adolescents who may require parental assistance as they plan to go to college. Educational materials such as encyclopedias or computers in the home are also pivotal. Finally, the educational attainment of parents and siblings (their human capital) may affect a developing child. The greater the number of siblings, the less resources that parents can divest to their children, whether they are social, intellectual, cultural, or economic. Under resource dilution principles, birth order does not make a difference but the number of siblings and the spacing between them does. As the family grows by adding children, there should be fewer resources to allocate to any given child. If parents space children widely, then they have time to recover from financial and other obligations before the next child arrives.

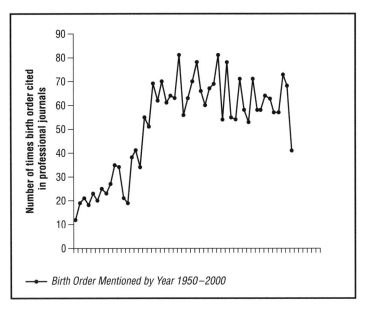

Table 1. *Scholarly research that includes birth order.* CENGAGE LEARNING, GALE.

Despite vast evidence to support the resource dilution hypothesis, scholars remain uncertain about whether or not children's development is negatively affected under all conditions. Whereas the number of siblings is usually found to be a detriment to children and adolescents when it comes to educational progress, there are noteworthy exceptions in which family size has either no impact or a positive one (Guo & Van Wey, 1999; Steelman et al., 2002). Under conditions in which families have abundant resources, the size of the sibling group may not count so much. Nonetheless, the majority of studies suggest that sibship size dampens educational progress. On other outcomes such as teamwork and sociability that remain less studied, the number of siblings may be beneficial (Steelman et al., 2002). Indeed the literature on siblings is often critiqued for its overwhelming concentration on just those outcomes that typically favor small families over large ones, such as educational advancement. It might be more than coincidence that as the birth rate has declined, the proportion of students going to college has generally increased.

Whether one takes a confluence model or a resource dilution perspective, as the number of siblings increases the amount of educationally lucrative resources that any given child can depend on relative to their counterparts seemingly decreases after other interrelated variables are statistically controlled (Downey, 1995; Steelman, 1985). Correspondingly, the number of siblings is almost invariably considered in attempts to understand where and when children and adolescents are privileged or are shortchanged as they mature. It is noteworthy that the number

of siblings has become a standard factor included in models of status attainment. Such models attempt to understand how family dynamics over time work to either encourage or discourage educational and occupational successes.

Birth order intersects with the interests of life course scholars because firstborn children are the sole ones who at least for some interval enjoy uninterrupted time with parents and, as such, may or may not profit from their position in the age rankings (see Conley, 2004). Correspondingly, later-born children more likely arrive at the time in the life course when parents reach their earnings peak and, because they are last, may profit from mentoring by older siblings.

Another sibling structure variable that meshes nicely with resource dilution and confluence model is child spacing. Brian Powell and Lala Carr Steelman (1993) observe that the more closely spaced consecutively born children are, the less likely parents can prepare for college and other advantages to pass on to their children. Nonetheless, there also may be payoffs to close spacing, such as greater opportunities to interact with brothers and sisters as age mates in ways that in no small measure shape interests, hobbies, and who children pick as friends. Siblings may also form coalitions with each other to challenge parental authority. Nonetheless, a widely spaced child may profit in a different way. Depending on birth order position, large age differences between siblings may set the stage for either giving or receiving help in school and other activities. As an example, the older sibling may help mentor the younger sibling in school activities and in so doing learn more himself or herself. Additionally, wider spacing allows parents to recover from financing older children's training just in time to assist their younger children.

Not surprisingly, many contemporary child advice books actually encourage families to consider such spacing strategies to benefit their children. Given the constraints of the childbearing years for women and the often accidental pregnancy, parents surely cannot completely control when they have children how their children will be spaced, nor can parents control the genders of the children they bear. All male or all female sibling groups may experience a lifelong difference in the way that parents rear them—the former focusing more on "masculine" domains, the latter on "feminine" ones. Still, there is no consistent support for any gender composition effect, and scholars remain baffled by these discrepancies observed across a number of studies (Conley, 2000; Steelman et al., 2002).

Notably, siblings may matter for different reasons as they mature. Social scientists have begun to broaden research on the sibling group to include alternatives to the traditional nuclear family. Specifically, scholars in the 21st century are concentrating on the addition or subtraction of biological siblings, half-siblings, step-siblings, or a combination of these that alter the overall character of the sibling group over time (Tillman, 2007; Zill, 1996). To a great extent the child's experiences are linked both to their siblings and their parents, who decide or accidentally are responsible for the structure of the sibling group. Birth order continues to tantalize laypeople despite the fact that many claims of early researchers that firstborns are superior to their later-born counterparts in terms of intellect have been seriously challenged. Most commonly, but not always, observed is that as the number of siblings increases, educational and occupational opportunities decline. So studies of educational and occupational attainment routinely include the number of siblings as a predictor but are much less likely to include birth order.

FUTURE RESEARCH

Research about siblings should and probably will move in many exciting and needed directions. For example, scholars need to collect direct behavioral observations of siblings as they interact. Such a study clearly would animate if and how siblings make a difference in each other's lives. A major flaw in existing sibling influence research is the lack of data from whole families. Whereas the earliest studies of siblings more or less presumed the presence of both biological parents, the need to expand research to encompass alternatives family forms that in turn affect the character of the sibling group is now acknowledged.

Step-, half-siblings, or siblings that may be in the custody of another parent may have different effects than do siblings who always live with one another (McLanahan & Sandefur, 1994; Zill, 1996). Across the life course young children and adolescents may experience the entrance or exit of siblings as parents divorce and remarry at high rates. To comprehend family dynamics requires investigating families across time.

Note also that the way in which children are arranged may infringe on the extent and manner in which parents may invest in and treat children. In contrast to living with both biological parents, children who abide with single parents, stepparents, and cohabiting partners are becoming more common. These variations also affect the character of the sibling group and pose a daunting challenge to researchers concerned with understanding sibling influence. Children who do not grow up with both biological parents are much more likely to live with step- and/or half-siblings and to experience fluctuations in the siblings with whom they reside. Consequently, contemporary research about the impact of siblings ordinarily must take into consideration types of families in which sibling groups are embedded. Co-resident siblings, for example, tend to take a negative toll on academic achievement during adolescence and early adulthood (Blake, 1989; Downey, 1995, 2001; Steelman et al., 2002) as does close spacing between siblings. It is unclear if non-coresident siblings have a similar impact. Young children may have to share parental attention with closely born siblings, often found in blended families, so much so that it has deleterious consequences for them. As adolescents approach independence, how economic resources are distributed in families with multiple types of siblings may even more profoundly affect educational expectations than is ordinarily observed in traditional nuclear families (Steelman et al., 2002).

In the early 21st century, scholars have begun to recognize how family change in the sibling group may have various consequences for children across the life cycle. A sibling can mean many different things in contemporary society. It is a status that may or may not include residential and nonresidential half- and step-siblings, as well as other relatives reared as veritable siblings such as cousins, foster children, or even the children of older siblings. All of these represent outstanding challenges to define what a sibling is. Grappling with this issue, although not easy, is a necessary step in future research.

Another direction involves evolutionary theory, which urges scholars to view siblings through a different prism. These theories posit that although siblings have a genetic connection with one another (they share 50% of their genes), their greatest gene share is with oneself. Proponents of an evolutionary psychology position would argue that genetic connectedness prods children to protect their siblings against external threats. Concern for the welfare of brothers and sisters emanates from the innate tendency to protect one's genetic heritage. Paradoxically, it also accounts for sibling rivalry. The genes that a person most wants to protect are his or her own. Hence siblings compete with each other to garner familial resources to advance an even larger portion (100%) of their genes into the next generation. Evolutionary theories also explain why biological siblings should be more vested than step-siblings (whose genes do not coincide with the other sibling) and why identical twins are the most protective of each other among the various types of siblings.

Future research on siblings must also move away from reliance on snapshot data of children taken at one point in time toward the examination of the role of siblings as they interact across the life span. So far sibling research is just not complete enough to capture the full essence of family life. Moreover, the need to collect and analyze data on whole families over the life cycle cannot be understated. More studies that analyze longitudinal data are warranted and may help reconcile the many inconsistencies already documented. Through these techniques researchers can assess the growing family forms and understand changes in the sibling group that occur through divorce, remarriage, and single-parent families. Finally, sibling research should devote more attention to outcomes other than those such as educational success that typically favor small families. The extent to which children cooperate and share, for example, could be a by-product of maturation in a large family. If scholars move in the recommended directions reviewed above, then a greater appreciation of the extent to which siblings affect one another across the life cycle could become a reality.

SEE ALSO Volume 1: *Identity Development; Family and Household Structure, Childhood and Adolescence; Parent-Child Relationships, Childhood and Adolescence;* Volume 2: *Sibling Relationships, Adulthood;* Volume 3: *Sibling Relationships, Later Life.*

BIBLIOGRAPHY

Blake, J. (1989). *Family size and achievement.* Berkeley: University of California Press.

Cicirelli, V. G. (1995). *Sibling relationships across the life span.* New York: Plenum Press.

Conley, D. (2000). Sibship sex composition: Effects on educational attainment. *Social Science Research, 29,* 441–457.

Conley, D. (2004). *The pecking order: Which siblings succeed and why.* New York: Pantheon.

Downey, D. B. (1995). When bigger is not better: Family size, parental resources and children's educational performance. *American Sociological Review, 60,* 746–761.

Downey, D. B. (2001). Number of siblings and intellectual development: The resource dilution explanation. *American Psychologist, 56,* 497–504.

Dunn, J. (1985). *Sisters and brothers (the developing child).* Cambridge, MA: Harvard University Press.

Galton, F. (1874). *English men of science.* London: Macmillan.

Guo, G., & Van Wey, L. K. (1999). Sibship size and intellectual development: Is the relationship causal? *American Sociological Review, 64,* 169–187.

McLanahan, S. S., & Sandefur, G. (1994). *Growing up with a single parent: What hurts, what helps.* Cambridge, MA: Harvard University Press.

Powell, B., & Steelman, L. C. (1993). The educational benefits of being spaced out: Sibship density and educational progress. *American Sociological Review, 58,* 367–381.

Steelman, L. C. (1985). A tale of two variables: A review of the intellectual consequences of sibship size and birth order. *Review of Educational Research, 55,* 353–386.

Steelman, L. C., Powell, B., Werum, R., & Carter, S. (2002). Reconsidering the effects of sibling configuration: Recent advances and challenges. *Annual Review of Sociology, 28,* 243–269.

Tillman, K. H. (2007). Family structure pathways and academic disadvantage among adolescents in stepfamilies. *Sociological Inquiry, 77,* 383–424.

Zajonc, R. B., & Markus, G. B. (1975). Birth order and intellectual development. *Psychological Review, 82,* 74–88.

Zill, N. (1996). Family change and student achievement: What we have learned, what it means for schools. In A. Booth & J. Dunn (Eds.), *Family-school links: how do they affect educational outcomes?* Mahwah, NJ: Lawrence Erlbaum Associates.

Lala Carr Steelman
Pamela Ray Koch

SINGLE PARENT FAMILIES

SEE Volume 1: *Family and Household Structure, Childhood and Adolescence.*

SOCIAL CAPITAL

Social capital refers to social relationships that have the capacity to enhance the achievement of one's goals. Social relationships are viewed as investments, whether the investment is conscious or unconscious. The fundamental insight of the idea of social capital is that life chances are influenced by the social resources available through social networks. This fundamental insight, coupled with the generality of the definition, has contributed to social capital's widespread appeal across a variety of research areas, including research on child and adolescent development. Social capital is believed to aid in the achievement of both individual goals, such as educational attainment and achievement, and collective goals, such as safe neighborhoods that promote healthy psychosocial development.

ORIGINS AND CONCEPTUAL DEVELOPMENT OF THE SOCIAL CAPITAL

The theoretical roots of social capital can be traced to the work of sociologists Pierre Bourdieu (1930–2002) and James Coleman (1926–1995) during the late 1980s. Bourdieu (1986) introduced social capital as a part of his larger project aimed at understanding the social reproduction of inequality, or how social class is passed from one generation to the next. He defined social capital as "the aggregate of the actual or potential resources which are linked to the possession of a durable network of more or less institutionalized relationships of mutual acquaintance and recognition" (1986, p. 248). The amount of social capital to which an individual has access depends on both the quantity of his or her connections and the amount of capital (e.g., financial, human, or cultural) that each network member possesses. Hence, membership in various social groups (such as gender, race, and social class) provides differential access to social capital. Bourdieu portrayed the accumulation of social capital as the result of conscious and unconscious long-term investment strategies designed to establish or maintain relationships of perceived obligations that can be accessed on some future occasion. The solidarity and resources provided by the social capital of the dominant class enable it to maintain its position.

Coleman's purpose in introducing social capital was to bridge the gap between sociological and economic explanations of social action by showing how explanations based on rational action could fit into a framework that also emphasized the importance of social context. Coleman wrote that social capital is a "variety of entities with two elements in common: They all consist of some aspect of social structures, and they facilitate certain actions of actors—whether persons or corporate actors—within structures" (1988, p. S98).

According to Coleman, social relationships can provide three major types of resources. The first, obligations and expectations, occurs when a person can count on their favors being returned. The reciprocity that allows one to collect "credit slips" for future use depends on the trustworthiness of the social environment. Information is the second type of resource. For instance, social relations are important sources of information about job openings and how to apply to college. The third type of resource is

norms and effective sanctions. Coleman argued, for example, that strong and enforceable pro-school norms within a community facilitate the job of schools and teachers. A social network with a high level of closure, when a group's members are highly interconnected and not very connected to persons outside of the group, can more effectively maintain compliance to norms because members can act collectively to restrain undesirable behavior. In the context of child development, Coleman argued that *intergenerational closure*, when parents know the parents of their children's friends, is particularly important. Because intergenerational closure enhances communication among parents, it facilitates norm enforcement and parental control over their children's behavior.

Bourdieu and Coleman's definitions share important similarities. Both identify social capital as the resources that exist in social relationships rather than as something that is tangibly possessed by individuals. In addition, in both definitions it is clear that investment in interpersonal relationships is what creates social capital. However, the Bourdieu and Coleman formulations differ in important ways. Consistent with their different purposes in introducing the concept, they diverged on the phenomena that they used social capital to explain. Bourdieu used it to explain the reproduction of inequality. In contrast, Coleman used the concept in a more general sense to explain a wider range of outcomes, from dropping out of high school to the successful organization of social movements, and did not emphasize the nature of groups that act in ways to concentrate resources within their boundaries. Also, while both theorists viewed social capital as something that can exist in collectivities, there is a crucial difference in their formulations at this level. In his discussion of social capital at the group level, Bourdieu refers to homogeneous groups that have a unified set of interests. Coleman's discussion suggests that social capital can serve as a public good in large groups, such as neighborhoods, that have a number of subgroups within them. His work on social capital did not consider inequality in access to social capital within such aggregates.

While Bourdieu's and Coleman's work on social capital has been highly influential, it was political scientist Robert Putnam's (b. 1941) work on social capital that galvanized its appeal across almost every social scientific discipline and even into popular discourse. Putnam drew from Coleman's work on the nature of social capital as a public good and defined social capital as "features of social organization, such as trust, norms and networks, that can improve the efficiency of society by facilitating coordinated actions" (1993, p. 167). He argued, for example, that regional levels of social capital account for why the identical local government structure functions so much better in the regions of northern Italy than in

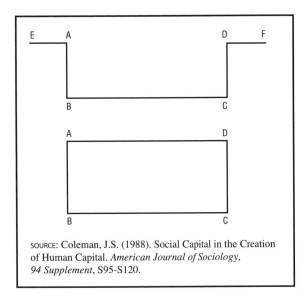

SOURCE: Coleman, J.S. (1988). Social Capital in the Creation of Human Capital. *American Journal of Sociology*, 94 Supplement, S95–S120.

Intergenerational Closure. *Network involving parents (A, D) and children (B, C) without (top) and with (bottom) intergenerational closure.* **CENGAGE LEARNING, GALE.**

the regions of southern Italy (1993). He has also attributed the decline in American political engagement, as reflected by voter turnout, to a decline in social capital (2000).

Coleman and Putnam's definitions of social capital are so broad that they have resulted in a proliferation of social phenomena labeled as social capital, leading some commentators—Charles Kadushin (2004) and Alejandro Portes (1998), for example—to question the utility of the concept. Despite this concern, the use of social capital continues to flourish. As Putnam (2000) notes, social capital is not just one thing, much like physical capital is not just one thing. It is useful to consider that there are different forms of social capital without extending the "notion of social capital beyond its theoretical roots in social relations and social networks" (Lin, 2001, p. 28). While there is no shortage of alternative definitions of social capital, research on child and adolescent development has tended to follow Coleman's framework by focusing on norm enhancing social relationships or, to a lesser extent, Bourdieu's ideas by focusing on the role of social capital in social reproduction.

RESEARCH ON SOCIAL CAPITAL AND CHILD AND ADOLESCENT DEVELOPMENT

Multiple social contexts, including the family, friendship network, school, and community, are important in the lives of children; as such, adolescents and most social capital research on children's outcomes has focused on one or more of these contexts. Following Coleman, early

research on social capital and child development focused on parent-child relationships. With some exceptions, research generally finds that living in a two-parent family, having fewer siblings, parental monitoring, discussion of school with parents, and parental expectations that that their child will go to college are associated with lower chances of dropping out of high school, higher chances of attending college, higher grades, and higher achievement on test scores. (For a comprehensive review of social capital studies in the area of education through 2001, see Sandra Dika's and Kusum Singh's *Applications of Social Capital in Educational Literature* [2002].) Research on the effects of parent involvement in school has found inconsistent results; some studies find positive associations with school achievement, whereas others find negative associations, likely because problems in school often prompt parent involvement.

Social capital research has also examined other important relationships. Friends can influence each other's academic outcomes in several ways. First, they act as agents of socialization through which pro-school norms can be encouraged and enforced, which can translate into more successful academic outcomes. Having academically successful friends provides an adolescent with access to educational resources. In addition, friends can act as role models. R. Crosnoe, S. Cavanagh, and G. H. Elder. (2003) observed that having friends with a higher grade point average and higher school attachment decreases having trouble at school.

Still other research has focused on school and community social capital. R. D. Stanton-Salazar and S. M. Dornbusch (1995) argued that schools can provide important social resources for low-income students by providing guidance in the college application process, for example. Furthermore, a good relationship between the family and the school enhances the likelihood of obtaining valuable resources at the school. Y. M. Sun (1999) found that in communities with higher levels of adolescent participation in religious activities and sports groups and where parents belong to more organizations, standardized tests scores were higher. In addition, higher community levels of intergenerational closure and being able to work together were related to higher achievement on test scores. Not only did these forms of community social capital enhance educational achievement, but they also explained a part of the association between neighborhood poverty and lower educational achievement.

Although most examinations of social capital concentrate on its positive effects, it can easily have negative consequences as well. Coleman (1988) himself recognized that a form of social capital that is useful for one desired end might be disadvantageous to the achievement of another goal. For example, community norms that discourage delinquency might also hinder creativity. In addition, not all social networks are characterized by pro-social norms. Moreover, some relationships might not provide helpful social support. For example, in a sample of African-American parents, M. O. Caughy, P. J. O'Campo, and C. Muntaner (2003) found that in poor neighborhoods, knowing many neighbors was associated with more behavioral problems among their children. In nonpoor neighborhoods, having neighborhood ties was associated with fewer child behavioral problems. Similarly, in a study of poor families, F. F. Furstenberg (1993) found that in the most disadvantaged and disordered neighborhoods, avoiding neighbors and activating ties to resources outside of the neighborhood proved to be a more successful parenting strategy than forging close ties with neighbors.

An important insight to take from these findings is that the value of social capital may depend on the situation. For example, is social capital more important and therefore effective for disadvantaged youths than for middle-class and affluent youths because it may offset the lack of other types of resources? Or, does social capital exacerbate inequality by being more effective for already advantaged youths? In one study, parental access to time and financial assistance from friends was associated with higher educational attainment for the offspring of high-income families, but not low-income families, suggesting that lower-income families are less able to access help that provided educational resources (Hofferth, Boisjoly, & Duncan, 1998). D. H. Kim and B. S. Schneider (2005) found the effects of college visits and matching educational expectations by the student and parent on attendance at a selective 4-year college or university were larger when parents had higher levels of education. In contrast, parental participation in college and financial aid programs at the high school pays off more for children of lower educated parents, which is consistent with Stanton-Salazar's and Dornbusch's (1995) argument that school resources are particularly important for working and lower-class families. Sun (1999) found that the negative effects of high levels of nontraditional families in the community on school achievement can be partially offset by high levels of parent participation in organizations and intergenerational closure, suggesting that community social capital can mitigate disadvantageous neighborhood circumstances. Future research will undoubtedly continue to examine the contexts in which social capital is more and less useful.

GAPS IN CURRENT KNOWLEDGE

Two decades of research on social capital have produced significant and important findings, but there is still more to learn. First, much of the research on social capital and child development has focused on educational outcomes. With some notable exceptions, relatively less is known about whether and how social capital influences outcomes such as social adjustment, behavioral problems, violence, and psychosocial resources such as an internal locus of control. S. De Coster, K. Heimer, and S. M. Wittrock (2006) found that intergenerational closure does not appear to curtail delinquent behavior. This is surprising in light of the association of intergenerational closure with a host of educational outcomes that is purportedly explained by the ability of parents to enforce norms, and, accordingly, more research is needed to understand the mechanisms that explain the relationship between closure and school outcomes.

Second, there is more to learn about the relative advantages of *bonding* versus *bridging* social capital. Bonding social capital is characterized by networks rich in strong and densely connected ties, usually of socially similar persons, whereas bridging social capital is characterized by dispersed networks that access disparate social worlds. In the context of education, bonding social capital might be beneficial because it is conducive to norm enforcement and the social support that adolescents need to be successful. Yet as S. L. Morgan and A. B. Sørensen (1999) argue, bridging social capital might be beneficial because it accesses diverse pools of information that create greater opportunities for learning. Rather than one form of social capital being generally more advantageous than the other, it may be the case that bonding social capital is more important for some types of educational outcomes, such as grades and school retention, whereas bridging social capital may be more advantageous to outcomes such as going to college because it is more likely to provide beneficial information. Thus far, research has not been able to adequately assess this debate.

Finally, as the field moves forward, measures will continue to be refined in ways that more closely tap the conceptual definitions. Certainly, much progress has already been made. Early research on social capital relied heavily on proxies, or very indirect and approximate measures, for social capital such as two-parent families, number of siblings, and number of residential moves. While these factors may influence the availability of family and community social capital, there are other factors, such as unmeasured aspects of socioeconomic status or family background, that could account for their relationships with better child outcomes. Many of the later studies on social capital and adolescent outcomes were based on data from the National Educational Longitudinal Study, a survey that includes several somewhat better measures of social capital, but is still limited in its information about social networks. A more recent survey, the National Longitudinal Study of Adolescent Health, has made it possible to measure the content and structure of school-based adolescent friendship networks.

However, there are still aspects of social capital that cannot be adequately measured with a widely available survey. For example, no survey has collected information about adolescent relationships outside of the immediate family and the school. In addition, the parental social networks are likely quite important in terms of the information they can provide about parenting techniques and the educational system, for example, and yet little information exists on them. Collecting more detailed and direct measures of social capital will allow researchers to address important issues, such as the relative benefits of bridging and bonding social capital, as well as to better understand the contexts in which various forms of social capital are more or less useful.

SEE ALSO Volume 1: *Coleman, James; Cultural Capital; Data Sources, Childhood and Adolescence; Friendship, Childhood and Adolescence; Human Capital; Intergenerational Closure; Mentoring; Parental Involvement in Education; Peer Groups and Crowds; Socialization;* Volume 2: *Parent-Child Relationships, Adulthood.*

BIBLIOGRAPHY

Bourdieu, P. (1986). The forms of capital. In J. G. Richardson (Ed.), *Handbook of theory and research for the sociology of education* (pp. 241–258). New York: Greenwood Press.

Caughy, M. O., O'Campo, P. J., & Muntaner, C. (2003). When being alone might be better: Neighborhood poverty, social capital, and child mental health. *Social Science & Medicine, 57*(2), 227–237.

Coleman, J. S. (1988). Social capital in the creation of human capital [Suppl.]. *American Journal of Sociology, 94*, S95–S120.

Crosnoe, R., Cavanagh, S., & Elder, G. H. (2003). Adolescent friendships as academic resources: The intersection of friendship, race, and school disadvantage. *Sociological Perspectives, 46*(3), 331–352.

De Coster, S., Heimer, K., & Wittrock, S. M. (2006). Neighborhood disadvantage, social capital, street context, and youth violence. *Sociological Quarterly, 47*(4), 723–753.

Dika, S. L., & Singh, K. (2002). Applications of social capital in educational literature: A critical synthesis. *Review of Educational Research, 72*(1), 31–60.

Furstenberg, F. F., Jr. (1993). How families manage risk and opportunity in dangerous neighborhoods. In W. J. Wilson (Ed.), *Sociology and the public agenda* (pp. 231–258). Newbury Park, CA: Sage.

Hofferth, S. L., Boisjoly, J., & Duncan, G. J. (1998). Parents' extrafamilial resources and children's school attainment. *Sociology of Education, 71*(3), 246–268.

Kadushin, C. (2004). Too much investment in social capital? *Social Networks, 26*(1), 75–90.

Kim, D. H., & Schneider, B. S. (2005). Social capital in action: Alignment of parental support in adolescents' transition to postsecondary education. *Social Forces, 84*(2), 1181–1206.

Lin, N. (2001). *Social capital: A theory of social structure and action.* Cambridge, UK: Cambridge University Press.

Morgan, S. L., & Sørensen, A. B. (1999). Parental networks, social closure, and mathematics learning: A test of Coleman's social capital explanation of school effects. *American Sociological Review, 64*(5), 661–681.

Portes, A. (1998). Social capital: Its origins and applications in modern sociology. *Annual Review of Sociology, 24*(1), 1–24.

Portes, A., & Landolt, P. (1996). The downside of social capital. *The American Prospect, 94*(26), 18–21.

Putnam, R. D. (1993). *Making democracy work: Civic traditions in modern Italy.* Princeton, NJ: Princeton University Press.

Putnam, R. D. (2000). *Bowling alone: The collapse and revival of American community.* New York: Simon & Schuster.

Stanton-Salazar, R. D., & Dornbusch, S. M. (1995). Social capital and the reproduction of inequality: Information networks among Mexican-origin high school students. *Sociology of Education, 68*(2), 116–135.

Sun, Y. M. (1999). The contextual effects of community social capital on academic performance. *Social Science Research, 28*(4), 403–426.

Jennifer L. Glanville

SOCIAL DEVELOPMENT

Peer relations are a critical aspect of youngsters' social development (Hartup, 1996). Children and adolescents place considerable importance on their peer relations and are influenced strongly by the attitudes and behaviors of their peers. Peers also provide a source of emotional support that rivals and, during adolescence, even exceeds support from parents.

Studying the development of peer relations in childhood and adolescence is essential for understanding the foundations of youths' friendships and romantic relationships into adulthood and across the lifespan. Children and adolescents who encounter serious difficulties in their peer relations early on are at risk for the development of later interpersonal and psychological difficulties, including delinquency, marital problems, substance use, and depression. Thus, efforts to understand and promote positive peer relations in youth is an important mental health goal. This entry describes the types of peer relations that are important for children and adolescents. Theory and measurement issues also are described.

CHILDREN'S PEER RELATIONS

During elementary school, children typically spend most of the day in self-contained classrooms with a specific group of classmates. In this context, children's peer status—or degree of acceptance from the peer group (i.e., classmates)—is a salient aspect of their peer relations. Peer acceptance provides children with a sense of belonging and inclusion.

Children's social status has been defined by the degree to which they are accepted by their peers. Specifically, social status groups are determined by ratings of peer acceptance (liking) and peer rejection (disliking) as follows: popular (high on liking, low on disliking), rejected (low on liking, high on disliking), neglected (low on liking and disliking), and controversial (high on liking and disliking). Popular children often have positive social skills and personal competencies, whereas rejected children display a number of interpersonal, emotional, and academic difficulties, such as aggressive, disruptive, and inattentive behaviors and feelings of loneliness (Coie, Dodge, & Kupersmidt, 1990).

In addition to peer acceptance, peer friendships (close supportive ties with one or more peers) provide children with a sense of intimacy, companionship, and self-esteem. Children's close friendships typically are with same-sex peers and vary in terms of quantity and quality. Most children have a close friend in school, although girls are more likely to have a best friend than are boys and to have more peers whom they rate as close friends. Common positive qualities of children's friendships include sharing activities, affection, receiving help, trust, and sharing. Intimacy and emotional support also are evident in children's friendships and increase during adolescence (La Greca & Prinstein, 1999). Friendships also have negative aspects, such as conflict and betrayal, although most children report more positive than negative qualities in their close friendships.

ADOLESCENTS' PEER RELATIONS

As children make the transition to middle school, their peer networks expand considerably. Most adolescents have a peer network that includes their best friends, other close friends, larger friendship groups or cliques, peer crowds, and often romantic relationships (Furman, 1989; Urberg, Degirmencioglu, Tolson, & Halliday-Scher, 1995).

Peer Crowds Adolescents interact with a large number of peers and need to find their place in the larger social system. In this context peer crowds càn be viewed as an outgrowth of the social status groups observed among children (La Greca & Prinstein, 1999). Peer crowds are distinct from smaller friendship cliques in that they are much larger, and peer crowd members may not know or be friends with one another.

Peer crowds reflect adolescents' peer status and reputation as well as the primary attitudes and behaviors by which adolescents are known to their peers. Although peer crowds and their labels vary, the most common crowds are jocks (or athletes), populars (or elites, preps, hot shots), brains (or nerds), burnouts (or druggies, dirts), alternatives (or nonconformists, goths, freaks), loners, and special-interest groups (e.g., dance, music; Brown, 1990; La Greca, Prinstein, & Fetter, 2001).

Peer crowds provide opportunities for social activities, friendships, and romantic relationships and provide a sense of belonging and identity. This is particularly important in middle school because peer crowds may help younger adolescents define their identity and reputation and provide norms for behavior and "fitting in." The importance of peer crowds peaks in high school, followed by a decline as adolescents focus their attention on close friends and romantic relationships. At the same time, peer crowds are evident among college-age youth and gay men, suggesting that crowd membership can be salient even into young adulthood (Urberg et al., 1995).

Friendships Adolescents spend more time talking to peers than they spend in any other activity. In the current technological age, adolescents' use of mobile phones and text messaging to connect with friends is widespread. Adolescents' interactions with friends occur primarily in the context of dyadic interactions with close friends or romantic partners and in friendship cliques.

Cliques are friendship-based groupings that may vary in size (usually from five to eight members), density (the degree to which each member regards others in the clique as friends), and tightness (the degree to which cliques are open or closed to outsiders). Cliques are the primary basis for adolescents' interactions with peers, and an adolescent's best friends are likely to participate in the same clique or group (Urberg et al., 1995). As adolescents begin to date, same-sex cliques may transition to mixed-sex groups. Over time cliques often are replaced by adolescents' interactions with close friends, romantic partners, or smaller friendship-based groups (La Greca & Prinstein, 1999).

Adolescents' close friendships have many of the same qualities as children's friendships (companionship, trust, and so on) but are characterized by intimacy, that is, the sharing of personal, private thoughts and feelings and knowledge of intimate details about friends. As with children, adolescent girls report having more close friends than do boys and report more intimacy in those relationships. However, adolescent boys have more open friendship groups than girls do and are more willing than girls to let others join ongoing interactions. During adolescence having close "other-sex" friends also becomes common and may set the stage for the development of romantic relationships.

The quality of adolescents' close friendships is linked with psychological adaptation. For example, adolescents who report more positive qualities in their best friendships also report less social anxiety than do their peers. Moreover, high levels of negative interactions in adolescents' close friendships are related to greater social anxiety and depressive symptoms.

SOCIALIZATION VERSUS SELECTION

As discussed above, peer relations serve many important developmental functions, providing support, a sense of belonging, opportunities for social interaction, and a means of developing identity. The literature on peer relations is compatible with the theory of homophily. Homophily (Kandel, 1978) posits that individuals with similar interests and characteristics tend to cluster together and seek one another out (selection). Further, peers reward and reinforce similar attitudes and behaviors among group members (socialization).

Children often gravitate toward others who are similar to themselves; this is consistent with selection theory. In fact, children select friends on the basis of similarities in observable characteristics such as sex, race, and preference for engaging in certain activities rather than on less observable factors such as personality, attitudes, and self-esteem. Friendships typically occur among youth of the same age and gender, although this changes in adolescence (Aboud & Mendelson, 1996).

Socialization is also important in children's peer relationships. For example, low-achieving 5th-, 6th-, and 7th-graders who affiliate with higher-achieving friends perform better academically than do those who affiliate with low-achieving peers. Thus, although children gravitate toward similar peers to form friendships, those friendships also socialize children and influence their behaviors and development.

Selection and socialization processes also are relevant for adolescents (Aboud & Mendelson, 1996). For example, adolescents may affiliate with certain peer crowds because of perceived similarities of interests, behaviors, or skills, reflecting a selection process. Smart adolescents may affiliate with the brains; athletes may affiliate with the jocks. However, socialization also may occur as adolescents feel real or perceived pressure to conform to or comply with the attitudes and behaviors that are prevalent in their peer crowd. For example, adolescents affiliating with the burnouts may feel pressure to drink or smoke; adolescents affiliating with the brains may be encouraged to take academic work seriously. Research also suggests that adolescents' aspiration to belong to a particular crowd may lead them to engage in behaviors they believe to be typical of that crowd (Brown, 2004).

Selection and socialization processes also occur within adolescents' friendships. Like children, adolescents gravitate toward others who share similar qualities, interests, and personal characteristics, although selection does not fully explain the formation and maintenance of adolescents' friendships. Socialization processes in friendships are also important and may influence adolescents' academic achievement. Although selection and socialization processes are difficult to disentangle, they provide a framework for understanding the importance of adolescents' friendships and peer crowds.

MEASUREMENT ISSUES

Peer relationships are challenging to study because they can be temporary and dynamic. Longitudinal studies are particularly revealing because they may capture the shifting nature of peer relationships and help clarify selection and socialization processes. Overall, self-report measures are used widely to capture youths' reports of close friendships and membership in a peer crowd; however, peer nominations, observations of group/dyadic interactions, and ethnographic observations also are important (Brown, 2004).

In assessing peer relationships, it is important to consider youngsters' ethnic/cultural background to understand the larger context in which relationships develop. Friendships may be defined and valued differently in different cultures; friendship selection also may vary by culture (Brown, 2004). In fact, research suggests that African American adolescents select friends on the basis of ethnic identity rather than academic orientation or substance use, whereas the opposite pattern may be prevalent for Asian and European American adolescents (Hamm, 2000). This may be because social concerns, such as racism, are more widely experienced by African American teens, making their ethnicity/race a more salient attribute for friendship selection than is the case for other adolescents. As another example, the high value placed on academic achievement in the Asian American culture may foster adolescent friendships that develop around academic pursuits.

Peer nominations are the most common method of assessing children's peer acceptance and friendships (Coie et al., 1990). Peer nominations require each child within a classroom or group to nominate up to three peers that he or she "likes the most" and "dislikes the most" (or "likes the least"), used for evaluating peer acceptance/rejection, or to nominate up to three "best friends," used for evaluating mutual friendships. Those nominations are used to categorize children into popular, rejected, neglected, and controversial social status groups to identify mutual friendships. Teachers' and parents' reports of peer acceptance and rejection may be useful when peer nominations are not possible. Peer networks, or mapping out patterns of relationships within the context of a school or other setting, are also commonly used in research to examine how adolescents form friendships and romantic relationships. Examining these networks facilitates the understanding of how adolescents transmit ideas, values, and also disease among their peers. In addition, self-report measures such as the Friendship Quality Questionnaire (Parker & Asher, 1993) have been used to evaluate the number and quality of children's close friendships. Overall, it is best to obtain several measures by using multiple approaches.

Although peer nominations are used widely with children, they are employed less with adolescents, whose peer networks are larger and are not limited to a particular school or classroom. Further, it is important to assess multiple relationships among adolescents, such as close same-sex and other-sex friendships and romantic relationships. Self-report measures such as the Network of Relationships Inventory (Furman & Buhrmester, 1985) have been used to assess the number and quality of adolescents' friendships and romantic relationships.

Although some investigators have obtained peers' perceptions of crowd affiliations (Sussman et al., 1990), self-report measures are used most commonly to assess adolescents' peer crowd affiliations. Measures such as the Peer Crowd Questionnaire (La Greca & Harrison, 2005; La Greca et al., 2001) describe and label each of the common peer crowds (e.g., jocks, populars, brains) and ask adolescents if those crowds are present in their school and, if so, what they are called. Adolescents then select the crowd they identify with most closely. Adolescents also may rate their degree of identification with each peer crowd (using a Likert scale) to obtain scores for multiple peer crowds. Sometimes additional items may be added, depending on the particular research questions. For example, adolescents have been asked how peers would classify them, which peer crowds they would like to belong to most, and the peer crowd affiliations of their three best friends or romantic partner. Overall, measures assessing peer crowd affiliation have been found to be reliable, have good construct validity and interrater reliability, and show good correspondence between adolescents' self-identification and peers' assignment to crowds.

FUTURE RESEARCH

Peer relations play a significant role in the social and emotional development of children and adolescents and provide an important source of social support. The quality and type of friendships change as children develop into adolescents. Future research might address cultural variations in peer relationships as well as the role of changing technology, such as the Internet and cell phones, on these relationships. Because friendships and peer relationships are so crucial for the social and emotional development of children and adolescents and have

implications for interpersonal functioning across the lifespan, efforts to understand and promote competence in youngsters' peer relationships should be an important target of developmental research, school and public policy, and mental health promotion and intervention.

SEE ALSO Volume 1: *Dating and Romantic Relationships, Childhood and Adolescence; Friendship, Childhood and Adolescence; Peer Groups and Crowds; Socialization.*

BIBLIOGRAPHY

Aboud, F. E., & Mendelson, M. J. (1996). Determinants of friendship selection and quality: Developmental perspectives. In W. M. Bukowski, A. F. Newcomb, & W. W. Hartup (Eds.), *The company they keep: Friendship in childhood and adolescence* (pp. 87–112). Cambridge, U.K.: Cambridge University Press.

Brown, B. B. (1990). Peer groups and peer cultures. In S. S. Feldman & G. R. Elliott (Eds.), *At the threshold: The developing adolescent* (pp. 171–196). Cambridge, MA: Harvard University Press.

Brown, B. B. (2004). Adolescents' relationships with peers. In R. M. Lerner & L. Steinberg (Eds.), *Handbook of adolescent psychology* (2nd ed., pp. 363–394). Hoboken, NJ: Wiley.

Coie, J. D., Dodge, K. A., & Kupersmidt, J. B. (1990). Peer group behavior and social status. In S. R. Asher & J. D. Coie (Eds.), *Peer rejection in childhood* (pp. 17–59). Cambridge, U.K.: Cambridge University Press.

Furman, W. (1989). The development of children's social networks. In D. Belle (Ed.), *Children's social networks and social supports* (pp. 151–172). New York: Wiley.

Furman, W., & Buhrmester, D. (1985). Children's perceptions of the personal relationships in their social networks. *Developmental Psychology, 21,* 1016–1022.

Hamm, J. V. (2000). Do birds of a feather flock together? The variable bases for African American, Asian American, and European American adolescents' selection of similar friends. *Developmental Psychology, 36*(2), 209–219.

Hartup, W. W. (1996). The company they keep: Friendships and their developmental significance. *Child Development, 67*(1), 1–13.

Kandel, D. B. (1978). Homophily, selection, and socialization in adolescent friendships. *American Journal of Sociology, 84,* 427–436.

La Greca, A. M., & Harrison, H. M. (2005). Adolescent peer relations, friendships, and romantic relationships: Do they predict social anxiety and depression? *Journal of Clinical Child & Adolescent Psychology, 34*(1), 49–61.

La Greca, A. M., & Prinstein, M.J. (1999). Peer group. In W. K. Silverman & T. H. Ollendick (Eds.), *Developmental issues in the clinical treatment of children* (pp. 171–198). Boston: Allyn & Bacon.

La Greca, A. M., Prinstein, M. J., & Fetter, M. D. (2001). Adolescent peer crowd affiliation: Linkages with health-risk behaviors and close friendships. *Journal of Pediatric Psychology, 26*(3), 131–143.

Parker, J. G., & Asher, S. R. (1993). Friendship and friendship quality in middle childhood: Links with peer group acceptance and feelings of loneliness and social dissatisfaction. *Developmental Psychology, 29*(4), 611–621.

Sussman, S., Dent, C. W., Stacy, A. W., Burciaga, C., Raynor, A., Turner, G. E., et al. (1990). Peer-group association and adolescent tobacco use. *Journal of Abnormal Psychology, 99*(4), 349–352.

Urberg, K. A., Degirmencioglu, S. M., Tolson, J. M., & Halliday-Scher, K. (1995). The structure of adolescent peer networks. *Developmental Psychology, 31*(4), 540–547.

Eleanor Race Mackey
Annette M. La Greca

SOCIAL INTEGRATION/ ISOLATION, CHILDHOOD AND ADOLESCENCE

SEE Volume 1: *Bullying and Peer Victimization; Friendship, Childhood and Adolescence; Peer Groups and Crowds.*

SOCIALIZATION

The concept of and approaches to socialization have a long history in sociology, psychology, and anthropology. Although life course sociologist Glen H. Elder Jr. (1994) has argued that socialization has declined as a research paradigm and become part of other paradigms, especially the life course perspective, theory and research on childhood and adolescent socialization in the 1980s and 1990s have revitalized the field. These new approaches stress the importance of the agency of children and youth. To say children have agency means they are active participants in their own socialization rather than being simply shaped or molded by adults. Such approaches also focus on how children collectively produce "peer cultures," which play a key role in their life trajectories and transitions as they become active members of adult culture. There has also been a convergence in theories of human development and socialization that is complementary to important theory and research on the life course (Corsaro & Fingerson, 2003).

THEORETICAL APPROACHES TO HUMAN DEVELOPMENT AND SOCIALIZATION

Theories of human development in psychology are primarily concerned with the individual child's acquisition of skills and knowledge and general adaptation to the environment. Sociologists, by contrast, normally use the term *socialization* when discussing human development. Their definitions of socialization highlight the ways in which the individual learns to fit into society. Some sociologists give more agency to children than others, but all emphasize interaction and collective processes when discussing socialization and argue that it is a life-long process. Gerald Handel, Spencer Cahill, and Frederick Elkin (2007) defined socialization as "*the processes by which we learn and adapt to the ways of a given society or social group so as to adequately participate in it*" (pp. 2–3).

In the 1980s and 1990s, important changes occurred in the conceptualization of human development and socialization in psychology and sociology. In general, these changes involved more of a focus on agency in the socialization process, more concern for the importance of social context, and agreement that experiences beyond the early years in the family (especially interactions and experiences with peers) are in need of careful theoretical development and empirical research.

RECENT TRENDS IN PSYCHOLOGICAL THEORIES OF HUMAN DEVELOPMENT

As noted earlier, psychologists use the term *human development* rather than *socialization*. Psychological theories of human development vary regarding (a) their perception of individuals as active or passive; (b) the importance they place on biological factors, the social environment, and social interaction; and (c) their conception of the nature of development or change. Three theoretical approaches have important implications for sociological approaches to socialization because they recognize the importance of the collective activities and processes of children and youth in social context and over time.

Cognitive Developmental Theory Since the early 1990s, work in cognitive development theory as centered around refinements and extensions of Jean Piaget's (1950) theory of intellectual development, which advocates an active view of the child who constructs his or her own place in the social and physical world. Several theorists argue that early interpretations of Piaget's work concentrate on the details of stages in cognitive development at the expense of an understanding of the theory they were intended to illustrate. Geoffrey Tesson and James Youniss (1995) argued that although Piaget described a series of stages of cognitive development he believed all children progress through, he did see these stages as the central focus of his theory. In his later work Piaget investigated the interrelationship between the logical and the social qualities of thinking. Tesson and Youniss argued that Piagetian intellectual operations enable children to make sense of the world as a set of possibilities for action, and thereby they can build a framework within which these possibilities may be envisioned. Thus Piaget attributes agency to children. Piaget also believed that children's peer relations were more conducive to the development of intellectual operations than the authoritative relations with adults, which primarily involved parental constraint over children, rather than a mutually influential and equal relationship.

Systems Theories of Human Development An excellent example of dynamic systems theory can be seen in the work of Esther Thelen and Linda Smith (1998), who criticized studies of human development that strive to discover invariants, that is, the programs, stages, structures, representations, schemas, and so on that underlie performance at different ages. Thelen and Smith argued that this approach uses the metaphor of a machine and that "knowledge is like the unchanging 'innards' of the machine, and performance subserves the more permanent structure" (p. 568).

Thelen and Smith (1998) offered instead the image of a mountain stream to capture the nature of development. They noted that there are patterns in a fast-moving mountain stream: Water flows smoothly in some places, but nearby there may be a small whirlpool or turbulent eddy, whereas in other parts of the stream there may be waves or spray. These patterns may occur for hours or even days, but after a storm or a long dry spell, new patterns may emerge. The mountain stream metaphor captures development as something formed or constructed by its own history and systemwide activity. In this approach there is a direct focus on processes, and outcomes are important primarily as part of further developing processes. The key strength of Thelen and Smith's systems approach is that it captures the complexity of real-life human behavior in physical, social, and cultural time and context.

Sociocultural Theories Sociocultural theorists refine and extend central concepts in the work of the Russian psychologist Lev Vygotsky (1978). According to Vygotsky, human activity is inherently mediational in that it is carried on with language and other cultural tools. A significant portion of children's everyday mediated activities take place in the zone of proximal development: "the distance between the actual developmental level as determined by independent problem solving and the level of potential development as determined through problem solving under adult guidance or in collaboration with more capable peers" (p. 86).

Building on Vygotsky's (1978) work, Barbara Rogoff (1996) argued that changes or transitions in children's lives can be best examined by asking how children's involvements in the activities of their community change, rather than by focusing on change as resulting from

individual activity. To capture the nature of children's involvements or changing participation in sociocultural activities, Rogoff suggested they be studied on three different planes of analysis: the community, the interpersonal, and the individual. In line with this view of change, Rogoff introduced the notion of "participatory appropriation" by which she means that "any event in the present is an extension of previous events and is directed toward goals that have not yet been accomplished" (p. 155). Thus, previous experiences in collectively produced and shared activities are not merely stored in memory as schema, plans, goals, and such and called up in the present; rather, the individual's previous participation contributes to and prepares or primes the event at hand by having prepared it.

Rogoff (1996) demonstrated this concept in a study of preadolescent girls' participation in the sale and delivery of Girl Scout cookies. The mothers of the girls initially managed and directed the interrelated tasks involved in this activity such as taking orders, calculating prices, and actually delivering the cookies. Over the course of time, however, the girls became more centrally involved and in some cases actually took over tasks (such as calculating prices) and played key roles (such as giving their mothers driving directions) as the mothers helped the girls deliver the cookies.

SOCIOLOGICAL THEORIES OF SOCIALIZATION AND THE SOCIOLOGY OF CHILDHOOD

As noted earlier, when discussing human development, sociologists normally use the term *socialization*. Since the 1980s, however, there has been a movement to refine or even replace this term in sociology because it has an individualistic and forward-looking connotation that is inescapable (Corsaro, 2005; Thorne, 1993). Some offer instead interpretive-reproductive theories that present a new sociology of children and childhood in which children's own cultures are the focus of research, not the adults they will become. These new approaches are refinements of earlier theoretical work on socialization in sociology.

Macro-Level Approaches to Socialization The major spokesperson of the macro-level view of socialization is the functionalist theorist Talcott Parsons, who envisioned society as an intricate network of interdependent and interpenetrating roles and consensual values. Parsons and Robert Bales (1955) likened the child to "a pebble 'thrown' by the fact of birth into the social 'pond'" (pp. 36–37). The initial point of entry—the family—feels the first effects of this "pebble," and as the child matures the effects are seen as a succession of widening waves that radiate to other parts of the social system. In a cyclical process of dealing with problems and through formal training to follow social norms, the child eventually internalizes the society. The influence of functionalist theorists has waned given their overconcentration on outcomes of socialization, deterministic views of society, and underestimation of the agency of social actors.

A recent and innovative macro-perspective of childhood can be seen in the work of Jens Qvortrup (1991), whose approach was based on three central assumptions: (a) Childhood constitutes a particular structural form, (b) childhood is exposed to the same societal forces as adulthood, and (c) children are themselves co-constructors of childhood and society. By childhood as a social form, Qvortrup means it is a category or a part of society such as social class, gender, and age groups. In this sense children are incumbents of their childhoods. Because childhood is interrelated with other structural categories, the structural arrangements of these categories and changes in these arrangements affect the nature of childhood. In modern societies, for example, changes in social structural arrangements of categories such as gender, work, family, and social class have resulted in many mothers working outside the home and their children both taking on more household work and spending more of their time in institutional settings, such as day-care centers and after-school programs, that did not exist in the past.

At a more intermediate level, analysis of socialization processes can be seen in work on social structure and personality and the life course. This work often escapes the deterministic nature of traditional macro-theories by documenting how specific features of social structure affect interaction in various contexts of socialization. For example, Elder (1994) argued that transitions in the life course are always embedded in trajectories that give them a distinct form and meaning. The life course approach, thus, overcomes the static nature of cross-sectional studies and captures the complexity of socialization across generations and key historical periods.

Interactionist Approaches to Socialization Interactionist approaches stem primarily from the social philosophy of George Herbert Mead (1934). Mead saw the genesis of self-consciousness as starting with the child's attempts to step outside him or herself by imitating others and reaching completion when the child, through participation in games with rules, acquires the ability to take on the organized social attitudes of the group. In Mead's stages in the genesis of self, however, children acquire more than a sense of self; they also appropriate conceptions of social structure and acquire a collective identity.

Surprisingly there has been little research by symbolic interactionists on early socialization. In one exception, Norman Denzin (1977) studied early childhood

and argued that socialization "from the standpoint of symbolic interactionism, represents a fluid, shifting relationship between persons attempting to fit their lines of action together into some workable, interactive relationship" (p. 2). From this perspective, Denzin studied the worlds of childhood in the preschool and family. In the family, for example, Denzin documented how parents incorporated their young children into shared discourse by responding to their gestures and vocalizations as meaningful communications. However, there has been no real research tradition or theoretical innovation on children and childhood from Denzin's work.

Other symbolic interactionists have been more persistent in theoretical and empirical work on preadolescents. Gary Fine (1987), for example, studied Little League baseball and identified how, over the course of a season, boys through their collective activities taught each other about morals, emotional expression and control, and language routines only indirectly related to the sport. Patricia Adler and Peter Adler (1998) identified clear status groups and cliques among elementary school children and how the clique dynamics varied across groups. Somewhat surprisingly, the members of the popular clique were visible but not always well liked, and the competition within the group worked against close friendships.

Interpretive Approaches to Children's Socialization and the New Sociology of Childhood Central to the interpretive view of socialization is the appreciation of the importance of collective, communal activity—how children negotiate, share, and create culture with adults and each other. In line with these assumptions regarding interpretive collective activity, William Corsaro (2005) offered the notion of *interpretive reproduction*. The term *interpretive* captures innovative and creative aspects of children's participation in society. Children produce and participate in their own unique peer cultures by creatively appropriating information from the adult world to address their own peer concerns. The term *reproduction* captures the idea that children do not simply internalize society and culture but also actively contribute to cultural production and change. The term also implies that children are, by their very participation in society, constrained by the existing social structure and by social reproduction.

Interpretive reproduction views children's evolving membership in their culture as reproductive rather than linear. According to the reproductive view, children and youth strive to interpret or make sense of the adult culture, and in the process they come to produce their own peer cultures (Corsaro, 2005; Eder, 1995). Appropriation of aspects of the adult world is creative in that it both extends or elaborates peer culture (transforms information from the adult world to meet the concerns of the peer world) and simultaneously contributes to the reproduction of the adult culture.

Corsaro (2003) pointed to children's dramatic role play as an instructive example of interpretive reproduction. In role play, children do not simply imitate the adult model but appropriate and embellish it to meet the interests, values, and concerns of their peer cultures. Corsaro documented complex differences in the dramatic role play of the upper-middle class compared to that of economically disadvantaged preschool children. The upper middle class children created an ice cream store in their private preschool where they portrayed themselves as owners of the store, invented a type of ice cream that would not melt in the hot sun, and decided to donate some of the money from their store to help sick kids pay their hospital bills.

The economically disadvantaged kids (attending a Head Start center) created a role-play event in which they pretended to be their mothers having a telephone conversation about the difficulties of parenting in poverty. They discussed how their children wanted them to take them to the store when there were no large grocery stores in the neighborhood and to the park. The girls talked about the difficulties of transportation to faraway stores and safe parks when the local bus system was slow and demanded transferring from one bus to another to get to the desired destination—a problem few middle-class families face. Although the complexities of both role-play events were impressive in regard to the children's cognitive and communicative skills, it was clear that the predispositions of the middle-class children were optimistic and confident, whereas those of the economically disadvantaged children displayed a sobering recognition of their challenging futures.

SEE ALSO Volume 1: *Developmental Systems Theory; Elder, Glen H., Jr.; Identity Development; Interpretive Theory; Piaget, Jean; Social Development; Socialization, Gender; Socialization, Race;* Volume 2: *Roles*

BIBLIOGRAPHY

Adler, P. A., & Adler, P. (1998). *Peer power: Preadolescent culture and identity.* New Brunswick, NJ: Rutgers University Press.

Corsaro, W. A. (2003). *"We're friends, right?" Inside kids' culture.* Washington, DC: Joseph Henry Press.

Corsaro, W. A. (2005). *The sociology of childhood.* (2nd ed.). Thousand Oaks, CA: Pine Forge Press.

Corsaro, W. A., & Fingerson, L. (2003). Development and socialization in childhood. In J. Delamater (Ed.), *Handbook of social psychology* (pp. 125–155). New York: Kluwer Academic/ Plenum.

Denzin, N. K. (1977). *Childhood socialization.* San Francisco, CA: Jossey-Bass.

Eder, D. (with Evans, C. C., & Parker, S.). (1995). *School talk: Gender and adolescent culture.* New Brunswick, NJ: Rutgers University Press.

Elder, G. H., Jr. (1994). Time, human agency, and social change: Perspectives of the life course. *Social Psychology Quarterly, 57,* 4–15.

Fine, G. A. (1987). *With the boys: Little League baseball and preadolescent culture.* Chicago: University of Chicago Press.

Handel, G., Cahill, S., & Elkin, F. (2007). *Children and society: The sociology of children and childhood socialization.* Los Angeles, CA: Roxbury.

Mead, G. H. (1934). *Mind, self, and society.* Chicago: University of Chicago Press.

Parsons, T., & Bales, R. F. (1955). *Family, socialization, and interaction process.* Glencoe, IL: Free Press.

Piaget, J. (1950). *The psychology of intelligence,* trans. M. Piercy & D. E. Berlyne. London: Routledge & Paul.

Qvortrup, J. (1991). *Childhood as a social phenomenon: An introduction to a series of national reports* (Eurosocial Rep. No. 36). Vienna: European Centre for Social Welfare Policy and Research.

Rogoff, B. (1995). Observing sociocultural activity on three planes: Participatory appropriation, guided participation, and apprenticeship. In J. Wertsch, P. Del Rio, & A. Alvarez (Eds.), *Sociocultural studies of mind* (pp. 139–164). New York: Cambridge University Press.

Rogoff, B. (1996). Developmental transitions in children's participation in sociocultural activities. In A. J. Sameroff & M. M. Haith (Eds.), *The five to seven year shift: The age of reason and responsibility* (pp. 273–294). Chicago: University of Chicago Press.

Tesson, G., & Youniss, J. (1995). Micro-sociology and psychological development: A sociological interpretation of Piaget's theory. *Sociological Studies of Children, 7,* 101–126.

Thelen, E., & Smith, L. (1998). Dynamic systems theories. In W. Damon (Series Ed.) & R. M. Lerner (Vol. Ed.), *Handbook of child psychology: Vol. 1. Theoretical models of human development* (5th ed., pp. 563–634). New York: Wiley.

Thorne, B. (1993). *Gender play: Girls and boys in school.* New Brunswick, NJ: Rutgers University Press.

Vygotsky, L. (1978). *Mind in society.* Cambridge, MA: Harvard University Press.

William A. Corsaro

SOCIALIZATION, GENDER

Individuals must learn the *culture* of the society in which they live—the ways of life, beliefs, values, behaviors, and symbols—to have the shared knowledge necessary to interact with others meaningfully. Using a life course perspective, culture is acquired through the ongoing process of socialization during which people first learn broad elements of the culture such as values and beliefs from the family and continue to learn the expectations and appropriate behaviors associated with particular social settings such as school and work from peers and coworkers. However, the socialization experience is not the same for everyone and the specific information transmitted to individuals through socialization can differ according to their race, socioeconomic status, and gender.

DEFINING SOCIALIZATION

The contemporary notion of socialization is a combination of ideas from psychology, sociology, anthropology, and human development. Some of these academic orientations, such as sociology, have traditionally emphasized the actions of external groups in the socialization process and stress the perpetuation of culture from one generation to another through social institutions such as the family, educational system, religion, and government. Other orientations, such as psychology, emphasize the individual's reaction to the information received through socialization. At this individual level, socialization leads to three outcomes. First, people internalize the values of their society that define what is good and desirable. Secondly, *social norms*, the rules that guide behavior, are learned. And third, people learn appropriate *roles*: the patterns of behavior attached to recognized positions in society such as son, daughter, husband, wife, or boss.

These dissimilar emphases on either the external processes or internal outcomes have actually served to make socialization an increasingly generic and ambiguous conception. So many different processes have been discussed under the heading of socialization that the concept has been critiqued for lacking a concise and authoritative meaning. Noting this ambiguity, socialization can generally be defined as both the external procedures by which the beliefs and appropriate behaviors of a society are conveyed to individuals and the internal processes resulting in the internalization of societal values, norms, and roles.

SOCIALIZATION AND GENDER

Gender is a very important component in the socialization process. Throughout the entire life course, people experience differential treatment and expectations depending on their sex. Boys and girls are encouraged by their parents, teachers, and friends to exhibit the behaviors and traits traditionally considered to be gender-appropriate. Eventually, individuals develop self-concepts based on the social meaning associated with their sex.

Whereas *sex* is usually used to describe the biological distinction between males and females determined by sex organs and genes, *gender* refers to a cultural distinction having to do with the attributes and traits thought to be related to each sex. A child's sex develops in the womb before birth, but gender is learned in a social context. Depending on the level of analysis, gender can involve both *gender roles*: society's notion of appropriate masculine or feminine behavior and attitudes, and *gender identity*: individuals' subjective feelings of themselves as male or female.

Gender roles are passed down from one generation to another through socialization and many of the attributes and behaviors considered to be innately male or female are largely determined by the society in which one lives. All societies have *gender stereotypes*: widely held sets of beliefs regarding the personal traits of men and women (Williams & Best, 1990). In U.S. society, males are commonly ascribed instrumental traits and are considered to be assertive, aggressive, independent, competent, and logical, whereas females are characterized as passive, warm, caring, submissive, and emotional.

Gender stereotypes can vary between societies depending on if perceived differences between the sexes are emphasized or minimized. Margaret Mead (1963) conducted a classic and pioneering study illustrating cultural variations in gender roles with three tribes in New Guinea. Among the members of the Arapesh tribe, both men and women exhibit qualities that U.S. society would define as being traditionally feminine: warmth, cooperation, and nurturance. In the Mundugumor tribe, both sexes have the traditionally masculine characteristics of anger and aggression and in the Tchambuli tribe, traditional gender roles are reversed with females being dominant and males being submissive and emotional. Even though the accuracy of some of Mead's conclusions have been questioned, her early research has inspired studies to focus on the social definitions of gender-appropriate behavior.

Just as gender stereotypes can differ between societies, variability can also exist within the same society depending on factors such as sex, race, and social class. Past research has found that boys have stronger gender-stereotypical views than girls as children and are more likely to conform to their gender role. African-American adults have been found to endorse greater gender equality than whites. Additionally, middle-class adults have more flexible gender-stereotypical views than individuals from the working class (for a more in-depth discussion of gender and a review of research findings, see Wharton, 2005).

MAJOR THEORETICAL APPROACHES TO STUDYING SOCIALIZATION

Socialization is a concept that bridges the gap between social experience and the development of individual personality. Three major theoretical approaches exist for studying socialization. Scholars with more social orientations have drawn largely from structural functionalism to highlight the structural dimensions and processes related to how people are socialized by other individuals and social groups. Structural functionalist approaches generally hold that societies are composed of complex social structures, such as families, the government, and the economy that function interdependently to promote stability and ensure continuation of the society (i.e., Parsons, 1951). The purpose of socialization is to transmit components of culture to new generations and to train individuals for their social roles. Therefore, conforming to the norms of society is necessary for the stability and the effective functioning of society.

Conversely, scholars with more individualistic orientations have used symbolic interaction theories to emphasize personal experiences in the socialization process and how social values and norms become part of identity. Symbolic interaction approaches focus much more on the individual and assume that people actively construct their own realities and self-concepts through interactions with others in specific social settings (i.e., Mead, 1962 [1934]). The self, therefore, develops through social activity and socialization. Symbolic interactionism stresses that socialization is a dynamic interactive process in which individuals adapt information to their specific needs (for an accessible overview of both theories see Ritzer & Goodman, 2004).

A third, developmental approach to studying socialization encompasses numerous different theories such as psychoanalytic, social learning, and cognitive development to explain how early social experiences influence personality. The theories included in this approach are the most focused on individual processes. The psychodynamic theory of Sigmund Freud (1856–1939)—and the extensions by Erik Erikson (1902–1994), Harry Stack Sullivan (1892–1949), and Alfred Adler (1870–1937)—address how the components of personality form and illustrate the specific mechanisms used to internalize social norms and expectations. Additionally, the constructivist theories of Jean Piaget (1896–1980) and Lev Vygotsky (1896–1934) deal with how individuals develop higher mental functioning by actively engaging their environment and interacting with others. Many of these theories have greatly informed the study of socialization and have been used to explain the development of gender identity; however, they are too varied to be adequately discussed at length in this entry (for more information see Miller, 2002).

AGENTS AND PROCESSES OF SOCIALIZATION

Numerous social experiences and interactions with others have the potential to affect beliefs and behaviors. However, certain groups in a society are much more influential than others. These *agents of socialization* are those groups that teach the important social norms and values that eventually become part of individual self-concept, including what is considered to be gender-appropriate behavior.

Agents socialize individuals through numerous processes. At the societal level, the *processes of socialization* are activities and actions of external groups that transmit societal norms and values to an individual such as direct instruction, modeling the appropriate behavior, rewarding proper behavior, and discouraging unacceptable behavior. However, socialization processes can also be analyzed at the individual level and can include individuals' behavioral responses and internal reactions to the socialization they receive such as attending to the instruction, imitating, remembering the lesson, and accepting or rejecting the value.

The Family The family is widely considered to be the most important agent of socialization. Because parents are more likely to follow what is considered to be appropriate parental roles for the culture in which they live, families serve to transmit the values of the society as a whole through their parental practices (Arnett, 1995). Additionally, each family is characterized by a specific ethnicity, income level, social class, religious belief, and political orientation that help to determine how children will be raised.

Differential treatment by parents can establish different expectations for behavior and teach young children much about gender. However, a review of the research conducted in North America has found that the only consistent difference in parental treatment between boys and girls is the encouragement of sex-typed play activities such as offering dolls to girls and balls to boys (Lytton & Romney, 1991). Of course, any differences in treatment between boys and girls in a specific family could also depend on factors such as the race and social class of the parents as well as the parents' ideas about appropriate gender roles.

School and Work Schools are formally responsible for teaching children the knowledge and skills needed to participate in society. However, the belief system of the larger society greatly influences the structure and culture of schools. Aside from academic lessons, schools also teach the norms, values, and socially appropriate behaviors of the society. Children learn numerous unintended lessons through this *hidden curriculum*. For instance, because students are counted as tardy if they are not in class by a certain time, they learn that punctuality is important. Additionally, the competitive nature of activities such as spelling bees, science fairs, and extra-curricular sporting events teach children to value success.

Teachers also continue the instruction about gender-appropriate behavior that children first receive from their parents. Boys are given more attention and praise than girls (Sadker & Sadker, 1994). Additionally, teachers are more likely to praise boys for their knowledge, but praise girls for obedience and following the rules. These differences in treatment further reinforce separate expectations for boys and girls and establish different standards of acceptable behavior.

In adulthood, the work setting replaces school. Instead of a teacher, most people work under the authority of a boss and hard work is rewarded with money instead of grades. Workers learn to adopt a strong work ethic and can eventually see their job as an important part of their identity.

The Peer Group A peer group is a collection of individuals who regularly interact, share similar interests, and are approximately the same age. For younger children, peer groups normally include nearby neighborhood playmates. Later, peer groups are composed of more distant acquaintances as preteens and adolescents make friends in school and join peer groups according to similar interests and concerns about social status. Whereas relationships with family members remain important, the significance of peers increases during the school years and become especially significant during adolescence.

Unlike the family or the school, the peer group provides children with a much more interactive context in which the meaning of gender is learned and reinforced. However, peers generally expect conformity to gender stereotypes and discourage others from nonappropriate behavior. Eleanor Maccoby (1998) details how boys and girls segregate themselves into same-sex play groups at an early age and learn different styles of interaction. Boys are more physical, competitive, and physically aggressive. Girls play in smaller groups, tend to be more cooperative, engage in taking turns during play, and express greater self-disclosure to each other. The different styles of interaction reinforced by each group help further socialize boys to be more independent and dominant whereas girls learn the importance of maintaining relationships.

Mass Media The mass media are different forms of communication directed to a very large audience and includes television, movies, radio, magazines, and the Internet. Public exposure to the mass media is considerable and has been increasing since the beginning of the 20th century due largely to technological innovations in the middle and latter parts of that century (such as the

wide sales of televisions after World War II, ca. 1946–1955), and the development of the World-Wide Web). Even if not intended by the producers, numerous messages presented by the media can teach values, beliefs, and ideologies.

Studies suggest that gender role perceptions and attitudes are affected by television programs and commercials that generally portray men in positions of authority and women as subordinate. Erving Goffman (1979) found that in magazine and newspaper advertisements, men were positioned in front of women or pictured as being taller than women whereas women were more likely to be in positions of subordination such as lying down or being seated on the floor. More recent research has supported Goffman's findings (i.e., Cortese, 1999), and even newer forms of media continue this traditional portrayal of men and women. For example, video games under-represent female characters and often portray them as sex objects.

Cohort Differences in Socialization Societies can be conceptualized as collections of various stratified groups based on factors such as gender, race, and socioeconomic status. However, societies are also composed of a number of *cohorts*: groups of people born during approximately the same time period. Values and beliefs can differ among cohorts depending on the socialization they received at a specific time. But as individuals in a cohort grow older, they also share a unique set of historical life experiences that have the potential to alter the initial beliefs and values learned through socialization (Settersten & Martin, 2002). This is especially true when the effects of a particularly important generation-defining historic event are experienced (e.g., the Great Depression in the 1930s, World War II with U.S. involvement from 1941–1945, or the counterculture movement of the 1960s). In turn, the new values and beliefs of the cohort will be transmitted to the next generation through socialization.

For example, a woman born in the 1920s would have experienced very different gender socialization than a woman born in the 1970s. Whereas significant differences in gender socialization could be expected depending on variables such as social class and race, many women born in the late 1920s would have probably been told that a woman's place was in the home, would have been discouraged from seeking employment, and would have been expected to get married and have children. Yet in their late teen years, this cohort would have experienced America's entrance into World War II during the early 1940s and witnessed the large numbers of women who contributed to the wartime effort by working in previously male-dominated professions. As a result of living through and possibly even participating in these events, some members of this cohort may have changed their ideas about the capabilities of women and subsequently taught their children an expanded notion of the appropriate behaviors and attitudes for women.

The children of this late 1920s cohort would have been born after World War II (the so-called Baby Boom). In turn, this new cohort would have experienced the Vietnam War and the counterculture movement. The gender role socialization that the Baby Boom generation provided to their children would probably be significantly affected by the women's movement of the 1960s and the resulting legislative changes banning sexual discrimination such as Title IX (outlawing sexual discrimination in any educational program or activity that receives federal funds). Additionally, the children of the Baby Boomers, born beginning around the mid-1970s (Generation X), would witness more and more of their mothers seeking employment outside of the house. Seeing this expansion of the homemaker role for women could have an influence on the messages about gender that this generation provides to their children. This greatly abbreviated review of the major historical events occurring over just three generations illustrates how radically different gender socialization could be for an individual born in 1930 and a member of Generation X born in the 1970s.

FUTURE DIRECTIONS

A review of the main theoretical approaches to socialization and previous research findings illustrate socialization as a multifaceted process. Gender socialization, in particular, is especially complex given the potential influence of numerous social and temporal variables. Trends in socialization research indicate that scholars are noting the influence of societal change on socialization patterns and are envisioning individuals as more active agents in their own socialization. Future research can be expected to continue along these trajectories with more of a focus on the individual and the effects of social change.

Societies are not static and at least some degree of change is inevitable. Whereas many theorists and researchers have stated the importance of the family as an agent of socialization, the nature of the traditional family is changing. Numerous couples divorce, more children are being born into cohabiting unions, more mothers are working, and more single women are the heads of their own households. Future research could investigate if witnessing these changes in the traditional homemaking compared with breadwinning roles is related to children's notions about the appropriate gender roles of men and women. Additionally, research could

also explore if witnessing these changes affect girls and boys similarly.

Research can also be expected to continue focusing on individuals as more active participants in their own socialization. With increasingly interactive forms of media such as the Internet and with so many entertainment choices offered by the multitude of channels on cable and satellite television, people can choose media that supports their already established notions of appropriate gender roles. Future research may explore the processes and mechanisms involved in how people choose the media they consume instead of the effects of the media on individual conceptions of gender.

SEE ALSO Volume 1: *Gender and Education; Identity Development; Media Effects; Peer Groups and Crowds; Social Development; Socialization, Race;* Volume 2: *Body Image, Adulthood; Parent-Child Relationships, Adulthood;* Volume 3: *Cohort.*

BIBLIOGRAPHY

Arnett, J. J. (1995). Broad and narrow socialization: The family in the context of a cultural theory. *Journal of Marriage and the Family, 57,* 617–628.

Cortese, A. J. (1999). *Provocateur: Images of women and minorities in advertising.* Lanham, MD: Rowan & Littlefield.

Goffman, E. (1979). *Gender Advertisements.* New York: Harper Colophon.

Lytton, H., & Romney, D. M. (1991). Parents' differential socialization of boys and girls: A meta-analysis. *Psychological Bulletin, 109,* 267–296.

Maccoby, E. (1998). *The two sexes: Growing up apart, coming together.* Cambridge, MA: Belknap.

Mead, G. H. (1962). *Mind, self, and society: From the standpoint of a social behaviorist.* Chicago: University of Chicago Press. (Original work published 1934.)

Mead, M. (1963). *Sex and temperament in three primitive societies.* New York: William Morrow.

Miller, P. (2002). *Theories of developmental psychology.* (4th ed.). New York: Worth Publishers.

Parsons, T. (1951). *The social system.* Glencoe, IL: Free Press.

Ritzer, G., & Goodman, D.J. (2004). *Sociological theory.* (6th ed.). New York: McGraw-Hill.

Sadker, M., & Sadker, D. (1994). *Failing at fairness: How our schools cheat girls.* Toronto: Simon and Schuster.

Settersten, R. A., & Martin, L. (2002). The imprint of time: Historical experiences in the lives of mature adults. In R. A. Settersten & T. J. Owens (Eds.), *Advances in life course research: Vol 7. New frontiers in socialization* (pp. 471–497). Oxford, U.K.: Jai.

Wharton, A. S. (2005). *The sociology of gender: An introduction to theory and research.* Oxford, U. K.: Wiley-Blackwell.

Williams, J. E., & Best, D. L. (1990). *Measuring sex stereotypes: A multi-nation study* (Rev. ed.). Newbury Park, CA: Sage.

Kurt Gore

SOCIALIZATION, RACE

Socialization is the lifelong process through which norms and expectations are transmitted among people in order to create stability in society through consensus. Agents of socialization can be individuals, groups, or institutions. These agents are constrained by cultural conditions and political forces such that socialization generally supports the status quo. In a certain sense then, socialization is conservative in nature because deviance from the status quo is discouraged. The content of socialization messages transmitted relates directly to the developmental stages (e.g., child, youth, adolescent, young adult, or adult), roles (e.g., son, teacher, student, mother, professor, or politician), and ascribed and achieved statuses (e.g., race, gender, or socioeconomic status) of individuals being socialized.

Race socialization is the lifelong process through which the social meaning and consequences of race and racism are transmitted (Brown, Tanner-Smith, et. al, 2007). Through this process, individuals learn about physical differences, history and heritage, identity politics, and/or prejudice and discrimination (Chesire, 2001; Hughes, Rodriguez, Smith, et. al, 2006; Lesane-Brown, 2006). Children and adolescents of color are the central focus in the literature because families of color are the agents of race socialization predominantly studied. More specifically, most studies focus on Black families. This is so because race socialization research started with attempts to explain how Black children could form and maintain positive self-concepts despite their in-groups' marginalized status in the U.S. racial hierarchy (Hughes et al, 2006). Familial race socialization continues to be deemed essential for the healthy development of Black children because it initiates a watershed experience after which Black children develop racial identity and become conscious regarding the nature of racial intolerance. Racial identity and consciousness are necessary for success in contexts such as the United States, where non-white physical features remain an impediment to social and upward mobility. Along those same lines, recognition that the lives of other U.S. racial minority groups are similarly shaped by racial hierarchy began during the 1990s (Hughes et al, 2006). In response, scholars began asking whether racial minorities other than Blacks socialize their children regarding race and racism (Aboud, 1988; Brown et al, 2007).

Parents raising White children teach them about race and racism too, yet are virtually neglected in discussions regarding the race socialization process because Whites define the norm. In addition, the goals of White families' teachings may differ from the goals of parents raising children of color. One study suggests that in White families the main objective of race socialization is to promote White children's tolerance of diversity (Katz & Kofkin, 1997). In contrast, another study finds that most White parents are indifferent toward socializing their children regarding interracial relations (Hamm, 2001). Although it probably has differential meaning within non-White and white families, race socialization occurs in all families rearing children and thus is universal.

TERMINOLOGY

Hughes, Rodriguez, Smith, Johnson, Stevenson, and Spicer (2006) suggest using the term *ethnic-racial socialization* when referring to the broad research literature. Unfortunately, some scholars use the term *ethnic socialization* when referring to, for example, *both* Blacks and African Americans, confounding race with ethnicity. In response, the present authors prefer the term *race socialization* when describing socialization processes involving U.S. racial group members of various ethnicities (e.g., Whites, Blacks, Hispanics, Asians, Native Americans, or multiracial individuals) and *ethnic socialization* when describing U.S. ethnic groups of various races (e.g., Irish Americans, Cuban Americans, Jamaican Americans, Jews, Mexican Americans, Japanese Americans, or Hopi Indians). Clearly race socialization and ethnic socialization might overlap, but in some cases one is present without the other. For reasons of clarity and parsimony, the authors use the term *race socialization* throughout the remainder of this entry.

WHY SOCIALIZE CHILDREN TO RACE

Brown, Tanner-Smith, Lesane-Brown, and Ezell (2007) theorize that there are three reasons families tend to socialize young children to race and racism: (a) the context in which the child lives exposes him or her to diversity; (b) some families are facile at communicating with children regardless of topic; and (c) the child is being prepared for membership in a marginalized racial minority group. An additional reason is that: (d) families desire to pass down traditions and customs. The first reason suggests that everyday life exposes the child to opportunities to learn about racial diversity. Indeed, research indicates that children, some as young as six months, recognize and react to differences in physical features (Katz & Kofkin, 1997). Second, in families where warmth characterizes interactions, children feel loved and probably discuss many topics with family members, race and racism among them (Aboud, 1988). The third reason is that race socialization prepares children of color to live successfully in an often hostile context (Lesane-Brown, 2006). The fourth reason deals with preservation of collective memory as some families

Lunchtime. *Muslim schoolchildren select from traditional Middle Eastern foods.* © GIDEON MENDEL/CORBIS.

pass in-group traditions and customs down to the next generation (Chesire, 2001).

THEORETICAL APPROACHES

There are five theoretical approaches that may be helpful for understanding the meaning and significance of the race socialization process: (a) social cognitive learning theory, (b) the life course perspective, (c) ecological theory, (d) racial identity theory, and (e) the social capital perspective. Connections among these theories are reflected in the centrality each assigns to meaning-making during interpersonal and group interactions.

Social cognitive learning theory is useful for understanding ways families teach children about race and racism. Broadly, this theory predicts how children come to think and behave in normative fashion at particular developmental stages, asserting that socially acceptable behaviors are learned through discussions, observation, modeling, vicarious reinforcement, and imitation of significant others (Bandura, 1977). Through these mechanisms, children learn about race and racism, and how to respond in race-related situations. For example, Chesire (2001) describes common practices (e.g., listening, tell-

ing, watching, and observation) related to transmission of cultural knowledge among American Indian women and their families. Children play an active role in this process because they must actively attend to their parents' messages, encode what they are hearing or observing, reject or store this new information in memory, and then retrieve this information at some later time (Bandura, 1977).

Life course perspective (Alwin, 1995; Elder, 1994) is useful because it describes how individuals' lives are shaped by social change. This perspective links agency, social conflict and change, individual development, and biography across various life domains including work, family, and health. In terms of the race socialization process, the life course perspective suggests that family communication about race and racism are contoured by historical time. The perspective acknowledges that parents' race socialization practices must adapt to fluctuations in cultural conditions and political forces in anticipation of when their children will come of age. Consequently, messages transmitted to one generation of children may be different than those transmitted to another generation (Brown & Lesane-Brown, 2006).

Ecological systems theory (Bronfenbrenner, 1979) defines several concentrically and temporally arranged levels of the environment (i.e., microsystem, mesosystem, exosystem, macrosystem, and chronosystem) that simultaneously interact to influence social development. Positioned at the center of these levels of the environment is a child who is influenced by and proactively influences the surrounding environment. In part, these environments determine when, how, and why parents racially socialize their children (Lesane & Brown, 2006). Ecological systems theory represents the causative layering of levels. More specifically, the theory addresses how multiple levels and socializing agents interrelate to influence individuals, and is truly social psychological in formulation (i.e., suggesting reciprocity between macro and micro levels).

Racial identity theory (Katz & Kofkin, 1997) views race socialization as the requisite precursor to development of racial identity. Racial identity represents a sense of attachment to a collective and imagined community of similar others. The sense of attachment provides purpose and an inter-subjective definition of the situation. Scholars theorize that race socialization is therefore psychologically affirming to children of color because it concretizes their emergent self-schemas—defined as organized, affective, coherent, and integrated generalizations about self and relationship of self to others in the in-group and out-group (Oyserman, et. al, 2003).

Social capital perspective (Putnam, 2000) defines social capital as an intangible resource that facilitates certain outcomes for actors within a social structure or an intangible resource that creates value through meaningful social relations within a social network. Scholars have neglected how the race socialization process creates two forms of social capital. First, race socialization encourages "bridging" social capital, formed of inclusive networks that encompass people across diverse social cleavages (e.g., region, socioeconomic status, or historical time) including members of one's racial group (Katz & Kofkin, 1997; Putnam, 2000). Second and simultaneously, race socialization encourages "bonding" social capital because it reinforces exclusive identities and homogeneous groups, framing racial identity as distinct from other identities (Oyserman et al, 2003).

EMPIRICAL RESEARCH

Most race socialization research focuses on the *content* of parental messages transmitted (Hughes et al, 2006). In these studies the research question has typically been: What are the specific things that parents say to teach their children about race and racism? Message content is usually divided into four broad categories: (a) racial pride and teachings about cultural heritage, traditions, and values; (b) awareness of and preparation for racism and

discrimination; (c) equality and coexistence among racial groups; and (d) the importance of life skills, citizenship, and individualism.

Prevalence assesses the proportion of parents that transmit race socialization messages. Research indicates that a larger proportion of parents raising non-White children practice race socialization as compared to the proportion of parents raising White children (Brown et al., 2007).

Frequency, which includes information such as persistence and accumulation that is obscured in studies of prevalence alone, assesses how routinely race socialization occurs. The frequency in which parents transmit race socialization messages may vary depending upon the content of the message. For instance, child and adolescent studies conducted since the 1980s have found that Black parents transmit messages regarding racial pride and heritage, and equality among racial groups more frequently than any other message (Hughes et al., 2006).

Several *predictors*—related to the child (e.g., age, gender, and race), parent (e.g., gender, education, income, marital status, discrimination experience, relationship with child, and racial identity), and situation (e.g., child school characteristics, neighborhood demographics, and region of residence)—are associated with whether (prevalence) and how often (frequency) parents transmit race socialization messages as well as the content of messages transmitted. For example, a nationally representative study comparing preschool-aged children found that families living in the West and families rearing American Indian children are likely to frequently teach children about their ethnic/racial heritage (Brown et al, 2007). Historical time is also emerging as an important situational predictor of race socialization, in the sense that appropriate race socialization depends upon cultural conditions and political forces (Brown & Lesane-Brown, 2006).

Regarding *outcomes,* several studies support the hypothesis that receiving race socialization messages in general and receiving group membership and pride messages in particular promote affirming racial identity structures, protection from internalizing negative racial stereotypes, positive academic orientations and outcomes, and positive psychosocial outcomes (Lesane-Brown, Brown, Caldwell, et. al, 2005). In addition, race socialization may act as a buffer against perceived experiences of discrimination. In contrast, messages emphasizing awareness of and preparation for racism and discrimination transmitted to children of color show inconsistent effects. Some studies suggest such messages are associated with positive psychosocial and academic outcomes, whereas others suggest these messages lead to negative academic outcomes, poor psychosocial functioning, and distrust of individuals outside one's racial group (Lesane-Brown, 2006). Little is known regarding implications of receiving messages that emphasize equality among racial groups or that deemphasize race and racism while touting individualism.

MEASUREMENT

Disagreement exists about what constitutes a reliable and valid measure of race socialization (Lesane-Brown et al, 2005). Some measures use open-ended questions to assess message content, whereas others use close-ended questions. Some measures are considered multidimensional and scalable, whereas others are not (Lesane-Brown et al, 2005). Some measures include few message content categories, whereas others include several.

Throughout the 1980s to late-2000s, race socialization measures that assess message content and frequency have dominated the literature, but there is progress toward more comprehensive assessment. In fact, Lesane-Brown, Brown, Caldwell, and Sellers (2005) developed an inventory that moves the field in that direction. In addition to capturing message content and frequency, their inventory, the Comprehensive Race Socialization Inventory (CRSI), examines neglected components of the race socialization process such as: onset and recency, the most useful message, multiple sources (i.e., agents of socialization), anticipatory socialization messages, and socializing behaviors.

FUTURE DIRECTIONS

The United States is becoming a more diverse and arguably more divided nation. Long-standing divisions relate to immigration, region of residence, religion, socioeconomic disparity, language, political orientation, and so on. What is interesting about these divisions is the extent to which they are racialized and to which individuals claiming specific racial backgrounds tend to represent particular positions. Given this, it is no wonder the race socialization process intrigues many scholars. To improve scholarship on the topic, it is particularly important to understand the history of race socialization research and where current research trends are headed. If one anthropomorphizes the race socialization literature, then she is currently an adolescent, experiencing excitement, amazing growth, and turmoil.

The following suggestions are offered to help facilitate her transition from adolescence to young adulthood. First, the race socialization literature needs studies that examine dynamism in the race socialization process (e.g., demonstrating whether message content is internalized or rejected). Second, because parents are but one of numerous socialization agents at work over the life course, scholars must examine how messages from multiple sour-

ces (including family, friends, the media, schools, or religious organizations) interact with one another to shape paradigms of children, youth, adolescents, young adults, and adults (Lesane-Brown et al, 2005). Third, long-term longitudinal, life course–inspired research designs are needed to reveal how early-life race socialization experiences link to older adults' conceptions regarding race and racism. Fourth, because there is little consensus regarding measurement of race socialization, studies that incorporate multiple measures and compare results are to be welcomed. In addition, rather than relying solely on self-report measures, scholars should employ multi-method approaches (such as use of observational measures, diaries, or vignettes) to investigate the race socialization process. Finally, more multi-generational studies of families are needed. These are particularly informative in terms of triangulating perspectives of socialization agents with those that they are intending to socialize.

SEE ALSO Volume 1: *Identity Development; Media Effects; Oppositional Culture; Parent-Child Relationships, Childhood and Adolescence; Peer Groups and Crowds; Racial Inequality in Education; Socialization, Gender;* Volume 2: *Ethnic and Racial Identity; Parent-Child Relationships, Adulthood.*

BIBLIOGRAPHY

Aboud, F. (1989). *Children and prejudice.* New York: Blackwell.

Alwin, D. F. (1995). Taking time seriously: studying social change, social structure, and human lives. In P. Moen, G. H. Elder, Jr., & K. Lüscher (Eds.), *Examining lives in context: Perspectives on the ecology of human development* (pp. 211–262). Washington, DC: American Psychological Association.

Bandura, A. (1977). *Social learning theory.* Englewood Cliffs, NJ: Prentice Hall.

Bronfenbrenner, U. (1979). *The ecology of human development.* Cambridge, MA: Harvard University Press.

Brown, T. N., & Lesane-Brown, C. L. (2006). Race socialization messages across historical time. *Social Psychology Quarterly, 69,* 201–213.

Brown, T. N., Tanner-Smith, E. E., Lesane-Brown, C. L., & Ezell, M. E. (2007). Child, parent, and situational correlates of familial ethnic/race socialization. *Journal of Marriage and Family, 69,* 14–25.

Chesire, T. C. (2001). Cultural transmission in urban American Indian families. *American Behavioral Scientist, 44,* 1528–1535.

Elder, G. H., Jr. (1994). Time, human agency, and social change: Perspectives on the life course. *Social Psychology Quarterly, 57,* 4–15.

Hamm, J. V. (2001). Barriers and bridges to positive cross-ethnic relations: African American and White parent socialization beliefs and practices. *Youth & Society, 33,* 62–98.

Hughes, D., Rodriguez, J., Smith, E. P., Johnson, D. J., Stevenson, H. C., & Spicer, P. (2006). Parent's ethnic-racial socialization practices: a review of research and directions for future study. *Developmental Psychology, 42,* 747–770.

Katz, P. A., & Kofkin, J. A. (1997). Race, gender, and young children. In S. A. Luthar, J. A. Burack, D. Cicchetti, & J. R. Weisz (Eds.), *Developmental psychopathology: Perspectives on adjustment, risk, and disorder* (pp. 51–74). New York: Cambridge University Press.

Lesane-Brown, C. L. (2006). A review of race socialization within Black families. *Developmental Review, 26,* 400–426.

Lesane-Brown, C. L., Brown, T. N., Caldwell, C. H., & Sellers, R. M. (2005). The Comprehensive Race Socialization Inventory. *Journal of Black Studies, 36,* 163–190.

Oyserman, D., Kemmelmeier, M., Fryberg, S., Brosh, H., & Hart-Johnson, T. (2003). Racial-ethnic self-schemas. *Social Psychology Quarterly, 66,* 333–347.

Putnam, R. D. (2000). *Bowling alone: The collapse and revival of American community.* New York: Touchstone.

Tony N. Brown
Chase L. Lesane-Brown

SOCIOECONOMIC INEQUALITY IN EDUCATION

Family background is consistently found to be related to educational outcomes such as grades, test scores, school dropout, and educational-degree attainment. Researchers often study the relationship between socioeconomic status (SES) and education. The term *socioeconomic status* often is used to describe the hierarchical arrangement of individuals and families on the basis of the social and economic factors that are rewarded in society. Although there is no consensus about the best way to measure SES in empirical studies, individual or family SES often is measured by three attributes: parents' income, education, and occupation. Studies using these individual-level measures of SES find that students from higher socioeconomic backgrounds tend to have higher test scores, lower rates of school dropout, and higher rates of postsecondary school enrollment and completion. Research on socioeconomic inequality also emphasizes the characteristics of a child's school and/or neighborhood, suggesting that it is not just the SES of individuals that matters but also the average socioeconomic level of all the individuals in the community. Schools with lower percentages of students eligible for free or reduced-price lunches have higher proportions of high school graduates, as do schools in districts with lower poverty rates. Because education matters for individuals' life chances, socioeconomic disparity in educational outcomes is a topic of concern to researchers and educators. In sum, family SES during childhood affects education, which in turn affects later SES as well as other life course outcomes.

This entry describes general trends in the relationship between SES and educational outcomes in the United States, then discusses the processes of social stratification in education, highlighting key mechanisms in the family and schools that reproduce social and economic disparities in schooling. It concludes with a brief discussion of current and future research on socioeconomic inequality in education.

TRENDS IN SOCIOECONOMIC INEQUALITY IN EDUCATION

The Coleman Report (Coleman et al., 1966) remains one of the most influential studies on educational inequality. One of its key findings is that family background has a greater influence on student academic achievement than do school resources; it is the characteristics of students and their families, not those of schools, that account for differences in students' test scores. Sociologists interested in social stratification have examined the ways in which family SES is transmitted.

Peter Blau and Otis Duncan's (1967) *The American Occupational Structure* marked the beginning of the predominance of research on status attainment in sociological studies of social stratification. Using data on adult males in the United States in the early 1960s, those researchers found that a father's education and occupation positively influence his son's educational attainment and occupational status. Although those authors found direct effects on both outcomes, the relationship between family background and a son's occupational status was mediated by the son's educational attainment. This means that although most men end up in occupations that are similar to those of their fathers, education is the main pathway for those from lower-SES backgrounds to move into occupations that place them in higher social strata than their parents. After Blau and Duncan, researchers studying status attainment processes documented the relationship between family SES and educational and occupational attainment.

Although traditional models of status attainment have been successful in documenting and describing the relationship between social origins and individuals' educational and occupational attainment, the Wisconsin model of status attainment developed by William Sewell, Archibald Haller, and Alejandro Portes in 1969 takes a more social psychological approach by including cognitive ability, educational aspirations, and the influence of significant others in the process of educational attainment. This model is distinguished from prior status attainment models by its concern with mediating factors and understanding of the way family background affects the attainment process. Though the Wisconsin study has been replicated and refined by others, the results of later studies generally support the original findings.

Socioeconomic inequality in educational achievement and attainment has persisted over time and across countries and is expected to continue to do so in the future (Gamoran, 2001). More recent studies have confirmed the persistence of the relationship between socioeconomic background and academic achievement over time. In Karl White's (1982) analysis of various studies that examine the relation between SES and academic achievement, a moderate correlation ($r = .343$) between SES and student academic achievement was found in studies conducted before the 1980s. Selcuk Sirin (2005) conducted a similar analysis of studies done between 1990 and 2000 and found that the average correlation between SES and academic achievement remained moderate ($r = 0.299$).

This persistent effect of family SES on educational outcomes has larger implications for many other life course outcomes and the reproduction of social and economic inequality. Academic achievement and educational attainment are important determinants of later social and economic success. Educational attainment is positively related to labor force participation, occupational prestige, income, and wealth in adulthood. Not only are SES differences reproduced in terms of later educational, financial, and occupational outcomes, initial SES differences in education can lead to later inequality in adult well-being. Education has been shown to have positive effects on both physical and mental health, and these relationships mediate some of the association between initial SES and health.

Figure 1 shows that the percentage of individuals reporting to be in good or excellent physical health increases with educational attainment. In addition, individuals with higher levels of educational attainment demonstrate more knowledge of current events and higher levels of political participation. Many of these relationships are not due only to differences in social origin; the studies described here have shown that after accounting for differences in family background, the effects of education on later social and economic outcomes still exist.

Having identified the role of education in maintaining the privileged status of those in higher social strata as well as its role in making it possible for those from lower strata to be upwardly mobile, researchers turned to the task of understanding how family SES affects academic achievement and educational attainment. This research has demonstrated that differences in financial resources, parental involvement, social and cultural capital, and school organization and quality all contribute to socio-

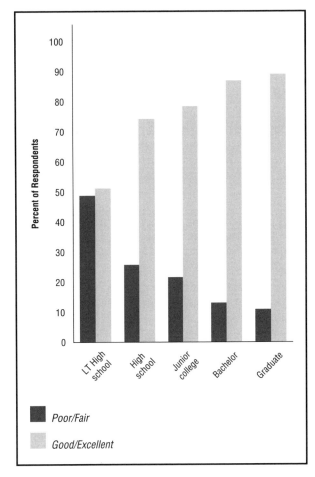

Figure 1. *Self–reported health condition by educational degree attainment, 2006 General Social Survey.* **CENGAGE LEARNING, GALE.**

economic differences in educational achievement and attainment.

PROCESSES OF SOCIAL STRATIFICATION IN EDUCATION

In the most basic sense, higher-SES parents have resources—human, financial, social, and cultural capital—that translate into advantages for their children in the educational system. One of the most obvious ways in which family SES matters for children's educational outcomes is the tangible resources money can provide. Children from higher-SES families have greater access to resources that enhance academic achievement: books, computers, tutors, trips to libraries and/or museums, and involvement in extracurricular activities. These families also have access to reliable transportation and flexible schedules that make it possible for their children to attend enrichment activities (Lareau, 2003).

Learning in the Summer Months Resource advantage also is reflected in opportunities for learning during the summer months, when school is not in session. Research by Barbara Heyns (1978) and Doris Entwisle, Karl Alexander, and Linda Olsen (1997) has attributed much of the disparity in scores on achievement tests between high- and low-SES children to the summer learning gap. Children with disadvantaged socioeconomic backgrounds make smaller increases in achievement during the summer than do their high-SES peers. This, combined with a slower rate of learning during the school year, leads to a widening gap in achievement over time. Some of the differences in summer learning have been attributed to differences in summer activities that are related to cognitive development, such as trips to the library, attendance in summer school, and even summer vacations. Research by David Burkham, Douglas Ready, Valerie Lee, and Laura LoGerfo (2004), however, showed that even controlling for these activities, the summer learning gap between high- and low-SES youth remains. Therefore, future research should focus not only on the frequency of participation in summer activities but also on the nature of activities that can foster higher levels of achievement.

Parental Involvement and Child Rearing The availability of those resources also allows higher-SES parents to be more involved in their children's schooling. According to Annette Lareau (1987, 2003), although both middle-class and working-class parents value education and want their children to do well academically, the two groups employ different strategies to help their children in this area. Middle-class parents are more involved in their children's education. They engage in supplemental activities at home, maintain established relationships with teachers and administrators, and are more comfortable intervening on behalf of their children. Working-class parents, in contrast, are more distanced from their children's experiences at school, relying on the schools to educate their children. The interactions they have with school personnel often are marked by a sense of distrust, powerlessness, and ineffectiveness.

Aside from these differences in direct involvement in school activities, there are social class differences in child rearing. Middle-class culture views parenting as requiring a conscious effort to develop children's skills and talents. To that end middle-class parents undertake a process of "concerted cultivation" (Lareau, 2003), emphasizing involvement in extracurricular activities, independence, and developing children's social and intellectual skills through reasoning and discussion. Working-class culture, in contrast, is more focused on what Lareau called the "accomplishment of natural growth." Parents are more concerned with providing children with the necessities—

food, clothing, shelter, safety—and believe that children will develop naturally if those needs are met. Working-class parents stress obedience, are less likely to engage in discussion or negotiation with their children, and expect their children to organize their own time.

Because the actions of middle-class parents are valued by schools, middle-class children reap the benefits of their parents' efforts in the form of higher levels of comfort in school and other formal settings and better academic performance. Taken together, the financial, human, and cultural capital that middle-class parents have at their disposal allows them to prepare their children better for academic success throughout their educational careers.

Social Capital Children's life chances also are influenced by their access to social capital. James Coleman (1990) defined social capital as resources that accrue to individuals as a result of their relationships with others (their social networks). Coleman (1990) argued that when parents know their children's friends' parents (intergenerational closure), they are in a better position to work in tandem with the other parents to monitor and evaluate their children's behavior because information, norms, and values can be shared through the network. Having parents whose social networks are characterized by higher levels of intergenerational closure leads to better educational outcomes for children. In addition, connections to others provide parents with information and other resources that they can use to improve their children's lives, including their educational experiences. For example, in an in-depth study of social class differences in family life, Lareau (2003) found that middle-class parents have social network connections that provide them with information they can use to help their children navigate the education system. Because social networks are stratified by social class, parents of children in lower strata are less likely to have connections that provide the same kinds of advantages.

Tracking and School Sector Although the studies cited above describe processes within the home, the nature and organization of schooling also contribute to socioeconomic differences in educational outcomes. For example, school tracking is considered one of the most important mechanisms in the social reproduction process. Tracking originated in efforts to allow teachers to modify their methods of instruction to suit the abilities of their students. However, Maureen Hallinan (1994) found that track assignment is influenced by a variety of factors other than academic ability, such as resource limitations, organizational constraints, parental influence, and teacher recommendations. Because of these factors, more advantaged families may be better able to manipulate the

Poverty. *Uniontown Elementary School Principal Ora Cummings helps first grade students read at the school in Uniontown, AL. Their desire to learn is not dulled by the fact that recent U.S. Census estimates show that more than one in three of the students in the mostly black Perry County school system, which includes Uniontown, live below the poverty level.* **AP IMAGES.**

system in their favor, resulting in the overrepresentation of lower-SES students in lower tracks and/or ability groups (Oakes, 1985). This is consequential for socio-economic inequality in eventual educational attainment because students in college preparatory tracks (also referred to as regular, advanced, or honors course levels) receive more and better academic instruction and have teachers with better skills who present them with more complex material and "high-status" knowledge. In light of the fact that lower tracks provide poorer-quality education, have less qualified teachers with lower expectations, and socialize youth differently, tracking systematically limits the educational opportunities of disadvantaged students.

School sector is another factor in the way SES affects educational outcomes. Public schools are managed by a public entity, generally the state, whereas private schools are not, although they sometimes are affiliated with and/or run by religious organizations. Private schools often require students to go through an extensive application process and charge tuition fees, making them much less accessible to disadvantaged families. Socioeconomic dif-

ferences in access to private schools are important because attending private school is associated with several positive educational outcomes, including student achievement, retention, and college attendance, although the causal effects of private school attendance have been debated. According to Caroline Persell, Sophia Catsambis, and Peter Cookson (1992), students who attend private schools not only are more likely to attend college but also are more likely to go on to elite private colleges and highly selective public universities. Elite private schools have connections to admissions offices at selective private colleges, making it easier for students to gain admission to those schools. With the increased importance placed on the selectivity of the college/university for one's life chances, the ability to gain admission to selective colleges and universities will be very beneficial for advantaged students and their families, allowing them to maintain their privileged position.

Access to Resources The mechanisms through which SES influences education extend beyond an individual and his or her family. As Coleman et al. (1966) demonstrated, the family backgrounds of a student's schoolmates also influence academic achievement. In a study of high school students in Louisiana, Stephen Caldas and Carl Bankston (1997) found that the social status of schoolmates' families has a significant effect on academic achievement and that this effect is only slightly smaller than that of a student's own family background. Other research suggests that socioeconomic inequality in education also is due to unequal resources at the school level. Because a large proportion of school funding is linked to local property taxes, there is considerable variation in per-pupil expenditures across school districts. Thus, students who live in areas with lower property values often attend schools with less funding. This disparity in funding is reflected in the quality of the schools. Although some research has suggested that school resources and spending do not have as large an effect on student achievement and later income inequality as does family background, other research has shown that unequal spending has led to disparate outcomes in academic achievement. For instance, Dennis Condron and Vincent Roscigno (2003) found that schools with higher per-pupil expenditures are in better physical condition and have a higher degree of classroom order, factors that are associated with higher student attendance rates, higher teacher quality, and higher average test scores.

CURRENT AND FUTURE RESEARCH ON SOCIOECONOMIC INEQUALITY IN EDUCATION

The studies described above focus primarily on socioeconomic inequality in primary and/or secondary education. However, as more individuals participate in postsecondary schooling, more research has been devoted to socioeconomic inequalities in higher education. Social reproduction theorists such as Pierre Bourdieu and Jean-Claude Passeron (1977) view the continued existence of socioeconomic inequality in education as resulting from the fact that privileged members of society use education to maintain their status and thus pass that privilege on to their children. Thus, as one level of education expands to become accessible to those in the lower strata, the threshold of inequality shifts upward. Most recently, the emphasis has shifted to postsecondary education, and those in the higher social strata have been able to maintain their advantage with higher rates of college attendance and completion. Recent institutional initiatives enacted by several public and private universities (e.g., the University of North Carolina, Brown University, and Stanford University) have attempted to reduce and in some cases eliminate financial barriers to higher education for lower-income and middle-class families. For example, in February 2008 Brown University announced that it would eliminate loans for students whose families earn less than $100,000, and

Stanford University approved a plan to waive tuition for students with family incomes under $100,000. Although these initiatives are intended to increase access to education, the effects of these efforts remain to be seen.

College access and enrollment has long been a concern, but research has shifted to studying access to and enrollment in different types of higher education (two-year vs. four-year, private vs. public) and college trajectories. Therefore, it is not only a matter of whether individuals attend college but also how they do so. Postsecondary perceptions and experiences vary by social class. For instance, Robert Bozick and Stephanie DeLuca (2005) found that low-income students are more likely than their higher-SES peers to delay college entry, resulting in a lower likelihood of college completion. Sara Goldrick-Rab (2006) found that college students with lower socioeconomic backgrounds are more likely to experience interrupted pathways through college, which prevent or delay degree completion. Bozick (2007) found that college students with fewer economic resources are more likely to be enrolled only part-time to keep tuition costs down or work to afford college expenses. This can lead to only a partial investment in or disengagement from school, which can lead to dropping out.

In addition to a greater focus on higher education, with the current emphasis on standards and accountability, researchers increasingly have focused on how education policies and reform such as No Child Left Behind (NCLB) alleviate or exacerbate socioeconomic differences in education. Although the explicit goals of reforms such as NCLB are to combat differences in education caused by family background, it remains to be seen whether they have been successful in doing so. Pamela Walters (2001) suggested that although educational reforms have increased educational attainment among disadvantaged groups, they have not closed or reduced social class gaps in education. The same might be true for NCLB. High standards and testing are intended to make school more difficult. This may result in higher rates of dropout, and researchers also are interested in how the effects of accountability and high-stakes testing on academic outcomes such as achievement and school dropout vary by SES.

Another educational reform that is associated with socioeconomic inequality is the school choice movement. Parents, educators, policy makers, and communities are interested in providing alternatives to large public schools. Charter schools allow parents and students to choose schools that better suit their needs. Although charter schools often are geared toward racial/ethnic minority groups or those of lower SES, as with the standards and accountability movement, it is not known whether they really are increasing levels of academic achievement and if there are social and economic differences in which families choose and attend these schools.

Although most gender differences in educational achievement and attainment have been eliminated and racial/ethnic differences are declining, SES continues to exert a stable moderate to strong influence on educational outcomes. Those from higher social strata are in an advantaged position in terms of financial resources and social and cultural capital and are able to use their resources to pass on that advantage to their children. This relationship between SES and educational outcomes has been shown to occur in a variety of settings, particularly in families, schools, and communities, and is evident at all levels of education (primary, secondary, and postsecondary). Education will also continue to mediate the effect of SES on later life outcomes such as occupation and income, political involvement, and physical and mental health. As Adam Gamoran (2001) noted, family background is likely to continue to influence children's educational achievement and attainment, making it an area of continued interest and importance for researchers, educators, and policy makers.

SEE ALSO Volume 1: *Academic Achievement; Cognitive Ability; Coleman, James; College Enrollment; Cultural Capital; Human Capital; Parental Involvement in Education; Policy, Education; Poverty, Childhood and Adolescence; Racial Inequality in Education; School Readiness; School Tracking; Segregation, School; Social Capital;* Volume 2: *Social Class.*

BIBLIOGRAPHY

Blau, P. M., & Duncan, O. D. (1967). *The American occupational structure.* New York: Wiley.

Bourdieu, P., & Passeron, J.-C. (1977). *Reproduction in education, society, and culture,* trans. R. Nice. London: Sage.

Bozick, R. (2007). The role of students' economic resources, employment, and living arrangements. *Sociology of Education, 80*(3), 261–284.

Bozick, R., & DeLuca, S. (2005). Better late than never? Delayed enrollment in the high school to college transition. *Social Forces, 84*(1), 531–554.

Burkham, D. T., Ready, D. D., Lee, V. E., & LoGerfo, L. F. (2004). Social-class differences in summer learning between kindergarten and first grade: Model specification and estimation. *Sociology of Education, 77,* 1–31.

Caldas, S. J., & Bankston, C. L. III. (1997). Effect of school population socioeconomic status on individual academic achievement. *Journal of Educational Research, 90*(5), 269–277.

Coleman, J. S. (1990). *Foundations of social theory.* Cambridge, MA: Belknap Press of Harvard University Press.

Coleman, J., Campbell, E., Hobson, C., McPartland, J., Mood, A., Weinfeld, F., et al. (1966). *Equality of educational opportunity.* Washington, DC: U.S. Department of Health, Education, and Welfare.

Condron, D. J., & Roscigno, V. J. (2003). Disparities within: Unequal spending and achievement in an urban school district. *Sociology of Education, 76*(1), 18–36.

Entwisle, D. R., Alexander, K. L., & Olsen, L. S. (1997). *Children, schools, and inequality.* Boulder, CO: Westview Press.

Gamoran, A. (2001). American schooling and educational inequality: A forecast for the 21st century. *Sociology of Education, 34,* 135–153.

Goldrick-Rab, S. (2006). Following their every move: An investigation of social-class differences in college pathways. *Sociology of Education, 79*(1), 61–79.

Hallinan, M. T. (1994). Tracking: From theory to practice. *Sociology of Education, 67*(2), 79–84.

Heyns, B. (1978). *Summer learning and the effects of schooling.* New York: Academic Press.

Jencks, C. L., Smith, M., Acland, H., Bane, M. J., Cohen, D. K., Gintis, H., et al. (1972). *Inequality: A reassessment of the effect of family and schooling in America.* New York: Basic Books.

Lareau, A. (1987). Social class differences in family–school relationships: The importance of cultural capital. *Sociology of Education, 60*(2), 73–85.

Lareau, A. (2000). *Home advantage: Social class and parental intervention in elementary education* (2nd ed.). Lanham, MD: Rowman & Littlefield.

Lareau, A. (2003). *Unequal childhoods: Class, race, and family life.* Berkeley: University of California Press.

Oakes, J. (1985). *Keeping track: How schools structure inequality.* New Haven, CT: Yale University Press.

Persell, C. H., Catsambis, S., & Cookson, P. W., Jr. (1992). Family background, school type, and college attendance: A conjoint system of cultural capital transmission. *Journal of Research on Adolescence, 2*(1), 1–23.

Sewell, W. H., Haller, A. O., & Portes, A. (1969). The educational and early occupational attainment process. *American Sociological Review, 34*(1), 82–92.

Sirin, S. R. (2005). Socioeconomic status and academic achievement: A meta-analytic review of research. *Review of Educational Research, 75*(3), 417–453.

Walters, P. B. (2001). The limits of growth: School expansion and school reform in historical perspective. In M. T. Hallinan (Ed.), *Handbook of the sociology of education* (pp. 241–261). New York: Kluwer Academic/Plenum.

White, K. R. (1982). The relation between socioeconomic status and academic achievement. *Psychological Bulletin, 91*(3), 461–481.

Shelley L. Nelson
Jennifer Catherine Lee

SPECIAL EDUCATION

SEE Volume 1: *Learning Disability; School Tracking.*

SPORTS AND ATHLETICS

A primary goal of life course sociology is to document the way in which institutions and interpersonal interactions within those institutions influence peoples' lives. For the study of childhood and adolescence, this focus often leads scholars to highlight the influence of traditional institutions such as the family, education, and religion. Yet research shows that sport is another significant field for understanding development from childhood to adulthood. Although the extent to which it achieves this goal is debated, sport is framed as a domain of activity with the potential to shape a broad range of characteristics and values, such as teamwork, following rules, and discipline, that are central to the development of a prosocial child and future adult.

SPORT VERSUS INFORMAL PLAY

Although seemingly obvious, clarifying what is meant by the term *sport* is necessary to understanding how it contributes to youth development. This definition often is made by distinguishing sport from what it is not—informal play.

Informal play is a free-flowing activity structured through the mutual agreement of the participants. Norms may exist, but they derive from the interaction of the participants and are open to modification. Sport, on the other hand, is defined as "institutionalized competitive activities that involve rigorous physical exertion" (Coakley, 2001, p. 20). The key distinctions between sport and play are: Sport inherently involves physical activities, sport is based on a goal of competitively establishing a clear winner and loser, and the rules and boundaries of sport are codified in institutions with regulatory powers. Unlike informal play, sport is an institution, meaning it exists without the presence or action of any particular participants. Sports are formalized with written and enforced rules and guidelines. The objective of sports always is predetermined (e.g., to get a basketball through a hoop the most times), minimizing the ability for participants to construct their own goals or outcomes.

For the purpose of understanding the place of sports in youth development, demarcating exactly which activities are sport and which are play is not nearly as important as studying how the interaction with and between participants and adults constructs activities to be more like one or the other. Gary Fine's (1987) study of Little League baseball showed that the balance between sport and play was a source of tension among adults and youth. Coaches and parents stressed the importance of Little League being "fun" and "enjoyable" while concurrently calling for "structure" and "rules" to teach the children important life lessons. Even within the same league, coaches took very different approaches to practice and games. For example, one coach allowed players to play any position they wanted and did not discipline players for playing makeshift games in the dugout. For the youth on this team, baseball would seem very much like informal play. Other coaches, however, held rigorously structured practices that emphasized learning uniform techniques with the ultimate goal of winning each game. Studying the impacts of these two models is a prime area for research in sports participation as it speaks directly to broader issues of learning and human development.

WHO GETS TO PLAY?

Beyond concerns about what the qualities of sports are and what they should be, there are pragmatic issues concerning how the current system of athletic participation plays out in the lives of youth. The majority of research in this area focuses on organized athletic participation, with primary emphasis on participation in school-sponsored sports teams. One of the unique aspects of studying sports participation, in contrast to involvement in other institutions such as education or the family, is that not all youth are equally likely to be involved. Thus, to understand the impact of being involved in sports, it is necessary to account for the factors that may affect the chances that one will participate in the first place.

A major element influencing the likelihood that a youth will participate in organized sports is one's physical stature. Although very few sports have physiological requirements, most have de facto ideal types, such as being tall in basketball or being petite in female gymnastics. Children may see these ideal types as necessary conditions for participation, and if they do not perceive themselves as fitting the mold, they may be less likely to try out for a team. At more competitive levels, youth who do not match the preferred physical types may be less likely to make the team even if they do try out, either because of stereotypical biases on the part of coaches or because of real physical limitations.

In addition to pure physical structure, cognitive and personality traits may lead certain youth to be more likely to participate in sports or in particular types of sports. For example, children who are more social and extroverted may be drawn more toward participating in sports in general and team sports more specifically than comparatively more introverted or shy children. Similarly, youth who are aggressive may enjoy contact sports such as football or wrestling, and children with high levels of hand–eye motor coordination will find success in sports that reward this ability, such as tennis or baseball. Sociologists of sports, however, recognize that none of these

Little League Baseball. *Little league players celebrate after a victory. There is a debate over whether sports have a positive influence on development or increase propensities toward anti–social and delinquent behaviors.* **AP IMAGES.**

patterns naturally occur; rather, sports are designed in such a way as to reward certain attributes over others. However, given the reality of the current sporting system, certain inherent physical and cognitive traits will continue to influence young people's likelihood of participation.

Social and structural factors also contribute to the ability of a child or adolescent to be involved in athletics and the likelihood of such involvement as well. One of the most prominent of these factors is a child's family's socioeconomic standing. Youth in families on the lower end of this hierarchy are more likely to live in poorer neighborhoods, which are less likely to have the economic and physical resources to fund organized athletic leagues (Pedersen & Seidman, 2005). Moreover, the move toward private athletic participation, such as club teams, requires families to invest more to keep children involved. From buying equipment and paying for specialized camps to taking time off for trips to tournaments, the cost of playing organized sports has continued to rise, further inhibiting youth with fewer economic resources from participating.

Perhaps one of the most polarizing attributes surrounding sports participation is gender. Title IX, passed in 1972, was intended to provide proportionate opportunities and funding to women in all public institutions. Surprisingly, most of the research and controversy in this area has focused on opportunities for female athletes at the collegiate level, with much

less attention paid to equality in secondary education. Although evidence shows that the number of females participating at all levels has increased since the passage of Title IX, their numbers still do not equal their representation of the population in high schools (National Women's Law Center, 2007). Furthermore, high schools, unlike colleges, are not federally required to even report on their achievement of Title IX requirements, creating a lack of accountability and enforcement of the goals of Title IX at this level of participation. Because of this dearth of attention and emphasis, one of the most needed avenues for future research is to examine the way in which Title IX has affected—or not affected—the athletic experience for female children and adolescents.

SPORT'S IMPACT ON YOUTH

Without question the most researched and contested topic concerning sports and youth is the debate over whether sports have a positive influence on development or increase propensities toward antisocial and delinquent behaviors. Many researchers hold that participation in athletics helps young children learn important life lessons, such as understanding the consequences of decisions, learning from constructive criticism, and accepting failure. More directly, sports are thought to promote prosocial skills and positive affect. Children who play sports learn how to work with others toward a common goal, gain respect for and acceptance of rules and punishments, and develop an increased sense of self-worth. In support of these claims, James McHale et al. (2005) found that sports participants had higher levels of self-esteem and were rated as more sociable by their teachers than nonparticipants.

For adolescents, research tends to focus more on sports' ability to minimize deviant behavior and increase academic performance. Travis Hirschi's (1969) social control theory posits that adolescents who are more attached to norm-enforcing institutions and people are less likely to commit deviant acts than those who are less attached. According to this perspective, sports are posited to limit deviance for several reasons. First, the majority of adolescents who play sports do so for school-sponsored teams, which have requirements for participation. Acting in deviant ways, such as committing crimes, skipping school, or failing in the classroom, would violate these regulations and result in a dismissal from the team, stripping the adolescent of a valued opportunity. Not surprisingly, the majority of research has shown that athletes are less likely to drop out of high school (Mahoney & Cairns, 1997) and, on average, achieve higher grades than nonparticipants (Eccles, Barber, Stone, & Hunt, 2003).

Second, participating in sports requires an extensive time commitment, which reduces the time available to be involved in deviant activities. Third, sports introduce adolescents to positive adult role models and mentors who can directly influence adolescents' choices as well as serve as another psychological control measure (i.e., a valued tie that may be broken by participating in deviant activities). Finally, as for younger children, sports are thought to engender a certain moral framework that disapproves of deviant behaviors.

Despite evidence demonstrating the positive influence of sports, many scholars argue that sports have detrimental consequences for all youth, participants and nonparticipants alike. Three primary areas in which sports are thought to have the most negative impacts are creating and supporting harmful status hierarchies in schools, maintaining and promoting gender inequality, and increasing delinquent behavior and encouraging substance use.

Donna Eder's (1995) ethnographic study of junior high students showed how the limited nature of athletic participation combined with the increased visibility of those who make school-sponsored teams led to status stratification in schools. Students who played sports, especially football for boys and cheerleading for girls, were attributed the highest status in the school, whereas those who were not on teams were socially marginalized. Frequently this stratification increased bullying, both physical and verbal, of the lower-status students, contributing to issues of self-esteem and depression among these students. Eder also contended that school officials were more lenient with the higher-status athletes, which only increased their status and facilitated their negative behaviors.

In addition to producing status hierarchies, sports have been identified as a significant source in the creation and maintenance of traditional gender stratification. Michael Messner (2002) claimed that sports are structured in a way that legitimizes and promotes male dominance. From the time youth begin playing sports, teams are segregated by gender, and even at very young ages children can see a highly unequal distribution of power and resources. Positions of authority in both male and female leagues are virtually all filled by men, and male teams receive disproportionate funding and resources (Eitzen, 2003). Beyond the structural segregation of gender, numerous studies have demonstrated that sports are a site of hegemonic masculine socialization. Often femininity is invoked as a threat and put-down by male coaches. For example, in his observations of a preschool soccer team, Messner found that when the boys were not paying attention the coach would suggest that he would bring "in the Barbies to play them." Similarly, Fine (1987)

observed numerous Little League baseball coaches chastise their teams for "playing like a bunch of girls." This type of language teaches children not only that boys and girls are different but also that boys are supposed to be tough and aggressive, that weakness and failure is a feminine quality, and that by and large males are superior to females.

The emphasis placed on stereotypical masculine characteristics in sports also has been linked to increased violence and aggression on the part of participants. Robert Hughes and Jay Coakley (1991) argued that involvement in sports creates *positive deviance*, meaning that athletes' commitment to the values and norms of sports actually increase the likelihood of deviant behavior. For example, participants in particular sports may come to think that violence is an acceptable means to solve problems. Coaches often use contests of aggression as a way to settle within-team disputes, and in certain sports (e.g., hockey), fighting is a tolerated aspect of the game itself. These alternative norms result in socially unacceptable behavior outside of sports. This theory is supported by Derek Kreager's (2007) analysis of a nationally representative sample of adolescents, in which he found that participants in contact sports were more likely to get into physical fights with peers than were nonparticipants, whereas involvement in noncontact sports, such as baseball or tennis, showed no influence on fighting.

The theory of positive deviance helps explain further the seemingly paradoxical finding that students involved in athletics drink alcohol more and use particular drugs more frequently than do nonparticipants (Hoffman, 2006). The idea that one must win at all costs quite clearly encourages athletes' higher use of steroids, and the value of pushing oneself to the extreme could explain collegiate athletes' higher rates of binge drinking. Additionally, unquestioning devotion to the team may lead high school athletes to be more influenced by fellow athletes who drink. Athletics themselves therefore may imbue adolescents with values that make them more inclined to commit deviant acts, as well as providing a social peer environment that supports these behaviors. Still some studies have found no differences between participant and nonparticipant levels of delinquency (e.g., see Miller, Melnick, Barnes, Sabo, & Farrell, 2007, for a review of mixed findings). Thus, despite the numerous studies that have examined this connection, the relationship between athletic participation and deviant behaviors is an area in which further research is needed.

CONCLUSION

For those interested in education policy, the question concerns the role of sports as an institution given these consequences. Some research points to the mere presence or inherent nature of sport as problematic, but most scholars conclude that it is the organization of formalized sports programs that leads to negative social consequences. For example, one study found that a program designed to teach Little League baseball coaches positive-reinforcement instruction strategies resulted in an increase in the players' feelings of self-worth over the course of the season (Barnett, Smoll, & Smith, 1992). As noted, the theories underlying arguments both for and against sports participation rest on the idea that youth learn particular values in sports. Thus, life course scholars must continue to study exactly what values and behaviors are being transmitted in the current system of sports as well as designing structures to help promote prosocial characteristics. Although the debate over whether sports are "good" or "bad" seems endless, the question of how sports could be designed in ways to promote gender empowerment, norms of nonviolence, and equality seems to be quite open and worthy of greater study.

SEE ALSO Volume 1: *Academic Achievement; Activity Participation, Childhood and Adolescence; Drinking, Adolescent; Drug Use, Adolescent; Health Behaviors, Childhood and Adolescence; Interpretive Theory; Peer Groups and Crowds; Self-Esteem; Social Development; Socialization, Gender.*

BIBLIOGRAPHY

Barnett, N. P., Smoll, F. L., & Smith, R. E. (1992). Effects of enhancing coach–athlete relationships on youth sport attrition. *The Sport Psychologist, 6,* 111–127.

Coakley, J. (2001). *Sport in society: Issues and controversies* (7th ed.). Boston: McGraw-Hill.

Eccles, J. S., Barber, B. L., Stone, M., & Hunt, J. (2003). Extracurricular activities and adolescent development. *Journal of Social Issues, 59,* 865–889.

Eder, D. (with Evans, C. C., & Parker, S.). (1995). *School talk: Gender and adolescent culture.* New Brunswick, NJ: Rutgers University Press.

Eitzen, D. S. (2003). *Fair and foul: Beyond the myths and paradoxes of sport.* (2nd ed.). Lanham, MD: Rowman & Littlefield.

Fine, G. A. (1987). *With the boys: Little League baseball and preadolescent culture.* Chicago: University of Chicago Press.

Hirschi, T. (1969). *Causes of delinquency.* Berkeley: University of California Press.

Hoffman, J. P. (2006). Extracurricular activities, athletic participation, and adolescent alcohol use: Gender-differentiated and school-contextual effects. *Journal of Health and Social Behavior, 47,* 275–290.

Hughes, R., & Coakley, J. (1991). Positive deviance among athletes: The implications of overconformity to the sport ethic. *Sociology of Sport Journal, 8,* 307–325.

Kreager, D. A. (2007). Unnecessary roughness? School sports, peer networks, and male adolescent violence. *American Sociological Review, 72,* 705–724.

Mahoney, J. L., & Cairns, R. B. (1997). Do extracurricular activities protect against early school dropout? *Developmental Psychology, 33,* 241–253.

McHale, J. P., Vinden, P. G., Bush, L., Richer, D., Shaw, D., & Smith, B. (2005). Patterns of personal and social adjustment among sport-involved and noninvolved urban middle-school children. *Sociology of Sport Journal, 22,* 119–136.

Messner, M. A. (2002). *Taking the field: Women, men, and sports.* Minneapolis: University of Minnesota Press.

Miller, K. E., Melnick, M. J., Barnes, G. M., Sabo, D., & Farrell, M. P. (2007). Athletic involvement and adolescent delinquency. *Journal of Youth and Adolescence, 36,* 711–723.

National Women's Law Center. (2007, June). *The battle for gender equity in athletics in elementary and secondary school.* Retrieved May 15, 2008, from http://www.nwlc.org/pdf

Pedersen, S., & Seidman, E. (2005). Contexts and correlates of out-of-school activity participation among low-income urban adolescents. In J. L. Mahoney, R. W. Larson, & J. S. Eccles (Eds.), *Organized activities as contexts of development: Extracurricular activities, after-school, and community programs* (pp. 85–109). Mahwah, NJ: Lawrence Erlbaum.

Kyle C. Longest

STAGES OF SCHOOLING

This entry contains the following:

I. OVERVIEW AND INTRODUCTION

In the United States, education is a system that has a collection of interrelated parts working together to produce some intended outcome—in this case, a well-trained, well-rounded youth who can contribute to the nation's economy and civic institutions. Like most systems, the American educational system originated in a simple form—the one-room, multigrade schoolhouse comes to mind—but slowly diversified into an increasingly complex form, typified by the large, diverse, departmentalized high schools of the modern era. Perhaps the most striking example of this diversification is the gradual reorganization of public education into the three distinct stages of schooling—elementary, middle, and high school—described in the following entries.

These stages emerged through a constellation of sociohistorical forces, including (a) an increasing supply of students because of population growth and the abolition of child labor laws, which meant children were available to attend school during the day; (b) demands introduced by the restructuring of the labor market; and (c) evolving ideas about the developmental needs of children and adolescents. Thus, the challenge of effectively educating increasing numbers of students with widening age ranges in developmentally appropriate ways to supply labor for the industrial (and then post-industrial) economy required that the educational system be divided into more manageable, age-homogenous subsystems.

Elementary schools are typically smaller than middle or high schools. Children move across classes in groups, with nonspecialized classes and few extracurricular activities. This more personalized, generalized approach to education is considered appropriate to the developmental status of young children. High schools emerged to answer concerns about the need to tailor education to adolescents. Larger, with more differentiated curricula and an increase of choice in coursework and activities, these schools are less about fostering learning in a supportive environment and more about preparing students in practical ways to enter adult roles. Middle schools (also referred to as junior high schools) emerged somewhat later as a bridge between these two models of education. Growing too old for the protective, common curriculum of elementary school but not yet socially or psychologically mature for the impersonal, individualized structure of high school, young people in or around the pubertal transition are educated in schools that are supposed to represent a middle ground in size, instruction, and course and activity offerings.

Progress through these stages is cumulative. What is learned in one stage influences what is taught and learned in the next. In this way, achievement trajectories diverge and demographic inequalities compound across stages. This cumulative nature undermines the ability of the system to serve as a vehicle for social mobility for historically disadvantaged groups because children from these groups tend to enter the system with less developed skills and, as a consequence, lose ground over time. In theory, transitions between stages could be turning points, allowing some students to start fresh or change directions. In reality, however, transitions are more likely to be points of disruption that increase performance disparities. Policies to manage transitions and to counteract the cumulative nature of education in general, therefore, are important to making the system work more effectively and better serve society.

SEE ALSO Volume 1: *Academic Achievement; High-Stakes Testing; Policy, Education; School Culture; School Tracking; School Transitions; School Violence; Socioeconomic Inequality in Education.*

BIBLIOGRAPHY

Angus, D. L., & Mirel, J. E. (1999). *The failed promise of the American high school: 1890–1995.* New York: Teachers College Press.

Cremin, L. A. (1970, 1980, 1988). *American education: The metropolitan experience, 1876–1980* (3 vols.). New York: Harper & Row.

Powell, A. G., Farrar, E., & Cohen, D. K. (1985). *The shopping mall high school: Winners and losers in the educational marketplace.* Boston: Houghton Mifflin.

Robert Crosnoe

II. ELEMENTARY SCHOOL

Walking into any elementary school in the United States, one sees brightly colored pictures on the walls, frequently the work of students. Stars, writing samples, photographs, and science projects adorn the bulletin boards, and children move quickly and not always quietly through the halls. Classrooms look much they way they did in the mid-20th century, which is not surprising because many schools were built in the 1950s and 1960s to accommodate the large baby boom cohort. In the early 21st century, schools are more diverse ethnically and racially. Unfortunately, children living in poverty still generally attend underresourced schools with less qualified teachers, whereas children from more affluent homes attend schools with computers, libraries, and a stable and well-qualified teaching staff. Nevertheless, the structure of elementary schools is very similar throughout the United States (Stigler & Hiebert, 1999).

ORGANIZATION OF SCHOOLS

Within a school district, typically headed by a superintendent, each elementary school has a principal and, frequently, an assistant principal. The principal is considered the educational leader of the school and is responsible for all aspects of the students' education, including who their teacher is going to be and which curriculum will be used. The principal is also the lead disciplinarian. Most principals take the achievement of their students very seriously and work hard to make sure every child gets a good education. This level of accountability is mandated in the No Child Left Behind (NCLB) Act of 2001, which holds districts and schools, and thus superintendents and principals, responsible for students' achievement. This represents an important change in policy away from using school resources as a measure of quality (input) and toward a focus on student learning (outcomes; Cohen, Raudenbush, & Ball, 2003).

For the most part, children attend elementary school from kindergarten through fifth or sixth grade. Although research suggests that a kindergarten through 8th grade model may be associated with stronger student achievement overall, this organization is not as popular. Sometimes children will share classrooms with children one grade level above or below (e.g., a class might be composed of first and second graders), but this is usually a way to stretch resources. Typically students begin kindergarten when they are 6 years old and are required to stay in school until the age of 16 to 18, depending on the specific state's law.

In contrast to middle and secondary (or high) schools, elementary schools are organized to be more protective and nurturing of young children. Generally schools are smaller and children have one teacher throughout the day (although they may change for math or reading). Classes are kept relatively small, and the quality of the teacher–child relationship is an important predictor of students' academic success.

Class size varies from state to state and even from district to district. Most children will attend classes that have between 20 and 30 children. Some states, such as California, Tennessee, and Florida, have mandated class size reductions in the early grades, based on research conducted within the Tennessee school systems. This research shows that in classes of fewer than 17 students, children, especially children living in poverty, generally learned better than did children in larger classes. Most classes include both boys and girls.

A highly qualified teacher is the heart of the U.S. educational system. Indeed, the NCLB Act requires that all students be taught by highly qualified teachers. Unfortunately, deciding what "highly qualified" means is not clear. Qualifications that are typically associated with being a capable teacher do not always predict how well students will learn. For example, years of experience, having a formal teaching credential, and years of education are not consistently associated with students' achievement (Morrison, Bachman, & Connor, 2005). Research does show, however, that when teachers are warm and care about their students, are highly knowledgeable about the subject they teach, and are masterful classroom managers so that children spend most of their day in meaningful activities, children learn more and are happier.

Increasingly, the content of instruction is dictated by state standards. Leading up to the passage of the NCLB Act, states were required to develop rigorous standards for what students were expected to learn at each grade level. These standards vary from state to state. There are no national standards, which is a feature that is fairly unique to the United States among the major countries of the world. Many countries, such as England and Korea, have national standards. Individual state standards (which are available on the World Wide Web) include content for reading, language arts, writing, science, social studies, mathematics, and other subjects. The topics and expectations are frequently provided by grade or by early versus late elementary grades (e.g., kindergarten through second vs. third through fifth). Updating these state standards can be a highly contentious process as stakeholders argue about what should and should not be included. For example, when Florida updated its math and science standards in the winter of 2008, there was controversy as to whether the scientific theory of evolution should be part of the science standards.

A number of states, including California, Texas, and Florida, provide a short list of acceptable curricula in each of the subject areas. Schools and districts may select curricula only from the approved list. In many states, however, the curriculum is a district- and sometimes school-level decision. Increasingly, states, districts, and schools are requiring that the curricula be founded on solid research. Part of the reason for this is that under the NCLB Act, districts, schools, and teachers are held accountable for whether their students learn or not, based on the results of mandated tests and a number of other indicators. By insisting on research-based curriculums, teachers hope to have more effective teaching materials at their fingertips.

Another result of the NCLB Act's focus on student outcomes is more frequent formal assessment of students' reading, math, and science skills. The benefits of this policy are highly contested, but most likely such tests will become a permanent part of the elementary school experience. Fortunately, more effective ways to judge students' learning are being developed. For example, value-added scores recognize that a large part of how students perform on tests is related to the skills with which they entered school (Morrison et al., 2005). If children begin 1st grade with weak reading scores, at the end of the year—even if they make very good progress—they will still not achieve scores that are as high as students who began first grade with very strong reading skills. Value-added scores do not penalize schools where many children begin the year with weak reading and math skills—for example at schools where many students come from families living in poverty. Rather, these scores

School Resources. *Elementary school students gather around school librarian Kathy Leonardis's desk at Lincoln elementary school in Edison, NJ. The library has two classrooms in it and is shared by 4 teachers. Unfortunately, children living in poverty still generally attend under-resourced schools.* © ED MURRAY/STAR LEDGER/CORBIS.

focus on children's progress from the beginning to the end of the school year and take this into account when judging school and district performance.

HISTORICAL ROOTS OF THE ELEMENTARY SCHOOL MODEL

The roots of the elementary school model that prevails in the United States in the early 21st century can be traced back to the Revolutionary War. Although the colonial settlers founded schools and instituted mandatory education and charity schools, these schools followed European traditions (Jeynes, 2007). With the onset of the Revolutionary War, the founding fathers, and Thomas Jefferson in particular, believed that a new nation deserved a new way of educating their children. Sons no longer attended Oxford and Cambridge but rather went to Harvard, Princeton, or Yale. Many children were schooled at home or in church schools. Prior to the revolution, the education of children was a local endeavor and varied depending on the town and how much the citizens and churches valued education. Most schooling focused on literacy so that children could read the Bible.

With political independence came an increasing focus on the importance of education and the belief that schools improved children intellectually and morally. Private schools flourished. In 1789, the year George Washington became president, Massachusetts passed the Massachusetts Education Act. Towns of 200 or more citizens were required to educate all children (boys and girls) at public expense through elementary school. This act implemented equal education for girls and formed the foundation of American schools. Many of these schools were church-run charity schools, which some credit with establishing many of the characteristics of contemporary U.S. elementary education, including the professionalization of teaching, a focus on developing moral character, and the implicit belief that education can help children escape poverty and become good citizens. These early schools provided the blueprints for the U.S. public education system of the early 21st century.

CURRENT EDUCATIONAL CHALLENGES

In many respects, the U.S. public school system is a model for providing high-quality free K–12 education to all children. However, the system can be improved and, indeed, faces significant challenges that have important implications for the future. These include closing the achievement gap, individualizing instruction, and monitoring students' response to interventions so that all children receive an effective education that prepares them for the demands of the modern world.

Closing the Achievement Gap Finding ways to close the achievement gap between children who are poor and their more affluent peers is one of the most important goals of the policy and research endeavors of the early 21st century (Jencks & Phillips, 1998). How can instruction be made more effective, especially for the nation's most vulnerable children? Converging evidence is showing that high-quality early intervention, although important (Reynolds, Temple, Robertson, & Mann, 2002), is not an inoculation against academic underachievement. Instead, the impact of instruction across grades appears to be additive and cumulative. This means that students are left further behind in math and reading each year they receive ineffective instruction. It has been argued that closing the achievement gap would do more to promote social justice than any other policy (Morrison et al., 2005). Much of the NCLB Act was designed to do this. States and schools must report achievement test data by ethnic group and poverty levels and must show that all students' performances are improving. Moreover, a move to outcomes rather than resources to hold schools accountable may, in the long run, help close the achievement gap (Cohen et al., 2003). With this in mind, accumulating research on new instructional strategies is presented next: individualized instruction and response to intervention.

Individualizing Instruction/Response to Intervention The amount and type of literacy instruction children receive in the classroom consistently and systematically predicts literacy growth. Increasingly, research studies have moved away from the "either/or" days of the reading wars (phonics vs. whole language; Ravitch, 2001). Research shows a combination of methods incorporating code-focused strategies with more holistic and meaning-focused strategies is more effective in supporting students' literacy skill growth (Connor, Morrison, Fishman, Schatschneider, & Underwood, 2007; National Reading Panel, 2000). Code-focused strategies are those that are designed to help children learn that letters and sounds go together and combine to make words. Code-focused skills include the alphabet, phonological awareness, phonics, and letter and word fluency. Meaning-focused strategies are designed to teach children how to extract and construct meaning from text. Activities include reading aloud together, discussion, comprehension strategies,

and writing. Generally, as long as interventions are sustained, intensive, and balanced with at least some time spent in explicit code-focused instruction (especially for children with weaker skills), educational differences across interventions that differ in pedagogical theory are small.

New research is finding that instruction that is tailored to the individual needs of children is more effective than high-quality instruction that is not individualized (Connor et al., 2007). Using software that computed recommended amounts of code- and meaning-focused instruction for first graders based on their vocabulary and reading test scores, many teachers were able to individualize reading instruction. The software recommended substantial amounts of time in teacher-managed code-focused instruction (e.g., phonics) for children who were struggling with reading. For children who were strong readers, however, the software recommended more time in independent meaning-focused instruction (e.g., reading books, writing). The more precisely the teachers provided the recommended amounts of each type of instruction, the stronger was their students' reading skill growth. Further, when teachers fully individualized instruction, they were able to close the achievement gap between children with strong and weak vocabulary skills (Connor et al., 2007). This finding was encouraging because children living in poverty often start school with weaker vocabulary skills compared to their more affluent peers.

Another way to individualize instruction for those children who are truly struggling to learn to read is *response to intervention*. There is a vital need for effective early identification and interventions for students who struggle to learn reading and mathematics. Remediating these problems becomes increasingly difficult over time. Once students fall behind, even the most powerful remedial interventions are often unable to help them (Torgesen, 2005). On average, children who have weak reading skills at the end of second grade almost never acquire average reading skills by the end of elementary school.

THE ROLE ELEMENTARY SCHOOLS PLAY IN THE LIVES OF STUDENTS AND IN SOCIETY AS A WHOLE

School and classroom environments are among the strongest influences on children's early life experiences. Thus, children who attend good schools have a lifelong advantage over children who attend underresourced and ineffective schools. Effective schools share common characteristics (Taylor & Pearson, 2002; Wharton-McDonald, Pressley, & Hampston, 1998), including a safe and orderly environment, strong leadership, high expectations for student achievement, an emphasis on academics, uninterrupted time devoted to literacy and mathematics instruction, the use of assessment to evaluate student progress and guide instruction, and good classroom management.

The climate of the classroom and the social and emotional support provided by effective teachers is important. Children generally learn better and develop important social skills in classrooms that provide high levels of social and emotional support as well as instructional support. These learning environments appear to be especially important for children who have weaker social skills in kindergarten. Students' interest in learning, or *engagement*, and their motivation to learn also contribute to their success in school. Thus the more teachers understand about how to engage and motivate their students to learn while spending time in meaningful instruction activities that take into account children's skills, the more effective classrooms and schools will be.

Effective schools are important for students because early success builds a foundation for a lifetime of achievement. Children who achieve well in school become contributing members of society (Shonkoff & Phillips, 2000; Snow, Burns, & Griffin, 1998). At the same time, school failure is costly for students and society alike. For example, children who cannot read well are more likely to be retained a grade, to enter special education, and to drop out of school, and ultimately are more likely to enter prison and be unemployed. Elementary schools serve a crucial role in American democracy.

SEE ALSO Volume 1: *Child Care and Early Education; Learning Disability; School Transitions.*

BIBLIOGRAPHY

Cohen, D. K., Raudenbush, S. W., & Ball, D. L. (2003). Resources, instruction, and research. *Educational Evaluation and Policy Analysis, 25,* 119–142.

Connor, C. M., Morrison, F. J., Fishman, B. J., Schatschneider, C., & Underwood, P. (2007). The early years: Algorithm-guided individualized reading instruction. *Science, 315,* 464–465.

Jencks, C., & Phillips, M. (1998). *The Black–White test score gap.* Washington, DC: Brookings Institution Press.

Jeynes, W. H. (2007). *American educational history: School, society, and the common good.* Thousand Oaks, CA: Sage.

Morrison, F. J., Bachman, H. J., & Connor, C. M. (2005). *Improving literacy in America: Guidelines from research.* New Haven, CT: Yale University Press.

National Reading Panel. (2000). *National Reading Panel report: Teaching children to read: An evidence-based assessment of the scientific research literature on reading and its implications for reading instruction* (NIH Publication No. 00-4769). Washington, DC: U.S. Department of Health and Human Services, Public Health Service, National Institutes of Health, National Institute of Child Health and Human Development.

Ravitch, D. (2001). It is time to stop the war. In T. Loveless (Ed.), *The great curriculum debate: How should we teach reading and math?* (pp. 210–228). Washington, DC: Brookings Institution Press.

Reynolds, A. J., Temple, J. A., Robertson, D. L., & Mann, E. A. (2002). Age 21 cost–benefit analysis of the Title I Chicago child–parent centers. *Educational Evaluation and Policy Analysis, 24,* 267–303.

Sanders, W. L., & Horn, S. P. (1998). Research findings from the Tennessee Value-Added Assessment System (TVAAS) database: Implications for educational evaluation and research. *Journal of Personnel Evaluation in Education, 12,* 247–256.

Shonkoff, J. P., & Phillips, D. A. (Eds.). (2000). *From neurons to neighborhoods: The science of early childhood development.* Washington, DC: National Academy Press.

Snow, C. E., Burns, M. S., & Griffin, P. (Eds.). (1998). *Preventing reading difficulties in young children.* Washington, DC: National Academy Press.

Stigler, J. W., & Hiebert, J. (1999). *The teaching gap: Best ideas from the world's teachers for improving education in the classroom.* New York: Free Press.

Taylor, B. M., & Pearson, P. D. (Eds.). (2002). *Teaching reading: Effective schools, accomplished teachers.* Mahwah, NJ: Lawrence Erlbaum.

Torgesen, J. K. (2005). Remedial interventions for students with dyslexia: National goals and current accomplishments. In S. O. Richardson & J. W. Gilger (Eds.), *Research-based education and intervention: What we need to know* (pp. 103–124). Baltimore, MD: International Dyslexia Association.

Wharton-McDonald, R., Pressley, M., & Hampston, J. M. (1998). Literacy instruction in nine first-grade classrooms: Teacher characteristics and student achievement. *Elementary School Journal, 99,* 101–128.

Carol McDonald Connor

III. MIDDLE SCHOOL

Middle schools play an important role in American society. Because early adolescence is a time of rapid cognitive, emotional, and social change, middle schools need to be places that promote this growth while fostering a sense of belonging. Contemporary research offers a clear understanding of the importance of teacher caring and a sense of belonging for students' academic and social well-being, especially at the middle school level. Yet many middle schools are currently not organized to promote belonging and a sense of community.

THE HISTORY AND ORGANIZATION OF MIDDLE SCHOOLS

Initially, the middle school concept was developed to deal with the multiple changes and wide range of maturing students undergo between the sixth and eighth grades. Middle schools replaced junior high schools that typically covered seventh through ninth grades, as well as the kindergarten through eighth grade model. Educators emphasized project-based learning within teams of 2 to 5 teachers. The advantage of project-based learning for this age group is that students can work on different aspects of the same project, reflecting their different levels of development. The initial concept also emphasized learning communities where students gained a sense of belonging by participating in decisions about topics of study and the social operation of the classroom.

The ways in which most middle schools are organized does not correspond with this initial concept. Rather than team-based instruction, most schools are organized with students taking courses from different teachers. According to Anthony Jackson and Gayle Davis, authors of *Turning Points 2000: Educating Adolescents in the 21st Century* (2000), the most common form of curriculum organization is tracking, which assigns students to different classes based on their actual or perceived capacity to learn. In this way middle schools—like the junior high schools that preceded them—are really "mini" high schools, lacking an organization that meets the special needs of this age group.

Educators continue to seek organizational changes in line with the initial middle school concept. When they describe aspects of effective schools, they continue to emphasize the importance of small learning units, team teaching, and other aspects that contribute to a sense of community and belonging. Jackson and Davis cite evidence from numerous studies showing that small schools lead to better educational outcomes and advocate dividing students in large schools into "houses." While they recommend houses numbering no larger than 250 students, there is evidence of especially positive student outcomes with houses of 120 students or fewer. The houses should be diverse—reflecting the ability levels, ethnicities, and social classes of the school population—given the problems that have been identified with dividing students on the basis of perceived or actual ability.

Entwined with the concept of houses is the concept of team teaching, which groups teachers covering the core subjects of math, science, social studies, and language arts. Ideally, these teams of teachers would stay together for a minimum of 5 years. Still other organizational aspects of effective schools include keeping students with the same team of teachers for more than one grade and using differentiated curriculum, which varies both by content and process according to different levels of readiness, interest, and learning styles.

THE ROLE OF MIDDLE SCHOOL AND SENSE OF BELONGING

Middle schools have social value as well as academic value for early adolescents. Schools are most successful when students see their schools as communities and experience a sense of belonging. This may happen naturally in rural areas where middle schools are often the focus of community life. It also occurs in affluent suburban areas that have the resources to create exemplary schools with low teacher-student ratios. In other cases, administrators and teachers have adopted the model of smaller houses with project-based learning to create definable learning communities.

Research on middle schools has focused on students' sense of belonging, drawing upon the theories of John Dewey (1859–1952) and Lev Vygotsky (1896–1934). Both theorists view education as a social process that is most successful when students are part of meaningful groups. There is now considerable evidence that teacher support, peer support, and a sense of belonging are important for students academically.

Students who perceive that their teachers support them have higher expectations for academic success, value education more, and often have higher academic motivation than do other students. Furthermore, when teachers were perceived as supportive, as promoting interaction with peers, and as promoting mutual respect among students, students showed an increase in motivation and engagement. However, when teachers were perceived as promoting academic comparisons and competition with other students, students experienced a decrease in motivation and engagement.

Other studies reveal the ways in which teacher caring influences students' social and emotional welfare. One study found that teacher caring was linked to students' goals for altruistic behavior and social responsibility. In another study, the school was shown to have more impact on students' emotional distress and acts of violence than the family at the middle school level (Resnick et al., 1997). In her review of research on a sense of school belonging, Karen Osterman (2000) notes that while an experience of belonging is important at all age levels, it is especially important during middle school.

Given the importance of teacher and peer support at this age level, some studies have examined the degree to which different students feel supported by their teachers and peers. While a sense of belonging led to positive outcomes for all students, African-American and Hispanic students in low-income schools have much less sense of belonging in school than do White students in middle-income schools (Goodenow & Grady, 1993). Even though peer support is often important for a male student's academic well-being, males in one study had more negative relationships with classmates (Wentzel & Caldwell, 1997). Osterman (2000) reports that secondary schools, on average, are less supportive than elementary schools. She indicates that tracking adds to this trend, with students in lower tracks expressing much lower levels of perceived teacher caring and support.

PEER CULTURES WITHIN MIDDLE SCHOOL

Although not studied as extensively as school belonging, another theme in the research on middle schools is the study of peer culture. In their 1995 book, *School Talk: Gender and Adolescent Peer Culture*, Donna Eder, Catherine Evans, and Stephen Parker found that students in the middle school they studied defined popularity as being well-known by other students and not necessarily as being well-liked. Because of the limited number of extracurricular activities, cheerleading and certain male sports became very important for gaining visibility in this school. The large size of this school (more than 750 students), combined with the limited opportunities for visibility, meant that many students reported feeling like social failures. While high schools also provide students with some visibility through extracurricular activities, the fact that high schools offer a much wider range of activities means many more students find ways to become socially connected with their peers than do middle school students.

Other research suggests that high-status groups, such as preps, achieve prestige because they reproduce mainstream values. In their study of cheerleading and school board policy, Pamela Bettis and Natalie Adams (2003) found that a school's policy to reduce race and social class bias in the composition of its cheerleading squads failed to be effective 5 years after being implemented. They found that cheerleaders tend to be selected on the basis of their petite appearance and ability to mask feelings with a smile, characteristics that were more common in the middle-class and White student culture.

Eder and her colleagues (1995) also found that the social isolates in the school they studied were frequent victims of bullying. Because students often take out their social anxieties on the most vulnerable students in the school, those who lacked a friendship group were targets of most of the bullying. While these students were initially isolated, in part, because of their appearance, lower intelligence, or atypical gender behavior, it was even more difficult for them to enter a peer group once they became the target of bullying.

Much of the research on peer culture focusing on peer harassment indicates that bullying occurs most often between the sixth and eighth grades. While there is little variance across urban or rural settings, the targets of harassment vary depending on the racial composition of the school. Sandra Graham (2004) found that the majority ethnic groups were perceived to be the aggressors (in this case, blacks and Hispanics), whereas the minority ethnic groups were perceived to be the victims (in this case, other racial groups as well as Whites). As in her other research, the bullies in this study had high social standing, while the victims reported more loneliness and more social anxiety than non-victims. Interestingly, when classrooms were ethnically balanced, students felt less lonely and less socially anxious.

The research on peer status and peer harassment cannot be explored fully here, but it is important to consider aspects of the middle school environment that could contribute to an emphasis on social rankings and bullying, such as the limited number of extracurricular opportunities leading to visibility for a few students in many schools. Also, when adults rank and label students through tracking assignments, they could be inadvertently promoting the ranking and labeling of students by each other. These larger organizational aspects can have a strong impact on peer culture as well as classroom factors such as those mentioned above.

DIRECTIONS FOR FUTURE RESEARCH

Middle school researchers know more about social psychological processes, such as the influence of a sense of belonging, than the influence of peer and teacher cultures on middle school students. Studies, such as the one on cheerleading policy, are a reminder that unless policy decisions take students' culture into account, they will meet with limited success (Bettis & Adams, 2003). Had the policy makers paid attention to the way cheerleading selection was shaped by middle-class and White peer culture, they might have crafted a more effective policy for limiting biases in selection. Currently most studies of peer culture are at the high school and elementary level, so that little is known about how peer cultures vary depending on the backgrounds of students and the organization of the school.

There is even less research on teacher culture at the middle school level. Some researchers have suggested that teachers may react to negative elements in peer culture by assuming that adolescents are naturally unkind. More research is needed to better understand the nature of teacher culture in diverse school settings.

Also needed is a better understanding of the impact of organizational changes beyond the classroom. Studies of the process and outcomes of school changes, such as James Rourke's *Breaking Ranks in the Middle* (2006), will benefit the understanding of these broader influences. In particular more research is needed comparing the influence of organizing students in diverse houses as compared to homogenous tracks, as well as comparing schools with intramural activities to those with extracurricular activities in terms of their influence on peer rankings and bullying. It is likely that by providing more students with opportunities for visibility and participation through intramural activities, students might focus less on peer status and isolates might have an easier time joining peer groups.

The impact of middle school teacher training is another subject deserving of attention. Currently most states do not require specialized preparation at this level, using instead the elementary versus secondary model. This may explain, in part, why many middle schools are structured like high schools, as most middle school teachers are trained along with high school teachers, with many using middle schools as their starting point on the career ladder.

Given the needs identified with this age group, middle school training might best employ what has been called a suppositional, as opposed to a propositional, approach to teaching. In this approach teachers are encouraged to invite student input as part of exploring a topic that might go in many directions. Teacher training should also include knowledge about organizational and cultural influences on students' experiences along with cognitive and psychological ones. By comparing teachers with specialized middle school training versus those with secondary training, one can better assess the impact of specialized training on teachers' and students' experiences.

In summary, middle schools were designed to address the special needs of and wide range of maturity in students between sixth and eighth grade. Unfortunately, many schools in the early 21st century are organized more like high schools, with the common use of curriculum tracking and the focus on extramural versus intramural activities. This may be due, in part, to the lack of specialized training for the middle school level. It may also be due to a lack of awareness of the cultural and organizational influences within middle schools. Until there is a better understanding of all the organizational and cultural factors at play, schools are not likely to foster the sense of belonging that research shows is so important at this period in the life course.

SEE ALSO Volume 1: *Academic Achievement; Bullying and Peer Victimization; Human Capital; Peer Groups and Crowds; Policy, Education; School Tracking; School Transitions; Segregation, School.*

BIBLIOGRAPHY

Bettis, P., & Adams, N. (2003). The power of the preps and a cheerleading equity policy. *Sociology of Education, 76*(2), 128–142.

Eder, D., Evans, C., & Parker, S. (1995). *School talk: Gender and adolescent peer culture.* New Brunswick, NJ: Rutgers University Press.

Goodenow, C., & Grady, K.E. (1993). The relationship of school belonging and friends' values to academic motivation among urban adolescent students. *Journal of Experimental Education, 62*(1), 60–71.

Graham, S. (2004). Ethnicity and peer harassment during early adolescence: Exploring the psychological benefits of ethnic diversity from an attributional perspective. In T. Urdan & F. Pajares (Eds.), *Educating adolescents: Challenges and strategies.* Greenwich, CT: Information Age Publishing.

Jackson, A., & Davis, G. (2000). *Turning points 2000: Educating adolescents in the 21ˢᵗ century.* New York: Teachers College Press.

Osterman, K. (2000). Students' need for belonging in the school community. *Review of Educational Research, 70*(3), 323–367.

Resnick, M., Bearman, P. S., Blum, R. W., Bauman, K. E., Harris, K. M., Jones, J., et al. (1997). Protecting adolescents from harm. *Journal of the American Medical Association, 278,* 823–832.

Rourke, J. (2006). *Breaking ranks in the middle: Strategies for leading middle level reform.* Providence, RI: Education Alliance.

Wentzel, K. R., & Caldwell, K. (1997). Friendships, peer acceptance, and group membership: Relations to academic achievement in middle school. *Child Development, 68*(6), 1198–1209.

Donna Eder

IV. HIGH SCHOOL

In the United States, high school is undoubtedly one of the most important institutions in an individual's life course. The high school plays two primary roles in the lives of students and society, the importance of which cannot be overestimated. The first role of the high school is that of sorting and selecting students for differentiated life tracks and opportunities. The second role is that of socializing students to community norms, roles, and behaviors. The ways in which students experience these processes holds profound implications for college-going, access to middle-class jobs, and quality of life more generally.

HISTORY AND ORGANIZATIONAL STRUCTURE

The American high school has its roots in the nation's earliest years. The high school developed in response to economic stimulus and cultural inspiration. An increasingly diversified, industrial economy, coupled with urbanization and modernization, created the need for specialized and advanced forms of training. At the same time, early educational reformers—responsible for putting a system in place that largely remains in the 21st century—found their philosophical footing in four cultural currents. As much as economic demand, the American high school was born from the social needs of early citizens, with the public (common) school system fueled by four primary cultural facts: the democratization of politics; the struggle to maintain social equality; changes in conceptions of the individual and society; and the rise of nationalism (Cremin, 1951). Educational founders and reformers, working from within government bureaucracies, felt confident that a public, specialized school system would meet demand for labor, guard against the creation of a caste system, and prepare youth for civic participation. Universal schooling would also Americanize increasing numbers of immigrants arriving in the 1830s and 1840s on matters pertaining to the primary roles and values of the new nation.

Early reformers such as Horace Mann (1796–1859) agreed that education should be free and available to all and that a curriculum should include essentials for everyday living, including reading, writing, spelling, and arithmetic, as well as instruction on moral adequacy and responsible citizenship. Free public education was first put into legislation following a report to the Philadelphia legislature in 1829, which stated that "all children, regardless of means, should have access to a liberal and scientific education." The same bill called for the establishment of high schools, based on a union of agricultural and mechanical instruction with literary and scientific instruction (Cremin, 1951).

Early high schools followed the village model, and were largely structured and run by local authorities. As cities and towns grew, and with increasing migration to urban centers in the mid- 19th century, high schools underwent important structural and governance changes. Tension emerged between a model of local governance and bureaucratic control. A struggle between community control and professionalism was waged openly until later in the century when two important trends took the country by force. As the industrial revolution advanced, the distinction between work and home, and the character of interactions, changed appreciably. Simultaneously, with the expansion of science, individuals relied less on folk knowledge and public schools lifted up rationalism and science-based expertise. These trends favored professionalism and the systematizing of U.S. high schools, both in terms of curricular decisions and with its embrace of corporate, centralized governance. Schools developed different tracks and specializations, and high schools began to occupy an increasingly important position in the life course, as knowledge gained ground on culture and social pedigree. This period of time came to be known as the Progressive Era (1890–1920), which was associated with efficiency, expertise, and the public service of disinterested elites. It was also an era that resonates strongly with contemporary educational reforms.

In the 20th century, high schools experienced massive expansion and growth and went through various structural reforms. In general, high schools became larger and more complex, and more sophisticated and ambitious in their goals with increased course offerings and specializations. This offered, in some part, changing notions of intelligence, and the emergence of psychometrics. Furthermore, with time, schools played an increasingly important role in students' lives, due in part to the passage of child labor laws, in addition to increasing demands from employers for certified credentials—a holdover from the Progressive Era. As schools began to matter more, the system of sorting and selection, as well as principal teaching philosophies developed dramatically. The latter part of the 20th century through the present day is defined most notably by its increasing expansion, adoption of desegregation laws in the 1960s and 1970s, and a decline in vocational training in favor of liberal education in the service of the emerging national standard of college for all.

STRATIFICATION AND SOCIALIZATION FUNCTIONS

Since its foundation, the high school has been a selective institution. During its earliest years only selected populations (primarily White males) were considered to be candidates, despite its availability "to all" per law. When high school attendance became compulsory in the 20th century, sorting and selection processes occurred within the schools. In spite of early Americans' resistance to a caste system of economic and social stratification, the United States has developed a distinct class system where some individuals occupy privileged positions and others remain at the bottom of the class ladder. Contrary to the position of social Darwinists, social hierarchies do not happen naturally but rather are built and maintained by a series of social institutions.

Writing in the mid-20th century, Harvard sociologist Pitirim Sorokin (1889–1968) summarized, "Within a stratified society, there seems to exist not only channels of vertical circulation, but also a kind of a 'sieve' within these channels which sifts the individuals and places them within society" (Sorokin, 1959 [1927], p. 182). Taking his cues from Max Weber's (1864–1920) classic writings on the interchange between modern democracy, bureaucracy, and education, Sorokin equates this "sieve" with social institutions such as the church, the army, the family, professional organizations, and, importantly, the school. In scholarship that examines the sorting functions of education, there is a tension among scholars about whether schools serve to advance social mobility or reinforce existing social hierarchies. This debate is discussed below.

High School Fashion Show. *High School students standing in a hallway wait to participate in a fashion show. High School socializes students to community norms, roles, and behaviors.* © **ARISTIDE ECONOMOPOULOS/STAR LEDGER/CORBIS.**

In addition to its sorting and selecting role, the high school is one of the most important socializing institutions in the individual life course. Writing in the latter part of the 19th century, Émile Durkheim (1858–1917) was among the first to acknowledge this role: "The aim of education, is, precisely, the socialization of the human being; the process of education, therefore, gives us in a nutshell the historical fashion in which the social being is constituted" (Durkheim, 1982 [1895], p. 6). Schools are not merely places where students sit, isolated from each other, in dialogue with words on a page. On the contrary, schools are complex social sites where students interact with teachers, and, not insignificantly, with their peers, during a very formative developmental period. To dismiss the social elements of the high school experience is to come away with an incomplete understanding of its role in the lives of students and society.

Adding significantly to knowledge on schools as sites of social activity, where certain behaviors are learned, rewarded, and sanctioned, was sociologist James Coleman (1926–1995). Writing first in 1961, Coleman published *The Adolescent Society*, which documented the importance of peer relationships for student learning. Later in the decade, Coleman led a research team to assess the impact of desegregation in schools. The subsequent report, *Equality of Educational Opportunity* (1966), commonly known as *The Coleman Report*, focused again on peer effects. The Coleman Report stirred up the world of educational research with an examination of desegregation. Most notable of his findings was the assertion that White schools were outperforming Black schools, even with equal funding. For Coleman, the difference in student achievement between and within high schools was influenced by students' communities, particularly by the cultural and social capital of parents and caregivers, and the degree to which parents and communities maintained meaningful contact and intergenerational social closure—a concept Coleman defined as social capital.

Another high profile finding in his study was that Black students benefited from being in integrated schools. Coleman's work led to the introduction of school busing policies, which were intended to diminish the lingering impact of school segregation, rooted in residential segregation. Following a decade of research on White flight and busing in the 1970s, during which Coleman retracted his commitment to busing policies, he launched yet another major strand of inquiry that compared student achievement in Catholic, private, and public high schools. Holding economic and racial status constant, Coleman concluded that students in Catholic high schools outperformed their peers in public schools. He attributed the advantages to greater parent engagement, better school discipline, and a strong sense of school community.

The importance of the high school as a socializing institution is being increasingly taken up by social scientists that have grown increasingly concerned with violence and disorder in schools, as well as with increasing incarceration rates of young adults that the public has associated with school inadequacies (e.g., 73% of adults believing that poor quality of schools is a critical or very important feature of increasing crime [Gallup, 1994, p. 184–185]). In particular, sociologists of education have argued that public school capacity to socialize youths requires the moral authority of school actors, which has been undermined both by adversarial legal challenges to school disciplinary practices as well as by policies that have limited educator's professional discretion by requiring mandatory zero-tolerance punishments (Arum, 2003).

RESEARCH DIRECTIONS

Within research on the U.S. high school, there is broad consensus about its importance in the life course. Students' high school experiences and performance have profound implications for quality of life and opportunity. Apart from this general agreement, however, researchers tend to fall into two different camps about how high schools impact student life chances. One group tends to view high schools as truly democratic places where students gain access to social mobility, and where special examinations allow for talented, hardworking students to rise to the top, regardless of social background. This is not to say that schools produce equality; on the contrary, according to Sorokin, "even in the most democratic school, open to everybody, if it performs the task properly, is a machinery of 'aristocratization' and stratification of society, not of 'leveling' and 'democratization'(Sorokin, 1959 [1927], pp. 189–190.). However, while the winners in this system constitute a new privileged stratum, schools themselves do not hinder social mobility (Weber, 1946). Other researchers, however, have shown that rather than promote social mobility, high schools themselves actually work to maintain and reinforce social hierarchy. Exemplary of research in this trend is Adam Gamoran's 1992 study, *Is Ability Grouping*

Equitable?, where he demonstrates that the common high school practice of ability grouping reinforces divisions that contribute to inequality through the separation of students from different racial, ethnic, and social backgrounds.

In line with Coleman's examination of school environment and peer effects, contemporary research on high schools is largely focused on the relationship between schools and communities. Unlike previous work, however, current research takes a closer look at what is meant by a school community. At present and going forward, researchers are working to identify and define the political, institutional, and network dimensions of school-community relationships. In particular, scholarship on schools is emerging in the neo-institutional tradition. Leaders in this field argue that schools are embedded not only in local communities, but are also situated in larger organizational communities, or what neo-institutional scholars refer to as *organizational fields*. Organizational fields include organizations directly related to schools, such as regulating bodies, union organizations, and professional schools, or that share similar structural attributes, such as other schools in the same school district. Researchers are currently working to demonstrate how these organizational relationships affect school institutional practices and subsequently are thus implicated in the shaping of life course trajectories of individuals. The field of institutions and individuals subject to such studies is vast: 3,303,000 U.S. high school students are on track to graduate in the 2007–2008 school year, 2,988,000 from public schools and 315,000 from private schools. These numbers are consistent with steady growth over time in percentages of the adult population completing high school. The percentage of the American adult population (ages 25 years and over) who have completed high school (including both diplomas and high-school equivalency exams) climbed from 82% to 86% from 1998 to 2008.

SEE ALSO Volume 1: *Academic Achievement; Coleman, James; High School Organization; Human Capital; Policy, Education; School Transitions; Segregation, School; Social Capital; Socialization.*

BIBLIOGRAPHY

Arum, R. (2000). Schools and communities: Ecological and institutional dimensions. *Annual Review of Sociology, 26*(1), 395–418.

Arum, R. (2003). *Judging school discipline: The crisis of moral authority.* Cambridge, MA: Harvard University Press.

Blau, P. M., & Duncan, O. D. (1967). *The American occupational structure.* New York: Wiley.

Blossfeld, H. P., & Shavit, Y. (Eds). (1993). *Persistent inequality: Changing educational attainment in 13 countries.* Boulder, CO: Westview Press.

Coleman, J. S. (1961). *The adolescent society: The social life of the teenager and its impact on education.* Glencoe, IL: Free Press.

Coleman, J. S. (1966). *Equality of Educational Opportunity.* Washington, DC: U.S. Department of Health, Education, and Welfare.

Coleman, J. S., Hoffer, T., & Kilgore, S. (1982). *High school achievement: Public, Catholic, and private schools compared.* New York: Basic Books.

Cremin, L. A. (1951). *The American common school: An historic conception.* New York: Teachers College Press.

Durkheim, E. (1982). *The rules of the sociological method.* (W. D. Halls, Trans.) New York: Free Press. (Original work published 1895.)

Gallup, G. (1994.) *The Gallup Poll: Public Opinion 1993.* Wilmington, DE: Scholarly Resources.

Gamoran, A. (1992). Is ability grouping equitable? *Educational Leadership, 50*(2), 11–17.

Oakes, J., Gamoran, A., & Page, R. N. (1992). Curriculum differentiation: Opportunities, outcomes, and meanings. In P. W. Jackson (Ed.), *Handbook of research on curriculum: A project of the American Educational Research Association.* New York: Macmillan.

Sorokin, P. (1959). *Social and Cultural Mobility.* New York: Free Press. (Original work published 1927.)

Tyack, D. B. (1974). *The one best system: A history of American urban education.* Cambridge, MA: Harvard University Press.

Weber, M. (1946). The rationalization of education and training. In H. H. Gerth & C. W. Mills (Eds. & Trans.), *From Max Weber: Essays in sociology.* New York: Oxford University Press.

Abby Larson
Richard Arum

STEPFAMILIES

SEE Volume 1: *Family and Household Structure, Childhood and Adolescence;* Volume 2: *Remarriage.*

STRESS IN CHILDHOOD AND ADOLESCENCE

SEE Volume 1: *Mental Health, Childhood and Adolescence; Resilience, Childhood and Adolescence;* Volume 2: *Risk.*

SUICIDE, ADOLESCENCE

DEFINING SUICIDE AND SUICIDALITY

Suicide refers to causing one's own death. This act may be pro-active or "positive," when one actively takes one's own life, or "negative," when one does not take actions to save one's own life when in danger. Distinctions are also made between direct and indirect suicide. *Direct suicide* refers to actions that one knows will almost certainly lead to death, while *indirect* refers to actions that one may knowingly take that could potentially lead to death, such as in entering a very risky situation like using potentially lethal drugs.

Scholars of suicide emphasize that suicidal behavior involves a continuum. At one end is *suicidal ideation,* which is talking or thinking about ending one's life. The next point on the continuum is developing a plan for how one would do this, followed by an attempt to kill oneself, generally referred to as a "suicide attempt." The final point of the continuum is an actual completed suicide. The term *suicidality* is sometimes used to refer to the entire continuum and also is used to describe attitudes and behaviors that may indicate that a person is at risk of committing suicide.

When the general public and scholars discuss teen suicide, they generally refer to the teenage years, from 13 to 19. However, because government data are almost always reported in 5-year age ranges, statistical reports of teen suicide usually involve the age range of 15 to 19 and, occasionally, 10 to 14 years of age.

VARIATIONS IN TEEN SUICIDE

Teens are not at equal risk of suicide. Studies throughout the years have noted differences by developmental status or age; by demographic characteristics, especially race and gender; and over historical time periods. Rates also vary between countries. International variations in suicide rates of teens generally mirror cross-cultural variations in suicide rates of adults.

Developmental Trends Suicide rates are higher for older teens than for younger teens. For instance, in 2005 suicide rates for youth age 15 to 19 in the United States were 12.1 per 100,000 for males and 3.0 per 100,000 for females. For 10- to 14-year-olds, the rates were 1.9 for males and 0.7 for females (Sourcebook of Criminal Justice Statistics Online, 2007). Changes that accompany adolescence, often involving concerns with body changes, emotions, and evolving sexuality, are associated with these developmental differences in suicide rates. Studies also indicate that factors related to the risk of suicide may vary developmentally, with some risk factors having a more powerful influence on suicide attempts in the younger teen years than the older teen years.

Demographic Variations Demographic variables are strongly related to completed suicides and suicide attempts. As noted above, completed suicides are much higher for males than for females. This pattern appears at all age ranges and cross-culturally. In contrast, non-fatal suicide attempts are more common among females than among males. Data from the Centers for Disease Control, for a representative sample of students in grades nine through twelve in 2005, indicate that 10.8% of females and 6.0% of males reported one or more attempts of suicide in the year preceding the survey (CDC, 2006).

Pronounced differences exist between race/ethnic groups in suicide deaths and attempts. Suicide rates are higher for Whites than for Blacks at all age groups and for both sex groups. For instance, for 15- to 19-year-olds, the rates for Whites are usually about twice as high as those for Blacks (13.2 per 100,000 versus 7.2 for males, 3.1 versus 1.4 for females in 2005) (Sourcebook, 2007). Given smaller population sizes, the data for other race and ethnic groups are somewhat less reliable. Consistently, however, American Indian/Alaskan Native young people have rates that are much higher than those of other race or ethnic groups. In addition, data from surveys of adolescents indicate that gay, lesbian, and bisexual teens and young adults have higher rates of both suicide ideation and suicide attempts than heterosexual teens (Silenzo, Pena, Duberstein, Cerel & Knox, 2007).

Historical Trends Suicide rates have fluctuated over historical time for people in all age groups. Within the United States, rates are often higher during times of economic distress, such as the era of the Great Depression in the 1930s. Changes in rates of suicide for teens have often paralleled the changes for the total population. For instance, the suicide rate for 55- to 59-year-olds in 1930 was 38.2 per 100,000, but by 1960 was substantially lower at 23.6. The rates for teens also declined, although not as dramatically: from 4.4 per 100,000 in 1930 to 3.6 per 100,000 in 1960. Life course scholars and demographers refer to changing patterns over time as a "history" or "period" effect.

For many years the suicide rates for adults were substantially higher than the rates for teens, reaching their highest levels at older ages, usually among men ages 65 and older. Life course scholars use the term *age effects* to describe these differences between age groups. The dramatic differences between the ages in the risk of suicide was noted by the earliest observers of suicide rates, the "moral statisticians" of 19th-century Europe, and were long thought to be immutable.

This traditional pattern of age effects changed quite dramatically, however, in recent decades. For instance, in 1930, suicide rates for adults in their late 50s were more than eight times as high as those for teens 15 to 19. By 1960 the ratio had declined slightly, to 6.6. But by 1995, adults in this age group had suicide rates that were only 1.2 times as high as teens. This resulted from both a substantial decline over time in the rates of adults (to 12.9/100,000 for 55- to 59-year-olds in 1995) and a very sharp rise in the rates of teens (to 10.5/100,000 for 15- to 19-year-olds in that year). These changing ratios between youth and adult suicide have persisted through the start of the new millennium. Some scholars suggest that these changes reflect birth cohort effects, patterns of suicide that differ between groups of people born in different eras. They suggest that more recent birth cohorts are much more prone to suicide at younger ages than were earlier cohorts (Stockard & O'Brien, 2002a, 2002b).

THEORETICAL EXPLANATIONS OF TEEN SUICIDE

Explanations of teen suicide are found in both psychological and sociological literature.

Psychological Explanations An important influence on all psychological explanations of suicide, both teen and adult, has been the work of Sigmund Freud, who wrote from the late 19th century through the 1930s. Many of these writings focused on aggressive urges and motivations, but, during his lifetime, Freud's views on the sources of aggression and suicide changed. In his earlier writings Freud suggested that suicide involved aggression turned inward, toward one's self, an action that is more likely to happen when an individual is anxious and fears punishment by others (Freud, 1957/1917). In his later years Freud rejected this formulation and suggested that aggression, including that against oneself, reflected a "death instinct" rather than simply a reaction against frustration (Freud, 1955/1920). One of the most notable scholars to follow in Freud's footsteps was Karl Menninger, also a psychoanalyst. Menninger suggested that every act of suicide includes three elements: hate, or the wish to kill; guilt, or the wish to be killed; and hopelessness, or the wish to die. Like Freud, he concluded that suicide is "disguised murder," motivated by an unconscious wish to kill another (Menninger 1938, p. 55).

Other scholars, especially those influenced by empirical psychology and tenets of learning theory, built upon Freud's earlier analyses, dismissing the notion of a "death instinct." The most influential work in this tradition is the frustration-aggression hypothesis developed by John Dollard and his colleagues in the late 1930s. They originally contended that "aggression is always a consequence of frustration" (Dollard, Miller, Doob, Mowrer et al. 1939, p. 27), but later modified the hypothesis to state that aggression can be one of a number of responses to frustration (Miller, Sears, Mowrer, Doob et al, 1941).

Following the early Freudian tradition, they suggested that the usual target of one's aggression would be the perceived source of the frustration, but that at times the aggression could be displaced onto others, including oneself. Thus, like previous scholars, they generally saw suicide as displaced homicide, but also suggested that suicide could occur when the self was seen as the source of frustration. Their analysis strongly influenced the work of later clinicians, as well as several sociologists.

Common to the work of all of the scholars working within this broadly defined Freudian heritage is a concern with the strength of the superego or conscience. A strong superego can impel individuals to check and control aggressive intentions toward others. At the same time, an excessively strong superego may prompt individuals to turn aggressive actions against the self. While few contemporary social scientists use an explicitly Freudian, or even neo-Freudian, approach, many are concerned with the circumstances related to the development of the conscience and self-control. In addition, some sociological work on lethal violence that appeared in the mid-20th century often built, at least implicitly, upon the neo-Freudian assumption that suicide (and homicide) are related to frustration, but translated this understanding into a macrolevel analysis of rates (Whitt, 1994).

Sociological Writings Psychological explanations of suicide focus on characteristics of individuals and, at times, their relations with others, generally trying to understand why one individual rather than another might choose to end his or her life. In contrast, sociological explanations focus on characteristics of societies and sub-populations of societies and the reason that some groups have higher rates of suicide than others.

The most important contributor to sociological analyses of suicide was Emile Durkheim, a French scholar writing in the late 19th and early 20th centuries and now considered one of the founders of the discipline. His book *Le Suicide*, published in 1897, has long been considered a classic contribution. Durkheim delineated four different types of self-destructive behavior, and contended that each type of suicide was related, in different ways, to varying amounts of social regulation and integration. He gave relatively little attention to two of these categories, altruistic and fatalistic suicides, which he suggested reflected reactions to very high levels of integration and regulation. Altruistic suicide may involve killing one's self for the good of society, whereas fatalistic suicide occurs when one sees no hope in the future, such as in the case of slaves who are confined by societal rules. The bulk of his writings and statistical analyses, as well as those of later generations of sociologists, concentrated on the impact of too little social integration and regulation

and increases in what Durkheim termed egoistic and anomic suicide.

Despite Durkheim's lengthy attempts to distinguish these two varieties of suicide, sociologists have long noted that his distinctions are both conceptually confusing and inconsistent. Most contemporary discussions tend to blur the distinction and focus on the two underlying causal variables posited by Durkheim, generally suggesting that they are strongly related. *Social integration* refers to the ties of individuals to the society, and the strength of one's relationships with others. *Social regulation* refers to social control that a group holds over its members. The basic hypothesis stemming from this work is that suicide rates will be higher among populations that have weaker ties with others within the society (lower levels of integration) and/or are subject to less social control (lower levels of regulation). Empirical analyses of suicide have consistently supported Durkheim's contention that suicide rates are higher in situations with lower levels of social integration and regulation. For example, parents and married persons have lower suicide rates than childless persons and unmarried persons, given their presumably higher levels of social integration.

Durkheim gave very little attention to the issue of youthful suicide, apart from noting its rarity. In perhaps his only reference to youthful suicide, he cites the analyses of Henry Morselli, one of the moral statisticians. Morselli, using data from the mid-and late-19th century in Paris, London, Petersburg, Vienna, and Berlin, documented the very low incidence of suicide among children in general but a striking increase among youth within cities. As Morselli put it, "In fact, it is in the great centres that the number of suicides amongst the young rises so extraordinarily high" (Morselli 1975/1882, p. 222). Durkheim concurred with Morselli's conclusions and explained the high incidence of youthful suicide in these regions as stemming from social factors:

> It must be remembered that the child too is influenced by social causes which may drive him to suicide. Even in this case their influence appears in the variations of child-suicide according to social environment. They are most numerous in large cities. Nowhere else does social life commence so early for the child.... Introduced earlier and more completely than others to the current of civilization, he undergoes its effects more completely and earlier. This also causes the number of child-suicides to grow with pitiful regularity in civilized lands. (Durkheim 1951, p. 101)

In one of the few sociological analyses to examine teen suicide, Peter Bearman (1991) describes how the anomic situation of modern society may affect young people. He notes that the "teen today is often a member of two separate societies, the family of origin and the peer group" (p. 517). The teen is integrated into these two "societies," but these social worlds are usually independent of one another and produce conflicting social demands. "The normative dissonance experienced by the teen is the same as anomie," according to Bearman.

Explaining Cohort Effects Sociological work on teen suicide has used a Durkheimian perspective and focused on the late-20th-century changes in the age distribution of suicide that were noted above. These scholars have examined ways in which social regulation and integration vary across birth cohorts and can account for the upswing in teen suicide relative to that of adults. Supported by a variety of complex statistical models, their results demonstrate the importance of cohort variations in family structure and relative size, both of which they take to reflect the integration and regulation that young people receive. Cohorts born in the 1980s and 1990s have higher suicide rates relative to adults because they have experienced less integration and regulation through their childhood years (Stockard & O'Brien, 2002a, 2002b).

Using cross-cultural data, researchers also have shown that the impact of cohort effects can vary from one social context to another. In the late 20th and early 21st centuries many modern societies have had similar risks of low levels of integration and regulation for their youth, stemming from changing family structures and numbers of youth relative to adults. Yet only some of these societies have experienced the dramatic upturn in teen suicides relative to adults that occurred in the United States and other nations. Research suggests that an important protective element is social policies that provide greater support to families and children. Teens in societies with these traditions have been much less likely to experience higher rates of suicide (Stockard & O'Brien, 2002a; Stockard, 2003).

FUTURE DIRECTIONS FOR THE STUDY OF TEEN SUICIDE

Undoubtedly, the most striking development in the area of teen suicide is the dramatic increase of youthful rates relative to adult rates. This issue is much more than scholarly in nature, for the deaths that the increased rates reflect have no doubt brought enormous emotional pain to many people and produced substantial losses in human and social capital for communities throughout the nation.

One of the most important areas for future research on teen suicide will be continued study of the relative rates of youth and adults. To what extent will the narrowed gap between teen and adult rates persist into the future? Do the factors that the recent cohort-related

analyses have uncovered continue to predict the age-period variations in suicide?

Prevention efforts will, of course, continue to be very important. Much prevention work concentrates on mental health efforts. These efforts often involve suicide prevention campaigns, mental health treatment for individuals who appear at risk, and training for adults working with youth to help them identify danger signals. Some scholars refer to these efforts as "downstream" policies, those that try to address issues of suicidality after they have already appeared. While these efforts are, of course, absolutely necessary and important, it is also possible to focus change efforts "upstream" and address factors more distally related to suicide risk, such as those identified in the analyses of cohort effects. Using this perspective, upstream approaches to diminishing teen suicide could focus on promoting social integration and regulation for teens through methods such as greater support for families and youth throughout childhood and for the society as a whole (Stockard, 2003).

SEE ALSO Volume 1: *Bullying and Peer Victimization; Freud, Sigmund; Gays and Lesbians, Youth and Adolescence; Mental Health, Childhood and Adolescence;* Volume 2: *Durkheim, Émile; Suicide, Adulthood;* Volume 3: *Suicide, Later Life.*

BIBLIOGRAPHY

Bearman, P. (1991). The social structure of suicide. *Sociological Forum, 6,* 501–524.

Centers for Disease Control (CDC). (2006). National Youth Risk Behavior Survey: 2005, National results by subgroup. Retrieved January 31, 2008 from http://www.cdc.gov/HealthyYouth/yrbs/subgroup.htm

Dollard, J., Miller, N. E., Doob, L. W., Mowrer, O.H., & Sears, R. R. (1939). *Frustration and aggression.* New Haven, CT: Yale University Press.

Durkheim, Emile. (1951). J. A. Spaulding (Trans.) & G. Simpson (Ed. and Trans.), *Suicide: a study in sociology.* New York: Free Press. (Original work published 1897.)

Freud, S. (1957). Mourning and melancholia. In J. Strachey (Ed. and Trans.), *The standard edition of the complete psychological works of Sigmund Freud,* Vol. 14 (pp. 237–258). London: Hogarth Press. (Original work published 1917.)

Freud, S. (1955). Beyond the pleasure principle. In J. Strachey (Ed. and Trans.), *The standard edition of the complete psychological works of Sigmund Freud,* Vol. 18 (pp. 3–66). London: Hogarth Press. (Original work published 1920.)

Menninger, K. A. (1938). *Man against himself.* New York: Harcourt, Brace, and Company.

Miller, N. E., Sears, R. R., Mowrer, O.H., Doob, L. W., & Dollard, J. (1941). The frustration-aggression hypothesis. *Psychological Review, 48,* 337–342.

Morselli, H. (1975). *Suicide: an essay on comparative moral statistics.* New York: Arnon Press. (Original work published in 1882.)

Silenzio, V. M. B., Pena, J. B., Duberstein, P. R., Cerel, J., Knox, K. L. (2007). Sexual orientation and risk factors for suicidal ideation and suicide attempts among adolescents and young adults. *American Journal of Public Health, 97* (11), 2017–2019.

Sourcebook of Criminal Justice Statistics Online. Table 3.137:2005. (2007). Retrieved March 17, 2008 from http:// www. albany.edu/sourcebook/pdf/t31372005.pdf

Stockard, J. 2003. Social science, social policy, and lethal violence: looking for upstream solutions: 2003 Presidential Address to the Pacific Sociological Association. *Sociological Perspectives, 46*: 291–308.

Stockard, J., & O'Brien, R. M. (2002a). Cohort effects on suicide rates: international variations. *American Sociological Review, 67*: 854–872.

Stockard, J., & O'Brien, R. M. (2002b). Cohort variations and changes in age-specific suicide rates over-time: explaining variations in youth suicide. *Social Forces, 81*: 605–642.

Whitt, H. P. (1994). Old theories never die. In N. Pr. Unnithan, L. Huff-Corzine, J. Corzine, & H. P. Whitt (Eds.), *The currents of lethal violence: An integrated model of suicide and homicide* (pp. 7–34). Albany: State University of New York Press.

Jean Stockard

T

TECHNOLOGY USE, ADOLESCENCE

SEE Volume 1: *Media and Technology Use, Childhood and Adolescence.*

TEEN PREGNANCY

SEE Volume 1: *Transition to Parenthood.*

THEORIES OF DEVIANCE

Crime, *deviance*, and *delinquency* are often used interchangeably. However, these terms refer to different behaviors. *Crime* refers to an act or a failure to act that is a violation of criminal law. *Delinquency* is a crime that only juveniles can commit. *Deviance* refers to violating social standards or norms about what is collectively considered conventional behavior. Deviance is often antisocial behavior that is typically condemned by the general public. Scholars who study theories that explain why youths commit different acts, whether these acts are truly criminal or merely deviant, overlap. In other words, a theory that explains deviance usually explains crime and delinquency as well. Several contemporary theories that explain why youth commit crime and deviance fall under three broad categories: psychological, sociological, and biological theories of deviance.

Psychological and sociological theories are by far the most commonly used to explain deviance and crime. Whereas psychological theories tend to focus on individual-level factors such as mental health and intelligence as influences on youthful deviance, sociological theories emphasize the role of the social structure and social processes that may impact a youth's life chances. For instance, sociological theories focus on race/ethnicity, gender, sexuality, and social class as the backdrop for understanding why an individual might commit a crime or deviant act. Few theories in the early 21st century are purely sociological or purely psychological. Instead, most of the tested theories of deviance emphasize a social-psychological approach, or a blended approach to understanding why people commit deviance.

Although the specific hypotheses generated by different social-psychological theories vary, these perspectives do share some common themes that may predict the social environments that generate youthful deviance. These themes include:

1. living in an impoverished area,

2. having poor quality relationships with parent(s), which are characterized by inconsistent discipline, abuse, and/or lack of parental affection,

3. poor academic performance, attendance problems, and/or dropping out of school,

4. access to deviant peers or parents who reinforce deviant behaviors (Giordano, Deines, & Cernkovich, 2006).

Four of the most influential social-psychological theories merit detailed discussion: differential association theory, social control theory, labeling theory, and life-course perspective. Differential association theory has the most longevity and has influenced other theories of deviance for nearly a century. Social control theory has had the most powerful impact on policy makers' strategies for how to reduce juvenile delinquency. Although such strategies have failed to produce the results promised, the conservative nature of social control theory has helped it remain a popular explanation of deviant behavior. Labeling theory began a revolutionary way of looking at delinquency as a problem created and maintained by the juvenile justice system itself. No other theory of lawmaking has been as widely accepted or integrated into policy as labeling theory. Finally, the life-course perspective is a theoretical integration, borrowing the best from various theories and helping to explain deviant behavior over longer periods of the life span than any other theory covers.

DIFFERENTIAL ASSOCIATION THEORY

Prior to the 1930s, most American theories of crime and deviance focused on biological causes, psychological malformations, or economic social conditions. Criminality was either an innate characteristic of an individual or crime was the product of abstract socioeconomic forces in an individual's life. Differential association was the first theory to challenge the existing positions by bridging sociological and psychological theories of behavior. Posited by Edwin H. Sutherland (1883–1950), differential association suggests that all crime and deviance are learned behaviors. Learning to commit crime occurs through associations or relationships with deviant peers or close family members. This theory was a revolutionary idea for its time and has come to be one of the most longstanding and valued theories of crime and deviance.

Differential association theory suggests that individuals learn to become criminal through interactions with immediate family and peers. Sutherland (1947) detailed this process of social learning of deviant behavior in nine postulates:

1. Criminal behavior is learned.

2. Criminal behavior is learned in interaction with other persons in a process of communication.

3. The principal part of the learning of criminal behavior occurs within intimate personal groups.

4. When criminal behavior is learned, the learning includes (a) techniques of committing the crime, which are sometimes very complicated, sometimes very simple; (b) the specific direction of motives, drives, rationalizations, and attitudes.

5. The specific direction of the motives and drives is learned from definitions of the legal codes as favorable or unfavorable.

6. A person becomes delinquent because of an excess of definitions and associations favorable to violation of law over definitions and associations unfavorable to violation of law. This is the principle of differential association.

7. Differential association may vary in frequency, duration, priority, and intensity. This means that associations with criminal behavior and also associations with anticriminal behavior vary in those respects.

8. The process of learning criminal behavior by association with criminal and anticriminal patterns involves all of the mechanisms that are involved in any other learning.

9. While criminal behavior is an expression of general needs and values, it is not explained by those general needs and values since noncriminal behavior is an expression of the same needs and values. Thieves generally steal in order to secure money, but likewise honest laborers work in order to secure money. The attempts to explain criminal behavior by general drives and values such as the money motive have been, and must continue to be, futile, since they explain lawful behavior as completely as they explain criminal behavior. They are similar to respiration, which is necessary for any behavior, but which does not differentiate criminal from noncriminal behavior. (pp. 6–7)

These postulates are generally understood by scholars to mean that criminal behavior is learned from deviant peers and close family members through talking about criminality. Not only must the how-to's of crime be learned by individuals, the norms, rationales, definitions, and attitudes that reinforce the deviant behavior also must be learned. It is these definitions that individuals make based on their associations with deviant peers and family that allow them to commit deviant acts.

From the 1930s through 1947, Sutherland revised this theory several times. Since 1947, however, differential association theory has remained relatively unchanged and has been extensively tested and criticized by scholars. Most research that evaluates differential association has focused on juvenile delinquency, partly because delinquency is considered a group phenomenon in which peer influences are probably the strongest (Vold, Bernard, & Snipes, 2002).

No academic studies have found that delinquency is caused by transmitting definitions and techniques of crime through delinquent peers and family members. However, several researchers have found some support

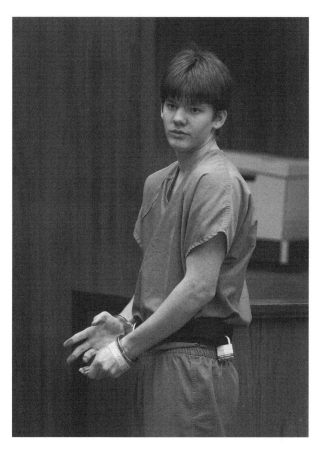

Teen Violence. *James Watson at the end of his sentencing hearing where he was sentenced to 9 to 12 months for the beating death of Shane Farrell in New Smyrna, FL.* AP IMAGES.

Abbot, Catalano, and Hawkins (1998) tested differential association theory focusing on gangs. The findings showed that the strongest and most consistent factor that predicts delinquency is prior delinquency. However, associating with peers who engage in delinquency plays a strong role in predicting delinquency as well.

Despite the fact that there is mixed support for differential association, most scholars who study deviance, delinquency, and crime still pay close attention to the issues that Sutherland raised in the early twentieth century. Ronald L. Akers (1998) and others added to the original tenets of differential association by integrating specific psychological learning processes such as operant conditioning. These additions are generally called *social learning theories* as they specify the process of actual learning better than Sutherland's original work. Most testing of social learning theories have shown evidence for the ideas proposed in differential association such as delinquent peer association and learning criminal behavior from close relationships with peers and family.

The influence of differential association extends far beyond social learning theories. The original concepts of delinquent peers or close family members are tested in most major theoretical tests of delinquency in the 21st century, even if the central purpose is not to test differential association. No other criminological theory has been as influential or as enduring.

SOCIAL CONTROL THEORY

Differential association is readily comparable to the next theory of deviance: social control theory. Both theories focus on explaining juvenile offenses more than adult crime, and both theories consider similar influences, but the order in which the influences act vary across the two theories. For differential association theory, an individual first has deviant peer and/or family associations. Next, the individual learns to become a criminal from those peers and family associations. Finally, the individual commits the criminal or deviant act. Conversely, social control theory is based on the assumption that deviant peers are more of a byproduct of being a deviant. In colloquial terms, "birds of a feather flock together": Like-minded, deviant juveniles congregate because they share a common interest in deviance (Glueck & Glueck, 1950, p. 164). According to most social control theorists, first an individual commits a criminal act; the individual then finds other deviants or criminals to associate with after already committing deviance.

More important than peers, however, is the role of social conformity in this theory. In 1957 Jackson Toby argued all young people have "stakes in conformity," meaning that juveniles are aware they might lose something by breaking the law. Toby's ideas on social

for the general spirit of Sutherland's differential association. For example, Ross L. Matsueda (1982, 1988) found that delinquent youth are more likely to have delinquent peers than are nondeviant youth. Matsueda (1988) also found that just having delinquent peers is not enough to increase risk of deviance. Youth must also have learned definitions favorable to committing crime. In the late 1990s, better statistical techniques and data allowed differential association to be further tested. Costello and Vowell (1999) reanalyzed Matsueda's results with more sophisticated statistical models, and his initial findings were not as strongly supported.

Other studies have shown some support for differential association. For instance, Jang (1999) found that there are direct effects of delinquent peers and school on delinquency in that having delinquent peers is related to committing more delinquency. This relationship, however, is age-limited. The effects of delinquent peers on one's own delinquency tend to increase from early to middle adolescence, reach a peak at the age of mid-13 to mid-15, and then decline. In addition, Battin, Hill, conformity were expanded and are now most often associated with the work of Travis Hirschi (1969), which begins with some unique assumptions compared with other theories of deviance. First, social control theories focus on the question: Why do people conform? The answer according to this perspective is because social control prevents individuals from committing crimes.

According to Hirschi (1969), there is no specific criminal motivation, but instead all individuals have a natural motivation to commit crime. Hirschi argued that deviance results from an individual having a weak or broken bond to society. Society in social control theory is specified for youth as family, school, church, and other conventional activities. Hirschi proposed that four primary bonds to society are what prevent juveniles from committing deviant acts. These bonds are: attachment, commitment, involvement, and belief (Hirschi, 1969).

Attachment refers to affectional ties that a youth has toward others, predominantly parents. Attachment is the building block necessary for all humans to internalize values and norms and thus be socially controlled (Vold et al., 2002). Unlike differential association theory, an individual can have affectional ties to deviant family members or peers and not commit deviance. Attachment to anyone should foster conforming behaviors in people.

The next social bond, *commitment*, refers to the level of investment individuals have in conventional society and in their personal long-term goals. The more an individual feels invested in schools and future careers, the less likely that individual will commit any deviant behaviors that may threaten society. *Involvement* is the actual tangible activities individuals participate in to show their commitment to the social structure. Moreover, involvement in conventional activities such as sports, extracurricular school activities, or religious activities leaves little or no time for deviant activities. The last bond, *belief*, refers to how much individuals believe in the legitimacy of the social structures around them. Hirschi (1969) argued that if individuals believe that they should obey the law and rules of society, they usually will.

Many scholars have tested Hirschi's (1969) social control theory over the past 40 years. Most research shows support for two of the four bonds: attachment and commitment (Costello & Vowell, 1999). There is no empirical evidence showing that belief and involvement are related to delinquency, as social control theory predicts. In the original study Hirschi conducted during the 1960s to develop social control theory, he could not find support for involvement. Many studies have found the opposite to be true: Heavy involvement in conventional activities, including extracurricular sports and having a part-time job, predict delinquency (Begg, Langley, Moffitt, & Marshall, 1996; Ploeger, 1997). Moreover,

tests that focus on the belief bond also have not supported social control theory (Marcos, Bahr, & Johnson, 1986; Massey & Krohn, 1986). Once delinquent peers are accounted for in these tests of social control, the impact of all four bonds decreases (Empey, Stafford, & Hay, 1999).

From the time social control theory emerged through the mid-1990s, it was an extremely popular explanation of deviance with policy makers. The policy implications of social control are fairly easy to implement and were appealing to the designers of tough-minded crime policies. The idea is to provide more opportunities to conventionally bond to legitimate social structures for young people. However, many of the programs implemented in the name of social control theory did not reduce deviant behavior among youth (Empey et al., 1999).

Social control theory has since waned in popularity partially because it is considered by many academic scholars to be too conservative (Greenberg, 1999). In addition, social control assumes an innate criminal nature of youth and adults that is disconcerting to some academics who believe that people are not inherently bad or criminal. Many scholars who study crime refer back to Sutherland's (1947) work on differential association in which he stated, "People are not inherently antisocial. If young people violate the law, it is because they have learned to do it" (p. 6).

LABELING THEORY

The theories discussed so far have focused on explaining deviant and law-breaking behavior. The next theory, labeling theory, became popular in the 1960s and focuses on lawmaking. According to the labeling perspective, peers, social control, and the other factors previously discussed are not relevant to explaining the true problem with delinquency. This perspective contends that the most important factor to study when focusing on crime is the formal label a youth receives as a result of being caught committing a delinquent act. The actual deviant behavior is secondary in importance. The labeling theorists instead ask: Who applies a deviant or criminal label? What determines when and to whom that label is applied? (Vold et al., 2002). It is the label itself that causes further criminal behavior in a process called *deviance amplification*.

Labeling theorists begin with a set of assumptions about the social structures and the nature of people that are quite different from those used by social control and differential association theorists. Labeling theorists believe that society is characterized by social conflicts. The so-called "winners" in these social conflicts are the people who are in the most powerful positions in terms of social class, status, wealth, and power. This powerful group, the elites, tends to have the most influence on laws about criminality and deviance. The elites tend to protect their own interests by defining lower-class behaviors as deviant or criminal. As a consequence, labeling theorists believe that laws and the persons involved in the legal system are responsible for generating crime by labeling the least powerful people as criminal (Empey et al., 1999).

Labeling theory emerged predominantly from the works of two theorists and over a 30-year period beginning in the 1930s (Vold et al., 2002). Frank Tannenbaum (1938) argued that when a child becomes enmeshed with the juvenile justice system, those experiences turn him or her from the occasional misbehaver to the active delinquent. Tannenbaum detailed how adult reactions become stronger and the perceptions of mischief become perceptions of evil. The child begins to accept the evil label by adults, further amplifying the problem.

Edwin Lemert (1951) added two additional concepts to labeling theory: primary and secondary deviance. Lemert argued that the first delinquent act may go undetected and unpunished, and this is common among youth from all social backgrounds (Empey et al., 1999). Those youths who commit deviant acts and are not caught, punished, or labeled usually desist their mischievous behaviors. However, juveniles who are caught are stigmatized by the label "delinquent." Usually the newly labeled delinquent is watched more closely and with stricter rules that may be too difficult to follow. Thus the youth will likely adapt to the new label by committing secondary deviance: additional crimes that result directly from the label.

Both Lemert (1951) and Tannenbaum (1938) believed that this labeling process was a gradual one that occurred over a period of time. A primary deviance would occur, followed by an initial labeling, the acquisition of a delinquent identity, and then secondary deviance. The deviance would be further amplified each time the youth was caught and relabeled by the formal juvenile justice system. Moreover, the labeling process would eventually lead to the juvenile accepting the label as a master status or a powerful label that may overshadow all other aspects of his or her identity. The master status as a delinquent would eventually lead to a life-long criminal career.

This process of deviance amplification following the acquisition of a formal deviant label does not affect all juvenile offenders equally. Howard Becker (1963) argued that juvenile justice administrators use labeling selectively. The least powerful youth in American society, such as the impoverished and racial and ethnic minorities, were more likely to be labeled as deviants than others.

By the 1960s and 1970s, when crime among youth was a top national concern for the first time, there was initial excitement about labeling theory. Empey and colleagues (1999) described the excitement: "Were it not for punitive reactions to mischievous children, entangling them in a self-fulfilling prophecy, there would be few career criminals" (p. 260). Labeling theorists outlined several reforms for the juvenile justice system to decriminalize, divert, and deinstitutionalize juvenile offenders to prevent the application of stigmatizing labels and summarily end deviance amplification. Many of these programs were implemented in the late 1960s and remain in existence at the beginning of the 21st century (Empey et al., 1999). However, the excitement over labeling theory faded quickly because of some theoretical problems.

The first problem is that scholars could find only limited support that the label was the problem causing deviance amplification. Although evaluations of labeling theory did find support for the idea that, once labeled, a youth was likely to commit more deviant acts, other factors were more important. In all of the tests of labeling theory, prior deviance was the largest predictor of secondary deviance, not the label (Empey et al., 1999).

One of the most influential tests of labeling theory was conducted by Smith and Paternoster (1990). These researchers found that juvenile offenders referred to juvenile courts were more likely to be referred again to the court for further offenses, which initially appeared to be consistent with labeling theory. However, a closer inspection of the data showed that it was not the juvenile court referral that caused the additional delinquency; rather, the juveniles who were referred to court were considered to be higher risk than those not referred because of the seriousness of prior convictions, the lack of capable guardianship in their homes, and other socioeconomic factors.

Labeling theory has not been abandoned outright, despite the scholarly evidence against it. Instead, labeling theory has been revamped by such academics as John Braithwaite (1989), who posits a theory of reintegrative shaming in lieu of the stigmatizing labels. In general, Braithwaite's theory has received some support and is an area in which the basic legacy of labeling theory is still used by scholars and policy makers in the early 21st century. However, labeling theory does not appear to be well supported independent of other socioeconomic factors that better predict youths' propensity toward deviant behavior.

LIFE-COURSE PERSPECTIVES ON DEVIANCE

The life-course perspective provides one of the newer frameworks for understanding deviance and crime. In general, life-course perspectives contain theoretical tenets about how changes in an individual's life determine the probability of that person becoming a criminal or ending a criminal career. Life-course perspectives integrate social, psychological, and environmental factors, which makes these theories more comprehensive and inclusive than other theories of crime. Researchers who study life-course perspectives typically follow a cohort or group of youths over a period of time, which allows scholars to document how changes in the life course affect youths' future opportunities and likelihood of committing deviant acts.

The term *perspective* is used instead of *theory* because one theoretical model does not unify the wide breadth of ideas that fall into this general category. Indeed, if one asked a sociologist and a psychologist about life-course theories of deviance, one may get surprisingly different answers. In general scholars who study life-course perspectives believe that a wide range of factors contribute to deviance at different points in an individual's life. Yet there is not a general agreement about which specific factors are most important to study. For these reasons, life-course perspectives should be divided into biosocial developmental issues and sociological issues.

Biosocial Life-Course Theories Biosocial theories suggest that some biological factors may predispose some youth to commit crime in addition to the social factors that are known to be related to crime. These biological factors usually capture some quality that inhibits normal biological development. For example, levels of hormones such as testosterone, physiological development indicators such as puberty onset, nutrition levels during formative years, or brain chemistry during different developmental ages have been linked to delinquent behavior.

Biosocial life course: Pubertal development. One of the most important biosocial areas of study related to life-course perspectives focuses on pubertal development and deviance. In fact, researchers have long noted that the onset of puberty for boys often coincides with the onset of delinquency (Felson & Haynie, 2002). Scholars have tested whether or not that coincidence indicates that delinquency and puberty are related to each other. For boys, pubertal development may be as important as peers and school performance in predicting deviant behaviors. Felson and Haynie showed that puberty predicts violent crime, property crime, illicit substance use, and precocious sexual behavior. However, this study and others cannot accurately determine the exact social or biological mechanisms that make this relationship apparent.

The relationship between puberty and deviance is different for girls and boys. Girls who enter puberty earlier than average are more likely to engage in deviant behaviors than girls who begin puberty at an average or older age. Girls who attend coeducational or mixed-sex schools are even more likely to be affected by early puberty, as boys tend to have an aggravating effect on girls' deviance (Caspi, Lynam, Moffitt, & Silva, 1993).

Biosocial life course: Neurological development. The widely known research of Terrie Moffitt (1993) is often credited for beginning a new and distinct discussion of life-course perspective and deviance. Moffitt created a taxonomy, or classification system, for two distinct types of offenders: adolescence-limited deviance and life-course persistent deviance. The first group, adolescence-limited, contains the majority of youthful offenders. These juveniles commit delinquent acts during their teen years and quickly end their criminal careers before adulthood. The offenders who fall into this group are for the most part normal youth who do not commit serious crimes.

Life-course persistent offenders, on the other hand, are far more problematic deviants. Moffitt (1993) suggested that individuals who have an early age of onset, or in other words begin committing antisocial acts during early childhood, will continue through adulthood committing more serious crimes than the other group. She argued that a neuropsychological impairment is the root cause of delinquency. Neuropsychological refers to "anatomical structures and physiological processes within the nervous system [that] influence psychological characteristics such as temperament, behavioral development, cognitive abilities, or all three" (Moffitt, 1993, p. 681). These impairments can be inherited or caused by maternal alcohol or drug use while in utero, poor prenatal nutrition, brain injury, or exposure to toxins such as lead. Moffitt, Lynam, and Silva (1994) argued that neuropsychological impairments lead to poor verbal skills and low self-control, which in turn causes delinquency.

Moffitt's (1993) theory has been the subject of many scholarly debates and research. The age of onset of criminal activities as well as desistance are the two factors that have been examined most frequently. Although Moffitt's own tests and other psychological research have shown some support for her theory (Donnellan, Ge, & Wenk, 2000), sociological studies show very limited support without factoring in sociological variables. For instance, one study found that, irrespective of the age of onset, criminal trajectories tended to specialize (Piquero, Paternoster, Mazerolle, Brame, & Dean, 1999). This finding suggests that there is no clear difference between adolescent-limited and life-course persistent offenders. Others found that parental and school reactions to deviant behavior are associated with long-term changes in children's behavior (Simons, Johnson, Conger, & Elder, 1998). Thus, social factors during development are more important in determining delinquency trajectories than neuropsychological traits.

The mixed support for this theory may reflect the variety of perspectives of the researchers who are interested in Moffitt's (1993) taxonomy. Sociologists are unlikely to believe in the more deterministic elements of her work, whereas psychologists may not be as willing to accept some of the social factors Moffitt addresses. Because Moffitt's (1993) theory is relatively new, future studies are sure to elaborate and refine her original work. Sociological studies of the life course are already working on refining her psychological ideas in the contexts of the social world.

Sociological Life-Course Theories Although not necessarily in agreement with Moffitt's (1993) pivotal work, sociological life-course researchers acknowledge the many contributions of her work. The most widely cited sociological life-course work has been conducted by Robert Sampson and John Laub (1993, 2005). Sampson and Laub's studies of juvenile delinquency through adult criminality begin with a direct discussion of Moffitt's work.

In a 2005 publication, Sampson and Laub outlined why they agree and disagree with Moffitt's ideas. Sampson and Laub moved away from biosocial determinism in favor of a sociological explanation of why people begin, persist, and desist their criminal careers or trajectories. They also see criminal careers as well as the life course as dynamic. Changes that happen during individuals' lives were more important to Sampson and Laub than any early childhood or neonatal factors that Moffitt emphasized.

Sampson and Laub (1993) presented an age-graded theory of informal social control. They argued that adolescents tend to not commit deviant acts when informal social controls are working. If a youth is properly supervised by fair and loving parents, that child is not likely to engage in deviance. Conversely, a lack of informal social controls, including poor parenting, low supervision, erratic or harsh discipline, and weak ties to schools, are all linked to predicting juvenile delinquency. Sampson and Laub's ideas were based in part on the tenets of Hirschi's (1969) social control theory, discussed earlier.

Sampson and Laub (1993) also maintained that informal social controls change over the life course. As adolescence ends and adulthood begins, social controls also change. Social capital now becomes an important form of social control. The more social capital individuals have, the less likely they are to commit a deviant act that will threaten their social capital. The most important forms of social capital according to this theory are marriage and job stability. Thus, securing adult milestones such as employment and marriage are linked to the ending of adolescent crime because individuals do not want to risk losing these social gains.

Adolescents can fail to make a positive transition to adulthood if their lives are marked by negative turning points or milestones. Sampson and Laub (1993) argued that prolonged incarceration, heavy drinking, and subsequent job instability during the transition to young adulthood will exacerbate criminal careers.

Numerous theoretical tests of Sampson and Laub's work (1993) that show support for their version of life-course theory have been published in criminology and sociology journals. Nagin, Farrington, and Moffitt (1995) studied Moffitt's (1993) adolescence-limited offenders along with life-course persistent offenders under Sampson and Laub's age-graded model of social control. This study followed offenders until age 32. The results show that adolescent-limited offenders do cease most serious types of crimes throughout adulthood. These offenders committed only minor infractions between the ages of 18 and 32, such as petty theft from employers and minor substance use. They also tended to have greater job stability and less family disruption than the chronic offenders in the study. The adolescence-limited offenders were careful to avoid committing crimes with a high risk of conviction that might jeopardize their work. These offenders did not engage in spousal abuse that might harm familial relationships. Thus, the theory of adolescence-limited offenders "aging out" of crime because of functioning informal social controls is well supported.

Several other tests show support for Sampson and Laub's (1993) theory in regard to adult criminal behavior. One study found that for adult convicted felons, living with a wife is associated with lower levels of offending, whereas living with a girlfriend is associated with higher levels of offending (Horney, Osgood, & Marshall, 1995). Another found that social capital is more important in terms of preventing adults from offending than threat of legal punishment (Nagin & Paternoster, 1994). Individuals in this study who were more concerned about losing their social capital also ranked high on scales measuring forethought and other negative consequences of criminal behavior.

Future Directions of Life-course Perspectives Whether sociological, psychological, or biosocial, it is clear that life-course perspectives will remain at the forefront of theories of deviance. Many life-course theoretical questions remain unanswered in part because of data limitations; there just are not enough longitudinal studies that follow children through adulthood to test all the competing theories of deviance. As data become more plentiful, theoretical development in this perspective is likely to continue to integrate the ideas of multiple disciplines.

SEE ALSO Volume 1: *Aggression, Childhood and Adolescence; Bandura, Albert; Crime, Criminal Activity in Childhood and Adolescence; Juvenile Justice; Peer Groups and Crowds; Puberty; School Violence;*

Volume 3: *Crime, Criminal Activity in Adulthood; Incarceration.*

BIBLIOGRAPHY

Akers, R. L. (1998). *Social learning and social structure: A general theory of crime and deviance.* Boston: Northeastern University Press.

Battin, S. R., Hill, K. G., Abbott, R. D., Catalano, R. F., & Hawkins, J. D. (1998). The contribution of gang membership to delinquency beyond delinquent friends. *Criminology, 36,* 93–115.

Becker, H. S. (1963). *Outsiders: Studies in the sociology of deviance.* New York: Free Press.

Begg, D. J., Langley, J. D., Moffitt, T., & Marshall, S. W. (1996). Sport and delinquency: An examination of the deterrence hypothesis in a longitudinal study. *British Journal of Sports Medicine, 30,* 335–341.

Braithwaite, J. (1989). *Crime, shame, and reintegration.* Cambridge, U. K.: Cambridge University Press.

Caspi, A., Lynam, D., Moffitt, T. E., & Silva, P.A. (1993). Unraveling girls' delinquency: Biological, dispositional, and contextual contributions to adolescent misbehavior. *Developmental Psychology, 29,* 19–30.

Costello, B. J., & Vowell, P. R. (1999). Testing control theory and differential association: A reanalysis of the Richmond Youth Project Data. *Criminology, 37,* 815–842.

Donnellan, M. B., Ge, X., & Wenk, E. (2000). Cognitive abilities in adolescent-limited and life-course-persistent criminal offenders. *Journal of Abnormal Psychology, 109,* 396–402.

Empey, L. T., Stafford, M. C., & Hay, C. H. (1999). *American delinquency: Its meaning and construction.* (4th ed.). Belmont, CA: Wadsworth.

Felson, R. B., & Haynie, D.L. (2002). Pubertal development, social factors, and delinquency among adolescent boys. *Criminology, 40,* 967–988.

Giordano, P. C., Deines, J. A., & Cernkovich, S. A. (2006). In and out of crime: A life course perspective on girls' delinquency. In K. Heimer & C. Kruttschnitt (Eds.), *Gender and crime: Patterns of victimization and offending* (pp. 17–40). New York: New York University Press.

Glueck, S., & Glueck, E. T. (1950). *Unraveling juvenile delinquency.* New York: Commonwealth Fund.

Greenberg, D. F. (1999). The weak strength of social control theory. *Crime and Delinquency, 45,* 66–81.

Hirschi, T. (1969). *Causes of delinquency.* Berkeley: University of California Press.

Horney, J. D., Osgood, W., & Marshall, I. H. (1995). Criminal careers in the short-term: Intra-individual variability in crime and its relation to local life circumstances. *American Sociological Review, 60,* 655–673.

Jang, S. J. (1999). Age-varying effects of family, school, and peers on delinquency: A multilevel modeling test of interactional theory. *Criminology, 37,* 643–685.

Lemert, E. M. (1951). *Social pathology: A systematic approach to the theory of sociopathic behavior.* New York: McGraw-Hill.

Marcos, A. C., Bahr, S. J., & Johnson, R. E. (1986). Test of bonding/association theory of adolescent drug use. *Social Forces, 65,* 135–161.

Massey, J. L., & Krohn, M. D. (1986). A longitudinal examination of an integrated social process model of deviant behavior. *Social Forces, 65,* 106–134.

Matsueda, R. L. (1982). Testing control theory and differential association: A causal modeling approach. *American Sociological Review, 47,* 489–504.

Matsueda, R. L. (1988). The current state of differential association theory. *Crime and Delinquency, 34,* 277–306.

Moffitt, T. E. (1993). Adolescence-limited and life-course-persistent antisocial behavior: A developmental taxonomy. *Psychological Review, 100,* 674–701.

Moffitt, T. E., Lynam, D. R., & Silva, P. A. (1994). Neuropsychological tests predicting persistent male delinquency. *Criminology, 32,* 277–300.

Nagin, D. S., Farrington, D., & Moffitt, T. (1995). Life-course trajectories of different types of offenders. *Criminology, 33,* 111–139.

Nagin, D. S., & Paternoster, R. (1994). Personal capital and social control: The deterrence implications of a theory of individual differences in criminal offending. *Criminology, 32,* 518–606.

Piquero, A., Paternoster, R., Mazerolle, P., Brame, R., & Dean, C. W. (1999). Onset age and offense specialization. *Journal of Research in Crime and Delinquency, 36,* 275–299.

Ploeger, M. (1997). Youth employment and delinquency: Reconsidering a problematic relationship. *Criminology, 35,* 659–676.

Sampson, R. J., & Laub, J. H. (1993). *Crime in the making.* Cambridge, MA: Harvard University Press.

Sampson, R. J., & Laub, J. H. (2005). A life-course view of the development of crime. *Annals of the American Academy of Political and Social Science, 602,* 12–45.

Simons, R. L., Johnson, C., Conger, R.D., & Elder, G. (1998). A test of latent trait versus life-course perspectives on the stability of adolescent antisocial behavior. *Criminology, 36,* 217–244.

Smith, D. A., & Paternoster, R. (1990). Formal processing and future delinquency: Deviance amplification as a selection artifact. *Law and Society Review, 24,* 1109–1131.

Sutherland, E. H. (1947). *Principles of criminology.* (4th ed.). Philadelphia: J. B. Lippincott.

Tannenbaum, F. (1938). *Crime and the community.* Boston: Ginn.

Toby, J. (1957). Social disorganization and stake in conformity: Complementary factors in the predatory behavior of hoodlums. *Journal of Criminal Law, Criminology, and Police Science, 48,* 12–17.

Vold, G. B., Bernard, T. J., & Snipes, J.B. (2002). *Theoretical criminology.* (5th ed.). New York: Oxford University Press.

Gini R. Deibert

THOMAS, W. I.
1863–1947

William Isaac Thomas was one of the most prolific and progressive sociologists of the early 20th century. The son

W. I. Thomas. COURTESY OF THE AMERICAN SOCIOLOGICAL ASSOCIATION.

tions of situations are arrived at collectively and are mediated by culture and history). By contrast, a positivistic reading of the Thomas theorem suggests a type of personal or collective delusion; the individual erroneously defines a situation as real even in the face of objective evidence to the contrary. In this vein, Robert Merton (1957) links the Thomas theorem to "self-fulfilling prophecies" and suggests that a "false" definition of a situation could become real if collectively shared. So, for example, the belief that the stock market is about to crash could cause an otherwise robust economy to fall into a recession. In a sense, a positivistic interpretation of the theorem implies that people sometimes fall in a sort of collective trance that causes them to believe a falsehood and act toward it as if it were true.

A close examination of Thomas's work suggests that he took a position somewhere between extreme objectivism and subjectivism. Indeed, the paragraph immediately following the theorem in *The Child in America* reads as follows: "The total situation will always contain more and less subjective factors, and the behavior reaction can be studied only in connection with the whole context, i.e., the situation as it exists in verifiable, objective terms, and as it has seemed to exist in terms of the interested persons" (1928, p. 572). Also, it is unlikely that Thomas intended for the theorem to apply to a limited class of human interactions (e.g., explaining how errors in judgment occur on a mass scale). Rather, as indicated in the excerpt below, the theorem simply articulates his previous research on what he believed to be a universal law of human interaction and personality.

> the human personality is both a continually producing factor and a continually produced result of social evolution, and this double relation expresses itself in every elementary social fact; there can be for social science no change of social reality which is not the common effect of pre-existing social values and individual attitudes acting upon them. (Thomas & Znaniecki, 1918, p. 5)

The phrase "every elementary social fact" is particularly significant in this context. It implies that it is not just a certain class of social facts (i.e., false perceptions) that fall under the Thomas theorem, but the theorem is a fundamental principle that applies to a wide range of social actions. In the case of the economic example, the Thomas theorem could also point to more foundational questions, such as how people come to define economic prosperity in terms of competitive trade among large corporations.

Interestingly, Thomas attributes the origins of this unique methodology to a happy accident, which is cited in Paul Baker's 1973 essay on Thomas:

> It was, I believe, in connection with *The Polish Peasant* that I became identified with "life history"

of a Protestant minister, Thomas was born in Virginia. He received his B.A. at the University of Tennessee and went on to hold teaching positions at several academic institutions, including the University of Chicago where he completed his Ph.D. in sociology. Thomas has made significant contributions to the field of sociology and the study of the life course in particular. He is best known for the *Thomas theorem*, which states, "If men define situations as real, they are real in their consequences" (Thomas & Thomas, 1928, p. 572). The book in which the theorem first appears, *The Child in America*, was coauthored by a graduate student, Dorothy Swaine (1899–1977), Thomas's second wife.

The Thomas theorem was likely inspired by his exposure to German philosophy and phenomenology—Thomas was an avid reader of German philosophy and studied in Germany for a year. The theorem can be interpreted in different ways. If one were to emphasize its phenomenological roots, the Thomas theorem means that reality is always subjective, contextual, and constructed via social interaction. In this context, the plural noun *men* connotes the collective endeavor that underlines the social construction of reality (i.e., defini-

and the method of documentation. ... I trace the origin of my interest in the document to a long letter picked up on a rainy day in the alley behind my house, a letter from a girl who was taking a training course in a hospital, to her father concerning family relationships and discords. It occurred to me at the time that one would learn a great deal if one had a great many letters of this kind. (p. 250)

Since its publication, *The Polish Peasant* has been criticized for its negative portrayal of Polish immigrants and for generalizations that did not necessarily grow out of the empirical data at hand. Despite these shortcomings, *The Polish Peasant* remains a seminal study that showed (a) identity is not an innate, stable trait but a fluid and malleable social form; and (b) individual identity is mediated by the broader social forces of culture and history. Thomas and Znaniecki's (1918) focus on how the individual fares in society over time continues to be the central theme of the study of the life course and social psychology in general.

The Polish Peasant and Thomas's other works challenged gender and ethnic stereotypes of marginal identities. For example, in *Sex and Society*, Thomas (1907) states that "the psychological differences of sex seem to be largely due, not to differences of average capacity, nor to difference in type of mental activity, but to differences in the social influences brought to bear on the developing individual from early infancy to adult years" (p. 438). Thomas was similarly critical of what is now referred to as objectification of women:

Woman was still further degraded by the development of property and its control by man, together with the habit of treating her as a piece of property, whose value was enhanced if its purity were assured and demonstrable. As a result of this situation man's chief concern in women became an interest in securing the finest specimens for his own use, in guarding them with jealous care from contact with other men, and in making them together with the ornaments they wore, signs of his wealth and social standing. (pp. 460–461)

Thomas's views on ethnic minorities were equally progressive. For example, in *Old World Traits Transplanted*, he and his coauthors warn against "demanding from the immigrant a quick and complete Americanization through the suppression and repudiation of all the signs that distinguish him from us" (1921, p. 281). They go on to note that the demand for the "destruction of memories" for immigrants is an oppressive totalitarian impulse shared by conservative Americans and radical Communists. In this sense, Thomas's reverence for memory and its role in identity formation is another hallmark in the study of the life course.

Thomas's professional career was eclipsed by a sex scandal in 1918. The 55-year-old Thomas was arrested along with a 24-year-old woman who was married to a U.S. Army officer serving in France. The couple was charged with registering under a false name at a hotel and later tried at the Chicago's Morals Court for disorderly conduct and other charges. It is rumored that Thomas's personal politics, his radical views on gender relations, and his wife's leadership of the Woman's Peace Party played a role in the Federal Bureau of Investigation's aggressive approach to his case. Thomas and the woman with whom he was having an affair were eventually cleared of the charges. Nonetheless, Thomas was dismissed from his position at the University of Chicago. Additionally, as a result of the scandal, the Carnegie Foundation, which was funding Thomas's research on immigration at the time, insisted that his name be removed from a forthcoming publication. That manuscript, *Old World Traits Transplanted*, instead gave writing credit to Robert Park (1864–1944) and Herbert Miller (1875–1951), both of whom contributed only minor parts to the book.

BIBLIOGRAPHY

Baker, P. J. (1973). The life histories of W. I. Thomas and Robert E. Park. *American Journal of Sociology, 79*(2), 243–260.

Thomas, W. I. (1907). *Sex and society: Studies in the social psychology of sex.* Chicago: University of Chicago Press.

Thomas, W. I., & Znaniecki, F. (1918). *The Polish Peasant in Europe and America: Monograph of an immigrant group.* Chicago: University of Chicago Press.

Thomas, W. I., Park, R. E., & Miller, H. A. (1921). *Old world traits transplanted.* New York: Harper.

Thomas, W. I., & Thomas, D. S. (1928). *The child in America: Behavior problems and programs.* New York: Knopf.

Amir Marvasti

TIME USE, CHILDHOOD AND ADOLESCENCE

SEE Volume 1: *Activity Participation, Childhood and Adolescence; Media and Technology Use, Childhood and Adolescence; Sports and Athletics.*

TRANSITION TO MARRIAGE

At the turn of the 21st century Americans were entering first marriages at later ages than at any other time in the previous hundred years. Combined with increasing rates of cohabitation and a growing proportion of births occurring to unmarried mothers, this suggests a diminished role for marriage in the lives of many young adults. However, it is important to study the transition to marriage for a number of reasons. Demographers estimate that the vast majority of Americans will marry, and most people continue to report that they value marriage highly. Marriage tends to be more stable than other types of intimate unions and confers greater legal rights and protections than does nonmarital cohabitation. Marriage also is associated with positive outcomes such as improved economic well-being, reduced risk-taking behaviors, and relatively higher levels of child well-being. The timing of the transition to marriage is important, with very early marriages associated with some negative outcomes. With a focus on the experience of young adults, this entry reviews trends and differentials in the transition to marriage, patterns of family formation outside marriage, and existing knowledge about the association between marriage and the well-being of adults and children.

SHIFTING PATTERNS
OF MARRIAGE ENTRY

The timing of the transition to marriage has changed considerably over recent decades. Although ages at first marriage decreased during the 1950s, they increased rapidly in the decades that followed. In 2006 the median age at first marriage in the United States reached 27.5 years for men and 25.5 years for women (U.S. Census Bureau, 2007), the latest marriage ages at any time in the previous century. These trends have reshaped the family lives of young adults dramatically. Whereas 64% of women age 20 to 24 had ever been married in 1970, the same was true of only 31% in 2000. Among men age 20 to 24, roughly 55% had ever been married in 1970, compared with only 21% in 2000 (U.S. Census Bureau, 1991, 2008). Similar patterns of marriage delay also are observed in many other industrialized countries. Average ages at first marriage throughout much of Western Europe are as high as or higher than those observed in the United States (Kiernan, 2000).

However, at least in the United States these patterns seem to be more about marriage delay than about eschewing marriage altogether. Demographers estimate that nearly 90% of recent cohorts of U.S. women will marry (Goldstein & Kenney, 2001). The vast majority of U.S. high school seniors surveyed in the late 1990s reported that having a good marriage and a happy family life was extremely important (Thornton & Young-DeMarco, 2001). The importance of marriage in the United States also is reflected in a strong social movement demanding marriage rights for same-sex couples and in the sometimes heated and emotional debate about the nature of marriage that this movement has generated (Hull, 2006). However, the fraction of people expected

to marry is lower in many European countries than in the United States, perhaps indicating the relatively lower salience of marriage in these countries. For example, fully 40% of Swedish women born in 1965 are expected to be never married at age 50, as are roughly 30% of such women in Norway, Finland, and France (Prioux, 2006).

FAMILIES FORMED
OUTSIDE MARRIAGE

Although young adults in the United States are less likely to marry than they were in the past, many are choosing to live with a romantic partner. In the last two decades of the 20th century and the first decade of the 21st rates of nonmarital cohabitation (living with a partner outside marriage) increased substantially. Whereas only 11% of first marriages formed between 1965 and 1974 were preceded by cohabitation, the same was true of 56% of first marriages formed in the early 1990s (Bumpass & Lu, 2000, Bumpass & Sweet, 1989). At the beginning of the 21st century, nearly one-third of all current unions among women age 19 to 24 were nonmarital cohabitations and 38% of these young women had ever cohabited (Bumpass & Lu, 2000). High levels of nonmarital cohabitation also are observed in many other parts of the world, including regions of Northern and Western Europe as well as many Central American and Caribbean nations (Fussell & Palloni, 2004, Kiernan, 2000). For example, 83% of women are expected to enter a nonmarital cohabitation before age 45 in France, as are roughly two-thirds of women in Sweden (Heuveline & Timberlake, 2004).

Nonmarital cohabitation in the United States tends to be relatively short-lived, with half of these unions lasting no more than about one year and only 1 in 10 lasting five years or longer (Bumpass & Lu, 2000). Roughly half of U.S. cohabiting unions eventually end in marriage (Bumpass & Lu, 2000). Cohabitation tends to be most common among those with relatively less education, although it is widespread among all educational groups. For example, 59% of women with less than 12 years of schooling had ever cohabited in the mid-1990s, compared with 37% of women with at least 4 years of college (Bumpass & Lu, 2000). Cohabitation also tends to be relatively more common among individuals from economically disadvantaged families and those who lived apart from at least one parent during childhood (Bumpass & Sweet, 1989). Although reasons for cohabitation are varied, individuals with relatively fewer economic resources may feel less able to afford the perceived economic responsibilities associated with marriage.

In addition to losing its dominance as a setting for intimate relationships, marriage has lost some of its dominance as a setting for childbearing. Roughly 37% of all births in the United States in 2005 were to unmarried mothers, compared with 11% in 1970 (Martin, Hamilton, Sutton, Ventura, MeNacker, Kirmeyer, et al., 2007, Ventura & Bachrach, 2000). The fraction of births occurring outside marriage is even greater among relatively younger women, representing more than half of all births to women age 20 to 24 and more than 80% of births to women under age 20 (Martin et al., 2007). These trends are driven by a number of factors, including declining rates of marriage in young adulthood, increased rates of childbearing among unmarried women, and declining rates of childbearing among married women. Also, women are considerably less likely to marry in response to a pregnancy than they were in the past. In the early 1950s more than half of all women age 15 to 29 married before a first premaritally conceived birth in the United States, compared with less than one-quarter in the early 1990s (Ventura & Bachrach, 2000).

Delayed entry into marriage, combined with a loosening of the connections between marriage and life course events such as entry into committed romantic relationships, entry into parenthood, school completion, and leaving the parental home, has led a number of scholars to argue that marriage has become less important as a marker of the transition to adulthood (Corijn & Klijzing, 2001, Furstenberg, Kennedy, McLoyd, Rumbaut, & Settersten, 2004). One study finds that whereas well over 90% of Americans believe that completing one's education, achieving financial independence, and working full-time are at least somewhat important to being considered an adult, only 55% say the same thing about marriage (Furstenberg et al., 2004). Other studies indicate that people in their teens and twenties place less emphasis on marriage and greater emphasis on qualities such as "accepting responsibility for one's self" and "making independent decisions" in self-identifying as adults (Arnett, 2000, p. 473). Contemporary cohorts of young people arguably tend to view marriage and parenthood as life choices rather than prerequisites for becoming an adult (Furstenberg et al., 2004).

The last decades of the 20th century and the first decade of the 21st brought delayed entry into marriage, growth in the proportion of young adults living with an unmarried partner, and an increase in the proportion of births occurring to unmarried parents. Although marriage is less dominant than it once was over the organization of family life, many family scholars argue that it remains important in American society. Sociologist Andrew Cherlin argues that marriage "has been transformed from a familial and community institution to an individualized choice-based achievement" (Cherlin 2004, p. 858). Recent shifts in marriage patterns likely result from a complex and deeply intertwined set of social and economic factors, many of which have deep historical roots. These include shifts in the labor market and educa-

tional opportunities of women and men, changing attitudes toward sex and childbearing outside marriage, improved control over the link between sex and reproduction, and growth in the importance individuals place on personal fulfillment. The fact that many people view school attendance and marriage as incompatible provides a nice twist on the life course theme of "linked lives." Although recent patterns of transition into marriage are well documented, there is less agreement about what marriage will look like in the future. The likely course of future change in marriage patterns is an important area of study and debate among social scientists.

DIFFERENTIALS IN EARLY ADULTHOOD MARRIAGE

A number of background factors are associated with the likelihood of becoming married before age 25. For example, there are substantial racial and ethnic differences in the likelihood of early marriage. Although 63% of White women and 61% of Hispanic women had transitioned to marriage by age 25 in 1995, the same was true of only 37% of non-Hispanic Black women and 44% of non-Hispanic Asian women (Bramlett & Mosher, 2002). Race differences in marriage extend into the later life course, with over 90% of recent cohorts of White women expected to ever marry, compared with roughly two-thirds of recent cohorts of Black women (Goldstein & Kenney, 2001).

Religion and region of residence also are associated with differences in early adulthood marriage patterns. Women who have no religious affiliation or who report that religion is not important to them are less likely than others to marry by age 25. Women in the Northeast are less likely to marry by age 25 than are women in other regions of the country. Although regional variation in the proportion ever married largely levels off by age 30, differentials by religiosity persist into older age groups (Bramlett & Mosher, 2002). Several of the same factors that predict a greater overall probability of marriage during one's lifetime are associated with a reduced probability of marrying during the early adult years: A number of studies demonstrate that people with relatively higher levels of education are more likely to marry over the course of their lives than are those with relatively less schooling, yet having completed more years of education is associated with a reduced probability of marrying by age 25. Estimates suggest that whereas roughly two-thirds of women with no more than a high school degree marry by age 25, the same is true of only half of women with relatively more schooling.

Variation in patterns of family formation across social class and racial/ethnic groups has been of particular interest to social scientists. Although better-educated women increasingly delay both marriage and parenthood, less-educated women delay marriage but not parenthood (Ellwood & Jencks, 2004). Black Americans are both less likely to marry than non-Hispanic Whites and more likely to bear children outside marriage. These patterns lead to considerable differences in the context in which children are born, with a much larger proportion of births occurring within marriage among Whites and the most educated than among Blacks and the least educated. The causal factors underlying social class and racial/ethnic differences in patterns of family formation, however, are not well understood. Sociologists William J. Wilson and Kathryn Neckerman (1987) argue that high rates of unemployment and incarceration among men in poor urban areas reduces the number of attractive male marriage partners. Yet most studies find only a relatively modest contribution of the availability of "marriageable" men to group differences in marriage patterns. Variation across groups in attitudes toward marriage is similarly unable to fully explain racial and socioeconomic gaps in marriage behavior (Edin & Kefalas, 2005; Ellwood & Jencks, 2004). The search for alterative explanations for these patterns remains an important area of future research.

MARRIAGE, HEALTH, AND WELL-BEING

Marriage is associated with a number of beneficial outcomes for men and women, including higher family income, improved health, and reduced risk-taking behavior (Waite, 1995). Although most children fare well regardless of the structure of their families, growing up with two biological parents who are married to each other rather than with a single parent or two cohabiting parents is associated with relatively higher child well-being (Amato, 2005). Scholars disagree, however, about the extent to which marriage causes these positive outcomes as opposed to people with higher preexisting levels of health, resources, and well-being being more likely to marry in the first place. Also, relatively little is known about whether these potential benefits of marriage extend to individuals who marry during the earliest years of adulthood. Attempts to answer this question are complicated by the fact that individuals who marry early in life tend to come from relatively more disadvantaged backgrounds than individuals who delay marriage. Marriage before age 20 also is associated with other negative outcomes for individuals, including a higher risk of experiencing a subsequent divorce than is the case for individuals who marry relatively later in life (Martin & Bumpass, 1989). Many explanations have been offered to explain this relationship. For example, relatively young people may be less able to predict what kind of partner would make the most compatible long-term match or may have fewer economic or emotional resources to draw on in support of their marriages.

In considering the potential benefits of marriage in the contemporary United States it is important to keep in mind that marriage brings certain rights and legal protections that are less easily available or not available to unmarried couples. Laws give special recognition to married couples in a number of domains, including their emotional attachment, parental status, and economic organization and resources (Hull, 2006). For example, the Family and Medical Leave Act gives spouses but not unmarried partners the right to take time off from work to care for a partner with a serious health condition. Married individuals generally can receive health insurance and other benefits through a spouse's employer and are subject to special rules governing inheritance and taxation. In court proceedings married partners receive protection from testifying against a spouse. Such legal and economic differences in the status of married versus unmarried couples have received considerable attention in debates about the extension of marriage rights to same-sex couples.

FUTURE DIRECTIONS

Although Americans increasingly are delaying the transition to marriage, most young people eventually marry and the desire to marry is widespread. However, young adults have more latitude in choosing what kinds of romantic partnerships to enter, when to enter those relationships, and the relationship context in which to bear and raise children. Although scholars know a considerable amount about patterns of marriage, much remains to be learned. For example, causal factors underlying differences in marriage patterns across educational and racial/ethnic groups are not well understood. Also, although marriage is associated with many positive outcomes, the processes underlying those outcomes are not well understood. How much of the relatively greater physical, emotional, and socioeconomic well-being of married versus unmarried people can be attributed to causal effects of marriage per se and how much is due to preexisting differences between people who enter marriage versus those who do not? Also, the extent to which any potential benefits of marriage extend to young adults is unclear.

There is much disagreement about the likely direction of change in future patterns of marriage. Will age at first marriage continue to increase over time? Will there be a meaningful reduction in the proportion of individuals who marry over the course of their lives? Answers to these questions are an important focus of ongoing efforts to understand the meaning and nature of marriage in the 21st century.

SEE ALSO Volume 1: *Dating and Romantic Relationships, Childhood and Adolescence; Family and Household Structure, Childhood and Adolescence; Sexual Activity, Adolescent; Transition to Parenthood;* Volume 2: *Cohabitation; Marriage.*

BIBLIOGRAPHY

Amato, P. R. (2005). The impact of family formation change on the cognitive, social, and emotional well-being of the next generation. *Future of Children, 15*(2), 75–96.

Arnett, J. J. (2000). Emerging adulthood: A theory of development from the late teens through the twenties. *American Psychologist, 55*(5), 469–480.

Bramlett, M. D., & Mosher, W. D. (2002). *Cohabitation, marriage, divorce, and remarriage in the United States.* Hyattsville, MD: U.S. Department of Health and Human Services, Centers for Disease Control and Prevention, National Center for Health Statistics.

Bumpass, L., & Lu, H-H. (2000). Trends in cohabitation and implications for children's family contexts in the United States. *Population Studies, 54*(1), 29–41.

Bumpass, L. L., & Sweet, J. A. (1989). National estimates of cohabitation. *Demography, 26*(4), 615–625.

Cherlin, A. J. (2004). The deinstitutionalization of American marriage. *Journal of Marriage and Family, 66*(4), 848–861.

Corijn, M., & Klijzing, E. (2001). Transitions to adulthood in Europe: Conclusions and discussion. In M. Corijn & M. Klijzing (Eds.), *Transitions to adulthood in Europe* (pp. 313–340). Dordrecht, Netherlands: Kluwer Academic Publishers.

Edin, K., & Kefalas, M. (2005). *Promises I can keep: Why poor women put motherhood before marriage.* Berkeley, CA: University of California Press.

Ellwood, D. T., & Jencks, C. (2004). The uneven spread of single-parent families: What do we know? Where do we look for answers? In K. M. Neckerman (Ed.), *Social inequality* (pp. 3–77). New York: Russell Sage Foundation.

Furstenberg, F. F. Jr., Kennedy, S., McLoyd, V. C., Rumbaut, R. G., & Settersten, R. A. Jr. (2004). Growing up is harder to do. *Contexts, 3*(3), 33–41.

Fussell, E., & Palloni, A. (2004). Persistent marriage regimes in changing times. *Journal of Marriage and the Family, 66*(5), 1201–1213.

Goldstein, J. R., & Kenney, C. T. (2001). Marriage delayed or marriage forgone? New cohort forecasts of first marriage for U.S. women. *American Sociological Review, 66*(4), 506–519.

Heuveline, P., & Timberlake, J. M. (2004). The role of cohabitation in family formation: The Untied States in comparative perspective. *Journal of Marriage and Family, 66,* 114–1230.

Hull, K. E. (2006). *Same-sex marriage: The cultural politics of love and law.* Cambridge, U.K., and New York: Cambridge University Press.

Kiernan, K. E. (2000). European perspectives on union formation. In L. J. Waite and C. Bachrach (Ed.), *The ties that bind: Perspectives on marriage and cohabitation* (pp. 40–58). New York: Aldine de Gruyter.

Martin, J. A., Hamilton, B. E., Sutton, P. D., Ventura, S. J., Menacker, F., Kirmeyer, S., et al. (2007). *Births: Final data for 2005. National Vital Statistics Reports, 56*(6). Hyattsville, MD: National Center for Health Statistics.

Martin, T. C., & Bumpass, L. L. (1989). Recent trends in marital disruption. *Demography, 26*(1), 37–51.

Prioux, F. (2006). Cohabitation, marriage, and separation: Contrasts in Europe. *Population and Societies, 422,* 1–4.

Thornton, A., & Young-DeMarco, L. (2001). Four decades of trends in attitudes toward family issues in the United States: The 1960s through the 1990s. *Journal of Marriage and the Family, 63*(4), 1009–1037.

U.S. Census Bureau. (1991). *Marital status and living arrangements: March 1990. Current Population Reports,* Series P-20, No. 450. Washington, DC: U.S. Government Printing Office.

U.S. Census Bureau. (2007). *Estimated median age at first marriage, by sex: 1890 to the present.* Retrieved June 15, 2008, from http://www.census.gov/population/socdemo/hh-fam/ms2.xls

U.S. Census Bureau. (2008). *1990 census of population and housing.* Summary tape file 3, Table PCT7, Retrieved June 15, 2008, from http://www.factfinder.census.gov

Ventura, S. J., & Bachrach, C. A. (2000). *Nonmarital childbearing in the United States, 1940–99. National Vital Statistics Report, 48*(16). Hyattsville, MD: National Center for Health Statistics.

Waite, L. J. (1995). Does marriage matter? *Demography, 32*(4), 483–507.

Wilson, W. J., & Neckerman, K., M. (1987). Poverty and family structure: The widening gap between evidence and public policy issues. In *The truly disadvantaged: The inner city, the underclass, and public policy* (pp. 63-92). Chicago: University of Chicago Press.

Megan M. Sweeney

TRANSITION TO PARENTHOOD

The transition to parenthood is typically defined as the event of becoming a biological mother or father. Early transitions to parenthood in the United States have been a topic of concern to scholars, policy makers, and the general public since the early 1970s. This concern has been fueled by the growing proportion of young mothers who are single and by the putative consequences of early and single childbearing. For instance, women who have a birth during their teens are more likely than their counterparts who postpone childbearing to drop out of high school or become dependant on welfare. Although early parenthood is a relative concept, considerable attention has been paid to transitions during the teen years (i.e., 15 to 19). Basic trends in adolescent and young adult transitions to parenthood, the causes and consequences of these trends, and directions for future research are discussed here.

HISTORICAL TRENDS

One of the more noted trends in the transition to parenthood is its postponement. As evidence of this, the proportion of women who have their first child before the age of 25 has decreased since the 1960s, whereas the proportion of women who have their first child beyond the age of 25 has increased. The postponement of childbearing has been concentrated mainly among women with college degrees

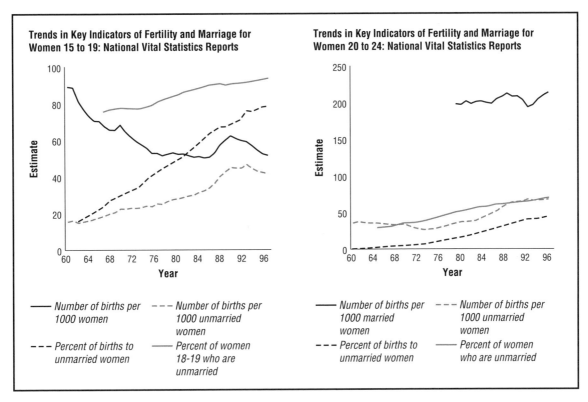

Trends in Key Indicators of Fertility and Marriage. CENGAGE LEARNING, GALE.

and reflects increases in their returns to higher education. Delayed childbearing on the part of more educated women has led to a divergence between educational groups in rates of early childbearing. In spite of this divergence, scholars documenting trends in early parenthood have focused mainly on the dramatic increase in the fraction of early births occurring to unmarried women.

Among women of all reproductive ages, changes in the proportion of births to unmarried women, termed the *nonmarital fertility ratio*, have been driven by declines in the proportion of women marrying, increases in nonmarital birthrates, and decreases in marital birthrates. Trends in these statistics, which are presented in Figure 1, are more frequently and fully documented for teen women than for woman of other ages, and they are rarely presented for men. The proportion of births occurring to unmarried teen women increased from 0.148 to 0.785 between 1960 and 1998. During this same period, the proportion of teen women who were unmarried increased from 0.719 to 0.936. Whereas the birthrate for married teen women (calculated as the number of births per 1,000 married women ages 18 to 19) declined from 530.6 to 322.1, the birthrate for unmarried teen women (the number of births per 1,000 unmarried women ages 15 to 19) increased from 15.3 to 41.5. The nonmarital birth ratio for women ages 20 to 24 (presented in Figure 2) also increased considerably

between 1960 and 1998 (i.e., from 4.8 to 47.7) but remained considerably lower than the ratio for teen women throughout this time period; the nonmarital birth ratio for teen women increased from 0.148 to 0.785.

Demographers attribute substantially more of the increase in nonmarital fertility ratio to changes in marriage than to changes in fertility. Although pregnancy rates among unmarried women have declined since the 1960s, the proportion of younger women who are single has increased considerably. Delays in marriage have dramatically increased women's exposure to the risk of having a nonmarital pregnancy. Furthermore, women are decreasingly likely to marry in response to a nonmarital conception. Consequently, birthrates among unmarried women have increased over time. Changes in marriage have been more dramatic for Black women than for White women and have driven relatively more of their increase in the nonmarital birth ratio. Changes in the measurement of race and ethnicity make it difficult to document and analyze changes in the nonmarital fertility ratio for Hispanic women.

Although the birthrates for unmarried teen women have increased, birthrates for teen women as a whole have fallen considerably since the 1960s (with some fluctuation over time). For instance, the teen birthrate declined from 89.1 to 51.1 (births per 1,000 women) between 1960 and 1998. Until the early 1990s, Blacks had the highest teen

Mobile Pregnancy Clinic. *Evelyn Flores holds her one-month-old daughter Hailey Interiano after their checkup at the mobile health clinic in Garland, TX. The clinic is just one way of helping teen mothers in Texas, which has the fifth-highest teen pregnancy rate in the nation.* AP IMAGES.

birthrates, followed by Hispanics and then by non-Hispanic Whites. Because of the large decline in teen birthrates among non-Hispanic Blacks during the 1990s, Hispanics have surpassed Blacks in their teen birthrates. As of 2000, rates for non-Hispanic Whites, non-Hispanic Blacks, and Hispanics were 32.6, 79.2, and 87.3 (births per 1,000 women), respectively.

Other developed countries have similarly experienced large declines in teen birthrates since the 1960s. Declining birthrates in developed countries have been attributed to increased education and increased knowledge of contraception and access to it. The decline in birthrates for teens in the United States has been considerably smaller than the decline observed in many other developed countries. Consequently, teen birthrates in the United States are high in comparison to those of Canada and European countries. The relatively higher birthrates among teens in the United States are only partly attributable to their greater socioeconomic disadvantage. Within all socioeconomic groups, U.S. youth tend to have higher teen birthrates than their counterparts from other developed countries. Nor are the higher birthrates on

the part of U.S. teens attributable to greater levels of sexual activity. U.S. youth begin having sex around the same age as youth in other developed countries. U.S. youth are said to have higher teen birthrates because of other factors: a weaker consensus that childbearing is only an adult behavior, less accepting attitudes toward teen sex, less access to family planning services, and greater inequality.

A handful of studies have attempted to explain the decline in teen motherhood by examining the relative magnitude of changes in the "proximate determinants" of fertility (i.e., sexual activity, contraception, abortion, and pregnancy). Declines in teen birthrates during the late 1990s have been linked to declines in pregnancy. A small fraction of the decline in pregnancy rates has been attributed to increases in sexual abstinence, but a large share of the decline has been attributed to increases in contraceptive use. Researchers have yet to explain the decline in teen parenthood over a broader time frame.

PERSPECTIVES

Explanations for the timing of transitions to parenthood are typically informed by rational choice perspectives.

According to these perspectives, individuals weigh the expected rewards of having a child against its anticipated costs (or disadvantages). Because women have historically been the primary caretakers of children and caretaking is time intensive, these perspectives highlight the opportunity costs of having children for women. Child rearing and childbearing interfere with the development of important educational and labor market experiences. In contrast, men have been more responsible for the financial support of children. The transition to fatherhood is likely to impede schooling (at younger ages) and increase men's participation in the labor force.

Studies of early parenthood focus not only on the role of opportunity costs but also on socialization, social control, stress, and risk preferences. Socialization perspectives emphasize the importance of adults and peers as role models for adolescent behavior, whereas social control perspectives highlight the importance of parental monitoring and supervision. Stress perspectives suggest that stresses accompanying major life events are critical to adolescent behavior to the extent that they change parenting practices or reduce an adolescent's sense of control. Deviance perspectives propose that a variety of risk-taking behaviors, including sexual behavior, are linked because of their dependence on a number of factors, including self-esteem, expectations about the future, and peer affiliations.

Studies have drawn on each of these perspectives to explain the influence of various factors on the risk of early motherhood. Although a growing number of studies are examining transitions to early fatherhood, studies typically focus on women because of concerns about men's representation in surveys and the reliability of their reports of fertility. Studies of early parenthood often highlight the role of a single variable or set of variables in measuring the various theoretical constructs. For example, growing up in a single-parent family is thought to be associated with lower levels of monitoring, a greater number of stressful life events, and more exposure to single motherhood.

Several patterns repeatedly emerge in studies of teen motherhood. Reflecting their higher birthrates, Black and Hispanic women are more likely than their White counterparts to become teen mothers; differences by race are explained only partly by family background variables. Women born to a teen mother are more likely to become teen mothers themselves than their counterparts born to an older mother, and women living with a single mother or a stepparent in adolescence are more likely to have a teen birth than their counterparts who live with two biological parents. Women are also more likely to become teen parents if they have parents who are less educated or economically advantaged, or if they themselves have lower academic achievement or aspirations. Women living in more disadvantaged communities (e.g., communities with larger percentages of impoverished families) are at greater risk of teen parenthood, regardless of their own characteristics. Women are more likely to have a teen birth if they reside in states with weaker child support enforcement, but their chances of having a teen birth increase only slightly, if at all, as the level of welfare benefits in their state increases. Yet women growing up in a household receiving Aid to Families with Dependent Children (AFDC) are much more likely to have a teen birth than women whose parents did not receive welfare. Finally, women are more likely to become teen mothers if they experience changes in family structure and income. Studies suggest that the influence of family background factors on teen parenthood does not differ markedly by cohort and is similar in other developed countries (e.g., England).

Rational choice perspectives assume that individuals with greater opportunities for advancement delay fertility in order to avoid incurring early parenthood. These perspectives suggest that the association between disadvantaged family background and the transition to parenthood will weaken or reverse at some point in the life course. Studies that examine how the influence of different variables on the transition to parenthood differs by age offer mixed evidence about if and when this turning point occurs. Some studies suggest that the association between background characteristics and fertility is stronger for teens than for young adults, whereas other studies fail to find any considerable differences.

CONSEQUENCES

Early childbearing continues to be an issue of public concern not only as a consequence of increases in the nonmarital fertility ratio but also because it is associated with negative outcomes for mothers and children. To consider just a few examples, teen mothers are less likely than their counterparts without an early birth to complete high school and to marry, and the children of teen mothers are less likely than the children of older mothers to receive prenatal care. More recently, scholars have argued that some of the negative outcomes associated with early childbearing are not a consequence of having a teen birth but a reflection of conditions that existed prior to the birth. According to this argument, women whose life chances are more precarious are less motivated to avoid early parenthood in the first place; these women would have had comparable experiences even if they had postponed a birth. Such an argument emphasizes the selection of women with more precarious circumstances into early motherhood, rather than the influence of early motherhood on women. Studies have attempted to identify the causal influence of early childbearing through the use of statistical models (e.g., fixed effects or random effects) that take into account family characteristics that are not directly measured by researchers and through the use of natural experiments, such as miscarriages, that reveal how women predisposed to early pregnancy fare when they delay childbearing. Taken together, the results of these studies suggest that some but not all of the negative outcomes associated with early childbearing are attributable to selection. Consequently, the experience of being a young mother has some negative influences on women. Researchers also suggest that the specific age at which women transition to parenthood is critical; becoming a teen mother at the age of 15 is more consequential than becoming a teen mother at 19.

DIRECTIONS FOR FUTURE RESEARCH

Although social scientists' understanding of early childbearing has broadened considerably, several important issues remain to be addressed. Studies of early parenthood and the events that lead to it often fail to consider more than one type of factor, focusing exclusively on the effects of demographic, socioeconomic, or policy variables. For instance, some studies concerning the effects of policies fail to control for socioeconomic factors such as male and female wages. Studies also need to take into account within-group variation by measuring physical attributes such as pubertal development, physical attractiveness, and body size and enduring attributes such as cognitive ability.

Studies also need to consider the family processes by which different factors influence parenthood and its proximate determinants. For instance, studies examining the effects of family structure on such outcomes typically measure mechanisms such as parental monitoring only indirectly. Consequently, it is not clear how family processes mediate family structure, family changes, and the intergenerational association in the timing and circumstances of parenthood. Research is additionally needed on how social-psychological processes (e.g., attitudes) mediate the effects of different factors. Validating opportunity-cost perspectives, qualitative studies reveal that some youth view early parenthood positively because other avenues to success are restricted to them and because they do not expect to live very long. Studies that consider "nothing to lose" attitudes about the future (e.g., expectations about graduating from college or surviving into adulthood) have shown that factors such as neighborhood context influence the transition to parenthood through their influence on expectations about the future. More work in this tradition is needed.

Studies typically examine union dynamics and fertility in isolation, even though pregnancies and births occur within the context of some form of relationship. Studies need to examine how different factors influence the formation of romantic and sexual relationships, in addition to the timing and sequencing of behaviors within these relationships. Related to this, studies need to examine the consistency of influences on the proximate determinants of early births. For example, if a particular welfare policy has no effect on pregnancy and no effect on abortion, then it would be inconsistent to find that it has a significant effect on births. Similarly, if a policy reduces pregnancies and births but has no effect on abortion, then this is evidence that the pathway of influence is through either less sex or more use of contraception. This approach allows researchers to make inferences about the pathways of influence and identify the policy levers most effective in delaying parenthood.

Because previous studies have focused simply on teen or premarital childbearing, it is not clear whether factors associated with having an early birth are also associated with the relationship context of the birth. Statistics based on mothers of all ages reveal that the number of births that occur to cohabiting women is growing and substantial. Changes in the relationship context of teen births beg several questions. Among women who become teen mothers, do factors such as economic disadvantage reduce the chances of having a cohabiting birth, as opposed to a single or marital birth? Do consequences of having an early birth differ by its relationship context? To the extent that children in cohabiting relationships tend to experience subsequent changes in family structure, having a child within a cohabiting relationship may not necessarily be more beneficial than having a child while single.

Information that teen mothers report about fathers on birth certificates suggests that teen fatherhood is much less common than teen motherhood. It is estimated that only about a third of the partners of teen women are teens themselves, and about a fifth of them are 25 or older. Premarital births appear to have important consequences for men, and assessments of the quality of male birth reports conducted in the early 21st century identify aspects of surveys that improve these estimates.

SEE ALSO Volume 1: *Family and Household Structure, Childhood and Adolescence; Sexual Activity, Adolescent; Transition to Marriage;* Volume 2: *Birth Control; Childbearing.*

BIBLIOGRAPHY

Furstenburg, F. F., Jr. (2003). Teenage childbearing as a public issue and private concern. *Annual Review of Sociology, 29,* 23–39.

Hamilton, B. E., Sutton, P. D., & Ventura, S. J. (2003). Revised birth and fertility rates for the 1990s and new rates for Hispanic populations, 2000 and 2001: United States. *National Vital Statistics Reports, 51*(12). Hyattsville, MD: National Center for Health Statistics.

Hynes, K., Joyner, K., Peters, H. E., & Yang, F. (2008). The transition to early fatherhood: National estimates based on multiple surveys. *Demographic Research, 18*(12), 337–376. Retrieved May 2, 2008, from http://www.demographic-research.org/Volumes

Rindfuss, R., Morgan, S. P., & Offutt, K. (1996). Education and the changing age pattern of American fertility: 1963–89. *Demography, 33*(3), 277–290.

Ventura, S. J., & Bachrach, C. A. (2000). Nonmarital childbearing in the United States, 1940–99. *National Vital Statistics Reports, 48*(16). Hyattsville, MD: National Center for Health Statistics.

Ventura, S. J., Mathews, T. J., & Hamilton, B. E. (2001). Births to teenagers in the United States, 1940–2000. *National Vital Statistics Reports, 29*(10). Hyattsville, MD: National Center for Health Statistics.

Kara Joyner

TWIN STUDIES
SEE Volume 1: *Genetic Influences, Early Life.*

V–Y

VALUES

SEE Volume 1: *Academic Achievement; Civic Engagement, Childhood and Adolescence; Employment, Youth; Moral Development and Education; Religion and Spirituality, Childhood and Adolescence.*

VIRGINITY

SEE Volume 1: *Sex Education/Abstinence Education; Sexual Activity, Adolescent.*

VOCATIONAL TRAINING AND EDUCATION

Vocational education, broadly speaking, is a type of course intended to train students for particular jobs. The extent to which an individual student is exposed to such courses can vary greatly, from a student taking a single vocational education course, to whole high schools being organized around vocational preparation. Most often, vocational education courses are intended to prepare students who are unlikely to attend college for the jobs they are likely to obtain. In so doing, vocational education attempts to retain students in high school by providing instruction that "at-risk" students see as useful and/or interesting. The rationale is that vocational involvement can benefit students by offering skills that will be of value in local labor markets. Although certainly plausible, certain fundamental questions remain unanswered. Who is placed into vocational education? Is the process relatively neutral, or are certain subgroups more likely to be vocationally steered despite actual achievement and expectation levels? And finally, do these processes play a part in creating labor market inequalities and concentration patterns by class, race, and/or gender?

THE HISTORY OF VOCATIONAL EDUCATION AND RESULTING STRATIFICATION

Federal funding of vocational education predated more general educational funding by about 40 years and had as its primary intent student preparation for occupational roles through differential socialization. Consequently, class, gender, and racial disparities in occupational status have a history of being reproduced through vocational programs. The recreation of inequality was sometimes implicit, although at other moments was quite explicitly mandated. Congress, for instance, had two goals for vocational training in the 1930s: first, to create programs that reflected the local labor market segmentation in terms of race and gender and, second, to reduce unemployment by matching workers to available blue-collar jobs (Werum, 2002).

As part of this process, federal vocational programs placed top priority on agricultural job training. For example, in 1936, two-thirds of the funding for vocational education targeted the agriculture industry. This

509

focus provided the greatest benefits to southern agrarian elites who had their workforce trained at the federal government's expense. The marginalization of vocational training available to females resulted from this overemphasis on agriculture, an industry dominated by men. Both the Smith-Hughes Act of 1917 and the George-Ellzey Act of 1934 left underfunded vocational education intended for females by limiting such training to home economics courses. In short, there is a history of highly gendered policies dating back to the beginnings of the vocational education movement (Werum, 2002).

Whereas the history of vocational education consistently demonstrates disadvantages faced by females, the race-related history is mixed. Dating back to the Smith-Hughes Act of 1917, states' rights arguments emphasized state control of education and prevented equitable distribution of resources across racial groups. Such policy once again benefited southern agrarian elites, because states and local school boards had discretion over how federal dollars would be spent. Although Black high schools were finally established in the South, their primary intent was to train workers. For their part, Black southern leaders preferred an emphasis on liberal arts education, yet the available external funding placed serious constraints on the curriculum. Note that race and class divisions also shaped the type of vocational training females received: Whereas middle-class White females were trained to be "household managers," immigrant and Black females were prepared to be "household workers." On a related point, Black males typically trained for "substance" farming, whereas White males were taught "scientific" farming techniques.

Historical cases such as these illustrate the overarching influence of local labor market conditions and elite decision making on racial and gender patterns of participation in vocational education and, consequently, segregation and differential opportunity in local labor markets. Moreover, such cases suggest quite clearly that vocational education may mean very different things to different groups in the stratification hierarchy or, at the very least, may influence social groups in very distinct ways.

For example, in a qualitative study of vocational education students, Royster (2003) argues that getting a job is driven not by qualifications but by who knows about the job openings and who has the "weak ties" that help them secure the position. Despite coming from similar backgrounds and performing similarly in school, Whites in Royster's study experienced far greater occupational success than the Blacks she interviewed. Specifically, Whites earned higher wages and experienced less unemployment than Blacks. Moreover, Black men were more likely to find work outside the trades they trained for in high school. In short, Whites had much more productive social networks beyond school than Blacks. As a result, rather than a meritocratic sorting of workers into jobs, racial inequality was reproduced through the social networks of working-class Whites and their links to the vocational education system.

BENEFITS, RISKS, AND COSTS OF VOCATIONAL PARTICIPATION

Considering the overall literature on vocational education, results suggest some contradictory patterns. Much of the research suggests that involvement in vocational education reduces the risk of dropping out of high school. Specifying this relationship, Rasinski and Pedlow (1998) argue that the protection from dropping out is generally indirect (working through performance), although the influence of certain types of vocational education (e.g., agricultural) may be more direct. Ainsworth and Roscigno (2005), by contrast, found that students taking large numbers of blue-collar vocational education classes were actually more likely to drop out of high school—a finding that calls into question the central reason vocational education exists.

Beyond high school persistence, other benefits have also been demonstrated. High school vocational education, for instance, has been linked to reductions in teen pregnancy among at-risk teenagers, reduced unemployment after high school, increases in one's earning potential, and the likelihood that one will work in a skilled position. Finally, some work suggests that graduates of vocational education programs are more satisfied with their jobs than are comparable students from other high school programs.

There nonetheless appear to be additional risks, if not costs, of vocational participation. For instance, students in vocational programs do not perform as well in basic academic skills such as mathematics, science, and reading compared to those enrolled in more general academic programs. When it comes to the school-to-work transition, Gamoran (1998) found that academic course work improved students' ability to find a job and enhanced occupational mobility relative to vocational course work. This benefit appears to increase over time. Consistent with this point, Rosenbaum (2001) suggested that most employers are more interested in basic academic skills and trainability rather than specific vocational skills. Along with such costs to occupational mobility, vocational training significantly reduces the chances of attending four-year college and limits opportunities for garnering a professional or managerial position (Ainsworth & Roscigno, 2005; Arum & Shavit, 1995).

High School Student Welding. © BOB ROWAN; PROGRESSIVE IMAGE/CORBIS.

VOCATIONAL EDUCATION AS TRACKING

By placing students into college preparatory or vocational tracks, schools act to stratify students—a process that differentially prepares some students for college attendance and others for work. This type of tracking could be interpreted as unproblematic, particularly when the channeling it entails is assumed to be based solely on merit. This, however, may not be the reality. Instead, micropolitical evaluation, including the frequent and questionable use of "neutral" standardized tests, may act to reproduce inequality at almost every turn. Thus, students from advantaged families (in terms of class, race, and gender) are more likely to be placed in college preparatory tracks.

But how and why might differential tracking of groups occur? As high school students form goals and aspirations, the adults and peers in their lives easily influence them. If teachers and counselors disproportionately encourage certain students (e.g., minorities, the poor, those who have parents with less education) to enroll in vocational courses, these students may come to believe that they are neither suited for nor capable of success in college preparatory courses. Moreover, low-income and minority parents rarely challenge such school-level decision making, including tracking decisions, because they themselves were tracked in a similar manner, because they trust the judgments of educators, or because they are politically disengaged or feel powerless relative to the educational process. Alternatively, middle-class parents are intimately familiar with how the educational system works and often act to place their children in an advantaged position. Such processes play out both individualistically, as a parent successfully negotiates for his or her child's placement into an advanced class, or collectively, as a group of parents may lobby for curricular development in a particular direction. Questions surrounding the sorting of students into vocational courses or tracks is a relatively understudied topic in the vocational education literature and represents a fruitful avenue for future research.

CONCLUSION

The literature suggests that higher status students avoid taking vocational education classes. Alternatively, disadvantaged students are funneled into vocational education classes, and this involvement increases the likelihood that they will drop out of high school and decreases the chances they will attend college. In short, vocational education programs promote very real advantages and disadvantages in the schooling process. They shape which students persist into college and the types of jobs available to them in the local labor market following their exit from high school.

Despite the potential for inequality reproduction, some researchers continue to advocate vocational education as the best way to motivate students to behave and work hard in school. The assumption here is that incentives—in this case, incentives for eventual job placement—are necessary to motivate students (Rosenbaum, 2001). To be sure, this viewpoint is consistent with the original purpose of tracking: to encourage students to resign themselves to their appropriate position in society, to garner marketable skills, and to see their placement as just. This viewpoint, however, is based on an assumption of a fair and meritocratic educational sorting process. The literature on vocational education suggests that this assumption is questionable at best.

SEE ALSO Volume 1: *Employment, Youth; High School Organization; Racial Inequality in Education; School Tracking; Socioeconomic Inequality in Education;* Volume 3: *Lifelong Learning.*

BIBLIOGRAPHY

Ainsworth, J. W., & Roscigno, V. J. (2005). Stratification, school–work linkages, and vocational education. *Social Forces, 84,* 257–284.

Arum, R., & Shavit, Y. (1995). Secondary vocational education and the transition from school to work. *Sociology of Education, 68,* 187–204.

Gamoran, A. (1998). The impact of academic course work on labor market outcomes for youth who do not attend college: A research review. In A. Gamoran (Ed.), *The quality of vocational education* (pp. 133–175). Washington, DC: National Institute on Postsecondary Education, Libraries, and Lifelong Learning.

Rasinski, K. A., & Pedlow, S. (1998). The effect of high school vocational education on academic achievement gain and high school persistence: Evidence from NELS:88. In A. Gamoran (Ed.), *The quality of vocational education* (pp. 177–207). Washington, DC: National Institute on Postsecondary Education, Libraries, and Lifelong Learning.

Rosenbaum, J. E. (2001). *Beyond college for all: Career paths for the forgotten half.* New York: Russell Sage Foundation.

Royster, D. A. (2003). *Race and the invisible hand: How White networks exclude Black men from blue-collar jobs.* Berkeley: University of California Press.

Werum, R. E. (2002). Matching youth and jobs? Gender dynamics in New Deal job training programs. *Social Forces, 81,* 473–503.

James W. Ainsworth

WELFARE REFORM
SEE Volume 1: *Policy, Child Well-being.*

YOUTH CULTURE

The study of youth culture and its formation is diffuse and expansive, charted across a variety of disciplines that include but are not limited to sociology, anthropology, psychology, education, history, literature, and cultural studies. Two central questions have directed the study of youth culture. What are the conditions for the emergence of a mass youth culture? In other words, what forces led to the formation, proliferation, and concentration of youth cultures as they are understood in the early 21st century? How can researchers define and identify a culture or cultures of youth as being distinct from the adult culture or cultures in a particular society?

DEFINING YOUTH CULTURE

Images of punks, burnouts, mods and rockers, suburban wiggas, goths, cheerleaders, graffiti writers, teenyboppers, skater kids, beats, hippies, zootsuiters, b-boys, DIY kids, lesbian zine-writers, and ravers come to the fore when one thinks about what youth culture is and the groups that constitute it. However, the precise meanings of the terms *youth* and *culture* have been difficult to establish. In response to the ambiguous and historically shifting boundaries that mark age categories—distinct youth cultures are developing earlier in the life course, whereas transitional markers representing adulthood such as marriage increasingly are delayed—and the broad use of the word *culture*, much scholarly discussion has involved matters of definition.

Those who conduct youth studies recognize youth cultures as expressive, emerging in and through a set of shared practices and collective meanings and organized by young people as they navigate a range of material, historical, and ideological forces. In this regard there is both a symbolic basis and a material basis to youth culture. Youth culture constitutes itself as a peer group yet is not homogeneous. An endless proliferation of splintering groups can found among the ranks of contemporary youth. Cultures of youth may be identified in terms of symbolic resources (sneakers, cell phones, iPods, hair extensions, slang, and argot), rituals and practices (car cruising, car racing, body piercing and tattooing, surfing, graffiti writing), and scenes (streets, raves, parking lots, proms, schools, skate parks, arcades).

Generally, scholars agree that youth culture should be regarded as semiautonomous in nature. Youth culture is not something young people create in isolation from the currents of the mainstream or dominant adult culture of a particular historical moment. Youth culture also is not fully independent of what often is referred to as parent culture, which typically is differentiated from the mainstream by race, ethnicity, and class. For example, the Brown Berets, a cohesive group of working-class Chicano and Chicana students in southern California in the late 1960s, positioned themselves against the White mainstream as they mobilized in support of Chicano studies in universities and improved educational opportunities for Chicano high school students and condemned the oppressive conditions responsible for racial profiling, violence, and poverty in their communities. At the same time they embraced an alternative language of identity from their parents, referring to themselves with proud defiance as Chicano instead of Hispanic. In doing that, they distinguished themselves from the mainstream culture and the parent culture simultaneously (Chávez, 1998).

Many scholars have investigated youth culture as an oppositional culture while recognizing that many of its defining features are not oppositional to the mainstream adult culture or the parent culture. Youth cultures can be part of the mainstream, originating from school-sanctioned activities (e.g., jocks) or being commercially developed (the MTV generation). Youth cultures can be defiant, embracing an ethos of irreverence and abandon, as concern about questions of mobility is suspended temporarily—as in the case of the college panty raids that weakened college administrators' role of being *in loco parentis* in the 1950s—yet also can be squarely mainstream in their orientation to the future.

CONDITIONS FOR THE EMERGENCE OF YOUTH CULTURE

The changing economic and social reality of childhood and adolescence in North America in the late 1900s set in motion swift changes in the leisure activities of young adults, the spaces they occupied, their activities, and the collective and individual selves they imagined. In the process it created a situation ripe for the emergence of a semiautonomous youth culture. Though a burgeoning youth culture can be traced to the early 1800s with evidence of young groups of men gathering in public settings for festivals and parades, most scholars agree that a mass youth culture emerged in the later part of the 19th century and was firmly a part of the cultural landscape by the mid-20th century.

Most youth historians agree that a confluence of forces combined to form a mass youth culture. In the United States, where a distinct youth culture is most visible, the growing freedom and independence of youth from family life that followed urban and industrial expansion in a modern capitalist nation-state set the stage

for increasing school attendance, the emergence of professional psychology, and public campaigns of youth advocacy that together played a role in shaping the boundaries of a mass youth culture.

High levels of age segregation are recognized as a necessary precondition for the emergence of a distinct youth culture. Institutions of mass socialization such as public schools conceived in terms of their potential as a socially integrative force served to cement a society that had been segmented by age. Before the 20th century, few youths attended high school; most worked, often beside adults, in factories and fields. The movement of youth into school meant that young people's lives played out in an institutional setting largely apart from adults, although there is some evidence of distinct youth cultures having taken shape through work before the mass ushering of youth into school. One example is the newsies, indigent boys in their early adolescence who sold newspapers on the streets of New York City, often lived in shared housing, and collectively organized for improved work conditions in the late 1800s.

The character and place of a semiautonomous youth culture was the subject of concern among educators, parents, and social reformers for most of the twentieth century. Youths' growing diversity with the mass influx of immigrants in the late 1800s and the shifting place of young people in a rapidly changing social landscape in the early 1900s fueled anxieties among a generation for whom the ability to supervise all aspects of young people's cultural life was increasingly beyond their grasp, giving rise to moral panics and widespread worry about youth deviance and delinquency.

Changes in the way young people had been culturally identified and constructed as a distinct age group also played a significant role in the formation of a mass youth culture. Beginning in the mid-nineteenth century, young people began to be seen by scholars for the first time as a cohort with distinctive habits and traits. By the 1930s the idea that adolescence is a distinct stage in the life course was entrenched in both the American cultural imagination and scholarly literature. At the threshold of adulthood, adolescence increasingly was treated as a tumultuous stage in the life course, a period of uncertainty and angst over the status and stability of the self. That characterization was influenced by the popularity of professional psychology, with its growing concern for adolescents' "normal" moral and psychological development and departures from that norm. The preoccupation with normal development, combined with widespread worry about youth delinquency and the fact that young adults' lives increasingly were dedicated to activities outside the home, led to concerted efforts to socialize young people inside and outside school, paving the way for the development of a mass youth culture (Palladino, 1996).

A variety of organized nonschool activities for youths sprang up to ensure their development as morally sound citizens and guard against youth complacency and delinquency, such as President Franklin Delano Roosevelt's youth workers' program. By the mid-20th century, teen canteens, school dances, and sock hops were a mainstay of cultural life for young adults, although race, religion, class, and sex played an influential role in participation in those activities. William Graebner (1990) showed how the sock hop was an event largely attended by White middle-class girls and was intended to protect that group of teenagers from activities organized within working-class, Black, and male youth cultures. Its conservative undertones, especially the selection of music played, kept many young people whose musical tastes were rooted in the emerging youth subcultures and not in the adult-sanctioned youth culture away.

School-sanctioned activities such as proms, student government, and after-school sports in the post–World War II period were also central to an evolving mass youth culture. Shaped by an uneasy cold war climate at home, school-based activities, much like school itself, were organized to create a more cohesive, assimilated, and homogeneous youth cohort. Willingness to participate in mainstream youth cultural activities reflected one's patriotic duty and democratic commitment and also was seen as an expression of essential teenhood by the 1950s.

The expansion of a middle class, its growing prosperity after World War II, rapid suburbanization, and rising consumerism also helped transform the leisure lives of young people and establish a mass youth culture. The growing presence of a consumer market radically transformed the leisure activities of youth as teens, a group that for the first time had disposable income (their incomes were less likely to be directed to family needs and allowance became more common) and thus emerged as a distinct consumer category (Palladino, 1996). The leisure of young adults was reinvented radically around consumption as entire markets developed to produce and sell distinct teen commodities (Palladino, 1996), distinguishing this age group from others.

THE PROLIFERATION OF CONTEMPORARY YOUTH CULTURES

Since the mid-1940s, leisure and consumption have been central means by which young adults constitute themselves as belonging to distinct cultures, distinguishing themselves not only from other age-based groups but also from one another. The consumer market has played a large role in the expansion of a mass youth culture and the proliferation and splintering of contemporary youth cultures. The consumer market has been a powerful force in organizing the social spaces and social activities of young people, appropriating signs and symbols that already register as repositories for youth culture and style. However, much research on youth culture, rather than seeing that culture as flowing from the market, has recognized that youth cultural groups use the objects offered by a consumer market for their own ends: to construct identities, express in-group solidarity, and define themselves apart from adults and other youth groups. For instance, Sarah Thornton (1996), in an investigation of ravers (those who participate in an underground dance scene emerging in the early 1990s, held in empty warehouses as an alternative to dance clubs), demonstrated how youth develop subcultural distinctions and symbolic boundaries to distinguish the hip from the unhip mainstream within club culture.

The focus on the creative means by which youth used objects and forms available through the mass market may be the legacy of the Centre for Contemporary Cultural Studies (CCCS) of Birmingham and its abiding interest in symbolic resistance by youth, especially young working-class men. CCCS scholars in the 1970s and 1980s attempted to understand the formation of oppositional youth cultures and subcultural styles as expressions of marginalized young people's oppressive structural location and the moral panics that developed around them (Hall & Jefferson, 1976; Hebdige, 1979). Blending ethnography and semiotics (a type of analysis concerned with understanding systems of signs and meanings) to reveal the contested nature of youth cultures, scholarly attention was given to young people's use of resources provided by a consumer market to establish subcultural boundaries and respond to the mainstream. Dick Hebdige's 1979 study of the mods, who used discarded market goods (clothes, records, clubs, hairstyles) as identity markers to craft an alternative youth culture that was defined against other youth cultures belonging to the mainstream, is an important example.

CURRENT AND FUTURE DIRECTIONS FOR RESEARCH

Although a great deal of research on the formation of oppositional youth cultures has focused primarily on rituals of resistance formed through consumption, a much smaller group of youth scholars has focused on the ways in which collective forms of transformative social action have emerged within youth culture. Musical and (maga)zine-based movements such as the Riot Grrrls have served as important conduits for girls to resist commodification, forge an alternative gender and sexual order, and articulate a feminist political agenda. Cultural jammers—young activists who protest the corporate control of everyday life—have been successful in disrupting the flow of information to the mass audience.

Far greater attention has been paid to the dynamic ways in which race, disability, ethnicity, sexuality, and gender have intersected in the formation of youth cultural groups in the last decades of the 20th century and the first decade of the 21st. Early youth cultural research focused almost exclusively on male youth cultures but with little attention to gender. The British scholar Angela McRobbie (1991) demonstrated that teenyboppers, who are by definition girls, consumed images of teen idols, courting them in the private space of their bedrooms, where the negotiation of feminine sexuality was a little less uncomfortable because girls were unlikely to have an audience there. In 2006 Amy Best examined how young Asian men struggle to assert their place as men among men and forge a pan-Asian identity in the underground import racing scenes in California against efforts by young men aligned with American muscle cars to discredit them as feminine.

Researchers increasingly have acknowledged that the boundaries of contemporary youth culture occur within a social landscape that has been transformed by globalization, the acceleration of production and consumption in late capitalism, the hypermobility of communication systems, and increasingly sophisticated media. The explosion of the Internet and people's movement into a digital age as "citizen consumers," the expansion of global markets, and the proliferation of global and corporate mediascapes have influenced the complex and historically distinct formations recognized as contemporary youth cultures. Technologies have reordered the social organization, expressive forms, and in-group communication that help constitute distinct youth cultures. Those cultures play out in a virtual world where elaborate cyber networks are forged and new publics are conceived. Important examples include the social utility Facebook and the blogs and online forums dedicated to topics and issues that are resonant with members of youth cultures. This direction is likely to be pursued in future research.

An increasingly global economy and the steady stream of people, cultural ideas, and cultural objects across ever-shifting borders has produced many changes in American youth cultures, impelling scholars of youth studies to examine the formation of youth cultures in a global context. As commercial forms of European and American youth culture from MTV to *Beverly Hills 90210* have been transmitted along global mediascapes, distinct and often hybridized youth cultures have emerged in settings where young people rarely were seen as a group apart from adults, pointing out the idea that youth cultures are the products of historical and cultural change. Another example can be seen among second-generation American youth of the Indian diaspora who combine traditional forms of Bhangra dance with the sampling elements of American hip-hop to forge a distinct club culture and construct race and gender identities as they negotiate a highly commercial American youth culture on the one hand and the nostalgic constructions of a India frozen in time by the parent culture on the other hand (Maira, 2002). Global considerations are also likely to inform future research. Because the social experiences of being young are thought to influence life outcomes in adulthood, youth cultural participation is quite relevant to the narratives that prevail as scholars try to make sense of the life course and understand different life trajectories.

SEE ALSO Volume 1: *College Culture; Dating and Romantic Relationships, Childhood and Adolescence; Friendship, Childhood and Adolescence; Identity Development; Interpretive Theory; Media and Technology Use, Childhood and Adolescence; Media Effects; Peer Groups and Crowds; School Culture; Social Development.*

BIBLIOGRAPHY

Best, A. L. (2000). *Prom night: Youth, schools, and popular culture.* New York: Routledge.

Best, A. L. (2006). *Fast Cars, cool rides: The accelerating world of youth and their cars.* New York: New York University Press.

Chávez, E. (1998). Birth of a new symbol. In J. Austin & M. N. Willard (Eds.), *Generations of youth: Youth cultures and history in twentieth century America.* New York: New York University Press.

Graebner, W. (1990). *Coming of age in Buffalo: Youth and authority in the postwar era.* Philadelphia: Temple University Press.

Hall, S., & Jefferson, T. (Eds.). (1976). *Resistance through rituals: Youth subcultures in post-war Britain.* London: Hutchinson.

Hebdige, D. (1979). *Subculture: The meaning of style.* London: Methuen.

Maira, S. M. (2002). *Desis in the house: Indian American youth culture in New York City.* Philadelphia: Temple University Press.

McRobbie, A. (1991). *Feminism and youth culture: From Jackie to just seventeen.* Boston: Unwin Hyman.

Palladino, G. (1996). *Teenagers: An American history.* New York: Basic Books.

Thornton, S. (1996). *Club cultures: Music, media, and subcultural capital.* Hanover, NH: University Press of New England.

Amy L. Best